Fundamentals with Elements of Algebra

FOURTH EDITION

Patricia J. Cass | Elizabeth O'Connor

Australia • Brazil • Japan • Korea • Mexico • Singapore • Spain • United Kingdom • United States

Fundamentals with Elements of Algebra
FOURTH EDITION

Patricia J. Cass l Elizabeth O'Connor

Executive Editors:
Michele Baird
Maureen Staudt
Michael Stranz

Project Development Manager:
Linda deStefano

Senior Marketing Coordinators:
Sara Mercurio
Lindsay Shapiro

Senior Production / Manufacturing Manager:
Donna M. Brown

PreMedia Services Supervisor:
Rebecca A. Walker

Rights & Permissions Specialist:
Kalina Hintz

Cover Image:
Getty Images*

ISBN-13: 978-0-7593-1000-1

ISBN-10: 0-7593-1000-9

Cengage Learning
5191 Natorp Boulevard
Mason, Ohio 45040
USA

Cengage Learning is a leading provider of customized learning solutions with office locations around the globe, including Singapore, the United Kingdom, Australia, Mexico, Brazil, and Japan. Locate your local office at:
international.cengage.com/region

Cengage Learning products are represented in Canada by Nelson Education, Ltd.

For your lifelong learning solutions, visit **custom.cengage.com**

Visit our corporate website at **cengage.com**

Printed in the United States of America

We dedicate this book to *our families, friends, and colleagues who have supported our efforts, but mostly to the many students who have taken the time to tell us that our books have really helped them. Thanks to you all!*

Contents

Preface

Looking around any developmental math classroom, you will see students of different ages, genders, and races. However, all these students have one trait in common: they have all been less than successful in previous math courses. The reasons for their lack of success are numerous and complex, but the result is that they now have some educational, career, or personal goals that they cannot attain unless this math deficiency is remedied. The desire to reach their goals will motivate most of these students, but, along with bad memories and experiences about math, attaining their goals creates a great deal of stress, anxiety, and self-doubt.

We have written this text with these factors in mind and with the understanding that many students fear and dislike math. Most of them are willing to work to be successful. They need good, caring instructors like you, support services, and resources that will help them learn the concepts in their own individual style and at their own speed.

Features

We have written the text in an informal, almost conversational style but have not minimized mathematical integrity. We have included extensive explanations and several detailed examples of each concept. When students leave campus and are studying with no access to resources they can find in this book explanations and examples to clarify the topic and to help them with the assignment. There are Self-Checks with almost every concept so they can quickly monitor their progress along the way. Students can test their preparation by doing the chapter review exercises or the chapter test. To review previous material, they can do the Cumulative Review exercises, which are grouped according to type so that a student can see the connections among, for example, addition of whole numbers, fractions, decimals, signed numbers, and expressions.

This edition of the text has several features that we believe will be interesting and helpful to students. First, we have included a Study Skills Module before each chapter. In these modules, we discuss a variety of topics. Each discussion is followed by several "Your Point of View" questions and activities that help students formulate an opinion about what they have read and put some of the ideas into practice. These modules are not designed to be done in any particular sequence; they are all independent of the chapters, so instructors and students can pick and choose according to their needs.

Another feature is an appendix on general calculator usage. Students often come to us with questions about even the simplest functions. We have kept this appendix as generic and basic as possible, but we have included examples using the "a b/c" key because these calculators are now affordable for most students.

The text contains a large number of examples that include detailed explanations for students to study when they are away from their instructor or tutor. The text is a "teacher they can take home."

In decimal equation, the method of eliminating decimal places involves multiplying by 10, 100, 1,000, and so forth, depending on the greatest number of decimal places in any term of the equation. Review the shortcut for doing this multiplication in Section 5.3.

EXAMPLE 5 Solve: $4.2B + 0.86 - 3.1B = 1.41$

Solution Multiply by 100 to eliminate the decimals.

$$100(4.2B) + 100(0.86) - 100(3.1B) = 100(1.41)$$
$$420B + 86 - 310B = 141$$
$$110B + 86 = 141$$
$$\underline{-86} \quad \underline{-86}$$
$$110B = 55$$
$$\frac{110B}{110} = \frac{55}{110}$$
$$B = 0.5$$

Check

$$4.2B + 0.86 - 3.1B = 1.41$$
$$4.2(0.5) + 0.86 - 3.1(0.5) = 1.41$$
$$2.1 + 0.86 - 1.55 = 1.41$$
$$2.96 - 1.55 = 1.41$$

arest hundredth: $5.6M + 0.4$

ıber equation to solve, mul-
greatest number of decimal

100.

r variable term, $240M$.

.

ıtely equal)

B. The following application problems will apply geometry concepts that were introduced in Section 2.3 and reviewed in Chapters 3 and 4. Required formulas will be given in the problem. You may want to try estimating your answer first. Simplify all fraction answers to lowest terms.

31. A room measures 12 3/4 feet by 9 1/2 feet wide. Find the area of the ceiling of that room. Use $A = LW$.

32. Katie is landscaping her backyard and needs to determine how many square yards of sod to order. The yard's shape is shown in Figure 4.7. How many square yards of sod should she buy? Use $A = 1/2\ h(b^1 + b^2)$

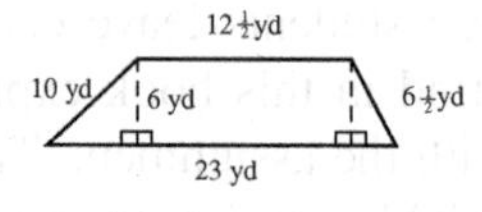

FIGURE 4.7

33. In the room described in Problem 31, the height of the walls (from floor to ceiling) is 8 feet. Find the area to be painted on the four walls, disregarding any window and door openings. (Hint: Two walls are $12\frac{3}{4} \times 8$ and two walls are $9\frac{1}{2} \times 8$.) Use $A = LW$ for each wall.

34. How much edging will Katie need to enclose her backyard if its dimensions are as shown in Figure 4.7? (Hint: Find the perimeter by adding all four sides togethers.)

35. The Parks and Recreation Department wants to develop a triangle-shaped flower bed in front of its building. How many feet of edging will be needed to go around the equilateral triangle with sides each 12 3/4 feet in length? If bricks that are 2/3 foot long are laid end to end to edge the flower bed, how many bricks will be needed on each side of the triangle? Use $P = 3a$.

36. Which has greater volume: a box that is 6 1/2 inches by 8 inches by 4 3/4 inches or one that is 5 5/8 inches by 2 3/4 inches by 16 inches? Use $V = LWH$.

4.6 CHAPTER REVIEW

We hope you are feeling more confident about working with fractions. As you prepare for the test for this chapter, remember:

1. Fractions can be added or subtracted only if they have the same denominator.
2. Find the least common denominator by inspection or by the least common multiple method (Section 2.6), and then write equivalent fractions.
3. Mixed numbers do not have to be changed to improper fractions when you are adding or subtracting.
4. If borrowing is required in subtraction, borrow 1 from the whole number, write the 1 as the common denominator over itself, add together the fraction parts of the numerator, and then subtract.
5. Apply the Order of Operations Agreement to fraction problems containing more than one operation.
6. To raise a fraction to a power or to take the square root of a fraction, you must apply the operation to both the numerator and the denominator.
7. To compare fractions, write them as equivalent fractions (fractions that have the same denominator), and then compare the numerators.

Chapter review sections contain a detailed summary of important points and procedures in each chapter.

Cumulative Review, Chapters 1–11

Find the answer to each of the following problems. Show all your work. Simplify all fraction answers to lowest terms.

In Exercises 1 through 5, add.

1. $11.56 + 3.8 + 18$
2. $11\ 4/9 + 7\ 2/3$
3. $5x + 8y + x + 3y$
4. $(-15) + (8) + (-3)$
5. $29 + (-17)$

In Exercises 6 through 10, subtract.

6. $(-15) - (-12)$
7. $0 - 8 - (-5)$
8. $4\ 1/4 - 2\ 5/6$
9. $18 - 3.56$
10. $12a - 3a - 18a$

In Exercises 11 through 15, multiply.

11. 0.176×100
12. (–14)(
13. $5\ 2/3 \times 2\ 1/8$
14. Find 4
15. $(-2)(3)(-1)(2)$

In Exercises 16 through 20, divide.

16. $(-32) \div (-8)$
17. $\frac{1.68}{0.04}$
18. $2\ 5/8 \div (-3)$
19. $\frac{-54}{9}$
20. $0.8\overline{)24}$

Simplify Exercises 21 and 22 using the
ations Agreement.

21. $3 + 2[6 - 2(4 \div 2)]$
22. $\frac{25}{36} \div \left(\frac{2}{3}\right)^2 \div \frac{5}{9}$

Find the value for each of the expressions in Exercises 23 through 25 by substituting the given values.

23. $ab^2 - ac$ when $a = 2$, $b = -1$, and $c = -3$
24. $\frac{2bc}{b^2 - c^2}$ when $a = 2$, $b = -1$, and $c = -3$
25. $2ab - 4c + 3ac$ when $a = 2$, $b = -1$, and $c = -3$

In Exercises 26 through 28, write an algebraic expression for each of the English phrases.

26. The difference between a number and six
27. seven more than twice a number
28. the product of a number and eiht

In Exercises 29 and 30, write an algebraic equation

Cumulative review exercises are arranged by operation to illustrate the commonality of procedures.

STUDY SKILLS MODULE

13 Working in Groups

More and more students in classrooms on college campuses are being given assignments where they must work on a group project. With the variety of class, study, and work schedules that students have, the difficulty of getting even a small group of students together outside of class seems almost insurmountable. You are fortunate that there are very few outside group projects assigned in math classes. However, instructors do ask students to work in small groups to solve problems in math classes. Why this push for group work? Why can't individual students be left alone to study and work on their own? Why do you need to learn to work together? The answer is simple. Most work situations in our society today are too complex for an individual to handle on his or her own. Employers are looking for workers who can work together to solve problems, make presentations, and complete projects. Small-group work in a math class is a relatively safe environment in which to develop good collaborative skills.

Several qualities characterize well-functioning groups:

1. A leader surfaces who can help to guide the group.
2. Someone acts as record-keeper who can help keep the group on-task.
3. Each member of the group makes a contribution to the discussion or project.
4. Members of the group discuss the problem in a supportive manner, listening to each person's opinion.
5. The group produces a final product in which each group member shares ownership.

In the case of group work in a math class, the instructor may assign a "part" to each student or allow the group itself to divide up the work. When students work in groups to solve math problems, it is very important that each person in the group understand how the solution is reached. It is the responsibility of each group member to try to help all the other group members. Personality differences need to be addressed on a respectful and constructive level. Learning how to resolve differences on this level can be an invaluable tool in an employment situation. If you work in an ongoing math group, try each of the different "jobs" in the group. Even if you don't see

521

Study skills modules, which precede each chapter, contain a brief discussion of a given topic, usually list practical hints, and end with "Your Point of View" questions that actively involve the student.

Exercise sets contain a unique blend of questions, including visual aplications from data, as well as geometry, calculator, exploration, error analysis, written response, and estimation practice.

For Exercises 98 and 99, refer to Figure 6.23, where ΔKJN is similar to ΔKLM, $KL = 32$ yards, $AM = 40$ yards, $KJ = 20$ yards, and $JN = 15$ yards.

98. Find the length of $\overline{KN}$.

99. Using the information in Exercise 98, find the length of $\overline{LM}$.

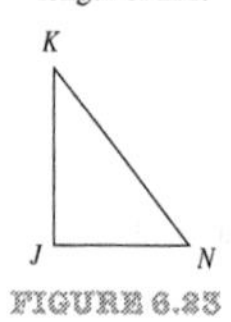

FIGURE 6.23

100. Change 50 miles per hour to feet per second using 1 mile = 5,280 feet, 1 hour = 60 minutes, and 1 minute = 60 seconds.

101. Change 25 miles per hour to kilometers per second using 1 mile = 1.61 kilometers, 1 hour = 60 minutes, and 1 minute = 60 seconds.

CHAPTER TEST

A. In Exercises 1 through 5, write each comparison as a ratio in simplest form.

1. 28 to 63
2. 7.2 to 0.12
3. 2/3 : 5/9
4. 2 dimes to 3 quarters

B. In Exercises 12 through 16, solve for the unknown. If necessary, round answers to the nearest hundredth.

12. $\frac{5}{9} = \frac{35}{W}$
13. $\frac{12}{C} = \frac{15}{32}$
15. $\frac{A}{14.5} = \frac{1.2}{0.5}$

contractor's building plan states
eet. What is the actual length of a
res 2 1/2 inches on the drawing?

ive 128 miles on 5 gallons of
many gallons will she need to
?

, with $QR = 19$ meters, $XY = 21$
meters, and $PR = 20$ meters. Find
Z and $\overline{XZ}$.

Equation Cost of Venetian beans plus Cost of Colombian beans equals Cost of Deluxe Mix.

$$10x + 4(40) = 6(x + 40)$$

Solution

$$10x + 4(40) = 6(x + 40)$$
$$10x + 160 = 6x + 240$$
$$-6x \quad -6x$$
$$4x + 160 = 240$$
$$-160 \quad -160$$
$$4x = 80$$
$$\frac{4x}{4} = \frac{80}{4}$$
$x = 20$ pounds of Venetian beans

Check

$$10(20) + 4(40) = 6(20 + 40)$$
$$200 + 160 = 6(60)$$
$360 = 360$ True.

EXAMPLE 11 Craft and Company is preparing a new shredded cheese mixture. How many pounds of \$4.30-per-pound romano cheese should be combined with 20 pounds of \$2.25-per-pound mozzarella to produce a pizza cheese mix costing \$3.80 per pound?

Sketch An organizational technique that you might prefer to use with mixture problems is to make a sketch. (see Figure 12.4).

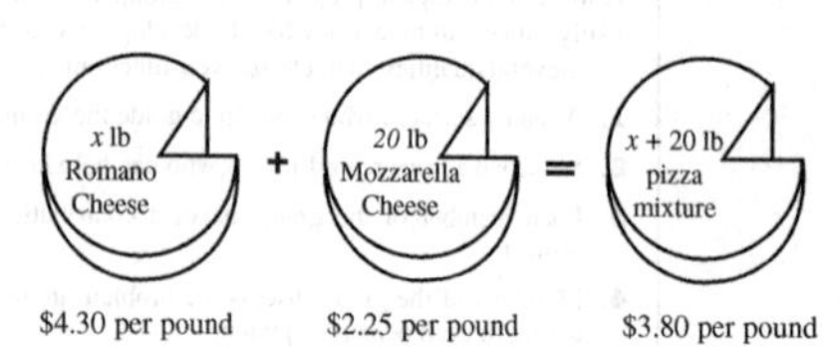

FIGURE 12.4

Chart Or you could record the information that is given in a box, and compare your box with Chart 12.5.

CHART 12.5

Item	Cost per Pound	Number of Pounds of Each	Total Cost of Each
Romano	4.30	x	$4.30x$
Mozzarella	2.25	20	2.25(20)
Mixture	3.80	$x + 20$	$3.80(x + 20)$

Word problem emphasis is on organization of information rather than on the type of problem presented. Many examples show two methods for organizing the same information.

This revised edition contains more emphasis on estimation, interpreting data, and critical thinking. We have included many exercises in which students only describe what they would do and never actually find an answer. Error analysis questions appear at the end of many chapters. In addition, we have increased our emphasis on real-world applications and have included more multi-step exercises. These more complex problems are usually included at the end of the chapter so that students can find success with the exercises as they progress through them. We also decided to allow as much flexibility for calculator usage as possible. As before, we use the calculator icon to indicate exercise sections specifically designed for calculator use, but we have removed most other calculator references.

The arithmetic chapters form a basis for students' future work in algebra. Little time was spent working with whole numbers, but topics such as the properties of whole numbers, prime factorization, least common multiple, and greatest common factor were developed in depth because of their application to algebra. The fraction and decimal chapters (Chapters 3, 4, and 5) will help students brush up on their basic skills so that they will be able to solve and check some algebraic equations. Chapter 6, Ratios and Proportions, and Chapter 7, Percents, introduce the idea of equation-solving without much algebra language so that the students can grow through this connection between arithmetic and algebra. The treatment of percents in Chapter 7 is formula-oriented. Although many instructors may consider this a new approach, students must be able to work with formulas in their future courses, and this has proven to be a good way to introduce them to these concepts. The text is arranged so that instructors wishing to teach percents by the proportion method can skip Section 7.3, the percent formula section, and use Section 7.4 instead. We have incorporated geometry through the arithmetic chapters so that the students improve their ability to work with formulas and become more familiar with the vocabulary of geometry.

Because of students' anxiety and self-doubt about their ability to learn algebra, the algebra topics are presented in a nontraditional order. We introduce the ideas of formulas, evaluating, equation solving, and word problems before introducing signed numbers. This approach gives the students success with algebra concepts before complicating problems with signed numbers. There is ample practice in and after Chapter 11 to perfect the signed-number operations.

Applications are presented throughout the text. Because critical-thinking and decision-making processes are being developed through this course, extensive practice with word problems is essential. We have included a wide variety of problems, knowing that teachers will select the ones most appropriate for their students. Emphasis is placed on organizing the information, not on memorizing types of problems and solutions.

The presentation of polynomial operations is fairly standard. Chapter 14, on factoring, gives students a good foundation, using both trial-and-error and product-sum methods. Chapter 15, on graphing, provides explanation and practice not only in graphing equations and inequalities, but also in writing equations from given information.

Should time not allow coverage of all chapters in this text, the later chapters are a good resource for students going on to a typical Introduction to Algebra course. We believe that once students have completed our material, they can progress into more traditional algebra courses and successfully build on their new-found skills. The emphasis on application problems and critical thinking carries throughout all the features.

Acknowledgments

We would like to thank our new team from Cengage Publishing—Denise Bosma and Jon Hughes Fuller. Our working relationship has been a pleasure. Special thanks to Janet Kerekes who first discovered us and introduced us to the world of publishing, and to Todd Rupp who went far beyond his sales rep duties to help us through this latest project.

Special thanks to all our students, who have taught us how to be better developmental educators, and to our colleagues at Columbus State, who share their ideas and opinions about improving these materials.

Patricia J. Cass
Elizabeth R. O'Connor

To the Student

To succeed in this or any other math course, you must start from the same point as every other student in your class and use the same vocabulary and the same rules. Because each of your past math experiences has been different, we need to begin with a few basics to ensure that everyone is starting with a solid foundation. It is likely that many of the topics in these first few chapters will be familiar to you. Nevertheless, please read each section as if the material were new. Pay attention to details. Underline the important statements and outline each topic. If you study all the sections, you will begin to fill in the gaps that exist in your own background. You will see how topics are related, and you can progress with confidence. The topics that are truly review material for you will bring you quick success. The weaker areas in your math skills will be repaired and reinforced and will soon become sources of confidence as well.

In spite of your past experiences with math texts, we urge you to read this one. It is written with you, the adult learner, in mind. It doesn't matter whether you are an adult of 17 or an adult of 45. You can pick things up quickly if they are explained clearly. You don't want to waste your time. You want to learn the facts and how to apply them so that you can move on to a successful working knowledge of math. To help you accomplish this, we have included many example exercises with detailed explanations. Chapter objectives are stated clearly at the beginning of each chapter and are summarized in the chapter review. Many practice problems in a variety of types and real-life applications are included, with the odd-numbered answers at the back of the book. Ask your instructor what learning and tutoring resources are available on your campus.

Whether or not you have had difficulty with mathematics courses in the past, we urge you to read and use the study skills modules that appear before each chapter. They are included to help you develop sound study behaviors that will aid you in becoming a more successful math student. Your instructor will decide which modules will be included in your course; there is no set order in which these units need to be completed. If topics that you feel you need are not included in your course, then pursue those modules on your own.

We wish you success in your mathematical endeavors and hope that this text helps you accomplish your goals.

Patricia J. Cass
Elizabeth R. O'Connor

STUDY SKILLS MODULE

1 Rationale for Study Skills Modules

Preceding each chapter in this book you will find an insert called a Study Skills Module. These modules are included to help you become a better, more successful math student. The level to which this improvement happens will depend on your commitment and effort with these materials. It will require that you do more than just read the pages and respond to the exercises. You will need to make a concerted effort to change your behaviors and, as you work through each chapter or prepare for each test, actually put into practice the strategies that are suggested. We suggest that you consciously apply the given strategies for several chapters, and then modify them to meet your needs and learning style.

It is never easy to break a habit—ask people who have tried to improve their eating habits or to add more exercise activities to their "old" lifestyle. Making substantial changes in how you study and do math will not be easy. But, as with the changes in eating and exercise, the effort will be beneficial to you.

Some of the topics included in these modules concern studying and homework tips; test preparation, test taking, and test anxiety; and time management and use of class time. The order of the topics in the book is not important, and your instructor will decide which topics to include and in what order, which to exclude, and perhaps which other topics to add. You do not even have to wait for an assignment to work on one of the modules. If you know that you are weak in one of the areas covered by a Study Skills Module, then work on that module on your own, right now!

Your most immediate goal in this course is probably to do well on the first test. After that, you might look ahead and think about doing things that will help you do well in the whole course. Such short-term goals are good, and using the Study Skills Modules should help you attain them. Your ultimate goal, however, for survival in a rapidly changing world, needs to be that of becoming a lifelong learner. The days when a person found a job and worked at it for 25 or 30 years are gone. Because of developing technology, jobs are being continually eliminated and new ones created. The skills required to do these new jobs are also changing rapidly, so that people entering the work force today are going to change the kind of work they are doing five or six times during their career years. In order to make those changes, workers must be willing and able to learn new skills. The company or the government will provide some of the training, but workers will also have to keep up with the changes that are occurring in their area of expertise. In order to hold onto their position and advance within the field, they will need to gain new knowledge and skills even before it becomes essential.

All of this means that you will need to know how to read and understand manuals, journals, and books pertaining to your job. You may find it necessary to find and study these materials on your own, just to stay current with the changes in your area. You might need to take courses or attend workshops or seminars that upgrade your skills or that teach you new skills. To be successful, you will need to know and apply good study skills techniques.

Your Point of View

1. After reading these pages, do you think that emphasis on improving study skills should be a part of this course? Explain your answer.
2. List several study skills topics with which you feel you need some help. Now look at the Contents page and see if those topics are included. If they are not, or if you are not sure they are part of one of the topics covered, ask your instructor to include the topic in his or her presentation. If it is important to you, it is probably important to others as well.
3. Describe the kind of job you would like to have after you reach your educational goal. Have the knowledge base or skills for that kind of work changed in the past ten years? What changes do you think will occur in the next ten years? How will you prepare for those changes?
4. List three things that you would like to change about your personal situation so that you could study more. (For example, work less hours, have a baby who take an hour-long nap, etc.)

CHAPTER

1 About Whole Numbers

LEARNING OBJECTIVES

When you have completed this chapter, you should be able to:

1. Read, write, and round whole numbers.
2. Perform the four basic mathematical operations (addition, subtraction, multiplication, and division) using whole numbers.
3. Estimate the answers to calculation and application problems.
4. Read application problems and solve them by using one or more of the four basic operations.
5. Apply the definitions of perpendicular lines and complementary, supplementary, and vertical angles.
6. Find the mean, median, mode, and range of given data.
7. Interpret data represented on a graph.

In this chapter, we cover many topics that are very elementary (maybe even boring). Adding, subtracting, and multiplying whole numbers are skills that most students take for granted. Here you are expected only to review and rebuild these skills. Long division, however, is a topic many adult students find difficult. Consequently, we will spend time going through division in detail. You will also begin work with word problems in this chapter. These application problems are designed to make mathematics real. The step-by-step approach introduced in this chapter will be used for word problems throughout the text.

1.1 Reading, Writing, and Rounding

It is not often that one writes a check for an amount as large as $4,208. Check writing, however, is one important activity where we write numbers in both word form and number form. Look at the following sample check and note where the word description and the numeric (number) designation appear on it.

No. 263

7/4 2002 25-2/440

PAY TO THE ORDER OF *Tri-City Car Sales* $ 4,208.00

Four Thousand Two Hundred Eight and 00/100 DOLLARS

THE FIRST NATIONAL BANK
Columbus, Ohio 43260

MEMO *Jane L. Edwards*

I: 044000024 I: 03898851946 II'

The ability to translate from word to number form is directly related to **place value,** so that is where we must begin.

Writing and Reading

Look at the chart in Figure 1.1. In this chapter, you will use only whole numbers. Whole numbers are often thought of as the counting numbers, or the natural numbers and zero. The **whole numbers,** then, are 0, 1, 2, 3, 4, and so on. Fractions, decimals, and all numbers less than zero are not included among the whole numbers. In the example of the check, there were no cents involved, and only zeros followed the decimal point. For now, we will look only at digits that fall to the left side of the decimal point: only whole dollars in the check example, only whole numbers in the other problems. Note that when we read from the right side of the number, the names for place value fall naturally into groups of three. Each of these groups of three is called a **block or group.** Each group is separated from the other groups by a comma. The three place values in the thousands block, for example, are all grouped together. It is the same for millions, billions, trillions, quadrillions, quintillions, and so on. This is how knowledge of place value makes writing numbers in numeric form possible.

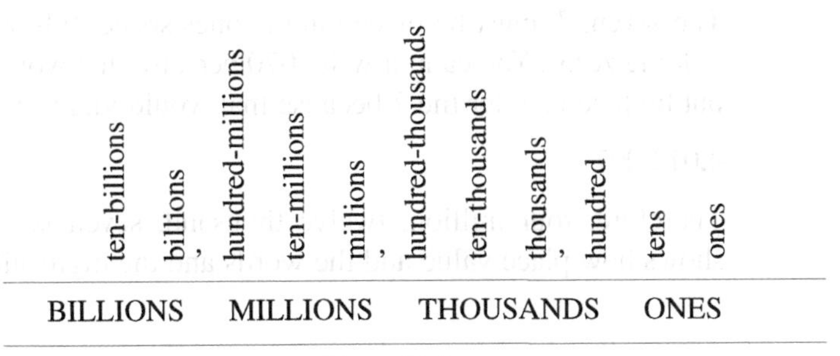

Place Value

FIGURE 1.1

EXAMPLE 1 Write eight thousand, two hundred sixteen in digits.

Solution Because you know you need eight thousand, write 8 in the thousands place and list the remaining three spaces:

8, — — —

Fill in two hundred sixteen, 216, in the blank spaces:

8,216

Note that the commas in the verbal statement and the commas in the number statement are written in the same place. Counting from right to left, commas are placed after each group of three digits. The commas separate the groups, or blocks of numbers, in both word and numeric form. Figure 1.2, where each digit is written under the correct column in the place value chart, also shows how the place value names correspond to placement of the digits in the number.

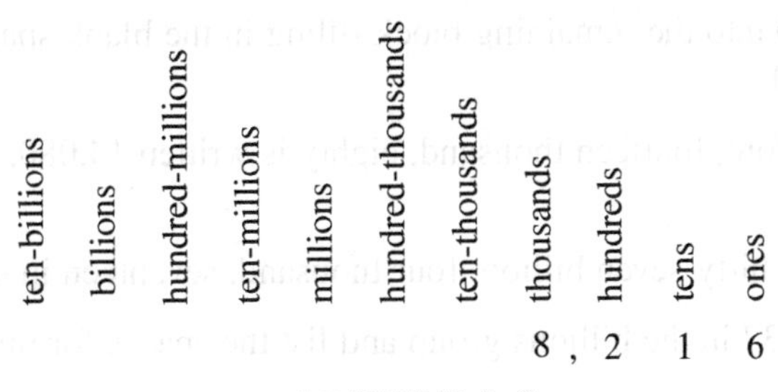

FIGURE 1.2

EXAMPLE 2 Write four million, twelve thousand, seven in digits.

Solution You need two periods of three spaces each to the right of the millions place. Write 4 in the millions place and indicate the other places that need to be filled:

4, — — — , — — —

Twelve thousand must be positioned in the thousands block. You have only two digits plus an extra space, so you will fill that space with a placeholding zero. This zero needs to go before the 12. If the zero were placed between the 1 and the 2, the number in the thousands group would be 102. If the zero were placed after the 12, it would be 120. Position of the placeholding zeros is very important.

4,012, — — —

The seven, 7, must be placed in the ones space. Fill in the blank spaces with placeholding zeros. You cannot write 070 because that would be seventy. And you cannot put both zeros after the 7 because that would yield seven hundred.

4,012,007

Therefore, four million, twelve thousand, seven is written 4,012,007. Figure 1.3 shows how place value and the words and the digits all fit together for this number.

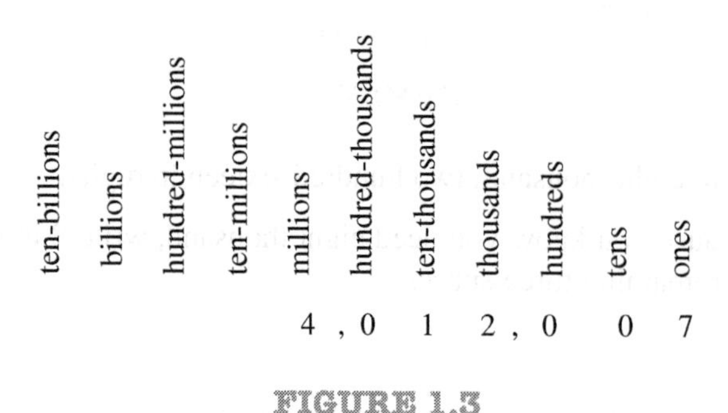

FIGURE 1.3

Always go back and read the number as you have written it to make sure that you wrote what was asked for in the problem. That is an easy way to check your answer.

EXAMPLE 3 Write fourteen thousand, eighty in digits.

Solution Write 14 in the thousands block follow it with the remaining three blanks:
14, __ __ __

Put 80 into the remaining block filling in the blank space with a placeholding zero:
14,080

Therefore, fourteen thousand, eighty is written 14,080.

EXAMPLE 4 Write thirty-seven billion, four thousand, seventeen in digits.

Solution Write 37 in the billions group and list the spaces for millions, thousands, and ones:

37, __ __ __, __ __ __, __ __ __

Write 4 in the thousands place and 17 in the ones block:

37, __ __ __, __ __ 4, __ 17

Fill in the blanks with placeholding zeros.

37,000,004,017

Therefore, thirty-seven billion, four thousand, seventeen is written 37,000,004,017.

Most numbers, of course, are not so large, but the length of the number actually makes no difference. Now that you understand how the groups or blocks are used, you can break every number down into groups of at most three digits.

The reverse process—going from numbers to words—is also possible if you break the number down into groups of three digits and use the *block of three* idea. But before you can correctly write the numbers in words, you need to know which numbers are hyphenated. All numbers between twenty and one hundred that are read as a combination of two single numbers are written with a hyphen. For instance, twenty-one, forty-three, and ninety-six are all hyphenated. Study the following examples to see how this works.

EXAMPLE 5 Write 3,264,581 in words.

Solution The first group on the left, millions, has only one digit, a 3, so it is read "three million"

The next group, thousands, uses all three spaces and is read

"two hundred sixty-four thousand"

The third block, ones, uses all three spaces and is read

"five hundred eighty-one"

Therefore, 3,264,581 is written

"three million, two hundred sixty-four thousand, five hundred eighty-one"

◀

Note that we do not use the word *and* in reading whole numbers. *And* stands for the decimal point, which separates the whole numbers from values that are less than 1. For example, $26.17 is read "twenty-six dollars *and* seventeen cents." Because we are interested only in whole numbers right now, *and* has no place here.

EXAMPLE 6 Write 75,009 in words.

Solution The first group on the left side, thousands, uses only two digits, 7 and 5. It is read "seventy-five thousand." The second group, ones, has only one digit, 9, and that digit is located in the ones place. It is read "nine." Therefore, 75,009 is read "seventy-five thousand, nine." Remember, no *and.*

◀

EXAMPLE 7 Write 203,014,085 in words.

Solution Two hundred three million, fourteen thousand, eighty-five

◀

Rounding

Rounding numbers also requires knowledge of place value. When we round in everyday situations, $19.89 becomes $20, and a gallon of gasoline for 92^9 cents becomes 93 cents a gallon. When we round in a mathematical sense, however, we must follow certain rules. Consider this problem:

Round 24,325 to the nearest *thousand.*

The question really is asking, "Is 24,325 closer to 24,000 or to 25,000?" Asked that way, it is easy to answer. It is closer to 24,000.

To determine the answer to this type of problem, it is helpful to follow these steps:

Rounding Whole Numbers

1. Mark the digit that occupies the place to which you are to round.
2. Underline the number to its right.
3. If the underlined number is less than 5, leave the digit in question unchanged and replace all the digits that follow it with zeros.
4. If the underlined number is 5 or more, increase the digit in question by one and replace all the digits that follow it with zeros.

EXAMPLE 8 Round 24,325 to the nearest *thousand.*

Solution

24,325 (mark under 4)	4 is the digit in the thousands place. Mark it.
24,3̲25 (mark under 4)	3 is the digit to the right of that 4. Underline it.
24,000	Because 3 is less than 5, leave the digit in the thousands place as it is and replace all the remaining digits with zeros.

The answer you found by following the steps agrees with the answer you determined before "by inspection."

◀

EXAMPLE 9 Round 22,876 to the nearest *hundred.*

Solution

22,876 (mark under 8)	8 is the digit in the hundreds place. Mark it.
22,87̲6 (mark under 8)	7 is the digit to the right of that 8. Underline it.
22,900	Because 7 is greater than 5, increase the 8 by one and replace all the remaining digits with zeros.

◀

EXAMPLE 10 Round 1,054,839 to the nearest *hundred-thousand.*

Solution

1,054,839 (mark under 0)	0 is the digit in the hundred-thousands place. Mark it.
1,05̲4,839 (mark under 0)	5 is the digit to the right of that 0. Underline it.
1,100,000	Because the underlined number is 5, increase the 0 by one and replace all the remaining digits with zeros.

◀

EXAMPLE 11 Round 186,925 to the nearest *hundred.*

Solution

186,925 (mark under 9)	9 is the digit in the hundreds place. Mark it.
186,92̲5 (mark under 9)	2 is the digit to the right of that 9. Underline it.
186,900	Because 2 is less than 5, leave the 9 as it is and replace all the remaining digits with zeros.

◀

EXAMPLE 12 Round 19,832 to the nearest *thousand*.

Solution

19,832 (^ under 9)	9 is the digit in the thousands place. Mark it.
19,832 (^ under 9, 8 underlined)	8 is the digit to the right of that 9. Underline it.
20,000	Because 8 is more than 5, increase the 19 by one and replace all the remaining digits with zeros. **Note that when a 9 is the number in question, and it needs to be increased, the change also involves the digit to the left of the 9.**

EXAMPLE 13 Round 23,971 to the nearest *hundred*.

Solution

23,971 (^ under 9)	9 is the digit in the hundreds place. Mark it.
23,971 (^ under 9, 7 underlined)	7 is the digit to the right of that 9. Underline it.
24,000	Because 7 is more than 5, increase the 239 by one to 240 and replace all the remaining digits with zeros.

EXAMPLE 14 Round $27,392,086 to the nearest *ten-thousand*.

Solution Rounding dollar amounts is no different from any other rounding.

$27,392,086 (^ under 9)	9 is the digit in the ten-thousands place. Mark it.
$27,392,086 (^ under 9, 2 underlined)	2 is the digit to the right of that 9. Underline it.
$27,390,000	Because 2 is less than 5, leave the 9 as it is and replace all the remaining digits with zeros.

1.1 EXERCISES

A. Write each of the following expressions in digits.

1. eight thousand, two hundred sixty-five
2. nine thousand, four hundred seventy-two
3. six million, fourteen thousand, three hundred eighty-five
4. eight million, twenty-six thousand, one hundred seventeen
5. forty-eight million, two thousand, eighty-five
6. seventy-three million, nineteen thousand, forty-seven
7. eight billion, four thousand, seven
8. nine billion, seven million, twelve
9. twenty-nine thousand, four hundred sixty
10. fifty-six thousand, twenty-nine
11. forty-six million, nine hundred eighty thousand, two
12. ninety million, thirty-six thousand, seven hundred four
13. seventy-two million, four hundred eighty thousand
14. fourteen billion, six hundred twenty-four thousand
15. eight million, six thousand, forty-three
16. twenty million, five thousand, two hundred twenty-four

B. Write each of the following numbers in words.

17. 8,914
18. 5,423
19. 2,009
20. 8,017

21. 70,070
22. 56,060
23. 1,234,567
24. 9,876,543
25. 14,008
26. 51,007
27. 12,014,235,006
28. 21,415,006,140
29. 35,000,000,008
30. 47,000,080,000

C. Round each of these numbers to the place requested.

31. 734 to the nearest ten
32. 967 to the nearest ten
33. 849 to the nearest hundred
34. 251 to the nearest hundred
35. 1,763 to the nearest thousand
36. 5,415 to the nearest thousand
37. 8,468 to the nearest hundred
38. 4,549 to the nearest hundred
39. 3,597 to the nearest ten
40. 59,264 to the nearest ten
41. 14,276 to the nearest hundred
42. 37,581 to the nearest hundred
43. 1,526 to the nearest thousand
44. 8,568 to the nearest thousand
45. 32,999 to the nearest hundred
46. 57,999 to the nearest hundred
47. 42,396 to the nearest ten
48. 92,598 to the nearest ten
49. 19,876 to the nearest thousand
50. 39,712 to the nearest thousand
51. 864,379 to the nearest ten
52. 752,663 to the nearest ten
53. 864,379 to the nearest hundred
54. 752,663 to the nearest hundred
55. 864,379 to the nearest thousand
56. 752,663 to the nearest thousand
57. 864,379 to the nearest ten-thousand
58. 752,663 to the nearest ten-thousand
59. 864,379 to the nearest hundred-thousand
60. 752,663 to the nearest hundred-thousand
61. 37,065,214 to the nearest ten-thousand
62. 86,396,021 to the nearest ten-thousand
63. 219,875 to the nearest hundred-thousand
64. 476,810 to the nearest hundred-thousand

D. Round each of these dollar amounts to the place requested.

65. $4,376 to the nearest hundred
66. $1,263,545 to the nearest thousand
67. $36,264 to the nearest thousand
68. $915,337 to the nearest hundred
69. $27,543,260 to the nearest million
70. $19,620,249 to the nearest million

E. Read and respond to the following exercises.

71. Describe in words the process used to round a number to the nearest thousand.
72. Describe in words the process you would use to write "fourteen million, two" in digits.
73. Explain how rounding 7,963 to the nearest hundred differs from rounding 7,863 to the nearest hundred.
74. Late in 1995, the national debt was reported to be approximately $4,988,882,588,134. How would you write that number in words? How might you round that number so it would be easier to discuss and still convey the size of the number?
75. Bring in a newspaper or magazine article that includes large whole numbers. Describe in words how you might round those numbers to make the main points of the story easier to understand.

1.2 Three Whole Number Operations

The use of calculators has made doing computations "in your head" or with paper and pencil almost a thing of the past. Many people have become so dependent on their calculators that they have almost lost their own computational skills. You could do the calculations in this section using only your head and paper and pencil. You may certainly use a calculator if your instructor allows you to do so. Once you have found the answer to a problem, check your answers in the answer key. Most of the problems in this chapter are not so difficult that a calculator will be necessary. Furthermore, there are many situations where calculators are of little benefit, and your personal skills are faster. If you have not used your own calculation skills in a long time, you will be amazed at how quickly you can work the rust off them with a little serious effort.

One thing you should notice in this section is the mention of estimation in many of the problems. This is an important step to include when you are doing mathematics. Generally, **estimation** should include "reasonable" numbers (those easy to work with and remember), and, whenever possible, an estimate should be computed using only mental math. The whole process of estimating an answer, getting a "ballpark" idea, can most often be done in your head. Forming an idea about the size of the quantities involved (numeracy) and the processes to be used is certainly a mental activity. Calculator usage is often most appropriate when you need to find the exact answer.

Addition

Given enough fingers and toes, anyone can do an addition problem successfully. Addition is the most basic of the mathematical operations. There is nothing wrong with using your fingers to help you add, but it is easier and faster to do it in your head. The only requirement for addition is that quantities must be combined with *like quantities:* thousands with thousands, hundreds with hundreds, ones with ones, and so on. If you always line up the numbers to be added (called **addends**) so that the ones values are under each other, the rest is easy. Study these examples for your review.

EXAMPLE 1 Add 2,304; 87; and 426.

Solution Notice that the three numbers are to be added: 2,304 + 87 + 426. A good approach in problem solving is first to estimate the answer by combining easy-to-add numbers such as 2,300, 100, and 400 that are close to the actual numbers. When you do this addition, you get an estimate of 2,800. Your actual answer should be close to that.

This estimation step is important even when you are using a calculator because it is easy to make an error when inputting a value and your calculator display will then show an incorrect result. Estimating first helps you catch that type of mistake whether you are solving with paper and pencil or using a calculator.

Now let's solve to find the actual answer.

$$\begin{array}{r} 2{,}304 \\ 87 \\ +\ 426 \\ \hline \end{array}$$

Line up the numbers vertically so that the digits in the ones place (4, 7, and 6) are under each other.

$$\begin{array}{r} {\scriptstyle 11} \\ 2{,}304 \\ 87 \\ +\ \ 426 \\ \hline 2{,}817 \end{array}$$

Add each column, beginning with the ones and carrying numbers over to the next column as required.

The estimate of 2,800 was very close to the actual answer of 2,817.

Check

$$\begin{array}{r} {\scriptstyle 11} \\ 2{,}304 \\ 426 \\ +\ \ \ 87 \\ \hline 2{,}817\ \surd \end{array}$$

You can check an addition problem that has three or more addends by arranging them in a different order. You can rewrite the problem, as is shown here. Or, if you normally add from top to bottom, you can add from the bottom to the top.

EXAMPLE 2 Find the sum of 91, and 384, and 78.

Solution An estimate might be 100 + 400 + 100, or 600.

$$\begin{array}{r} {\scriptstyle 21} \\ 91 \\ 384 \\ +\ \ \ 78 \\ \hline 553 \end{array}$$

The word **sum** indicates addition. Line up the digits in the ones place: 1, 4, and 8. Add, and carry to the next place as necessary.

Check

$$\begin{array}{r} 384 \\ 78 \\ +\ \ \ 91 \\ \hline 553\ \surd \end{array}$$

Check by rearranging the addends and adding again.

EXAMPLE 3 Find the total of 19; 3,804; and 198.

Solution

$$\begin{array}{r} {\scriptstyle 112} \\ 19 \\ 3{,}804 \\ +\ \ 198 \\ \hline 4{,}021\ \surd \end{array}$$

The word **total** indicates addition. Line up the 9, 4, and 8. Add, and carry as required.

Check

$$\begin{array}{r} 3{,}804 \\ 198 \\ +\ \ \ 19 \\ \hline 4{,}021\ \surd \end{array}$$

Check by rearranging the addends and adding again.

You could have first estimated the answer by adding 20, 3,800, and 200. The result would have been 4,020, which is every close to the actual answer, 4,021.

Subtraction

Subtraction also requires keeping the place values lined up properly. Begin with the ones place and line up the digits. Often when subtracting, you will need to **borrow** or regroup. Borrowing will be discussed in the examples where it is necessary. You can check subtraction problems by adding, because addition and subtraction are opposite, or inverse, operations.

EXAMPLE 4 Subtract 623 from 8,897.

Solution Estimate an answer by subtracting 600 from 9,000 to get 8,400.

```
  8,897
−   623
```

The number to be subtracted *from* goes on top. Line up the 7 and the 3.

```
  8,897
−   623
  8,274
```

Subtract, beginning with the ones column and moving to the left:
3 from 7 is 4.
2 from 9 is 7.
6 from 8 is 2.
Nothing from 8 is 8.

Check

```
  8,897
−   623
+ 8,274
  8,897 √
```

Check the solution by adding 623 and 8,274. 8,897 is the same as 8,897, so the answer is right.

As you consider Example 5, notice the detailed explanation of borrowing or regrouping in the third step. You will follow this same process every time the number you are subtracting from is not large enough. Borrowing is often a source of difficulty for students, particularly when borrowing from a zero, as in the second step in Example 6.

EXAMPLE 5 Find the difference between 17,824 and 9,460.

Solution

```
  17,824
−  9,460
```

The word **difference** indicates subtraction. Line up the 4 and the 0.

```
  17,824
−  9,460
       4
```

Subtract from right to left. 0 from 4 is 4.

```
     71
  17,824
   9,460
      64
```

Because 6 cannot be taken from 2, borrow 1 from the hundreds column. Now there are only 7 hundreds left. Since 1 hundred = 10 tens, regroup the 10 tens with the 2 tens already there, and now there are 12 tens, or 12 in the tens column. Six from 12 is 6.

```
     71
  17,824
−  9,460
   8,364
```

Continue to subtract:
4 from 7 is 3.
9 from 17 is 8.

Check

```
  17,824
−  9,460
+  8,364
  17,824 √
```

Check the subtraction answer by adding.

EXAMPLE 6 Subtract 438 from 2,004.

Solution You could estimate an answer by subtracting 400 from 2,000. The actual answer should be close to 1,600.

```
  2,004
-   438
```

Line up the digits 4 and 8 in the ones column.

```
  1 9 9 1
  2,004
-   438
```

In this step, you need to borrow from 0. You cannot subtract 8 from 4, so you must borrow 1. Since we cannot borrow from zero, we will regroup and borrow 1 from 200 tens. This leaves 199 tens and 1 ten (or 10 ones) that can be combined with the 4 ones in the problem. Subtract 8 ones from 14 ones.

```
  1 9 9 1
  2,004
-   438
  1,566
```

Continue to subtract:
8 from 14 is 6.
3 from 9 is 6.
4 from 9 is 5.
Nothing from 1 is 1.

Check

```
  2,004
-   438
+ 1,566
  2,004 √
```

Check by adding,

EXAMPLE 7 Subtract 2,684 from 30,000.

Solution

```
  30,000
-  2,684
```

Line up the digits 0 and 4 in the ones column.

```
  29 991
  30,000
-  2,684
```

Yon cannot subtract 4 from 0, so you must borrow 1 from 3,000 because you cannot borrow 1 from 0. When you borrow 1 from 3,000, the number 2,999 remains.

```
  29 991
  30,000
-  2,684
  27,316
```

Continue to subtract:
4 from 10 is 6.
8 from 9 is 1.
6 from 9 is 3.
2 from 9 is 7.
Nothing from 2 is 2.

Check

```
  30,000
-  2,684
+27,316
  30,000 √
```

Check by adding.

EXAMPLE 8 Find the difference between 70,165 and 18,476.

Solution and Check

$$\begin{array}{r} 70{,}165 \\ -\ 18{,}476 \\ \hline +\ 51{,}689 \\ \hline 70{,}165 \end{array} \quad \surd$$

Self-Check

1. 17,246 + 958 + 4,006
2. 18,007 – 12,989
3. Find the sum of 14,293; 20,484; and 8,094.
4. Find the difference between 9,072 and 6,485.

Answers:

1. 22,210 2. 5,018 3. 42,871 4. 2,587

Multiplication

Multiplication of whole numbers demands a firm knowledge of the basic multiplication facts. You must know without hesitation that 6 × 7 = 42 and that 9 × 4 = 36. You may not have used these basic facts for a while. Take the short multiplication review check that follows and see if you can answer these problems easily, quickly, and correctly. If you cannot, go to the Skills Check (Appendix A of this text) and do the exercises suggested to help you rebuild this skill. It will not take a lot of time, but the effort you invest *now* will payoff in the rest of your math work.

Self-Check Try to do all the problems in 2 minutes, and then check your answers against the answers that follow. Be honest with yourself and work on this skill now if you need to brush up.

4 × 7 =	9 × 3 =	0 × 5 =	11 × 10 =	8 × 5 =
6 × 4 =	7 × 8 =	4 × 8 =	3 × 9 =	7 × 6 =
12 × 6 =	2 × 7 =	2 × 6 =	4 × 1 =	11 × 1 =
0 × 4 =	12 × 9 =	10 × 3 =	7 × 3 =	4 × 0 =
7 × 9 =	5 × 7 =	5 × 9 =	11 × 12 =	8 × 3 =
6 × 3 =	12 × 8 =	10 × 7 =	1 × 7 =	8 × 9 =
3 × 5 =	6 × 7 =	4 × 3 =	9 × 5 =	0 × 9 =
12 × 5 =	6 × 0 =	11 × 4 =	6 × 8 =	4 × 9 =

Multiplication answers:

28	27	0	110	40
24	56	32	27	42
72	14	12	4	11
0	108	30	21	0
63	35	45	132	24
18	96	70	7	72
15	42	12	45	0
60	0	44	48	36

Multiplication, like addition, requires paying careful attention to place value and to carrying. The **factors** (the numbers that are multiplied) are lined up vertically so that the ones values are under each other, just as in addition. Most students prefer to put the larger factor on top, but that is not necessary. It usually makes the multiplication easier, though, because you are multiplying by fewer digits. Study these examples.

EXAMPLE 9 Find the product of 473 and 85.

Solution

```
  473
× 85
```

The word **product** indicates multiplication. Line up the ones values, 3 and 5.

```
   31
  473
 × 85
 2365
```

Multiply by the 5. Because 5 is located in the first column, the first "partial product" begins in the first column. Carry as required.

$5 \times 3 = 15$. Write 5, carry 1.
$5 \times 7 = 35 + 1 = 36$. Write 6, carry 3.
$5 \times 4 = 20 + 3 = 23$. Write 23.

```
   52
  473
× 85
 2365
3784
```

Now multiply by the 8. It is located in the tens place, so this partial product begins in the second column. Carry as needed.

$8 \times 3 = 24$. Write 4, carry 2.
$8 \times 7 = 56 + 2 = 58$. Write 8, carry 5.
$8 \times 4 = 32 + 5 = 37$. Write 37.

```
   473
×   85
  2365
+ 3784
 40205
```

Finally, add the partial products. Carry in the addition as required.

5 + nothing = 5
$6 + 4 = 10$. Write 0, carry 1.
$1 + 3 + 8 = 12$. Write 2, carry 1.
$1 + 2 + 7 = 10$. Write 0, carry 1.
$1 + 3 = 4$. Write 4.

$473 \times 85 = 40{,}205$

There are two ways to check a multiplication problem. One is to rearrange the factors and multiply again. The partial products will be different, but the final product will be the same. Because multiplication and division are inverse (opposite) operations, there is a second method for checking. Divide the final product by one of the factors. The **quotient** (answer) in the division should be the other factor. This division should be easier than most long division problems because you already have an idea of which numbers should appear in the quotient. Both methods are shown here. Of course, you could use your calculator, instead of paper and pencil, to do these checking calculations.

$$\begin{array}{r} 85 \\ \times\ \underline{473} \\ 255 \\ 595 \\ \underline{340} \\ 40205\ \surd \end{array} \qquad \begin{array}{r} 473\ \surd \\ 85\overline{)40205} \\ \underline{-340} \\ 620 \\ \underline{-595} \\ 255 \\ \underline{-255} \end{array}$$

EXAMPLE 10 Multiply 2,483 by 79.

Solution You could estimate the answer by multiplying 2,500 by 80. Multiply 25 times 8 then attach three zeros to the answer (one from the 80 and two from the 2,500). Now you know that your answer should be close to 200,000.

$$\begin{array}{r} {\scriptstyle 4\ 72} \\ 2,483 \\ \times\ \underline{79} \\ 22347 \end{array}$$

Line up the digits 3 and 9 in the ones column. Multiply by 9, beginning the answer in the first column. Carry as required.
9 × 3 = 27. Write 7, carry 2.
9 × 8 = 72 + 2 = 74. Write 4, carry 7.
9 × 4 = 36 + 7 = 43. Write 3, carry 4.
9 × 2 = 18 + 4 = 22. Write 22.

$$\begin{array}{r} {\scriptstyle 3\ 52} \\ 2,483 \\ \times\ \underline{79} \\ 22347 \\ \underline{+17381} \\ 196157 \end{array}$$

Now multiply by 7. Because 7 is in the second column, the answer begins in that column.
7 × 3 = 21. Write 1, carry 2.
7 × 8 = 56 + 2 = 58. Write 8, carry 5.
7 × 4 = 28 + 5 = 33. Write 3, carry 3.
7 × 2 = 14 + 3 = 17. Write 17.
Add the partial products. Carry as required.

2,483 × 79 = 196,157

The actual answer of 196,157 is very close to the estimate of 200,000.

Check Check your answer. Two methods are shown. The checks should be the same if you are using a calculator.

```
       79              2483 √
  × 2483           79)196157
      237            - 158
     632                381
    316               - 316
   158                   655
   196157 √            - 632
                          237
                        - 237
```

◀

EXAMPLE 11 Find the product 9,371 and 407.

Solution

```
     24
   9371
 ×  407
  65597
```

The word **product** indicates multiplication. Line up the digits that occupy the ones place. Multiply by 7 first.

When you multiply by 0, the result is always 0. You can indicate this product in two different ways. You can insert a whole row of zeros, or you can just put a zero in the second column and then multiply by the 4, beginning your partial product in the *third* column. Both ways are shown.

```
     12            12
     24            24
   9371          9371
 ×  407        ×  407
  65597         65597
  0000         374840
37484         3813997
3813997
```

9,371 × 407 = 3,813,997

◀

EXAMPLE 12 Find the product of 48 and 206.

Solution You could estimate your answer by finding the product of 50 and 200. Multiply 5 by 2 and attach three zeros. The approximate answer is 10,000.

```
    2
    4
  206
 × 48
 1648
 824
 9888
```

The word **product** indicates multiplication. As the checking process in Examples 9 and 10 shows, you can line up the factors with either number on top, but it is easier to multiply twice instead of three times. Remember, 8 × 0 = 0.

48 × 206 = 9,888

◀

EXAMPLE 13 Find the product of 2,615 and 3,009.

Solution You could estimate the answer to be 7,800,000 by multiplying 2,600 by 3,000. In this problem, you need to be careful to place the partial products correctly when multiplying by the zeros. Again, two methods are shown.

```
   1 1                 1 1
   514                 514
  2615                2615
× 3009              × 3009
 23535               23535
 0000               784500
0000               7868535
7845
7868535
```

$2,615 \times 3,009 = 7,868,535$

EXAMPLE 14 From the sum of 1,289 and 3,046 subtract 986.

Solution This problem requires two steps. First you need to add the numbers 1,289 and 3,046 and then subtract 986 from that answer.

```
  1,289        4,335
+ 3,046      −   986
  4,335        3,349
```

The answer is 3,349.

EXAMPLE 15 Marc has a $287 car payment each month. Explain, in words, how he could estimate how much money he needs to include in his annual budget for this year's car payments.

Solution Since we are asked for an *annual* amount we need to include 12 monthly payments. For estimation purposes, we could use $300 for the monthly payment amount. To estimate how much Marc will need this year we could multiply 12 months by $300 and estimate $3,600 for this year's payments.

1.2 EXERCISES

A. Perform the indicated operations. Whenever practical, check your answers by the methods suggested in this section.

```
1.   245        2.   372
   + 566              29
                   + 495

3.    59        4.   614
     178           +  92
   +  35

5.  3,125       6.   801
  −   256          −  45

7.    776       8.  1,254
  −   125         −   363

9.    856      10.   579
  ×    37          ×  58

11. 1,892      12.   297
  ×    92          ×  46
```

13. 725 − 88

14. 8,921 + 726

15. 149 × 8

16. 497 × 6

17. $$\begin{array}{r} 20{,}007 \\ -\ 8{,}564 \\ \hline \end{array}$$

18. $$\begin{array}{r} 4{,}098 \\ \times\ 306 \\ \hline \end{array}$$

19. $$\begin{array}{r} 6{,}078 \\ 847 \\ +\ 4{,}651 \\ \hline \end{array}$$

20. $$\begin{array}{r} 2{,}056 \\ -1{,}778 \\ \hline \end{array}$$

21. $$\begin{array}{r} 4{,}003 \\ -\ \ 875 \\ \hline \end{array}$$

22. $$\begin{array}{r} 486 \\ \times\ \ 39 \\ \hline \end{array}$$

23. $$\begin{array}{r} 1{,}756 \\ 326 \\ +\ \ 28 \\ \hline \end{array}$$

24. $$\begin{array}{r} 895 \\ 3{,}046 \\ +\ \ 57 \\ \hline \end{array}$$

25. $$\begin{array}{r} 7{,}631 \\ -\ \ 987 \\ \hline \end{array}$$

26. $$\begin{array}{r} 15{,}080 \\ -8{,}795 \\ \hline \end{array}$$

27. $$\begin{array}{r} 348 \\ 49 \\ +\ 806 \\ \hline \end{array}$$

28. $$\begin{array}{r} 1{,}006 \\ \times\ 278 \\ \hline \end{array}$$

29. Find the difference between 1,587 and 849.

30. What is the product of 2,016 and 83?

31. Find the total of 2,985, 384, and 1,912.

32. From the sum of 3,548 and 2,095 subtract 3,296.

33. Multiply the sum of 1,483 and 982 by 14.

34. From the product of 48 and 32 subtract 893.

35. Find the difference between 54,008 and 27,519.

36. Find the product of 431; 5; and 22.

37. Subtract 578 from 2,113.

38. Find the difference between 62,000 and 8,747.

39. Find the product of 387 and 78.

40. Subtract 5,376 from 10,002.

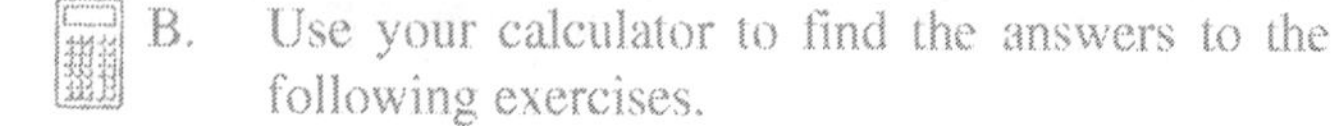

B. Use your calculator to find the answers to the following exercises.

41. $$\begin{array}{r} 204{,}965 \\ 371{,}508 \\ +\ 734{,}918 \\ \hline \end{array}$$

42. $$\begin{array}{r} 16{,}243{,}002 \\ -\ \ 9{,}875{,}799 \\ \hline \end{array}$$

43. 17,395 × 2,645

44. 20,034 × 857

45. $$\begin{array}{r} 40{,}000{,}031 \\ -\ 29{,}998{,}754 \\ \hline \end{array}$$

46. $$\begin{array}{r} 1{,}836{,}752 \\ 39{,}098 \\ +\ \ 157{,}489 \\ \hline \end{array}$$

47. Find the sum of 17,384 and 209,195.

48. Find the difference between 198,000 and 99,493.

49. Find the product of 16,453 and 975.

50. From the product of 2,493 and 26,548 subtract 43,821.

C. Read and respond to the following exercises. In each one you are asked to explain, in words, how you would find the estimate. You are not asked to do the calculations.

51. Galia manages a 28-unit apartment complex. Each unit rents for $415 per month. Describe how you would calculate how much rent she should collect each month.

52. Jeremy is in charge of buying the fuel for his company's fleet of small planes. Each plane uses approximately 10,003 gallons of fuel in a month's time. How could he estimate how much fuel one plane will use in a year?

53. Galia is paid $25 per month per unit for managing the complex. (See Exercise 51.) How would she estimate her monthly income as manager?

54. If Jeremy's company (see Exercise 52) has 7 small planes, what process could he use to decide how much fuel he should plan to purchase each month?

55. Celeste charges $12 per day for a single child in her day care and $20 per day for two children in the same family. How would you find the amount Celeste would charge each day when caring for the children of 3 families with one child each and 2 families with two children each?

56. Eduardo's electric bills for the summer months were $56 in June, $83 in July, and $78 in August. Explain how you would find the amount of decrease from July's bill to August's bill.

57. If Eduardo's average monthly electric bill for the year was $44, how would you find his average annual cost for electricity?

58. Explain how to multiply 30 by 700 without using long multiplication on paper and without using a calculator.

59. The Golden Trippers Club is considering a trip to Gatlinburg, Tennessee. The bus fare will be $137 per person, and the lodging costs for two nights will be $91 per person. Explain how you could estimate the approximate cost per person for transportation and lodging.

60. Describe, in detail, the process for borrowing in a whole number subtraction problem.

61. Find the product of 123,456 and 789,012 using your calculator. What does the display say? How do you explain what is happening? What might you do to find this product?

62. Find the sum of 23,456,789,015 and 86,472,389,158 using your calculator. What does the display say? Describe what you might try to do to find the sum. Compare your method with a classmate's method.

1.3 Whole Number Division

Division can be indicated in one of four ways:

$6\overline{)4{,}521}$ $\quad 4{,}521 \div 6 \quad \dfrac{4{,}521}{6}$ "Find the quotient of 4,521 divided by 6."

The number that does the dividing is called the **divisor**. The number into which the divisor goes is called the **dividend**. The answer that you get when you divide is called the **quotient.** If the division does not come out evenly, there is a **remainder**.

$\text{divisor}\overline{)\text{dividend}}$ with quotient remainder above $\quad$ dividend ÷ divisor $\quad \dfrac{\text{dividend}}{\text{divisor}}$

Dividing by a single digit is the easiest to do. Carefully studying one of these example problems will serve as a review of the basics of division.

EXAMPLE 1 Divide 4,521 by 6.

Solution You could estimate an answer by dividing 4,800 by 6, because 6 divides easily into 48 and then you attach two zeros. The actual answer should be close to 800.

```
  753
6)4521     6 will not go into 4, but 6 will go into 45.
 -42       6 × 7 = 42. Place the 7 over the 5.
   32      45 – 42 = 3. Bring down the next digit, 2.
  -30      6 × 5 = 30. Place the 5 over the 2.
    21     32 – 30 = 2. Bring down the next digit, 1.
   -18     6 × 3 =18. Place the 3 over the 1.
     3
```

Caution: If you are using a calculator to do this problem, the display will show 753.5 because calculators do not display whole number remainders. Read Appendix B to learn more about calculator division.

There are no more digits, so there is a remainder of 3. Whenever there is a remainder, it must be written as part of the quotient. Otherwise, someone looking at the answer might think that the number divided exactly (with a zero remainder).

4,521 ÷ 6 = 753 R3

The actual answer is, in fact, close to the estimate, 800.

Check To check the division, multiply 753 by 6 and add 3.

753 × 6 = 4,518 4,518 + 3 = 4,521 √

EXAMPLE 2 Find the quotient of 3,652 divided by 8.

Solution

```
   456 R4
8)3652
 -32
   45
  -40
    52
   -48
     4
```

Check

```
  456
×   8
 3648
+   4
 3652 √
```

EXAMPLE 3 Divide 21, 341 by 7.

Solution To estimate an answer divide 21,000 by 7. Twenty-one divided by 7 is 3 and we need to attach three zeros; 3,000.

```
   3048 R5
7)21341
 -21
   03
  -00
   34
  -28
    61
   -56
     5
```

This problem has a new situation. When you bring the 3 down, the divisor, 7, will not divide into 3. Before you continue, show that 03 ÷ 7 has no natural number answer by placing a zero in the quotient above the 3, multiplying 0 by 7, and writing down the product, 00. Subtract, getting 3. Then you can bring down the next digit, 4, continue to divide 7 into 34, and so on. When you have run out of numbers in the dividend in a whole number division problem, you are finished. If the final remainder is not zero, your answer must include that remainder.

Check

```
  3048
×    7
 21336
+    5
 21341 √
```

Dividing by a number with more than one digit follows the same pattern as dividing by a single digit, but the calculations are longer. Try to do the following example on your own paper first. Then follow the worked-out steps and see if your process agrees with these steps.

EXAMPLE 4 Divide 3,815 by 26.

Solution You could estimate an answer by dividing 3,900 by 30. This shows that the answer in the original problem should be close to 130.

```
     1
26)3815
  -26
   121
```

26 goes into 38 one time. Write 1 over the 8.
1 × 26 = 26
38 – 26 = 12. 12 is less than 26. This step is extremely important. Each time you subtract, be sure to check to see that *the number you get is less than the divisor*. If it is not, erase the number in the quotient and try a larger number. 12 is less than 26. OK. Bring down the next digit, 1.

```
    14
26)3815
  -26
   121
  -104
    17
```

You can guess that 26 goes into 121 four or five times because 30 × 4 = 120 and 30 × 5 = 150.
Try 5. 5 × 26 = 130. Too big.
Try 4. 4 × 26 = 104. 121 – 104 = 17.
17 is less than 26. OK.

```
    146 R19
26)3815
  -26
   121
  -104
    175
   -156
     19
```

Bring down the 5.
You know that 6 × 30 = 180, which is close to 175, so try 6 × 26 = 156. Place 6 above the 5 and see if, when you subtract 175 – 156, your remainder is less than the divisor, 26. 175 – 156 = 19. OK. Write the remainder in the quotient. The answer, 146 R19, is close to the estimate of 130.

Check

```
   146
 ×  26
   876
  292
  3796
 +  19
  3815 √
```

Check the exact answer by multiplying. Remember to add on the remainder, 19.

Therefore, 3,815 divided by 26 equals 146 R19.

EXAMPLE 5 Find the quotient of 6,204 divided by 73.

Solution Estimate an answer: 6,300 ÷ 70 = 90

```
      8
73)6204
  -584
    36
```

The word **quotient** indicates division. 73 will not go into 6. 73 will not go into 62. But 73 will go into 620.
70 × 8 = 560, so we can try 8.
8 × 73 = 584, which is less than 620.
620 – 584 = 36. 36 is less than 73. OK.

```
     84  R72
73)6204        Bring down the 4.
  -584         70 × 5 = 350, so we can try 5.
    364        73 × 5 = 365. 365 is bigger than 364.
   -292        73 × 4 = 292. 364 – 292 = 72.
     72        72 is less than 73. OK.
```

Write the remainder in the answer. The estimate, 90, is close, so 84 R72 is a reasonable answer to the problem.

Check

```
    84         Check by multiplication. Add the remainder.
  × 73
   252
  588
  6132
 +  72
  6204 √       Therefore, the quotient of 6,204 divided by 73 is 84 R72.
```

EXAMPLE 6 A software conversion project will require 1,440 labor hours to complete. Describe, in words, how you would calculate the number of hours to be assigned to each member of the 6-person conversion team.

Solution We know the total number of hours but we need to split the project up among 6 people. We need to find *each* person's share. We will divide 1,440 hours by 6 people to find the *hours per person.*

As you do the four exercises that follow, try estimating the answer before you divide. Check your answers against those given below the exercises. If you have made an error, see if you can find it and correct it. It may help to know that the four most common errors in long division problems are

1. Placing the first answer over the wrong digit
2. Subtracting incorrectly
3. Not spotting a remainder that is larger than the divisor (correcting this error requires that you increase the digit in the quotient and multiply again; be sure to check your remainder each time)
4. Not placing a zero in the quotient *once you have brought a digit down* and realized that the divisor is larger than the number created

HINT: When you do division problems on a test, it always a good idea to check your answer by multiplying the divisor by the quotient and adding if there is a remainder.

Self-Check 1. 9,876 ÷ 42 2. 56)61,152 3. $\frac{68,452}{18}$

4. Find the quotient of 147,729 divided by 68.

Answers:

1. 235 R6 2. 1,092 3. 3,802 R16 4. 2,172 R33

What about division and zero? What happens there? Look at the following examples of some division problems that involve zero. See if you can figure out any rules for division with zero.

$6\overline{)0}$? $35\overline{)0}$? $114\overline{)0}$?

Remember, to check division, you multiply the answer times the divisor. These division problems might be worded another way:

"What number times 6 = 0?"

"What number times 35 = 0?"

"What number times 114 = 0?"

The answer to all these questions is zero: 6 × 0 = 0, 35 × 0 = 0, 114 × 0 = 0. Therefore, *whenever you divide zero by any non-zero number, you get zero*. This is always true.

What happens when you divide *by zero?* Look at these examples:

$0\overline{)7}$? $0\overline{)42}$? $0\overline{)215}$?

Put another way,

"What number times 0 equals 7?"

"What number times 0 equals 42?"

"What number times 0 equals 215?"

The answer is that you cannot multiply 0 by any number and get anything but zero. There is no number that will answer these questions. Division by zero cannot be done. It is **undefined**. When you encounter a problem that involves division by zero, your answer should be "Undefined" or "No Solution." This is because division by zero has no answer.

1.3 EXERCISES

A. Divide as indicated in each of the following exercises.

1. 2,136 ÷ 8
2. 3,213 ÷ 7
3. 4,032 ÷ 6
4. 17,810 ÷ 5
5. 42,728 ÷ 7
6. 74,754 ÷ 9
7. 8,246 ÷ 38
8. 15,687 ÷ 27
9. 26,187 ÷ 43
10. 1,435 ÷ 8
11. 4,735 ÷ 9
12. 6,381 ÷ 7
13. $\frac{63,015}{61}$
14. $\frac{82,196}{34}$
15. $\frac{77,458}{29}$
16. $62\overline{)24,159}$
17. $77\overline{)53,215}$
18. $36\overline{)11,614}$
19. Find the quotient of 116,248 divided by 287.
20. Find the quotient of 320,064 divided by 215.
21. What is the result when 3,458 is divided by 88?
22. What is the result when 7,083 is divided by 45?
23. Divide 5,092 by 41.
24. Divide 8,176 by 64.
25. Find the quotient of 76,368 divided by 37.
26. What is the result when 51,815 is divided by 43?
27. Divide 8,020 by 26.
28. Find the quotient when 65,060 is divided by 13.

B. Use your calculator to find the answers to the following exercises.

29. 72,430,140 ÷ 20,036
30. 6,368,670 ÷ 16,542
31. $\frac{41,382,302}{2,351}$

32. $\frac{39{,}135{,}200}{5{,}680}$

33. $7{,}895\overline{)15{,}813{,}685}$

34. $3{,}006\overline{)26{,}975{,}844}$

35. Find the quotient of 18,293,658 divided by 5,001.

36. What is the result when 2,118,825 is divided by 3,225?

37. What is the result when 2,950,992 is divided by 8,019?

38. Find the quotient of 48,650,625 divided by 6,975.

39. Divide 1,056,784 by 1,028.

40. Divide 17,538,458 by 8,743.

C. Read and respond to the following exercises.

41. The weekly fuel bill for a fleet of 8 delivery vans was $368. How would you find the cost of fuel per van?

42. A 76-hour project is to be divided among four workers. How would you find how many hours each person is expected to spend working on this project?

43. A group of 7 coworkers held a winning lottery ticket. Describe how you could calculate each person's share if the ticket was worth $89,635.

44. The liability insurance premium for a 28-unit apartment complex is $350 each month. What would the manager have to do to calculate each unit's share of this bill?

45. Explain how to check a division problem that has a remainder in the quotient.

46. Explain why 173 ÷ 0 has no answer.

47. Explain what must be done when the remainder in a division problem is greater than the divisor.

48. How might you estimate the population of the United States if you read that each person's share of the $4,988,882,589,000 national debt is approximately $19,375?

49. The local income tax collected in Centerburg last year was $404,010. Estimate how much each of Centerburg's 1,206 working citizens paid in local tax if everyone paid the same amount.

    a. Tell what values you could use to estimate this answer, and explain why you chose those values.

    b. Calculate the answer using the estimated values.

    c. Compare your estimated answer to those of your classmates. Calculate the actual answer. Did different methods of estimation produce answers closer to the actual answer?

50. First use your calculator to find the answer to each of the following exercises. Then, explain, in words, how you would prove that your answers for parts a through d are correct.

    a. 1,320 ÷ 24

    b. 1,205 ÷ 5

    c. 280 ÷ 7

    d. 0 ÷ 18

    e. 27 ÷ 0

1.4 Applications

There is probably no more dreaded term in a mathematics classroom than "applications" or "word problems." Students, as a whole, dislike them and are uncomfortable trying to solve them. This really is strange because application problems are much more practical than plain calculation problems. They are the real-world part of math for most people. What probably makes them so feared is that they force students to combine two kinds of thinking. You need to *read and interpret* the words and then apply the *mathematical operation* necessary to reach a solution. Having to do two things in one problem hardly seems fair, does it? But once you understand the cause of your dislike and discomfort, application problems will be easier for you to do.

This is the approach that you should follow when faced with an application.

Suggested Approach to Application Problems

1. *Read* the problem carefully once to see what is going on in the problem. What is it about?
2. *Read* it very carefully again to see if you can determine which mathematical operation you should use. Sometimes there are key words or phrases that will help you to decide.
3. *Estimate* what your answer should be. You just need to be close. What kind of an answer would make sense?
4. *Perform* the mathematical operations(s) carefully, making sure you have copied the numbers correctly.
5. *Answer* the question that has been asked, and be sure to label your answer. Because word problems have a context, their answers are not just numbers, but numbers that mean something—for example, 12 feet, 57 people , $42.15, and so on.
6. *Reread* the problem, putting your answer in to see if it does, in fact, makes sense. Is the exact answer reasonably close to the estimate?

To help determine which mathematical operation to use, study the following list of key words. These (and many other words) can tell you when to add, subtract, multiply, or divide.

Addition	**Subtraction**	**Multiplication**	**Division**
sum	difference	product	quotient
total	less	times	per
altogether	fewer	part of	split
increase	decrease	twice	each
more	remain	double	shared
in all	larger	area	average
combined	-er words	total	ratio

Sometimes, getting the basic idea of what is happening in a problem will help you decide what to do. Sometimes a key word or idea will stand out and get you started. Remember, there are only four basic operations. If you are really stumped, estimate a reasonable answer and try all four operations. See which answer comes closest to your estimate and makes the most sense.

The best way to become successful with applications is to do lots of them! The more you practice and follow the patterns, the more successful you become. Practice really helps. The first time you went into the deep end of a swimming pool, the result was probably not too graceful. Once you had gone swimming a few more times, you became more comfortable in the water, and your strokes became smoother and more effective. This can happen with applications too, but it takes effort and persistence on your part. If you have been working a problem, and it just doesn't make sense to you, skip it and try another one. Spending a lot of time stewing over one or two problems does not help you learn. It only makes you frustrated. And sometimes if you leave a confusing problem and return to it later, you find you are suddenly able to solve it.

In each of the following examples, the six steps listed at the beginning of this section are given in detail. You should try to follow these steps each time you do a word problem. You do not always have to write everything down, but the pattern should be there in your thinking.

General Applications

EXAMPLE 1 Margot weighed 153 pounds on March 15 and weighed 129 pounds on September 15. How much weight had she lost in those 6 months?

Steps 1 and 2: Read to see what is happening. Though there is no key word, you are asked to find the difference between two quantities. Subtract.

Solution For estimating purposes, use 150 and 130: 150 – 130 = 20

Step 3: You could use 155 and 130 just as well to get close to the answer.

153 – 129 = 24

Step 4: Subtract and borrow carefully.

24 pounds

Step 5: Be sure to label the answer as pounds.

Margot lost 24 pounds in that 6-month period of time.

Step 6: 24 pounds seems reasonable and it is close to the estimate of 20 pounds.

◀

EXAMPLE 2 How far can you drive on your first day of vacation if you plan to drive for 8 hours and can average 55 miles per hour?

Step 1 and 2: You will need to know the number of hours and the rate per hour to find the distance by multiplying.

Solution 8 hours × 60 mph = 480 miles

Step 3: For estimation, it is easier to multiply by a number ending in zero.

8 × 55 = 440

Step 4: Multiply carefully.

440 miles

Step 5: Distance will be measured in miles because the speed is in miles per hour.

You can plan to travel 440 miles during the first 8-hour day.

Step 6: This seems possible, and it is close to the estimate of 480 miles.

◀

EXAMPLE 3 How many pounds of grass seed can you buy if the seed costs $2 per pound and you have $8 in your wallet?

Step 1 and 2: You have $8 in your wallet and you are going to split it up into $2 units. To find the number of parts, you will divide.

Solution $8 divided by $2 = 4

Step 3: Both quantities are easy to work with, so you can use them as they are. There is no need to estimate.

$$2\overline{)8}\quad 4$$

Step 4: Perform the division.

4 pounds

Step 5: Be sure to label.

You can buy 4 pounds of seed for $8 if it costs $2 per pound.

Step 6: This answer makes sense and agrees with the estimate.

◀

EXAMPLE 4 Jose and Maxine went to the store to buy the supplies that they needed for their son's school project. They bought 2 packages of pencils for 99 cents each, 2 packages of index cards for 59 cents each, a bag of pipe cleaners for 79 cents, and a box of tissues for 85 cents. How much did they spend?

Step 1 and 2: You need to find the total amount spent at the store, so you will add. Because they bought more than one package of the pencils and index cards, you multiply to find the cost of these items.

Solution

2 times 100 cents = 200 cents
2 times 60 cents = 120 cents
1 times 80 cents = 80 cents
1 times 90 cents = 90 cents
Approximate cost = 490 cents, or $4.90

Step 3: *Approximate* the item costs to the nearest dime, or even the nearest dollar or half dollar if you like, just to get an idea of the answer.

2×99 cents = 198 cents
2×59 cents = 118 cents
1 at 79 cents = 79 cents
1 at 85 cents = 85 cents
Total = 480 cents, or $4.80

Step 4: Perform the multiplication with the actual costs, and then add to find the total.

They spent $4.80 for the school items at the store.

Step 5: Give the answer to the problem, labeling it as money.

Considering the amount of each item and the fact that they bought only 6 items, the total seems to make sense.

Step 6: Reading the problem again, $4.80 seems like a reasonable answer and it is close to the estimate, $4.90

◀

You certainly do not need to write out this many steps to do a simple application problem. That would be ridiculous. But you should be going through the steps mentally each time, particularly the estimating and checking steps that enable you to tell if your answer is reasonable. If it does not come out the way you think it should (or if it does not agree with the answer key), check your calculations first, or use a calculator, and then decide if perhaps you used the wrong operation.

Geometry Applications

You may be wondering what this section about geometry is doing in the middle of an arithmetic review and beginning algebra text. That's a good question and one that has several answers. But before we answer it, you need to understand what is usually included in a geometry course.

There are two main functions of a beginning geometry course. The first is to acquaint students with the basic figures, definitions, and relationships that are part of plane geometry. The second and most important aspect of such a course is the development of formal proof. This is the backbone of any complete geometry course, and it is one of the means used to develop critical thinking in mathematics students.

The definitions and relationships in geometry are the practical nuts and bolts used in applications; they are also the concepts that appear on many standardized tests. Another reason geometry is included in basic math courses is that many students are visual learners, and applying arithmetic processes to pictures can help these students better understand the math concepts. In the geometry sections, you will study definitions and relationships and work with practical situations.

Because there are a number of definitions and formulas in the geometry sections, you should ask your instructor which, if any, you need to memorize.

Lines and Angles

To begin the study of geometry, we must consider the most basic of all geometric quantities, a point. From this simple concept, we can build everything that we need in practical geometry. If you were to take a straight pin and poke it through a piece of paper, making a hole in the paper, you would have produced a representation of a point. There are other, more scientific ways to describe a point, but you get the idea. A **point** is represented by a single dot or period. If you were to put a number of points next to each other in a row, stretching out from that first point in exactly opposite directions, and if you continued that process indefinitely, you would have drawn a representation of a straight line. A **straight line** looks like a piece of string stretched out indefinitely with no bends or curves. If you were to place a number of straight lines right next to each other on a flat surface and if you extended that process indefinitely, you would have a representation of a surface called a plane. A **plane** looks like a flat surface made up of points, extending indefinitely in all directions, with no spaces in between.

This indefinite and infinite language makes these concepts difficult to understand. Point, line, and plane are in fact *undefined* terms, so you need to have an idea in your mind of what they look like. Lines, unless otherwise described, are straight, and planes are flat surfaces (see Figure 1.4)

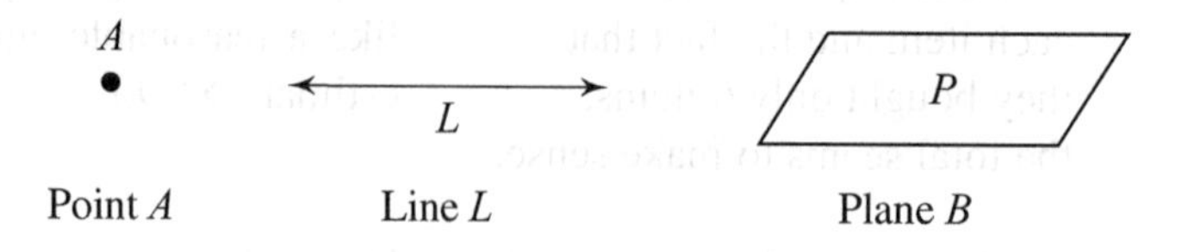

FIGURE 1.4

Try thinking of the paper you write on as a plane; the pencil you use can present the line, and the pencil tip is a point. Remember, however, that lines and planes extend forever. They don't really have edges or ends.

Two lines in a plane **intersect** (meet at some point), or they are **parallel** (never meet and always are the same distance apart), or they **coincide** (one lies on top of the other). The rails in a straight stretch of railroad track are parallel. The distance between the rails is always the same, and they never intersect. Two streets that cross one another on a map intersect at that point. The notation for parallel lines is $L_1 \parallel L_2$, and the notation for intersecting lines is $L_1 \cap L_2$.

$L_1 \parallel L_2$ says that line 1 is parallel to line 2.

$L_1 \cap L_2$ says that lines 1 and 2 intersect at some point.

A section of line is called a **line segment** (Figure 1.5). It has two endpoints. A section of a line that has one endpoint and extends forever in the other direction is called a **ray**. Note that the endpoint of a ray is named first. Be sure to use the arrow on one end of the ray to show that it continues.

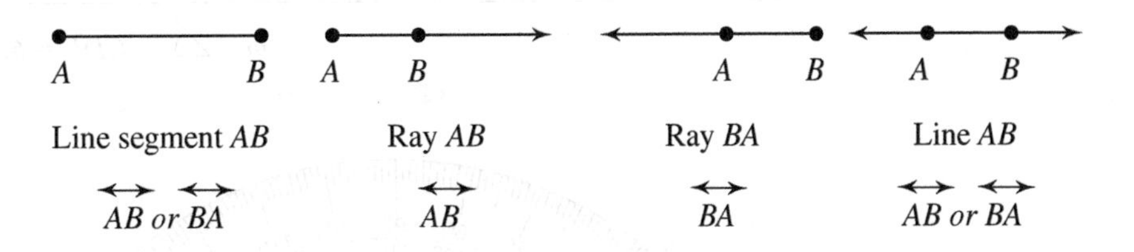

FIGURE 1.5

When you connect three or more line segments, you make geometric figures. We will examine some of those in Section 2.3. When you join two rays at their endpoints you have formed an **angle.** ∡ is the symbol used to name an angle. Figure 1.6 shows an angle. In naming an angle with three letters, be sure to indicate the **vertex**, or the corner of the angle, as the middle letter. You can also use the single vertex letter to name the angle when there is no possibility of it being confused with another angle. Remember that the sides of an angle are rays that extend forever.

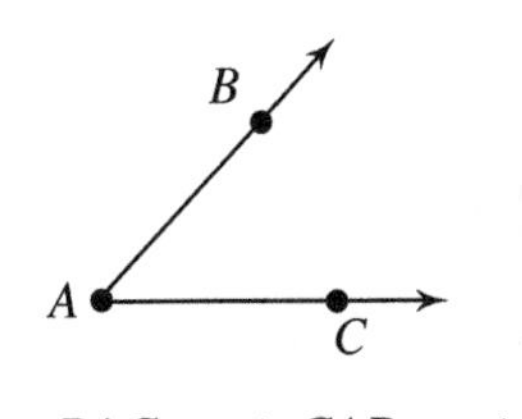

∡ *BAC* or ∡ *CAB* or ∡ *A*
FIGURE 1.6

Before we define different kinds of angles, you need to know how they are measured. The instrument used to measure angles is called a **protractor** (Figure 1.7). For an angle to be measured on a protractor, its sides must be extended out past the scale marks on the protractor. Make sure that the center of the protractor is on the vertex of the angle and that one side of the angle is on the 0° (zero degree) mark of the protractor. Figure 1.8 shows that ∡*CDE* has a measurement of 62°. This could be written m ∡ *CDE* = 62°. Figure 1.9 shows how to measure ∡ *XYZ*.

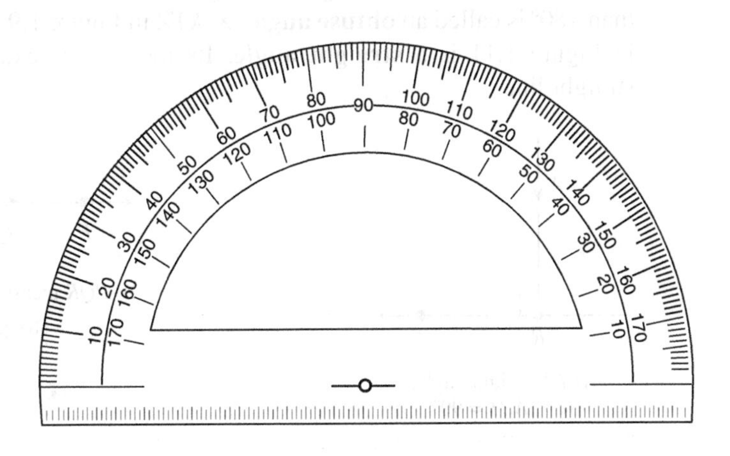

FIGURE 1.7

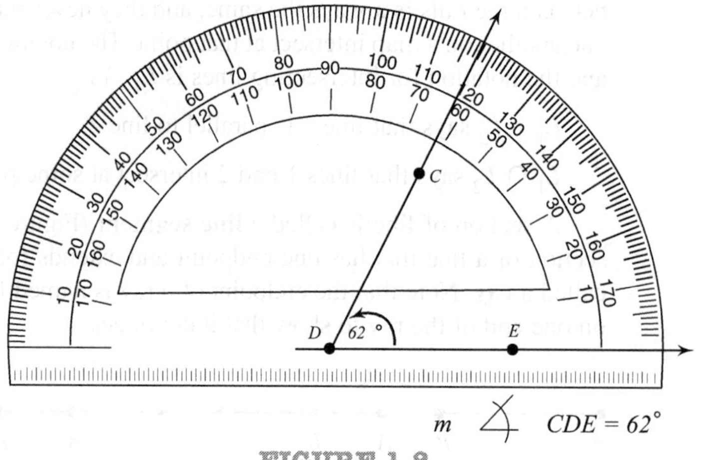

FIGURE 1.8

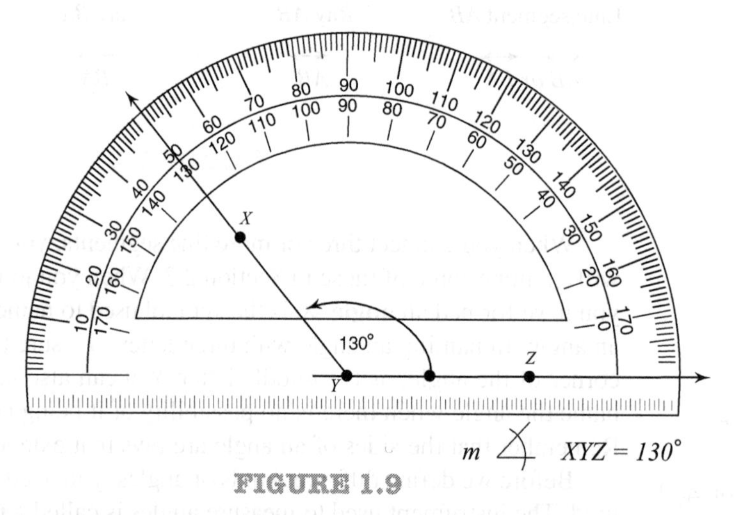

FIGURE 1.9

There are several kinds of angles that you need to learn about and be able to recognize. The most widely known angle is a **right angle**. Its sides form a square corner. They are said to be **perpendicular** (⊥), and the measurement of a right angle is 90º. (See Figure 1.10, where the symbol ⌉ inside the angle shows that it is a right angle.)

Angles that are greater than 0º but less than 90º are called **acute angles**. ∡ *CDE* in Figure 1.8 is an acute angle. An angle whose measure is greater than 90º but less than 180º is called an **obtuse angle**. ∡ *XYZ* in Figure 1.9 is an obtuse angle. ∡ *PQR* in Figure 1.11 is a **straight angle**. Its measure is exactly 180º, and it forms a straight line.

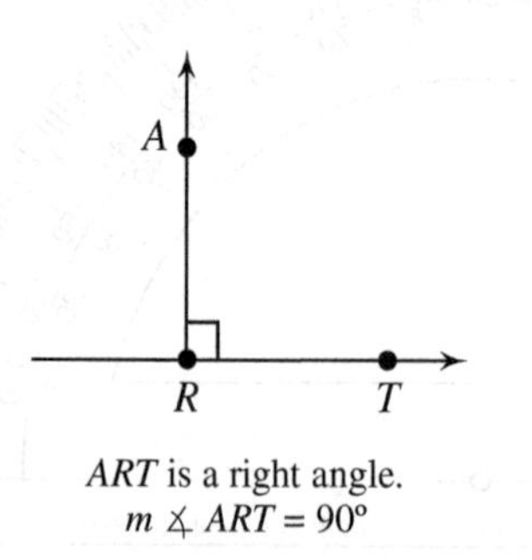

ART is a right angle.
m ∡ *ART* = 90º

FIGURE 1.10

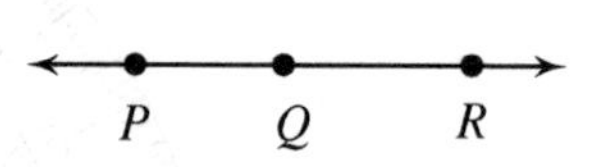

∡ *PQR* measures 180º and is a straight angle.

FIGURE 1.11

Angles whose measures are equal are said to be **congruent.** $\cong$ is the symbol for congruence. If $m \measuredangle DEF = 82^\circ$ and $m \measuredangle KLM = 82^\circ$, then $\measuredangle DEF \cong \measuredangle KLN$. Their sides could seem to be of different lengths because they are rays and extend indefinitely. However, if the measures of the angles between the rays are the same, the angles are said to be congruent. In Figure 1.12, $m \measuredangle KLM = 130^\circ$ and $m \measuredangle PLN = 130^\circ$. Because the two angles have the same measure, $\measuredangle KLM \cong \measuredangle PLN$.

There are a few other special relationships between angles that you should know. Two angles are said to be **complementary** if their sum is 90°. Two angles are said to be **supplementary** if their sum is 180°. Angle measures can be calculated without a protractor or diagram if certain relationships are given.

If $m \measuredangle ABC = 43^\circ$ and $m \measuredangle RST = 47^\circ$, $\measuredangle ABC$ and $\measuredangle RST$ are complementary angles because $43^\circ + 47^\circ = 90^\circ$.

If $m \measuredangle ABC = 43^\circ$ and $m \measuredangle GHI = 137^\circ$, $\measuredangle ABC$ and $\measuredangle GHI$ are supplementary angles because $43^\circ + 137^\circ = 180^\circ$.

EXAMPLE 5 Angles U and V are complementary. The $m \measuredangle V$ is 38°. Find the measure of $\measuredangle U$.

Solution Because angles U and V are complementary, their sum must be 90°. If the measure of $\measuredangle V$ is 38°, the measure of $\measuredangle U$ must be $90^\circ - m \measuredangle V$, or $m \measuredangle U = 90^\circ - 38^\circ = 52^\circ$.

The opposite angles formed by two intersecting lines are called **vertical** angles. Two angles that share a common side and vertex are called **adjacent** angles. Figure 1.12 shows that $\measuredangle KLM$ and $\measuredangle PLN$ are vertical angles. (Can you pick out and name the other pair of vertical angles?) These intersecting lines form four pairs of adjacent angles. $\measuredangle KLM$ and $\measuredangle MLN$ are adjacent angles. $\measuredangle KLM$ is also adjacent to $\measuredangle KLP$. What do you know about the sum of the measures of $\measuredangle KLM$ and $\measuredangle MLN$?

EXAMPLE 6 If $m \measuredangle KLM$ in Figure 1.12 is 130°, find the measure of

a. $\measuredangle MLN$ b. $\measuredangle NLP$ c. $\measuredangle PLK$

Solution

a. $\measuredangle KLM$ and $\measuredangle MLN$ are adjacent angles, and $\measuredangle KLN$ is a straight angle because $\measuredangle KLN$ is a straight line. Therefore,

$m \measuredangle KLM + m \measuredangle MLN = 180^\circ$ Definition of straight angle.

$130^\circ + m \measuredangle MLN = 180^\circ$ Substitute; $m \measuredangle KLM = 130^\circ$.

$m \measuredangle MLN = 50^\circ$ Subtract 130° from 180°.

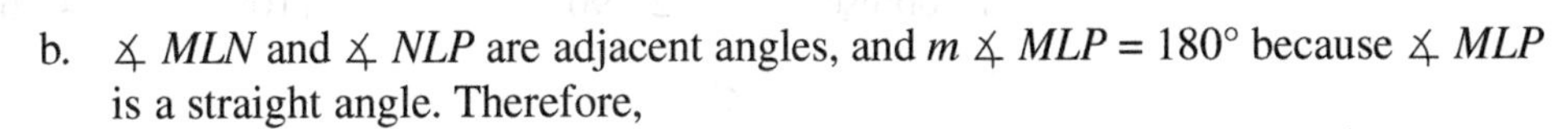

b. $\measuredangle MLN$ and $\measuredangle NLP$ are adjacent angles, and $m \measuredangle MLP = 180^\circ$ because $\measuredangle MLP$ is a straight angle. Therefore,

$m \measuredangle MLN + m \measuredangle NLP = 180^\circ$ Definition of straight angle.

$50^\circ + m \measuredangle NLP = 180^\circ$ From part (a).

$m \measuredangle NLP = 130^\circ$

FIGURE 1.12

c. $\measuredangle NLP$ and $\measuredangle PLK$ are adjacent angles, and $\measuredangle NLK$ is a straight angle with a measure of 180°. Therefore,

$m \measuredangle NLP + m \measuredangle PLK = 180^\circ$ Definition of straight angle.

$130^\circ + m \measuredangle PLK = 180^\circ$ From part (b).

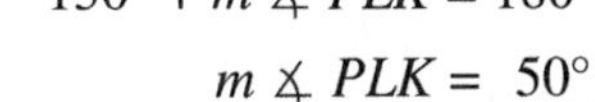

$m \measuredangle PLK = 50^\circ$

From Figure 1.12, you can see that the intersection of two lines produces four angles. There are four pairs of adjacent, supplementary angles. There are two pairs of vertical angles, and those vertical angles are congruent because their measures are equal: $m \measuredangle KLM = m \measuredangle NLP$, and $m \measuredangle MLN = m \measuredangle KLP$. This is always true and can be proved using formal geometric proofs that are similar to the way these facts were proved in Example 6.

EXAMPLE 7 Lines $\overleftrightarrow{AB}$ and $\overleftrightarrow{CD}$ intersect at point E. $\overleftrightarrow{AB}$ is perpendicular ($\perp$) to $\overleftrightarrow{CD}$.

a. Sketch a figure for this problem.

b. Find the measure of $\measuredangle AEC$.

c. Find the measure of $\measuredangle CEB$.

d. Find the sum of the measures of $\measuredangle CEA$ and $\measuredangle DEA$.

e. Angles CEA and DEA are _____, _____, and _____.

Solution

a. Figure 1.13 shows a possible sketch for this problem. Because $\overleftrightarrow{AB}$ and $\overleftrightarrow{CD}$ are perpendicular, they must intersect in a right angle (make a square corner).

b. Because $\overleftrightarrow{AB}$ and $\overleftrightarrow{CD}$ are $\perp$, the angles at their intersection are right angles and $\measuredangle AEC$ is a right angle. The measure of $\measuredangle AEC = 90°$.

c. The measure of $\measuredangle CEB$ is also 90°.

d. $m \measuredangle CEA + m \measuredangle DEA = 90° + 90° = 180°$.

e. Angles CEA and DEA are congruent, adjacent, and supplementary. The measures of angles CEA and DEA are equal.

Self-Check

1. Because the measure of $\measuredangle PQR$ is 102°, $\measuredangle PQR$ is a(n) _____ angle.

2. The measure of a right angle is _____ degrees.

3. $\measuredangle C$ and $\measuredangle D$ are supplementary angles. Find the measure of $\measuredangle D$ if the measure of $\measuredangle C$ equals 70°.

4. $\measuredangle RST$ is vertical to $\measuredangle USV$. Find the measure of $\measuredangle RST$ if the measure of $\measuredangle USV$ equals 63°.

Answers:

1. obtuse 2. 90 3. 110° 4. 63°

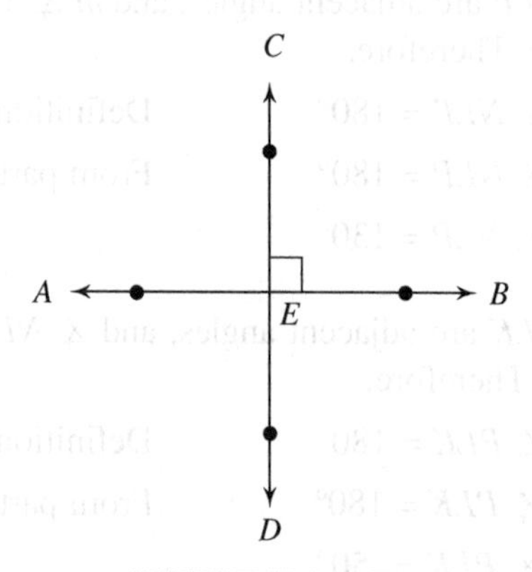

FIGURE 1.13

1.4 EXERCISES

A. Estimate an answer to each of the following exercises. Remember that when you estimate, the numbers should be reasonable to work with so you can use mental math.

1. Max and Yvette decided to open a restaurant. They figured that it would cost them $2,250 for rent, $890 for taxes, $1,200 for utilities, $1,950 for supplies, and $2,400 for salaries for the first 2 months of operation. How much money would they need to cover expenses for those first 2 months?
2. Bentley had some unexpected car expenses. His faithful old car developed some serious problems. He was told that the repair bill would be $126 for new brakes, $82 for new shocks, $26 for an alignment, $87 for a rebuilt carburetor, and $17 for a new headlight. The garage will charge $220 for labor for all of these repairs. What will his total bill be?
3. Don planted 67 acres of spring wheat and 39 acres of corn. How many more acres did he plant with wheat?
4. Suzanna and Luis figured that it would cost them $418 a month to continue to rent their apartment. They could own their own home for $527 a month. How much more would they have to pay each month if they bought a home?
5. Aunt Millie died and left her estate to her 4 nieces and nephews. The total amount left to her 4 heirs was $81,452. How much was each heir to receive?
6. The Sun Valley Travel Association is planning a trip for the Senior Citizens Club from Marshall City. Hotel expenses will be $285 per person, and airfare will cost each traveler $487. How much will the trip cost if 18 members decide to go?
7. At holiday time, the Hill family's home is decorated with many strands of colored lights. The Hills' normal electric rate is $1 per day, but it jumps to $3 per day with the extra lights. How much extra will the Hills pay for the 3-week holiday season (21 days) when they use more electricity?
8. John, the manager of Noah's Restaurant, orders 36 12-ounce packages of cream cheese to make cheesecakes for the week. If each cheesecake uses 8 ounces of cream cheese, how many cheesecakes can he make?

B. For Exercises 9 through 16, find the actual answer to each of Exercises 1 through 8. Compare your actual answer with the estimate.

9. See Exercise 1.
10. See Exercise 2.
11. See Exercise 3.
12. See Exercise 4.
13. See Exercise 5.
14. See Exercise 6.
15. See Exercise 7.
16. See Exercise 8.

C. Find the answer to each of the following exercises. You might estimate an answer first. Be sure to label your answers with the appropriate units.

17. A pair of glasses with a scratch-resistant surface costs $23 more than a pair without that surface. What will a pair of $78 glasses cost if you add the scratch-resistant surface?
18. Farmer McDonald has a herd of 93 cattle. If each animal eats 2 pounds of grain at each of 3 feedings per day, how many pounds of feed will he need per day for his herd?
19. Traveling along U.S. Highway 36, the distance from Hannibal, Missouri, to St. Joe, Missouri, is 189 miles. Continuing on, the distance from St. Joe to Norton, Kansas, is 275 miles. How far is it along U.S. 36 from Hannibal to Norton?
20. In Indianapolis, Indiana, the average temperature in June is 75 degrees. In January the average temperature is 28 degrees. On the average, how much colder is Indianapolis in January than in June?
21. In Exercise 19, how much farther is it from Norton to St. Joe than from Hannibal to St. Joe?
22. Cosmetic tint for glasses costs $19. If Marlene's new glasses cost $83 and she decides to add a tint, what will her total bill be?
23. Mr. Hutton's estate, valued at $13,486,835, is to be divided equally among his 5 heirs. How much money will each person inherit?

24. A Boeing 747 jet flew 2,348 miles on Monday, 1,456 miles on Tuesday, 1,089 miles on Wednesday, and 4,576 miles on Friday. How many total miles did it fly during those 4 days?
25. Your new car averages 19 miles per gallon. How far can you drive on 16 gallons of gas?
26. A flight from Winnipeg, Canada, to Honolulu is approximately 3,824 miles. If the flight takes 8 hours, how many miles does the plane fly in 1 hour?
27. Upholstery fabric for new sofa cushions will cost approximately $7 a yard. The trim for each cushion will be $6 a yard.Each cushion pattern calls for 3 yards of fabric and 2 yards of trim. What will it cost to make 4 of these large pillows?
28. The total bill for computer desks for the new lab is $4,296. If there are to be 24 workstations in the lab, how much does each desk cost?

D. For each of the following statements, fill in the missing word, words, or symbols.

29. Two angles whose sum is 180° are _____ .
30. An angle less than 90° is a(n) _____ angle.
31. A line consists of _____ points. (how many?)
32. A(n) _____ is a part of a line that has one end-point and extends indefinitely in the opposite direction.
33. Two angles whose sum is 90° are _____.
34. A(n) _____ is a part of a line that has two end-points.
35. An angle whose measure is between 90° and 180° is a(n) _____ angle.
36. Angles that share a common vertex and a common side are called _____.
37. Opposite angles formed by two intersecting lines are called _____ angles.
38. An angle of exactly 180° is a(n) _____ angle.
39. The device used to measure angles is called a(n) _____.
40. An angle of exactly 90° is a(n) _____ angle.
41. Angles are measured in units called _____.
42. $\parallel$ is the symbol that means the two lines are _____.
43. $\cap$ is the symbol that means that two lines are _____.
44. A plane has _____ edges. (how many?)
45. If $\measuredangle A$ and $\measuredangle B$ are complementary, and $m\measuredangle A = 29°$, what is the measure of $\measuredangle B$?
46. If $\measuredangle P$ and $\measuredangle R$ are supplementary angles, and the measure of $\measuredangle R = 97°$, what is the measure of $\measuredangle P$?
47. The measure of $\measuredangle K = 107°$. $\measuredangle K$ and $\measuredangle L$ are supplementary angles. What is the measure of $\measuredangle L$?
48. If the measure of $\measuredangle C$ is 43°, and $\measuredangle C$ and $\measuredangle D$ are complementary, find the measure of $\measuredangle D$.

E. Solve each of the following exercises using Figure 1.14

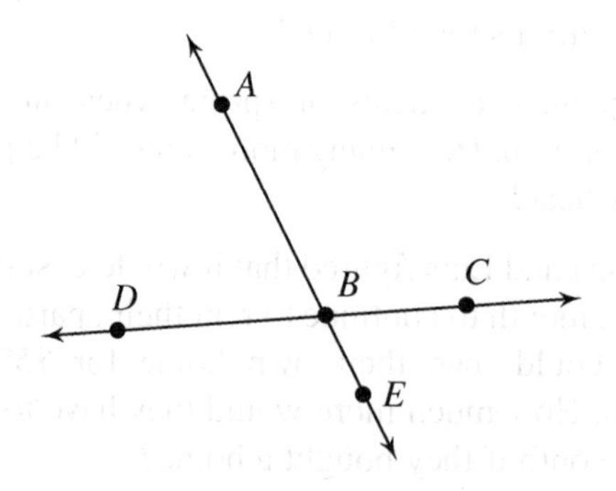

FIGURE 1.14

49. Find the measure of $\measuredangle DBE$ if $m\measuredangle ABC = 127°$.
50. Find the measure of $\measuredangle ABD$ if $m\measuredangle ABC = 127°$.
51. Find the measure of $\measuredangle DBE$ if $m\measuredangle CBE = 62°$.
52. Find the measure of $\measuredangle EBC$ if $m\measuredangle ABD = 43°$.

F. Draw two lines $\overleftrightarrow{PQ}$ and $\overleftrightarrow{RS}$ intersecting at point T. Let the measure of $\measuredangle PTR$ be 52°.

53. Find the measure of $\measuredangle STQ$.
54. Find the measure of $\measuredangle PTS$.
55. Find the measure of $\measuredangle PTQ$.
56. Find the measure of $\measuredangle RTQ$.
57. $\measuredangle PTS$ and $\measuredangle RTQ$ are _____ and _____.
58. $\measuredangle STQ$ and $\measuredangle QTR$ are _____ and _____.

G. Read and respond to the following exercises.

59. Explain in two or three sentences what you can learn by studying application problems.

60. Write an application problem that could be solved by other students in your class. Be sure it is clear and easy to understand but also interesting and realistic.

61. Write a real-life geometry problem involving the measurement of angles. Ask a question and then calculate the answer to the question.

62. Bring in a magazine, newspaper article, or advertisement that you could use as the basis for a mathematical application problem. Write the problem and prepare a solution. Exchange the problem with a classmate.

1.5 Interpreting Data

Often in work or classroom situations, you will be expected to look at a collection of numeric facts (data) and see a pattern or relationship. Perhaps you will need to organize and display the data in an easy-to-interpret manner. This kind of work requires the use of graphs to display and organize the data and the use of statistics to interpret the data.

The basic statistical quantities that you will compute in this section are the mean, the median, the mode, and the range. By providing these four pieces of information, you are telling someone the average value (mean), middle value (median), the most often repeated value (mode), and the spread between the highest and lowest values (range). You have provided your audience with key information that should enable them to understand and make sound decisions.

Mean

The most common statistical term used with a set of values is the **mean**, or average. What if your grades on mathematics tests for this semester were 75, 84, 68, 93, 44, 86, 80, 70, and 84, and you would like to know what your average grade is in the class? To find the mean score for your tests, add all of the scores together and divide that sum by the number of grades.

$$\text{Mean} = \frac{\text{Sum of the grades}}{\text{Number of grades}}$$

$$\text{Mean} = \frac{75 + 84 + 68 + 93 + 44 + 86 + 80 + 70 + 84}{9}$$

$$\text{Mean} = \frac{684}{9} = 76$$

The average score, or mean score, then, is 76.

Median

The **median** of a set of scores is the score that is in the middle position *when the scores are arranged in order of size.*

93, 86, 84, 84, 80, 75, 70, 68, 44

The number in the middle of the nine scores is 80. Therefore, the median of these scores is 80. If the list of data had had an even number of entries, there would have been no single middle value. The median for such data is the average of the two middle values.

Mode

The **mode** of a set of scores or data is the most often repeated value. Sets of data can have one mode, more than one mode, or no mode (if none of the values is repeated). For the grades on these mathematics tests, the only repeated score is 84, so that is the mode.

Range

Another meaningful statistic is the range. The range shows how far apart the highest and lowest values are. The range for these test scores is the difference between 93 and 44. It is calculated as 93 – 44, or 49 points.

EXAMPLE 1 The daily low temperatures for Indianapolis last week were 58°, 49°, 46°, 38°, 35°, 38°, and 44°. Find the mean, median, mode, and range for these temperatures,

Solution

$$\text{Mean} = \frac{\text{Sum of temperatures}}{\text{Number of temperatures}}$$

$$= \frac{58 + 49 + 46 + 38 + 38 + 35 + 44}{7}$$

$$= \frac{308}{7} = 44^\circ$$

Median: Middle value when the elements are arranged in order of size:
58°, 49°, 46°, 44°, 38°, 38°, 35°
The middle value, the median, is 44°.

Mode: Most often repeated temperature
Because 38° appears twice, it is the mode.

Range: Difference between the highest and lowest temperatures:
58° – 35° = 23°
The range is 23°.

EXAMPLE 2 For the past six months, Antonio's monthly gasoline bills have been $49, $30, $29, $32, $40, and $36. Find the mean, median, mode, and range for these data.

Solution

$$\text{Mean} = \frac{49 + 30 + 29 + 32 + 40 + 36}{6}$$

$$= \frac{216}{6} = \$36$$

The mean, or average, gasoline bill was $36.

Median: Arrange the values in order of size:

49, 40, 36, 32, 30, 29

There is an even number of entries, so the median is the average of the middle two values.

$$\frac{36 + 32}{2} = \frac{68}{2} = 34$$

Therefore, the median is \$34.

Mode: No value is repeated, so there is no mode.

Range: \$49 – \$29 = \$20

The range of these bills is \$20.

◀

Self-Check Liam makes the bank deposits each evening for the restaurant where he works. The amounts for the last five days were \$1,200, \$1,050, \$775, \$1,325, and \$1,150. Find the a) mean, b) median, c) mode, and d) range for these deposit amounts.

Answers:

a) \$1,100 b) \$1,150 c) none d) \$550

Although the mean, median, mode, and range provide a great deal of information about a set of numbers, they don't show the relationship among the values. Graphs are used to quickly show how values may be related or what trends or patterns exist.

A **bar graph** is used to compare several items. One of the scales is labeled to show the units represented by each interval. The other scale indicates the items being compared. The height or length of a bar indicates the value for that item.

In Figure 1.15, the vertical scale is labeled "Number of Students," and each increment in that scale represents 4,000 students. The horizontal scale is used to represent four different years in which the autumn quarter enrollments are being compared. Reading across the top of the 1997 bar to the vertical scale shows that the enrollment that year was 12,000 students. The 1999 enrollment was about 14,000 students.

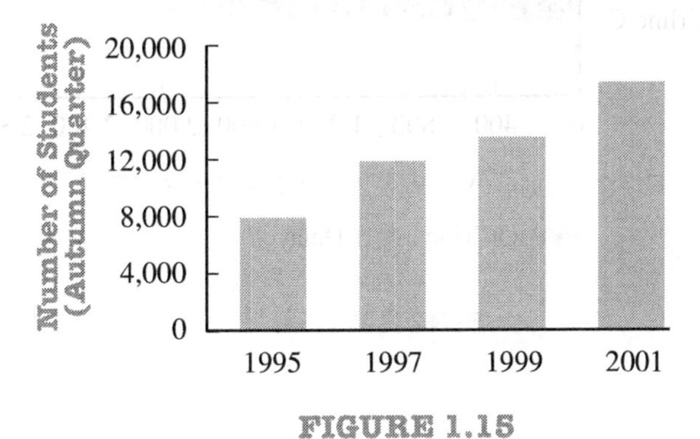

FIGURE 1.15

EXAMPLE 3 Use Figure 1.15 to answer the following questions.

a. What was the enrollment in 1995?

b. In what year was the enrollment 18,000 students?

c. How many more students enrolled in 2001 than in 1995?

d. Is the sum of the 1995 and 1997 enrollments greater than or less than the enrollment in 2001?

e. In what year did the enrollment show the least increase over the previous year studied?

Solution a. 8,000 students

b. 2001

c. 18,000 – 8,000 = 10,000 more students

d. 8,000 + 12,000 = 20,000

20,000 is greater than 18,000

e. The 1995–1997 increase was from 8,000 to 12,000 or 4,000 students.
The 1997–1999 increase was 2,000 students.
The 1999–2001 increase was 4,000 students.
The increase was the least in 1999.

Bar graphs can also be arranged horizontally (Figure 1.16).

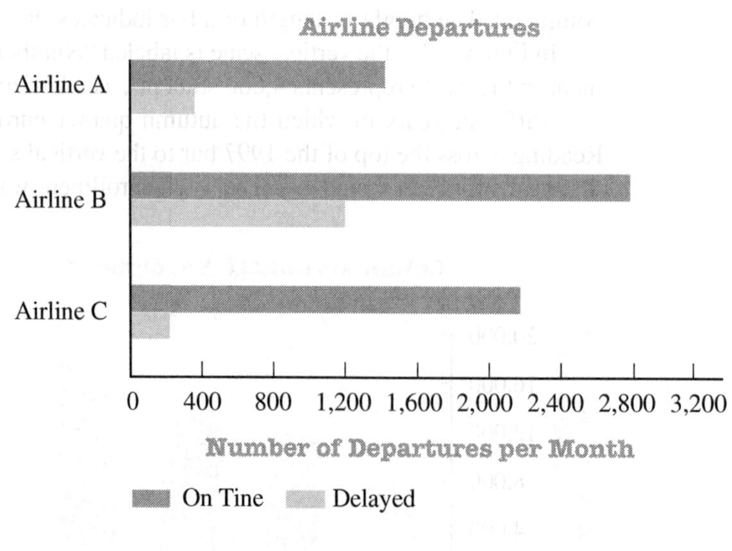

FIGURE 1.16

EXAMPLE 4 Use Figure 1.16 to answer the following questions.

a. How many delayed flights did Airline B have during the month?

b. Which airline flew the greatest number of flights during this period? How many did it fly?

c. How many more on-time departures did Airline B have than Airline C?

d. Is the sum of the delays for Airlines A and C greater than, less than, or equal to the number of delays for Airline B? By how much?

e. If all other factors are the same, which airline would you prefer to fly? Explain your answer.

Solution

a. Airline B had 1,200 delayed flights.

b. Airline B flew 2,800 + 1,200 = 4,000 flights.

c. Airline B had 2,800 and Airline C had 2,200. Therefore, Airline B had 600 more on-time departures.

d. Airlines A and C had a total of 600 delays. Airline B had 1,200. Therefore, Airline B had 600 more delays than Airlines A and C combined.

e. One possible response would be to fly Airline C because it has a "middle" number of on-time flights but experiences lots fewer delays.

◀

Some bar graphs use rows of pictures in place of the bars (Figure 1.17). The pictures suggest the items being represented. These bar graphs are called **pictographs** or **pictograms**. Each symbol represents a stated number of objects. And a part of a symbol represents a fractional part of the stated number.

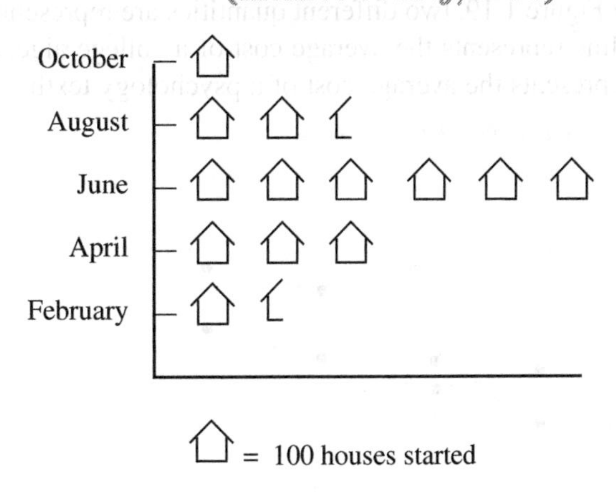

FIGURE 1.17

EXAMPLE 5 Use Figure 1.17 to answer the following questions.

a. How many houses were started in April?

b. In what month were 250 houses started?

c. How many more houses were started in June than in February?

Solution

a. 300 houses were started in April.

b. August had 250 housing started.

c. 600 houses were started in June and only 150 in February. Therefore, there were 450 more houses started in June than in February.

The data from Figure 1.15 might also be represented by **line graph**. In such a graph, points rather than bars represent the values (Figure 1.18). A line connecting the points shows the change in enrollment over the period represented. The pattern, or trend, is evident in a line graph.

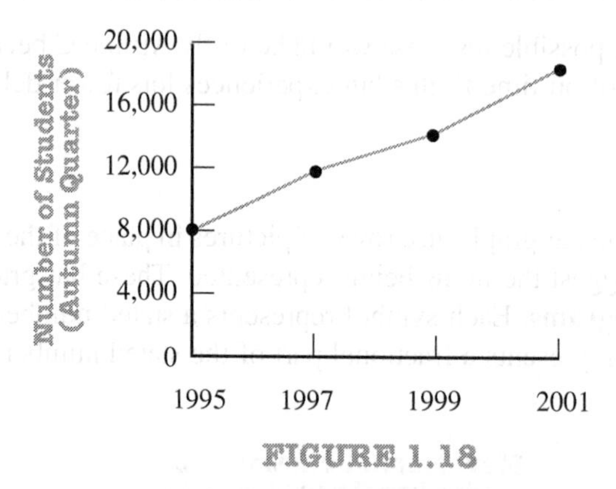

FIGURE 1.18

In Figure 1.19, two different quantities are represented in the same line graph. A solid line represents the average cost of a college algebra textbook, and the broken line represents the average cost of a psychology textbook.

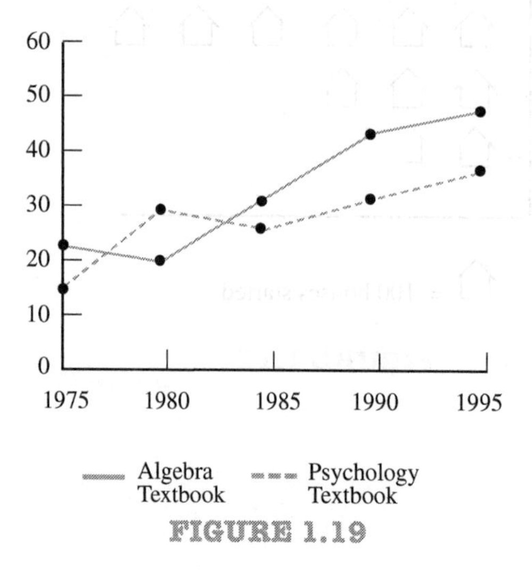

FIGURE 1.19

EXAMPLE 6 Use Figure 1.19 to answer the following questions.

a. How much more did the psychology textbook cost than the algebra textbook in 1980?

b. In what 5-year interval did the psychology text increase the most?

c. Which book had a cost of $30 in 1985?

d. Which book price experienced a decrease in price of $5 in a 5-year interval?

Solution a. In 1980, the algebra text cost $20 and the psychology text cost $30:

$30 – $20 = $10 more

b. The increase in the price of the psychology text was the greatest between 1975 and 1980.

c. The algebra text cost $30 in 1985.

d. The price of the psychology text decreased by $5 between 1980 and 1985.

1.5 EXERCISES

A. Find the answer to each of the following exercises.

1. The following data represent the number of hours per week that Carolyn worked at her part-time job during the last 7 weeks. Find the requested statistical information for this set of values:

   13 hours, 19 hours, 26 hours, 19 hours, 10 hours, 21 hours, and 18 hours

   a. mean b. median
   c. mode d. range

2. The following data represent the number of minutes used for long-distance calls on Taria's last phone bill. Find the requested statistical information for this set of values:

   12 minutes, 14 minutes, 18 minutes, 22 minutes, 12 minutes, 24 minutes, and 17 minutes

   a. mean b. median
   c. mode d. range

3. The following data represent the monthly food bill totals that Clyde has spent each month during the last 6 months. Find the requested statistical information for this set of values:

   $128, $218, $176, $204, $228, and $204

   a. mean b. median
   c. mode d. range

4. The following data represent the number of minutes that Eloise has spent exercising per week during the last 6 weeks. Find the requested statistical information for this set of values:

   358 minutes, 313 minutes, 358 minutes, 332 minutes, 307 minutes, and 330 minutes

   a. mean b. median
   c. mode d. range

B. Using the graphs and tables that are given, find the answers to each of the following exercises. For Exercises 5 through 8. Use Table 1.1.

TABLE 1.1

	Scores on French tests			
	1	2	3	4
Maria	86	75	91	84
Igor	78	84	84	86
Chris	68	82	76	66

5. Find the mean, median, mode and range of Maria's grades.

6. Find the mean, median, mode, and range of Chris's grades.

7. Which of the three students has the widest range of scores? State each student's score range.
8. Which student has a mode score?

For Exercises 9 and 10, use Figure 1.20.

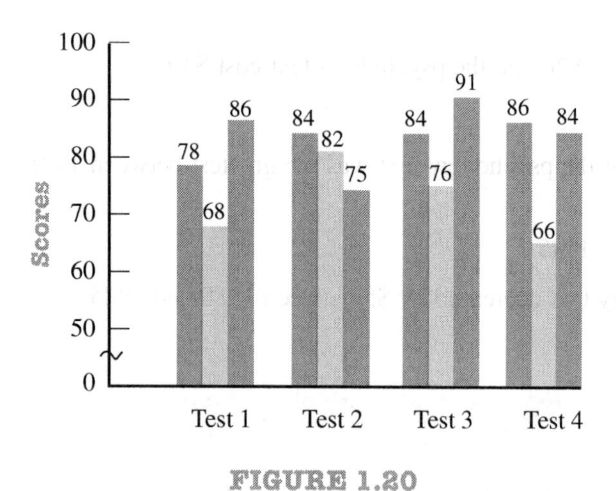

FIGURE 1.20

9. The data from Table 1.1 are displayed in Figure 1.20. However, the key that tells which student's scores are depicted by each bar has been left off. Whose scores do the bars on the far right represent?
10. The data from Table 1.1 are displayed in Figure 1.20, but the key that tells which student's scores are depicted by each bar has been left off. Whose scores do the middle bars represent?

For Exercises 11 through 16, use Figure 1.21.

11. What was the monthly high temperature in Portland in April?
12. Which city had a high temperature of 85 in June?
13. How much warmer was Nassau than Portland on the warmest day in December?
14. How much cooler was Portland than Nassau on the warmest day in July?
15. During what month(s) was the high temperature in Portland closest to the high temperature in Nassau?
16. Between what months does the high temperature in Portland undergo the greatest change?

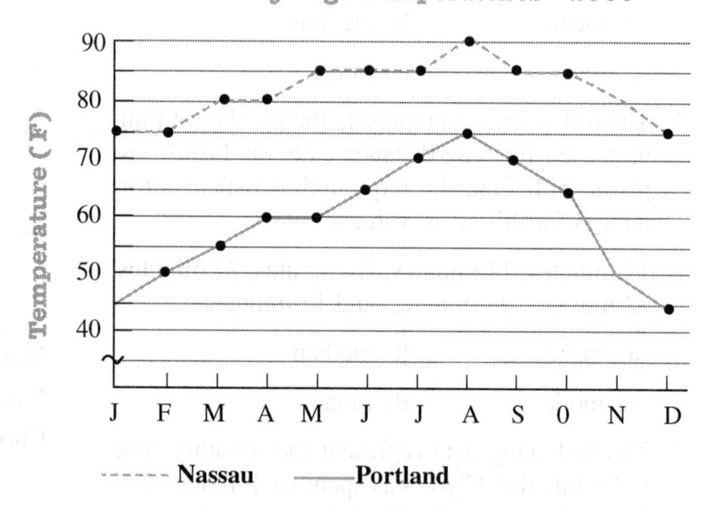

FIGURE 1.21

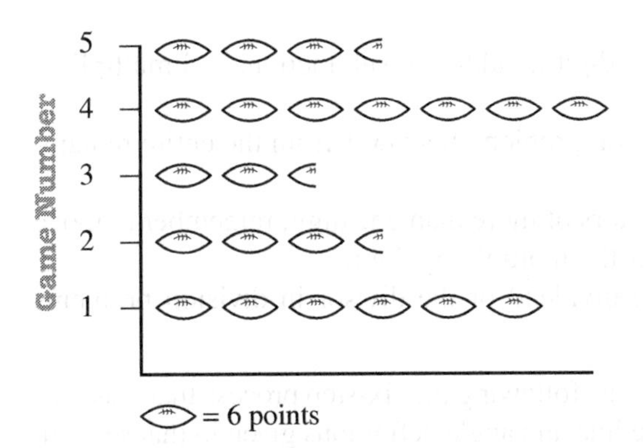

FIGURE 1.22

Use Figure 1.22 to answer Exercises 17 through 24.

17. During which game did the Flyers score 15 points?
18. What was the Flyers' score in game 4?
19. How many more points were scored in the fourth game than in the third game?
20. Is the sum of the scores in game 2 and game 3 greater than, less than, or equal to the score in game 1?
21. What is the mean score for the five games?
22. What is the median of the scores?
23. What is the mode of the scores?
24. What is the range of the scores?

C. Read and respond to the following exercises.

25. A fellow classmate is having difficulty remembering the meanings of mean, median, mode, and range. Give her some suggestions for ways to remember them.
26. Bring in a copy of a graph from a magazine or newspaper and explain what information you would need to notice in order to understand what the graph is describing.
27. Write four questions based on the graph that you have brought in to class. (See Exercise 26.)
28. Describe a situation where you would use a pictograph. Explain what information you would include on the graph and what "picture" you would use and why.

1.6 CHAPTER REVIEW

The purpose of this introductory chapter is to give you many of the basics you will need to study mathematics. Because whole numbers are the most basic numbers, that is the place to begin. Most mathematical operations depend on an understanding of place value. The practical side of basic calculations is the solving of application problems. You will be successful with them if you practice often.

When you have finished studying this chapter, you should be able to:

1. Explain the following terms: place value, placeholding zeros, hyphenated numbers, rounding, sum, addends, difference, product, factors, quotient, divisor, dividend, remainder, estimation, undefined answer.
2. Read and write whole numbers using commas and hyphens correctly.

3. Do addition, subtraction, multiplication, and division problems without errors and by using the following hints:
   a. Remember to line up the ones digit in addition, subtraction, and multiplication problems.
   b. When borrowing in a subtraction problem, borrow 1 from the entire remaining quantity.
   c. In multiplication involving factors of more than one digit, remember to move the partial product over, under the multiplying digit.
   d. Zeros appearing in either the dividend or the divisor in division problems require great care.
4. Read and solve application problems, following the six-step process that was outlined in Section 1.4, and apply the line and angle definitions given in that section.
5. Interpret data from a graph.
6. Find the mean, median, mode, and range of given data.

REVIEW EXERCISES

A. Name the digit in the indicated place in each of the following exercises.

1. 238,195
   a. tens place
   b. thousands place
   c. hundred-thousands place
2. 18,743,092
   a. thousands place
   b. hundreds place
   c. millions place
3. 5,264,381
   a. hundred-thousands place
   b. ones place
   c. millions place
4. 20,381,467
   a. ten-millions place
   b. ten-thousands place
   c. hundreds place

B. Write each of the following quantities in numeric form.

5. eighteen million, four thousand, seventeen
6. six billion, four hundred thousand, twelve
7. nine thousand, forty-two
8. twelve million, eighty-one thousand, six
9. seventeen billion, nine thousand, fifty-five
10. four hundred twenty-three thousand
11. eight hundred forty-six thousand
12. eight thousand, four hundred thirty-nine

C. Write each of the following numbers in words.

13. 32,189
14. 27,693
15. 891,407
16. 375,682
17. 1,424,856
18. 3,427,895
19. 77,923
20. 98,137
21. 492,108
22. 648,000
23. 23,011
24. 62,039

D. Round each number to the place indicated.

25. 2,655 to the nearest ten
26. 7,983 to the nearest hundred
27. 34,549 to the nearest hundred
28. 62,004 to the nearest ten
29. 184,563 to the nearest thousand
30. 22,761 to the nearest thousand
31. 358,901 to the nearest hundred
32. 26,511 to the nearest thousand
33. 117,241 to the nearest ten-thousand
34. 267,189 to the nearest ten
35. $92,341 to the nearest hundred
36. $1,457,832 to the nearest million
37. $389,726 to the nearest thousand
38. $4,223,971 to the nearest hundred
39. $175,306 to the nearest thousand
40. $214,925 to the nearest thousand

E. Perform the mathematical operations indicated.

41. $\begin{array}{r}1{,}289\\ \times\ \ 35\\ \hline\end{array}$

42. $14\overline{)8{,}223}$

43. $\begin{array}{r}247\\ 1184\\ +\ 936\\ \hline\end{array}$

44. $\begin{array}{r}4{,}005\\ -\ \ 428\\ \hline\end{array}$

45. $\dfrac{14{,}582}{46}$

46. $\begin{array}{r}16{,}236\\ -9{,}277\\ \hline\end{array}$

47. $\begin{array}{r}9{,}982\\ \times\ \ 48\\ \hline\end{array}$

48. $\begin{array}{r}159\\ 3{,}803\\ +\ \ 547\\ \hline\end{array}$

49. $\begin{array}{r}6{,}004\\ -3{,}847\\ \hline\end{array}$

50. $\begin{array}{r}2{,}468\\ \times\ \ 15\\ \hline\end{array}$

51. $\begin{array}{r}5{,}080\\ \times\ \ 25\\ \hline\end{array}$

52. $106\overline{)21{,}518}$

53. $400\overline{)312{,}000}$

54. $\begin{array}{r}7{,}009\\ \times\ \ 34\\ \hline\end{array}$

55. Find the difference between 82,066 and 37,848.
56. Compute the quotient of 11,396 divided by 28.
57. Find the product of 4,509 and 38.
58. Find the sum of 53,472 and 3,906.
59. Compute the quotient of 10,962 divided by 27.
60. Find the difference between 7,060 and 2,843.
61. Find the sum of 329 and 1,047.
62. Find the product of 4,042 and 205.
63. Compute the product of 2,007 and 304.
64. Compute the quotient of 16,192 divided by 32.

F. Estimate an answer to each of the following exercises. Then calculate the actual answers. Be sure to label your answers.

65. Martin swam 18 laps on Monday, 23 laps on Tuesday, 19 laps on Friday, and 32 laps on Saturday. How many laps did he swim on those 4 days?
66. How many feet of copper tubing will Eduardo need to purchase at the hardware store if he has to run three lines of 17 feet each?
67. Kelly and two of her friends split the cost of dinner at Rocco's Spaghetti House. If the total bill was $42, how much did each of them pay?
68. As a fund-raiser, students at the Edgehill School are selling flats of the bedding plants to raise money for a new freezer for the school cafeteria. The school makes $8 on each flat of plants sold. If the freezer will cost about $4,400, how many flats will the students need to sell to pay for the freezer?
69. The population of Columbus, Indiana, is approximately 30,800. The population of Columbus, Ohio, is about 566,100. How much larger is Columbus, Ohio?
70. The area of the state of Alaska is 586,412 square miles. The area of the state of Rhode Island is 1,214 square miles. How much larger is Alaska? Estimate your answer to the nearest hundred square miles.
71. Find the total cost of buying computers for the new lab if there are 24 workstations in the lab and each computer costs $1,078.

72. Textbook for the next term are being sold in the campus bookstore. Santiago needs an economics book ($54), a literature book ($47 used), and a mathematics book ($72). What will his textbook bill be for these three courses?

73. Gabrielle went on a 4-month diet to lose some extra pounds. She weighed 153 pounds when she started the diet and weighed 129 pounds at the end of the four months. How much weight did she lose?

74. Janet and Sam want to buy a new home. They need $3,600 more for a down payment. If they are saving $240 a month, how many more months will it be before they will have all the money saved?

75. In Exercise 73, how many pounds was Gabrielle's average loss each month during her diet?

76. Alexander has a part-time job cutting grass for 9 of his neighbors during the summer season. He charges $15 per yard to cut the grass each week. At this rate, how much money can he make each week?

77. Marc, a salesman for Rayburn Electronics, travels an average of 372 miles each week. On the average, how many business miles does he put on his car each year during the 49 weeks that he works?

78. The air distance from San Francisco to Tokyo is 5,410 miles. The distance from San Francisco to London is 5,980 miles. How much closer to San Francisco is Tokyo?

G. Fill in the blank in each of the following statements.

79. If $m \measuredangle ABC$ is 76°, $\measuredangle ABC$ is a(n)_______angle.

80. If $m \measuredangle PQR$ is 102°, $\measuredangle PQR$ is a(n)_______ angle.

81. The measure of $\measuredangle XYZ$ is 180°, so $\measuredangle XYZ$ is a(n)_______angle.

82. $\measuredangle ABC$ and $\measuredangle CBD$ are complementary angles, so $m \measuredangle ABC + m \measuredangle CBD =$ _______.

83. $\measuredangle PRS$ and $\measuredangle SRT$ are supplementary angles, so $m \measuredangle PRS + \measuredangle SRT =$ _______.

84. Two angles are_______ if their measures are exactly the same.

H. Use Figure 1.23 to answer each of the following exercises.

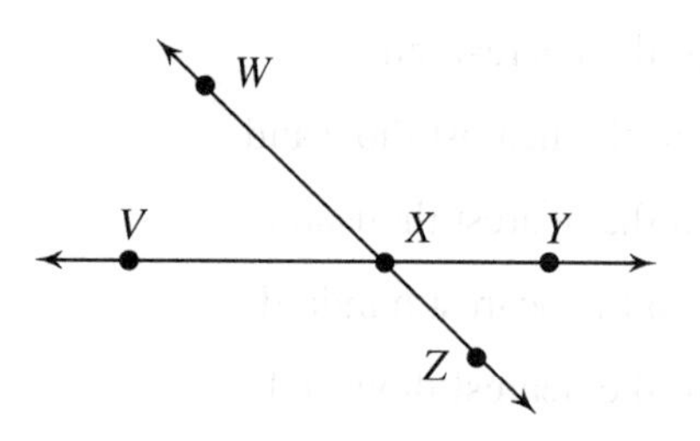

FIGURE 1.23

85. Find the measure of $\measuredangle XYZ$ if $m \measuredangle WXY = 122°$.

86. Find the measure of $\measuredangle VXZ$ if $m \measuredangle WXY = 122°$.

87. $\measuredangle WXY$ and $\measuredangle VXZ$ are _______ angles.

88. $\measuredangle WXV$ and $\measuredangle VXZ$ are _______ angles.

89. $m \measuredangle WXV + m \measuredangle VXZ =$ _____.

I. Use Figure 1.24 to answer the following questions.

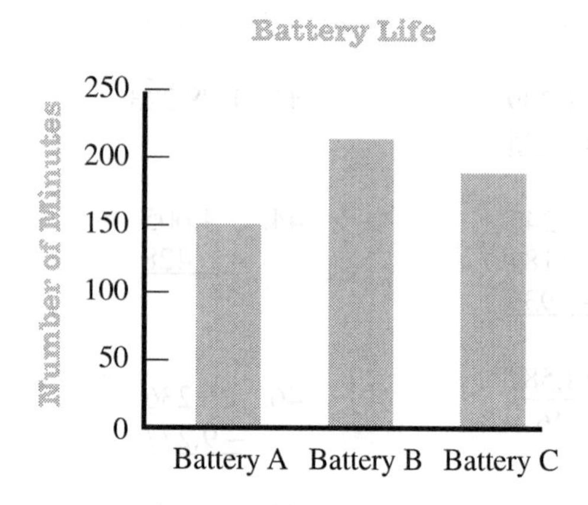

FIGURE 1.24

90. How many more minutes did Battery C last than Battery A?

91. Which battery lasted exactly 150 minutes?

92. How long did Battery B last?

93. How many more minutes did Battery B last than Battery C?

J. Answer the following questions by using the information in Figure 1.25.

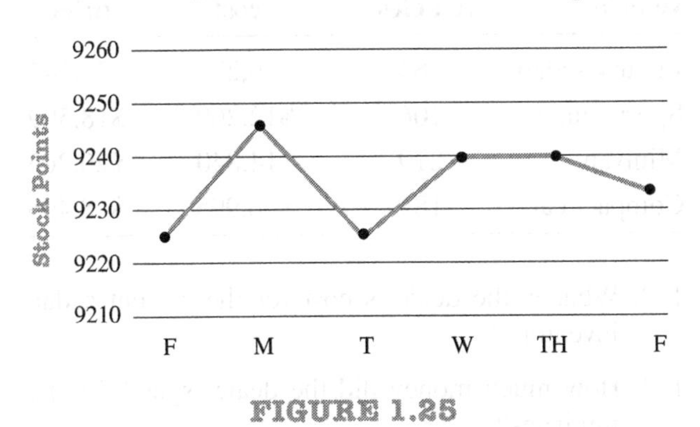

FIGURE 1.25

94. What was the mean for the stock market closings from Friday through Friday?

95. What was the median for the stock market closings from Friday through Friday?

96. What was the mode for this time period?

97. What was the range in closings for this time period?

98. Between which two days was there the greatest decrease?

99. Between which two days was there the greatest increase?

100. What was the overall change from the first Friday to the next Friday?

101. Was this a good week for the stock market or not? Explain your answer.

K. Each exercise that follows has an error in the solving process. State what the error is; then work the problem correctly.

102. Write six million, eighteen thousand, four hundred in digits.
Solution: 6,018,004

103. Write 8,061,025 in words.
Solution: eight million, sixteen thousand, twenty-five

104. Find the sum of 338; 1,245; and 47.

Solution:
```
  338
 1245
+  47
 9325
```

105. Subtract 498 from 2,001.

Solution:
```
   8
 1 9 9 1
 2,0 0 1
 -  498
   1403
```

106. Multiply 5,027 by 46.

Solution:
```
   5027
 ×   46
  30762
  20508
 235842
```

107. Find the product of 5,027 and 26.

Solution:
```
    193 R9
 26)5027
   -26
    242
   -234
     87
    -78
      9
```

L. Read and respond to the following exercises.

108. Write down an estimated answer (approximation) for each of the following problems. Then do each of the problems using your calculator. How good were your estimates? What could you have done to make them better?

a. 11,206 + 5,437 + 2,861

b. 20,105 − 11,768

c. 32,638 × 1,726

d. 485,616 ÷ 1,608

109. Explain why you think it is just as important to estimate an answer when using a calculator as it is when calculating by hand?

110. Explain two ways in which you could multiply 48 times 54.

111. What reasons can you think of to study some geometry in a fundamentals course?

112. Bring in a chart or a graph from a magazine or a newspaper and write four mathematical questions that you could ask from the data shown.

M. Using the information given, answer the following questions. Considering the size of the numbers involved, a calculator would be very helpful.

The Thurness Mega Auto Mall has four different basic styles of vehicles in its one location. The styles are a luxury sedan, a sport coupe, a minivan, and a compact car. There are two dollar values associated with each vehicle: the dealer's cost (what he pays the factory for the car) and the selling price (what the consumer pays the dealer). As you can imagine, the amount of money involved in an operation of this size is very large. During a recent inventory, the totals shown in Table 1.2 were reported. The report also included the average dealer's cost and the average selling price for each style of car.

TABLE 1.2

Style of vehicle	Number of vehicles	Average dealer cost	Average selling price
Luxury sedan	84	$19,275	$27,345
Sport coupe	106	$12,260	$18,509
Minivan	219	$14,840	$20,205
Compact car	184	$ 6,995	$ 8,495

113. What is the dealer's cost for the current sedan inventory?

114. How much money did the dealer spend for the minivans?

115. What is the difference between the average selling prices for a sport coupe and a compact car?

116. What is the total dealer cost of the current inventory?

117. If a nearby auto dealer wanted to buy 7 sport coupes at the average dealer cost, how much money would he have to pay to the Thurness Company?

118. In order to cover his growing expenses and still make a profit, the dealer needs to increase the average selling price for each vehicle by $1,095. For how much does he hope to sell his inventory of compact cars?

CHAPTER TEST

1. Write in digits: forty-seven million, six thousand, eight hundred four.
2. Write 30,206,011 in words.
3. Round 176,852 to the nearest hundred.
4. Round 347,826 to the nearest hundred-thousand.
5. Find the product of 347 and 29.
6. Find the quotient of 149,036 divided by 37.
7. What is the difference between 40,082 and 27,995?
8. From the sum of 3,827 and 1,046 subtract 2,785.
9. 0 ÷ 143 = _____.
10. 2,642 ÷ 0 = _____.
11. Tim and Julia went to the laundromat to wash several large blankets and area rugs. Each large washer required 8 quarters, and each oversized dryer used 5 quarters. How many quarters will they need to wash and dry five large loads?
12. The United Campaign predicts that it will receive $17 from each employee working at American Limited. American Limited currently has 463 employees. How much money can the United Campaign expect to receive from this group?

13. Projected monthly expenses for Le Cafe are as follows: rent, $1,200; utilities, $486; insurance, $389; salaries, $7,385; taxes, $243; and telephone, $157. *Estimate*, to the nearest hundred dollars, the total amount needed to pay these expenses.

14. According to its advertising material, Jillian's new car can travel 392 miles on its 14-gallon gas tank. Using these figures, how many miles per gallon of gasoline should she expect from this car?

15. $\measuredangle ABC$ and $\measuredangle CBD$ are supplementary angles. If m $\measuredangle ABC$ is 46°, find the measure of $\measuredangle CBD$.

16. If two angles are complementary and congruent, what is the measure of each angle?

17. The problem that follows has an error in the solving process. State what the error is; then work the problem correctly.

$$\begin{array}{r} 286 \text{ R } 12 \\ 127\overline{)36234} \\ \underline{-254} \\ 1083 \\ \underline{-1016} \\ 774 \\ \underline{-762} \\ 12 \end{array}$$

18. Malcolm's scores on his biology tests for this semester were

    82, 76, 93, 84, 100, 76, and 84

    Find the following statistical information.

    a. mean
    b. median
    c. mode
    d. range

19. Two angles are complementary and you are given the measure of one angle. Explain, in words, how you would find the measure of the other angle.

20. Explain, in words, how you would multiply 27,000 by 40 using what you know about multiplying numbers that end in zero.

STUDY SKILLS MODULE

Attitude Toward Math

Psychologists, coaches, and parents often say that "you need to have a positive attitude in order to be successful." Does this principle apply to doing math? You bet it does!

Many students, especially those who fear or hate math the most, have attempted to take as few math classes as possible. In those few courses they were forced to take, they fixed their goal on just getting through, just passing and getting out. Other students felt they had no control over how well they did in their math courses. They blamed their lack of success on factors other than their own effort—bad teachers, poor facilities, distracting classmates, no time to study, and so forth. And still others strongly believed that they were "too dumb" to learn mathematics. They expected to do poorly or to fail, and they probably lived up to those expectations. Perhaps you were one of those students. Now these students find themselves with educational or career goals that require successful completion of several math courses. If these students are going to reach proficiency in their required mathematics courses, they are going to have to undergo a serious attitude adjustment.

First, students must recognize that they do have a considerable amount of control over their own success. Dr. Benjamin Bloom, an educational researcher, has found that a person's IQ and math background account for only 50% of a student's academic achievement. The quality and effectiveness of instruction can account for another 25%, and the student's affective characteristics will account for the rest.

The affective characteristics are factors such as anxiety, motivation, study and test-taking skills, and self-esteem. These factors are ones over which you have control. The Study Skills Modules in this book are aimed at helping you improve in some of those areas.

The first step in taking control of your math destiny is to set a goal for yourself—a goal to complete this course with a good grade. If this seems too long range and overwhelming to you, focus on the more immediate goal of getting a good grade on the weekly quiz or on the next unit test. Once you find success with your short-term goals, you will gain confidence and be able to work toward the broader goal of earning a good grade in the course.

To accomplish the task that you set for yourself, you must schedule adequate time for study, work to improve your test-taking and study skills, get help to overcome math anxiety, and convince yourself that you can be a successful math student.

Put the bad experiences and memories behind you because you are a different person now than you were in junior or senior high school, when you probably took your last math course. Think about other areas in your life in which you are successful. If you have been able to grow in those areas, you can also grow in your mathematical ability. In fact, you have probably come a long way already without realizing it.

To find success, you must first begin with a math course that is at the true level of your math skills. You will struggle terribly or be very bored if your placement is not correct. Master the material in the course that is right for you and move on to the next, knowing that you now have the prerequisite skills to be successful in that next course. Start off correctly on the road to math achievement by committing time and effort toward reaching understanding of the material and take control of and responsibility for your success.

Your Point of View

1. Describe a situation in your life in which having a positive attitude has helped.
2. Which of the affective characteristics listed is likely to be a hindrance to your academic achievement if you do not do something about it? Why?
3. Do you believe that you have a significant amount of control in obtaining success in this math course? Why?
4. Describe your attitude toward math as you begin this course.
5. List three specific steps that you will take this week to improve your attitude about this course. Even if you have a good attitude about math, things can always be better.

CHAPTER

Whole Number Concepts

LEARNING OBJECTIVES

When you have completed this chapter, you should be able to:

1. Simplify expressions involving exponents and roots.
2. Simplify problems using the Order of Operations Agreement.
3. Answer questions that require the use of a formula.
4. Apply triangle, quadrilateral, and solid figure definitions and formulas to the solution of geometry applications.
5. Identify and make use of the following number properties:
   a. Commutative Properties of Addition and Multiplication
   b. Associative Properties of Addition and Multiplication
   c. Properties of Zero and One
   d. Distributive Principle
6. Recognize prime numbers.
7. List all the factors of a given whole number.
8. Use the divisibility tests for 2, 3, and 5.
9. Find the prime factorization of a given number.
10. Find the least common multiple of a group of numbers.
11. Find the greatest common factor of a group of numbers.

This chapter covers a number of interesting and useful topics. Some of them will follow each other very closely. Others won't seem to be related at all to the previous or following sections. For this reason, the chapter may seem confusing to you. Don't be discouraged. Take it one step at a time. Each section introduces a skill or an approach you will need in the chapters to come and in later courses.

2.1 Exponents and Roots

Before going any further, you need to understand what exponents and roots are. The expression 7^2 is read "7 to the second power" or "7 squared." It means 7×7, or 49. The 7 is referred to as the **base**, and the ^{2} is referred to as the **exponent** or **power**. 4^3 means $4 \times 4 \times 4$, or 64. The 4 is the base, and the ^{3} is the exponent or power. The exponent tells how many times the base is used as a factor. The expression 3×4 means 4 will be used as an *addend* 3 times: $4 + 4 + 4$. The expression 4^3 means 4 will be used as a *factor* 3 times: $4 \times 4 \times 4$. An exponent can be any number, but the most commonly used exponents are 2 and 3. 4^2 means 4×4, or 16. 5^3 is read "5 to the third power" or "5 cubed"; it means $5 \times 5 \times 5$, or 125. Exponents greater than 3 have no special names. The expression 9^5 is read "9 to the fifth power."

To review,

6^2 means $6 \cdot 6$, or 36.

2^3 means $2 \cdot 2 \cdot 2$, or 8.

8^2 means $8 \cdot 8$, or 64.

2^8 means $2 \cdot 2 \cdot 2 \cdot 2 \cdot 2 \cdot 2 \cdot 2 \cdot 2$, or 256.

The last two examples make it clear that interchanging the base and power, when those numbers are not the same, produces a different answer. You can see that 8^2 and 2^8 are not the same thing.

$\sqrt{49}$ is read "the square root of 49." The square root means the number that, when squared, gives 49. Therefore, $\sqrt{49} = 7$. $\sqrt{36}$ is read "the square root of 36," and it represents the number that, when multiplied by itself, gives 36. Therefore, $\sqrt{36} = 6$. The root symbol, $\sqrt{}$, is called a **radical symbol**, and the quantity inside the symbol is called the **radicand**. Problems involving exponents and radicals are easy to do once you understand what the symbols mean. Look at the pattern of numbers in Table 2.1, which gives the most commonly used squares, cubes, square roots, and cube roots. In the second column, the radicands are all squared numbers (sometimes called **perfect squares**)—numbers produced when a quantity is "squared," or used as a factor two times.

TABLE 2.1

Squares	Square Roots	Cubes	Cube Roots
$1 \times 1 = 1$	$\sqrt{1} = 1$	$1^3 = 1$	$\sqrt[3]{1} = 1$
$2 \times 2 = 4$	$\sqrt{4} = 2$	$2^3 = 8$	$\sqrt[3]{8} = 2$
$3 \times 3 = 9$	$\sqrt{9} = 3$	$3^3 = 27$	$\sqrt[3]{27} = 3$
$4 \times 4 = 16$	$\sqrt{16} = 4$	$4^3 = 64$	$\sqrt[3]{64} = 4$
$5 \times 5 = 25$	$\sqrt{25} = 5$	$5^3 = 125$	$\sqrt[3]{125} = 5$
$6 \times 6 = 36$	$\sqrt{36} = 6$	$6^3 = 216$	$\sqrt[3]{216} = 6$
$7 \times 7 = 49$	$\sqrt{49} = 7$	$7^3 = 343$	$\sqrt[3]{343} = 7$
$8 \times 8 = 64$	$\sqrt{64} = 8$	$8^3 = 512$	$\sqrt[3]{512} = 8$
$9 \times 9 = 81$	$\sqrt{81} = 9$	$9^3 = 729$	$\sqrt[3]{729} = 9$
$10 \times 10 = 100$	$\sqrt{100} = 10$	$10^3 = 1{,}000$	$\sqrt[3]{1000} = 10$
$11 \times 11 = 121$	$\sqrt{121} = 11$	$11^3 = 1{,}331$	$\sqrt[3]{1331} = 11$
$12 \times 12 = 144$	$\sqrt{144} = 12$	$12^3 = 1{,}728$	$\sqrt[3]{1728} = 12$

Try these exercises and see if your answers agree with the ones that follow. If possible, don't look up the values given in the table. Do the calculations on your own!

Self-Check Find the value of each of the following expressions.

1. 3^2 2. 12^2 3. $\sqrt{64}$
4. 2^4 5. $\sqrt{121}$ 6. $\sqrt{4}$
7. 3^3

Answers:

1. 9 2. 144 3. 8
4. 16 5. 11 6. 2
7. 27

What happens when the number inside the square root symbol, $\sqrt{}$, is not a squared number? For instance, what is the value of $\sqrt{20}$? To answer this question, you may want to use a square root table or a calculator. Table 2.2 is a portion of a square root table. The corresponding square root values, except for the squared number, are rounded to three decimal places.

TABLE 2.2

Number	Square root	Number	Square root
14	3.743	21	4.583
15	3.873	22	4.690
16	4	23	4.796
17	4.123	24	4.899
18	4.243	25	5
19	4.359	26	5.099
20	4.472	27	5.196

The values given in Table 2.2, except 16 and 25, are decimal approximations. When squared, each gives a number very close to the number whose square root you wish to find. In order to find the value of $\sqrt{20}$, it helps to realize that 20 is between 16 and 25. Because $\sqrt{16} = 4$ and $\sqrt{25} = 5$, $\sqrt{20}$ should have a value between 4 and 5, somewhat closer to 4 than to 5. Looking up the square root of 20 in the table reveals that $\sqrt{20} \doteq 4.472$. (The symbol $\doteq$ means "is approximately equal to." Some texts use $\approx$ to mean the same thing.) To see if this answer is correct, multiply 4.472 by itself.

$4.472 \times 4.472 = 19.998784$

Remember that the values in the table are approximations. Because 20 is not a squared number, it has no exact whole number square root. A more comprehensive square root table is located on the inside back cover of your text. Use the values from that table to find the following square roots.

Self-Check Find an approximate value for each of the following expressions. Use the square root table on the inside back cover.

1. $\sqrt{18}$ 2. $\sqrt{7}$ 3. $\sqrt{39}$
4. $\sqrt{42}$ 5. $\sqrt{11}$

Answers:

1. 4.243 2. 2.646 3. 6.245
4. 6.481 5. 3.317

Radical or root problems do not always involve square roots. $\sqrt[3]{8}$ is asking you to find the cube root, or the third root, of 8—that is, the number that, when cubed (used as a factor 3 times), gives 8. That number is 2 because $2 \times 2 \times 2 = 8$. Therefore, $\sqrt[3]{8} = 2$. The raised 3 on the outside of the root symbol is called its **index**. A raised 4 on the outside of the radical is asking you to find the fourth root of the radicand. $\sqrt[4]{81} = 3$ because $3 \times 3 \times 3 \times 3 = 81$. Are you wondering where the raised 2 is in square root problems? $\sqrt{25}$ could indeed be written $\sqrt[2]{25}$, but in writing square roots there is no need to include the index. You may assume you are working with a square root unless there is an index to tell you differently.

Use your own calculation abilities or Table 2.1 to simplify the following problems.

Self-Check Find the value of each of the following expressions.

1. $\sqrt[3]{125}$ 2. $\sqrt[4]{16}$
3. $\sqrt{64}$ 4. $\sqrt[3]{1000}$

Answers:

1. 5 2. 2
3. 8 4. 10

2.1 EXERCISES

A. Simplify each of the following expressions involving exponents. Do the exercises without using the tables.

1. 3^3
2. 4^2
3. 5^2
4. 6^3
5. 1^5
6. 0^4
7. 2^8
8. 1^7
9. 0^3
10. 3^5

B. Find the value of each of the following expressions involving square roots. When necessary, refer to the table on the inside back cover.

11. $\sqrt{35}$
12. $\sqrt{49}$
13. $\sqrt{17}$
14. $\sqrt{27}$
15. $\sqrt{36}$
16. $\sqrt{85}$
17. $\sqrt{52}$
18. $\sqrt{100}$
19. $\sqrt{16}$
20. $\sqrt{96}$

C. Find the value of each of the following expressions involving radicals. When necessary, refer to the table on the inside back cover.

21. $\sqrt[3]{64}$
22. $\sqrt[4]{1}$
23. $\sqrt[4]{16}$
24. $\sqrt[3]{729}$
25. $\sqrt[5]{1}$
26. $\sqrt[2]{100}$
27. $\sqrt[3]{1000}$
28. $\sqrt[5]{0}$
29. $\sqrt[2]{121}$
30. $\sqrt[3]{216}$
31. $\sqrt[2]{64}$
32. $\sqrt[3]{8}$
33. $\sqrt[3]{125}$
34. $\sqrt[2]{71}$
35. $\sqrt[2]{23}$
36. $\sqrt[3]{1331}$
37. $\sqrt[3]{1728}$
38. $\sqrt[3]{1}$
39. $\sqrt[2]{84}$
40. $\sqrt[2]{65}$

D. Answer each of the following questions.

41. What is the square root of 196?
42. What is the third root of 125?
43. What is the third root of 512?
44. What is the square root of 144?
45. Is 289 a squared number? Why or why not?
46. Is 169 a squared number? Why or why not?
47. Is 343 a cubed number (third power)? Why or why not?
48. Is 64 a cubed number (third power)? Why or why not?
49. What number used as a factor four times is 256?
50. What number used as a factor five times is 243?

E. Use your calculator to find the value for each of the following expressions. See Appendix B.

51. 8^4
52. 7^5
53. $\sqrt[5]{32}$
54. $\sqrt[5]{3,125}$
55. $\sqrt[4]{6,561}$
56. 12^4
57. 16^5
58. $\sqrt[3]{4,913}$
59. $\sqrt[3]{1,000,000}$
60. $\sqrt[4]{50,625}$

F. Read and respond to the following exercises.

61. Describe the relationship between squared numbers and square roots.
62. Do the following problems using the exponent and root key on your calculator. Report the answer that your calculator shows.
    a. 6^0 b. 0^8 c. $\sqrt[0]{8}$
    d. $\sqrt[6]{0}$ e. 9^1
63. Complete the following pattern:

    $10^1 =$
    $10^2 =$
    $10^3 =$
    $10^4 =$
    $10^5 =$

    Based on what you have observed, what would 10^9 be?
64. Key the following expressions into your calculator and report what the display shows:

    $1^4 =$
    $1^9 =$
    $1^{12} =$
    $1^{17} =$

    Formulate a rule based on what you have observed.

2.2 Order of Operations and Formulas

Order of Operations Agreement

In an example in Chapter 1, you calculated the total needed to pay for an order at the school supply store. To figure that total, you added after doing some multiplication because the order included more than one of some items. Often in mathematical situations, you need to do more than one operation within a problem. In application problems such as the one at the store, the order in which the operations must be done is obvious. In some calculation problems, however, that is not the case. Consider this simple problem:

$25 - 4 \times 3$

It would seem perfectly normal to do the problem from left to right because that is the direction in which you read. In some other countries, however, people read from top to bottom because their alphabet and communication customs are different from our own. Because mathematics is a universal language, it became necessary, centuries ago, for mathematicians to reach an agreement about how certain problems would be done. The **Order of Operations Agreement** used today is a result of that effort. All people throughout the world do mathematics by following these same steps. As you will see, there are four steps in the Order of Operations Agreement, and you must **complete each step**, if applicable, **before going on to the next.**

Order of Operations Agreement

When simplifying a mathematical expression involving more than one operation, follow these steps:

1. Simplify within parentheses or other grouping symbols, using steps 2, 3 and 4 below in order.
2. Simplify any expression involving exponents or square roots.
3. Multiply or divide **in order from left to right.**
4. Add or subtract (combine) **in order from left to right.**

Let's look again at the problem $25 - 4 \times 3$. This problem involves subtraction and multiplication. The agreement says to multiply before subtracting. Underlining each step as it is to be done helps you to keep on track. It may be easier to keep your work organized if you simplify one step at a time and sift your way down through the problem rather than work "sideways." Therefore,

$25 - \underline{4 \times 3}$ Multiply before subtracting.

$= \underline{25 - 12}$ Subtract from left to right.

$= 13$

Grouping symbols, such as parentheses, are used to indicate an order other than the normal operations agreement. When parentheses appear, we work *within* them before following the normal procedure. If the problem had been written $(25 - 4) \times 3$, we would have done the operation within the parentheses first and then the multiplication.

$\underline{(25-4)} \times 3$	Work within parentheses first.
$= \underline{21 \times 3}$	Multiply from left to right.
$= 63$	

Here is another example where parentheses make a difference.

$\underline{28 \div 4} + 3$	Divide before you add.
$= \underline{7 + 3}$	Add from left to right.
$= 10$	

Now with the parentheses included in the problem, the order of the operations changes.

$28 \div \underline{(4 + 3)}$	Work within the parentheses first.
$= \underline{28 \div 7}$	Divide from left to right.
$= 4$	

Wherever you see a number next to a parenthesis, with no operation symbol between them, that indicates multiplication. Look at this problem:

$6\underline{(7-4)}$	Work within parentheses first.
$= \underline{6(3)}$	Multiply by 6.
$= 18$	

Here's another:

$20 - 4\underline{(8-6)}$	Work within parentheses first.
$= 20 - \underline{4(2)}$	Multiply before you subtract.
$= \underline{20 - 8}$	Subtract from left to right.
$= 12$	

Study each of these examples.

EXAMPLE 1

$\underline{25 - 17} + 3 - 5$	Combine (add or subtract) from left to right.
$= \underline{8 + 3} - 5$	Combine left to right.
$= \underline{11 - 5}$	Combine left to right.
$= 6$	

EXAMPLE 2

$3 \times 15 - \underline{(6-2)}$	Work in parentheses first.
$= \underline{3 \times 15} - 4$	
$= \underline{45 - 4}$	
$= 41$	

EXAMPLE 3

$\underline{12 \div 3} \times 4 \div 2$ — Multiply or divide from left to right.

$= \underline{4 \times 4} \div 2$

$= \underline{16 \div 2}$

$= 8$

EXAMPLE 4

$\underline{\sqrt{81}} + 24 \div 3 - 2$ — Square root.

$= 9 + \underline{24 \div 3} - 2$ — Divide before adding or subtracting.

$= \underline{9 + 8} - 2$ — Combine left to right.

$= \underline{17 - 2}$

$= 15$

EXAMPLE 5

$5 + \underline{(14 - 3)} \times 2$ — Work within parentheses.

$= 5 + \underline{11 \times 2}$ — Multiply before adding.

$= \underline{5 + 22}$ — Combine (add).

$= 27$

EXAMPLE 6

$\underline{(36 \div 3)} + (12 \times 2 - 6)$ — Divide in first parentheses.

$= 12 + (\underline{12 \times 2} - 6)$ — Multiply in second parentheses.

$= 12 + \underline{(24 - 6)}$ — Subtract in second parentheses.

$= \underline{12 + 18}$ — Combine (add).

$= 30$

EXAMPLE 7

$14 + \underline{12 \times 4} - 3$ — Multiply before adding or subtracting.

$= \underline{14 + 48} - 3$ — Add (or combine) from left to right.

$= \underline{62 - 3}$ — Subtract.

$= 59$

EXAMPLE 8

$\underline{5}^2 - 6 \times 3 + 4$ — Exponent.

$= 25 - \underline{6 \times 3} + 4$ — Multiply before subtracting.

$= \underline{25 - 18} + 4$ — Combine from left to right.

$= \underline{7 + 4}$ — Combine from left to right.

$= 11$

EXAMPLE 9

$3\underline{\sqrt{64}} - 4^2$ — Square root.

$= 3 \cdot 8 - \underline{4^2}$ — Exponent.

$= \underline{3 \cdot 8} - 16$ — Multiply.

$= \underline{24 - 16}$ — Subtract.

$= 8$

EXAMPLE 10

$42 - 3\underline{(8-6)} + 5^2$	Parentheses.
$= 42 - 3(2) + \underline{5^2}$	Exponent.
$= 42 - \underline{3(2)} + 25$	Multiply before adding or subtracting.
$= \underline{42 - 6} + 25$	Combine from left to right.
$= \underline{36 + 25}$	Combine.
$= 61$	

EXAMPLE 11

$\underline{(220 \div 10)}\ \underline{(18 + 5)}$	Inside parentheses.
$= \underline{(22)\ (23)}$	Multiply.
$= 506$	

EXAMPLE 12

$5 \times 9 - \underline{(4 + 8)} \div 2 - (7 \times 4)$

$= 5 \times 9 - 12 \div 2 - \underline{(7 \times 4)}$

$= \underline{5 \times 9} - 12 \div 2 - 28$

$= 45 - \underline{12 \div 2} - 28$

$= \underline{45 - 6} - 28$

$= \underline{39 - 28}$

$= 11$

In Examples 13 through 15, you will notice more than one pair of grouping symbols within each problem. This happens often in mathematics. The symbols [ ], called brackets, serve the same function as a set of parentheses. They indicate that whatever is to be done inside of them should be completed before moving on. They are like another set of parentheses but are shaped differently so that you don't get confused by a set of parentheses within a set of parentheses. Work within the innermost set of grouping symbols first. You may then use the quantity that you have found to complete the problem. In most problems involving more than one set of grouping symbols, the parentheses are inside the brackets. You just need to remember to begin with the innermost grouping symbols and work your way out.

EXAMPLE 13

$[2\underline{(5 \times 5)} - 7]\ [6 - \underline{(12 \div 3)}]$	Parentheses inside brackets.
$= [\underline{2(25)} - 7]\ [6 - (4)]$	Multiply before subtracting.
$= [\underline{50 - 7}]\ [\underline{6 - 4}]$	Inside brackets.
$= \underline{[43]\ [2]}$	Multiply.
$= 86$	

EXAMPLE 14

$8 + [13 - 5\underline{(6 - 4)}]$	Parentheses inside brackets.
$= 8 + [13 - \underline{5(2)}]$	Multiply before subtracting.
$= 8 + \underline{[13 - 10]}$	Brackets.
$= \underline{8 + 3}$	Combine.
$= 11$	

EXAMPLE 15

$7[21 - 8\underline{(5 - 3)}] - 5 \times 2$	Parentheses inside brackets.
$= 7[21 - \underline{8(2)}] - 5 \times 2$	Multiply before subtracting.
$= 7\underline{[21 - 16]} - 5 \times 2$	Subtract in brackets.
$= \underline{7[5]} - 5 \times 2$	Multiply.
$= 35 - \underline{5 \times 2}$	Multiply before subtracting.
$= \underline{35 - 10}$	Subtract.
$= 25$	

If there are no exponents or square roots within the problem, you can skip over step 2 of the Order of Operations Agreement. If there are no parentheses, you can begin with step 2. Remember, though, that it is important for you to check for each step *in order* in each problem. Order of operations means the order in which things are to be done. Don't be discouraged if it takes a while before you can do these problems correctly all the time. Start working the problems with the Order of Operations Agreement on a notecard right in front of you, and go through all four steps each time. When you learn the Agreement and can put it away, you will quickly become successful with these problems.

Self-Check Simplify each of these expressions using the Order of Operations Agreement.

1. $32 - 10 + 12 - 7$
2. $32 - (10 + 12) - 7$
3. $28 - 3(7 - 2) + 2^3$
4. $41 - 2[16 - 4\ (12 \div 6)]$

Answers:

1. 27 2. 3 3. 21 4. 25

Using Formulas

One application for the Order of Operations Agreement is solving problems that use a formula. **Formulas** are general statements that are always true and that can be applied to particular situations. Formulas are general statements that express relationships. For instance, if you were planning a trip, you might need to know how far you can travel in an 8-hour day if you can average 52 miles per hour. There is a formula that states that the distance (D) you can travel is equal to your rate (r), or speed, times the length of time (t) that you travel. Written as a formula, it is $D = rt$. Whenever two **variables**, or letters that stand for numbers, appear next to each other with no symbol in between, the operation called for is multiplication. A number next to a variable also means to multiply. This formula can be used to determine distance whether you are talking about a trip by car, by plane, or by space shuttle. The relationship among distance, rate, and time is always the same. The values for the letters change, but the formula remains the same. To determine how far you can travel in 8 hours on your trip, write the formula, substitute the given values, and perform the mathematical calculation that is called for.

Steps to Use with a Formula

When working with a formula, you are encouraged to follow these steps:

1. Write the formula.
2. Substitute the values you know
3. Perform the indicated calculations.
4. Check the answer in the original problem.
5. Label the answer, if appropriate.

EXAMPLE 16 How far could you travel in an 8-hour day if you could average 52 miles per hour? (Use $D = rt$.)

Solution

$D = rt$	Write the formula.
$D = (52)(8)$	Substitute the given values.
$D = 52 \times 8$	Two letters next to each other tell you to multiply the values.
$D = 416$	You could travel 416 miles at 52 miles per hour during an 8-hour day.

◀

When you are given a mathematical formula and the values to be substituted for the letters, you will often need to remember how to apply the Order of Operations Agreement to do the calculations. Here are two examples of formulas that do not have real application.

EXAMPLE 17 Using the formula $D = 7A^2 - 4BC$, solve for D when $A = 3$, $B = 2$, and $C = 5$.

Solution

$D = 7A^2 - 4BC$	Write the formula.
$D = 7(3)^2 - 4(2)(5)$	Substitute the given values.
$D = 7(9) - 4(2)(5)$	Exponent.
$D = 63 - 4(2)(5)$	Multiply from left to right.
$D = 63 - 8(5)$	Multiply before subtracting.
$D = 63 - 40$	Multiply before subtracting.
$D = 23$	

◀

EXAMPLE 18 Using the formula $L = M + K(N - P)$, solve for L when $M = 12$, $K = 4$, $N = 9$, and $P = 3$.

Solution

$L = M + \mathrm{K}(N - P)$	Write the formula.
$L = 12 + 4(9 - 3)$	Substitute the given values.
$L = 12 + 4(6)$	Work within the parentheses.
$L = 12 + 24$	Multiply before adding.
$L = 36$	Add from left to right.

◀

As you study further in this book, you will find a number of sections that deal with formulas. Some will be taken from geometry; others will involve interest or percents; some will be taken from science. The process and steps remain the same in all problems involving a formula.

Self-Check

1. Using the formula $W = XY - Z(Y - 2)$, find the value of W when $Y = 5$, $X = 7,1$ $Y = 4$, and $Z = 2$.
2. Using the formula $P = C + M$, find the price of an item (P) when the dealer's cost (C) is \$17 and the markup (M) is \$6.

Answers:

1. $W = 22$
2. $P = \$23$

2.2 EXERCISES

A. Simplify the mathematical expressions using the Order of Operations Agreement. Be sure to check for each of the four steps in every exercise.

1. $42 \times 3 \div 7$
2. $65 \div 5 + 8$
3. $42 \div (7 \times 3)$
4. $65 \div (5 + 8)$
5. $19 - (3 \times 4)$
6. $28 + (9 \div 3)$
7. $(12 \times 8) \div (4 \times 4)$
8. $(20 \times 9) \div (5 \times 3)$
9. $(5 + 2)(8 \div 2)$
10. $(3 \times 5)(4 \div 2)$
11. $30 - (2 + 1)(5 \times 2)$
12. $42 - (3 \times 2)(6 + 1)$
13. $18 \div 3 + 6$
14. $56 \times 2 - 12$
15. $18 \div (3 + 6)$
16. $56 - 12 \times 2$
17. $12 + 4 \times 5 - 6$
18. $10 + 5 \times 3 - 7$
19. $24 - 6 + 7 - 2$
20. $18 - 4 + 9 - 7$
21. $14 + 6(5)$
22. $12 + 7(3)$
23. $36 \div 4 \times 3$
24. $42 \div 6 \times 7$
25. $20 - 3(4 + 2)$
26. $24 - 4(2 + 1)$
27. $20 - (3 \cdot 4) + 2$
28. $24 - (4 \cdot 2) + 1$
29. $28 \div (4 + 3)$
30. $36 \div (9 + 3)$
31. $(28 \div 4) + 3$
32. $(36 \div 9) + 3$
33. $24 + 12 \div 6 + 3$
34. $21 + 14 \div 7 - 2$
35. $(24 + 12) \div 6 + 3$
36. $(21 + 14) \div 7 - 2$

B. Simplify completely.

37. $3^2 + \sqrt{25}$
38. $5^2 - \sqrt{49}$
39. $2^4 - \sqrt{16}$
40. $4^2 + \sqrt{36}$
41. $3 \cdot 4^2 \div 2^3$
42. $8 \cdot 2^4 \div 4^2$
43. $3\sqrt{49} - 3^2$
44. $4\sqrt{25} - 4^2$
45. $\sqrt{81} - \sqrt{9} + \sqrt{36}$
46. $\sqrt{121} - \sqrt{64} + \sqrt{16}$
47. $38 - 2\sqrt{25}$
48. $42 - 3\sqrt{9}$
49. $4 + [3(9 - 5)]$
50. $6 + [4(8 - 3)]$
51. $24 - 3[5 + (2 + 1)]$
52. $37 - 4[3 + (8 - 2)]$
53. $6^2 \div 4 - 5 + 2$
54. $4 + \sqrt{100} \div 2 - 8$
55. $5 + 4[20 - 2(7 - 2)]$
56. $4 + 2[18 - 3(5 - 2)]$

57. $[2 + 3(4)]\,[5 + 2(3)]$
58. $[4 + 2(3)]\,[7 - 2(2)]$
59. $3[60 \div 3 + 9] - 14 \times 2$
60. $48 \div 3[12 - 4 \times 2] - 4$

C. Exercises 61 through 64 use the distance formula $D = rt$, which says that the distance you can travel is equal to the product of the rate, or speed, at which you move and the length of time that you travel.

61. Find the distance you can travel at 48 mph in 8 hours.
62. Find the distance you can travel at 52 mph in 7 hours.
63. Find the distance an object can travel at 1,200 feet per second in 15 seconds.
64. Find the distance that an object can travel at 1,500 feet per minute in 8 minutes.

D. Solve Exercises 65 through 74 using the formulas given. (They have no specific application in the real world.)

65. Find P in $P = 2R + 3S$ when $R = 5$ and $S = 4$.
66. Find P in $P = 2R + 3S$ when $R = 4$ and $S = 5$.
67. Find K in $K = 3L - 2M$ when $L = 12$ and $M = 9$.
68. Find K in $K = 3L - 2M$ when $L = 10$ and $M = 12$.
69. Find Y in $Y = 3A + 2B - 4C$ when $A = 5$, $B = 4$, and $C = 3$.
70. Find Y in $Y = 3A + 2B - 4C$ when $A = 4$, $B = 5$, and $C = 2$.
71. Find R in $R = 4S - 3T + 2U$ when $S = 12$, $T = 8$, and $U = 5$.
72. Find R in $R = 4S - 3T + 2U$ when $S = 10$, $T = 4$, and $U = 8$.
73. Find A in $A = 3B^2 - 5C$ when $B = 4$ and $C = 7$.
74. Find A in $A = 3B^2 - 5C$ when $B = 6$ and $C = 12$.
75. Find L in $L = 2M - N^2$ when $M = 18$ and $N = 5$.
76. Find R in $R = 4P^2 - 5TV$ when $P = 5$, $T = 3$ and $V = 4$.
77. Find A in $A = B(E - C^2)$ when $B = 4$, $C = 3$ and $E = 14$.
78. Find J in $J = 4[F - G(H + K)]$ when $F = 51$, $G = 5$, $H = 4$ and $K = 2$.
79. Find M in $M = 3[N - 3P^2R]$ when $N = 53$, $P = 2$ and $R = 4$.
80. Find W in $W = Y(X - 4Z^2)$ when $X = 26$, $Y = 3$ and $Z = 1$.

E. Use your calculator to simplify the following expressions using the Order of Operations Agreement. Give your answer to three decimal places if necessary.

81. $5.2(8.1 - \sqrt{20})$
82. $17.9(8.76 - \sqrt{12})$
83. $(14.3)\,(19.72) - 6.4(8.9)$
84. $28.2[16.4 - 2.9(4.7)]$
85. $(17.3)^3 - (16.2)^2$
86. $(21.8)\,(14.7) - 9.2(12.45)$
87. $37.4[26.8 - 7.3(2.5)]$
88. $(23.5)^2 - (7.3)^3$
89. $(6.5)^2 - \sqrt{169}$
90. $(14.5)^2 - \sqrt{225}$

F. Solve the following exercises using your calculator.

91. How far can a rocket travel in 126 seconds at 62 feet per second? (Use $D = rt$.)
92. How far does an airplane travel in 47 minutes if it averages 12 miles per minute? (Use $D = rt$.)
93. If the interest on America's national debt grows at the rate of $23,132 per day, how much will it grow in September? (Growth = Rate × Days.)
94. How much will the interest in Exercise 93 grow in a year? (Use $G = RD$.)
95. If $A = 3B - 4D + 2C$, find A when $B = 124$, $C = 92$, and $D = 76$.
96. If $K = 4M - 2N + 5P$, find K when $M = 426$, $N = 195$, and $P = 28$.
97. Find R in $R = 5S^2 - 4T$ when $S = 27$ and $T = 32$.
98. Find W in $W = 7X^2 - 9Y$ when $X = 42$ and $Y = 26$.

G. Read and respond to the following exercises.

99. Write an application problem in which you would have to use two different mathematical operations. Is there more than one way to solve your problem? Does it matter which math operation you do first?

100. Write the mathematical problem that you would use to answer the following question. Then use the Order of Operations Agreement to find the solution.

Tyetta went shopping and charged a coat for $85; 4 shirts, priced at $18 each but on sale at $3 off; and 2 pairs of stockings marked 3 for $12. Her charge account was credited when she returned 2 pairs of shoes that cost $26 per pair but now each had a $2 restock charge. What was the overall effect on the balance of her charge account after this trip?

101. Do the following exercise on paper using the Order of Operations Agreement: $26 - 4 \times 5$. Now do the exercise using your calculator. Are both answers the same? Did you get 6 both times? If your calculator display showed 6, it has built-in order of operations. This feature is called **hierarchy**. If your calculator display showed 110, it is not programmed to follow the Order of Operations Agreement, and you must be careful when doing this type of problem with your calculator.

Do the following problem on paper.

$33 - 6 \times 4 \div 3 + 5 \times 2$

If your calculator has hierarchy (order of operations), key in the same problem and check the answer that you got when you worked the problem by hand. State your answer.

2.3 Geometry Applications

Suppose you are putting a fence around your backyard, and the yard has an unusual shape. You are putting a fence around the outside of the yard, so you need to calculate the perimeter of the yard. (Whenever you are looking for the *distance around* something, you are looking for **perimeter**. If you want to know how much space is *contained inside* the fence, you are looking for **area**.) Suppose the yard has five different sides. You could write the formula for the perimeter of any five-sided figure as $P = a + b + c + d + e$, where a, b, c, d, and e represent the lengths of the sides.

EXAMPLE 1 The measurements of the sides of your yard are 50 feet, 35 feet, 42 feet, 12 feet, and 47 feet. Use the formula to find the amount of fencing that you would need to buy.

Solution

$P = a + b + c + d + e$ — Write the formula.

$P = 50 + 35 + 42 + 12 + 47$ — Substitute the values. Perform the operation called for.

$P = 186$ — You would need 186 feet of fencing for the yard.

As you saw in Example 1, real-life situations make use of geometric ideas.

Geometry figures and terms are often used in books other than geometry texts because pictures make it easier to remember a concept. Geometry is, in part, the study of shapes. Geometric figures, such as rectangles, are appropriate to this study of formulas because the formulas apply no matter how large or small the rectangle. The formula for the area of a rectangle is $A = L \times W$, or $A = LW$. (Remember that two variables next to each other tell you to multiply their values.)

EXAMPLE 2 Find the area of a rectangle whose length is 27 yards and whose width is 14 yards (see Figure 2.1).

Solution

$A = LW$	Write the formula.
$A = (27)\,(14)$	Substitute the given values. Perform the indicated operation.
$A = 378$	The area of the rectangle is 378 square yards. (Area is always measured in *square* units.)

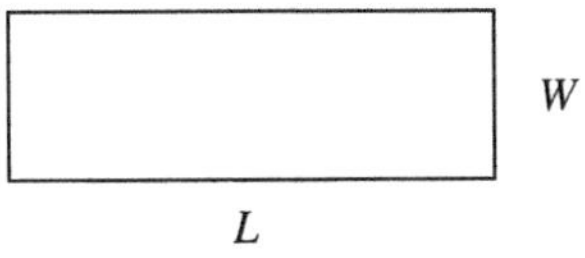

W

L

FIGURE 2.1

Steps to Use with Geometry Problems

Approaching geometry problems in a well-organized manner might make them more manageable. We suggest that you read the problem thoroughly and then do the following:

1. Make a sketch of the figure that is described and write on the sketch the measurements that are given.
2. Determine what is missing (what value you need to find).
3. Decide what formulas you could use to find that missing value and write them down.
4. Replace the letters (variables) in the formula with the measurements that are given and solve for the missing quantity.
5. Label the answer with the correct units of measure and check to see that the answer is reasonable.

EXAMPLE 3 Find the perimeter of the rectangle in Example 2.

Solution

$P = 2L + 2W$	Because perimeter is equal to the sum of the sides, and because a rectangle has two long sides called **lengths** and two short sides called **widths**, this is the formula to use.
$P = 2(27) + 2(14)$	Substitute the given values.
$P = 54 + 28$	Perform the indicated operations.
$P = 82$	The perimeter of the rectangle is 82 yards.

EXAMPLE 4 Find the perimeter of an **equilateral** (all sides equal) **triangle** if each of the three equal sides is 23 inches long.

Solution

$P = 3S$	The formula for perimeter of any triangle is $P = a + b + c$, where a, b, and c represent the lengths of the sides. In this case, however, all the sides are of equal length so $P = S + S + S$, or $P = 3S$ (see Figure 2.2).
$P = 3(23)$	Substitute the given values. Perform the indicated operation.
$P = 69$	The perimeter of this equilateral triangle is 69 inches.

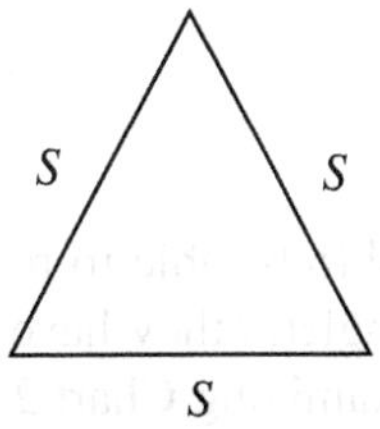

S S

S

FIGURE 2.2

EXAMPLE 5 Find the area of the square in Figure 2.3. Each of its equal sides is 17 inches long.

Solution

$A = S^2$	Because a square is a type of rectangle, use the formula for the area of a rectangle, $A = LW$. But in a square, L and W are the same; call them S. Therefore, $A = LW = S \times S = S^2$.
$A = 17^2$	Substitute.
$A = 289$	The area is 289 square inches.

S

S

FIGURE 2.3

Before going any further in these application problems, you need to learn some additional general information about geometric shapes and formulas. In these applications, you will work with perimeter of triangles, with area and perimeter of polygons, and with the volume of some three-dimensional figures called rectangular solids.

Triangles

When you use three line segments to join three points in a plane (on a flat surface) that are not all on the same straight line, you form a **triangle**. The symbol Δ is used to denote a triangle. A triangle, as its name implies, has three angles and, of course, three sides. Figure 2.4 shows samples of different kinds of triangles. The line segments are called **sides** and the "corners" are called **vertices**. We describe the sides by using line segment notation. That is, each of these triangles has three sides: $\overline{AB}$, $\overline{BC}$, and $\overline{AC}$. We denote the *lengths of these sides* by AB, BC, and AC without the line segment notation. Each triangle also has three vertices, A, B, and C. $\measuredangle A$ is the **angle opposite** side $\overline{BC}$. $\overline{AC}$ is the **side opposite** $\measuredangle B$. $\measuredangle C$ is the angle opposite $\overline{AB}$.

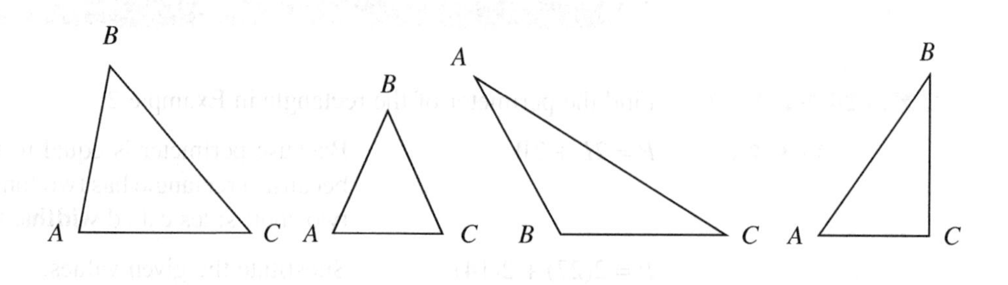

FIGURE 2.4

To show that two sides have the same length, you can say $AB = BC$ (the length of AB is the same as the length of BC), or you can say that $\overline{AB} \cong \overline{BC}$ (line segment $\overline{AB}$ is congruent to ($\cong$) line segment $\overline{BC}$). **Congruent to** means that one is an exact replica of the other. To show that two angles have the same measure, you could say

$m \measuredangle ABC = m \measuredangle DEF$

or you could say

$\measuredangle ABC \cong \measuredangle DEF$ (Angle ABC is congruent to angle DEF.)

There are several different types of triangles that you need to be able to recognize. Triangles are described in two ways: by the type(s) of angle(s) they have and by their sides. We shall first consider angle classification by examining Chart 2.1.

CHART 2.1 Classification by Angles

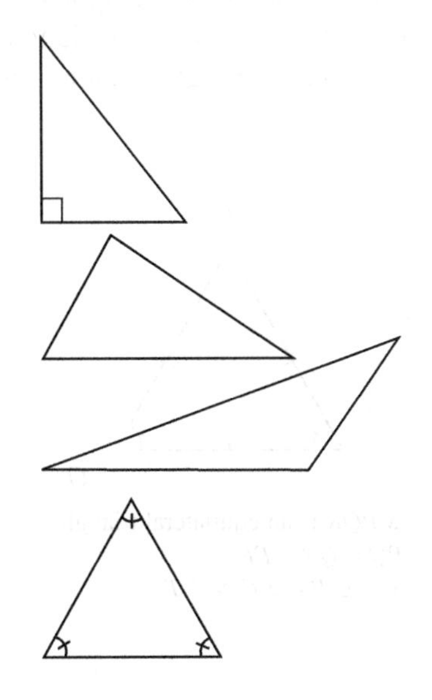

Right triangle: has one right angle; the other angles are acute; the side opposite the right angle is called the *hypotenuse*.

Acute triangle: all angles are acute (less than 90°).

Obtuse triangle: has one obtuse angle (greater than 90°); both other angles are less than 90°.

Equiangular triangle: all three angles are the same size—are congruent; the lengths of the sides are equal as well.

In this chart, angles with the same number of tick marks are congruent angles, or angles whose measures are equal.

Now let's use Chart 2.2 to consider classification by sides. The tick marks on the sides of the triangles indicate sides of the same length (same number of tick marks) or of different lengths (different numbers of tick marks).

CHART 2.2 Classification by Sides

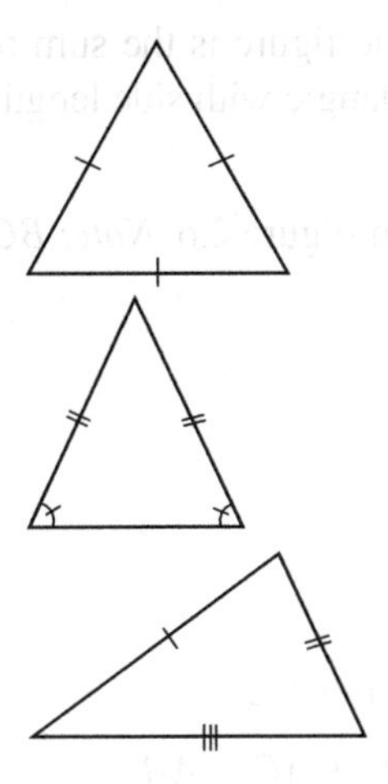

Equilateral triangle: three equal (or congruent) sides; the measures of the angles are equal as well.

Isosceles triangle: only two congruent sides; the angles opposite those sides are also congruent, and they are called **base angles**.

Scalene triangle: all three sides have different lengths; all three angles have different measures as well.

Triangles, therefore, can be described by using angle classification, or side classification, or both. You can talk about an obtuse isosceles triangle. This triangle has two congruent sides and one obtuse angle. A right isosceles triangle has two congruent sides and one right angle. An equilateral triangle is also equiangular. Figure 2.5 shows several different triangles and gives some information about their parts.

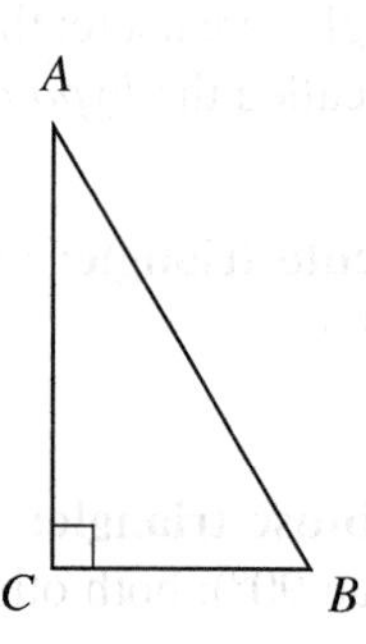

Δ *ABC* is a right triangle.
∡ *C* is a right angle.

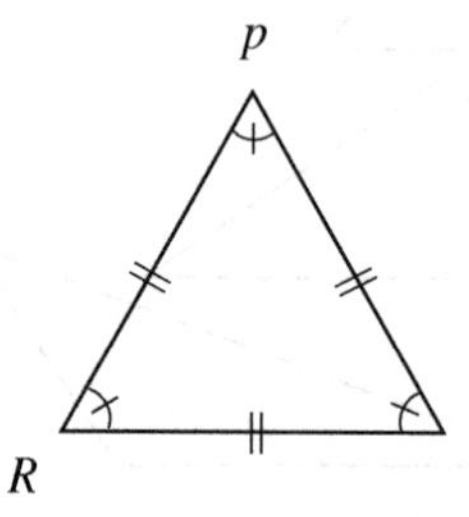

Δ *PQR* is an equilateral triangle.
PQ = *QR* = *PR*,
and ∡ *P* ≅ ∡*Q* ≅ ∡ *R*.

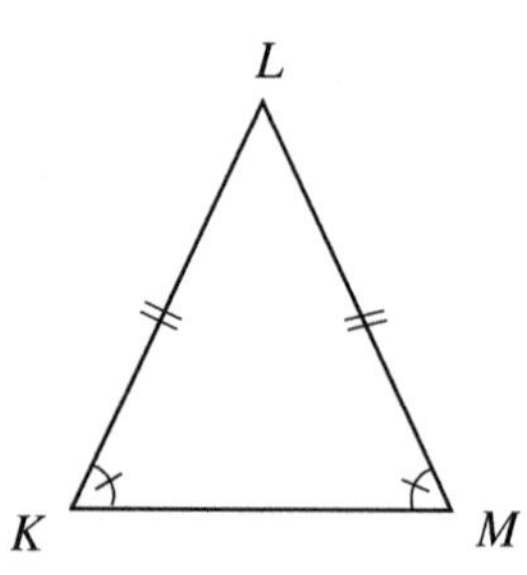

Δ *KLM* is an isosceles triangle.
KM = *LM*, ∡ *K* and ∡ *M* are base angles,
and ∡ *K* ≅ ∡ *M*.

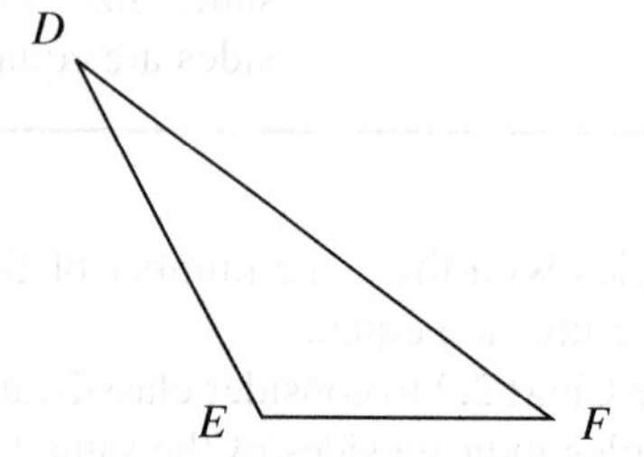

Δ *DEF* is an obtuse triangle.
∡ *E* is an obtuse angle.

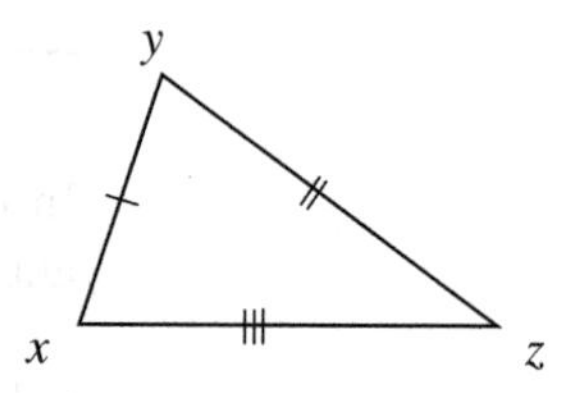

Δ *xyz* is a scalene triangle.
All sides and angles are different.

FIGURE 2.5

The perimeter of any geometric figure is the sum of the lengths of its sides. The formula for the perimeter of any triangle with side lengths *a*, *b*, and *c* is $P = \underline{a} + \underline{b} + \underline{c}$.

EXAMPLE 6 Find the perimeter of the triangle in Figure 2.6. *Note: BC* means the length of side $\overline{BC}$.

Solution

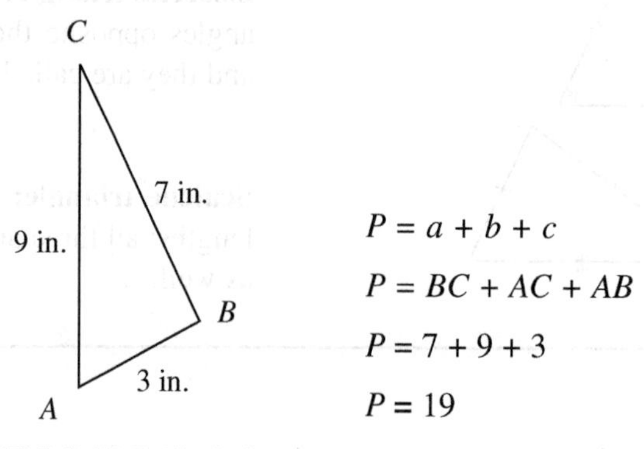

FIGURE 2.6

$P = a + b + c$

$P = BC + AC + AB$

$P = 7 + 9 + 3$

$P = 19$

Therefore, the perimeter of Δ*ABC* is 19 inches.

Sum of the Angles of a Triangle

The sum of the measures of the angles in a triangle is always equal to 180°.

EXAMPLE 7 Find the measure of $\measuredangle A$ in ΔABC in Figure 2.6 if $m\measuredangle B = 92°$ and $m\measuredangle C = 30°$.

Solution

$$m\measuredangle A + m\measuredangle B + m\measuredangle C = 180°$$
$$m\measuredangle A + 92° + 30° = 180°$$
$$m\measuredangle A + 122° = 180°$$
$$m\measuredangle A = 58°$$

Therefore, the measure of $\measuredangle A$ is 58 degrees.

EXAMPLE 8 Find the perimeter of ΔDEF in Figure 2.7.

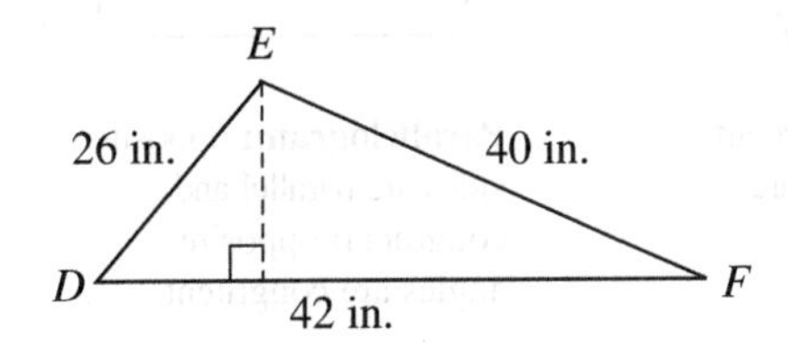

FIGURE 2.7

Solution We find the perimeter of a triangle by using the formula $P = a + b + c$.

$P = a + b + c$

$P = DE + EF + DF$

$P = 26 + 40 + 42$

$P = 108$

Therefore, the perimeter of triangle *DEF* is 108 inches.

Polygons

Four-sided plane geometric figures are called **quadrilaterals**. There are plane geometric figures that have more than four sides too, of course. This whole group of plane figures with straight-line sides is called **polygons**. A stop sign, for instance, is a type of polygon called an **octagon**. It has 8 sides. The Defense Department in Washington, D.C., is housed in a building shaped like a polygon with five equal sides that is called, for this very reason, the **Pentagon**. Triangles are also a special kind of polygon. If all the sides in a polygon are the same length, and all the angles have the same measure, such a polygon is called a **regular polygon**. A square is a regular polygon. An equilateral triangle is also a regular polygon. The stop sign we mentioned is a regular polygon. For the purposes of this section, we will study only four-sided polygons—quadrilaterals. Several special types of quadrilaterals are described in Chart 2.3. You need to be able to recognize all of them and to know what characteristics are particular to each type.

CHART 2.3

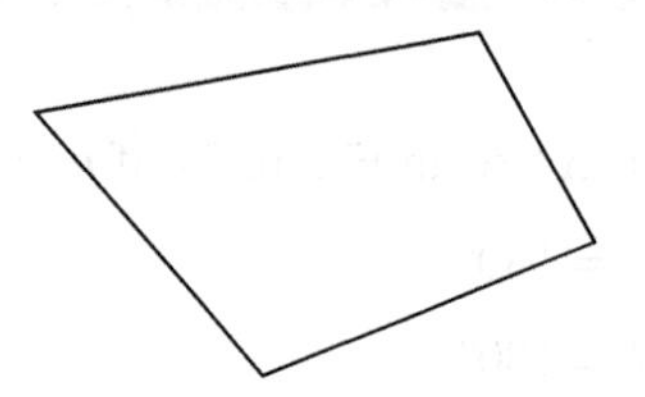

Quadrilateral: four sides.

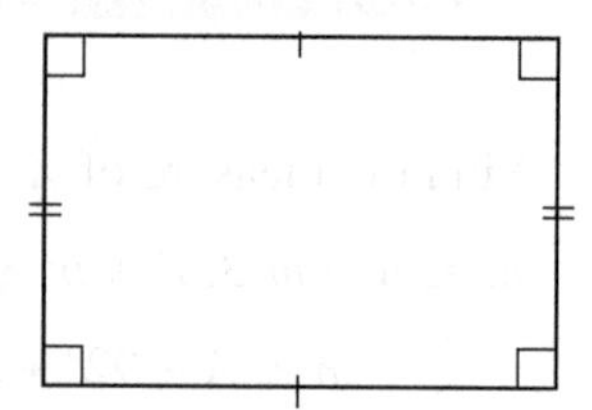

Rectangle: four sides, four right angles; opposite sides are parallel (||) and congruent ≅.

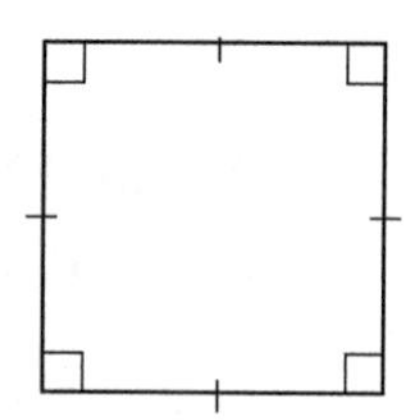

Square: four congruent sides, four right angles; opposite sides are parallel; all sides are congruent.

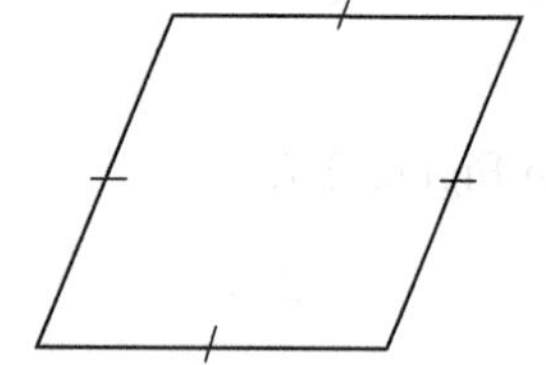

Rhombus: four congruent sides, opposite angles are congruent.

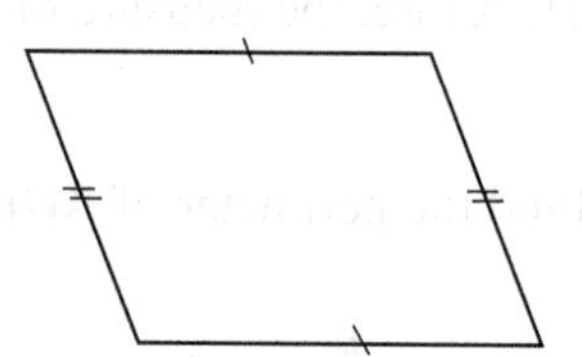

Parallelogram: opposite sides are parallel and congruent; opposite angles are congruent.

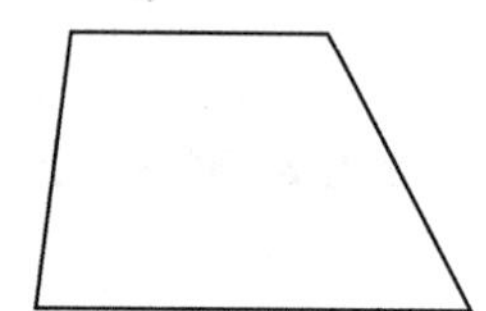

Trapezoid: only one pair of parallel sides, called bases.

EXAMPLE 9 Find the perimeter of the quadrilateral in Figure 2.8 if $a = 2$ feet, $b = 4$ feet, $c = 7$ feet, and $d = 6$ feet.

Solution

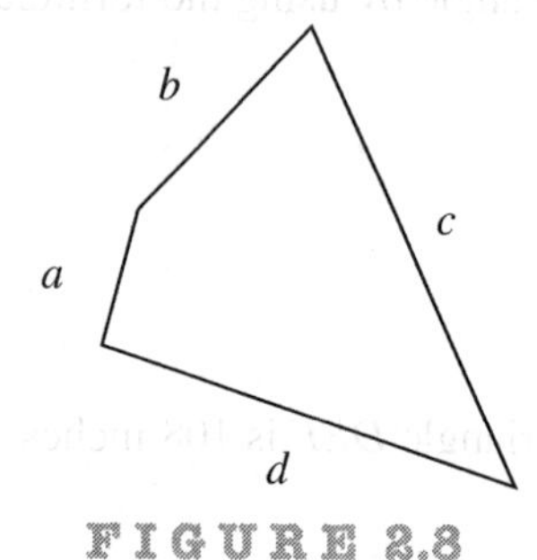

FIGURE 2.8

$P = a + b + c + d$

$P = 2 + 4 + 7 + 6$

$P = 19$

The perimeter is 19 feet.

Sum of the Angles in a Quadrilateral

The sum of the angles in any quadrilateral is 360°.

EXAMPLE 10 Find the measure of $\measuredangle B$ as seen in Figure 2.9 if $m \measuredangle A = 104°$, $m \measuredangle C = 102°$, and $m \measuredangle D = 96°$.

Solution

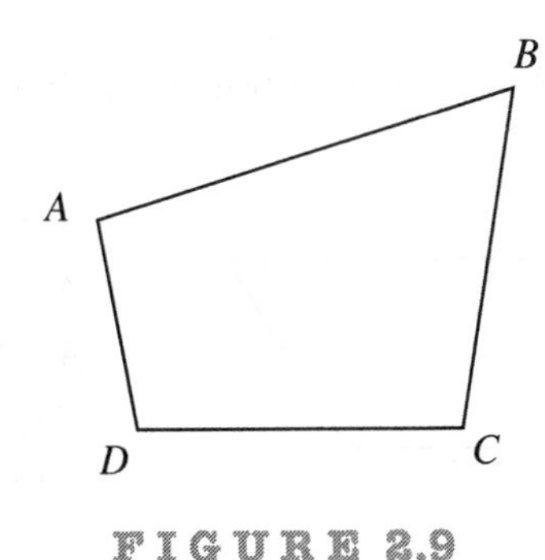

FIGURE 2.9

$$m \measuredangle A + m \measuredangle B + m \measuredangle C + m \measuredangle D = 360°$$
$$104° + m \measuredangle B + 102° + 96° = 360°$$
$$m \measuredangle B + 302° = 360°$$
$$m \measuredangle B = 58°$$

Therefore, the measure of $\measuredangle B$ is 58°.

EXAMPLE 11 Find the perimeter of a rhombus if one of its sides is 12 meters.

Solution First, sketch a rhombus similar to the one shown in Figure 2.10. A rhombus has 4 equal sides.

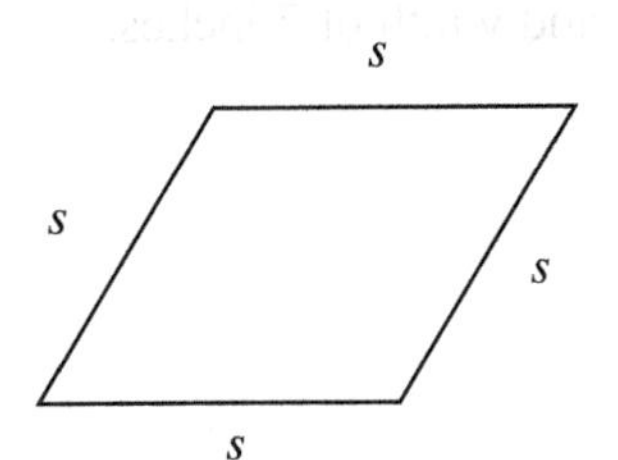

FIGURE 2.10

$P = s + s + s + s$ $\quad$ $P = 4s$

$P = 12 + 12 + 12 + 12$ $\quad$ or $\quad$ $P = 4(12)$

$P = 48$ $\quad$ $P = 48$

Therefore, the perimeter of this rhombus is 48 meters.

There is only one area formula for triangles, $A = 1/2\ bh$, but quadrilaterals have a different formula for each special shape. The area formulas for various quadrilaterals are given in Chart 2.4. Be sure to memorize these formulas and to remember that area is always given in square units.

CHART 2.4

Area Formulas for Quadrilaterals

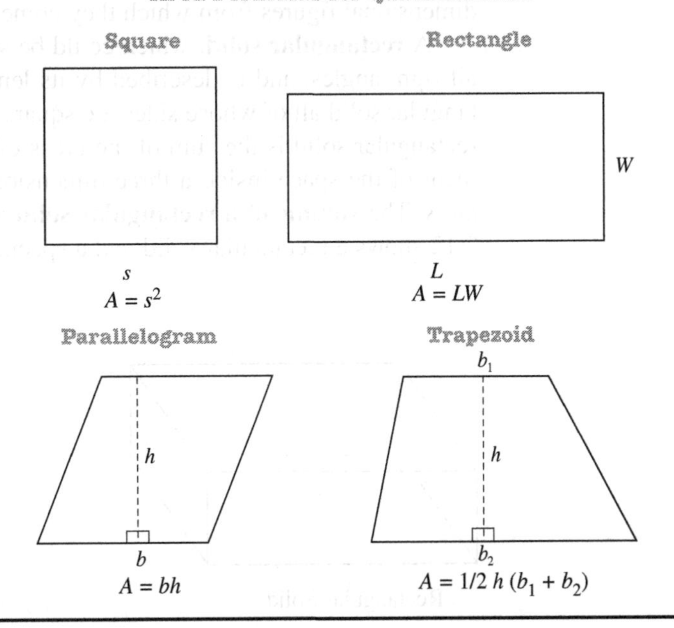

Note: In the formula for the area of a trapezoid, the two bases are denoted by b, but b_1 is a different base from b_2. Subscript notation is often used in mathematics.

EXAMPLE 12 Find the area of a parallelogram whose base is 16 feet and whose height is 5 feet. (See Figure 2.11.)

Solution

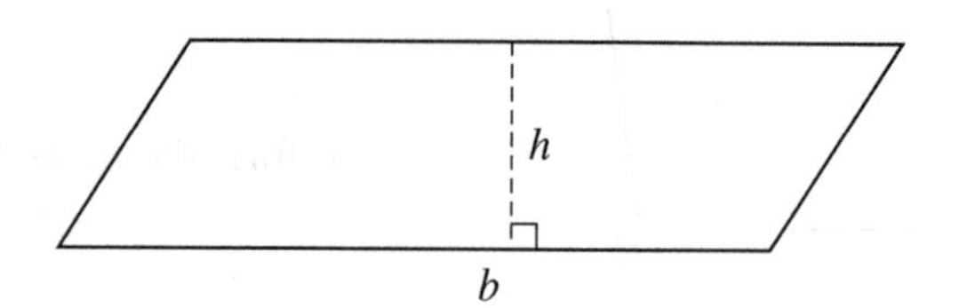

FIGURE 2.11

$A = bh$

$A = (16)(5)$

$A = 80$

The area of this parallelogram is 80 square feet.

Self-Check

1. Find the perimeter of a regular pentagon (5 sides) if one of its sides measures 6 centimeters.
2. Find the area of a rectangle with length of 10 inches and width of 7 inches.
3. Find the perimeter of the rectangle in Problem 2.

Answers:

1. 30 centimeters
2. 70 square inches
3. 34 inches

Three-Dimensional Figures

Up to this point, all the geometric figures you have studied could be drawn on a piece of paper—that is, on a plane. They were at most two-dimensional, having only length and width. Now we will consider three-dimensional figures—figures that have length, width, and height. Most of these figures get their names from the two-dimensional figures from which they come.

A **rectangular solid**, which could be shaped like a shoe box, has six sides, has all right angles, and is described by its length, width, and height. A **cube** is a rectangular solid all of whose sides are squares of the same size. The **surface area** of a rectangular solid is the sum of the areas of all its surfaces. Volume is the measurement of the space inside a three-dimensional figure. Volume is measured in *cubic units*. The **volume of a rectangular solid** is given by the formula $V = LWH$. Figure 2.12 shows a rectangular solid and a special type of rectangular solid called a cube.

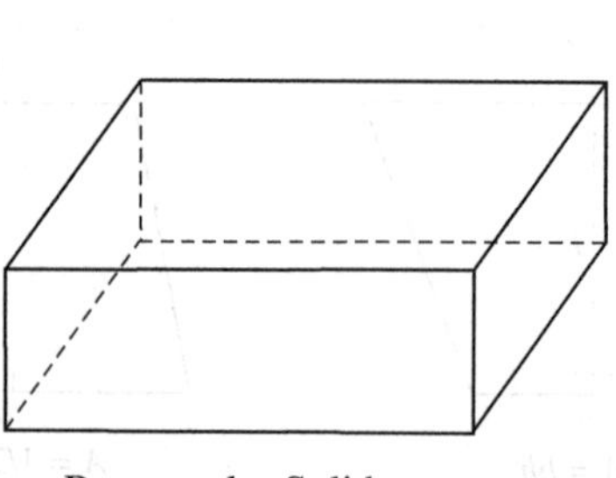

Rectangular Solid

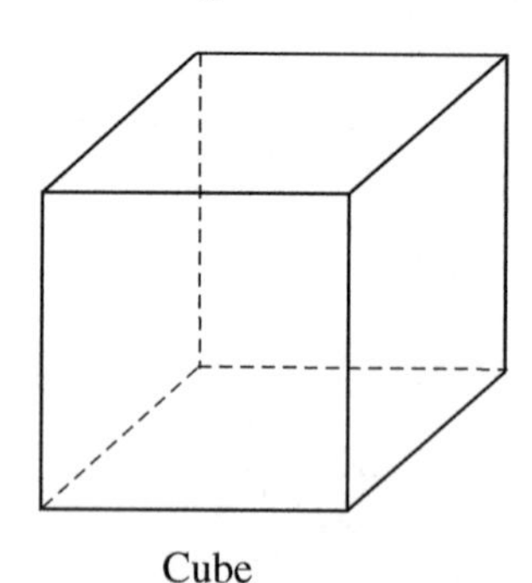

Cube

FIGURE 2.12

EXAMPLE 13 Find the volume of a rectangular solid whose length is 14 feet, height is 12 feet, and width is 3 feet.

Solution Because you are finding volume, use the formula $V = LWH$.

$V = LWH$

$V = (14)(3)(12)$

$V = (42)(12)$

$V = 504$

Therefore, the volume of this rectangular solid is 504 cubic feet.

EXAMPLE 14 Find the surface area of the rectangular solid in Example 13.

Solution In order to find the surface area, you will need to compute the area of the six surfaces. There will be three pairs of matching surfaces: a top and a bottom, a front and a back, and two sides. Draw a sketch similar to the one in Figure 2.13 to help you determine each area. Because each surface is a rectangle, you will use the area formula $A = LW$.

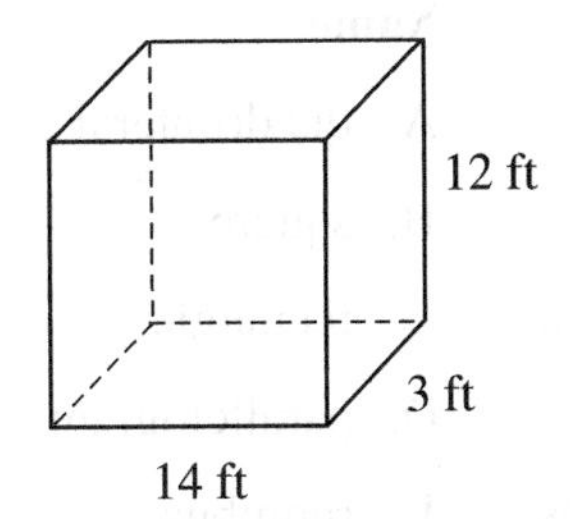

FIGURE 2.13

The area of the top and of the bottom is found using $A = (14)(3)$, so each of those two areas measures 42 square feet.

The area of each end is found using $A = (12)(3)$, so each measures 36 square feet.

The area of the front and back is found using $A = (14)(12)$, so each of the two measures 168 square feet.

Therefore, the total surface area is $42 + 42 + 36 + 36 + 168 + 168$, or 492 square feet.

2.3 EXERCISES

A. Complete the following sentences by filling in the blanks.

1. A(n) _____ triangle has three equal sides and three equal angles.
2. The two equal angles in an isosceles triangle are called _____ angles.
3. The corners in a triangle are called _____.
4. A triangle with one right angle is called a(n) _____ triangle.
5. A(n) _____ triangle has two equal sides.
6. The perimeter formula for a triangle whose sides have lengths a, b, and c, is _____.
7. A(n) _____ is a figure with 5 sides.
8. A(n) _____ and a(n) _____ each have four equal sides, but the _____ also has four right angles.
9. The sum of the measures of the angles in a triangle is always _____.
10. The sum of the measures of the angles in a quadrilateral is always _____.
11. If ΔABC and ΔDEF are exact duplicates of each other, ΔABC is said to be _____ to ΔDEF.
12. _____ is the symbol for "is congruent to."

B. Find the answer to each of the following exercises.

13. If in ΔPQR, $m\measuredangle P = 62°$ and $m\measuredangle Q = 54°$, find the measure of $\measuredangle R$.
14. If in ΔDEF, $\measuredangle E$ is a right angle and $m\measuredangle F = 36°$, find the measure of $\measuredangle D$.
15. In ΔKLM, $m\measuredangle K = 40°$ and $m\measuredangle L = 65°$. Find the measure of $\measuredangle M$.
16. In DEF, $m\measuredangle D = 26°$ and $m\measuredangle F = 22°$. Find the measure of $\measuredangle E$.

17. Triangle XYZ is a right triangle with $\overline{XY} \perp \overline{YZ}$. Find the measure of $\measuredangle Z$ if $m \measuredangle X = 30°$.
18. In ΔGHI, $\measuredangle G$ is a right angle and $m \measuredangle I = 18°$. Find the measure of $\measuredangle H$.

C. Find the answer to each of the following exercises.

19. Find the perimeter of a regular pentagon if each of its sides measures 17 inches.
20. Find the perimeter of a rhombus if one of its sides is 9 meters long.
21. Find the perimeter of a regular octagon if each of its sides measures 12 centimeters.
22. Find the measure of $\measuredangle B$ in quadrilateral $ABCD$ if $m \measuredangle A = 102°$, $m \measuredangle C = 45°$, and $m \measuredangle D = 110°$.
23. Find the perimeter of a parallelogram if two of its adjacent sides are 11 inches and 15 inches long.
24. Find the perimeter of ΔPQR in Figure 2.14.

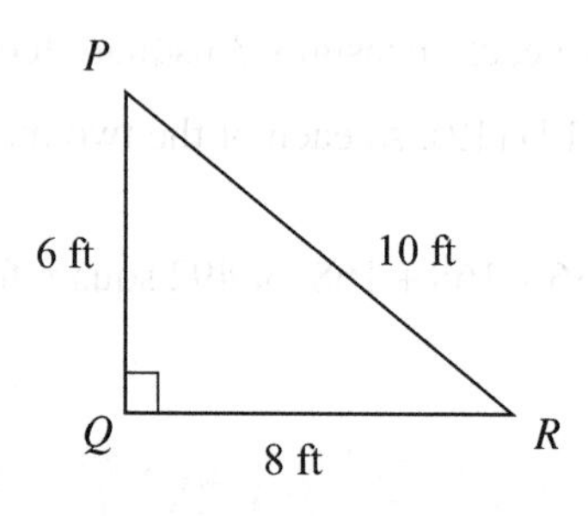

FIGURE 2.14

D. Exercises 25–30 refer to Figure 2.15. The following are measurements to be used: $PQ = 13$ yd, $QR = 15$ yd, $PR = 14$ yd, $PS = 5$ yd, and $SQ = 12$ yd.

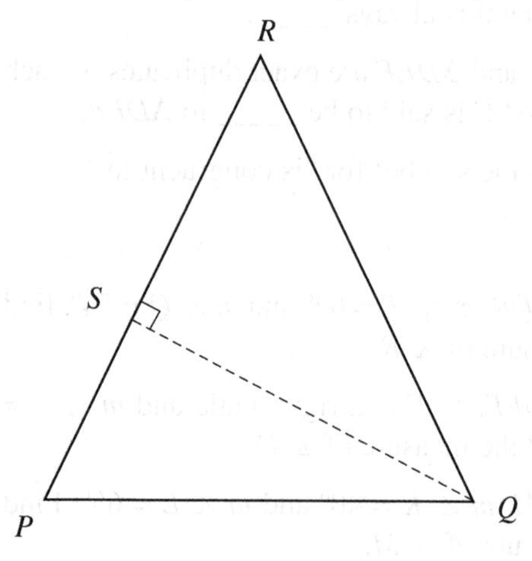

FIGURE 2.15

25. Find the perimeter of ΔPSQ.
26. Find the perimeter of ΔQRS.
27. Find the perimeter of ΔPQR.
28. If the $m \measuredangle P = 67°$ and $m \measuredangle R = 53°$, find the measure of $\measuredangle PQR$.
29. If the $m \measuredangle R = 53°$ and $\measuredangle QSR$ is a right angle as shown, find the measure of $\measuredangle SQR$.
30. Using Exercises 28 and 29, find the measure of $\measuredangle PQS$.

E. For each characteristic listed on the left, choose, from the list on the right, the names of all figures that always have that characteristic.

Condition	**Name**
31. four equal angles	A. quadrilateral
32. four equal sides	B. square
33. two pairs of parallel sides	C. rectangle
34. four sides	D. parallelogram
35. two pairs of equal sides	E. rhombus
36. one pair of parallel sides	F. trapezoid
37. no right angles	
38. four right angles	

F. Solve each of the following exercises.

39. Find the area of the rectangle in Figure 2.16.

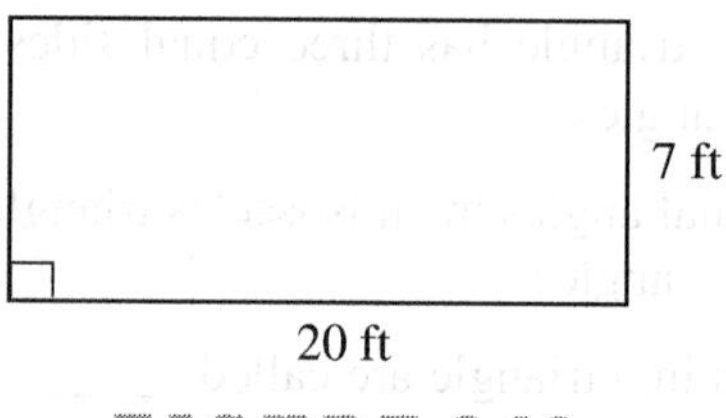

FIGURE 2.16

40. Find the perimeter of the rectangle in Figure 2.16.
41. Find the area of a square if one of its sides measures 18 inches.
42. Find the perimeter of a rhombus if one of its sides measures 23 meters.

43. Find the perimeter of the trapezoid in Figure 2.17.

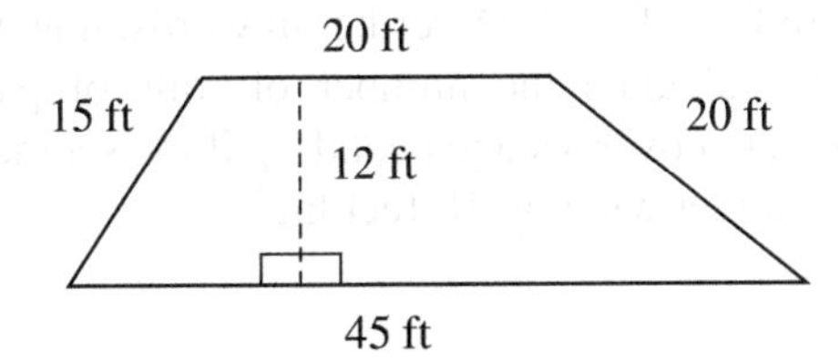

FIGURE 2.17

44. Find the perimeter of the parallelogram in Figure 2.18.

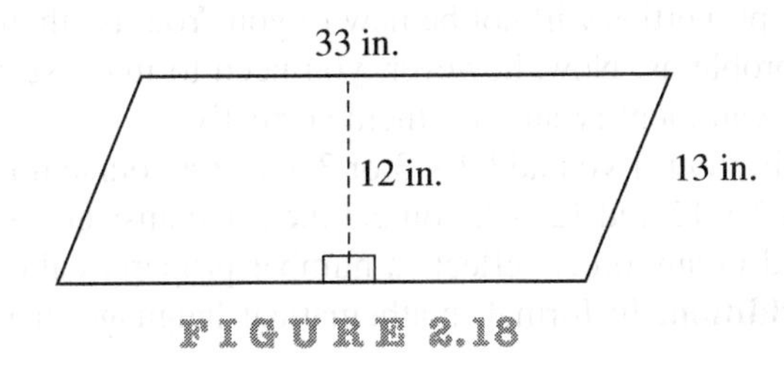

FIGURE 2.18

45. Find the area of the parallelogram in Figure 2.18.

46. Find the perimeter of the trapezoid in Figure 2.19.

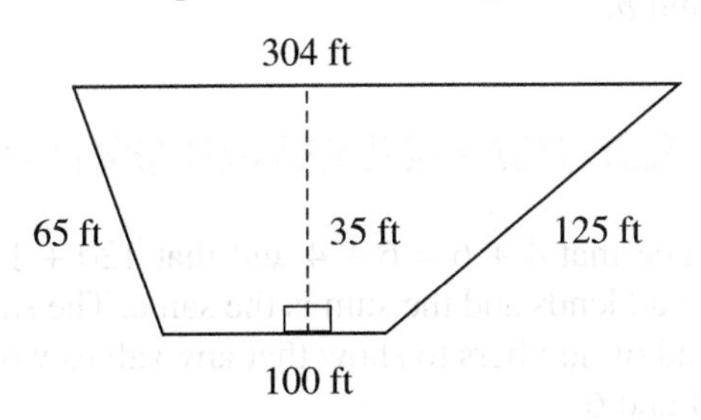

FIGURE 2.19

G. Respond to each of the following exercises.

47. Find the volume of a cube that is 6 cm wide.

48. Find the surface area of the cube in Exercise 47.

49. Find the surface area of a rectangular box that is 11 inches long, 8 inches wide, and 4 inches deep.

50. Find the volume of the box in Exercise 49.

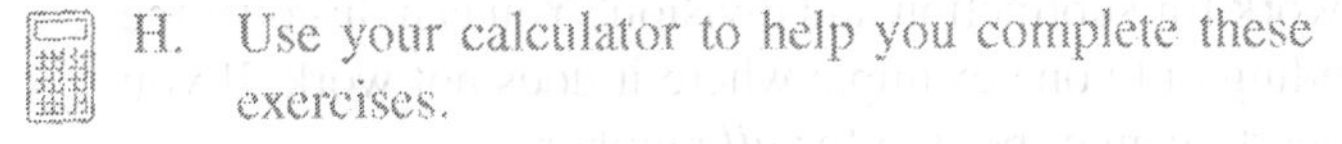

H. Use your calculator to help you complete these exercises.

51. Find the surface area of the carton in Figure 2.20.

52. Find the volume of the carton in figure 2.20.

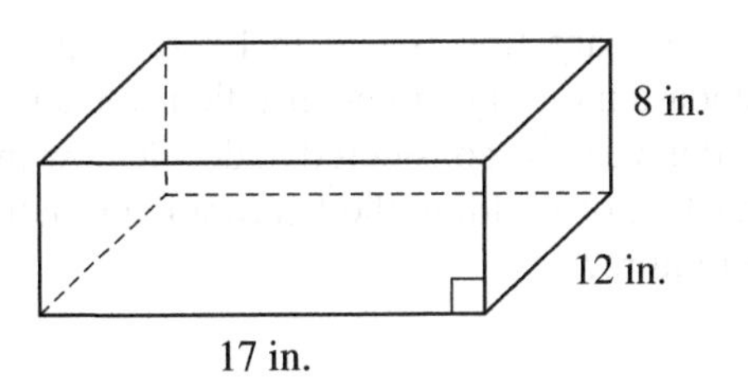

FIGURE 2.20

53. What is the surface area of the solid shown in Figure 2.21?

54. What is the volume of the solid shown in Figure 2.21?

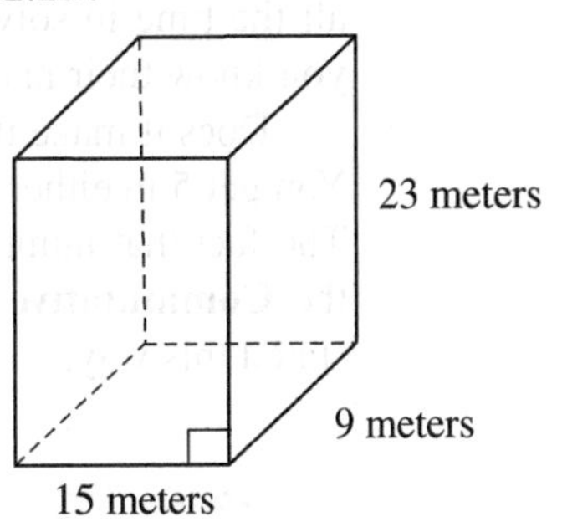

FIGURE 2.21

55. Find the cost to carpet a banquet hall whose dimensions are 27 yards by 24 yards if the cost of the carpet is $18 pr square yard.

56. Find the cost to lay parquet tile in an office entrance area if the dimensions of the entryway are 24 feet by 15 feet. The parquet tile sells for $7 a square foot.

I. Read and respond to each of the following exercises.

57. Your boss has given you the responsibility for planning the redecorating of the second floor in his office building. Explain what factors would have to be taken into consideration when planning your presentation on new floor covering to the Board of Directors and the Finance Committee.

58. Wright Rug is laying carpet in the Swenson's bedroom and adjoining hallway. Explain, in words, how you would decide how many square yards of carpeting are needed to cover rectangular areas that measure 14 feet by 11 feet and 4 feet by 12 feet.

59. After the carpeting is laid in Exercise 58, new baseboard molding will be installed. What things would have to be considered to decide how many linear feet of molding the bedroom and hallway will require?

60. A roll of wallpaper will cover approximately 35 square feet of wall. Describe, in words, how you would calculate the number of rolls of paper needed to cover a stage backdrop if the set measures 42 feet wide by 18 feet high.

2.4 Some Common Properties of Whole Numbers

When you are beginning any new project, it is important to review the basics and be sure you know the rules. It is difficult to follow any process if you don't understand what is going on from the very beginning. In this section, you are going to study some properties of numbers. These properties will not be new to you. You use them all the time in solving arithmetic problems. Now, however, you need to make sure you know their names and that you can identify and use them correctly.

Does it make the answer any different if you add 2 + 3 or 3 + 2? Of course not. You get 5 in either case. If you add 7 + 12 and 12 + 7, you get the same answer: 19. The fact that numbers can be added in any order reflects a number property called the **Commutative Property of Addition.** In formal, mathematical language, it is stated this way:

Commutative Property of Addition

For all numbers represented by a and b,

$a + b = b + a$

You know from your own experience that 4 + 6 = 6 + 4 and that 120 + 18 = 18 + 120. You can *commute* or move the addends and the sum is the same. The statement of the property uses a and b instead of numbers to show that any values would work, not just specific values such as 4 and 6.

The next property is like the first except that is applies to multiplication problems. What is 6 × 7 equal to? What is 7 × 6 equal to? Both answers are 42. What about 12 × 5 and 5 × 12? You get 60 each time. You can *commute* or move the factors and the product is the same. That is what this next property says.

Commutative Property of Multiplication

For all numbers represented by a and b,

$a \times b = b \times a$

Does this same property work for subtraction and division? You can discover the answer to that question by finding only one example where it does not work. If you find such an example, the property cannot be true for *all* numbers.

If you have \$20 in your checking account and write a check for \$8, how much money is left in your account? \$20 – \$8 = \$12. If, however, you have only \$8 in your checking account and write a check for \$20, what happens? \$8 – \$20 = an overdrawn account. You are \$12 in the red, in debt. From this example, you know that \$20 – \$8 is not the same as \$8 – \$20. Therefore, subtraction is *not* commutative.

The Commutative Property is easy to disprove for division as well. Suppose you made 8 pies for Thanksgiving and you are splitting them up equally for your two daughters to take home to their families. How many pies does each daughter get? (See Figure 2.22.)

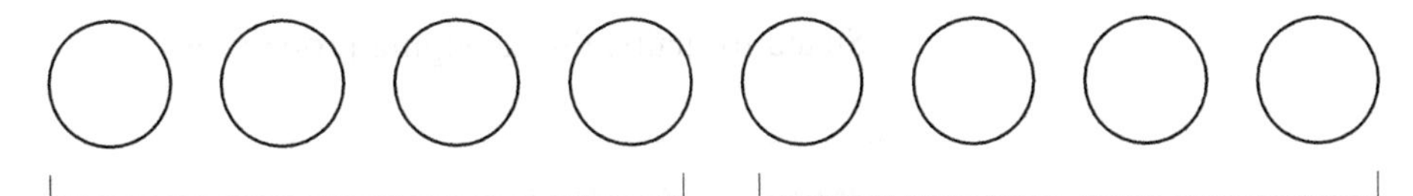

8 pies divided between 2 people

2 pies divided between 8 people

FIGURE 2.22

Eight pies divided by 2 daughters gives 4 pies to each family: $8 \div 2 = 4$. However, if you only have two pies, and there are 8 people at your dinner table, how much pie does each person get? $2 \div 8 = 1/4$. By this example, you can tell that $8 \div 2 = 4$ but $2 \div 8 = 1/4$. The results 4 and 1/4 are obviously not the same; the Commutative Property does not work for division.

When you are trying to learn about something as unexciting as "properties of numbers," it is a good idea to try to get a practical or even funny picture in your mind to help you remember the fact. It makes the property more interesting and easier to remember.

The next two properties, the **Associative Property of Addition** and the **Associative Property of Multiplication**, are easiest to show by example. Associates are people who group together to do something. Barnes and Associates might be a law firm or a consulting firm or a construction company. The Associative Properties of Addition and Multiplication talk about the way numbers are grouped together.

The way in which addends are grouped together does not change the answer in an addition problem. You know from your own experience that if you are doing a column addition problem, it does not matter whether you start from the top and add each succeeding number going down the column, or if you begin at the bottom of the column, getting sums as you go up the column. You can also *combine* certain numbers in the middle of a column to form 10s or 9s if you choose. The grouping does not matter in an addition problem. Look at this example. The parentheses, ( ), join quantities in a special way. The operations that appear inside the parentheses are always done *first.*

$3 + (7 + 9) = (3 + 7) + 9$ $3 + (16) = (10) + 9$ $19 = 19$	The order of the addends is the same, but they are grouped in different ways. However, both arrangements produce the same answer.

Associative Property of Addition

For all numbers represented by a, b, and c,

$(a + b) + c = a + (b + c)$

The same property holds true for multiplication. Look at this example:

$(5 \times 4) \times 8 = 5 \times (4 \times 8)$
$(20) \times 8 = 5 \times (32)$
$160 = 160$

The order of the factors is the same but the groupings are different.

Stated formally, this grouping property is:

Associative Property of Multiplication

For all numbers represented by a, b, and c,

$(a \times b) \times c = a \times (b \times c)$

Is there an Associative Property for Subtraction? Consider this example.

Is 12 – (8 – 2) the same as (12 – 8) – 2?

$12 - (8 - 2)$	?	$(12 - 8) - 2$
$12 - 6$	?	$4 - 2$
6	$\neq$	2

In order to disprove a property, you need to find only one example that will not work. This example proves that subtraction is not associative. See if you can disprove an Associative Property for Division. All you need is one situation where the property fails to be true. Neither subtraction nor division is associative.

Now let's consider zero. Zero is one digit that has some special properties that are not true for other numbers. What happens when you add zero to a number? Absolutely nothing. The identity of the number remains the same. For this reason, zero is called the *Identity Element for Addition.* You add $0 to your checking account and nothing happens to your balance: $5 + 0 = $5. 38 + 0 = 38. The **Addition Property of Zero** is just that simple.

Addition Property of Zero

For any number represented by a,

$a + 0 = a$

Zero is the Identity Element for Addition.

The **Multiplication Property of Zero** is easy to understand and to remember. You know from your own experience that zero times any number is equal to zero: $0 \times 8 = 0$, $12 \times 0 = 0$, and so on. That is all the property says.

Multiplication Property of Zero

For any number represented by a,

$a \times 0 = 0$

What about multiplying by the number 1? Does it have a special rule? You know that multiplying a number by 1 doesn't change a thing: $7 \times 1 = 7$, $25 \times 1 = 25$, and so on. Because the identity of the number multiplied by 1 does not change, 1 is called the *Identity Element for Multiplication.* That is exactly what the Multiplication Property of One says.

Multiplication Property of One

For any number represented by a,

$a \times 1 = a$

One is the Identity Element for Multiplication.

There is a corresponding property when you divide by 1.

Division Property of One

For any number represented by a, $a \div 1 = a$.

There are two special properties for division with zero. The rationale for these properties was discussed in Section 1.3. The two properties are listed formally here.

Division by Zero

For all real numbers represented by a,

$a \div 0$ is undefined

It has no real number answer.

Division into Zero

For all real numbers represented by a, $a \neq 0$,

$0 \div a = 0$

The last property we will study at this time is one that combines addition and subtraction operations with multiplication. The **Distributive Principle** is used throughout mathematics, and it is not difficult to understand. Consider the following example:

$6(5 + 4) = ?$ Because a number written just outside a set of parentheses means to multiply, the problem says to multiply 6 times the result of adding 5 and 4.

$6(9) = 54$ In order to do that, first add 5 and 4 to get 9, and then multiply 9 by 6.

Now, do the problem by finding the products first and then adding.

$6(5) + 6(4) = ?$ This time, find the product of 6 and 5 and the product of 6 and 4 and then combine those results.

$30 + 24 = 54$ You get 54 both times, so the two expressions must be equal.

$$6(5 + 4) = 6(5) + 6(4)$$
$$6(9) = 30 + 24$$
$$54 = 54$$

This example suggests that it doesn't matter whether you add first and then multiply the sum by the factor or you find the two products first and then add those results. This is, in fact, always true.

Distributive Principle

For all numbers represented by a, b, and c,

$a(b + c) = a(b) + a(c)$

and

$a(b - c) = a(b) - a(c)$

The following examples show that you get the same answer whether you add (or subtract) first and then multiply or you multiply first and then add (or subtract). See if you can understand each step of the "proof."

EXAMPLE 1 $5(9 + 3) = 5(9) + 5(3)$

$$5(12) = 45 + 15$$
$$60 = 60$$

EXAMPLE 2 $8(7 - 5) = 8(7) - 8(5)$

$$8(2) = 56 - 40$$
$$16 = 16$$

The Distributive Principle is a very powerful tool. To multiply 7 by 15 mentally, you could use the Distributive Principle and turn the problem into 7(10) + 7(5). That equals 70 + 35, which equals 105. If you wanted to multiply 23 × 14, you could do the multiplication in long form, using paper and pencil, or you could do the problem partly (or wholly) in your head by using the Distributive Principle.

$$\begin{aligned} 23 \times 14 &= 20(14) + 3(14) \\ &= 280 + 42 \\ &= 322 \end{aligned}$$

In this case, the multiplication by 14 is *distributed over* the sum of 20 and 3.

Sometimes in class, instructors amaze students with their ability to do complicated calculations in their heads. Perhaps they are using the Distributive Principle. Practice it and see if you can develop your own mental skills.

All of these properties are very useful for you to know. You have, in fact, been using most of them all along and may never have been aware of them. It is important for you to know each property and be able to recognize it when it is used.

2.4 EXERCISES

A. Match each of the statements in Column I with the appropriate property in Column II.

Column I	Column II
1. 7 × 0 = 0	a. Distributive Principle
2. 1 × 9 = 9 × 1	b. Multiplication Property of One
3. 1 + (2 + 3) = (1 + 2) + 3	c. Addition Property of Zero
4. 6(7 – 5) = 6(7) – 6(5)	d. Multiplication Property of Zero
5. 4 × (3 × 2) = (4 × 3) × 2	e. Commutative Property of Addition
6. 0 + 1 = 1 + 0	f. Associative Property of Addition
7. 1 + 0 = 1	g. Commutative Property of Multiplication
8. 19 × 1 = 19	h. Associative Property of Multiplication

B. Write a numerical example of each of the following properties and simplify each side to verify that the example is true.

9. Commutative Property of Multiplication
10. Commutative Property of Addition
11. Associative Property of Addition
12. Associative Property of Multiplication
13. Distributive Principle with Addition
14. Distributive Principle with Subtraction

C. Verify the Distributive Principle by simplifying each side of the equals sign. (See Examples 1 and 2.)

15. 8(12 + 2) = 8(12) + 8(2)
16. 25(10 – 7) = 25(10) – 25(7)
17. 7(9 – 3) = 7(9) – 7(3)
18. 13(8 + 3) = 13(8) + 13(3)
19. 4(5) + 4(7) = 4(5 + 7)
20. 6(9) – 6(4) = 6(9 – 4)
21. 14(10) – 14(6) = 14 (10 – 6)
22. 8(9) + 8(7) = 8(9 + 7)
23. (12 – 5)22 = 12(22) – 5(22)

24. 104(38) – 104(23) = 104(38 – 23)

25. 68(33) + 22(33) = (68 + 22)33

26. 47(53 + 27) = 47(53) + 47(27)

D. Read and respond to the following exercises.

27. Describe an activity that you do every day that could be described as commutative. Write a commutative property for the activity that shows how it is commutative.

28. Describe an activity that you do every day that could not be described as commutative and explain why it is not commutative.

29. Describe a real situation where the distributive property could apply.

30. Give a real-world example of a situation that shows why subtraction is not commutative.

31. Give a real-world example of a situation that shows why division is not commutative.

2.5 Factors

In the last chapter, addends were mentioned when the topic was addition, and factors were mentioned in the discussion of whole number multiplication. **Addends** are numbers that are added. **Factors** are numbers that are multiplied. You will encounter factors throughout your study of math. **Divisors** are numbers that divide into a given number. *Factors* are divisors whose remainders are zero. They can also be numbers that multiply together to give a certain number. This sounds pretty complicated, but it is not confusing once you get the idea. In this unit only whole number factors and divisors will be considered.

All Factors

EXAMPLE 1 Find all the factors of 30. (This problem is asking for all the whole numbers that divide into 30 without a remainder.)

Solution You will need some system for doing this because it asks for *all* the factors and you don't want to miss any of them.

What is the smallest whole number that is a factor of 30? 1, of course. Therefore, because $1 \times 30 = 30$, both 1 and 30 are factors of 30. The next number to try is 2. $2 \times 15 = 30$, so both 2 and 15 are factors of 30. And $3 \times 10 = 30$, so both 3 and 10 are factors of 30. There is no whole number to multiply by 4 to get 30, so 4 is not a factor of 30. $5 \times 6 = 30$, so both 5 and 6 are factors of 30. $6 \times 5 = 30$, but you already have 6 and 5, so you are finished. Let's look at this systematically.

$1 \times 30 = 30$

$2 \times 15 = 30$

$3 \times 10 = 30$

4 × *******

$5 \times 6 = 30$

6 × 5, which you already have

Therefore, 1, 2, 3, 5, 6, 10, 15, and 30 are all the factors of 30.

EXAMPLE 2 List all the factors of 28.

Solution $1 \times 28 = 28$

$2 \times 14 = 28$

3 × *******

$4 \times 7 = 28$

5 × *******

6 × *******

$7 \times 4 = 28$, which you already have

Therefore, 1, 2, 4, 7, 14, and 28 are the only whole numbers that divide evenly into 28. They are the only factors of 28. It is not necessary for you to write out the 3 × *****, the 5 × *****, and the 6 × ***** in your list as long as you remember to check all the numbers in numerical order until you come to a pair you already have.

EXAMPLE 3 List all the factors of 29.

Solution $1 \times 29 = 29$

That's it. Those are the only two whole numbers that are factors of 29.

Divisibility Tests

It would be helpful if you could tell very quickly whether a number will be divisible by, say, 2. If you think about it, you already know how to tell. Is 2 a factor of 35? No. Is 2 a factor of 66? Yes. What is the difference between 35 and 66 that lets you know that? 35 is an **odd number**—a number whose last digit is 1, 3, 5, 7, or 9. Therefore, 2 will not divide into it evenly and is not a factor of 35. 66 is an **even number**—a number whose last digit is 0, 2, 4, 6, or 8. Therefore, 2 is a factor of 66. This simple test is actually a *divisibility test*. There are several of these tests. You will need to learn three of them.

Divisibility Tests

1. A number is divisible *by 2 if it ends in 0, 2, 4, 6, or 8.* 2 is then a factor of the number.
2. A number is divisible *by 3 if the sum of its digits is divisible by 3.* 3 is then a factor of the number.
3. A number is divisible *by 5 if it ends in 0 or 5.* 5 is then a factor of the number.

The divisibility tests for 2 and for 5 make good sense, and you probably would have been able to come up with the test for 5 on your own, just as you did the test for 2. The test for 3 is a strange one, though. It needs some explanation, and using it takes some practice. You don't need to know why it works. But solving problems will be much easier if you remember how it works.

The reason the test for 3 is so convenient is that the sum of the digits is usually a fairly small number, and most people recognize the early multiples of 3 from having learned the multiplication facts. There are divisibility tests for some of the other numbers, but you won't need them very often. You will be using the tests for 2, 3, and 5 many times.

EXAMPLE 4 Is 3 a factor of 114?

Solution You could divide 114 by 3 and see that it divides evenly without a remainder. That would take some time. If you use the test for 3, you just add $1 + 1 + 4 = 6$. Because 3 is a factor of 6, you know that it is also a factor of 114.

EXAMPLE 5 Is 3 a factor of 57?

Solution $5 + 7 = 12$. 3 is a factor of 12. Therefore, 3 is a factor of 57.

EXAMPLE 6 Is 2 a factor of 58?

Solution 58 ends in 8, an even number, so 2 is a factor of 58.

EXAMPLE 7 Is 5 a factor of 86?

Solution 86 does not end in 0 or 5, so 5 is not a factor of 86.

EXAMPLE 8 Is 3 a factor of 46?

Solution $4 + 6 = 10$. Because 3 does not divide evenly into 10, 3 is not a factor of 46.

EXAMPLE 9 Decide if 2, 3, and 5 are factors of 180.

Solution 180 ends in 0, an even number, so 2 is a factor of 180.

$1 + 8 + 0 = 9$. 3 is factor of 9, so 3 is a factor of 180.

180 ends in 0. Therefore, 5 is also a factor of 180.

You know that these answers are correct because $2 \times 90 = 180$, $3 \times 60 = 180$, and $5 \times 36 = 180$.

Self-Check

1. Is 5 a factor of 120? How do you know?
2. Is 3 a factor of 3,021? How do you know?
3. Is 3 a factor of 128? How do you know?
4. Is 2 a factor of 527? How do you know?

Answers:

1. Yes; 120 ends in 0.
2. Yes; $3 + 0 + 2 + 1 = 6$ and 3 is a divisor of 6.
3. No; $1 + 2 + 8 = 11$ and 3 does not divide into 11 evenly.
4. No; 527 does not end in 0, 2, 4, 6, or 8.

Earlier in Example 3 we looked at the number 29 and found that it had only two factors, itself and 1. A number that has only two factors, itself and one, is called a **prime number** so 29 is a prime number.

Prime Number

A prime number is a whole number greater than 1 that has exactly two factors: itself and 1.

Numbers greater than 1 that are made up of more than two factors are called **composite numbers**. In Example 9, 180 is a composite number because it has factors of 2, 3, 5, and many more as well.

Composite Number

A composite number is a number greater than 1 that has more than two factors.

Let's look at the first few numbers and see which of them are prime numbers. To prove that a number is not a prime, you need only find one factor of that number other than the number itself and 1.

$0 \times 4 = 0$, $0 \times 12 = 0$. $0 \times$ any number is 0. Therefore, 0 *is not* a prime number.

$1 \times 1 = 1$. 1 only has one factor. 1 *is not* a prime number.

$2 \times 1 = 2$. Therefore, 2 *is* a prime number.

$3 \times 1 = 3$. Therefore, 3 *is* a prime number.

$2 \times 2 = 4$. Therefore, 4 *is not* a prime number. It is a composite number.

$5 \times 1 = 5$. Therefore, 5 *is* a prime number.

$2 \times 3 = 6$. Therefore, 6 *is not* a prime number. It is a composite number.

You should continue the list in this manner up through the number 30 and see how many prime numbers there are between 0 and 30. There is no pattern to the prime numbers. You will just need to find and remember which numbers they are. The list of prime numbers goes on indefinitely, but the ones you will use the most are the ones less than 30. See if your list includes 2, 3, 5, 7, 11, 13, 17, 19, 23, and 29. These are the prime numbers less than 30. Remember them!

Prime Factorization

The next step in this trip through new topics is to put together the concepts of prime numbers and factors and discuss a concept called prime factorization. As the name implies, **prime factorization** involves *writing a given number as a product of its* *prime factors.* Once again, if you follow a pattern, it's easy. Two methods will be shown. You will have an opportunity to try both, and then you can decide which one you prefer. The first example uses a technique called a **factor tree**.

EXAMPLE 10 Use a factor tree to find the prime factorization of 36.

Solution

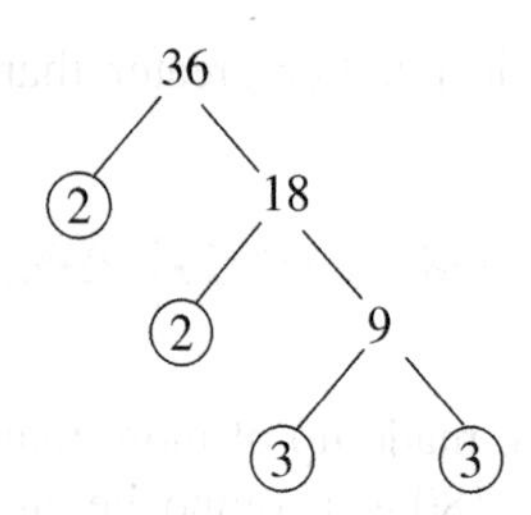

One way to make this tree is to begin with the smallest prime number that is a factor of 36. That number is 2. Write 36 as 2×18. Then take the smallest prime number that is a factor of 18. Next do the same for the 9 this step yields. That number is 3. Write 9 as 3×3. Because the remaining factor, 3, is already a prime number, you are finished. Therefore, $36 = 2 \times 2 \times 3 \times 3$.

Check This type of problem is easy to check. 2 and 3 are indeed prime numbers, and when you multiply all the factors you do get 36.

You do not have to begin to factor 36 by using the smallest prime. You could have begun with 6 times 6 or with 4 times 9. Then each factor would need to be factored again until each branch ended in a prime. Look at the following two trees used to factor 36. Be sure to circle each prime as you find it so that you don't lose any of them along the way. Note that the prime factorization is the same no matter which two factors you use first.

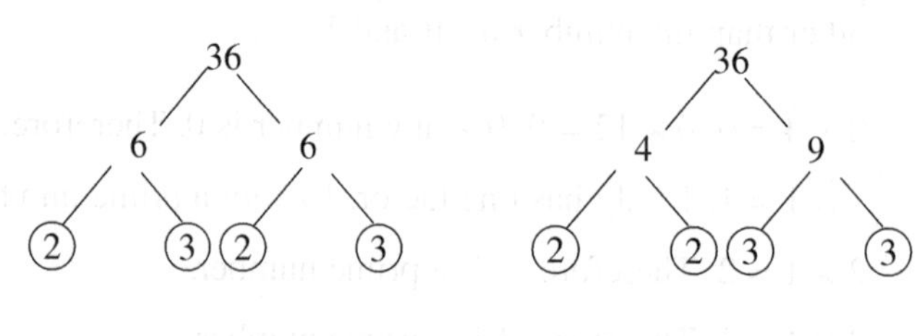

$36 = 2 \cdot 3 \cdot 2 \cdot 3$ $\qquad$ $36 = 2 \cdot 2 \cdot 3 \cdot 3$

Each answer could also have been written using exponent notation instead of expanded notation. In exponent notation, the prime factorization for 36 is written

$36 = 2^2 \times 3^2$ or $36 = 2^2 \cdot 3^2$

because 2 and 3 are both used as a factor twice.

EXAMPLE 11 Write 90 in its prime factored form. Use a factor tree.

Solution

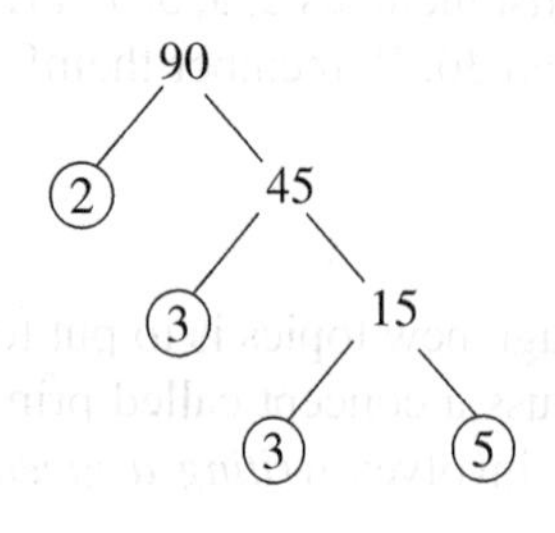

Because 3 and 5 are primes, you are finished. $90 = 2 \times 3 \times 3 \times 5$, or $90 = 2 \times 3^2 \times 5$.

Check This is correct because 2, 3, and 5 are prime numbers and when you multiply, you get 90.

EXAMPLE 12 Write the prime factorization of 189.

Solution

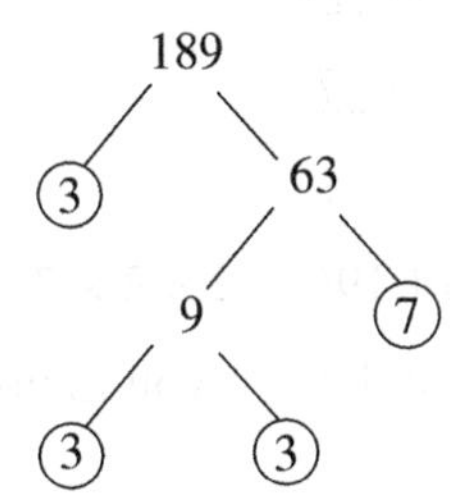

Because the sum of the digits is 18, 3 is a factor of 189. Now factor each branch until only primes remain. Remember to circle each prime factor as it appears.

Therefore, the prime factorization for 189 is $3 \cdot 3 \cdot 3 \cdot 7$ or $3^3 \cdot 7$.

Check 3 and 7 are primes, and when you multiply $3 \cdot 3 \cdot 3 \cdot 7$, you get 189.

EXAMPLE 13 Write 40 in its prime factored form. Use the method of **repeated division.**

Solution This time you will actually use the division process, but *each of your divisors must be a prime number.* You might begin with 2, the smallest factor that will divide without a remainder. Divide 2 into 40 and write the answer, 20, on the next line. Continue using 2 as the divisor until it no longer works. (The divisibility test will tell you this.) Then try 3, then 5, and so on through the list of prime numbers until the answer in the division is 1.

2)40	$40 \div 2 = 20$
2)20	$20 \div 2 = 10$
2)10	$10 \div 2 = 5$
5)5	2 is not a factor of 5 and neither is 3, but 5 is. $5 \div 5 = 1$.
1	

Therefore, the prime factorization of 40 is $2 \times 2 \times 2 \times 5$. It could also be written using exponent notation: $2^3 \times 5$.

Check All the factors are prime, and when you multiply, you get 40.

EXAMPLE 14 Write 975 in its prime factored form. Use the method of repeated division. It may be quicker to begin with 5 because we know from observation and the divisibility test that 5 is a factor of 975.

Solution

5)975	$975 \div 5 = 195$
5)195	$195 \div 5 = 39$
3)39	$39 \div 3 = 13$
13)13	$13 \div 13 = 1$
1	

Therefore, the prime factorization of 975 is $3 \cdot 5 \cdot 5 \cdot 13$ or $3 \cdot 5^2 \cdot 13$.

Check All the factors are prime numbers and, when you multiply, you get 975.

EXAMPLE 15 Find the prime factorization for 196. Use either method.

Solution

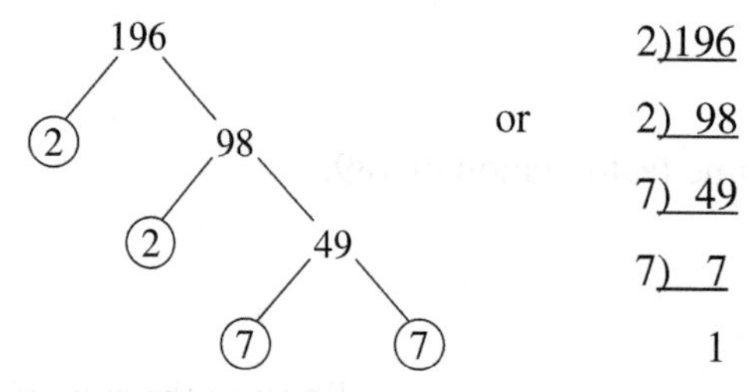

Therefore, the prime factorization of 196 is $2 \times 2 \times 7 \times 7$, or $2^2 \cdot 7^2$.

Check All the factors are prime numbers, and when you multiply, you get 196.

Self-Check

1. Write the prime factorization of 54.
2. Write the prime factorization of 84.
3. Write the prime factorization of 175.
4. Write the prime factorization of 450.

Answers:

1. $2 \cdot 3 \cdot 3 \cdot 3$ or $2 \cdot 3^3$
2. $2 \cdot 2 \cdot 3 \cdot 7$ or $2^2 \cdot 3 \cdot 7$
3. $5 \cdot 5 \cdot 7$ or $5^2 \cdot 7$
4. $2 \cdot 3 \cdot 3 \cdot 5 \cdot 5$ or $2 \cdot 3^2 \cdot 5^2$

These examples show why it is so helpful to know the divisibility tests, recognize the early prime numbers, know what a factor is, and be confident of your multiplication facts. You use all of those skills in the prime factorization process. We will use prime factorization to reduce fractions in the next chapter. It will make some of the toughest looking fractions reduce easily.

2.5 EXERCISES

A. List *all* the factors of the given number.

1. 8
2. 12
3. 21
4. 15
5. 7
6. 13
7. 32
8. 36
9. 18
10. 28
11. 45
12. 42
13. 70
14. 85
15. 120
16. 90
17. 150
18. 140
19. 200
20. 300

B. Answer the following questions without actually doing the division. Then explain you answer; that is, tell why or why not.

21. Is 14 divisible by 3?
22. Is 15 divisible by 2?
23. Is 216 divisible by 2?
24. Is 216 divisible by 3?

25. Is 216 divisible by 5?
26. Is 13,170 divisible by 2?
27. Is 13,170 divisible by 3?
28. Is 13,170 divisible by 5?
29. Is there a remainder when 525 is divided by 2?
30. Is there a remainder when 525 is divided by 3?
31. Is there a remainder when 525 is divided by 5?
32. Is 176 divisible by 3?
33. Is 2,142 divisible by 2?
34. Is 2,142 divisible by 3?
35. Is 2,142 divisible by 5?
36. Is 3 a factor of 153?
37. Is 2 a factor of 178?
38. Is 5 a factor of 270?
39. Is 3 a factor of 270?
40. Is 2 a factor of 270?

C. Find the prime factorization for each of the following numbers. You may use either a factor tree or the method of repeated division. You may write your answer in expanded form or use exponent notation.

41. 12
42. 18
43. 20
44. 30
45. 36
46. 50
47. 54
48. 42
49. 84
50. 98
51. 85
52. 100
53. 120
54. 66
55. 154
56. 132
57. 144
58. 225
59. 242
60. 250

D. Use your calculator to help you find the prime factorization for each of the following numbers.

61. 2,805
62. 3,325
63. 3,927
64. 644
65. 14,553
66. 6,006
67. 4,095
68. 13,013
69. 16,445
70. 34,307
71. 10,780
72. 10,626

E. Read and respond to the following exercises.

73. Using what you have learned about divisibility tests, write a divisibility test for 6.
74. Consider the list of the first 12 multiples of the number 9 (9, 18, 27, and so on.) Think about those numbers and write a divisibility test for 9. Test your new divisibility test by seeing if 9 is a divisor of 1,449.
75. Is your own words, explain the process you use to list all the factors of a given number.
76. Explain when it would be better to use the factor tree to find the prime factorization of a number and when the method of repeated division by primes might be the better choice.

2.6 Least Common Multiple

Least Common Multiple

The least common multiple (LCM) for two or more numbers is the smallest number into which each of those numbers divide evenly.

Try to find the least common multiple (LCM) for 8 and 12. The list of multiples of 12 begins

12, 24, 36, 48, . . .

The list of multiples of 8 begins

8, 16, 24, 32, 40, 48, . . .

You can see that 24 is the first common multiple for 8 and 12. There are many other numbers, such as 48, 72, and 96, that are also common multiples of 8 and 12, but 24 is the *least.* Therefore, 24 is the **least common multiple** (LCM) for 8 and 12.

You may remember that when you need to add or subtract fractions, you need a common denominator. Sometimes the process of finding a common denominator can be very difficult or require a lot of trial and error. In this section you will learn two methods for finding the least common multiple of several numbers. When those numbers are the denominators of fractions, these same methods will enable you to find the *least common denominator* for fractions. The method that looks so much like the prime factorization method is often the one that students prefer.

The approach of listing multiples works, but it is not very efficient. The other two methods you can use are less time-consuming. You need to remember that the name *least common multiple* explains exactly what you are trying to find. The examples that follow illustrate these two methods.

EXAMPLE 1 Find the least common multiple for 24 and 15 by using the method of repeated division by primes.

Solution This method looks a lot like the method for prime factorization described in the previous section. In fact, it is identical except that you divide into more than one number each time and multiply the factors together at the end. It is only necessary that the prime number you choose to divide by divides evenly into one of the numbers, although it may go into more than one of them. If it does not divide into a number, just bring that number down to the next line.

2)	24	15	Smallest prime into any is 2.
2)	12	15	Smallest prime into any is 2.
2)	6	15	Smallest prime into any is 2.
3)	3	15	Smallest prime into any is 3.
5)	1	5	Smallest prime into any is 5.
	1	1	All 1s across the bottom. Stop.

The least common multiple for 24 and 15 is $2 \times 2 \times 2 \times 3 \times 5 = 120$.

The LCM is 120.

EXAMPLE 2 Find the least common multiple for 4, 6, and 20.

Solution

2)	4	6	20	Smallest prime into any is 2.
2)	2	3	10	Smallest prime into any is 2.
3)	1	3	5	Smallest prime into any is 3.
5)	1	1	5	Smallest prime into any is 5.
	1	1	1	All 1s across bottom. Stop.

The LCM for 4, 6, and 20 is $2 \times 2 \times 3 \times 5 = 60$.

The LCM is 60.

EXAMPLE 3 Find the LCM for 14, 35, and 10.

Solution

2)	14	35	10	Smallest prime into any is 2.
5)	7	35	5	Smallest prime into any is 5.
7)	7	7	1	Smallest prime into any is 7.
	1	1	1	All 1s across the bottom. Stop.

The LCM for 14, 35, and 10 is $2 \times 5 \times 7 = 70$.

The LCM is 70.

There is another way to find the least common multiple (LCM). This method parallels the one you will use later in algebra. Find the prime factorization for each of the given numbers. List the primes under each other. Then bring down each prime number and multiply those primes together. That product is the LCM of the given numbers. Study Example 4, which is done using this method. It is exactly the same problem as Example 3, so the answer should be the same.

EXAMPLE 4 Find the LCM for 14, 35, and 10.

Solution Begin by finding the prime factorizations for 14, 35, and 10.

$14 = 2 \times 7$	List prime factors for 14.
$35 = \quad 7 \times 5$	Position 7 under 7 and start a new column for 5.
$10 = 2 \times \quad 5$	Position 2 under 2 and the 5 under the 5.
$\text{LCM} = 2 \times 7 \times 5 = 70$	Bring down the factor from each column. Their product is the least common multiple.

Compare this answer with that in Example 3.

EXAMPLE 5 Find the LCM for 24 and 15.

Solution

$24 = 2 \times 2 \times 2 \times 3$	$= 2^3 \times 3$
$15 = \qquad 3 \times 5$	$= \quad 3 \times 5$
$\text{LCM} = 2 \times 2 \times 2 \times 3 \times 5$	$= 2^3 \times 3 \times 5 = 120$

Compare this answer with that in Example 1.

Note that when you use exponential notation for the prime factorization, each factor is used to its *highest* power in the least common multiple.

EXAMPLE 6 Find the LCM for 12, 20, and 18.

Solution

$12 = 2 \times 2 \times 3$	$= 2^2 \times 3$
$20 = 2 \times 2 \quad\quad \times 5$	$= 2^2 \quad \times 5$
$18 = 2 \quad \times 3 \times 3$	$= 2 \times 3^2$
$\text{LCM} = 2 \times 2 \times 3 \times 3 \times 5$	$= 2^2 \times 3^2 \times 5 = 180$

Again, each factor in the prime factorization appears in the LCM raised to its highest power.

Self-Check

1. Find the Least Common Multiple for 6 and 10.
2. Find the Least Common Multiple for 12 and 18.
3. Find the LCM for 10 and 24.
4. Find the LCM for 9, 12, and 15.

Answers:

1. $\text{LCM} = 2 \cdot 3 \cdot 5 = 30$. The LCM is 30.
2. $\text{LCM} = 2 \cdot 2 \cdot 3 \cdot 3 = 2^2 \cdot 3^2 = 36$. The LCM is 36.
3. $\text{LCM} = 2 \cdot 2 \cdot 2 \cdot 3 \cdot 5 = 2^3 \cdot 3 \cdot 5 = 120$. The LCM is 120.
4. $\text{LCM} = 2 \cdot 2 \cdot 3 \cdot 3 \cdot 5 = 2^2 \cdot 3^2 \cdot 5 = 180$. The LCM is 180.

2.6 EXERCISES

A. Find the least common multiple (LCM) for each of the following groups of numbers. Use either method.

1. 6 and 8
2. 4 and 10
3. 7 and 12
4. 5 and 9
5. 4, 5, and 6
6. 7, 8, and 10
7. 2, 5, and 8
8. 3, 6, and10
9. 10 and 12
10. 12 and 20
11. 18 and 20
12. 15 and 12
13. 10, 15, and 18
14. 9, 12, and 20
15. 3, 4, 5, and 6
16. 3, 5, 8, and 10
17. 15, 20, and 24
18. 12, 18, and 20
19. 3, 7, 12, and 14
20. 5, 7, 10, and 21

B. Find the least common multiple for each of the following groups of numbers. You might find your calculator helpful.

21. 10, 12, 18, and 45
22. 16, 12, 20, and 15
23. 24, 30, 54, and 42
24. 60, 48, 108, and 36
25. 63, 45, and 25
26. 27, 45, and 12
27. 100, 60, and 75
28. 48, 30, and 72
29. 72, 180, and 108
30. 96, 72, and 120

C. Read and respond to the following exercises.

31. Describe in your own words what "least common multiple" means.
32. Which method do you prefer to use when finding the LCM? Why do you prefer it?

2.7 Greatest Common Factor

Another use for prime factorization is to find the greatest common factor for a group of numbers. The **greatest common factor** (GCF) is the largest factor (divisor) common to all the given numbers. To find the greatest common factor, you will find the prime factorization for each number and then list those factors in an orderly fashion. You can then identify the factors that need to be multiplied together to produce the greatest common factor. When there are no common prime factors, the GCF is 1. Numbers whose greatest common factor is 1 are said to be **relatively prime**.

EXAMPLE 1 Find the greatest common factor for 20 and 30.

Solution First, find the prime factorization for each number.

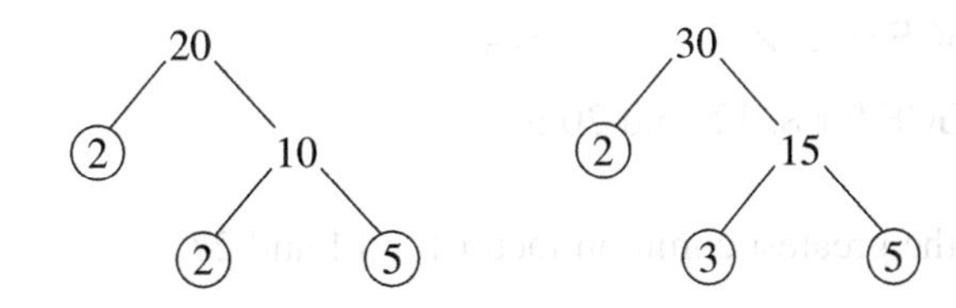

List the factors, with like factors under each other, leaving spaces for those that are missing. The greatest common factor (GCF) is the product of the factors that are *common to both 20 and 30.*

$$\begin{array}{rl} 20 = & \boxed{2} \times 2 \times \boxed{5} \\ 30 = & \boxed{2} \times 3 \times \boxed{5} \\ \hline \text{GCF} = & 2 \times 5 = 10 \end{array}$$

Therefore, 10 is the largest number that is a divisor (factor) of both 20 and 30.

EXAMPLE 2 Find the greatest common factor for 12, 24, and 30.

Solution

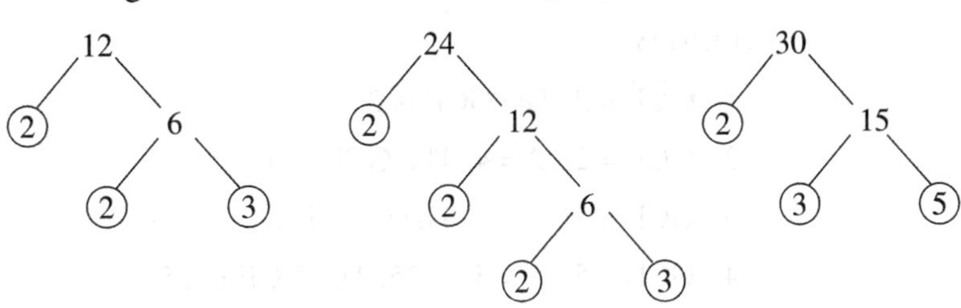

$$\begin{array}{rl} 12 = & \boxed{2} \times 2 \times \boxed{3} \\ 24 = & \boxed{2} \times 2 \times 2 \times \boxed{3} \\ 30 = & \boxed{2} \times \boxed{3} \times 5 \\ \hline \text{GCF} = & 2 \times 3 = 6 \end{array}$$

Line the factors up under each other in the same way that you did when finding the LCM. This time, however, you want the factors common to *all* of the numbers. Circle the 2 and 3 and bring them down. Their product is the greatest common factor.

The GCF for 12, 24, and 30 is 6.

EXAMPLE 3 Find the greatest common factor for 8, 12, and 20.

Solution

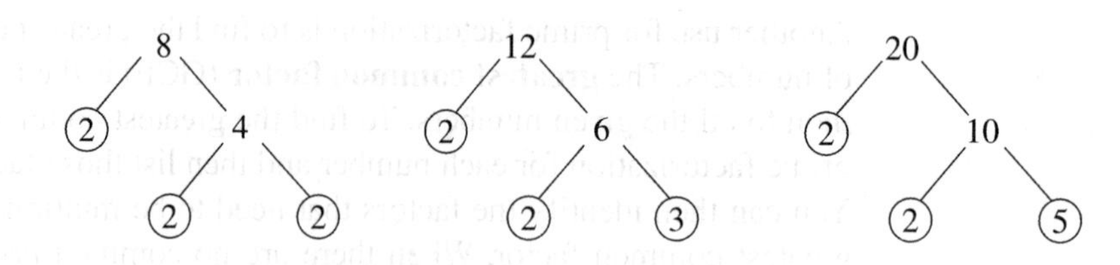

$$\begin{aligned} 8 &= 2 \times 2 \times 2 \\ 12 &= 2 \times 2 \times 3 \\ 20 &= 2 \times 2 \times 5 \\ \hline \text{GCF} &= 2 \times 2 = 4 \end{aligned}$$

The GCF for 8, 12, and 20 is 4.

EXAMPLE 4 Find the greatest common factor for 14 and 27.

Solution

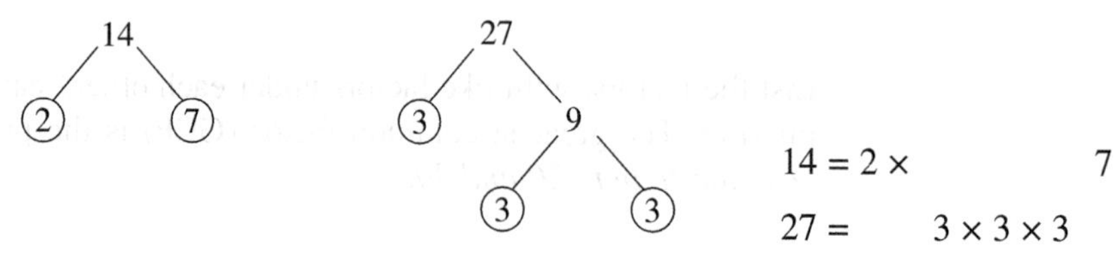

$14 = 2 \times 7$

$27 = 3 \times 3 \times 3$

Because there are no common prime factors for 14 and 27, their greatest common factor (GCF) is 1. These numbers are said to be *relatively prime* because their GCF is 1.

Self-Check

1. Find the Greatest Common Factor of 12 and 10.
2. Find the Greatest Common Factor of 8 and 12.
3. Find the GCF of 18 and 24.
4. Find the GCF of 50 and 75.

Answers:

1. GCF = 2. The GCF is 2.
2. GCF = $2 \cdot 2 = 4$. The GCF is 4.
3. GCF = $2 \cdot 3 = 6$. The GCF is 6.
4. GCF = $5 \cdot 5 = 5^2 = 25$. The GCF is 25.

As you practice finding the greatest common factors, your skills in using prime factorization and the divisibility tests will also improve.

2.7 EXERCISES

A. Find the greatest common factor (GCF) for each of the following sets of numbers. If the numbers are *relatively prime*, please include that in your answer.

1. 10 and 12
2. 15 and 20
3. 8 and 12
4. 8 and 20
5. 24 and 40
6. 24 and 60
7. 9 and 10
8. 12 and 35
9. 15 and 30
10. 12 and 48
11. 16, 18, and 20
12. 8, 12, and 20
13. 21 and 35
14. 28 and 42
15. 14 and 15
16. 15 and 28
17. 15, 20, and 30
18. 12, 16, and 20
19. 18, 30, and 42
20. 20, 30, and 40

B. Find the greatest common factor (GCF) for each of the following groups of numbers. You might find it helpful to use your calculator.

21. 72, 180, and 108
22. 42, 105, and 63
23. 210, 336, and 294
24. 700, 420, and 1,120
25. 216, 360, and 432
26. 288, 180, and 108
27. 252, 504, and 315
28. 160, 192, and 256
29. 120, 216, and 144
30. 270, 360, and 315

C. Read and respond to the following exercises.

31. Describe in your own words what "greatest common factor" means.
32. A classmate is having difficulty remembering GCF and LCM. He gets them confused. What would you say to help him?

2.8 CHAPTER REVIEW

This chapter has been a real mixture of topics and ideas. The difficult thing is not mastering each one separately but knowing what to do when all of them are used together. Be sure you have studied each topic throughly and are comfortable with each one. Stydy related topics together, and you will see the progression. Use the Order of Operations Agreement to simplify some formulas. Factors and prime numbers are both involved in prime factorization. And the use of prime factorization ties together least common multiples and great common factors.

When you have finished studying this chapter, you should be able to:

1. Recognize and discuss the following terms and expressions: exponent, base, square root, radical, radicand, perfect square, root, formula, perimeter, area, factors, prime numbers, LCM, GCF, and relatively prime.
2. Simplify and find approximate values for exponent and radical expressions by using tables.
3. Use the Order of Operations Agreement to simplify expressions that include multiple operations and grouping symbols.
4. Substitute values into a formula, and evaluate to find the answer to a question.
5. Apply the definitions and formulas for triangles, polygons, and rectangular solids.
6. Recognize and identify the number properties.
7. Beginning with the least and greatest factors of a number, find *all* the factors of the number.

8. Use the divisibility tests for 2, 3, and 5.
9. Find the prime factorization of any number by using a factor tree or repeated division by primes.
10. Find the least common multiple for any group of numbers by using repeated division by primes or by writing each factor to its highest power.
11. Find the greatest common factor for any group of numbers by writing each factor that is common to all the given values.

REVIEW EXERCISES

A. Find the value of each of the following exponent or radical expressions.

1. 5^3
2. $\sqrt{36}$
3. $\sqrt[3]{64}$
4. 3^5
5. $\sqrt{144}$
6. $\sqrt[3]{1}$
7. $\sqrt[4]{16}$
8. $\sqrt[4]{81}$
9. 1^4
10. 0^6

B. Use the Order of Operations Agreement to simplify each of the following expressions.

11. $4 + 9 - 3 + 2$
12. $8 \div 2 \times 4 \div 2$
13. $24 \div 3 \times 2 \div 4$
14. $14 + 6 - 5 + 3$
15. $32 - 4 \times 3 \div 2$
16. $45 - 6 \times 3 \div 9$
17. $18 - 2 + 4 \times 3$
18. $22 - 5 + 3 \times 6$
19. $(8 - 5) + (12 \div 3)$
20. $(9 - 2) + (16 \div 2)$
21. $7(9 + 12)$
22. $9(6 + 11)$
23. $2^3 + 9 \div 3 - 1$
24. $3^2 - 8 \div 2 + 1$
25. $\sqrt{16} - 3 + 1$
26. $\sqrt{36} - 4 + 2$
27. $5^2 - \sqrt{49} + 8 \times 2$
28. $7^2 - \sqrt{81} + 7 \times 4$
29. $34 - 3(8 - 4)$
30. $26 - 4(7 - 3)$
31. $(18 \div 3)(12 - 6)$
32. $(30 \div 6)(9 + 2)$
33. $40 + 5[8 - 2(6 - 3)]$
34. $36 + 4[9 - 2(5 - 3)]$

C. Solve for the indicated variable.

35. Find the value for A in $A = P + R(T - V)$ when $P = 8, R = 5, T = 7,$ and $V = 3$.
36. Find the value for A in $A = P + R(T - V)$ when $P = 6$, $R = 8$, $T = 9$, and $V = 4$.
37. Find the value of K in $K = 2L + 3M$ when $L = 8$ and $M = 4$.
38. Find the value of R in $R = 5W - 3Y$ when $W = 8$ and $Y = 4$.
39. Find the value of H in $H = K^2 - 3L$ when $K = 9$ and $L = 5$.
40. Find the value of D in $D = E^2 - 5F$ when $E = 6$ and $F = 4$.
41. Using $D = rt$, find out how far you can travel in 22 hours at 56 mph.
42. Find the perimeter of the form in Figure 2.23.

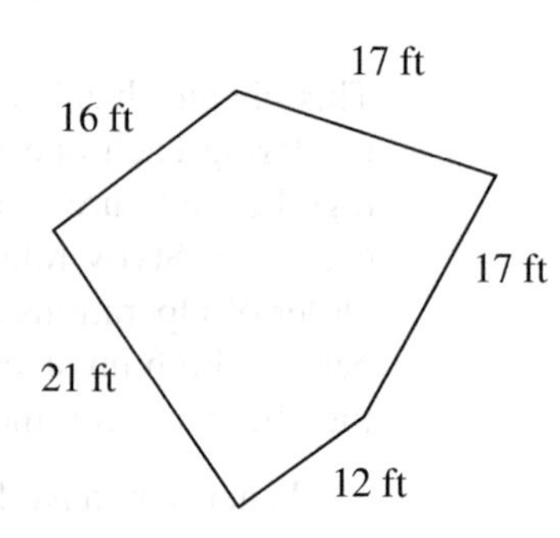

FIGURE 2.23

43. Find the area for the rectangle in Figure 2.24. ($A = LW$)

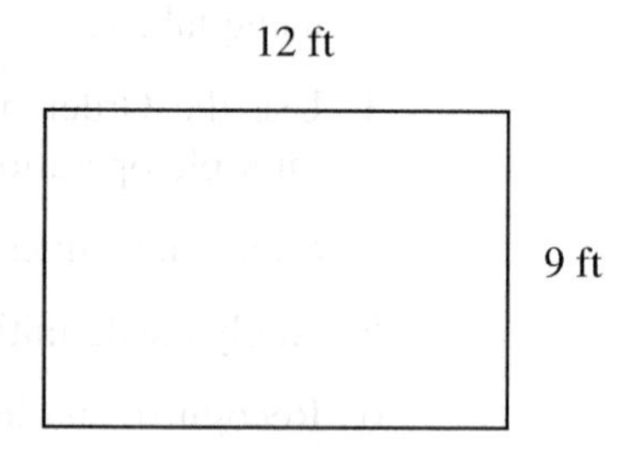

FIGURE 2.24

44. Using $D = rt$, see how far you can fly in 3 hours at 420 mph.

45. Find the perimeter of the polygon in Figure 2.25.

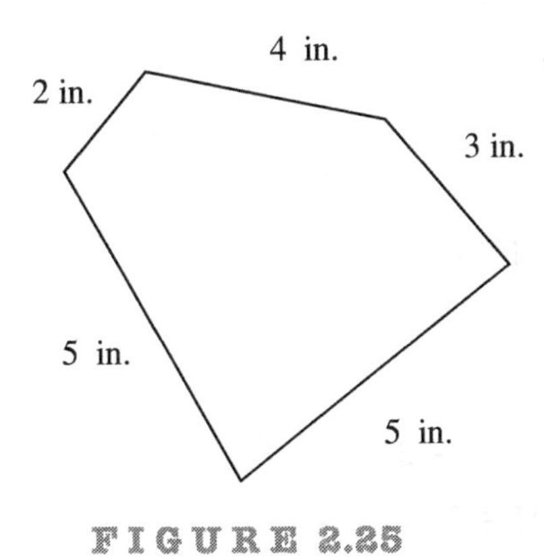

FIGURE 2.25

46. Find the area of the rectangle in Figure 2.26. $(A = LW)$

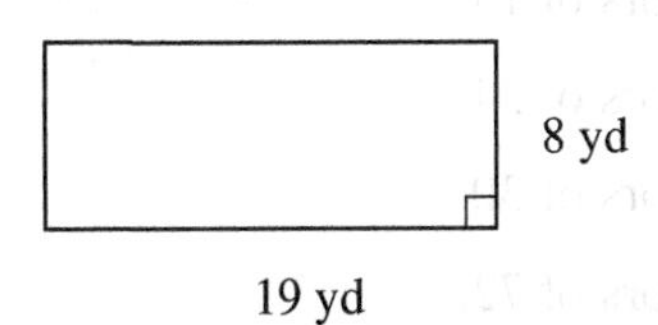

FIGURE 2.26

47. Figure *ABCD* is a quadrilateral wherein $m \measuredangle A = 25°$, $m \measuredangle C = 110°$, and $m \measuredangle D = 145°$. Find $m \measuredangle B$.

48. Find the area of a parallelogram if its base is 11 meters and its height is 8 meters.

49. Find the perimeter of a rectangle that is 14 yards long and 9 yards wide.

50. Find the area of the rectangle in Exercise 49.

51. Find the volume of a cardboard packing crate whose dimensions are 3 feet by 2 feet by 4 feet.

52. Find the surface area of the crate in Exercise 51.

D. Match each of the statements in Column I with the appropriate property in Column II.

Column I	Column II
53. 19 × 43 = 43 × 19	a. Distributive Principle
54. 173 + 0 = 173	b. Multiplication Property of One
55. 17(85) = 17(80) + 17(5)	c. Addition Property of Zero
56. 7 + (9 + 3) = (7 + 9) + 3	d. Multiplication Property of Zero
57. 276 × 1 = 276	e. Commutative Property of Addition
58. 92 + 79 = 79 + 92	f. Associative Property of Addition
59. (14 × 3)5 = 14(3 × 5)	g. Commutative Property of Multiplication
60. 0 = 46 × 0	h. Associative Property of Multiplication

E. Verify whether the Distributive Principle has been applied correctly by simplifying each side of the equals sign in these exercises.

61. 9(14 – 6) = 9(14) – 9(6)
62. 5(12) + 5(14) = 5(12 + 14)
63. 11(6) – 11(4) = 11(6 + 4)
64. 8(8 + 7) = 8(8) + 7(7)
65. 7(14 + 6) = 7(14) + 7(6)
66. 9(15 – 3) = 9(15) – 9(3)
67. 6(8) + 6(9) = 6(8 + 6)
68. 7(9) – 7(4) = 7(9 – 4)
69. 22(5 + 8) = 22(5) + 22(8)
70. 18(16 – 7) = 18(16) – 18(7)

F. Respond to each of the following.

71. List all the factors of 18.
72. List all the factors of 24.
73. List all the factors of 30.
74. List all the factors of 72.
75. Is 126 divisible by 2? Why or why not?
76. Is 147 divisible by 3? Why or why not?
77. Is 280 divisible by 5? Why or why not?
78. Is 117 divisible by 5? Why or why not?
79. Is 118 divisible by 3? Why or why not?
80. Is 215 divisible by 3? Why or why not?

G. Find the prime factorization for each of the following numbers.

81. 80
82. 84
83. 210
84. 90
85. 114
86. 150
87. 75
88. 126
89. 66
90. 85
91. 132
92. 242

H. Find the least common multiple (LCM) for each of the following groups of numbers. Use any method.

93. 12 and 20
94. 18 and 30
95. 8, 12, and 15
96. 45 and 60
97. 3, 14, and 12
98. 4, 15, and 18
99. 36 and 60
100. 42 and 70
101. 70 and 105
102. 12, 20, and 15

I. Find the greatest common factor (GCF) for each of the following groups of numbers.

103. 42 and 70
104. 20 and 21
105. 30 and 77
106. 105 and 70
107. 12, 18, and 30
108. 16, 40, and 24

J. Each problem that follows has an error in the solving process. State what the error is: then work the problem correctly.

109. $18 - 5(8 - 6)$

$18 - 5(2)$

$13(2)$

26

110. $26 + 18 - 12 + 7$

$44 - 19$

25

111. $\sqrt{16} + 2^4 - 15 \div 3$

$8 + 16 - 15 \div 3$

$8 + 16 - 5$

$24 - 5$

19

112. $(21 \div 3)(32 + 4)$

$(7)(8)$

56

113. Write the prime factorization of 200.

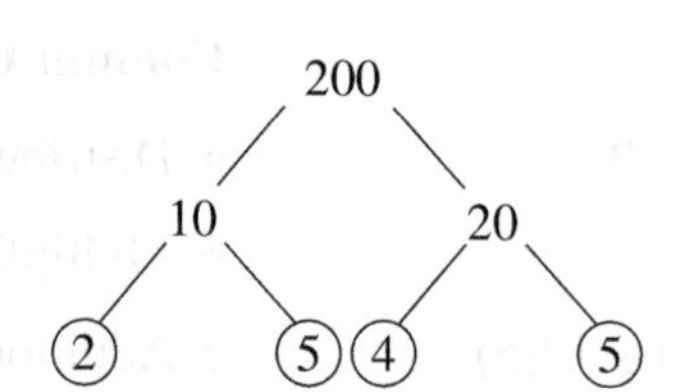

$200 = 2 \times 4 \times 5 \times 5$

114. Find the LCM of 20, 12, and 15.

$20 = 2 \cdot 2 \cdot 5$

$12 = 2 \cdot 2 \cdot \quad 3$

$30 = 2 \cdot \quad 5 \cdot 3$

LCM = 2

115. Find the area of ΔABC if the base is 18 in. and the height is 12 in.

$A = 1/2\ bh$

$A = 1/2(18)(12)$

$A = (9)(6)$

$A = 54$ square inches

K. Describe in words the process used to go from one step to the next.

116. $15 - 3(8 - 6) + 7$

$15 - 3(2) + 7$	1.
$15 - 6 + 7$	2.
$9 + 7$	3.
16	4.

117. $85 - 4[12 \div 6 \times 8 - 5 \times 2]$

$85 - 4[2 \times 8 - 5 \times 2]$	1.
$85 - 4[16 - 5 \times 2]$	2.
$85 - 4[16 - 10]$	3.
$85 - 4[6]$	4.
$85 - 24$	5.
61	6.

L. Read and respond to the following exercises.

118. Now that you have finished studying Chapter 2, what advice would you give to someone just beginning to study it? (Be encouraging, but realistic.)
119. Comment on whether you find a fragmented chapter, like Chapter 2, easier or harder to study than a chapter like Chapter 1, which deals with just one topic.
120. Describe the difference between the least common multiple (LCM) and the greatest common factor (GCF) of a given set of numbers.
121. Describe the process you would use to list *all* the factors of a given number.
122. Explain the difference(s) between "listing all the factors of a given number" and "finding the prime factorization of a given number."
123. List the information that you would include on a study card for the geometry included in this chapter.
124. What reasons can you think of for studying geometry in a fundamentals and algebra course?

M. Refer to the information and answer the following questions.

Maria is planning to redecorate her living room. She needs to replace the carpet, paint the ceiling, and put wallpaper on the walls. She also wants to put a border paper around the top of the walls. The diagram for the living room is given in Figure 2.27; the cost for various supplies is given below it.

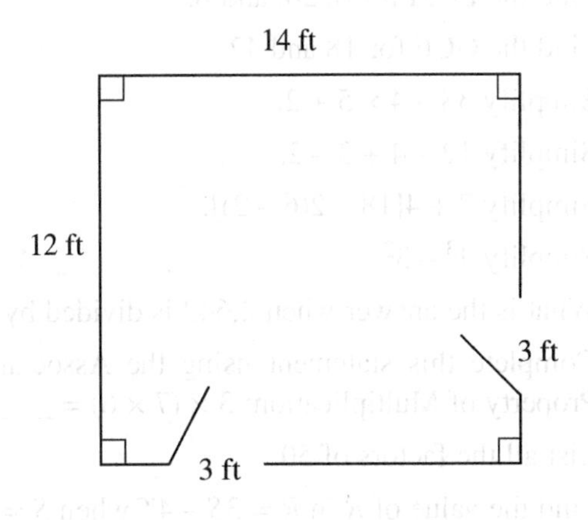

FIGURE 2.27

Carpet with pad: $14 per sq yard

Ceiling paint: $11 per gallon (1 gal covers 400 sq ft)

Border paper: $6 per roll (1 roll = 7 ft)

Wallpaper: $12 per roll (1 roll covers 75 sq ft)

125. How many square feet of carpet will she need?
126. How many square feet are in 1 square yard?
127. How many square yards of carpet will she need?
128. Maria needs to use two coats of paint for the ceiling. How many gallons will she need?
129. How many rolls of border paper must she buy?
130. How much border paper will she have left over?
131. How many square feet of wall space is there to be covered with wallpaper if the room has an 8-foot ceiling? (Don't worry about windows or door openings. A little extra wallpaper is good for repairs later on.)
132. How many rolls of wallpaper will she need to buy?
133. A professional paperhanger will charge $11 per roll to hang the wallpaper. How much will Maria save if she hangs the paper herself?
134. Approximately how much will the redecorating supplies cost for this project?

CHAPTER TEST

1. List all the prime numbers less than 25.
2. Is 3 a factor of 276? Why or why not?
3. Find the prime factorization for 630.
4. Find the LCM for 6, 20, and 8.
5. Find the GCF for 18 and 42.
6. Simplify $38 - 4 \times 5 + 2$.
7. Simplify $12 - 4 + 3 - 2$.
8. Simplify $7 + 4[18 - 2(6 - 2)]$.
9. Simplify $4^3 - 3^2$.
10. What is the answer when 2,642 is divided by 0?
11. Complete this statement using the Associative Property of Multiplication: $3 \times (7 \times 6) =$_____.
12. List all the factors of 50.
13. Find the value of R in $R = 3S - 4T$ when $S = 24$ and $T = 13$.
14. If $M = 3X^2 - 4X + 2$, find the value of M when $X = 4$.
15. Using the formula $A = LW$, find the area of a rectangle that is 18 feet long and 7 feet wide.
16. Find the perimeter of the rectangle in Exercise 15 by using the formula $P = 2L + 2W$.
17. Find the distance you can travel in 7 hours if you average 43 miles per hour. Use $D = rt$.
18. Find the area of a parallelogram that has a base of 16 yards and a height of 7 yards. Use $A = bh$.
19. Explain, in words, the steps you would follow to answer this exercise, and then find the answer: In quadrilateral $PQRS$, $m \measuredangle R = 110°$, $m \measuredangle S = 85°$, $m \measuredangle P = 120°$. Find the measure of $\measuredangle Q$.
20. Find the surface area of the box in Figure 2.28.

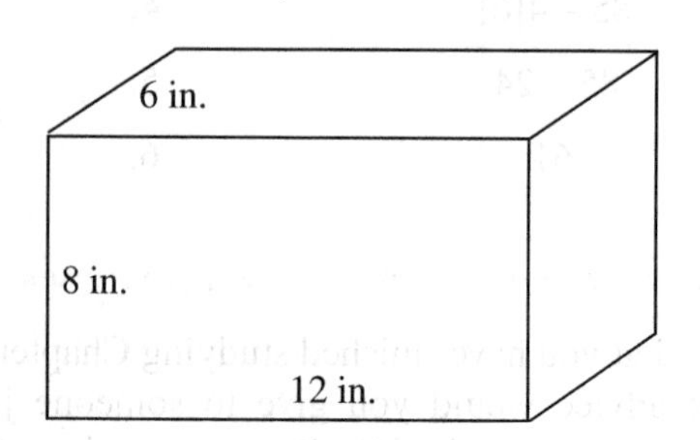

FIGURE 2.28

STUDY SKILLS MODULE

Math Anxiety

Math anxiety! You probably know it as that awful feeling you get in the pit of your stomach when a teacher announces a math test, or when fellow students start talking about all the math they know, and you can't remember a thing! Rest assured, you are not alone. At least one-fourth of all students suffer from math anxiety.

Is this disease terminal? Do you have it? How did you get it? What can you do about it? Would you do a lot better in your math courses if you didn't have it? All of these are the kinds of questions that might run through your mind when you think about math anxiety. Let's see if there isn't something that can be done about it.

First, you need to understand that math anxiety is a learned response. You were not born with it. You developed it as a result of something that happened in your life. And *because it is a learned response, you can unlearn it.* In order to unlearn math anxiety, it is helpful to try to figure out when in your own life this problem originated. In fact, there are "math inventories" and "math autobiographies" that you can complete to help you pinpoint when you first developed the problem and what situation originally caused you to experience math anxiety. Most often, students find out it was not something they themselves did or didn't do, but rather someone else who caused them to have a bad experience with math. Perhaps it was a teacher who embarrassed you, or a parent who pressured you to bring home "good math grades." Maybe a particular math topic has always caused you trouble, and now you are sure you will never be able to learn it—or any other math. In this module, the responses in Your Point of View address this issue. If you need more help, ask your instructor or the counseling center on campus for information about math inventories and math autobiographies. Once you identify where the problem started, you can begin to break its hold on you.

Second, math anxiety is just another type of stress. In these hectic times, life can be full of stress. You probably don't need another kind to add to your list. Actually, a little stress is good for you. When it is used in the right way, it can help you perform at your very best. It can keep you on your toes during a test so you don't become too complacent or too relaxed and make careless mistakes. It is a problem only when it becomes so powerful that it overtakes your whole person and makes you unable to function normally.

When you find yourself faced with a math situation, there are things you can do to help you deal with anxiety. As you go into a stress-causing situation, allow a few minutes to practice some relaxation techniques to get yourself prepared physically for the task ahead. Take a few deep breaths; think of something that you find very peaceful or soothing; concentrate on individual limbs of your body and try to relax each one; and so on. Once you are ready to begin, put a smiley face or a phrase like "I can do this!" on the top of your assignment, class notes, or test to make yourself feel good about your ability to do the task at hand. And don't allow yourself to talk negatively about your ability to do this work. Work through the homework or test problems carefully, doing the easiest ones first. This starts you off in a positive, comfortable frame of mind. When you finish studying and doing your math assignment, go back to a few problems that you feel good about and review those again. This leaves a positive impression for the next time you work on math.

The best way to overcome a serious problem with math anxiety is to study carefully and steadily all along. Keeping up with the course work and not cramming at the last minute is the strongest defense against anxiety. Believe in yourself and your ability to be successful. Don't expect an immediate cure from math anxiety, but each step taken will be a step toward unlearning the anxiety.

Your Point of View

1. Write a brief autobiography of your math history. Describe specific good and bad memories.
2. If you experience any level of math anxiety, try to think back to an incident that first made you fear and dread math. Describe that incident. That is probably where your cycle started. Make a list of behaviors that keep the cycle going. List some behavior changes that you will make to try to break the cycle.

CHAPTER

3 Multiplication and Division of Fractions

LEARNING OBJECTIVES

When you have completed this chapter, you should be able to:

1. Understand the terms: numerator, denominator, proper fraction, improper fraction, mixed number, equivalent fraction, reduced to lowest terms, and reciprocal.
2. Change improper fractions to mixed numbers, and vice versa.
3. Write fractions equivalent to other fractions.
4. Simplify fractions to lowest terms by prime factorization and by repeated division.
5. Multiply fractions and mixed numbers.
6. Divide fractions and mixed numbers.
7. Solve application problems that require multiplication or division of fractions and mixed numbers, including area and volume problems.

Here we are ready to start fractions and many of you may be having negative thoughts about your ability to learn fractions. We are going to try to turn that thinking around! Successful operations with fractions require only that you learn a few rules and that you use them carefully.

The material on fractions has been divided into two parts (Chapters 3 and 4) so you can learn and digest some of it before going on to the rest. This should make fractions more manageable.

Calculator usage will not be demonstrated in this chapter, but the exercises presented are appropriate for either manual or calculator solution. If you are going to use a calculator as you do this work, you may want to study the keying sequences described in Appendix B.

3.1 The Language of Fractions

You need to begin by learning the vocabulary of fractions. Then you will understand what is being explained when terms such as *numerator* and *improper fractions* are used by your instructor or appear in the text.

Every fraction indicates division. When you say 3/4, you mean "3 divided by 4," or "3 parts out of 4." The rectangle in Figure 3.1 is divided into 4 equal parts. Of these 4 parts, 3 are shaded; that is, 3/4 of the rectangle is shaded. In this example, the top number represents the number of parts you are interested in, and the bottom number represents the total number of parts the rectangle contains.

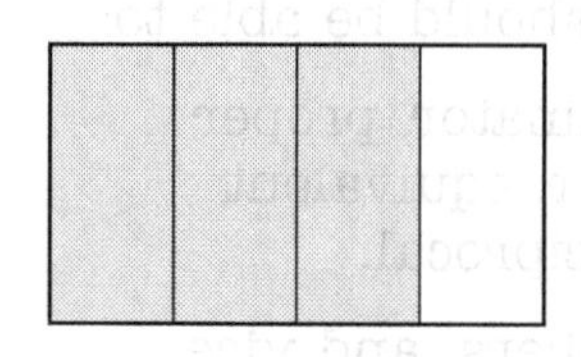

FIGURE 3.1

$$\frac{3}{4} \qquad \frac{\text{Number of parts that are shaded}}{\text{Total number of parts in the rectangle}}$$

The top number is called the **numerator** and the bottom number is the **denominator**.

$$\frac{3}{4} \qquad \frac{\text{Numerator}}{\text{Denominator}}$$

EXAMPLE 1 Write a fraction to represent each of the following statements.

a. Three students in this math class of 25 are absent today. What fraction of the class is absent?

b. This class of 25 people has 17 male students. What fraction of the class is male?

c. What fraction of the class is not male?

d. Another class contains 13 males and 19 females. What fraction of this class is female?

e. Seventeen minutes is what fractional part of an hour?

Solution

a. $\dfrac{\text{3 students absent}}{\text{25 students enrolled in class}}$ or $\dfrac{3}{25}$ are absent.

b. $\dfrac{\text{17 male students}}{\text{25 students enrolled in class}}$ or $\dfrac{17}{25}$ are male.

c. 25 total – 17 male = 8 not male students

$$\frac{8 \text{ not male students}}{25 \text{ students enrolled in class}} \text{ or } \frac{8}{25} \text{ are not male.}$$

d. You need the total number of students.

13 male + 19 female = 32 total

$$\frac{19 \text{ female students}}{32 \text{ total students}} \text{ or } \frac{19}{32} \text{ are female.}$$

e. The quantities must have the same units, and an hour contains 60 minutes.

$$\frac{17 \text{ minutes}}{60 \text{ minutes}} \text{ or } \frac{17}{60} \text{ of an hour.}$$

◀

Self-Check Neena's son has 21 toys in his toy box. Six are trucks, 11 are cars, and the rest are airplanes. Write each of the following:

a. The fraction of cars to the total number of toys.

b. The fraction of airplanes to the total number of toys.

Answers:

a. $\frac{11}{21}$ are cars.

b. $\frac{4}{21}$ are airplanes.

Any fraction whose numerator is less than its denominator is called a **proper fraction.**

$\frac{3}{4}, \frac{5}{18}, \frac{8}{9}, \frac{1}{2}, \frac{31}{64}$, and $\frac{10}{15}$ are all proper fractions.

All proper fractions have a value *less than* 1 because they involve only some of the total number of parts.

Any fraction whose numerator is greater than or equal to its denominator is called an **improper fraction.**

$\frac{5}{4}, \frac{19}{7}, \frac{5}{5}, \frac{12}{11}$, and $\frac{10}{10}$ are all improper fractions.

All improper fractions have a value of 1 or more. 5/4 means that all of one rectangle and 1/4 of another are shaded, as shown in Figure 3.2.

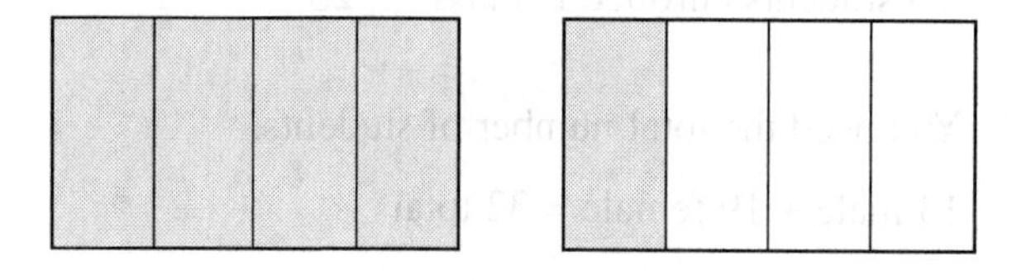

FIGURE 3.2

A number that is made up of a whole number and a fraction part is called a **mixed number**.

$1\frac{1}{2}$, $2\frac{4}{5}$, and $17\frac{1}{8}$ are mixed numbers.

Figure 3.2 not only illustrates 5/4 but also shows the mixed number 1 1/4, which is read "one and one-fourth." One whole rectangle (4 parts out of 4) and one-fourth (1 part out of 4) more were shaded.

You will often need to change from one form to the other: improper fraction to mixed number or mixed number to improper fraction. To change from an improper fraction to a mixed number, divide the denominator into the numerator. This answer becomes the whole number. Then write the remainder over the divisor (denominator). In 5/3, 3 divides into 5 one time (1) with 2 left over.

$\frac{5}{3} \to$ mixed number divisor $3\overline{)5}$ with quotient 1 (whole number), -3, 2 remainder $= 1\frac{2}{3}$

EXAMPLE 2 Change 17/5 to a mixed number.

Solution $\frac{17}{5} \to$ mixed number $5\overline{)17}$ with quotient 3, -15, 2 $= 3\frac{2}{5}$

It isn't necessary to write out these steps if you can do them in your head.

Self-Check Change these improper fractions to mixed numbers.

1. $\frac{5}{3}$ 2. $\frac{22}{7}$ 3. $\frac{65}{4}$ 4. $\frac{13}{9}$

Answers:

1. $1\frac{2}{3}$ 2. $3\frac{1}{7}$ 3. $16\frac{1}{4}$ 4. $1\frac{4}{9}$

To change from a mixed number to an improper fraction, you do the opposite operation from division: you multiply. Multiply the whole number times the denominator, and then add the numerator to this product and write the answer over the original denominator. This is not as complicated as it sounds. (Figure 3.3 shows that 3 1/4 is the same as 13/4.)

$3\frac{1}{4} \rightarrow$ improper fraction

$$\frac{\text{Whole number} \times \text{Denominator} + \text{Numerator}}{\text{Original Denominator}}$$

$$= \frac{3 \times 4 + 1}{4} = \frac{13}{4}$$

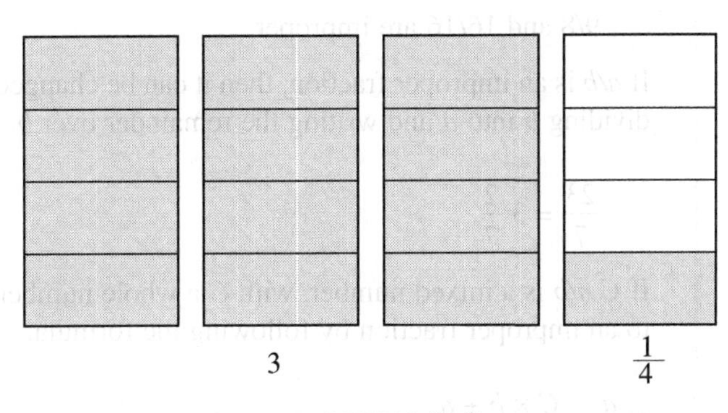

FIGURE 3.3

Self-Check Change these mixed numbers to improper fractions. (Again, you don't have to write the steps out if you can do them in your head.)

a. $6\frac{4}{5}$ b. $2\frac{5}{9}$ c. $1\frac{2}{3}$

Answers:

a. $6\frac{4}{5} = \frac{6 \times 5 + 4}{5} = \frac{34}{5}$

b. $2\frac{5}{9} = \frac{2 \times 9 + 5}{9} = \frac{23}{9}$

c. $1\frac{2}{3} = \frac{1 \times 3 + 2}{3} = \frac{5}{3}$

The definitions in this section can be stated in general form. That means they are written with letters in place of the numbers to show that the statements are true for *all* numbers.

Defining Fractions

A **fraction** is a number in the form *a/b*, where *a* is any number and *b* is any number except zero (division by zero is undefined).

$\frac{2}{3}, \frac{14}{9}$

The **numerator** of the fraction *a/b* is *a*, and the **denominator** is *b*.

In 2/3, 2 is the numerator and 3 is the denominator.

If *a* is less than *b* in *a/b*, then the fraction is **proper**.

2/3 is proper because 2 is less than 3.

If *a* is greater than or equal to *b* in *a/b*, then the fraction is **improper**.

9/8 and 16/16 are improper.

If *a/b* is an improper fraction, then it can be changed to a **mixed number** by dividing *b* into *a* and writing the remainder over *b*.

$$\frac{23}{7} = 3\frac{2}{7}$$

If *C a/b* is a mixed number, with *C* a whole number, then it can be changed to an improper fraction by following the formula:

$$C\frac{a}{b} = \frac{C \times b + a}{b}$$

$$5\frac{3}{8} = \frac{43}{8}$$

3.1 EXERCISES

A. Write a fraction to represent each statement.

1. An algebra class has 35 students enrolled. The number of female students is 13. Write
   a. The fraction of the total number of students that are female
   b. The fraction of the total number of students that are male
2. A button box contains 5 white buttons and 3 black buttons. Write
   a. The fraction of the total number of buttons that are white
   b. The fraction of the total number of buttons that are black
3. The Learning Center is open 13 hours each weekday. Write
   a. The fraction of the total hours in a day that the center is open
   b. The fraction of the total hours in a day that the Center is not open
4. The Ice Kreme Shoppe has 7 flavors that contain nuts and 15 flavors that do not. Write
   a. The fraction of the total number of flavors that do not contain nuts
   b. The fraction of the total number of flavors that do contain nuts

5. The office supply drawer contains 9 yellow legal pads and 11 white pads. Write
   a. The fraction of the total number of pads that are white
   b. The fraction of the total number of pads that are yellow
6. Sixty minutes of television time for a sporting event includes 11 minutes of commercials. Write
   a. The fraction of the hour that is not commercials
   b. The fraction of the hour that is commercials

B. Change these improper fractions to mixed numbers.

7. $\frac{21}{5}$
8. $\frac{13}{8}$
9. $\frac{7}{2}$
10. $\frac{3}{2}$
11. $\frac{5}{3}$
12. $\frac{11}{9}$
13. $\frac{29}{11}$
14. $\frac{6}{5}$
15. $\frac{67}{32}$
16. $\frac{101}{25}$
17. $\frac{9}{9}$
18. $\frac{24}{6}$

C. Change these mixed numbers to improper fractions.

19. $6\frac{2}{7}$
20. $2\frac{2}{3}$
21. $11\frac{3}{5}$
22. $14\frac{2}{3}$
23. $1\frac{3}{8}$
24. $9\frac{4}{9}$
25. $7\frac{5}{9}$
26. $3\frac{7}{12}$
27. $2\frac{4}{7}$

D. Consider the following set of numbers:

$$1\frac{1}{2}, \frac{5}{9}, \frac{19}{4}, \frac{8}{8}, 25, \frac{6}{5}, 4\frac{4}{9}, 3, \frac{1}{2}$$

List all the numbers among these that:

28. Are mixed numbers in the given form
29. Are improper fractions in the given form
30. Are proper fractions in the given form
31. Have a value greater than 1
32. Have a value less than 1
33. Have a value equal to 1
34. Are improper fractions or could be expressed as improper fractions
35. Are mixed numbers or could be expressed as mixed numbers

E. Read and respond to the following exercises.

36. Make a sketch of the fraction 2 4/7. Describe your drawing. (See Figure 3.3.)
37. Make a sketch of the fraction 11/9. Describe your drawing. (See Figure 3.3.)
38. When writing the fractions for Exercises 1–6, did it make a difference which number went in the denominator and which in the numerator? Why?

3.2 Equivalent Fractions

Write the fraction that is represented by the shaded portion of each rectangle in Figure 3.4.

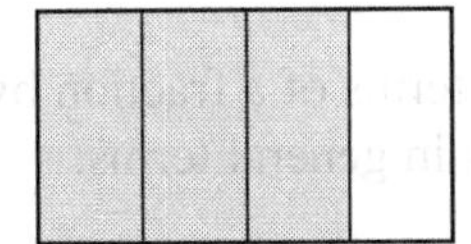
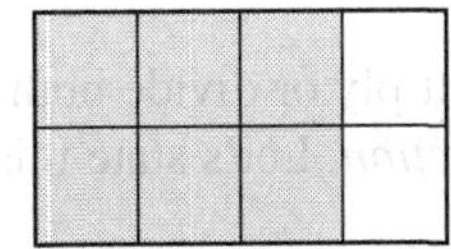
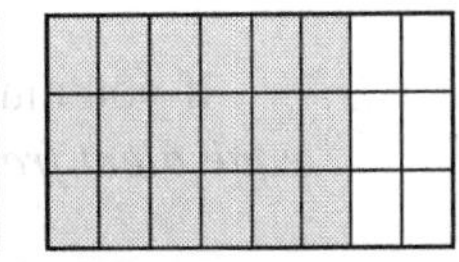
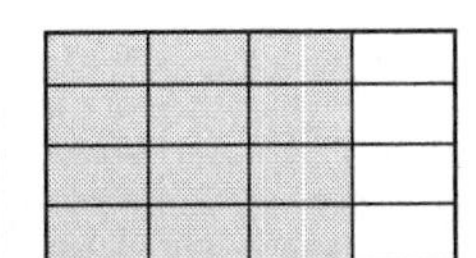

FIGURE 3.4

Did you get 3/4, 6/8, 18/24, and 12/16? Even though the fractions look different, is the shaded portion of the rectangle different? No. In fact, if you were able to place the rectangles on top of each other, all the shaded portions would match exactly.

The fractions 3/4, 6/8, 18/24, and 12/16 all represent the same amount of shading in these identical figures. The fractions are all equal to each other and are **equivalent fractions.**

$$\frac{3}{4} = \frac{6}{8} = \frac{18}{24} = \frac{12}{16}$$

If you took 3/4 and multiplied *both* the numerator and the denominator by the *same nonzero number*, 2, you would get 6/8, an equivalent fraction.

$$\frac{3}{4} \times 1 = \frac{3}{4} \times \frac{2}{2} = \frac{3 \times 2}{4 \times 2} = \frac{6}{8}$$

If you multiply both parts of the fraction 3/4 by 6 or by 4, you get other equivalent fractions.

$$\frac{3}{4} \times 1 = \frac{3}{4} \times \frac{6}{6} = \frac{18}{24} \qquad \frac{3}{4} \times 1 = \frac{3}{4} \times \frac{4}{4} = \frac{12}{16}$$

In each of these problems, the original fraction is being multiplied by a form of 1 because 2/2, 6/6, and 4/4 all equal 1. The new fraction looks different from the original, but its value is the same because when you multiply a quantity by 1, the product is the same as the original number. This is an example of the Multiplication Property of One.

If you took 18/24 and divided *both* the numerator and denominator by the *same number*, you would also get an equivalent fraction.

$$\frac{18}{24} \div 1 = \frac{18 \div 6}{24 \div 6} = \frac{3}{4} \qquad \frac{18 \div 3}{24 \div 3} = \frac{6}{8}$$

You are using forms of 1 (6/6 and 3/3) to divide, so the value does not change even though the appearance changes.

EXAMPLE 1 Are 3/4 and 9/12 equivalent fractions?

Solution They are if you can multiply 3 and 4 by the same factor and get 9 and 12.

Isn't $\frac{3 \times 3}{4 \times 3} = \frac{9}{12}$? Then $\frac{3}{4}$ is equivalent to $\frac{9}{12}$, or $\frac{3}{4} = \frac{9}{12}$.

EXAMPLE 2 Are 18/24 and 9/12 equivalent?

Solution They are if you can divide 18 and 24 by the same factor to get 9 and 12.

Isn't $\frac{18 \div 2}{24 \div 2} = \frac{9}{12}$? Then $\frac{18}{24}$ is equivalent to $\frac{9}{12}$, or $\frac{18}{24} = \frac{9}{12}$.

If you multiply or divide both terms of a fraction by the same factor, you get an *equivalent fraction.* Let's state this in general terms:

Equivalent Fractions

If a, b, and c are any number (with $b, c \neq 0$), then

$$\frac{a}{b} = \frac{a \times c}{b \times c} \quad \text{and} \quad \frac{a}{b} = \frac{a \div c}{b \div c}$$

EXAMPLE 3 Are 21/24 and 3/8 equivalent?

Solution No, because the only number that divides into both 21 and 24 is 3. But $\frac{21 \div 3}{24 \div 3} = \frac{7}{8}$ and $\frac{3}{8} \neq \frac{7}{8}$.

EXAMPLE 4 Write 8/9 as an equivalent fraction with a numerator of 72.

Solution This is asking you to complete $\frac{8}{9} = \frac{72}{}$ with the correct denominator to make an equivalent fraction. Since $8 \times 9 = 72$, then $\frac{8 \times 9}{9 \times 9} = \frac{72}{81}$.

EXAMPLE 5 Write $1\frac{2}{3}$ as an equivalent fraction with 12 as the denominator.

Solution $1\frac{2}{3}$ is the same as $\frac{5}{3}$. Since $3 \times 4 = 12$, multiply the fraction by $\frac{4}{4}$. $\frac{5 \times 4}{3 \times 4} = \frac{20}{12}$.

Or keep $1\frac{2}{3}$ as a mixed number and multiply $\frac{2}{3} \times \frac{4}{4}$.

$$1\frac{2 \times 4}{3 \times 4} = 1\frac{8}{12}$$

The process of *multiplying both terms* of the fraction by the same factor *gives an equivalent fraction with larger numbers* than the original fraction had. You will apply this procedure extensively when you add and subtract fractions in the next chapter. The process of *dividing both terms* of the fraction by the same factor *produces an equivalent fraction with smaller numbers* than the original fraction. This very important concept is called **reducing**, or simplifying. You will apply it to every fraction answer so that the answers are always simplified completely, or *reduced to lowest terms*.

When you solve a problem and get an answer of 3/4 and other students in the class get 12/16 or 9/12, how do you recognize that you all have the same answer? That is not a problem if all the answers are reduced to lowest terms. All those answers are 3/4 when reduced.

Simplifying all fraction answers to lowest terms is a universally accepted practice. Reducing requires that you divide out common factors that are in both the numerator and denominator.

When simplifying 24/28, write the numerator and denominator as the product of prime factors, using the processes you learned in Chapter 2.

$$\frac{24}{28} = \frac{2 \cdot 2 \cdot 2 \cdot 3}{2 \cdot 2 \cdot 7}$$

Divide out the two pairs of factors that are in both the numerator and the denominator:

$$= \frac{\overset{1}{\cancel{2}} \cdot \overset{1}{\cancel{2}} \cdot 2 \cdot 3}{\underset{1}{\cancel{2}} \cdot \underset{1}{\cancel{2}} \cdot 7}$$

Then multiply together the factors remaining in the numerator and those remaining in the denominator.

$$= \frac{2 \cdot 3}{7} = \frac{6}{7}$$

The fraction is in lowest terms because there are no common factors other than 1.

EXAMPLE 6 Simplify these fractions to lowest terms.

a. $\frac{14}{21}$ b. $\frac{18}{45}$ c. $\frac{21}{32}$ d. $\frac{44}{48}$

Solution

a. $\frac{14}{21} = \frac{2 \cdot \overset{1}{\cancel{7}}}{3 \cdot \underset{1}{\cancel{7}}} = \frac{2}{3}$ b. $\frac{18}{45} = \frac{2 \cdot \overset{1}{\cancel{3}} \cdot \overset{1}{\cancel{3}}}{\underset{1}{\cancel{3}} \cdot \underset{1}{\cancel{3}} \cdot 5} = \frac{2}{5}$

c. $\frac{21}{32} = \frac{3 \times 7}{2 \times 2 \times 2 \times 2 \times 2}$

$\frac{21}{32}$ is in lowest terms because there are no common factors.

d. $\frac{44}{48} = \frac{\overset{1}{\cancel{2}} . \overset{1}{\cancel{2}} . 11}{\underset{1}{\cancel{2}} . \underset{1}{\cancel{2}} . 2 . 2 . 3} = \frac{11}{12}$

◀

You may not recognize this procedure as the method you've used to simplify fractions, but it is a similar process described in a more mathematical format. If you reduce fractions by dividing a number into both the numerator and denominator, then you are dividing out common factors, even though you didn't list the prime factors to do this. Part d of the previous example can be done this way:

$$\frac{44}{48} = \frac{44 \div 2}{48 \div 2} = \frac{22 \div 2}{24 \div 2} = \frac{11}{12} \quad \text{or} \quad \frac{44}{48} = \frac{44 \div 4}{48 \div 4} = \frac{11}{12}$$

EXAMPLE 7 Simplify $\frac{48}{60}$ to lowest terms.

Solution The prime factors of 48 and 60 can be found by either the factor tree method or the repeated division method (see Section 2.5.).

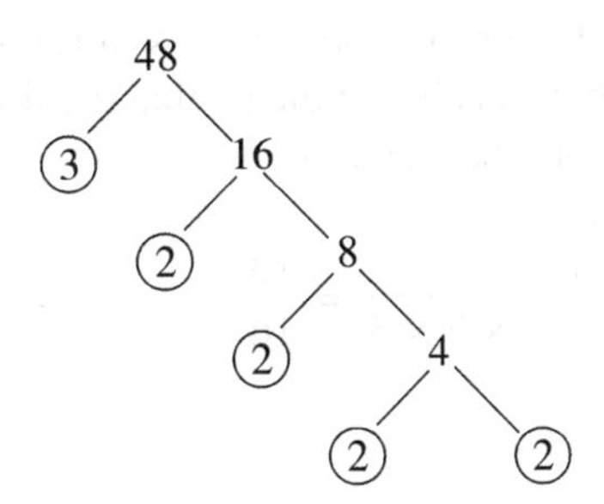

2) 60
2) 30
3) 15
5) 5
1

$$\frac{48}{60} = \frac{2 \times 2 \times 2 \times 2 \times 3}{2 \times 2 \times 3 \times 5}$$

$$= \frac{2}{2} \times \frac{2}{2} \times \frac{3}{3} \times \frac{2 \times 2}{5}$$

$$= \frac{4}{5}$$

◀

EXAMPLE 8 Simplify $\frac{312}{384}$ by reducing to lowest terms.

Solution Find the prime factors of 312 and 384.

2) 312	2) 384
2) 156	2) 192
2) 78	2) 96
3) 39	2) 48
13) 13	2) 24
1	2) 12
	2) 6
	3) 3
	1

List the primes in the numerator and denominator:

$$\frac{312}{384} = \frac{2 \times 2 \times 2 \qquad\qquad \times 3 \times 13}{2 \times 2 \times 2 \times 2 \times 2 \times 2 \times 2 \times 3}$$

Each factor of 2 in the numerator that matches a factor of 2 in the denominator can be written as a fraction with a value of 1. The pair of 3s also has a value of 1.

$$\frac{312}{384} = 1 \times 1 \times 1 \times \frac{\qquad}{2 \times 2 \times 2 \times 2} \times 1 \times \frac{13}{\quad}$$

Multiply together what remains in each part of the fraction:

$$\frac{312}{384} = \frac{13}{16}$$

Instead of rearranging and rewriting the factors several times, just divide out the pair of factors and write a 1 above each to indicate that the division has a value of 1. Then multiply together the factors that remain.

$$\frac{312}{384} = \frac{\overset{1}{\cancel{2}} \times \overset{1}{\cancel{2}} \times \overset{1}{\cancel{2}} \times \overset{1}{\cancel{3}} \times 13}{\underset{1}{\cancel{2}} \times \underset{1}{\cancel{2}} \times \underset{1}{\cancel{2}} \times 2 \times 2 \times 2 \times 2 \times \underset{1}{\cancel{3}}} = \frac{13}{16}$$

EXAMPLE 9 Using prime factorization, simplify 1,350/315 to lowest terms.

Solution

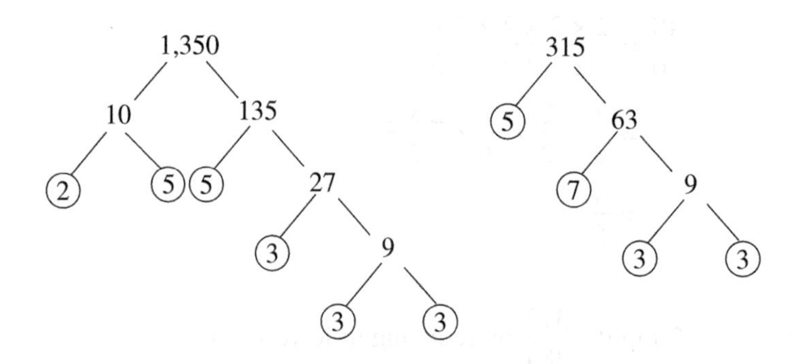

$$\frac{1{,}350}{315} = \frac{2 \times 3 \times 3 \times 3 \times 5 \times 5}{3 \times 3 \times 5 \times 7}$$

$$\frac{1{,}350}{315} = \frac{2 \times \overset{1}{\cancel{3}} \times \overset{1}{\cancel{3}} \times 3 \times \overset{1}{\cancel{5}} \times 5}{\underset{1}{\cancel{3}} \times \underset{1}{\cancel{3}} \times \underset{1}{\cancel{5}} \times 7}$$

Divide out pairs of matching factors, and then multiply together the factors that remain.

$$= \frac{30}{7} \text{ or } 4\frac{2}{7}$$

Self-Check a. Complete each of the following as an equivalent fraction:

1. $\frac{3}{8} = \frac{15}{\quad}$ 2. $\frac{4}{21} = \frac{\quad}{63}$

b. Reduce each fraction to lowest terms:

3. $\frac{27}{36}$ 4. $\frac{124}{240}$

Answers:

1. $\frac{15}{40}$ 2. $\frac{12}{63}$ 3. $\frac{3}{4}$ 4. $\frac{31}{60}$

3.2 EXERCISES

A. Are these pairs of fractions equivalent fractions? Tell why. See Examples 1, 2, and 3 in this section.

1. 5/8 and 25/40
2. 18/21 and 6/7
3. 24/30 and 4/5
4. 3/5 and 21/35
5. 2/3 and 12 /24
6. 10/12 and 20/24
7. 12/20 and 24/40
8. 45/54 and 5/9
9. 30/21 and 15/7
10. 11/64 and 33/128

B. Write an equivalent fraction with the denominator or numerator that is given by multiplying or dividing both the numerator and denominator by the same value.

11. $\frac{3}{4} = \frac{}{12}$
12. $\frac{1}{2} = \frac{}{10}$
13. $\frac{5}{8} = \frac{}{24}$
14. $\frac{3}{5} = \frac{}{15}$
15. $\frac{1}{2} = \frac{8}{}$
16. $\frac{2}{3} = \frac{10}{}$
17. $\frac{7}{1} = \frac{}{4}$
18. $2\frac{3}{4} = \frac{}{8}$
19. $3\frac{7}{8} = \frac{}{16}$
20. $\frac{4}{1} = \frac{}{9}$
21. $4\frac{1}{3} = \frac{}{12}$
22. $1\frac{7}{15} = \frac{}{45}$

C. Simplify to lowest terms.

23. $\frac{15}{18}$
24. $\frac{24}{36}$
25. $\frac{9}{15}$
26. $\frac{8}{16}$
27. $\frac{4}{16}$
28. $\frac{12}{14}$
29. $\frac{6}{8}$
30. $\frac{6}{9}$
31. $\frac{14}{21}$
32. $\frac{3}{6}$
33. $\frac{10}{16}$
34. $\frac{2}{4}$
35. $\frac{8}{15}$
36. $\frac{15}{22}$
37. $\frac{39}{21}$
38. $\frac{84}{36}$
39. $\frac{36}{80}$
40. $\frac{120}{200}$
41. $\frac{98}{144}$
42. $\frac{124}{236}$
43. $\frac{96}{124}$
44. $\frac{320}{480}$
45. $\frac{64}{84}$
46. $\frac{192}{336}$
47. $\frac{225}{300}$
48. $\frac{196}{320}$
49. $\frac{54}{126}$
50. $\frac{78}{130}$

D. Read and respond to the following exercises.

51. Describe, in words, how you would reduce 16/18 to lowest terms.
52. Describe, in words, how you would find the missing denominator in:

$$\frac{5}{} = \frac{15}{27}$$

53. Is 2 3/5 equivalent to 39/15? In words, describe your answer.
54. Is 42/16 equivalent to 2 3/8? In words, explain your answer.
55. Describe how you would simplify 24/39.
56. When simplifying 36/60, how will you know that the answer is in lowest terms?

3.3 Multiplying Fractions

If you packed one-half a box of books each day this week, how many boxes did you pack?

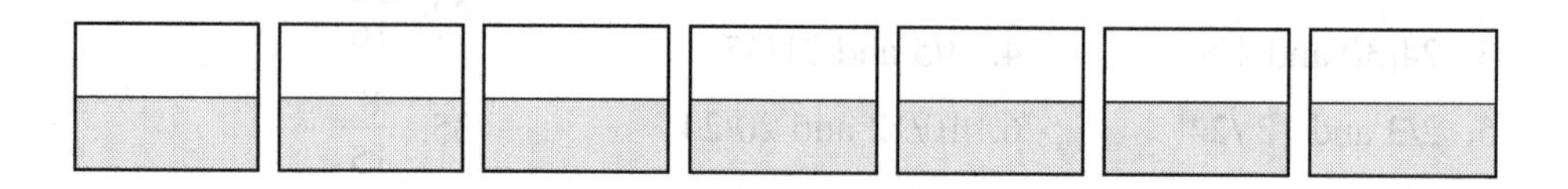

FIGURE 3.5

The diagram in Figure 3.5 indicates that 7 half boxes were packed. 1/2 + 1/2 + 1/2 + 1/2 + 1/2 + 1/2 + 1/2 is equal to "seven halves" and is expressed as 7/2. You could find this value by multiplying 7 times 1/2, or 7 × 1/2, because multiplication is just a short way of doing repeated addition. The same amount is packed each day for 7 days. Write 7 as 7/1 (you can write a whole number over 1 to put it in fraction form), then 7/1 × 1/2 will equal 7/2 if you multiply the numerators together and multiply the denominators together.

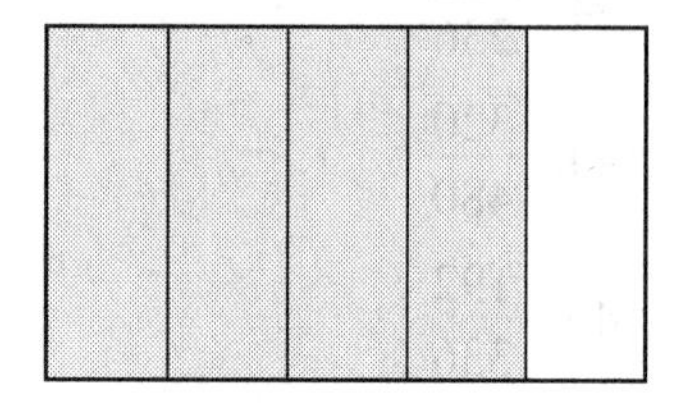

FIGURE 3.6

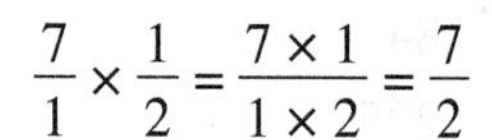

$$\frac{7}{1} \times \frac{1}{2} = \frac{7 \times 1}{1 \times 2} = \frac{7}{2}$$

This is the process used to multiply any fractions together.

Let's look at a pictorial representation of 4/5 × 2/3. The shaded part of Figure 3.6 represents the fraction 4/5. Now divide that figure into three equal parts (Figure 3.7). Because you are interested in two of those thirds, darken the shading on two of the rows of the fourth-fifth region (Figure 3.8). Figure 3.8 represents the answer to 4/5 × 2/3 as 8/15. Using mathematical process rather than pictures,

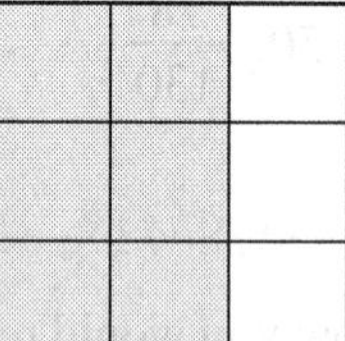

FIGURE 3.7

$$\frac{4}{5} \times \frac{2}{3} = \frac{4 \times 2}{5 \times 3} = \frac{8}{15}$$

FIGURE 3.8

Multiplication of Fractions

Given the fractions a/b and c/d (where b and d are not equal to zero), then

$$\frac{a}{b} \times \frac{c}{d} = \frac{a \times c}{b \times d}$$

EXAMPLE 1 Multiply 3/4 by 5/8.

Solution

$$\frac{3}{4} \times \frac{5}{8} = \frac{3 \times 5}{4 \times 8} = \frac{15}{32}$$

You don't have to write out the middle step.

EXAMPLE 2 Find the product of 2/15 and 4.

Solution **Product** means to multiply. Any whole number can be written in fraction form by writing it over 1. (Isn't 4/1 the same as 4 ÷ 1, which has a value of 4?)

$$\frac{2}{15} \times 4 = \frac{2}{15} \times \frac{4}{1} = \frac{8}{15}$$

In each of the previous examples the answer was already in lowest terms, but you need to check every answer to be sure that it is reduced.

EXAMPLE 3 Multiply 8/15 by 25/28.

Solution To estimate an answer, think of $\frac{8}{15}$ as about $\frac{1}{2}$ $(\frac{8}{16} = \frac{1}{2})$ and $\frac{25}{28}$ as about 1 $(\frac{28}{28} = 1)$.

Then, $\frac{1}{2} \times 1 = \frac{1}{2}$.

$$\frac{8}{15} \times \frac{25}{28} = \frac{8 \times 25}{15 \times 28} = \frac{200}{420}$$

This must be simplified, or reduced to lowest terms.

$$\frac{200}{420} = \frac{\overset{1}{\cancel{2}} \times \overset{1}{\cancel{2}} \times 2 \times \overset{1}{\cancel{5}} \times 5}{\underset{1}{\cancel{2}} \times \underset{1}{\cancel{2}} \times 3 \times \underset{1}{\cancel{5}} \times 7} = \frac{10}{21}$$

Isn't $\frac{10}{21}$ close to the estimated answer of $\frac{1}{2}$?

In Example 3, multiplying was done first, then reducing. A shortcut that can be used in multiplication of fractions is to reduce first and then multiply. Let's look at the previous example and write each number in factored form before doing any multiplying.

$$\frac{8}{15} \times \frac{25}{28}$$

$$= \frac{2 \times 2 \times 2}{3 \times 5} \times \frac{5 \times 5}{2 \times 2 \times 7}$$

Divide out matching factors, one in the numerator and one in the denominator. Their value is 1.

$$= \frac{\overset{1}{\cancel{2}} \times \overset{1}{\cancel{2}} \times 2 \times \overset{1}{\cancel{5}} \times 5}{3 \times \underset{1}{\cancel{5}} \times \underset{1}{\cancel{2}} \times \underset{1}{\cancel{2}} \times 7}$$

Multiply the factors that remain.

$$= \frac{10}{21}$$

Problems involving the multiplication of fractions can be done by reducing first and then multiplying, or by multiplying first and then reducing.

EXAMPLE 4 Find the product of 15/16 and 24/39.

Solution Multiplying first:

$$\frac{15}{16} \times \frac{24}{39}$$

Multiply 15 × 24 and 16 × 39.

$$= \frac{360}{624}$$

Find the prime factors of 360 and 624.

$$= \frac{2 \cdot 2 \cdot 2 \cdot 3 \cdot 3 \cdot 5}{2 \cdot 2 \cdot 2 \cdot 2 \cdot 3 \cdot 13}$$

Divide out pairs of like factors.

$$= \frac{\overset{1}{\cancel{2}} \cdot \overset{1}{\cancel{2}} \cdot \overset{1}{\cancel{2}} \cdot 3 \cdot \overset{1}{\cancel{3}} \cdot 5}{\underset{1}{\cancel{2}} \cdot \underset{1}{\cancel{2}} \cdot \underset{1}{\cancel{2}} \cdot 2 \cdot \underset{1}{\cancel{3}} \cdot 13}$$ Multiply the remaining factors.

$$= \frac{15}{26}$$

Reducing first:

$$\frac{15}{16} \times \frac{24}{39}$$ Write each number in prime factored form.

$$= \frac{3 \times 5}{2 \times 2 \times 2 \times 2} \times \frac{2 \times 2 \times 2 \times 3}{3 \times 13}$$

$$= \frac{3 \times 5 \times 2 \times 2 \times 2 \times 3}{2 \times 2 \times 2 \times 2 \times 3 \times 13}$$ Divide pairs of like factors.

$$= \frac{3 \times 5 \times \overset{1}{\cancel{2}} \times \overset{1}{\cancel{2}} \times \overset{1}{\cancel{2}} \times \overset{1}{\cancel{3}}}{2 \times \underset{1}{\cancel{2}} \times \underset{1}{\cancel{2}} \times \underset{1}{\cancel{2}} \times \underset{1}{\cancel{3}} \times 13}$$ Multiply the factors that remain.

$$= \frac{15}{26}$$

A popular technique for doing the reducing first lets you bypass the prime factoring step. You look mentally for pairs of factors (not necessarily primes). In this example you might "see" the pair of 3s that could be divided out of 15 and 39. There is also a pair of 8s—one in 16 and one in 24.

$$\frac{15}{16} \times \frac{24}{39} = \frac{\overset{5}{\cancel{15}}}{\underset{2}{\cancel{16}}} \times \frac{\overset{3}{\cancel{24}}}{\underset{13}{\cancel{39}}}$$

Because there are no more pairs of factors, multiply the remaining factors together:

$$= \frac{15}{26}$$

The reason this works is the same as in the earlier explanation about simplifying (Section 3.1). The pairs divided have a value of 1.

$$\frac{15}{16} \times \frac{24}{39} = \frac{5 \times 3}{2 \times 8} \times \frac{8 \times 3}{3 \times 13}$$

$$= \frac{5 \times 3 \times 8 \times 3}{2 \times 3 \times 8 \times 13}$$

$$= \frac{5}{2} \times 1 \times 1 \times \frac{3}{13}$$

$$= \frac{15}{26}$$

Remember that dividing out pairs of common factors in the numerator and denominator can be done only if the fractions are being multiplied. **Do not do this process if the fractions are linked by + , –, ÷, or = signs.**

EXAMPLE 5 Find the product: $\frac{3}{4} \times \frac{5}{9} \times \frac{12}{20}$

Solution First estimate an answer. The first fraction, $\frac{3}{4}$, is exactly halfway between $\frac{1}{2}$ and 1, so we will leave it as it is. $\frac{5}{9}$ is about $\frac{1}{2}$ ($\frac{5}{10} = \frac{1}{2}$) and $\frac{12}{20}$ is about $\frac{1}{2}$ ($\frac{12}{24} = \frac{1}{2}$ or $\frac{10}{20} = \frac{1}{2}$). So, multiply $\frac{3}{4} \times \frac{1}{2} \times \frac{1}{2} = \frac{3}{16}$.

By dividing common factors:

$$\frac{3}{4} \times \frac{5}{9} \times \frac{12}{20}$$

$$\frac{\overset{1}{\cancel{3}}}{\underset{1}{\cancel{4}}} \times \frac{\overset{1}{\cancel{5}}}{\underset{\underset{1}{\cancel{3}}}{\cancel{9}}} \times \frac{\overset{\overset{1}{\cancel{3}}}{\cancel{12}}}{\underset{4}{\cancel{20}}}$$

$$= \frac{1}{4}$$

The dividing out can be done in many different ways and still produce the same answer. In this example, 3 and 9 were divided by 3, 4 and 12 by 4, and 5 and 20 by 5. Then 3 and 3 were divided by 3.

By prime factorization:

$$\frac{3}{4} \times \frac{5}{9} \times \frac{12}{20}$$

$$= \frac{\overset{1}{\cancel{3}}}{2 \times 2} \times \frac{\overset{1}{\cancel{5}}}{\underset{1}{\cancel{3}} \times \underset{1}{\cancel{3}}} \times \frac{\overset{1}{\cancel{3}} \times \overset{1}{\cancel{2}} \times \overset{1}{\cancel{2}}}{\underset{1}{\cancel{5}} \times \underset{1}{\cancel{2}} \times \underset{1}{\cancel{2}}}$$

$$= \frac{1}{4}$$

Divide two pairs of 2s, two pairs of 3s, and one pair of 5s. Multiply what remains.

The answer 1/4 is close to the estimate answer of 3/16 (4/16 = 1/4).

When asked to multiply mixed numbers such as 2 and 3 1/5, you must realize that 2 is not multiplying just 3, but 1/5 as well. From the discussion of the Distributive Principle in Chapter 2, you know you could write this as

$$2 \times 3\frac{1}{5} = 2\left(3 + \frac{1}{5}\right) = 2 \times 3 + 2 \times \frac{1}{5}$$

$$= 6 + \frac{2}{5}$$

$$= 6\frac{2}{5}$$

The problem is more commonly calculated by converting the mixed number and the whole number to improper fractions before doing the multiplication.

$$2 \times 3\frac{1}{5}$$

$$= \frac{2}{1} \times \frac{16}{5}$$

$$= \frac{32}{5} \quad \text{or} \quad 6\frac{2}{5}$$

The following multiplication examples that involve mixed numbers all illustrate this second procedure, changing mixed numbers to improper fractions before multiplying.

EXAMPLE 6 Multiply the mixed numbers: $2\frac{3}{4} \times 1\frac{2}{3}$

Solution

$2\frac{3}{4} \times 1\frac{2}{3}$ Change to improper fractions.

$= \frac{11}{4} \times \frac{5}{3}$ Because there are no common factors to divide out, multiply the numerators and the denominators.

$= \frac{55}{12}$ or $4\frac{7}{12}$ Because 55/12 is reduced to lowest terms, it would have been all right to leave the answer as an improper fraction. But, if you prefer, change it to a mixed number.

In this a reasonable answer? Quickly calculate the product of the next larger whole numbers: 3 × 2 = 6. Is the actual answer close to 6? It is, so you know you have a reasonable answer.

EXAMPLE 7 Find the product: $7\frac{1}{7} \times 8 \times 6\frac{3}{10}$

Solution First estimate the answer.

$7 \times 8 \times 6 = 336.$ The answer should be about 336.

$7\frac{1}{7} \times 8 \times 6\frac{3}{10}$

$= \frac{50}{7} \times \frac{8}{1} \times \frac{63}{10}$

$= \frac{\overset{5}{\cancel{50}}}{\underset{1}{\cancel{7}}} \times \frac{8}{1} \times \frac{\overset{9}{\cancel{63}}}{\underset{1}{\cancel{10}}}$

$= \frac{360}{1} = 360$ This answer is close to the estimated answer.

EXAMPLE 8 Find 3/4 of 90.

Solution The instruction to find "**part of**" something indicates **multiplication**.

$\frac{3}{4}$ "of" 90 means $\frac{3}{4} \times 90$.

$\frac{3}{4} \times \frac{90}{1} = \frac{3}{\underset{2}{\cancel{4}}} \times \frac{\overset{45}{\cancel{90}}}{1} = \frac{135}{2}$ or $67\frac{1}{2}$

Self-Check a. $\frac{3}{8} \times \frac{5}{9}$ b. $1\frac{4}{5} \times 2\frac{6}{12}$ c. $\frac{9}{10}$ of 12

Answers:

a. $\frac{5}{24}$ b. $\frac{9}{2}$ or $4\frac{1}{2}$ c. $\frac{54}{5}$ or $10\frac{4}{5}$

EXAMPLE 9 Two-thirds of the students in this class are doing B or better work. If there are 27 students in the class, how many are doing B or better work?

Solution The problem could be restated by saying that 2/3 of the 27 students are doing B or better work. It is a "**part of**" problem, so multiply. Estimate your answer by reasoning that 2/3 of the class will be over half, but not all, of the 27 students.

$$\frac{2}{3} \text{ of } 27 = \frac{2}{3} \times \frac{27}{1}$$

$$= \frac{2}{\cancel{3}_1} \times \frac{\cancel{27}^9}{1} = 18$$

Therefore, 18 students are doing B or better work. This answer is reasonable because it is over half of 27 but not the whole class.

EXAMPLE 10 On the Stay Trim diet program, Carolyn averaged a weight loss of 2 1/2 pounds per month. If she stayed on the program for 6 months, how many pounds did she lose?

Solution 2 1/2 pounds every month for 6 months means that the same loss occurred six times.

First estimate the answer.

2 pounds per month for 6 months = 12 pounds or

3 pounds per month for 6 months = 18 pounds. The answer should be between 12 and 18 pounds.

$$2\frac{1}{2} \times 6 = \frac{5}{\cancel{2}_1} \times \frac{\cancel{6}^3}{1} = 15 \text{ pounds}$$

This is within the estimated range.

EXAMPLE 11 Five-eighths of the teachers at Mecca High School have master's degrees. If there are 160 teachers at the school, how many *do not* have masters degrees?

Solution Be careful and read this problem thoroughly. It is asking how many teachers *do not* have master's degrees. To do the problem is to find the "part of" the 160 teachers that do have the degree, then subtract that number from 160 to find the number who do not.

$$\frac{5}{8} \text{ of } 160 = \frac{5}{\cancel{8}_1} \times \frac{\cancel{160}^{20}}{1} = 100 \text{ teachers do have master's degrees.}$$

Then 160 – 100 = 60 teachers do not have master's degrees.

EXAMPLE 12 Cookbook directions say to cook a roast 1/4 of an hour for each pound of meat. How long should you cook a $4\frac{5}{8}$ pound roast?

Solution For each pound, you cook the meat for 1/4 of an hour. This gets done for each pound of the weight, so multiply.

$$\frac{1}{4} \times 4\frac{5}{8} = \frac{1}{4} \times \frac{37}{8} = \frac{37}{32} = 1\frac{5}{32} \text{ hours}$$

◀

EXAMPLE 13 Ray is ordering fence materials for his backyard. He knows that the length of the yard is 22 3/4 yards and that the width is 2/3 of the length. What is the width of the yard?

Solution The width is 2/3 "of" the length = 2/3 × 22 3/4

$$\frac{\overset{1}{\cancel{2}}}{3} \times \frac{91}{\underset{2}{\cancel{4}}} = \frac{91}{6} = 15\frac{1}{6} \text{ yards}$$

Leaving the answer to an application problem as an improper fraction is not meaningful, so change it to a mixed number. Remember to label the answer with the correct units.

◀

EXAMPLE 14 In the previous problem, Ray's rectangular backyard was found to have a length of 22 3/4 yards and a width of 15 1/6 yards. Use the formula $A = LW$ to help Ray find the number of square yards of sod needed to cover the yard.

Solution As you approach this geometry problem, you may want to use a method for organizing the information, such as the procedure described in Section 2.3.

1. Make a sketch and write in the given measurements.
2. Decide what is missing.
3. Decide which formula could be used to find the missing value and write it down.
4. Substitute the measurements into the formula and solve for the missing quantity.
5. Label the answer with the correct units and check to see that the answer is reasonable.

$22\frac{3}{4}$ yd

$15\frac{1}{16}$ ft

FIGURE 3.9

Figure 3.9 could represent Ray's backyard; the dimensions are written on the drawing. Because square yards of sod are needed, the formula for the area of a rectagle, $A = LW$, is the one to use.

Solution

$A = LW$	Formula for the area of a rectangle. (See Section 2.3.)
$A = 22\frac{3}{4} \times 15\frac{1}{6}$	Substitute the values for L and W.
$= \frac{91}{4} \times \frac{91}{6}$	Change to improper fractions.
$= \frac{8281}{24}$	Multiply, then simplify the answer.
$= 345\frac{1}{24}$ square yards of sod	

◀

EXAMPLE 15 Find the area of a triangular flower bed that has a base of 12 1/2 feet and a height of 7 3/4 feet. Use $A = 1/2\ bh$.

Solution The area formula for a triangle is $A = 1/2\ bh$, where b represents the length of one side, and h represents the length of the height drawn from the opposite vertex. Height is a line segment drawn from a vertex, perpendicular to the line containing the opposite side. Look at Figure 3.10.

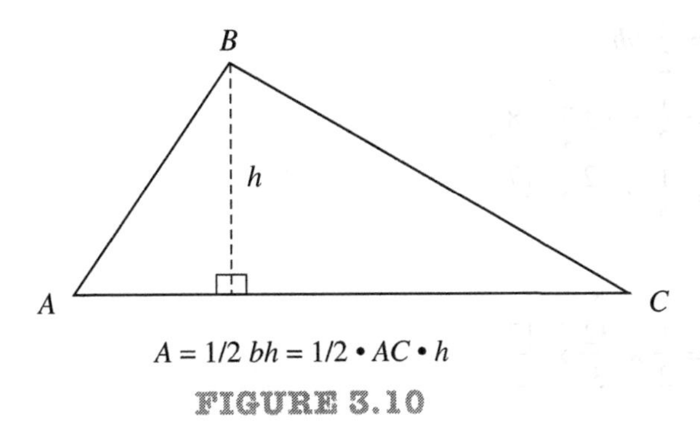

$A = 1/2\ bh = 1/2 \bullet AC \bullet h$

FIGURE 3.10

Base does not always have to be the side that the triangle seems to be sitting on; rather, any side can be a base if the altitude is drawn to that side. The base in this triangle would be the length of side $\overline{AC}$, and the height h is drawn from B perpendicular to $\overline{AC}$. Remember, perimeter is measured in linear units (inches, feet, meters) and area is measured in square units (square inches, square feet, square meters).

For this problem, the base (AC) is 12 1/2 feet and the height (h) is 7 3/4 feet. Substitute those values into the formula and simplify.

$$A = \frac{1}{2} bh$$

$$A = \frac{1}{2} \times 12\frac{1}{2} \times 7\frac{3}{4}$$

$$A = \frac{1}{2} \times \frac{25}{2} \times \frac{31}{4}$$

$$A = \frac{775}{16}$$

$$A = 48\frac{7}{16}$$

The area of the flower bed is 48 7/16 square feet.

EXAMPLE 16 Find the area of the triangle in Figure 3.11 if $QR = 10\ 2/3$ meters and $h = 8\ 1/2$ meters.

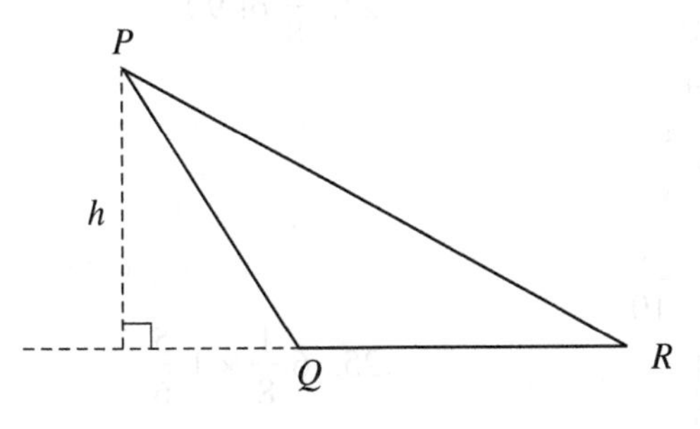

FIGURE 3.11

Solution The formula for the area of a triangle is $A = 1/2bh$.

Consider ΔPQR in Figure 3.11. $\overline{QR}$ is the base, and the height h is drawn perpendicular from vertex P to the extension of side $\overline{QR}$. In this case, the height is measured outside the triangle. (This happens in obtuse triangles.) To find the height in that situation, you extend the base side and measure the perpendicular distance from the vertex to that base. The base is only the measure from Q to R.

$$A = \frac{1}{2}bh$$

$$A = \frac{1}{2} \times 10\frac{2}{3} \times 8\frac{1}{2}$$

$$A = \frac{1}{2} \times \frac{32}{3} \times \frac{17}{2}$$

$$A = \frac{1}{\underset{1}{\cancel{2}}} \times \frac{\overset{\overset{8}{\cancel{16}}}{\cancel{32}}}{3} \times \frac{17}{\underset{1}{\cancel{2}}}$$

$$A = \frac{136}{3}$$

Therefore, the area of triangle PQR is 45 1/3 square meters.

Let's estimate the answer to see if our calculated answer is reasonable. Use 11 meters for the base and 8 meters for the height. Since $A = \frac{1}{2}bh = \frac{1}{2} \times 11 \times 8 =$ 44 square meters. This is close to the calculated answer.

3.3 EXERCISES

A. Find the product and simplify all answers to lowest terms.

1. $\frac{2}{3} \times \frac{5}{9}$
2. $\frac{4}{5} \times \frac{7}{9}$
3. $\frac{6}{7} \times \frac{2}{5}$
4. $\frac{1}{3} \times \frac{4}{9}$
5. $\frac{3}{4}$ of 18
6. $\frac{5}{6}$ of 24
7. $\frac{12}{18} \times \frac{27}{36}$
8. $\frac{8}{25} \times \frac{10}{24}$
9. $\frac{3}{8} \times \frac{12}{15} \times \frac{2}{3}$
10. $1\frac{2}{3} \times 3\frac{3}{5}$
11. $5\frac{2}{5} \times 2\frac{7}{9}$
12. $\frac{1}{2} \times \frac{5}{6} \times \frac{8}{10}$
13. $12\frac{2}{3} \times 7\frac{1}{2}$
14. $2\frac{7}{8} \times \frac{4}{5}$
15. $4 \times 5\frac{1}{2}$
16. $12\frac{2}{3} \times 9$
17. $1\frac{5}{9} \times 2\frac{1}{3}$
18. $\frac{13}{24} \times \frac{12}{39}$
19. $1\frac{3}{7}$ of 200
20. $2\frac{4}{9}$ of 36
21. $11\frac{5}{6} \times 3\frac{1}{2}$
22. $19\frac{1}{2} \times \frac{6}{7}$
23. $\frac{7}{8}$ of 90
24. $\frac{5}{16}$ of 84

B. Estimate an answer to each of the following exercises. Remember that when you estimate, the numbers should be reasonable so you can use mental math whenever possible.

25. $6\frac{1}{8} \times 1\frac{5}{6}$
26. $12\frac{1}{3} \times 4\frac{7}{8}$

27. $\frac{3}{8} \times 1\frac{5}{6}$

28. $21\frac{3}{4} \times 10\frac{1}{5}$

29. $12\frac{1}{3} \times \frac{3}{8}$

30. $\frac{7}{8} \times \frac{15}{16}$

C. Solve these application problems. We suggest that you first estimate the answer.

31. Two-thirds of a class of 27 are going on a field trip. How many students are **not** going?

32. Becky has delivered 3/4 of her 160 newspapers. How many have **not** been delivered?

33. Niko has eaten 5/6 of the carton of ice cream bars. If the box had contained 12 bars, how many has he eaten?

34. Five-sixths of the students in this class passed the last exam. If there are 42 students, how many passed?

35. Three-fifths of the textbooks sold in the bookstore were paperbacks. If 48,395 books were sold, how many were paperbacks?

36. Victor's raise in salary means that he will receive one and one-eighth times his previous pay. If he had been earning $120 per week, what is his new salary?

37. Dr. Garfield has computed the averages of two-thirds of his students. How many averages has he done if he has 165 students this quarter?

38. A national diet program advertises that customers can expect a weight loss of 2 3/4 pounds per month. What is the total loss expected for six months?

39. During a record-breaking storm in December 1995, Billings, New York, had 1 3/8 inches of snow fall every hour for a 24-hour period. What was the total accumulation during this period?

40. Sarah wants to make 1 1/2 times her recipe for chocolate chip cookies. Please help her figure the amounts of flour, sugar, and vanilla to use if the original amounts were 2 1/2 cups of flour, 3/4 cup of sugar, and 1 1/2 teaspoons of vanilla.

41. In the Cavaliers' last basketball game, two-ninths of their 108 points were scored with 3-point shots. How many points were scored this way?

42. Sue's new exercise plan calls for her to walk 3/4 mile for 8 days and then 1 1/4 miles for 8 days. What is the total distance she will walk during this period?

43. Mary addresses three-eighths of a box of envelopes each day. At this rate, how many boxes will she complete in 4 1/2 days?

44. If a candy bar contains 1 3/8 ounces of almonds, how many ounces of almonds are in 4 1/2 bars?

D. Use the given formula to find the solution to each of the following exercises.

45. A rectangle has a length of 14 1/3 meters and a width of 8 meters. Find the area using $A = LW$.

46. Find the area of a triangle with a base of 2/3 foot and a height of 5/6 foot. Use the formula $A = 1/2\ bh$.

47. The area of a parallelogram is found by using the formula $A = bh$, where b is the base and h is the height of the figure. Find the area of a parallelogram with a base of 5 2/3 inches and a height of 3 5/8 inches.

48. Find the volume of a rectangular box that is 5 1/2 inches wide, 11 inches long, and 8 3/4 inches high. Use $V = LWH$.

E. Read and respond to the following exercises.

49. Describe, in words, the process you would use to find out how many invoices have been processed by an accounts payable clerk if he has completed five-eighths of the 128 invoices.

50. Describe, in words, how you would estimate the answer to this problem: 4 5/8 × 2 1/3.

51. A method is shown in the fraction multiplication process where common factors in the numerator and the denominator are divided out before the multiplying is done. Can this process be used in other operations?

52. Write an application problem that requires multiplication of fractions to reach the answer. Base the problem on a situation from your work or home setting.

3.4 Dividing Fractions

Before learning how to divide fractions, you need to understand the concept of reciprocals. Look at these:

$$\frac{5}{8}\times\frac{8}{5}=\frac{\overset{1}{\cancel{5}}}{\underset{1}{\cancel{8}}}\times\frac{\overset{1}{\cancel{8}}}{\underset{1}{\cancel{5}}}=\frac{1}{1}=1 \qquad \text{or} \qquad \frac{5}{8}\times\frac{8}{5}=\frac{40}{40}=1$$

$$\frac{1}{6}\times 6=\frac{1}{\underset{1}{\cancel{6}}}\times\frac{\overset{1}{\cancel{6}}}{1}=1 \qquad \frac{11}{3}\times\frac{3}{11}=\frac{33}{33}=1$$

In all of these examples, the product is 1. *If you multiply two fractions together and the product equals 1*, then the fractions are **reciprocals** of each other.

$\frac{5}{8}$ is the reciprocal of $\frac{8}{5}$, and $\frac{8}{5}$ is the reciprocal of $\frac{5}{8}$.
$\frac{1}{6}$ is the reciprocal of 6, and 6 is the reciprocal of $\frac{1}{6}$.

The reciprocal of a fraction is written by inverting the fraction.

EXAMPLE 1 Write the reciprocals of 1/4 , 3/5, 7/6, and 2 5/8.

Solution The reciprocals are 4/1 (or 4), 5/3, 6/7, and 8/21. The reciprocal of 2 5/8 was written after changing the mixed number to an improper fraction, 21/8.

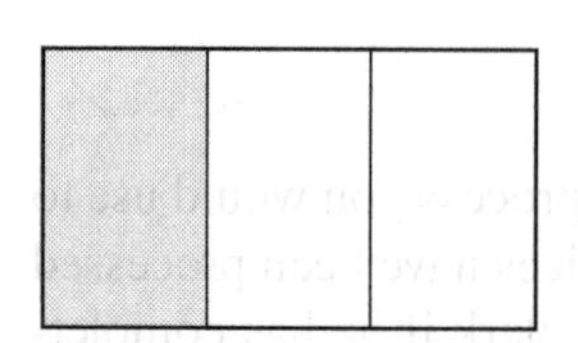

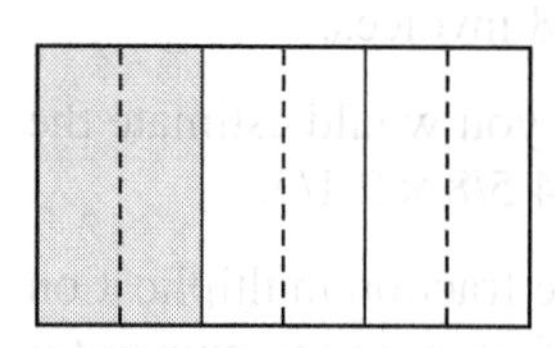

FIGURE 3.12

Reciprocals are necessary for dividing fractions. Let us look at an illustration of fraction division before learning rules for the operation. One-third of the rectangle in Figure 3.12 is shaded. If the shaded portion is divided into two equal parts, what portion of the whole figure does one of these parts represent? The drawings show that the answer is 1/6 of the whole rectangle. Therefore,

$$\frac{1}{3}\div 2=\frac{1}{6}$$

Here is another illustration. In Figure 3.13, the remaining piece of pizza, one-fourth of the whole pizza, must be split into 3 equal portions. What part of the whole pizza does one of these portions represent? Do you see that it would be 1/12 of the whole pizza? Therefore, you could say that

$$\frac{1}{4}\div 3=\frac{1}{12}$$

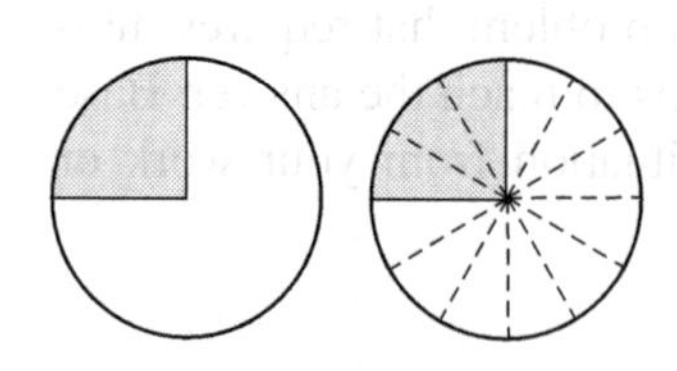

FIGURE 3.13

Look again at Figure 3.12. The only way that $\frac{1}{3}\div 2$ can equal $\frac{1}{6}$ is if $\frac{1}{3}$ is multiplied by $\frac{1}{2}$. Note that $\frac{1}{3}\div 2$ and $\frac{1}{3}\times\frac{1}{2}$ must be equivalent expressions because they give the same answer, $\frac{1}{6}$. Thus $\frac{1}{3}\div 2$ can be solved by multiplying $\frac{1}{3}$ by the reciprocal of the second fraction (divisor):

$$\frac{1}{3} \div 2 = \frac{1}{3} \div \frac{2}{1} = \frac{1}{3} \times \frac{1}{2} = \frac{1}{6}$$

The same pattern is true for Figure 3.13. The only way that $\frac{1}{4} \div 3$ can equal $\frac{1}{12}$ is if $\frac{1}{4}$ is multiplied by $\frac{1}{3}$. The correct answer to $\frac{1}{4} \div 3$ can be found by multiplying the first fraction (dividend) by the reciprocal of the second fraction (divisor):

$$\frac{1}{4} \div 3 = \frac{1}{4} \div \frac{3}{1} = \frac{1}{4} \times \frac{1}{3} = \frac{1}{12}$$

Division of Fractions

To divide fractions, multiply the dividend by the reciprocal of the divisor. In other words, multiply the first fraction by the reciprocal of the second fraction. For the fractions *a/b* and *c/d* (*b*, *c*, and $d \neq 0$),

$$\frac{a}{b} \div \frac{c}{d} = \frac{a}{b} \times \frac{d}{c} = \frac{ad}{bc}$$

EXAMPLE 2 Divide the fraction 2/3 by 5/6.

Solution $\frac{2}{3} \div \frac{5}{6} = \frac{2}{3} \times \frac{6}{5}$ Rewrite as multiplication by the reciprocal of the second fraction.

$\frac{2}{\cancel{3}_1} \times \frac{\cancel{6}^2}{5}$ Because it is now multiplication, we can divide out common factors.

$= \frac{4}{5}$

EXAMPLE 3 Divide 3/5 by 9.

Solution $\frac{3}{5} \div 9 = \frac{3}{5} \div \frac{9}{1} = \frac{3}{5} \times \frac{1}{9}$

$= \frac{\cancel{3}^1}{5} \times \frac{1}{\cancel{9}_3} = \frac{1}{15}$

EXAMPLE 4 Find the quotient of 2 1/2 divided by 8.

Solution $2\frac{1}{2} \div 8$ **Quotient means to divide.**

$= \frac{5}{2} \div \frac{8}{1}$ First change both numbers to improper fractions.

$= \frac{5}{2} \times \frac{1}{8}$ Now rewrite as multiplication by the reciprocal.

$= \frac{5}{16}$

EXAMPLE 5 Simplify $12 \div 4\frac{7}{8}$

Solution $12 \div 4\frac{7}{8}$ Change the numbers to improper fractions.

$= \frac{12}{1} \div \frac{39}{8}$ Rewrite as multiplication by the reciprocal of the second fraction.

$= \frac{12}{1} \times \frac{8}{39}$

$= \frac{2 \cdot 2 \cdot \overset{1}{\cancel{3}}}{1} \times \frac{2 \cdot 2 \cdot 2}{\underset{1}{\cancel{3}} \cdot 13}$

$= \frac{32}{13}$ or $2\frac{6}{13}$

EXAMPLE 6 Tony can drive 360 miles on 17 1/2 gallons of gas. How many miles does his car average on 1 gallon?

Solution Anytime you are asked for a rate of miles per 1 gallon or miles per 1 hour, you will need to divide the miles by the gallons or hours.

$360 \div 17\frac{1}{2}$ First estimate an answer to $360 \div 18 \doteq 20$ miles per gallon

$= \frac{360}{1} \div \frac{35}{2}$

$= \frac{\overset{72}{\cancel{360}}}{1} \times \frac{2}{\underset{7}{\cancel{35}}}$

$= \frac{144}{7}$

$= 20\frac{4}{7}$ miles per gallon

EXAMPLE 7 A 7 7/8-acre parcel of land is to be divided into nine equal-sized lots. What will be the size of each lot?

Solution First estimate the answer. The numbers are very close in size, so dividing 9 into a number that is nearly 8 should give an answer that is a little less than 1.

7 7/8 is to be *divided* into 9 equal parts:

$$7\frac{7}{8} \div 9 = \frac{63}{8} \div \frac{9}{1} = \frac{\overset{7}{\cancel{63}}}{8} \times \frac{1}{\underset{1}{\cancel{9}}} = \frac{7}{8} \text{ acre}$$

This agrees with the estimate.

◀

EXAMPLE 8 How many gasoline cans, each with a capacity of 2 1/2 gallons, can be filled from a tank holding 23 3/4 gallons?

Solution The total amount is 23 3/4 gallons. It is being divided into parts that are each 2 1/2 gallons in size.

Total amount ÷ Size of each part = Number of parts

To get an estimate answer, divide 24 by either 2 or 3. 24 ÷ 2 = 12 and 24 ÷ 3 = 8. The answer should be between 8 and 12 cans.

$$23\frac{3}{4} \div 2\frac{1}{2} = \frac{95}{4} \div \frac{5}{2} = \frac{\overset{19}{\cancel{95}}}{\underset{2}{\cancel{4}}} \times \frac{\overset{1}{\cancel{2}}}{\underset{1}{\cancel{5}}} = \frac{19}{2} \text{ or } 9\frac{1}{2} \text{ cans}$$

◀

A fraction whose numerator and/or denominator is a fraction is called a **complex fraction**. It's an appropriate name, because expressions like the ones that follow can be very confusing!

$$\frac{\frac{2}{3}}{\frac{3}{4}} \qquad \frac{8}{\frac{6}{5}} \qquad \frac{4\frac{2}{5}}{11}$$

Actually, they look much worse than they are. If you keep in mind that a fraction is an indicated division and use that concept, then you can rewrite complex fractions as the division of two fractions.

$$\frac{\frac{2}{3}}{\frac{3}{4}} \text{ becomes } \frac{2}{3} \div \frac{3}{4} = \frac{2}{3} \times \frac{4}{3} = \frac{8}{9}$$

EXAMPLE 9 Simplify: $\frac{8}{\frac{6}{5}}$

Solution $$\frac{8}{\frac{6}{5}} \text{ becomes } 8 \div \frac{6}{5} = \frac{\overset{4}{\cancel{8}}}{1} \times \frac{5}{\underset{3}{\cancel{6}}} = \frac{20}{3} \text{ or } 6\frac{2}{3}$$

◀

EXAMPLE 10 Simplify: $\frac{4\frac{2}{5}}{11}$

Solution $\dfrac{4\frac{2}{5}}{11}$ rewritten with the division symbol becomes

$$4\frac{2}{5} \div 11 = \frac{22}{5} \div \frac{11}{1} = \frac{\overset{2}{\cancel{22}}}{5} \times \frac{1}{\underset{1}{\cancel{11}}} = \frac{2}{5}$$

EXAMPLE 11 Simplify: $\dfrac{8\frac{1}{6}}{11\frac{2}{3}}$

Solution $8\frac{1}{6} \div 11\frac{2}{3} = \frac{49}{6} \div \frac{35}{3} = \frac{\overset{7}{\cancel{49}}}{\underset{2}{\cancel{6}}} \times \frac{\overset{1}{\cancel{3}}}{\underset{5}{\cancel{35}}} = \frac{7}{10}$

Self-Check Divide and simplify answers to lowest terms.

a. $\frac{6}{7} \div \frac{2}{3}$ b. $5\frac{1}{4} \div 1\frac{3}{4}$ c. $6 \div 2\frac{4}{9}$

Answers:

a. $\frac{9}{7}$ or $1\frac{2}{7}$ b. 3 c. $\frac{27}{11}$ or $2\frac{5}{11}$

3.4 EXERCISES

A. Write the reciprocal of each of the following.

1. $\frac{2}{3}$
2. $\frac{5}{8}$
3. 4
4. $\frac{4}{5}$
5. 12
6. $\frac{1}{6}$
7. $\frac{9}{7}$
8. 15
9. $\frac{13}{3}$
10. 2

B. Divide and simplify answers to lowest terms.

11. $\frac{2}{3} \div \frac{3}{4}$
12. $\frac{5}{8} \div \frac{2}{3}$
13. $\frac{5}{6} \div \frac{1}{3}$
14. $\frac{4}{5} \div \frac{2}{15}$
15. $8 \div \frac{5}{6}$
16. $1\frac{2}{3} \div 3\frac{1}{8}$
17. $5\frac{5}{6} \div 2\frac{1}{3}$
18. $4 \div \frac{2}{3}$
19. $\frac{7}{15} \div \frac{4}{5}$
20. $63 \div \frac{3}{8}$
21. $26 \div \frac{2}{9}$
22. $\frac{5}{6} \div \frac{2}{18}$
23. $\frac{4}{5} \div \frac{7}{8}$
24. $21\frac{3}{10} \div 7\frac{2}{5}$
25. $2\frac{1}{3} \div 3$
26. $2\frac{1}{3} \div 1\frac{1}{9}$
27. $\frac{2}{15} \div 32\frac{1}{5}$
28. $11\frac{3}{8} \div 6$

C. Simplify these complex fractions.

29. $\dfrac{\frac{2}{3}}{4}$

30. $\dfrac{8}{\frac{3}{4}}$

31. $\dfrac{1\frac{1}{3}}{8}$

32. $\dfrac{\frac{3}{5}}{\frac{5}{6}}$

33. $\dfrac{\frac{4}{9}}{\frac{2}{3}}$

34. $\dfrac{1\frac{7}{8}}{1\frac{1}{4}}$

35. $\dfrac{7\frac{7}{8}}{9}$

36. $\dfrac{5\frac{1}{3}}{8}$

37. $\dfrac{4\frac{1}{3}}{2\frac{8}{9}}$

D. Solve these application problems. We suggest that you first estimate an answer.

38. Divide 26 2/3 into eight equal parts.

39. If Lori evenly divides a 3 1/8-pound boneless roast among her nine dinner guests and herself, how much meat will each receive?

40. Each can of ready-to-eat soup contains 10 3/4 ounces. If the soup is divided into three equal servings, how many ounces will each person receive?

41. When at full production, a car assembly plant has a new car roll off the line every 2/3 hour. If the plant is operating around the clock, how many cars are produced in 24 hours?

42. A tanning booth is used for 12 hours a day. If each session is 3/4 hour long, how many sessions can be scheduled in a day?

43. If a plane flies 65 2/3 miles in 7 1/2 minutes, how far can it travel in one minute?

44. How many pieces of cotton cloth, each 5 3/8 yards long, can be cut from a bolt containing 64 1/2 yards?

45. Laura earned 38 1/2 dollars for working 7 hours. How much does she earn per hour? Express the answer in fraction form.

46. A 3 1/8-pound roast costs $10. How much per pound does it cost? Give the answer in fraction form.

47. Marilyn bought 7 1/2 yards of material to make scarves. How many can she make if each requires 3/4 of a yard?

48. If 297 bushels of corn were produced by a 6 3/4-acre farm, how many bushels were produced per acre?

49. At the Colonial Cheese Shoppe, a clerk used a 20 1/4-pound block of cheese to make small packages, each 3/8 pound in size. How many packages did she made?

50. In five hours, a car was driven 240 1/2 miles. How many miles per hour did the car average?

51. A land developer plans to divide 70 5/8 acres of land into 1 1/4-acre plots. How many plots will be available?

52. A spool of wire 28 1/2 feet long is cut into 19 equal pieces. How long is each piece?

53. How many servings, 5 3/8 ounces, are there in a 32 1/4-ounce can of fruit cocktail?

54. If a train travels 268 2/3 miles in 5 1/6 hours, how far can it travel in 1 hour?

E. Read and respond to the following exercises.

55. Describe, in words, what is meant by the reciprocal of a fraction.

56. Describe, in words, the meaning of "complex fraction."

57. Tell, in words, how to simplify 2 1/3 ÷ 1 1/9.

58. Describe, in words, the process used to go from one step to the next in the exercise below.

$3\frac{5}{8} \div 4$	
$= \frac{29}{8} \div 4$	1.
$= \frac{29}{8} \div \frac{4}{1}$	2.
$= \frac{29}{8} \times \frac{1}{4}$	3.
$= \frac{29}{32}$	4.

3.5 CHAPTER REVIEW

Your success on the test for this chapter will depend on your understanding the vocabulary of fractions and on your being able to apply a few rules to perform multiplication and division. By now you should know that:

1. The *numerator* is the top number in a fraction, and the *denominator* is the bottom number.
2. An *improper fraction* has a numerator that is greater than or equal to the denominator. A *proper fraction* has a numerator that is smaller than the denominator.
3. A *mixed number* has a whole number part and a fraction part.
4. *To change from a mixed number to an improper fraction, multiply the whole number times the denominator and add the numerator.* Write this answer over the denominator.
5. *To change from an improper fraction to a mixed number, divide the denominator into the numerator.* This answer is written as the whole number, and the remainder is written over the denominator for the fraction part.
6. *Equivalent fractions* are fractions that have the same value but different forms. An equivalent fraction can be obtained by multiplying or dividing both numerator and denominator by the same factor.
7. A fraction has been simplified completely, or *reduced to lowest terms*, when there is no factor other than 1 common to both the numerator and the denominator.
8. *To multiply fractions*, multiply the numerators together, multiply the denominators together, and then reduce the answer to lowest terms.
9. *To find the area of a regular geometric shape*, use the appropriate formula, substitute the given values, simplify, and label the answer.
10. *To divide fractions*, multiply the first fraction by the reciprocal of the second fraction.
11. *To multiply or divide mixed numbers*, first change them to improper fractions,
12. A fraction is a *reciprocal* of another fraction if their product is 1.
13. To simplify a *complex fraction*, rewrite it as one fraction divided by another.
14. All fractional answers *must be simplified to lowest terms*.

REVIEW EXERCISES

A. Change these improper fractions to mixed numbers, reduced to lowest terms.

1. $\frac{44}{15}$
2. $\frac{12}{7}$
3. $\frac{19}{3}$
4. $\frac{5}{2}$
5. $\frac{22}{12}$
6. $\frac{19}{2}$
7. $\frac{61}{5}$
8. $\frac{126}{10}$

B. Change these mixed numbers to improper fractions.

9. $3\frac{3}{8}$
10. $5\frac{1}{2}$
11. $12\frac{2}{3}$
12. $1\frac{5}{8}$
13. $54\frac{2}{5}$
14. $32\frac{3}{4}$
15. $10\frac{1}{8}$
16. $2\frac{2}{3}$

C. Simplify these fractions to lowest terms.

17. $\frac{27}{36}$
18. $\frac{15}{21}$
19. $\frac{8}{20}$
20. $\frac{10}{18}$
21. $\frac{3}{9}$
22. $\frac{7}{14}$
23. $\frac{32}{80}$
24. $\frac{24}{64}$
25. $\frac{24}{39}$
26. $\frac{18}{48}$
27. $\frac{54}{81}$
28. $\frac{56}{96}$
29. $\frac{96}{126}$
30. $\frac{140}{252}$
31. $\frac{224}{320}$
32. $\frac{156}{234}$

D. Find the correct numerator or denominator to make these fractions equivalent.

33. $\frac{2}{3} = \frac{\quad}{12}$
34. $\frac{1}{4} = \frac{\quad}{16}$
35. $\frac{3}{5} = \frac{\quad}{15}$
36. $\frac{5}{8} = \frac{15}{\quad}$
37. $\frac{24}{36} = \frac{8}{\quad}$
38. $\frac{15}{18} = \frac{\quad}{6}$
39. $\frac{3}{4} = \frac{\quad}{24}$
40. $\frac{12}{16} = \frac{6}{\quad}$
41. $\frac{8}{24} = \frac{\quad}{3}$
42. $\frac{16}{1} = \frac{\quad}{2}$
43. $\frac{11}{1} = \frac{44}{\quad}$
44. $\frac{63}{35} = \frac{\quad}{5}$

E. Multiply or divide as indicated and simplify answers to lowest terms.

45. $\frac{2}{3} \times \frac{5}{6}$
46. $\frac{4}{9} \times \frac{3}{8}$
47. $\frac{7}{8} \div \frac{1}{4}$
48. $\frac{3}{4} \div \frac{1}{8}$
49. $1\frac{2}{3} \times 4$
50. $8 \times 5\frac{1}{3}$
51. $6\frac{1}{4} \div \frac{5}{8}$
52. $4 \div 1\frac{3}{5}$
53. $\frac{3}{4}$ of 6
54. $2\frac{4}{9} \div \frac{3}{11}$
55. $\frac{2}{3} \div \frac{3}{4}$
56. $1\frac{5}{8}$ of 39
57. $\frac{3}{8} \times 1\frac{3}{5} \times 2\frac{2}{3}$
58. $1\frac{1}{2} \times 4\frac{2}{3} \times 1\frac{5}{6}$
59. $\frac{5}{6} \times 1\frac{2}{3} \times 3\frac{3}{8}$
60. $4\frac{1}{3} \times \frac{10}{39} \times 1\frac{1}{5}$
61. $\dfrac{2\frac{2}{3}}{1\frac{2}{4}}$
62. $\dfrac{\frac{3}{8}}{\frac{5}{6}}$
63. $\dfrac{\frac{5}{9}}{\frac{2}{3}}$
64. $\dfrac{1\frac{4}{5}}{3\frac{1}{3}}$

F. Find the solution to each exercise. Be sure to reduce all answers to lowest terms.

65. McArches Corporation processes beef for its hamburger restaurants. It forms 3/8-pound patties from a package of ground beef weighing 24 3/4 pounds. How many of these patties can be formed from each package?

66. Five-sixths of the students doing a problem will get it correct. If 48 students are doing the problem, how many will get the correct answer?

67. Sara hates to get up in the morning, but she has decided to try harder. Three-fourths of the last 24 school days she has gotten up on time. How many days did she get up on time?

68. A farmer died and left his land to his children. The 74 2/3 acres are to be divided evenly among his four children. How much land will each child inherit?

69. Five-eighths of the compact disks sold by Hi-Notes music store are by rock artists. If 1,280 disks were sold last month, how many were by rock artists?

70. Connie has read two-thirds of the book assigned for psychology class. How many of the 396 pages in the book has she read?

71. If a train travels 162 3/4 miles in 3 1/2 hours, what is its average speed in miles per hour?

72. At the Java Jungle Coffee Shop, 28 pounds of coffee beans are made into 7/8-pound packages. How many packages can be made?

73. How far has Manuella run if she has gone around the 5/8-mile track 2 1/2 times?

74. A tank holding 138 1/2 liters of kerosene is to be used to fill 2 1/4-liter cans. How many full cans will there be?

75. How many 10 3/4-ounce cans of soup can be filled from a large kettle containing 700 ounces of soup?

76. Each piece of flooring tile measures 1/9 square yard. How many pieces of tile will be needed to cover a floor whose area is 12 square yards?

77. A large display balloon that looks like a box of crackers needs to be inflated. How many cubic yards of helium are needed to fill the balloon, which is 5 1/2 yards high, 3 3/4 yards long, and 2 1/2 yards wide? Use $V = LWH$.

78. Find the area of a triangle if h = 11 1/3 centimeters, PQ = 14 1/2 centimeters, PR = 19 3/4 centimeters, and QR = 13 3/8 centimeters. Use $A = 1/2\ bh$. (See Figure 3.14.)

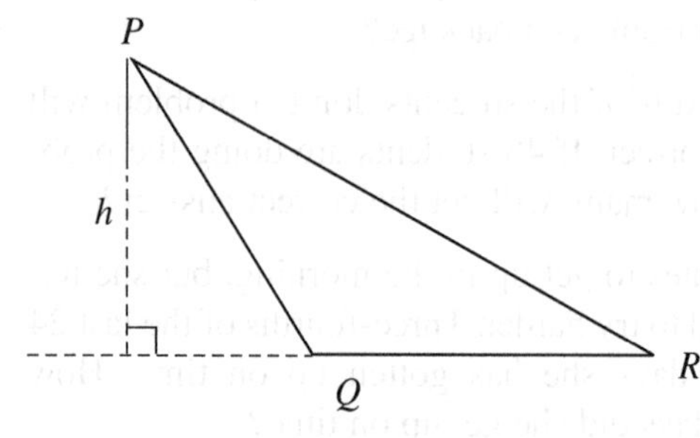

FIGURE 3.14

G. Each exercise that follows has an error in the solving process. State what the error is, then work the problem correctly.

79. $\frac{15}{16} \times \frac{4}{9}$

$= \frac{\cancel{15}^{5}}{\cancel{16}_{2}} \times \frac{\cancel{4}^{1}}{\cancel{9}_{3}}$

$= \frac{5}{6}$

80. Simplify: $\frac{5}{16} \div \frac{1}{4}$

$= \frac{16}{5} \times \frac{4}{1}$

$= \frac{64}{5}$ or $12\frac{4}{5}$

81. Find $\frac{1}{2}$ of $\frac{7}{8}$.

$= \frac{1}{2} \div \frac{7}{8}$

$= \frac{1}{\cancel{2}_{1}} \times \frac{\cancel{8}^{4}}{7}$

$= \frac{4}{7}$

82. Are $\frac{4}{9}$ and $\frac{16}{27}$ equivalent fractions?

Yes, because $\frac{4}{9} \times \frac{4}{3} = \frac{6}{27}$

83. Find $\frac{5}{8}$ of $2\frac{1}{2}$.

$= \frac{5}{8} \div 2\frac{1}{2}$

$= \frac{5}{8} \div \frac{5}{2}$

$= \frac{5}{8} \times \frac{2}{5}$

$= \frac{1}{4}$

H. Read and respond to the following exercises.

84. The following facts were found in a newspaper article describing yesterday's football game: The

Bengals scored 42 points in the game. Only twenty-seven players saw action in the game. Four-sevenths of the points scored came on passing plays. One-ninth of the players who saw action were rookies (first-year players). Six of the points scored came from kicking field goals. All other points came from running plays.

a. Describe, in words, how you would find the number of points scored with running plays.

b. What facts are given that are not needed to do the steps described in part a of this exercise?

85. In the rule for multiplication of fractions on page 120, why is the statement "where *b* and *d* are not equal to zero" included?

86. Some students have difficulty understanding that a fraction division problem can produce an answer that is a larger number than the original numbers in the problem. Using the following example, describe this "phenomenon" to such a puzzled student:

A 5 1/2-foot length of wire is to be cut into pieces that are each 5/6 of a foot in length. How many pieces will be produced?

87. Describe, in words, how to simplify this fraction:

$$\frac{\frac{4}{9}}{\frac{2}{3}}$$

I. The following problems might require two or more steps to reach a solution. Show your work.

88. How many cans of applesauce, each holding 48 3/8 ounces, will be needed to provide 40 servings, each 5 1/4 ounces?

89. Marina wants to paint one wall of her bedroom a contrasting color. The label on the paint can says that 1 quart covers 96 square feet. How many quarts will she need to buy if her wall is 11 3/8 feet long and 9 1/2 feet high? Use $A = LW$. Explain your answer.

90. Three students attempt to answer the following question:

To cover the cushions on a sofa, Nancy will need 1 1/4 yards of fabric for each cushion. Does she have enough fabric to cover 6 cushions if she has a piece of fabric that is 8 1/2 yards long?

The first student said to multiply 1 1/4 × 6 to see if the total amount of fabric needed is less than 8 1/2 yards.

The second student said to divide 8 1/2 by 6 to see if this gives at least 1 1/4 yards for each cushion.

The third student said to divide 8 1/2 by 1 1/4 to see if there is at least enough fabric for 6 cushions.

Which student is correct? Explain your answer.

CHAPTER TEST

1. Change 19/7 to a mixed number,
2. Change 4 5/8 to an improper fraction.
3. Reduce 36/80 to lowest terms.
4. Simplify 160/252 to lowest terms.
5. Find the correct numerator to make these fractions equivalent: 7/8 ≠ 32.
6. Write the reciprocal of 3/5.

Multiply or divide as indicated in the following exercises. Reduce all fractions to lowest terms.

7. $\frac{2}{3} \times \frac{5}{7}$

8. $\frac{5}{9} \div \frac{1}{5}$

9. $2\frac{3}{4} \times \frac{8}{9}$

10. Find $\frac{5}{6}$ of 42.

11. $\frac{3}{8} \times 2\frac{3}{5} \times 2\frac{2}{3}$

12. $3\frac{2}{3} \div 2\frac{4}{9}$

13. $\frac{\frac{5}{6}}{\frac{4}{9}}$

Solve the following application problems.

14. Danny has repaid two-thirds of the money that he borrowed from his mother to buy a car. If he borrowed \$2,970 for the car, how much has he repaid?

15. A 28 1/2-pound block of cheese is being cut into smaller pieces to be wrapped and sold. How many 4/5-pound packages can be made?

16. If Jan can drive 238 miles on 9 4/5 gallons of gas, what is her car's average number of miles per gallon?

17. A bow to decorate a wreath of dried flowers requires 4 3/8 feet of ribbon. If Marilyn has 9 wreaths to decorate, how much ribbon will she need?

18. Find the area of a parallelogram whose base is 12 2/3 yards and whose height is 5 5/8 yards. Use $A = bh$.

19. What is the volume of a box that is 3 1/2 feet by 2 1/3 feet by 2 1/4 feet? Use $V = LWH$.

20. Describe, in words, the process used to go from one step to the next.

$$4\frac{3}{8} \div 6\frac{1}{4}$$

$$= \frac{35}{8} \div \frac{25}{4} \qquad 1.$$

$$= \frac{35}{8} \times \frac{4}{25} \qquad 2.$$

$$= \frac{\overset{7}{\cancel{35}}}{\underset{2}{\cancel{8}}} \times \frac{\overset{1}{\cancel{4}}}{\underset{5}{\cancel{25}}} \qquad 3.$$

$$= \frac{7}{10} \qquad 4.$$

STUDY SKILLS MODULE 4

Reading a Math Textbook

Open this textbook to the first page of Section 4.1 and read that section for about ten minutes. Did the reading experience seem similar to the reading you do for pleasure (novels or magazines) or even the reading you do for nontechnical subjects (psychology or history)?

You probably noticed that the pages were not made up of paragraph after paragraph of just words. These pages contained drawings, colored boxes, bold-faced words, and example problems simplified with detailed solutions. Reading just the first and the last paragraphs of the section wasn't enough to allow you to understand most of the material presented in the section. Math and technical textbooks require different reading strategies than other kinds of textbooks.

Reading a math textbook is not quick and one-directional reading like reading a novel or a history textbook. When reading a math textbook, you need to have a pencil and paper available so you can jot down notes or try practice problems. You may need to go back in the chapter or to another chapter to clarify a term or a procedure. Sometimes, just reading the text and definitions does not produce understanding. Often you will need to go back and reread preliminary material and carefully follow example problem solutions to understand a particular topic. Once the topic does make sense, reread the explanation that you found difficult to be sure you understand all the details.

Generally, nontechnical textbooks present material by stating the concept several different ways, and on different levels; by relating the concept to previously learned material; and by using paragraph structure with clearly stated topic and summary sentences. Math and technical textbooks, however, offer less repetition of concepts, assume knowledge of previous concepts, and make little use of formal paragraph structure. The math author uses as few words as possible to convey a thought, and the language she or he uses has very precise meaning. Much of the teaching is done with example problems, charts, graphs, or diagrams. You will need to take time to study each of these items thoroughly when they are used in your textbook.

Students use three approaches in reading for a math assignment. Some students start from the beginning and read straight through, studying the example problems as they go; others start with the example problems and then read the parts of the text needed to understand the examples; and the third group starts doing homework problems and goes back and reads only what is needed to do those problems. The last procedure is the least desirable method because too much material may be skimmed over, and this may cause difficulties later in trying to understand more complex concepts. Unfortunately, the third method is the one most commonly practiced by math students.

If you are learning math topics for the first time or have had difficulty learning math, then you should try to use the first method just described. Before class, read for understanding of the material, closely following the descriptions of steps used in the example problems. Highlight important terms and explanations. Jot down things that are not clear to you. After hearing the lecture or using other instructional aids such as videotapes or computer programs, review the chapter and check for understanding of the chapter material and your notes. If your earlier list of unclear topics has not been resolved, then reread that material and seek help from your instructor or a math tutor. Use the book's resources—the index, glossary, table of contents, list of objectives, and chapter review—to find related material that might clarify any trouble areas. After rereading each section, try to write or state a summary of the concepts. If the vocabulary is unfamiliar, make your own glossary or use index cards to make vocabulary flash cards.

Don't expect to sit down and read quickly through math material from beginning to end. Instead, be prepared to read and reread, to search for related information in previous chapters or sections, to take notes and list new vocabulary, to work out the example problems, and then to solve the exercises in the book, working them out step by step. To understand and be successful with math, you will need to develop reading strategies that are unique to this type of material.

Your Point of View

1. You have probably already completed or read several chapters. Describe the procedure you have been using to read the textbook.
2. Can you describe several differences between pleasure reading and math textbook reading that were not pointed out here?
3. Describe how your reading strategy might differ if you were in a math class that is reviewing previously learned material instead of a class that is presenting all new material.
4. Active learning is preferable to just "soaking it in." List some active things that you can do while reading your math book that will help you stay focused and increase your understanding.

CHAPTER

Addition and Subtraction of Fractions

LEARNING OBJECTIVES

When you have completed this chapter, you should be able to:

1. Find the least common denominator of several fractions.
2. Add or subtract fractions.
3. Add or subtract mixed numbers.
4. Apply the Order of Operations Agreement to fraction problems.
5. Compare the sizes of several fractions.
6. Solve application problems involving all operations with fractions and mixed numbers.

As we continue with fractions, keep in mind that you have already mastered many of the concepts of fractions, and only a few remain. In this chapter, you will study the operations of addition and subtraction. Then you will work with all four operations in order of operations and application problems.

Calculator usage will not be demonstrated in this chapter, but the exercises presented are appropriate for either manual or calculator solution. If you are going to be using a calculator as you do this work, you may want to study the keying sequences described in Appendix B.

4.1 Adding and Subtracting Fractions

It would be difficult to combine the fractions represented by the shaded amounts in Figure 4.1 because each rectangle is divided into parts of a different size.

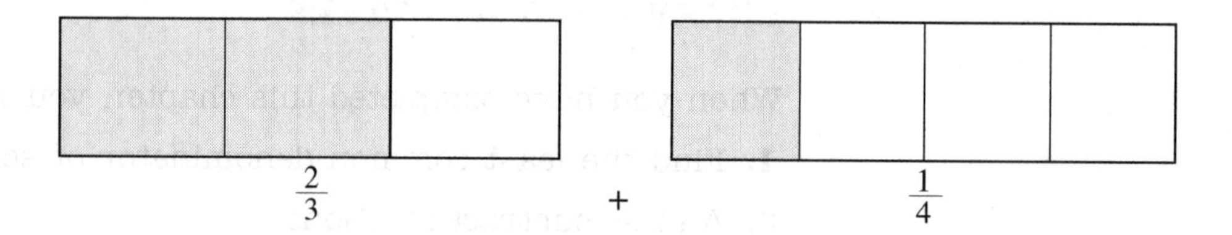

FIGURE 4.1

If equivalent figures were drawn—rectangles that had divisions of the same size, as in Figure 4.2—the problem would be easy to solve.

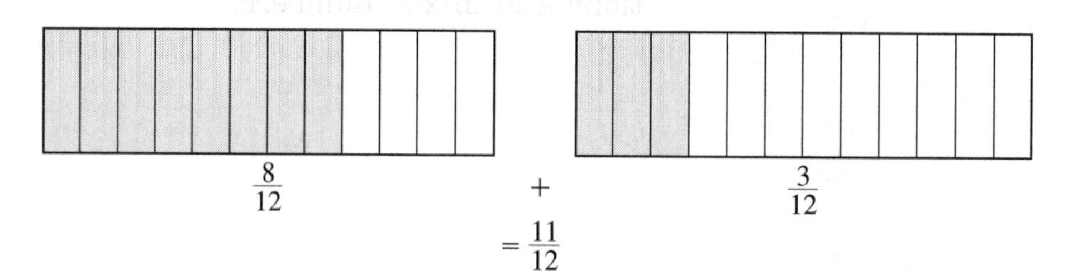

FIGURE 4.2

If the same drawings were used to show subtraction, as in Figure 4.3, the same difficulty would arise.

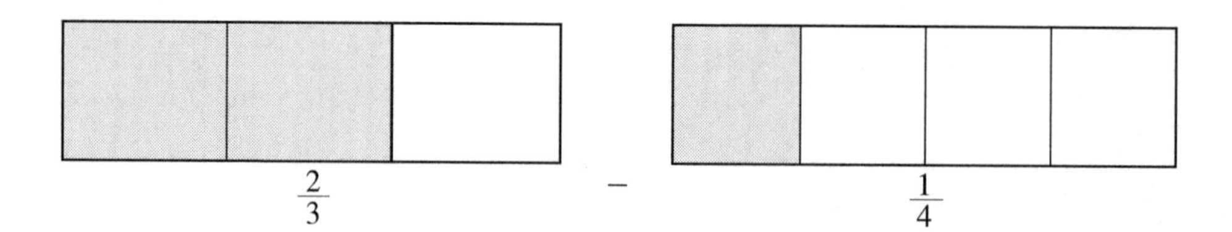

FIGURE 4.3

The difference can be seen only when the rectangles are divided into parts of the same size, as in Figure 4.4.

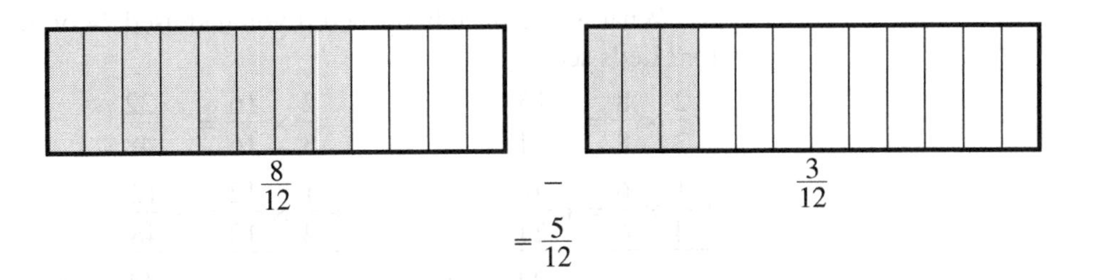

FIGURE 4.4

This is the principle that must be applied to every addition or subtraction problem involving fractions. The quantities can be combined only if they have the same denominators. This is similar to the concept used with whole number addition, where like place values were combined (that is, added or subtracted). You can only add or subtract like items. The following fractions can be added because they have the same denominator (they are divided into parts of the same size). Add the numerators together and write the answer over the common denominator.

$$\frac{4}{8}+\frac{1}{8}=\frac{5}{8}$$

The next two fractions can be subtracted because they have the same, or a common, denominator. Subtract the numerators and write the answer over the common denominator.

$$\frac{11}{16}-\frac{4}{16}=\frac{7}{16}$$

If the fractions that you are to add or subtract do not have the same denominator, then you must find a value that all denominators can divide into evenly. Equivalent fractions, with that new denominator, must be written before the addition or subtraction can be done. Remember, to change a fraction to an equivalent fraction, multiply it by a fraction whose value is 1 (4/4, 9/9, 2/2, and so forth; see Section 3.2).

Let's take the following problem; it is the same problem that we did visually using figures 4.1 and 4.2.

$$\frac{2}{3}+\frac{1}{4}$$

12 is a number into which 3 and 4 can divide evenly. Use it as the common denominator and write equivalent fractions.

You may find it easier to make the change to equivalent fractions if the problem is written vertically.

$$\begin{array}{rcr} \frac{2}{3} & = & \frac{8}{12} \\ +\frac{1}{4} & = & +\frac{3}{12} \\ \hline & & \frac{11}{12} \end{array}$$

Multiply: $\frac{2}{3}\times\frac{4}{4}=\frac{8}{12}$

Multiply: $\frac{1}{4}\times\frac{3}{3}=\frac{3}{12}$

What would have happened if you had used 24 or 48 as the common denominator? Let's see.

$$\frac{2}{3} \times \frac{8}{8} = \frac{16}{24} \qquad \frac{2}{3} \times \frac{16}{16} = \frac{32}{48}$$

$$+\frac{1}{4} \times \frac{6}{6} = +\frac{6}{24} \qquad +\frac{1}{4} \times \frac{12}{12} = +\frac{12}{48}$$

$$\frac{22}{24} = \frac{11}{12} \qquad \frac{44}{48} = \frac{11}{12}$$

All three gave the same answer, but you had to work with larger numbers in the last two, and you had to reduce the answer to lowest terms. Using the smallest common denominator, the **least common denominator**, gives you a problem that is a little easier to solve. Most of the time this value can be found by inspection. However, in problems with more than two denominators or with large denominators, the **least common multiple** methods (Section 2.6) can be used. A least common multiple (LCM) and a least common denominator (LCD) are the same thing.

Addition and Subtraction of Fractions

Fractions can be added or subtracted only if they have common denominators.

$$\frac{a}{c} + \frac{b}{c} = \frac{a+b}{c} \qquad \frac{a}{c} - \frac{b}{c} = \frac{a-b}{c}$$

EXAMPLE 1 Find the sum of 4/7 and 3/4.

Solution

$$\frac{4}{7} = \frac{16}{28}$$

The LCD, 28, is the product of the denominators 4 and 7 because 4 and 7 have no factors in common.

$$+\frac{3}{4} = +\frac{21}{28}$$

Write equivalent fractions, and then add the numerators and write that answer over 28.

$$\frac{37}{28}$$

Because 37/28 cannot be reduced, it can be left as the answer or changed to the mixed number.

$$\text{or } 1\frac{9}{28}$$

EXAMPLE 2 Find the difference between 19/24 and 1/2.

Solution

$$\frac{19}{24} = \frac{19}{24}$$

The common denominator is the larger denominator (24) because 2 also divides into 24.

$$-\frac{1}{2} = -\frac{12}{24}$$

Write equivalent fractions, and then subtract. Be sure the answer is in lowest terms, which it is.

$$\frac{7}{24}$$

EXAMPLE 3 Add: 3/4 + 5/12 + 2/3

Solution

$$\begin{array}{rcr} \frac{3}{4} & = & \frac{9}{12} \\ \frac{5}{12} & = & \frac{5}{12} \\ +\ \frac{2}{3} & = & +\ \frac{8}{12} \\ \hline & & \frac{22}{12} \\ & = & \frac{11}{6} \text{ or } 1\frac{5}{6} \end{array}$$

It is still easy to find the common denominator by inspection because 3 and 4 divide into 12.

Write equivalent fractions, and then add the numerators.

Reduce the answer to lowest terms.

EXAMPLE 4 Combine: $\frac{8}{15} - \frac{5}{18} + \frac{4}{9}$

Solution Find the LCD for 15, 18, and 9.

Here we will use a method that was demonstrated in Section 2.6, because these denominators have factors in common.

$$\begin{array}{l} 15 = \quad 3 \quad \times 5 \\ 18 = 2 \times 3 \times 3 \\ \ \ 9 = \quad 3 \times 3 \\ \hline \text{LCD} = 2 \times 3 \times 3 \times 5 = 90 \end{array}$$

$$\frac{8}{15} - \frac{5}{18} + \frac{4}{9}$$

Write equivalent fractions:

$$\frac{8}{15} \times \frac{6}{6} = \frac{48}{90} \qquad \frac{5}{18} \times \frac{5}{5} = \frac{25}{90} \qquad \frac{4}{9} \times \frac{10}{10} = \frac{40}{90}$$

$$= \frac{48}{90} - \frac{25}{90} + \frac{40}{90}$$

Follow Order of Operations and combine from left to right.

$$= \frac{23}{90} + \frac{40}{90}$$

$$= \frac{63}{90} = \frac{7}{10}$$

Reduce to lowest terms.

EXAMPLE 5 Combine: $\frac{5}{9} + \frac{11}{15} + \frac{7}{24}$

Solution The LCD is more difficult to find with these denominators. The methods for finding least common multiples (Section 2.6) can be used because the LCM and LCD are the same thing. In one of those LCM methods, the numbers were arranged horizontally and then divided by a prime number that divided into at least one of them (it didn't have to divide into all of them). This process was continued until there were all 1s across the bottom. Apply that method to the denominators in this example. List the denominators side by side with some space between them.

$$\begin{array}{r|lll} 2 & 9 & 15 & 24 \\ 2 & 9 & 15 & 12 \\ 2 & 9 & 15 & 6 \\ 3 & 9 & 15 & 3 \\ 3 & 3 & 5 & 1 \\ 5 & 1 & 5 & 1 \\ & 1 & 1 & 1 \end{array}$$

Begin dividing by a prime, and write the answers on the line below. If, a number is not divisible by the prime, just bring the number down. Once you get a 1 in a row, continue to bring it down.

$\text{LCD} = 2 \times 2 \times 2 \times 3 \times 3 \times 5 = 360$

Now multiply the primes together to get the LCD.

This means that 360 is the least common denominator for this problem. Write the equivalent fractions and add.

$$\frac{5}{9} = \frac{5}{9} \times \frac{40}{40} = \frac{200}{360}$$

$$\frac{11}{15} = \frac{11}{15} \times \frac{24}{24} = \frac{264}{360}$$

$$+\frac{7}{24} = \frac{7}{24} \times \frac{15}{15} = +\frac{105}{360}$$

$$\frac{569}{360} \text{ or } 1\frac{209}{360}$$

EXAMPLE 6 Simplify: $\frac{17}{21} - \frac{9}{15}$

Solution Find the LCD of 21 and 15.

$$\begin{array}{r|ll} 3 & 21 & 15 \\ 5 & 7 & 5 \\ 7 & 7 & 1 \\ & 1 & 1 \end{array}$$

$\text{LCD} = 3 \times 5 \times 7 = 105$

$$\frac{17}{21} \times \frac{5}{5} = \frac{85}{105}$$

$$-\frac{9}{15} \times \frac{7}{7} = -\frac{63}{105}$$

$$\frac{22}{105}$$

Is 22/105 in lowest terms? Yes, because $22/105 = (2 \times 11)/(3 \times 5 \times 7)$ and there are no factors common to both the numerator and the denominator.

Self-Check a. $\frac{4}{5} + \frac{7}{15}$ b. $\frac{15}{16} - \frac{7}{8} + \frac{3}{4}$

Answers:

a. $\frac{19}{15}$ or $1\frac{4}{15}$ b. $\frac{13}{16}$

EXAMPLE 7 Myra swam 7/8 of a mile on Monday, 3/4 mile on Wednesday, and 1/2 mile on Saturday. What was her total swimming distance for the week?

Solution First estimate an answer. Since 7/8 and 3/4 are close to 1, add 1 + 1 + 1/2. 2 1/2 miles is the estimate answer. You are looking for a total of several different quantities, so addition is indicated.

$$\frac{7}{8}+\frac{3}{4}+\frac{1}{2}$$

The LCD will be 8 since 4 and 2 can divide into 8.

$$=\frac{7}{8}+\frac{6}{8}+\frac{4}{8}$$

$$=\frac{17}{8}=2\frac{1}{8}\text{ miles}$$

For application problems, the improper fraction answer is not meaningful. It should be changed to a mixed number.

The answer is close to the estimate of 2 1/2 miles.

EXAMPLE 8 Using the information in Example 7, find how much farther Myra swam on Monday than on Saturday.

Solution When asked "how much farther (or shorter or whatever)," you need to subtract to find the difference. Be sure that the larger fraction is on top or to the left of the subtraction sign. (You may be able to tell which fraction is larger only when they have the same denominator.)

$$\frac{7}{8}=\frac{7}{8}$$

$$-\frac{1}{2}=\frac{4}{8}$$

$$\frac{3}{8}\text{ mile farther}$$

EXAMPLE 9 Find the perimeter of a triangle with sides that measure 3/8 yard, 4/5 yard, and 3/4 yard.

Solution As you studied in Section 2.3, the perimeter of any geometric figure is found by adding the lengths of all the sides together. The formula that should be used is $P = a + b + c$.

$$P = a + b + c$$

$$P=\frac{3}{8}+\frac{4}{5}+\frac{3}{4}$$

If you can't find the LCD for 8, 5 and 4 by inspection, use one of the LCM methods from Section 2.6.

$$=\frac{15}{40}+\frac{32}{40}+\frac{30}{40}$$

$$=\frac{77}{40}$$

$$=1\frac{37}{40}$$

Therefore, the perimeter is 1 37/40 yards.

4.1 EXERCISES

A. If the given numbers were denominators in a fraction addition or subtraction problem, what would be the least common denominator?

1. 4 and 3
2. 7 and 14
3. 8 and 2
4. 6 and 9
5. 6 and 4
6. 5 and 3
7. 2, 3 and 4
8. 15, 3 and 60
9. 30, 40 and 60
10. 3, 4 and 5
11. 42 and 15
12. 39 and 15
13. 32 and 18
14. 48 and 30
15. 72, 15 and 48
16. 12, 45 and 30
17. 28, 42 and 63
18. 4, 24, 15 and 12
19. 18, 24, 6 and 3
20. 4, 6, 32 and 48
21. 12, 32, 8 and 18

B. Add or subtract as indicated. Simplify all answers to lowest terms.

22. $\frac{1}{8}+\frac{3}{8}$
23. $\frac{2}{9}+\frac{5}{9}$
24. $\frac{8}{11}-\frac{5}{11}$
25. $\frac{6}{5}-\frac{3}{5}$
26. $\frac{7}{12}+\frac{5}{6}$
27. $\frac{7}{15}+\frac{3}{5}$
28. $\frac{3}{4}+\frac{2}{3}$
29. $\frac{5}{7}-\frac{3}{14}$
30. $\frac{7}{8}-\frac{1}{2}$
31. $\frac{5}{6}+\frac{7}{9}$
32. $\frac{5}{6}-\frac{3}{4}$
33. $\frac{3}{5}+\frac{2}{3}$
34. $\frac{27}{32}-\frac{5}{6}$
35. $\frac{15}{16}-\frac{4}{5}$
36. $\frac{11}{18}+\frac{3}{4}$
37. $\frac{1}{2}+\frac{2}{3}+\frac{3}{4}$
38. $\frac{2}{5}+\frac{2}{3}+\frac{5}{6}$
39. $\frac{5}{3}-\frac{3}{4}+\frac{1}{6}$
40. $\frac{2}{3}-\frac{1}{4}-\frac{2}{5}$
41. $\frac{11}{12}-\frac{7}{15}$
42. $\frac{7}{9}-\frac{4}{15}$
43. $\frac{7}{12}-\frac{5}{18}$
44. $\frac{11}{18}+\frac{19}{20}$
45. $\frac{7}{12}-\frac{7}{15}+\frac{7}{18}$
46. $\frac{7}{12}-\frac{11}{45}+\frac{13}{30}$
47. $\frac{3}{28}+\frac{5}{42}+\frac{11}{63}$
48. $\frac{3}{4}+\frac{11}{24}+\frac{8}{15}+\frac{5}{12}$
49. $\frac{5}{18}+\frac{11}{24}+\frac{5}{6}+\frac{1}{3}$
50. $\frac{4}{9}+\frac{3}{4}+\frac{5}{12}+\frac{4}{5}$

C. Solve the following application problems. Be sure the answers are reduced to lowest terms.

51. Carol is making her own mixed-nut combinations to give as holiday gifts. She has blended 7/8 pound of cashews, 3/4 pound of pecans, 1/2 pound of almonds, and 3/8 pound of pistachio nuts. How much "deluxe mix" does she have?
52. In replacing the floor in Jo's kitchen, the contractor used plywood 3/8-inch thick and insulation material 1/4-inch thick. What was the total thickness of these materials?
53. In Exercise 51, how many more pounds of pecans than pistachios were used?
54. In Exercise 52, how much thicker is the plywood than the insulation material?
55. Raoul spent 1/2 hour on Monday, 7/8 hour on Tuesday, and 3/4 hour on Wednesday doing math homework. What is his total math study time? (If he can't solve this problem, then he didn't spend enough time!)
56. Scott is working out to get in condition for baseball season. He exercised 1/2 hour on Monday, 3/4 hour on Wednesday, and 7/8 hour on Friday. What was his total number of hours of exercise?
57. In Exercise 55, how much more time did Raoul study on Tuesday than on Wednesday?
58. In Exercise 56, how much more time did Scott exercise on Friday than on Monday?
59. Dominique studied her Spanish vocabulary words for 1/2 hour on Monday, 2/3 hour on Tuesday, and 3/4 hour on Wednesday. Altogether, how much time did she spend studying the vocabulary words on those days?

60. In Exercise 59, how much longer did Dominique study on Wednesday than on Tuesday?

D. Find the perimeter of each of the following figures:

61. A triangle with sides that measure 5/8 meter, 3/4 meter, and 1/2 meter. Use $P = a + b + c$.

62. A rectangular piece of cloth with a length of 15/16 yard and a width of 5/8 yard. Use $P = L + W + L + W$.

63. A rectangular piece of picture frame matting that is 5/8 foot by 3/4 foot. Use $P = L + W + L + W$.

64. A triangle that has sides measuring 11/36 yard, 15/16 yard, and 7/8 yard. Use $P = a + b + c$.

E. Read and respond to the following exercises.

65. Describe the process you would use to find the least common denominator for fractions with denominators of 6, 12, and 20.

66. When adding 5/8 and 13/16, is 16 the only possible denominator that could be used? Explain your answer.

67. When the answer to a fraction addition problem is 25/24, must the answer be changed to 1 1/24? Explain your answer.

68. How is addition of fractions similar to addition of whole numbers?

4.2 Adding and Subtracting Mixed Numbers

When adding or subtracting mixed numbers, you can either work with those numbers or change them to improper fractions. You will continue to follow the rules of addition or subtraction of fractions and write equivalent fractions with common denominators. Both methods will be demonstrated in the problems that follow. After studying these solutions, you will be able to choose the method that works best for you.

In the following problem, the proper fractions were written as equivalent fractions with the LCD of 10. Then the fraction parts and whole number parts were added.

$$1\frac{4}{5} = 1\frac{8}{10}$$

$$+3\frac{7}{10} = +3\frac{7}{10}$$

$$4\frac{15}{10} = 4\frac{3}{2} = 4 + 1\frac{1}{2} = 5\frac{1}{2}$$

Note: You cannot leave an improper fraction in a mixed number; 4 15/10 and 4 3/2 are not acceptable. Someone looking quickly at the answer might think that the answer is a little more than 4 because they see 4 and a fraction. In fact, because 3/2 is 1 1/2, you can (and must) add 4 + 1 1/2 and get 5 1/2 as the answer.

Many mixed number subtraction problems proceed the same way as the previous problem. 14 7/8 – 5 2/3 can be done in this way.

$$14\frac{7}{8} = 14\frac{21}{24}$$

$$-5\frac{2}{3} = -5\frac{16}{24}$$

$$9\frac{5}{24}$$

Mixed number addition and subtraction problems can be done by changing to improper fractions, though the numbers may become very large and awkward to work with. Doing the previous problem with improper fractions looks like this:

$$14\frac{7}{8} = \frac{119}{8} = \frac{357}{24} \qquad \frac{119}{8} \times \frac{3}{3} = \frac{357}{24}$$

$$-5\frac{2}{3} = -\frac{17}{3} = -\frac{136}{24} \qquad \frac{17}{3} \times \frac{8}{8} = \frac{136}{24}$$

$$\frac{221}{24} \text{ or } 9\frac{5}{24}$$

Mixed number addition and subtraction exercises can be simplified as mixed numbers or as improper fractions. The mixed number method will be demonstrated first; then Examples 3-6 will be simplified using both methods.

EXAMPLE 1 Solve: $11\frac{2}{3} - 9\frac{1}{2} + 6\frac{2}{15}$

Follow Order of Operations and subtract before you add.

Solution

$$11\frac{2}{3} = 11\frac{20}{30}$$

$$-9\frac{1}{2} = -9\frac{15}{30}$$

$$2\frac{5}{30}$$

$$2\frac{5}{30} = 2\frac{5}{30}$$

$$+6\frac{2}{15} + 6\frac{4}{30}$$

$$8\frac{9}{30}$$

$$= 8\frac{3}{10}$$

EXAMPLE 2 Solve: $5\frac{3}{8} + 6\frac{3}{4}$

Solution

$$5\frac{3}{8} = 5\frac{3}{8}$$

$$+6\frac{3}{4} = +6\frac{6}{8}$$

$$11\frac{9}{8}$$

You cannot leave an improper fraction if it is part of a mixed number.

$$= 11 + 1\frac{1}{8}$$

$$= 12\frac{1}{8}$$

EXAMPLE 3 Solve: 5 1/3 – 2 3/4

Solution

$$5\frac{1}{3} = 5\frac{4}{12}$$

$$-2\frac{3}{4} = -2\frac{9}{12}$$

Once the equivalent fractions are written, a problem occurs in this mixed number subtraction because 9/12 cannot be subtracted from 4/12, When this happens, you must regroup. Borrow 1 from the whole number 5 and write it in the form most useful in this problem. Because 12 is the common denominator, write the 1 as 12/12 (12/12 = 1) and add that to the fraction already in the top number .

$$\begin{array}{llll} 5\frac{4}{12} = & \overset{4}{\not{5}}\frac{4}{12} + 1 = & \overset{4}{\not{5}}\frac{4}{12} + \frac{12}{12} & = 4\frac{16}{12} \\ \underline{-2\frac{9}{12}} = & \underline{-2\frac{9}{12}} \quad = & \underline{-2\frac{9}{12}} & = \underline{-2\frac{9}{12}} \\ & & & \quad 2\frac{7}{12} \end{array}$$

Regrouping is also necessary when a mixed number or fraction is subtracted from a whole number. In the following problem, 3/8 needs something from which to be subtracted. Borrow 1 from 7. Because 8 is the needed denominator, write the borrowed 1 as 8/8, Then the subtraction can be done.

$$\begin{array}{l} 7 \\ \underline{-5\frac{3}{8}} \end{array}$$

$$\begin{array}{llll} 7 & = 6 + 1 = & 6\frac{8}{8} \\ \underline{-5\frac{3}{8}} = & \underline{-5\frac{3}{8}} \quad = & \underline{-5\frac{3}{8}} \\ & & 1\frac{5}{8} \end{array}$$

The previous two examples could have been worked with improper fractions. Example 3 asked us to subtract 5 1/3 – 2 3/4. The solution with improper fractions is

$$\begin{array}{lll} 5\frac{1}{3} = & \frac{16}{3} \times \frac{4}{4} = & \frac{64}{12} \\ \underline{-2\frac{3}{4}} = & \underline{-\frac{11}{4} \times \frac{3}{3}} = & \underline{-\frac{33}{12}} \\ & & \frac{31}{12} \text{ or } 2\frac{7}{12} \end{array}$$

And improper fractions can be used as follows to solve 7 – 5 3/8.

$$7 - 5\frac{3}{8} = \frac{7}{1} - \frac{43}{8} \qquad \frac{7}{1} \times \frac{8}{8} = \frac{56}{8}$$

$$= \frac{56}{8} - \frac{43}{8} = \frac{13}{8} \text{ or } 1\frac{5}{8}$$

EXAMPLE 4 Subtract 121 7/8 from 146 9/16.

Solution Pay attention to the wording of this problem. The quantity we are subtracting from comes first. Using mixed numbers, we have

$$\begin{array}{rcrcrcr} 146\frac{9}{16} & = & 146\frac{9}{16} & = & \overset{145}{\not{1}46}\frac{9}{16}+\frac{16}{16} & = & 145\frac{25}{16} \\ -121\frac{7}{8} & = & -121\frac{14}{16} & = & -121\frac{14}{16} & = & -121\frac{14}{16} \\ \hline & & & & & & 24\frac{11}{16} \end{array}$$

As in the previous examples, after the equivalent fractions are written with common denominators, the need to borrow is evident. The numerator 14 cannot be subtracted from 9. Borrow 1 from the whole number 146 and write it as 16/16. Then regroup 16/16 with the 9/16 already there. Then the subtraction can be done.

Solution Using improper fractions, we write

$$\begin{array}{rcrcr} 146\frac{9}{16} & = & \frac{2345}{16} & = & \frac{2345}{16} \\ -121\frac{7}{8} & = & -\frac{975}{8}\times\frac{2}{2} & = & -\frac{1950}{16} \\ \hline & & & & \frac{395}{16} \text{ or } 24\frac{11}{16} \end{array}$$

You will need to decide which method is better for you to use. In some problems it may be easier to use the mixed numbers and borrow, if necessary, in order to avoid large numbers in the improper fractions. You can make your choice of method after you see the problem.

EXAMPLE 5 Solve: $5 - 2\frac{5}{9}$

Solution Using mixed numbers, we proceed as follows:

$$\begin{array}{rcr} 5 & = & \overset{4}{\not{5}}\frac{9}{9} \\ -2\frac{5}{9} & = & -2\frac{5}{9} \\ \hline & & 2\frac{4}{9} \end{array}$$

There has to be a fraction on top from which 5/9 can be subtracted. Borrow 1 from the 5 and write it as 9/9. Then subtract.

Solution Using improper fractions, we write

$$\begin{array}{rcrcr} 5 & = & \frac{5}{1}\times\frac{9}{9} & = & \frac{45}{9} \\ -2\frac{5}{9} & = & -\frac{23}{9} & = & -\frac{23}{9} \\ \hline & & & & \frac{22}{9} \text{ or } 2\frac{4}{9} \end{array}$$

EXAMPLE 6 Find the difference between 6 3/8 and 5 1/2.

Solution

$$6\frac{3}{8} = 6\frac{3}{8}$$
$$-5\frac{1}{2} = -5\frac{4}{8}$$

Until the fractions are written with the common denominator, you do not know if you have to borrow. You will borrow here because 4/8 cannot be taken from 3/8.

$$6\frac{3}{8} = \overset{5}{\cancel{6}}\frac{3}{8} + \frac{8}{8} = 5\frac{11}{8}$$
$$-5\frac{4}{8} = -5\frac{4}{8} \qquad = -5\frac{4}{8}$$
$$\frac{7}{8}$$

Borrow 1 from 6 and write it as 8/8. Add 3/8 and 8/8. Then subtract.

Self-Check: a. $4\frac{5}{8} + 9\frac{3}{16}$ b. $7 - 2\frac{2}{5}$ c. $28\frac{1}{4} - 13\frac{3}{8}$

Answers:

a. $13\frac{13}{16}$ b. $4\frac{3}{5}$ c. $14\frac{7}{8}$

EXAMPLE 7 A cookie recipe combines 3 1/2 cups of flour, 1 2/3 cups of sugar, and 3/4 cup of oatmeal. What is the total amount of these dry ingredients?

Solution You are looking for the *total amount* of several different quantities, so addition is indicated. First estimate, using convenient whole numbers: 4 + 2 + 1 = 7. The answer should be a little less than 7.

$$3\frac{1}{2} = 3\frac{6}{12}$$
$$1\frac{2}{3} = 1\frac{8}{12}$$
$$+\ \frac{3}{4} = +\frac{9}{12}$$
$$4\frac{23}{12} = 4 + 1\frac{11}{12} = 5\frac{11}{12} \text{ cups}$$

EXAMPLE 8 Some pizzas are 15 1/4 inches in diameter, and other are 12 1/2 inches. How much difference in diameter is there?

Solution The word *difference* means to subtract.

$$15\frac{1}{4} = 15\frac{1}{4} = \overset{14}{\cancel{15}}\frac{1}{4} + \frac{4}{4} = 14\frac{5}{4}$$
$$-12\frac{1}{2} = -12\frac{2}{4} = -12\frac{2}{4} = -12\frac{2}{4}$$
$$2\frac{3}{4} \text{ inches}$$

EXAMPLE 9 Find the amount of wallpaper border needed to go on the walls of a rectangular room that is 9 5/8 feet wide and 11 3/4 feet long.

Solution If we are measuring around a room, we are finding perimeter. The perimeter of the room will be the sum of the measures of the four walls. This can be written as $P = L + W + L + W$. Substitute values for L and W.

$$P = L + W + L + W$$

$$P = 11\frac{3}{4} + 9\frac{5}{8} + 11\frac{3}{4} + 9\frac{5}{8}$$

$$= 11\frac{6}{8} + 9\frac{5}{8} + 11\frac{6}{8} + 9\frac{5}{8}$$

$$= 40\frac{22}{8}$$

$$= 40 + 2\frac{6}{8}$$

$$= 42\frac{6}{8}$$

$$= 42\frac{3}{4}$$

You will need 42 3/4 feet of wallpaper border.

4.2 EXERCISES

A. Add or subtract as indicated. Be sure to simplify all answers to lowest terms.

1. $4\frac{5}{8} + 3\frac{1}{2}$
2. $7\frac{7}{8} + 1\frac{5}{6}$
3. $6\frac{2}{3} - 4\frac{1}{2}$
4. $17\frac{5}{8} - 12\frac{1}{3}$
5. $2\frac{1}{4} + 7\frac{5}{6}$
6. $10\frac{1}{3} - 4\frac{5}{16}$
7. $22\frac{17}{18} - 10\frac{3}{4}$
8. $13\frac{1}{12} + 4\frac{7}{10}$
9. $16 - 7\frac{5}{6}$
10. $12 - 5\frac{5}{7}$
11. $15\frac{14}{15} - 10\frac{2}{3}$
12. $30 - 17\frac{2}{3}$
13. $11\frac{1}{4} + 4\frac{3}{7}$
14. $18\frac{5}{9} - 17\frac{2}{30}$
15. $21\frac{2}{5} + 10\frac{5}{8}$
16. $4\frac{5}{8} - 3$
17. $11\frac{5}{9} - 5$
18. $6 - 2\frac{3}{8}$
19. $8\frac{2}{3} - 7\frac{5}{6}$
20. $126\frac{3}{4} - 125\frac{2}{3}$

21. $\begin{array}{r} 296\frac{11}{16} \\ -152\frac{7}{8} \\ \hline \end{array}$

22. 2 5/8 – 1 15/16

23. 26 2/3 – 21 3/4

24. 13 4/5 – 11 13/15

25. 7 4/9 – 3 5/12 + 5/6

26. 22 3/4 – 5 1/8 + 2 2/3

27. 12 3/5 + 9 1/2 + 13 2/3 + 6 5/6

28. 2 7/16 + 5 3/8 + 7 3/24 + 2 1/2

29. 6 3/5 + 11 + 4 3/4 + 7 5/12

30. 1 9/16 + 3 3/5 + 9 + 2 1/2

B. Find the solution to each application problem. Reduce all answers to lowest terms.

31. How much more is 138 5/8 than 117 3/4?

32. Find the difference between 10 5/6 and 8 2/9.

33. Find the total of 14 1/3, 11 3/4, and 12 5/8.

34. Find the sum of 162 1/2, 131 4/9, and 120 3/5.

35. The Nut Shoppe blends its own deluxe mixed nuts. If 5 3/4 pounds of cashews, 2 1/2 pounds of almonds, 7 1/4 pounds of pecans, and 1 1/3 pounds of hazelnuts are combined, how much deluxe mix will there be?

36. One package of ground beef weighs 3 7/10 pounds and another weighs 2 1/2 pounds. How much ground beef is there altogether?

37. Sue weighed 137 1/2 pounds when she started her diet. If she has lost 5 3/4 pounds, what is her new weight?

38. On three consecutive days of rain, Columbus, Ohio, had 1 1/2, 5/8, and 3/4 inches of rain. What was the total rainfall for that period?

39. After six weeks of dieting, Sue weighs 126 1/4 pounds. What is her total weight loss if she started at 137 1/2 pounds?

40. In Exercise 38, how much more rain fell on the first day than on the second day?

41. A cake recipe calls for 2 1/2 cups of flour and 1 2/3 cups of sugar. How much more flour than sugar is used?

42. Jim bought 10 gallons of paint. After painting his house, he had 1 3/4 gallons left. How much paint did he use to paint the house?

43. Jan works part-time in the library. If she worked 4 1/2 hours on Monday, 2 3/4 hours on Thursday, and 5 1/3 hours on Saturday, what were her total hours for the week? Try estimating the answer before solving this problem.

44. On the stock market report, a certain stock opened at 42 1/8 points and closed at 43 3/8. How much did the stock increase?

45. In Exercise 43, how many more hours did Jan work on Saturday than on Monday?

46. Mike took a long-distance bike trip. He rode 56 3/4 miles on the first day, rode 48 1/3 miles on the second day, and then biked home by the same route. Altogether, how many miles did he ride? Try estimating the answer before solving this problem.

47. Find the amount of wallpaper border needed to go around a room that is 8 5/8 feet by 10 9/16 feet. Use $P = L + W + L + W$.

48. For a remodeling job, Janina must replace the baseboard that goes around the entire room. The dimensions of the room are 9 3/16 feet by 11 1/2 feet. Disregarding doorways and other openings, how many feet of baseboard material will she need? Use $P = L + W + L + W$.

49. Cara wants to put wallpaper border around a window frame and a door frame in her bedroom. The window measures 5 3/16 feet by 3 1/2 feet. The doorway is 7 1/2 feet by 3 feet. How many feet of border material will she need? Use $P = L + W + L + W$.

50. The room in Exercise 48 has two door openings that won't require the baseboard material. One opening is 2 1/2 feet wide, the other is 3 3/8 feet wide. What is the minimum amount of baseboard material needed?

C. Read and respond to the following exercises.

51. Describe, in words, how you could estimate an answer to Exercise 47.

52. Describe, in words, how you could estimate an answer to Exercise 48.

53. How do the processes used to solve these two problems differ?

$$\begin{array}{r} 12 \\ -\ 7\frac{5}{8} \\ \hline \end{array} \qquad \begin{array}{r} 12\frac{5}{8} \\ -\ 7 \\ \hline \end{array}$$

54. Explain, in words, why fractions must have the same denominators before they can be added or subtracted.

55. Simplify the problem that follows. On a separate sheet of paper, write the original problem and a detailed description of your steps. Have a classmate try to simplify the problem doing only the steps that you have described. Was your description clear enough that the classmate was able to get the correct solution by following your steps?

51 3/8 – 22 5/9

4.3 Order of Operations in Fraction Problems

If asked to solve a fraction problem that has more than one operation, apply the **Order of Operations Agreement,** which was introduced in Section 2.2.

Order of Operations Agreement

When simplifying a mathematical expression involving more than one operation, follow these steps:

1. Simplify within parentheses or other grouping symbols.
2. Simplify any expressions involving exponents or square roots.
3. Multiply or divide in order from left to right.
4. Add or subtract (combine) in order from left to right.

In regard to step 2, when a fraction is raised to the power indicated by an exponent, both the numerator and the denominator are raised to that power. When you are taking the square root of a fraction, you must take the square root of both parts of the fraction.

$$\left(\frac{2}{5}\right)^2 = \frac{(2)^2}{(5)^2} = \frac{4}{25} \qquad \left(\frac{2}{3}\right)^3 = \frac{(2)^3}{(3)^3} = \frac{8}{27}$$

$$\sqrt{\frac{4}{9}} = \frac{\sqrt{4}}{\sqrt{9}} = \frac{2}{3} \qquad \sqrt{\frac{49}{25}} = \frac{\sqrt{49}}{\sqrt{25}} = \frac{7}{5}$$

EXAMPLE 1 Simplify: $\frac{3}{4} + \frac{3}{8} \times \frac{5}{9}$

Solution $\frac{3}{4} + \frac{3}{8} \times \frac{5}{9}$ Do multiplication before addition.

$$= \frac{3}{4} + \frac{\overset{1}{\cancel{3}}}{8} \times \frac{5}{\underset{3}{\cancel{9}}}$$

$= \frac{3}{4} + \frac{5}{24}$ Then find a common denominator and add.

$= \frac{18}{24} + \frac{5}{24} = \frac{23}{24}$

◀

EXAMPLE 2 Simplify: $\frac{3}{4}\left(\frac{2}{3} + \frac{1}{2}\right)$

Solution

$\frac{3}{4}\left(\frac{2}{3} + \frac{1}{2}\right)$ Work inside the parentheses first.

$= \frac{3}{4}\left(\frac{4}{6} + \frac{3}{6}\right)$ Find the common denominator and add.

$= \frac{3}{4}\left(\frac{7}{6}\right)$ 3/4 outside the parentheses says to multiply. Use the cancellation tool.

$= \frac{\overset{1}{\cancel{3}}}{4} \cdot \frac{7}{\underset{2}{\cancel{6}}} = \frac{7}{8}$

◀

EXAMPLE 3 Simplify: $\left(\frac{2}{3}\right)^2 + \sqrt{\frac{16}{81}}$

Solution

$\left(\frac{2}{3}\right)^2 + \sqrt{\frac{16}{81}}$ Remember that the exponent and square root apply to both the numerator and the denominator.

$= \frac{4}{9} + \frac{4}{9}$

$= \frac{8}{9}$

◀

EXAMPLE 4 Simplify: $\left(\frac{7}{9} - \frac{2}{3}\right)\left(\frac{1}{2} + \frac{4}{5}\right)$

Solution

$\left(\frac{7}{9} - \frac{2}{3}\right)\left(\frac{1}{2} + \frac{4}{5}\right)$ First do the operation within each grouping symbol.

$= \left(\frac{7}{9} - \frac{6}{9}\right)\left(\frac{5}{10} + \frac{8}{10}\right)$

$= \left(\frac{1}{9}\right)\left(\frac{13}{10}\right)$ Now multiply since the parentheses are side by side.

$= \frac{13}{90}$

◀

EXAMPLE 5 Simplify: $\left(\frac{5}{6}\right)^2 + 1\frac{1}{3} \times \frac{7}{8}$

Solution

$\left(\frac{5}{6}\right)^2 + 1\frac{1}{3} \times \frac{7}{8}$

$= \frac{25}{36} + 1\frac{1}{3} \times \frac{7}{8}$ Do exponents first; then multiplication; then addition.

$$= \frac{25}{36} + \frac{\overset{1}{\cancel{4}}}{3} \times \frac{7}{\underset{2}{\cancel{8}}}$$

$$= \frac{25}{36} + \frac{7}{6}$$

$$= \frac{25}{36} + \frac{42}{36}$$

$$= \frac{67}{36} \text{ or } 1\frac{31}{36}$$

EXAMPLE 6 Find the value of T in $T = 3R - W$ when $R = 3/4$ and $W = 5/12$.

Solution

$T = 3R - W$ Substitute the given values.

$T = 3 \times \frac{3}{4} - \frac{5}{12}$ Multiply first.

$T = \frac{9}{4} - \frac{5}{12}$ Then do subtraction.

$T = \frac{27}{12} - \frac{5}{12}$

$T = \frac{22}{12}$

$T = 1\frac{5}{6}$

Self-Check Simplify completely.

a. $\frac{5}{6} + \frac{3}{4} \times \frac{1}{9}$ b. $\sqrt{\frac{25}{36}} \div \left(\frac{2}{3}\right)^2$

Answers:

a. $\frac{11}{12}$ b. $1\frac{7}{8}$

4.3 EXERCISES

A. Simplify each exercise using the Order of Operations Agreement, and reduce answers to lowest terms.

1. $\frac{2}{3} + \frac{5}{6} \times \frac{1}{2}$
2. $\frac{3}{4} - \frac{5}{8} \times \frac{1}{2}$
3. $\frac{5}{9} \times \frac{3}{25} + \frac{7}{10}$
4. $\frac{9}{16} + \frac{3}{4} \times \frac{1}{2}$
5. $\frac{7}{8} - \frac{2}{3} \times \frac{3}{4}$
6. $\frac{6}{7} \times \frac{21}{24} + \frac{5}{8}$
7. $\sqrt{\frac{25}{36}} - \frac{2}{3}$
8. $\left(\frac{5}{8} + \frac{3}{4}\right)\left(\frac{5}{6} - \frac{3}{8}\right)$
9. $\frac{7}{8} \div \left(\frac{3}{4}\right)^2$
10. $\sqrt{\frac{16}{49}} - \frac{1}{3}$
11. $\left(\frac{3}{4} - \frac{2}{3}\right)\left(\frac{1}{2} - \frac{3}{8}\right)$
12. $\frac{5}{9} \div \left(\frac{2}{3}\right)^2$
13. $2\frac{1}{4} - 1\frac{1}{2} \times \frac{3}{4}$
14. $3\frac{1}{2} + 1\frac{1}{4} \times \frac{3}{5}$

15. $\left(\frac{3}{4}\right)^2 + \frac{1}{2} \times \frac{5}{6}$

16. $\sqrt{\frac{1}{9}} + \left(\frac{2}{3}\right)^2$

17. $\left(\frac{3}{4}\right)^2 + \sqrt{\frac{4}{9}}$

18. $\left(\frac{5}{6}\right)^2 + \frac{3}{4} \times 1\frac{5}{9}$

19. $3\frac{5}{6} - \left(1\frac{2}{3} + \frac{3}{4}\right)$

20. $4\frac{4}{9} - \left(2\frac{3}{5} + \frac{2}{3}\right)$

21. $\left(\frac{4}{5}\right)^2 \div \sqrt{\frac{64}{9}}$

22. $\left(1\frac{2}{3} + 3\frac{1}{2}\right)\left(\frac{5}{6} \times \frac{2}{3}\right)$

23. $1\frac{7}{15} \div 3\frac{2}{3} \times 1\frac{7}{16}$

24. $3\frac{5}{9} \div 2\frac{2}{3} \times 1\frac{3}{5}$

25. $\left(\frac{4}{9} \times \frac{5}{6}\right)\left(3\frac{1}{2} - 1\frac{3}{4}\right)$

26. $\sqrt{\frac{81}{49}} \div \left(\frac{2}{3}\right)^2$

B. Describe, in words, the process used to go from one step to the next.

27. $\frac{1}{8} + \frac{3}{4} \times \frac{5}{6}$

$= \frac{1}{8} + \frac{5}{8}$ 1.

$= \frac{6}{8}$ 2.

$= \frac{3}{4}$ 3.

28. $\frac{5}{6}\left(\frac{2}{4} + \frac{1}{4}\right)$

$= \frac{5}{6}\left(\frac{3}{4}\right)$ 1.

$= \frac{5}{\cancel{6}_{2}}\left(\frac{\cancel{3}^{1}}{4}\right)$ 2.

$= \frac{5}{8}$ 3.

29. $\sqrt{\frac{4}{9}} + \left(\frac{5}{6}\right)^2$

$= \frac{2}{3} + \left(\frac{5}{6}\right)^2$ 1.

$= \frac{2}{3} + \frac{25}{36}$ 2.

$= \frac{24}{36} + \frac{25}{36}$ 3.

$= \frac{49}{36}$ or $1\frac{13}{36}$ 4.

30. $\frac{3}{8}\left(3\frac{1}{4} - 1\frac{2}{4}\right)$

$= \frac{3}{8}\left(2\frac{5}{4} - 1\frac{2}{4}\right)$ 1.

$= \frac{3}{8}\left(1\frac{3}{4}\right)$ 2.

$= \frac{3}{8}\left(\frac{7}{4}\right)$ 3.

$= \frac{21}{32}$ 4.

C. Substitute the given values into the formula and simplify.

31. $P = 2(L + W)$ Find P if $L = 5\ 1/3$ feet and $W = 2\ 1/4$ feet.

32. $P = 2a + b$ Find P if $a = 4\ 5/8$ centimeters and $b = 3\ 2/3$ centimeters.

33. $m = 1/2\ a(b + c)$ Find m if $a = 1\ 3/5$, $b = 3\ 2/3$, and $c = 4\ 1/2$

34. $s = vt + 1/2\ at^2$ Find s if $v = 8$, $t = 3/4$, and $a = 32$.

35. $T = 1/2\ ch^2$ Find T if $c = 5/6$ and $h = 3/4$.

36. $m = 1/2a(b + c)$ Find m if $a = 3/8$, $b = 2\ 1/4$, and $c = 5/6$.

37. $W = vm + 1/2\ t^2$ Find W if $v = 2\ 1/2$, $m = 5/6$, and $t = 2/3$.

38. $Q = c(a - b)$ Find Q if $c = 4/5$, $a = 5$, and $b = 2\ 2/3$.

D. Read and respond to the following exercises.

39. Is $\sqrt{49/4} = 7/4$? Explain why or why not.

40. Find the value of $(\sqrt{25/16})^3$. Explain, in words, how you got your answer.

41. Is the Order of Operations Agreement you used with whole numbers different from the one you used with fractions? Would you expect the Agreement to differ if decimals are involved?

42. Is $(2/3)^3 = (2)^3/(3)^3$? Explain your answer.

4.4 Comparing Fractions

In order to determine whether 5/16 is larger than 13/32, we must write the fractions as equivalent fractions with the same denominator. Then the fraction with the larger numerator is the larger fraction. The symbol > means "is greater than," and < means "is less than." Suppose you were asked, "Which fraction is larger, 5/16 or 13/32?" You would need to change 5/16 to 32nds so that both fractions had the same denominator. You would reason that 5/16 = 10/32, and clearly 13/32 > 10/32. Therefore, 13/32 > 5/16.

EXAMPLE 1 Determine which symbol, >, <, or =, should be between 2/5 and 3/5.

Solution The fractions have the same denominator. Thus 2/5 is smaller than 3/5, because 2/5 contains one less "piece" than 3/5, and the pieces are the same size. Therefore, 2/5 < 3/5.

EXAMPLE 2 Which is larger, 2/3 or 3/4?

Solution The fractions must be changed to equivalent fractions with the denominator of 12 before they can be compared.

$$\frac{2}{3} = \frac{8}{12} \quad \text{and} \quad \frac{3}{4} = \frac{9}{12}, \quad \text{and} \quad \frac{9}{12} > \frac{8}{12}$$

Therefore, $\frac{3}{4} > \frac{2}{3}$.

EXAMPLE 3 Determine whether 5/12 or 7/16 is the larger fraction, and then subtract the smaller number from the larger one.

Solution

2) 12 16 — Find the LCD for 12 and 16 and write equivalent fractions.

2) 6 8

2) 3 4

2) 3 2

3) 3 1

1 1

LCD = 2 × 2 × 2 × 2 × 3 = 48

$$\frac{5}{12} = \frac{20}{48} \qquad \frac{7}{16} = \frac{21}{48}$$

Because $\frac{21}{48} > \frac{20}{48}$, subtract 20/48 from 21/48.

$$\frac{21}{48} - \frac{20}{48} = \frac{1}{48}$$

EXAMPLE 4 Arrange these fractions in order from smallest to largest: 3/8, 1/3, 5/16, 5/12.

Solution Find the LCD for 8, 3, 16, and 12.

$$3 = 3$$
$$8 = \quad 2 \times 2 \times 2$$
$$12 = 3 \times 2 \times 2$$
$$16 = \quad 2 \times 2 \times 2 \times 2$$
$$\text{LCD} = 3 \times 2 \times 2 \times 2 \times 2 = 48$$

Change the fractions to equivalent fractions with a denominator of 48.

$$\frac{3}{8} = \frac{18}{48} \qquad \frac{1}{3} = \frac{16}{48}$$
$$\frac{5}{16} = \frac{15}{48} \qquad \frac{5}{12} = \frac{20}{48}$$

Arrange the new fractions from smallest to largest.

$$\frac{15}{48}, \frac{16}{48}, \frac{18}{48}, \frac{20}{48}$$

Replace these with their original form for the answer.

$$\frac{5}{16}, \frac{1}{3}, \frac{3}{8}, \frac{5}{12}$$

Self-Check Arrange the fractions in order from *largest to smallest.*

a. $\frac{1}{2}, \frac{2}{3}, \frac{5}{8}$ b. $\frac{3}{8}, \frac{5}{12}, \frac{4}{9}$

Answers:

a. $\frac{2}{3}, \frac{5}{8}, \frac{1}{2}$ b. $\frac{4}{9}, \frac{5}{12}, \frac{3}{8}$

4.4 EXERCISES

A. Determine which symbol should be between the fractions: >, <, or = (greater than, less than, or equal to).

1. $\frac{3}{4}, \frac{2}{3}$
2. $\frac{5}{9}, \frac{7}{12}$
3. $\frac{5}{6}, \frac{3}{4}$
4. $\frac{7}{8}, \frac{49}{56}$
5. $\frac{1}{6}, \frac{2}{12}$
6. $\frac{10}{21}, \frac{3}{7}$
7. $\frac{4}{5}, \frac{5}{9}$
8. $\frac{11}{12}, \frac{2}{3}$
9. $\frac{4}{7}, \frac{5}{6}$
10. $\frac{11}{32}, \frac{2}{3}$
11. $\frac{1}{2}, \frac{4}{5}$
12. $\frac{15}{16}, \frac{11}{12}$
13. $\frac{48}{64}, \frac{3}{4}$
14. $\frac{6}{36}, \frac{24}{144}$
15. $\frac{19}{32}, \frac{7}{12}$

B. Arrange these fractions in order from smallest to largest.

16. $\frac{1}{2}, \frac{2}{5}, \frac{3}{4}$
17. $\frac{3}{8}, \frac{5}{6}, \frac{2}{3}$
18. $\frac{4}{5}, \frac{5}{6}, \frac{7}{8}$
19. $\frac{15}{16}, \frac{3}{4}, \frac{7}{8}$
20. $\frac{1}{4}, \frac{1}{2}, \frac{1}{6}$
21. $\frac{2}{9}, \frac{2}{3}, \frac{4}{5}$
22. $\frac{4}{5}, \frac{2}{3}, \frac{5}{8}$
23. $\frac{15}{16}, \frac{2}{3}, \frac{3}{4}$
24. $\frac{10}{12}, \frac{13}{16}, \frac{5}{8}, \frac{3}{4}$
25. $\frac{2}{3}, \frac{5}{8}, \frac{3}{4}, \frac{1}{2}$
26. $\frac{15}{16}, \frac{5}{6}, \frac{7}{8}, \frac{11}{12}$
27. $\frac{4}{7}, \frac{3}{8}, \frac{2}{3}, \frac{1}{2}$

28. $\frac{7}{12}, \frac{8}{15}, \frac{5}{9}, \frac{3}{4}$

29. $\frac{11}{24}, \frac{5}{8}, \frac{7}{16}, \frac{17}{32}$

30. $\frac{3}{32}, \frac{5}{64}, \frac{1}{8}, \frac{3}{16}, \frac{1}{4}$

C. Determine which fraction is larger and subtract the smaller fraction from it.

31. $\frac{2}{3}, \frac{3}{4}$

32. $\frac{5}{8}, \frac{5}{6}$

33. $\frac{11}{12}, \frac{15}{16}$

34. $\frac{4}{5}, \frac{3}{4}$

35. $\frac{5}{8}, \frac{9}{16}$

36. $\frac{5}{12}, \frac{1}{3}$

37. $\frac{17}{16}, \frac{13}{12}$

38. $\frac{19}{24}, \frac{5}{8}$

39. $\frac{29}{32}, \frac{7}{8}$

40. $\frac{11}{18}, \frac{17}{27}$

41. $\frac{54}{49}, \frac{31}{28}$

42. $\frac{11}{16}, \frac{17}{32}$

D. Read and respond to each exercise.

43. Is 5/6 – 1/3 > 1/9 + 8/27? Explain, in words, how you got your answer.

44. State whether 11/32 is greater than, less than, or equal to 35/96. Explain, how you determined your answer.

45. Find the value of the following expressions, and then arrange the answers in order from largest to smallest.

$\frac{5}{8} - \frac{3}{16} \qquad \frac{1}{4} + \frac{3}{5} \qquad \left(\frac{2}{3}\right)^2 \div \sqrt{\frac{4}{9}}$

Could you arrange the answers just by doing the addition, subtraction, or division indicated? Why?

4.5 Fraction Application Problems General Applications

Review the procedures and suggestions for solving application problems in Section 1.5. They apply to all types of word problems, including the fraction problems in this section. Remember to review, as well, the rules for fractions that follow.

Basic Rules for Fraction Operations

1. Fractions must have common denominators to be added or subtracted, but not to be multiplied or divided.
2. Mixed numbers must be changed to improper fractions to be multiplied or divided, but they do not need to be changed to improper fractions when added or subtracted.
3. The cancellation shortcut can be performed only when the fractions are being multiplied.
4. All fraction answers should be reduced to lowest terms. Improper fraction answers to application problems should be changed to mixed numbers.
5. Improper fractions cannot be left as part of a mixed number answer. The result needs to be simplified to a mixed number with a proper fraction.

The examples in this section involve all four operations with fractions, not just addition and subtraction. More than one step may be required to reach the solution. When determining what operation to use, remember that putting amounts together is addition or multiplication and splitting up amounts is subtraction or division.

EXAMPLE 1 Cans of soup normally have a net weight of 10 3/4 ounces. How many cans can be filled from a kettle containing 150 1/2 ounces?

Solution The large amount of soup is going to be split up into smaller amounts. Because the total amount is given (150 1/2 ounces) and the size of each part is given (10 3/4 ounces), division needs to be done. Remember to place the total amount before the division symbol. Estimate an answer by rounding to convenient whole numbers: 150 ÷ 10 = 15 cans.

$$150\frac{1}{2} \div 10\frac{3}{4}$$

$$= \frac{301}{2} \div \frac{43}{4}$$

$$= \frac{301}{\cancel{2}_1} \times \frac{\cancel{4}^2}{43}$$

$$= \frac{602}{43} = 14 \text{ cans}$$

This is close to the estimated answer

EXAMPLE 2 The value of one share of Neighborhood Cooperative stock went from $64\frac{3}{8}$ points to $71\frac{1}{6}$ points. How much did the stock increase?

Solution We are looking for the difference between the quantities, so we need to subtract.

$$71\frac{1}{6} = 71\frac{4}{24} = \overset{70}{\cancel{71}}\frac{4}{24} + \frac{24}{24} = 71\frac{28}{24}$$

$$-64\frac{3}{8} = -64\frac{9}{24} = -64\frac{9}{24} = -64\frac{9}{24}$$

$$6\frac{19}{24} \text{ points}$$

EXAMPLE 3 Two-thirds of the students in this college drive their own car to campus each day. If there are 9,600 students attending, how many drive their own car?

Solution 2/3 of the total students drive. The total number of students is 9,600. 2/3 of the 9,600 drive. "Of" connecting two numbers usually indicates multiplication.

To estimate an answer, 2/3 is between 1/2 and 1. 1/2 × 9600 = 4800 and 1 × 9600 = 9600. The answer should be between these two.

$$\frac{2}{\cancel{3}_1} 1 \times \frac{\overset{3,200}{\cancel{9,600}}}{1} = 6,400 \text{ students drive}$$

EXAMPLE 4 If a car can travel 364 miles on 15 3/4 gallons of gasoline, how far can it travel on 1 gallon?

Solution When looking for the amount for one unit, you will divide.

First estimate an answer by using 360 ÷ 18 = 20 miles per gallon.

$$364 \div 15\frac{3}{4}$$

$$= \frac{364}{1} \div \frac{63}{4}$$

$$= \frac{\overset{52}{\cancel{364}}}{1} \times \frac{4}{\underset{9}{\cancel{63}}}$$

$$= \frac{208}{9} = 23\frac{1}{9} \text{ miles per gallon}$$ This is close to the estimated answer.

EXAMPLE 5 A bolt of fabric contained 22 yards when it arrived at the store. After sales of 4 3/8 yards, 1 1/2 yards, and 3 3/4 yards, how much fabric remains on the bolt?

Solution The total amount of fabric that has been sold can be found by addition. Then "how much remains" suggests subtraction. Estimate the answer by using convenient whole numbers: 4 + 2 + 4 = 10 and 22 – 10 = 12 yards.

$$4\frac{3}{8} = 4\frac{3}{8}$$

$$1\frac{1}{2} = 1\frac{4}{8}$$

$$+\ 3\frac{3}{4} = +\ 3\frac{6}{8}$$

$$8\frac{13}{8}$$

$$= 8 + 1\frac{5}{8}$$

$$= 9\frac{5}{8} \text{ yards have been sold}$$

$$22 = 21\frac{8}{8}$$

$$-\ 9\frac{5}{8} = -\ 9\frac{5}{8}$$

$$12\frac{3}{8} \text{ yards remain on the bolt}$$

Geometry Applications

EXAMPLE 6 Find the area and the perimeter of the triangle in Figure 4.5. Use $A = 1/2\ bh$, $P = a + b + c$, and $h = 2\ 3/4$ meters.

Solution $A = 1/2\ bh$ Substitute given values.

First estimate. The answer as $A = \frac{1}{2} \times 14 \times 3 = 21$ square meters.

$$A = \frac{1}{2}(14)\left(2\frac{3}{4}\right)$$

$$A = \frac{1}{2}\left(\frac{14}{1}\right)\left(\frac{11}{4}\right)$$

$$A = \frac{1}{\cancel{2}_1}\left(\frac{\cancel{14}^7}{1}\right)\left(\frac{11}{4}\right)$$

$$A = \frac{77}{4}$$

$$A = 19\frac{1}{4}$$

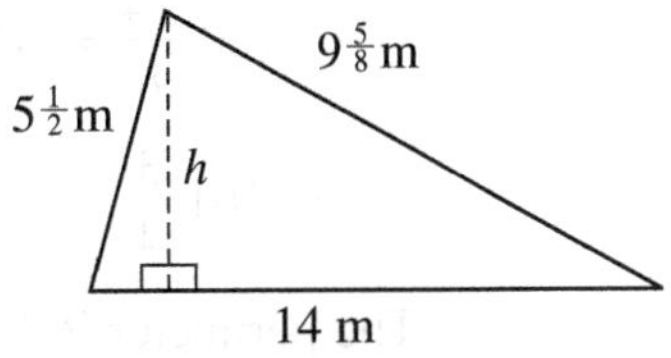

FIGURE 4.5

The area is 19 1/4 square meters.

The perimeter is the sum of the three sides of the triangle: $P = a + b + c$.

First estimate the answer as 14 + 5 + 10 = 29 meters.

or

14 + 6 + 10 = 30 meters

$$P = a + b + c$$

$$P = 14 = 5\frac{1}{2} + 9\frac{5}{8}$$

$$= 14 + 5\frac{4}{8} + 9\frac{5}{8}$$

$$= 28\frac{9}{8}$$

$$= 28 + 1\frac{1}{8} = 29\frac{1}{8}$$

The perimeter is 29 1/8 meters

EXAMPLE 7 Find both the perimeter and the area of the field that is shown in Figure 4.6. Use the formula $A = 1/2\ h(b_1 + b_2)$.

This shape is called a trapezoid. The b^1 and b^2 refer to the two parallel bases that have different lengths.

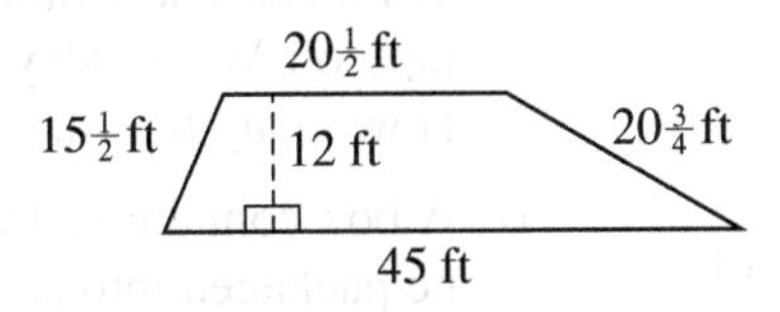

FIGURE 4.6

Solution The perimeter is the sum of the four sides.

$$P = a + b_1 + b_2 + c$$

$$P = 20\frac{3}{4} + 45 + 20\frac{1}{2} + 15\frac{1}{2}$$

$$= 20\frac{6}{8} + 45 + 20\frac{4}{8} + 15\frac{4}{8}$$

$$= 100\frac{14}{8} = 100 + 1\frac{6}{8}$$

$$= 101\frac{3}{4}$$

The perimeter is 101 3/4 feet.

In the area formula, the 1 and 2 beside the b's are just there for identification of the two bases (the parallel sides of the trapezoid). The 1 and 2 are not used in the calculation.

$$A = \frac{1}{2}h(b_1 + b_2)$$

$$A = \frac{1}{2} \times 12\left(20\frac{1}{2} + 45\right)$$ Apply Order of Operations.

$$= \frac{1}{2} \times 12\left(65\frac{1}{2}\right)$$

$$= \frac{1}{2} \times \frac{12}{1} \times \frac{131}{2}$$

$$= \frac{1}{\cancel{2}_1} \times \frac{\overset{3}{\cancel{\overset{6}{\cancel{12}}}}}{1} \times \frac{131}{\cancel{2}_1}$$

$$= 393$$

The area of the trapezoid is 393 square feet.

4.5 EXERCISES

A. Solve the following application problems, simplifying fraction answers to lowest terms. Remember that the problem can require any operation, and in some problems you will need more than one step to get the solution. You may find that estimating an answer first will help you decide which operation to use.

1. Sarah's cookie recipe requires 2 2/3 cups of flour. How much flour will she need if she triples the recipe?
2. Five-eighths of an economics class are female students. If there are 32 students enrolled in that class, how many are female?
3. In Exercise 1, how much flour will Sarah need if she makes only half of the recipe?
4. In Exercise 2, what fraction of the class is male? How many students are male?
5. Sandy is trying to build up her endurance as a swimmer. She swam 5 1/2 laps on Monday, 6 3/4 laps on Wednesday, and 7 1/2 laps on Friday. How many laps did she swim altogether?
6. A box containing 15 pounds of hard candy is to be packaged into small baskets each holding 3/4 of a pound. How many baskets can be filled? (Hint: see Example 1)
7. In Exercise 5, how much farther did Sandy swim on Friday than on Wednesday?
8. Sue has started another weight-loss program. If she lost 2 1/2 pounds the first week, 1 1/3 pounds the second week, and 3/4 pound the third week, how much has she lost so far?

9. In Exercise 5, if each lap is one-third of a mile in length, how many miles did Sandy swim on Monday?

10. In Exercise 8, if Sue's weight-loss goal is 10 pounds, how many more pounds does she need to lose to reach her goal?

11. If a car travels 95 1/2 miles in 2 3/4 hours, how far can it travel in 1 hour? (Hint: See Example 4)

12. Marilyn is making a matching skirt and blouse. The blouse requires 1 7/8 yards of material, and the skirt requires 2 1/4 yards. How much fabric must she purchase to make both?

13. A tank contains 26 3/4 gallons of maple syrup. It will be used to fill bottles that hold 1/8 of a gallon. How many bottles that hold 1/8 of a gallon-can be filled? (Hint: See Example 1)

14. In Exercise 12, how much fabric would Marilyn need if she decided to make three of the blouses and two of the skirts?

15. Mark has formed his own lawn service business in order to earn money for college. He has contracted to mow the Engels' yard each week. The first time he mowed the yard, it took him 1 1/4 hours, but later in the summer he was able to do it in 5/8 hour. How much less time did it take late in the summer?

16. In a small town in Pennsylvania, one-third of the voters are registered as Democrats and nine-sixteenths are registered as Republicans. What fractional part of the registered voters are neither Democrats nor Republicans? (Hint: Use 1 or 16/16 to represent the total number of voters)

17. One-eighth of the players on the college's football team are maintaining an A average in their class work, and three-sixteenths have a B average. What fractional part of the team is doing less than B work? (Hint: Use 1 or 16/16 to represent all the players on the team)

18. If the town in Exercise 16 has a population of 14,688 registered voters, how many are registered as Republican?

19. If there are 32 players on the football team in Exercise 17, how many are doing B work?

20. A piece of wire that is 5 1/4 feet long is divided into three equal pieces. How long is each piece?

21. Maria is making matching skirts and blouses for her ethnic dance class. Each skirt requires 2 1/4 yards of fabric, and each blouse requires 1 3/8 yards. How much more fabric does a skirt need than a blouse?

22. Three-eights of the children in a preschool class live in single-parent households. If there are 16 children in the class, how many live in single-parent households? (Hint: See Example 3)

23. If Maria is making blouses and skirts for five members of the dance class (see Exercise 21), how much fabric will she need altogether?

24. Forest rangers and volunteers have replanted trees on 16 3/4 acres of land that had been destroyed by a forest fire in Wilson State Park. If the area destroyed by the fire consists of 35 1/2 acres, how many acres remain to be replanted?

25. A pork roast is to be cooked 1/3 hour for each pound. If the roast weighs 5 1/6 pounds, how many hours should it be cooked? (Hint: Think of how much time would be needed for a 6 pound roast)

26. In Exercise 24, if an average of 2,420 trees is planted per acre, what is the total number of trees that have been planted so far in this recovery project?

27. How many volumes of an encyclopedia set will fit on a library shelf that is 34 inches long if each book is 2 1/4 inches thick?

28. A piece of pipe that is 7 1/3 feet long is to be cut into five equal pieces. Assume there is no waste and find how long each piece will be.

29. A length of upholstery fabric that is 26 7/8 yards long is to be cut into pieces that are each 5 1/4 yards in length. How many whole pieces will there be?

30. A cookbook states that a beef roast should be cooked one-fourth of an hour for each pound of meat. If the roast weights 4 7/8 pounds, how long should it be cooked?

B. The following application problems will apply geometry concepts that were introduced in Section 2.3 and reviewed in Chapters 3 and 4. Required formulas will be given in the problem. You may want to try estimating your answer first. Simplify all fraction answers to lowest terms.

31. A room measures 12 3/4 feet by 9 1/2 feet wide. Find the area of the ceiling of that room. Use $A = LW$.

32. Katie is landscaping her backyard and needs to determine how many square yards of sod to order. The yard's shape is shown in Figure 4.7. How many square yards of sod should she buy? Use $A = 1/2\ h(b^1 + b^2)$

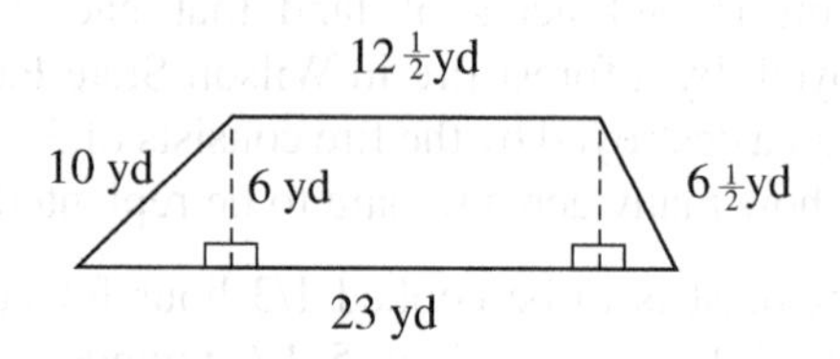

FIGURE 4.7

33. In the room described in Problem 31, the height of the walls (from floor to ceiling) is 8 feet. Find the area to be painted on the four walls, disregarding any window and door openings. (Hint: Two walls are $12\frac{3}{4} \times 8$ and two walls are $9\frac{1}{2} \times 8$.) Use $A = LW$ for each wall.

34. How much edging will Katie need to enclose her backyard if its dimensions are as shown in Figure 4.7? (Hint: Find the perimeter by adding all four sides together.)

35. The Parks and Recreation Department wants to develop a triangle-shaped flower bed in front of its building. How many feet of edging will be needed to go around the equilateral triangle with sides each 12 3/4 feet in length? If bricks that are 2/3 foot long are laid end to end to edge the flower bed, how many bricks will be needed on each side of the triangle? Use $P = 3a$.

36. Which has greater volume: a box that is 6 1/2 inches by 8 inches by 4 3/4 inches or one that is 5 5/8 inches by 2 3/4 inches by 16 inches? Use $V = LWH$.

4.6 CHAPTER REVIEW

We hope you are feeling more confident about working with fractions. As you prepare for the test for this chapter, remember:

1. Fractions can be added or subtracted only if they have the same denominator.
2. Find the least common denominator by inspection or by the least common multiple method (Section 2.6), and then write equivalent fractions.
3. Mixed numbers do not have to be changed to improper fractions when you are adding or subtracting.
4. If borrowing is required in subtraction, borrow 1 from the whole number, write the 1 as the common denominator over itself, add together the fraction parts of the numerator, and then subtract.
5. Apply the Order of Operations Agreement to fraction problems containing more than one operation.
6. To raise a fraction to a power or to take the square root of a fraction, you must apply the operation to both the numerator and the denominator.
7. To compare fractions, write them as equivalent fractions (fractions that have the same denominator), and then compare the numerators.

8. To solve fraction application problems, use the same steps described for application problems with whole numbers.

9. To calculate the area, perimeter, or volume of geometric figures: Write the formula, substitute the given values, simplify the expression using the Order of Operations Agreement. Label answers with appropriate units.

10. Simplify all fraction answers to lowest terms.

REVIEW EXERCISES

A. Solve each exercise, simplifying answers to lowest terms.

1. $\frac{3}{4}+\frac{5}{6}$
2. $\frac{9}{16}-\frac{3}{8}$
3. $\frac{5}{24}-\frac{1}{8}$
4. $\frac{5}{9}+\frac{1}{2}$
5. $2\frac{3}{4}+1\frac{5}{9}$
6. $3\frac{4}{5}-1\frac{1}{3}$
7. $7\frac{6}{7}-3\frac{3}{8}$
8. $4\frac{2}{3}+2\frac{5}{8}$
9. $3\frac{1}{2}-2\frac{3}{4}$
10. $1\frac{3}{4}+5\frac{2}{3}+7\frac{1}{6}$
11. $4\frac{1}{8}+2\frac{2}{5}+7\frac{3}{10}$
12. $\frac{4}{9}+\frac{11}{12}-\frac{3}{4}$
13. $\frac{5}{7}+\frac{3}{4}-\frac{5}{14}$
14. $9-2\frac{5}{8}$
15. $2\frac{1}{4}-\frac{3}{8}$
16. $8\frac{11}{16}+5\frac{3}{4}$
17. $20-11\frac{2}{3}$
18. $4\frac{1}{8}-\frac{7}{16}$
19. $24\frac{5}{8}+10\frac{1}{3}$
20. $5\frac{5}{9}-2\frac{3}{4}$
21. $116-84\frac{5}{9}$

B. Solve each exercise, applying the Order of Operations Agreement. Simplify answers to lowest terms.

22. $\frac{2}{3}-\frac{1}{2}+\frac{3}{4}$
23. $\left(\frac{3}{4}\right)^2+\frac{1}{2}\times\frac{5}{8}$
24. $\frac{2}{3}+\frac{3}{8}-\sqrt{\frac{4}{9}}$
25. $\frac{7}{10}+\sqrt{\frac{9}{25}}-\frac{4}{5}$
26. $\left(\frac{5}{6}\right)^2+\frac{3}{4}\times\frac{7}{9}$
27. $\frac{2}{3}-\frac{1}{2}\div\frac{3}{4}$
28. $1\frac{1}{2}+2\frac{1}{3}\times\frac{3}{4}$
29. $5\frac{1}{4}\div\frac{7}{12}\times\frac{1}{2}$

C. Estimate an answer to each of the following exercises. You do not have to simplify the problem.

30. $2\frac{7}{12}+3\frac{1}{3}$
31. $17\frac{5}{8}-15$
32. $129-112\frac{17}{32}$
33. $11\frac{3}{5}+9\frac{2}{9}$
34. $45\frac{13}{32}-37$
35. $\frac{4}{15}+\frac{19}{32}+\frac{7}{12}$
36. $\frac{5}{16}+\frac{17}{18}+\frac{11}{24}$
37. $2\frac{5}{8}+3\frac{1}{2}\times1\frac{3}{4}$
38. $2\frac{2}{3}\div1\frac{3}{5}\times\frac{3}{10}$

D. Arrange the fractions in order from *largest to smallest.*

39. $\frac{3}{4}, \frac{3}{8}, \frac{5}{16}$
40. $\frac{4}{9}, \frac{4}{5}, \frac{4}{15}$
41. $\frac{5}{8}, \frac{2}{3}, \frac{7}{9}$
42. $\frac{3}{16}, \frac{1}{4}, \frac{5}{12}$
43. $\frac{3}{4}, \frac{7}{12}, \frac{5}{6}$
44. $\frac{11}{18}, \frac{2}{3}, \frac{5}{9}$

E. Solve each exercise and simplify answers to lowest terms. Remember that any operation may be required, not just addition and subtraction. Try estimating your answer first.

45. If a man drives 264 miles in 5 1/2 hours, how far does he drive in one hour?

46. Carol promised herself that she would study algebra and do homework for 1 1/4 hours every night of the week (7 days). What is the total number of hours that she will spend on algebra each week?

47. A stock opens at 16 3/8 points and closes at 20. How much of an increase did it experience?

48. Greg worked 5 3/4 hours on Tuesday, 4 hours on Wednesday, and 5 1/8 hours on Saturday. What was the total number of hours he worked for the week?

49. Ann Marie will pay her babysitter at the end of the week. How much is due if the sitter earns $3 per hour and worked 7 7/8 hours Tuesday, 6 1/2 hours Wednesday, and 8 5/8 hours Saturday?

50. An airplane required 22 gallons of gasoline to travel 232 1/2 miles. How far does it travel on 1 gallon?

51. To make a batch of toffee, three-fourths of a pound of butter is needed. How much butter would be needed to make 5 batches?

52. Two-thirds of the students at a community college are on financial aid. If 10,200 students are enrolled, how many are on financial aid?

53. A 12 1/2-acre plot of land is to be divided into building lots, each 5/8 acre in size. How many lots will there be?

54. Five-sixteenths of the voters in the last election cast ballots choosing candidates of only one political party. If 32,496 people voted in the election, how many cast one-party ballots?

55. In a record snowfall on Mt. Washington in New Hampshire, an average of 1 7/8 inches of snow fell each hour for a 24-hour period. What was the accumulation for this period?

56. During the past week, Jamar walked on his treadmill for 3/4 hour on Monday, 1/2 hour on Tuesday, 1 hour on Thursday, 7/8 hour on Friday, and 1 1/4 hours on Saturday. What is the average or mean length of his workouts for this period?

57. If a pork roast is to be cooked 1/3 of an hour for each pound of weight, how long should a 3 4/5-pound roast cook?

58. A box containing 8 1/2 pounds of mints is to be used to fill gift baskets that each hold 2/3 of a pound of mints. How many baskets can be filled from the box?

F. The following application exercises will involve geometry. The formulas will be given in the problem.

59. Find the area of a triangle whose height is 13 2/3 meters and whose base is 20 meters. Use $A = 1/2\ bh$.

60. Find the area of a trapezoid with bases 10 1/2 yards and 14 3/8 yards when the height is 6 1/4 yards. Use $A = 1/2\ h(b_1 + b_2)$.

61. The other two sides of the triangle in Exercise 59 are 15 5/6 meters and 18 1/4 meters. Find the perimeter of the triangle.

62. An octagon is a figure that has eight equal sides. Find the perimeter of an octagon with a side measuring $6\frac{5}{8}$ centimeters.

63. Find the area and the perimeter of the top of a square coffee table that measures 2 2/3 feet on each side. Use $A = s^2$.

64. Find the area of a square with sides 5 5/6 inches. Use $A = s^2$.

65. Find the perimeter of a hexagon (a figure with six equal sides) with each side measuring $11\frac{1}{4}$ yards.

66. An octagon is a figure with eight equal sides. Find its perimeter if each side measures $4\frac{3}{8}$ centimeters.

G. Each of the simplifications that follow have an error in the process. Describe what should have been done and then simplify completely.

67.
$$\begin{array}{r} 16\frac{3}{8} \\ +\ 12\frac{3}{4} \\ \hline 28\frac{6}{12} \end{array} = 28\frac{1}{2}$$

68.

$$\begin{array}{rcrcr} 238\frac{2}{5} & = & 238\frac{8}{20} & = & \overset{7}{\cancel{238}}\frac{18}{20} \\ -145\frac{3}{4} & = & -145\frac{15}{20} & = & -145\frac{15}{20} \\ \hline & & & & 92\frac{3}{20} \end{array}$$

69.

$$\begin{array}{rcr} 84 & = & 84\frac{3}{3} \\ -76\frac{2}{3} & = & -76\frac{2}{3} \\ \hline & & 8\frac{1}{3} \end{array}$$

70.

$$= \frac{3}{8} + \frac{7}{8} \times \frac{2}{3}$$

$$= \frac{10}{8} \times \frac{2}{3}$$

$$= \frac{10}{\underset{4}{\cancel{8}}} \times \frac{\overset{1}{\cancel{2}}}{3}$$

$$= \frac{10}{12} = \frac{5}{6}$$

H. Read and respond to the following exercises.

71. $2\frac{2}{3} - 1\frac{5}{8} + \frac{5}{6}$

Describe, in words, the steps you would use to simplify this exercise. You do not have to do the work.

72. Describe, in words, the entire process of writing equivalent fractions with larger numbers. You might want to describe an actual problem, such as 3/4 = 15/20.

73. Why must fractions have the same denominator before being added or subtracted?

74. Describe, in words, the process of borrowing in a mixed number subtraction problem.

75. Write several pairs of equivalent fractions. In each pair, compare the products when the numerator of each fraction is multiplied by the denominator of the other. Can you make an assumption about equivalent fractions?

76. Is $\frac{1}{2} \times \frac{1}{2} \div \frac{1}{2} = \frac{1}{2} \div \frac{1}{2} \times \frac{1}{2}$? Could grouping symbols be added to the right side of the equals sign and still maintain equivalency?

77. Describe how you would find the total surface area (all six sides) of a rectangular box that is 6 1/2 inches by 12 1/2 inches by 8 3/4 inches. The formula for a rectangle is $A = LW$. You do not have to actually calculate an answer.

78. From the information given in Exercise 60, can you find the perimeter of that trapezoid? Explain your answer.

I. The following problems might require two or more steps to reach a solution. Show your work.

79. Using the formula $A = LW$, find the area to be painted in a room that has two walls that are 9 3/4 feet by 8 1/2 feet and two walls that are 7 5/8 feet by 8 1/2 feet. Do not include the area taken up by two windows. One window is 3 1/2 feet by 2 1/4 feet, and the other measures 2 1/2 feet by 2 1/4 feet. There are also two doorways, which are each 3 1/4 feet by 7 1/2 feet. If 1 gallon of latex semigloss paint can cover about 400 square feet, how many gallons of paint must be purchased to do one coat on the walls?

80. One-eighth of Joel's take home salary is spent on car expenses. Three-sixteenths goes toward clothing and the same amount goes toward entertainment. Five-sixteenths of the money is spent on rent. The rest of his salary goes to pay for food. What fractional part of his salary is spent for food?

81. How many square feet of ceiling would there be in the room described in Exercise 9? If 1 gallon of ceiling paint covers approximately 275 square feet, how many gallons must be purchased to do two coats of paint on the ceiling? Use $A = LW$.

82. Joel's monthly take home salary is \$1840. Referring to the information in Problem 80, how much does he spend for car expenses, how much for food, and how much for rent?

CHAPTER TEST

1. If 3, 15, and 8 are denominators of fractions in an addition or subtraction problem, what is the least common denominator?

Simplify as indicated. Reduce all answers to lowest terms.

2. $\frac{5}{6}+\frac{2}{3}$

3. $\frac{4}{5}-\frac{1}{2}$

4. $\frac{3}{8}+\frac{15}{16}$

5. $1\frac{2}{3}+4\frac{3}{4}$

6. $3\frac{3}{5}-2\frac{1}{3}$

7. $8-4\frac{3}{7}$

8. $\frac{7}{8}+\frac{2}{3}+\frac{5}{12}$

9. $\frac{7}{9}-\frac{7}{15}$

10. $4\frac{1}{4}-2\frac{2}{3}$

11. $7\frac{1}{3}-3\frac{4}{5}+2\frac{4}{15}$

12. $14\frac{2}{3}-11$

13. $\left(\frac{2}{3}\right)^2+2\frac{1}{3}\times\frac{1}{2}$

14. $\frac{3}{10}+\sqrt{\frac{9}{25}}-\frac{2}{5}$

15. $9-2\frac{3}{5}$

16. Describe, in words, the steps that you would use to get an answer to:

    Arrange from largest to smallest: 3/16, 3/8, and 1/4.

Solve the following application problems.

17. Laura worked 6 1/3 hours on Monday, 4 3/4 hours on Tuesday, and 7 5/8 hours on Wednesday. What was the total number of hours that she worked?

18. How much more volume is there in a box that is 8 3/4 inches by 7 1/2 inches by 9 inches than one that is 12 1/3 inches by 7 inches by 6 5/8 inches? Use $V = LWH$.

19. Marc has mowed two-thirds of a 1 3/4-acre field. How many acres have been mowed?

20. If Tom can drive 132 miles on 5 1/2 gallons of gas, how far can he travel on 1 gallon?

Cumulative Review, Chapters 1–4

Solve each of the following exercises. Show all your work. Simplify all fraction answers to lowest terms.

A. In Exercises 1 through 10, subtract.

1. $26 + 114 + 236$
2. $46{,}551 + 28{,}779$
3. $\frac{5}{8} + \frac{7}{8}$
4. $\frac{16}{25} + \frac{4}{5}$
5. $7\ 5/9 + 8\ 1/2$

B. In Exercises 6 through 10, subtract.

6. $729 - 455$
7. $30{,}004 - 2{,}965$
8. $\frac{8}{15} - \frac{2}{15}$
9. $2\ 1/2 - 1\ 5/6$
10. $124 - 52\ 3/8$

C. In Exercises 11 through 15, multiply.

11. 539×78
12. 45×100
13. $\frac{3}{4} \times \frac{9}{16}$
14. $1\ 2/3 \times 5/9 \times 12/25$
15. Find 7/8 of 90.

D. In Exercises 15 through 20, divide.

16. $429 \div 3$
17. $42\overline{)7{,}594}$
18. $\frac{5}{6} \div \frac{1}{3}$
19. $4\ 3/8 \div 2/5$
20. $\frac{\frac{20}{2}}{3}$

E. Find the answer to each of the following exercises.

21. Is 1,593 divisible by 2? by 3? by 5?
22. Find the greatest common factor for 28 and 56.
23. Find the least common multiply for 7, 28, and 56.
24. Find the prime factorization for 63.
25. Find the prime factorization for 240.
26. $2^5 =$
27. $\sqrt{81} =$
28. List *all* the factors of 48.
29. Find the value of $K = L + 2M$ when $L = 6$ and $M = 5$.
30. Find the value of K in $K = L + 2M$ when $L = 2\ 1/2$ and $M = 1\ 1/3$.

F. Write the letter of the statement that exemplifies each of the properties listed in Exercises 31 through 35.

31. Commutative Property of Multiplication
32. Multiplication Property of One
33. Associative Property of Multiplication
34. Distributive Principle
35. Zero Property of Multiplication

a. $6 \times 1 = 6$
b. $6 \times 0 = 0$
c. $6 \times 1 = 1 \times 6$
d. $2(6 \times 1) = (2 \times 6) \times 1$
e. $2(6 + 1) = 2(6) + 2(1)$

G. In Exercises 36 through 43, simplify each of the expressions using the Order of Operations Agreement.

36. $7 + 3 - 5 + 2$
37. $8 \div 2 \times 4 + 2$
38. $2^4 - 9 \div 3$
39. $\sqrt{81} + 2^3 \div 4$
40. $5/8 \times 2/3 + 1/2$
41. $\sqrt{\frac{4}{9}} \div 6 + \frac{4}{5}$

42. $7 + 3[8 - 2(6 - 3)]$

43. $2[4(18 \div 9 + 2)]$

H. Find the answer to each of the following application problems.

44. Three-fourths of the women students in this college are part-time students. If there are 6,400 women enrolled in the college, how many are part-time students?

45. In his job, Raoul drives an average of 115 miles a day. How many miles does he drive during a 5-day workweek?

46. Find the area of ΔCDE (Figure 4.8). Use $A = 1/2\ bh$.

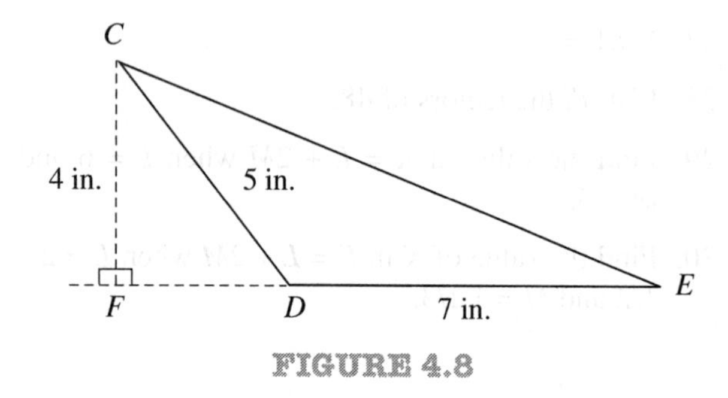

FIGURE 4.8

47. If Raoul can drive 228 miles on 12 gallons of gas, how far can he drive on 1 gallon?

48. An 8-pound block of cheese is to be cut into pieces that each weigh two-thirds of a pound. How many pieces can be cut from the block?

49. Find the difference between 26 2/3 and 14 3/8.

50. Find the mean of 76, 54, 83, and 63.

STUDY SKILLS MODULE 5

Using Class Time/ Listening Skills

Attending every class session is important, but you also want to get the maximum benefit out of the time you spend in class. That means you must come prepared; arrive on time or a little early; be mentally and physically alert; and actively participate in class by taking notes, answering questions, and working in-class problems.

Coming prepared means that you have studied the material covered in the last lecture and have done the homework problems related to those topics. It also means that you have read ahead and studied the material that is to be included in today's lecture. You have tried to master the new vocabulary and tried to follow the example problems. Perhaps you jotted down some questions about things you did not understand in this new material, and will wait to see if those questions are resolved by the lecture. You arrive in class a few minutes early and use that time to get your materials organized, to review your notes and the last textbook section, and perhaps even look over the material that is to be presented. You have ready the questions you wrote down concerning the material from the previous class or the homework assignment. Maybe you even have time to work a few problems that were not assigned, so that you refresh your memory.

Even deciding where you sit in class should be done with some consideration. Be sure that you are able to see the entire board or overhead screen from your seat and that you can clearly see the instructor. Sitting too far back or too far to the outside edges of the classroom can inhibit your view. Don't sit so far away that you canot hear the instructor clearly. If someone sitting near you has distracting habits, then look for another seat for the next class. Keep in mind that the nearer you are to the instructor, the fewer the number of distractions between the two of you.

To stay active during lectures, you should take notes. Don't try to write down everything that is said, but listen for and write the main ideas and anything that is stressed by the instructor. To know what to write and what not to write, you will have to be listening attentively and not let your mind wander. The general concepts and important issues can be determined only if you know (from listening and prereading) the overall content of the unit of instruction.

You must actively participate in any in-class activities because these assignments let you learn and practice new material under the guidance of the instructor. Your participation lets you and the instructor know if you understand the concepts before he or she goes on to build on those concepts.

As soon as possible after class, reread your notes and the textbook material. Fill in the gaps in your notes and use them and the book as you attempt a few homework problems. These steps will help you stabilize your knowlege so you will remember it later as you do your homework or study for a test.

If you must miss a class, at least read the material that would have been covered and take notes; then try some of the problems before the next class. Having a contact person or a "study buddy" in each class who will share notes if you are absent is an excellent resource to have arranged ahead of time.

Your Point of View

1. Where do you usually sit in your math classroom? What motivates you to choose that seat? What are the advantages and disadvantages of that location?
2. How much preparation do you do before you attend a math lecture? Do you see any benefits to incorporating some of the suggestions from this module in your future work?
3. Try to list three to five pieces of information that your instructor gave during the first five minutes of class today. How might your performance in this class have been affected if you were tardy and missed those facts?
4. Choose one suggestion from those made in this module and try it during your next class. Describe which activity you chose and comment on its benefit to you.

CHAPTER

5

Decimals

LEARNING OBJECTIVES

When you have completed this chapter, you should be able to:

1. Read, write, round, and rank decimal numbers.
2. Add, subtract, multiply, and divide decimals.
3. Change decimal numbers to their fractional equivalents and fractions to their decimal values.
4. Simplify expressions using the Order of Operations Agreement.
5. Estimate answers to problems.
6. Solve application problems involving one or more operations with whole numbers, fractions, and decimal numbers.
7. Solve geometry application problems involving triangles, quadrilaterals, solids, and circles.

With this chapter on decimal numbers, we will expand the type of numbers we are able to use. Mixed fractions have a whole number and a part; many decimal numbers also have a whole number and a part. The decimal point separates the whole number from the portion of the decimal that represents a value less than 1. We will also expand our knowledge of geometry to include the study of circles.

5.1 Reading, Writing, Rounding, and Ranking Decimals

Almost all numbers can be written in decimal form using numerals and a decimal point. The whole number 8 can be written 8.0. The fraction 3/4 is equivalent to 0.75. 5/8 is 0.625. Because our monetary system is based on the decimal system, people are usually more comfortable and confident with decimals than with their fractional equivalents.

Place Value

Decimal numbers have a place value system that is very similar to that of the whole number system. In fact, because the numbers to the left of the decimal point are whole numbers, the only new thing you need to learn is the place values to the right of the decimal point.

ten-millions
millions
,
hundred-thousands
ten-thousands
thousands
,
hundreds
tens
ones
. and
tenths
hundredths
thousandths
ten-thousndths
hundred-thousndths
millionths

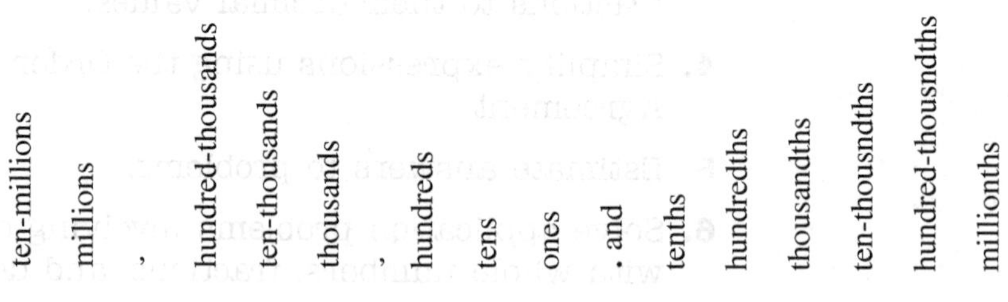

FIGURE 5.1

In Figure 5.1, you can see that the progression of place names is the same for the "part" side (to the right of the decimal point) as it is for the whole number side, except that there are no "oneths." Because "tenths" means a number divided into "10" parts, "oneths" would mean a number divided into "1" part. A number divided by "1" is a whole number and belongs on the left of the decimal point. Therefore, the progression of place values on the right of the decimal point begins with tenths. Note also that the "ths" ending signifies place value on the right side of the decimal point. Three hundred is written 300. Three-hundredths is written 0.03. If you think in terms of money, $300 is very different from $0.03. That "ths" ending is extremely important.

Self-Check To practice, identify the digit in the given place values in 24,578.9061. (See Figure 5.2.)

1. tenths place ________
2. hundreds place ________
3. ones place ________

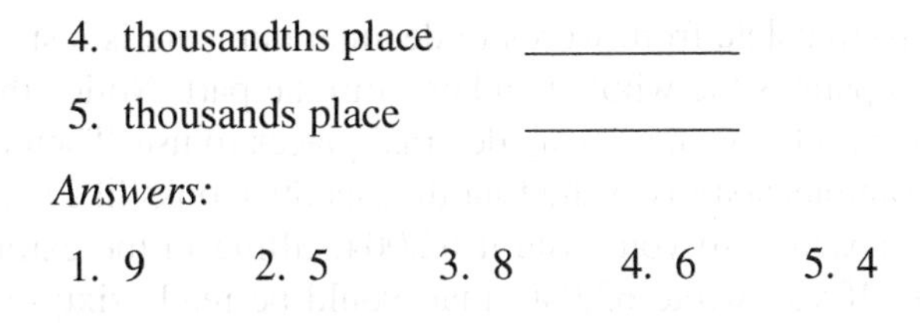

4. thousandths place __________
5. thousands place __________

Answers:

1. 9 2. 5 3. 8 4. 6 5. 4

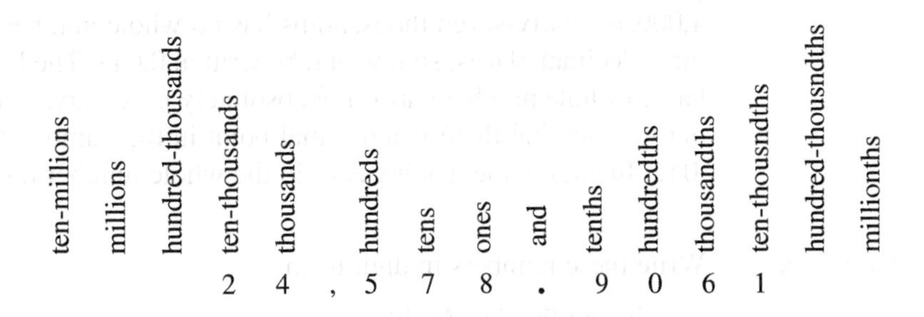

FIGURE 5.2

Reading and Writing Decimal Numbers

Place value plays an important role in the reading of decimal numbers. As an example, 1,624.038 would be read "one thousand six hundred twenty-four and thirty-eight thousandths." The whole number is read as discussed in Chapter 1; the decimal point is read "and." The part of the number to the right of the decimal point is read like a whole number and ends with the place value name of the last digit. Consider these examples:

14.23	fourteen and twenty-three hundredths
14.023	fourteen and twenty-three thousandths
14.0023	fourteen and twenty-three ten-thousandths

Self-Check Write these numbers in words.

1. 26.025 ____________________
2. 3.1456 ____________________
3. 76.421 ____________________
4. 147.05 ____________________
5. 0.0046 ____________________

Answers:

1. twenty-six and twenty-five thousandths
2. three and one thousand four hundred fifty-six ten-thousandths
3. seventy-six and four hundred twenty-one thousandths
4. one hundred forty-seven and five hundredths
5. forty-six ten-thousandths

To translate from words to decimal numbers is just as easy. Look for the "and" that separates the whole number from the part. Notice the last word in the number; it lets you know how many decimal places to use. Then play "fill in the blanks." As an example, sixty-two and four thousandths looks like 62.__ __ __ (thousandths uses three spaces), so you write it 62.004, filling in the extra spaces with placeholding zeros. If you wrote 62.040, that would be read "sixty-two and forty thousandths." 62.400 would be read "sixty-two and four hundred thousandths." Sixty-two and four thousandths must be 62.004. Four thousand and sixteen hundredths would be 4,000.16. Forty-seven thousandths has no whole number part, and thousandths uses three decimal places, so it would be written 0.047. The leading zero in a number that has no whole number part is not absolutely necessary. It does help, however, to point out the fact that there is a decimal point in the number. 0.047 is easier to read than .047. Including the leading zero in the whole number ones place is recommended.

Self-Check Write these numbers in digit form.

1. six and five hundredths ____
2. forty-three and fifteen thousandths ____
3. seventy-seven ten-thousandths ____
4. four hundred and four hundredths ____
5. six thousand two and two thousandths ____

Answers:

1. 6.05 2. 43.015 3. 0.0077 4. 400.04 5. 6,002.002

Rounding

Another noncomputational process involving decimals is rounding. The rule for rounding decimal numbers is almost identical to the rule used for whole numbers. The main difference is that the places to the right of the rounded place are dropped and not replaced by zeros.

Rules for Rounding Decimal Numbers

1. Mark the place value being considered.
2. Look at the number to its right.
   a. If that digit is less than 5, the place marked remains as it is, and all following places are dropped.
   b. If that digit is 5 or more, the place marked increases by 1, and all following places are dropped.

A good hint to remember is that when you round decimals, the final answer must have *exactly* the number of digits named.

For example, round 921.6387 to the nearest hundredth.

1. Mark the hundredths place: 921.63^87
2. Look at the number to the right. Because it is 8, which is larger than 5, the 3 increases to 4, and the 87 is dropped.
3. 921.6387 becomes 921.64, nine hundred twenty-one and sixty-four *hundredths.* Hundredths is the second decimal place, and the answer must have exactly two decimal places.

Consider the following completed examples.

EXAMPLE 1 Round 8.049 to the nearest hundredth.

Solution

8.04^9	9 is greater than 5, so the 4 increases by 1.
8.05	Drop all digits past hundredths.
8.05	Eight and five hundredths.

EXAMPLE 2 Round 16,047.346 to the nearest tenth.

Solution

16,047.3^46	4 is less than 5.
16,047.3	Drop all digits past tenths.
16,047.3	Sixteen thousand forty-seven and three tenths.

EXAMPLE 3 Round 3,426.795 to the nearest hundredth.

Solution

3,426.79^5	5 is equal to 5, and therefore the 79 increases to 80.
3,426.80	Keep the last zero so that the answer ends in hundredths.
3,426.80	Three thousand, four hundred twenty-six and eighty hundredths.

EXAMPLE 4 Round $83.925 to the nearest cent.

Solution

$83.92^5	Because the number to the right of the digit in the cents place is 5, the 2 increases by 1.
$83.93	Cents is two decimal places.
$83.93	Eighty-three dollars and ninety-three cents.

EXAMPLE 5 Round $47.884 to the nearest cent.

Solution

$47.88^4	4 is less than 5, so the 8 remains the same.
$47.88	Cents is always two decimal places.
$47.88	Forty-seven dollars and eighty-eight cents.

While you are studying rounding, it is good to learn about another common mathematical practice. You do not write the answer to a problem as 5, for instance, unless the calculation comes out to be exactly the whole number 5. The number 5.0 is not the same as 5, because 5.0 indicates that the answer is approximately equal to ($\doteq$) 5.0. It could be 5.036, for example, rounded to the nearest tenth.

Self-Check Round each decimal to the place indicated.

1. 3,276.4958 nearest hundredth
2. 2,543.2716 nearest tenth
3. 7,349.3335 nearest hundred
4. 861.0007 nearest thousandth
5. $327.8648 nearest cent

Answers:

1. 3,276.50 2. 2,543.3 3. 7,300 4. 861.001 5. $327.86

Ranking Decimals

Suppose you were asked to decide which of two decimal numbers is the larger or to arrange three or four decimal numbers in order from the smallest to the largest. You would need some basis for comparing the sizes of the numbers, such as writing the numbers with the same number of decimal places. In order for us to compare two or more quantities, they must be described in the same way. For example,

"Which is larger, 0.26 or 0.264?"

If you annexed a zero to 0.26 (twenty-six hundredths), it would be written 0.260 (two hundred sixty thousandths). Adding a zero does not change the value of the decimal. You can see this by comparing the numbers in their fraction forms:

$$0.26 = \frac{26}{100} \qquad 0.260 = \frac{260}{1000} = \frac{260 \div 10}{1000 \div 10} = \frac{26}{100}$$

You can see that 0.26 and 0.260 are equivalent numbers. Annexing zeros to the end of a decimal number does not change its value. You can now compare 0.264 and 0.260. Because two hundred sixty-four thousandths is larger than two hundred sixty thousandths, 0.264 is larger than 0.260, which is equivalent to 0.26.

The same process would work if you were ranking three decimal numbers from the smallest to the largest. Consider this example.

EXAMPLE 6 Rank 0.302, 0.31, and 0.306 in order from the *smallest to the largest.*

Solution Write each decimal number with three decimal places, since three is the most number of decimal places in any of the given values, and then order the numbers.

0.302 already has three decimal places.

0.31 becomes 0.310.

0.306 has three decimal places.

Therefore, the order would be

0.302 0.306 0.310

Going back to the original form, the order from smallest to largest is

0.302 0.306 0.31

EXAMPLE 7 Arrange 1.21, 1.1, and 1.11 from the *largest to the smallest.*

Solution Add zeros where necessary to write each number with two decimal places.

1.21 stays 1.21.

1.1 becomes 1.10.

1.11 stays as it is.

The order from largest to smallest is

1.21 1.11 1.10 or 1.21 1.11 1.1

EXAMPLE 8 Insert > or < between 3.26 and 3.267.

Solution The symbol > means **"greater than"** and < means **"less than."** In order to compare these two decimal numbers, you will need to add a zero to make both of them have three decimal places.

3.26 becomes 3.260.

3.267 stays as it is.

Because 3.260 is less than 3.267, the correct way to write the expression is 3.26 < 3.267.

EXAMPLE 9 Write 6.32, 6.305, 6.23, and 6.31 in order using the symbol >.

Solution First, write each number with three decimal places.

6.32 becomes 6.320.

6.305 stays as it is.

6.23 becomes 6.230.

6.31 becomes 6.310.

Because > means "greater than," begin with the largest number and rank the numbers down to the smallest.

6.320 > 6.310 > 6.305 > 6.230 or 6.32 > 6.31 > 6.305 > 6.23

Self-Check Rank each of the following in order from the largest to the smallest.

1. 3.026 3.1 3.101 3.01
2. 4.13 4.132 4.103 4.3
3. 0.026 0.26 0.206 0.21
4. 1.23 1.231 1.223 1.023
5. 16.3 16.33 16.03 16.303

Answers:

1. 3.101 3.1 3.026 3.01
2. 4.3 4.132 4.13 4.103
3. 0.26 0.21 0.206 0.026
4. 1.231 1.23 1.223 1.023
5. 16.33 16.303 16.3 16.03

Try the following problems using the reading, writing, rounding, and ranking skills demonstrated in this section.

5.1 EXERCISES

A. Look at the given number, and name the digit that is in the place requested.

1. 2,635.41
   a. tenths b. tens
   c. hundreds d. hundredths
2. 31.0569
   a. tenths b. tens
   c. thousandths d. ten-thousandths
3. 0.43762
   a. thousandths b. tenths
   c. ones d. hundredths
4. 143.567
   a. hundreds b. tenths
   c. hundredths d. ones

B. Write the following numbers in words.

5. 160.6
6. 8.008
7. 182.034
8. 200.2
9. 0.0456
10. 143.143
11. 3.017
12. 0.68
13. 45.45
14. 12.1507

C. Write the following numbers in digits.

15. six and seventy-four thousandths
16. fourteen and four tenths
17. one hundred seven thousandths
18. eighty-three ten-thousandths
19. one hundred sixteen thousand
20. twenty-seven hundred

D. Round each of the following numbers to the place requested.

21. 0.038 to the nearest hundredth
22. 0.826 to the nearest tenth
23. 1.452 to the nearest tenth
24. 3.005 to the nearest hundredth
25. 146.286 to the nearest ten
26. 275.188 to the nearest hundred
27. 146.286 to the nearest tenth
28. 275.188 to the nearest hundredth
29. 16.0997 to the nearest thousandth
30. 38.2998 to the nearest thousandth
31. 16.047; hundredths
32. 43.186; tenths
33. 167.352; tens
34. 346.008; hundreds
35. 0.0753; hundredths
36. 6.01352; thousandths
37. 14.2635; thousandths
38. 26.396; hundredths
39. 357.985; tenths
40. 0.0063; hundredths

41. $286.452; dollar
42. $149.685; dollar
43. $24.8556; cent
44. $38.8947; cent
45. 267.521; whole number
46. 39.499; whole number

E. Arrange the following numbers in order from the smallest to the largest.

47. 2.03, 2.2, and 2.12
48. 0.006, 0.01, and 0.02
49. 0.7, 0.635, and 0.67
50. 3.1, 3.095, and 3.09
51. 4.03, 4.003, and 4.304
52. 6.12, 6.001, and 6.102

F. Rank the numbers using > or < as requested.

53. 0.04, 0.4, 0.392, and 0.401; >
54. 6.07, 6.6, 6.76, and 6.067; <
55. 2.15, 2.123, 2.146, and 2.015; <
56. 0.28, 0.208, 0.21, and 0.218; >
57. 2.24, 2.004, 2.402, and 2.204; >
58. 7.18, 7.118, 7.081, and 7.108; <
59. 0.305, 0.31, 0.351, and 0.053; <
60. 0.451, 0.541, 0.405, and 0.514; >

G. Read and respond to the following exercises.

61. Explain in words how "rounding to the nearest hundred" and "rounding to the nearest hundredth" are the same and how they are different.
62. A fellow student says that 0.51 is smaller than 0.506. Explain in words why he is incorrect.
63. Explain in words why 0.7 and 0.70 represent the same value even though they appear to be different.
64. Explain in words the steps you would follow to arrange the following numbers in order from largest to smallest: 0.913 0.92 0.902 0.9

5.2 Adding and Subtracting Decimals

Consider the following mixed number problem:

12 3/10 + 36 6/10

```
 12 3/10
+36 6/10
 48 9/10
```

First, the whole numbers are added; then the parts (with like denominators) are combined. The format of the numbers—numerals indicating whole numbers and the fraction bar indicating the parts—tells which quantities to combine.

This same thinking applies to decimal addition and subtraction. The whole numbers are aligned vertically, the decimal point separates the wholes from the parts, and the part values are on the right side of the decimal point. Below is the same problem, done in decimal form.

```
 12.3
+36.6
 48.9
```

Separating the whole numbers and the decimal quantities is the key.

In the next examples, you will notice that the practice of attaching or annexing zeros to the "part" (right) side of the decimal point is used. You can understand that this is the correct process to follow if you look at the fraction equivalents of some decimal numbers.

.7 = 7/10
.70 = 70/100, which reduces to 7/10
.700 = 700/1,000, which reduces to 7/10

Therefore, 0.7 = 0.70 = 0.700, which are all equal to 7/10.

.18 = 18/100, which reduces to 9/50
.180 = 180/1,000, which reduces to 18/100 = 9/50
.1800 = 1,800/10,000, which reduces to 9/50

Therefore, 0.18 = 0.180 = 0.1800, which are all equal to 9/50.

Now consider these examples of decimal addition and subtraction.

EXAMPLE 1 Add: 16.3 + 1.96 + 37.43

Solution Annex a zero to three tenths, making it thirty hundredths (16.3 = 16.30), so you have like "denominations." Although this is not absolutely necessary, it really is a good idea.

```
 11
16.30
 1.96
37.43
-----
55.69
```

You could have estimated this answer first by adding 16, 2, and 37. The result, 55, would have been very close to the actual answer.

EXAMPLE 2 Add: 93.8 + 17 + 6.765

Solution

```
                    11
  93.8     =      93.800
  17       =      17.000
+  6.765   =   +   6.765
               ---------
                 117.565
```

Be sure to put 17 on the whole number side and to annex zeros as necessary. Any whole number is understood to have a decimal point to the right of the last digit.

EXAMPLE 3 Add: $2,063.80 + $149 + $13.65 + $0.40

Solution You might first estimate the answer by adding $2,060 + $150 + $14 + $1. This will let you know that your actual answer should be close to $2,225.

```
                     111
$2,063.80 =    $2,063.80
   149    =       149.00
    13.65 =        13.65
+    0.40 =   +     0.40
              ----------
               $2,226.85
```

You can see from these examples that annexing the zeros only serves to keep the number of decimal places the same in addition, subtraction, or comparison problems; it in no way affects the value of the number. One last example makes this principle easier to see: $35 = $35.00.

The process of using decimals in subtraction is no different from using them in addition problems. Just be careful to keep the whole numbers with the whole numbers and the parts with the parts.

EXAMPLE 4 Subtract 287.65 from 396.47.

Solution

```
 8¹5 1
 396.47
-287.65
+108.82
```

Check You can always check your subtraction answer by adding and you should form the good habit of checking your answer whenever possible.

```
 287.65
+108.82
 396.47 √
```

EXAMPLE 5 Find the difference between 83 and 17.86.

Solution First we need to write a decimal point and two zeros after the 83. Then we line up the numbers according to their place value.

```
 83.00
-17.86
```

We cannot subtract 6 from 0 so we need to borrow 1 from 30, making 29 and 10.

```
   2 9₁
 83.00
-17.86
   .14
```

Now subtract 6 from 10 and then 8 from 9.

```
 7 12 9₁
 83.00
-17.86
 65.14
```

Now borrow 1 from the 8 to make 7 and 12 and finish the problem.

Check

```
  65.14
+ 17.86
  83.00 √
```

Always add to check your answer.

EXAMPLE 6 Find the difference between 27.002 and 8.965.

Solution Subtracting from zeros is always tricky. In this problem, you might think of regrouping the numbers as 2, 700, and then 2. Borrow from 700, leaving 699. Regroup the borrowed number with the 2. Subtract 5 from 12 and continue. You learned how to do this with whole numbers in Section 1.2. It will help, too, to estimate an answer first by subtracting 9.000 from 27.000. You know then that the answer should be close to 18.000.

```
  6 99₁
 27.002
-  8.965
+ 18.037
  27.002 √
```

Self-Check

1. Find the sum of 26.8; 32; and 4.37.
2. Subtract 14.28 from 32.91.
3. Find the difference between 46 and 17.43.

Answers:

1. 63.17   2. 18.63   3. 28.57

EXAMPLE 7 Review what you have learned about geometric figures in Chapter 2 and use that information to find the perimeter of quadrilateral ABCD shown in Figure 5.3. Also calculate the measure of $\measuredangle D$ if the $m\measuredangle A = 82.4°$, $m\measuredangle B = 101.2°$, and $m\measuredangle C = 120.5°$.

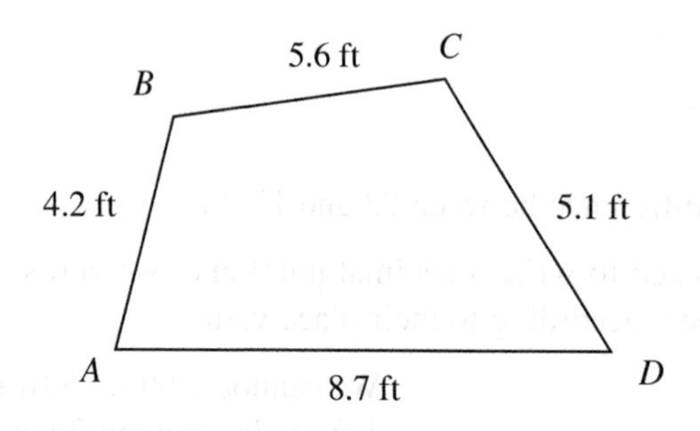

FIGURE 5.3

Solution The perimeter of any geometric figure is the sum of the lengths of its sides, so

$P = 4.2\text{ ft} + 5.6\text{ ft} + 5.1\text{ ft} + 8.7\text{ ft} = 23.6\text{ ft}$

The sum of the measures of the angles in any quadrilateral is 360°, so find the sum of the measures of the angles *A*, *B*, and *C* and subtract that sum from 360° to find the measure of $\measuredangle D$.

$$m\measuredangle A + m\measuredangle B + m\measuredangle C + m\measuredangle D = 360°$$
$$82.4° + 101.2° + 120.5° + m\measuredangle D = 360°$$
$$304.1° + m\measuredangle D = 360°$$
$$m\measuredangle D = 360° - 304.1°$$
$$m\measuredangle D = 55.9°$$

5.2 EXERCISES

A. Estimate an answer to each of the following exercises. Remember that when you estimate, the numbers should be reasonable to work with so you can use mental math whenever possible.

1. Find the sum of 22.9, 14.7 and 6.3.
2. Add: 16.3 + 29.07 + 8
3. Subtract 27.42 from 40.
4. Find the difference between 43.46 and 19.2.
5. From the sum of 20,324.86 and 6,921.4, subtract 18,003.9.
6. From the sum of 17,261.8 and 6,429.63, subtract 9,027.63.
7. How much change would you get back from a $20 bill if the cost of your meal was $8.26?
8. How much change would you get back from a $50 bill if you made purchases of $17.38 and $22.10?

B. Perform the indicated operations. Be sure to check your answers.

9. 48 – 15.26
10. 0.26 + 3.45 + 4.2
11. 1.7 + 2.035 + 6 + 11.5
12. 2.4 + 3.086 + 4 + 15.1
13. 8.269 – 4.837
14. 8.34 + 19 + 2.8
15. 17.343 – 11.565
16. 21.822 – 14.954
17. 143.2 + 11.68 + 0.07
18. 43.86 + 0.09 + 173.4
19. 12.83 – 9.964
20. 11.76 – 5.893
21. 27.6 – 17.923
22. 72 – 40.36
23. 64.8 – 37.06 + 9.7
24. 49.1 – 29.83 + 6.4
25. 51.02 – 17.8 + 8.23
26. 70.3 – 12.6 + 19.34
27. $176 + $8.32 + $9.57
28. $18.53 + $43.96 + $147
29. $87 – $49.73
30. $27 – $19.97

C. Find the answer to the following verbal exercises.

31. Find the difference between 37 and 19.265.
32. What is the sum of 18.3, 43, and 1.79?
33. Find the sum of 47.5, 82, and 17.59.
34. What is the difference between 72 and 56.549?
35. From the sum of 17.6 and 38, take 22.94.
36. From the sum of 82.4 and 66, take 109.55.
37. How much change would you get back from a $50 bill if your purchases totaled $37.59?
38. How much change would you get back from a $20 bill if your purchases totaled $14.83?
39. How much larger is 85.1 than 27.42?
40. How much larger is 106.4 than 87.58?

D. Use your calculator to find the answers to the following exercises.

41. 117.365 + 829 + 16.8
42. 214.38 + 789 + 37.95
43. 21,000 – 17,952.43
44. 38,000 – 29,452.8
45. 209,075 – 43,869.57
46. 312,004 – 93,768.29
47. What is the difference between 867.4 and 12.982?
48. Find the sum of 1,002, 19.458, and 183.2.
49. What is the sum of 120,458.9 and 331.43?
50. What is the difference between 21,854 and 7,738.6?

51. From the sum of 1,876 and 339.45, take 994.534.

52. From the sum of 3,217.45 and 39,851.6, take 221.98.

53. From the sum of 0.00985 and 0.00874, take 0.00032.

54. From the sum of 0.0876 and 0.000321, take 0.00287.

55. Find the difference between 0.00875 and 0.000941.

56. Find the difference between 0.1024 and 0.00983.

E. Find the answer to each of the following exercises.

57. Find the perimeter of $\triangle ABC$ if $AB = 11.2''$, $AC = 23.5''$, and $BC = 24.6''$.

58. Find the measure of $\measuredangle B$ in $\triangle ABC$ if $m\measuredangle A = 42.7°$ and $m\measuredangle C = 34.6°$.

59. Explain in words how you would find the measure of $\measuredangle K$ in quadrilateral *KLMN* if you were given the measures of $\measuredangle L$, $\measuredangle M$, and $\measuredangle N$.

60. Explain in words how you could find the amount of tape necessary to seal a box if the parcel delivery service allows one strap of tape around the length of the box and one strap around the width.

5.3 Multiplying Decimals

Multiplying decimal numbers is perhaps the least troublesome of the operations. You multiply the quantities, ignoring the decimal points for the moment. When writing whole numbers, the decimal point is not usually written, but it is "imagined" at the far right of the whole digits. Therefore, in order to determine where to place the decimal point in the product (the result of the multiplication), count the total number of decimal places in the factors and move the decimal point that number of places from right to left in the answer. For example, consider:

3.27×1.5

First perform the multiplication as if there were no decimal points in either number.

$327 \times 15 = 4905$

There are three decimal places altogether in the factors. Therefore, count three places from right to left, and position the decimal point between the 4 and the 9.

$3.27 \times 1.5 = 4.905$

Now let's look at 5.3×2.6.

$53 \times 26 = 1378$

There are two decimal places in the original factors, so your answer has two decimal places.

$5.3 \times 26 = 13.78$

Have you ever wondered where the multiplication rule for decimal placement comes from? If the rule says that 0.06 × 0.9 equals 0.054, try doing the problem in its fraction form.

0.06 = 6/100 0.9 = 9/10

$$\frac{6}{100} \times \frac{9}{10} = \frac{54}{1{,}000} = 0.054$$

hundredths × tenths = thousandths
(two places) + (one place) = (three places)

EXAMPLE 1 Multiply: 14.9 × 16

Solution You could first estimate the answer by multiplying 15 × 16. The actual answer should be close to 240.

```
  14.9
×   16
   894
  149
  2384
```

14.9 × 16 = 238.4

There is no decimal point in the whole number 16, so you have a total of one decimal place in the answer.

The actual answer is close to the estimate.

EXAMPLE 2 Multiply: 7.35 × 0.2

Solution We might first estimate the answer by multiplying 7 by 0.2. We can then tell that the actual answer should be close to 1.4.

```
  7.35
×  0.2
 1.470
```

There is a total of three decimal places in the factors, so there will be three decimal places in the product. It is standard practice to drop off ending zeros *after* the decimal point has been placed.

7.35 × .02 = 1.47

EXAMPLE 3 Multiply: 1.006 × 0.29

Solution To estimate this product we might choose to multiply 0.3 by 1. Our actual answer should, therefore, be close to 0.3.

```
  1.006
×  0.29
   9054
  2012
0.29174
```

Five decimal places in the factors require five decimal places in the answer (product).

It is common practice in this type of answer to place a zero to the left of the decimal point to call attention to the decimal point. Either way is certainly correct, but .29174 is not as clear as 0.29174.

1.006 × 0.29 = 0.29174

EXAMPLE 4 Multiply: 0.823 × 0.07

Solution You could estimate the answer by multiplying 0.8 by 0.07. Multiply 8 times 7 to get 56 and then you need three decimal places, so your estimate would be 0.056.

```
  0.823
×   .07
0.05761
```

In this example, you need to move the decimal point five spaces left in the product, but there are only four digits. Add a zero to the left of your answer, and then move the decimal point a total of five places. Note also the leading zero to the left of the decimal point. Again, it is optional.

0.823 × 0.07 = 0.05761

The answer, 0.05761, is reasonably close to the estimate, 0.056.

◀

Whenever you are doing a problem that involves money, it is customary to round the answer to the nearest cent, even though no specific rounding instructions are given. See how this is done in Examples 5 and 6.

EXAMPLE 5 Multiply: $83.48 × 0.055

Solution

```
  $83.48
× 0.055
   41740
  41740
$4.59140
```

Because your answer is money, you must round to the nearest cent, even though the problem says nothing about rounding.

$83.48 × 0.055 = $4.59

◀

EXAMPLE 6 Multiply: $143.84 × 1.7

Solution

```
 $143.84
×    1.7
  100688
  14384
$244.528
```

This problem is money once again, so you need to round to the nearest cent. 2 is in the cents place and 8 is more than 5, so you should round up to 53 cents.

$143.84 × 1.7 = $244.53

◀

Self-Check Do the following multiplication problems without a calculator. Pay careful attention to the decimal point placement.

1. 1.3 × 0.06
2. 0.002 × 41
3. 0.0035 × 0.4
4. $15.45 × 0.3

Answers:

1. 0.078 2. 0.082 3. 0.0014 4. \$4.635 = \$4.64

EXAMPLE 7 Multiply: 246.2 × 100

Solution

$$\begin{array}{r} 246.2 \\ \times \quad 100 \\ \hline \end{array}$$

When you multiply by a number ending in zeros, you can simply annex that number of zeros to the right of the number, multiply by the nonzero number(s), and position the decimal point.

$$\begin{array}{r} 246.2 \\ \times \quad 100 \\ \hline 24{,}620.0 \end{array}$$

Because 100 has two zeros, add two zeros to the other factor and move the decimal point one place to the left because there is only one decimal point in the factors being multiplied.

There is a shortcut that you can use when multiplying a decimal number by a power of 10. In other words, you can use it when one of the factors in the problem is 10, 100 or 1,000 ($10^1 = 10$, $10^2 = 100$, $10^3 = 1{,}000$), or another power of 10. You will use this shortcut often, especially when doing percent problems. Here it is:

Multiplication by Powers of Ten

To multiply by

10; move the decimal point one place to the right.

100; move the decimal point two places to the right.

1,000; move the decimal point three places to the right.

And so on.

In order to move the decimal point, you may need to add some placeholding zeros. The following examples illustrate this shortcut.

EXAMPLE 8 26.4 × 100 = 26.40. = 2,640

EXAMPLE 9 0.056 × 10 = 0.0.56 = 0.56

EXAMPLE 10 17 × 100 = 17.00. = 1,700

Remember that in a whole number, the decimal point is understood to be to the right of the last digit.

Decimals are also often used in geometry applications. Review what you learned about area and volume in Chapters 2, 3, and 4, and then study Examples 11 and 12.

EXAMPLE 11 Find the area and perimeter of a rectangle with length of 21.6 centimeters and width of 4.5 cm. (Use $P = 2(L + W)$ and $A = LW$.)

Solution Use the formula $P = 2(L + W)$ to find the perimeter. Follow the Steps for Using a Formula from Section 2.3.

$P = 2(L + W)$
$P = 2(21.6 + 4.5)$
$P = 2(26.1)$
$P = 52.2$ centimeters

To find the area, we will use $A = LW$.
$A = LW$
$A = (21.6)(4.5)$
$A = 97.2$ square centimeters
Therefore, the perimeter is 52.5 centimeters and the area is 97.2 square centimeters.

EXAMPLE 12 Find the perimeter of a regular hexagon (six-sided figure, all sides the same length) if one of its sides measures 7.4 inches.

Solution Perimeter means the sum of the sides and we have six sides of the same length. We can add 7.4 six times or we can multiply the length of one side by 6. The second approach is more efficient.

$P = 6s$
$P = 6(7.4)$
$P = 44.4$ inches
Therefore, the perimeter of this hexagon is 44.4 inches.

5.3 EXERCISES

A. Estimate an answer to each of the following exercises. Remember that when you estimate, the numbers should be reasonable to work with so you can use mental math whenever possible.

1. 26 × 3.9
2. 37 × 7.4
3. 84.9 × 16.7
4. 76.8 × 18.6
5. 0.00176 × 0.0046
6. 0.0045 × 0.000825

B. Multiply in each of the following exercises. Check your answers by using the answer key.

7. 17.42 × 2.8
8. 16.35 × 3.7
9. 4.6 × 0.009
10. 5.8 × 0.007
11. 0.07 × 0.5
12. 0.06 × 0.8
13. 3.48 × 10
14. 12.451 × 100
15. 248 × 11.6
16. 357 × 12.4
17. \$18.39 × 0.055
18. \$32.75 × 0.055
19. 23 × 100
20. 45 × 1,000
21. \$47 × 8.6
22. \$59 × 9.4
23. 632 × 14
24. 753 × 26
25. 3.82 × 0.035
26. 4.96 × 0.185
27. 0.59 × 100
28. 0.962 × 10
29. 0.0035 × 0.54
30. 0.00987 × 0.0032
31. Find the product of 21, 18.36, and 0.4.
32. Find the product of 0.9, 32, and 21.45.

C. Find the answer to each of the following exercises by using your calculator.

33. 0.296 × 14.834
34. 72.35 × 0.849
35. 0.279 × 0.0038
36. 0.821 × 0.0762
37. From the product of 897.5 and 38.6, subtract 4,005.4.

38. To the product of 117.9 and 0.376, add 215.9.
39. To the product of 376.91 and 1.038, add 476.1.
40. From the product of 737.4 and 93.5, subtract 29,879.42.
41. From the product of 0.35 and 0.086, subtract 0.0009.
42. To the product of 0.0095 and 0.765, add 0.0045.

D. Find the answer to each of the following exercises.

43. Find the perimeter for a rectangle if the length is 12.4 yards and the width is 5.8 yards.
44. Find the area of a rectangular practice field if its width is 32.5 feet and its length is 140.4 feet. ($A = LW$)
45. Find the area of the rectangle in Exercise 43. ($A = LW$)
46. Find the perimeter of the practice field in Exercise 44.
47. Find the perimeter of a triangle if its sides are 12.2 inches, 15.7 inches, and 19.5 inches.
48. Find the area of a triangle whose base is 18.4 centimeters and whose height is 9.3 centimeters. ($A = 1/2\ bh$)
49. Find the perimeter of a regular octagon (8 equal sides) if the measure of one of its sides is 6.3 inches.
50. Find the perimeter of a regular nonagon (9 equal sides) if the measure of one of its sides is 0.43 feet.
51. Find the area of one side of a cube if the cube is 7.2 cm high. ($A = s^2$)
52. Find the volume of a cardboard carton whose dimensions are 16.4 inches by 12.5 inches by 10.8 inches ($V = LWH$)
53. Find the volume of a cube that is 7.2 centimeters deep. ($V = LWH$)
54. Find the area of one end of the carton in Exercise 52. The ends are the smaller sides. ($A = LW$)
55. Find the total surface area of the cube in Exercise 51. (Use $A = s^2$ to find the area of each side.)
56. Find the total surface area of the carton in Exercise 52. (Use $A = LW$ to find the area of each side.)

E. Read and respond to the following exercises.

57. Explain the similarities between the ways decimal and fraction problems are solved. Include addition, subtraction, and multiplication.
58. Using your calculator, compute the answer 226,672 × 345,185.
    a. What does the display look like? What do you think it means?
    b. You could get an approximate answer if you multiplied 227,000 × 345,000. Try it on your calculator. What happens?
    c. Explain how you could get an approximate answer by using your calculator and applying some other things you know about multiplying by numbers that end in zero(s).
59. What happens when you try to multiply 0.0003762 by 0.000258 on your calculator? Can you think of a way to find the answer to this problem by using your calculator and applying some other things you know about decimal multiplication? Explain your ideas.
60. Explain to another student in your class why 0.003 × 0.017 = 0.000051 and what procedures you would follow to get that answer without using a calculator.

5.4 Dividing Decimals

If multiplication is one of the simplest decimal operations, division is the one that causes the most trouble. There is, however, a series of steps that, when followed in order, make the decimal division process more manageable.

First, some vocabulary is necessary. Study each of these forms:

$$\text{divisor}\overline{)\text{dividend}}^{\text{quotient}} \qquad 9\overline{)45}^{5}$$

$$\frac{\text{dividend}}{\text{divisor}} = \text{quotient} \qquad \frac{45}{9} = 5$$

$$\text{dividend} \div \text{divisor} = \text{quotient} \qquad 45 \div 9 = 5$$

Division with decimals, like division with whole numbers, often results in a remainder. This difficulty is usually handled by directions that state where the quotient should be rounded. These directions allow the division process to be halted before it runs off the page! For example, the quotient of 7 divided by 9 is

$$0.7777777777777777777\ldots, \quad \text{or} \quad 0.\overline{7}$$

The three dots following the last 7 mean that the 7s continue indefinitely. The line written over the 7 in $0.\overline{7}$ also indicates that the 7 repeats indefinitely. If the quotient is rounded to the nearest thousandth, the answer is 0.778.

Now let's see how the series of steps for division are applied in a simple problem. They are listed in the order in which they should be considered.

Divide 14 by 1.5 and round your answer to the nearest hundredth.

1. *Is there a decimal point in the dividend?* If not, put it at the end (on the right side) of the whole number.

   No, there is no decimal point, so place one at the right of the whole number.

   $1.5\overline{)14.}$

2. *Is the divisor a whole number?* If it is, then put the decimal point in the quotient right above the decimal point in the dividend. If it is not, swing the decimal point in the divisor to the right so that the divisor becomes a whole number. Then swing the decimal point in the dividend the same number of places to the right (multiplying the divisor and dividend by ten). Place the decimal point straight up in the quotient.

   You are dividing by a decimal number, so move the decimal point in both numbers one place to the right, writing 1.5 as 15 and 14 as 140. Place the decimal point in the quotient.

   $1.5.\overline{)14.0.}$

3. *Does the problem call for rounding?* If yes, annex enough zeros so that the quotient *is one place longer than requested.* If rounding is not called for, you may still need to annex zeros until the division ends with a zero remainder.

   Yes, it calls for hundredths, so you will add three zeros.

   $1.5.\overline{)14.0.000}$

4. Now, *divide.*

   $$\begin{array}{r} 9.333 \\ 1.5.\overline{)14.0.000} \end{array}$$

5. *Round* your quotient as requested. (Remember that ≐ means "is approximately equal to.")

   $9.333 \doteq 9.33$

6. You can *check* the answer by multiplication.

   $9.33 \times 1.5 = 13.995 \doteq 14.0$

   Because the quotient has been rounded, the check will be only approximately equal to the original dividend. You might find a calculator helpful when checking your answers.

Although the questions in these steps may seem cumbersome, they will become familiar very quickly and will make dividing with decimals more routine. All of the decimal point and rounding questions are resolved before the dividing even begins.

Follow these examples. Each of the question steps is numbered for you.

EXAMPLE 1 Divide 1.83 by 2 and round to the nearest hundredth.

Solution

1. Is there a decimal point in the dividend?

   Yes, 1.83 already has a decimal point.

   $2\overline{)1.83}$

2. Are you dividing by a whole number?

   Yes, 2 is a whole number. Therefore, place the decimal point straight up into the quotient.

   $2\overline{)1.83}$

3. Does the problem call for rounding?

   Yes, it calls for the nearest hundredth. You already have a digit in the hundredths place, so you need to add only one zero to have enough places for the rounding.

   $2\overline{)1.830}$

4. Now, do the division.

   $$\begin{array}{r} 0.915 \\ 2\overline{)1.830} \end{array}$$

5. Round to the nearest hundredth.

   $0.915 \doteq 0.92$

6. Check. Because of the relationship between division and multiplication, you can check the solution by multiplying the quotient by the divisor. In the case of a rounded answer, however, the product may only be close to the original dividend.

   $0.92 \times 2 = 1.84 \doteq 1.83$

EXAMPLE 2 What is the quotient when 0.367 is divided by 1.4? Round your answer to the nearest thousandth.

Solution

1. Is there a decimal point in the dividend?

   Yes, there is.

   $1.4\overline{)0.367}$

2. Is the divisor a whole number?

   No, it is not. Therefore, move the decimal point one place to the right to make the divisor 14. Do the same thing to the dividend, and then place the decimal point in the quotient.

   $1.4.\overline{)0.3.67}$

3. Does the problem call for rounding?

   Yes, you need to add two zeros to round to the thousandths place.

   $1.4.\overline{)0.3.6700}$

4. Do the division.

   $$\begin{array}{r} 0.2621 \\ 1.4.\overline{)0.3.6700} \end{array}$$

5. Round the quotient as requested.

   $0.2621 \doteq 0.262$

6. Check by multiplication.

   $0.262 \times 1.4 = 0.3668 \doteq 0.367$

EXAMPLE 3 Divide 18 by 0.32. Divide until the answer comes out even (has a remainder of zero).

Solution

1. Is there a decimal point in the dividend?

   No. Place it to the right of 18.

   $0.32\overline{)18.}$

2. Is the divisor a whole number?

   No, so move both decimal points two places to the right and position a decimal point straight up in the quotient.

   $0.32\overline{)18.00.}$

3. Are there any rounding instructions?

   The problem says to divide until the quotient comes out even and there is no remainder. You may have to add several zeros, or it may work out quickly. Add several zeros and continue dividing until the remainder is zero.

   $0.32.\overline{)18.00.000}$

4. Divide.

   $$\begin{array}{r} 56.25 \\ 0.32.\overline{)18.00.000} \end{array}$$

5. Round as requested.

   The answer is 56.25 and we had no rounding instructions, so we leave the answer as it is.

6. Check the answer by using multiplication.

   $56.25 \times 0.32 = 18$

   The check is exact because the quotient was not rounded.

EXAMPLE 4 Simplify $\frac{2.007}{1.5}$ by dividing until the answer comes out even.

Solution

1. Decimal point in dividend?

   Yes.

   $1.5\overline{)2.007}$

2. Divisor a whole number?

   No. Move both of the decimal points one place to the right.

   $1.5.\overline{)2.0.07}$

3. Divide until the answer comes out even.

   Add several zeros and see what happens.

   $1.5.\overline{)2.0.0700}$

4. Divide.

   $$\begin{array}{r} 1.338 \\ 1.5.\overline{)2.0.0700} \end{array}$$

5. Round as requested.

   The answer comes out even, so

   $$\frac{2.007}{1.5} = 1.338$$

6. Check by multiplication.

   $1.338 \times 1.5 = 2.007$

EXAMPLE 5 Divide 6 by 7. Round the answer to the nearest thousandth if it does not come out even before that time.

Solution

1. Decimal point in dividend? No.

   $7\overline{)6.}$

2. Dividing by a whole number? Yes. Place a decimal point straight up into the quotient.

   $$\begin{array}{r} . \\ 7\overline{)6.} \end{array}$$

3. Rounding? Yes, to the nearest thousandth, so add four zeros—three for thousandths and the fourth to round.

   $$\begin{array}{r} . \\ 7\overline{)6.0000} \end{array}$$

4. Divide.

   $$\begin{array}{r} .8571 \\ 7\overline{)6.0000} \end{array}$$

5. Round to the nearest thousandth.

   $0.8571 \doteq 0.857$

6. Check by using multiplication.

   $0.857 \times 7 = 5.999 \doteq 6$

Here is a summary list of the steps that can help you do long division problems successfully. You don't need to write the six steps out each time, but you should follow those steps, in order, when you are doing the problems.

Six Steps for Dividing Decimals

1. Is there a decimal point in the dividend?
2. Is the divisor a whole number?
3. Does the problem call for rounding?
4. Now, divide.
5. Round the quotient if required.
6. Check by multiplication.

Try the Self-Check exercises on some scrap paper and see if you can follow the six steps demonstrated in Examples 1 through 5. Check it see if your answers are the same as those given. If not, review the examples and try again. In case you need to ask your instructor or a friend for help, be sure to show all your work so that together you can find your errors.

Self-Check

1. Divide 0.8 into 3.67 and round the answer to the nearest hundredth.
2. Find the quotient of 0.3744 divided by 2.4, and divide until your answer comes out even.
3. $36 \div 9.3$ Round to the nearest hundredth.

4. Simplify 4/26 (divide 4 by 26) and round to the nearest thousandth.
5. Divide 70.5 by 28.2 and round to the nearest thousandth if the answer does not come out even before that time.

Answers:

1. $4.587 \doteq 4.59$
2. 0.156
3. 3.87
4. $0.1538 \doteq 0.154$
5. 2.5

There is a division shortcut that parallels the multiplication shortcut given in Section 5.3. When a quantity is divided by a power of 10 ($10^1 = 10$, $10^2 = 100$, $10^3 = 1{,}000$ and so on), the easiest approach is to move the decimal point to the left. The number of places in the move is determined by the number of zeros in the divisor.

Division by Powers of Ten

To divide by

10; move the decimal point one place to the left.
100; move the decimal point two places to the left.
1,000; move the decimal point three places to the left.

And so on.

Consider Examples 6, 7, and 8.

EXAMPLE 6 Divide: 146.5 ÷ 100

Solution Move the decimal point two places to the left.

146.5 ÷ 100 = 1.46.5 = 1.465

EXAMPLE 7 Divide: 0.082 ÷ 100

Solution Move the decimal point two places to the left.

0.082 ÷ 100 = 0.00.082 = 0.00082

EXAMPLE 8 Divide: 963 ÷ 10

Solution Move the decimal point one place to the left.

963 ÷ 10 = 96.3. = 96.3

Here are two examples that use decimals in formulas.

EXAMPLE 9 Given the formula $m = e/c^2$, find m when e = 13.14 and c = 3.

Solution

$M = e/c^2$	Write the formula.
$M = 13.14/3^2$	Substitute the given values.
$M = 13.14/9$	Simplify the denominator.
$M = 1.46$	Perform the division.
$M = 1.5$	The answer is to be rounded to the nearest tenth.

EXAMPLE 10 Given the formula $\pi = c/d$

Solution

$\pi = c/d$	Write the formula.
$\pi = 12.566/4$	Substitute the given values.
$\pi = 3.1415$	
$\pi = 3.142$	The answer is rounded to the nearest thousandth.

5.4 EXERCISES

A. Divide in each of the following exercises. Follow the six steps, positioning the decimal point first. Divide until the answer comes out even. (Each will eventually have a remainder of zero.) You should check your answers by multiplication.

1. $6 \div .03$
2. $8 \div 0.04$
3. $128 \div 0.4$
4. $327 \div 0.3$
5. $1.5 \div 6$
6. $2.8 \div 5$
7. $2.45 \div 100$
8. $98.2 \div 10$
9. $2{,}000 \div 5$
10. $1{,}400 \div 8$
11. $1.6\overline{)32}$
12. $2.4\overline{)120}$
13. $10\overline{)2.65}$
14. $1000\overline{)3.275}$
15. $0.9\overline{)0.081}$
16. $0.7\overline{)0.0035}$
17. $2.5\overline{)1.6}$
18. $3.2\overline{)2.4}$
19. $0.15\overline{)90}$
20. $0.18\overline{)9}$
21. $\frac{3.6}{8}$
22. $\frac{4.2}{12}$
23. $\frac{18.34}{100}$
24. $\frac{2.762}{100}$

B. Divide in each of the following exercises. Round your answer to the nearest thousandth if the answer does not come out even before that place. You might check your answers by multiplication or by using your calculator, but remember that the checks may not be exact if the answers have been rounded.

25. $13 \div 0.6$
26. $22 \div 0.7$
27. $2.6 \div 3$
28. $1.8 \div 8$
29. $0.23 \div 0.7$
30. $0.34 \div 0.6$
31. $12 \div 2.3$
32. $27 \div 2.6$
33. $26 \div 1{,}000$
34. $1{,}452 \div 100$
35. $0.028 \div 4$
36. $0.059 \div 9$
37. $2 \div 3.5$
38. $5 \div 5.4$
39. $0.38 \div 1.4$
40. $0.45 \div 2.3$
41. $0.556 \div 2.6$
42. $0.489 \div 3.2$
43. $66 \div 2.3$
44. $57 \div 1.8$
45. $5.935 \div 100$
46. $0.267 \div 1{,}000$
47. $0.023 \div 35$
48. $0.036 \div 63$
49. $0.007 \div 10$
50. $0.093 \div 100$

C. Find the answers to the following exercises by using your calculator. Round each to the place indicated.

51. $827.6 \div 0.0192$; hundredths
52. $769.8 \div 0.925$; thousandths
53. $18.276 \div 12.85$; ten-thousandths
54. $34.598 \div 18.37$; ten-thousandths
55. $1{,}000.5 \div 27.38$; thousandths
56. $24{,}015.9 \div 387.95$; thousandths
57. $356.2 \div 36.895$; tenths
58. $185.4 \div 18.692$; hundredths
59. $283.5 \div 0.824$; tenths
60. $369.5 \div 0.377$; hundredths

D. Find the value of K in each of the following formulas if $a = 1.2$, $b = 0.5$, $c = 2$, and $d = 0.15$.

61. $K = abc$
62. $K = 2a - 3b$
63. $K = 3a + b - 2/c$
64. $K = a^2 - 2b + d$
65. $K = 3b + 5c - a$
66. $K = a/b$
67. $K = ab^2/c$
68. $K = ac - bd$
69. $K = a + b + c + d$
70. $K = ab/d$

E. Read and respond to each of the following exercises.

71. Explain in your own words the six steps to be followed in a decimal division problem.

72. Using your calculator, compute the answer to this exercise and round your answer to the nearest hundredth.

$1.56\overline{)26.732}$

Which value needs to be keyed into the calculator first? Why?

73. Explain in words how you could write $\frac{18.4}{0.23}$ in an equivalent fraction form using only whole numbers.

74. Explain in words how you could estimate the quotient of 12 divided by 0.47.

5.5 Decimals and Fractions

The ability to change from a decimal form to fraction form and from fraction form to decimal equivalent is a very useful skill when you are solving problems that involve both fractions and decimals. For example, to figure the cost of 2 1/2 pounds of tomatoes at $0.47 a pound, you need to multiply 2 1/2 by $0.47. You can change 2 1/2 to 2.5, following the process outlined below, and then multiply 2.5 by $0.47, which equals 1.175. Because you are talking about money, round your answer to the nearest cent. The cost is $1.18. For this problem, the important first step was to change 1/2 to its decimal equivalent as follows:

Changing Fractions to Decimals

To change a fraction to its decimal equivalent, divide the numerator by the denominator and follow the rounding instructions if any are given.

Change 1/2 to a decimal by dividing 1 by 2.

$$2\overline{)1.0}\quad = 0.5$$

Let's change a few more fractions to their equivalent decimal form.

Change 3/4 to a decimal.

$$4\overline{)3.00}\quad 0.75 = 0.75$$

Change 5/7 to a decimal, rounded to the nearest thousandth.

$$7\overline{)5.0000}\quad 0.7142 \doteq 0.714$$

Change 2 5/12 to a decimal, rounded to the nearest hundredth.

2 5/12 = 29/12 $\qquad 12\overline{)29.000}\quad 2.416 \doteq 2.42$

or you can leave the 2 alone and change 5/12 to 0.416. Then 2 5/12 = 2.416 $\doteq$ 2.42.

The example with which we began—find the cost of 2 1/2 pounds of tomatoes at $0.47 a pound—could also be solved using fractions.

$0.47 = 47/100

so 2 1/2 × 47/100

= 5/2 × 47/100

= 235/200

= $1 7/40

Although this answer is not very useful, it is also a correct solution. $0.47 was changed to its equivalent fraction form by the following steps:

Changing Decimals to Fractions

To change a decimal number to its fraction equivalent,
Read, Write, and Reduce.

0.08 is 8 hundredths, 8 over 100, 8/100 = 2/25

1.473 is 1 and four hundred seventy-three thousandths, 1 and 473 over 1,000 = 1 473/1,000

24.6 is twenty-four and six tenths, 24 and 6 over 10, 24 6/10 = 24 3/5

It is useful to memorize a few of the most common fraction and decimal equivalencies to save time when doing problems. The following short list may be very helpful.

1/2 = 0.5	1/10 = 0.1	1/8 = 0.125
1/4 = 0.25	1/100 = 0.01	3/8 = 0.375
3/4 = 0.75	1/1000 = 0.001	5/8 = 0.625
1/5 = 0.2		7/8 = 0.875
2/5 = 0.4		
3/5 = 0.6		
4/5 = 0.8		

Self-Check

1. Write 3/7 as a decimal rounded to the nearest hundredth.
2. Write 3 5/9 as a decimal rounded to the nearest tenth.
3. Write 0.56 as a fraction reduced to lowest terms.
4. Write 3.008 as a fraction or mixed number reduced to lowest terms.

Answers:

1. 0.43 2. 3.6 3. 14/25 4. 3 1/125 or 376/125

Look at the following examples involving decimals and fractions. Does it seem to matter which way the problems are done? Generally not, if the fraction is one that changes exactly to decimal form. The solutions could be found using the fractions or their decimal equivalents. With 1/2, 1/4, 1/10, and so on, it is usually easier to do the problem with decimals. Fractions such as 1/3, 5/6, and 7/9, however, do not change to exact decimals. Instead, 1/3 = 0.333 . . . , 5/6 = 0.833 . . . , and 7/9 = 0.777 . . . . (The three dots following a repeating digit let you know that that digit continues to repeat.) Such quantities are more accurate to work with in fraction form. *Remember:* When a problem requires a money answer, the answer must be changed to decimal form even if the problem was worked using fractions.

EXAMPLE 1 Add: 1/2 + 0.8

Solution

Fractions	**Decimals**
1/2 + 8/10	0.5 + 0.8
= 5/10 + 8/10	= 1.3
= 13/10 or 1 3/10	

EXAMPLE 2 Subtract: 6.4 – 2 3/5

Solution

Fractions	**Decimals**
6 4/10 – 2 6/10	6.4 – 2.6
= 5 14/10 – 2 6/10	= 3.8
= 3 8/10 = 3 4/5	

EXAMPLE 3 Multiply: 5/8 × $1.70

Solution This problem could be worked either way but since it involves money, decimals are preferable. Again, we will round a money answer to the nearest cent.

Fractions	**Decimals**
5/8 × 1 7/10	0.625 × 1.7
= 5/8 × 17/10	= 1.0625
= 85/80 = 17/16	= $1.06
= $1 1/16	

EXAMPLE 4 Multiply 7/9 × 1.1

Solution

Fractions	**Decimals**
7/9 × 1 1/10	Should not be done using decimals because
= 7/9 × 11/10	7/9 = 0.77777 . . .
= 77/90	

5.5 EXERCISES

A. Write the following decimals in their fraction or mixed number form. Be sure to reduce where possible.

1. 0.14
2. 0.86
3. 2.003
4. 9.017
5. 17.8
6. 23.96
7. 0.072
8. 0.825
9. 36.125
10. 11.008

B. Change each of the following fractions or mixed numbers to their decimal form. Round to the nearest thousandth were necessary.

11. 3/11
12. 7/12
13. 2 5/6
14. 7 5/9
15. 1/12
16. 4/7
17. 21/2
18. 23/8
19. 16 4/19
20. 14 8/13

C. Do the following calculations. Change to the same type of numbers, either fractions or decimals.

21. 3/4 × 2.9
22. 5/8 × 3.4
23. 1.6 + 2 1/2
24. 3/10 + 2.8
25. 19.4 ÷ 1/2
26. 13.5 ÷ 1/3
27. 7 1/4 – 3.8
28. 9 3/4 – 6.9
29. 4 3/8 × 1.2
30. 8 3/4 × 2.4

D. Use your calculator to change the following fractions and mixed numbers to their decimal equivalents. Round your answers to the places indicated.

31. 35/76; thousandths
32. 84/79; hundredths
33. 3 95/112; hundredths
34. 7 44/53; thousandths
35. 2356/8429; thousandths
36. 3568/4579; hundredths
37. 9 56/67; hundredths
38. 4 67/74; thousandths
39. 128/3655; ten-thousandths
40. 359/5678; ten-thousandths

E. Perform the following calculations using your calculator. Round division answers to the nearest ten-thousandth.

41. 19 5/8 + 39.79
42. 17 3/4 + 84.92
43. 38 1/4 × 73.6
44. 126 – 78 9/10
45. 68.5 ÷ 18 7/8
46. 85.4 × 29 5/8
47. 284 – 97 3/5
48. 14 3/5 ÷ 7.8
49. 89.4 × 26 3/4
50. 37 5/8 – 19 3/4

F. Read and respond to the following exercises.

51. Using your calculator, change 2/3 to a decimal. Write down the exact answer that the calculator gives in the display and explain it.
52. Using your calculator, complete the following decimal equivalents and answer the questions.

    a. 1/9 =

       2/9 =

       3/9 =

       4/9 =

       5/9 =

       6/9 =

       7/9 =

    b. What do you suppose 8/9 will look like? Why?

    c. What do you suppose 11/9 will look like? Why?

    d. What do you suppose 18/9 will look like? Why?

53. Helene needs 8 2/3 yards of fabric for a dress she is making. The fabric costs $4.68 a yard. When you estimate the cost of the fabric for the dress, what numbers would you use and why?
54. If you were to calculate the actual cost of the fabric in Exercise 53, would you use fractions or decimals? Why? Try the problem both ways and compare your results.

5.6 Applications of Decimals

Before looking at application problems, it will be helpful for you to review the list of key words and expressions given in Section 1.4. Also, remember that you can use the following ideas to help you decide which type of problem you are being asked to do.

> **Identifying the Operation**
>
> When putting quantities together, you might expect to add or multiply. When separating quantities, you are likely to subtract or divide.

Read each of the following examples carefully and then, before you read the explanation that is given, see if you can decide what you would do. Estimating an answer first is usually a good idea so that you can tell if you have chosen the correct operation before actually doing the calculations.

General Applications

EXAMPLE 1 You are taking your daughter's Brownie troop to the skating rink. There are 11 little girls going. The tickets are $2.25 apiece. How much will it cost?

Solution You could add $2.25 + $2.25 + $2.25 + . . . + $2.25 up 11 times, or you could simply multiply 11 × $2.25. The result would be $24.75 in either case. If you had estimated the amount first in your head, using $2+ (a little more than $2) per Brownie, you would have known that 11 Brownies would cost slightly more than $22.

◀

EXAMPLE 2 You have $24 to spend at the grocery store on the hamburger for a cookout. Hamburger is on special for $1.60 a pound. How many pounds can you buy?

Solution You could begin with the $24 and subtract $1.60 to get $22.40, and then subtract $1.60 again to get $20.80, and continue until you ran out of money. The simpler way, of course, is to divide $24 by $1.60.

$$1.60.\overline{)24.00.00} \quad = 15.00$$

So you find out that you can buy 15 pounds of hamburger for $24.

◀

EXAMPLE 3 Fabric for the new jacket you are making costs $6.98 a yard. The pattern calls for 4 3/4 yards of fabric. How much will the fabric cost?

Solution Estimating, you reason that if the fabric cost $7 a yard and you need 5 yards, you would multiply $7 by 5 and get $35, so your actual answer should be about $35. Now that you have decided which operation to use, you need to multiply $6.98 by 4 3/4 yards. You are looking for cost, so a decimal answer is appropriate.

1. Change 4 3/4 to 4.75.
2. Multiply: 6.98 × 4.75 = 33.155
3. Round to the nearest cent, $33.16.

The fabric for the jacket will cost $33.16. The answer is reasonably close to the estimated cost of $35.

◀

EXAMPLE 4 A travel agent booking a tour to Orlando, Florida, for the 17 members of the Seniors Club needs to collect $137.80 for airfare and $214.95 for hotels from each member. How much money must the agent collect?

Solution There are two different ways to do this problem. One method is to multiply $137.80 by 17 to get the total amount due for airfare and to multiply $214.95 by 17 to get the total amount due for hotels. Then you would add $2,342.60 and $3,654.15 to get the total amount to be collected, $5,996.75.

The second method is to figure each person's individual total by adding $137.80 and $214.95. Then multiply that sum, $352.75, by 17 to get $5,996.75, the total amount due. The answer should be the same using either method.

◀

Example 4 was a two-step problem. Many times there is more than one way to arrive at an answer, or more than one calculation that must be done. Read each problem carefully and plan your approach before you begin to figure.

Geometry Applications

Circles are different from the plane figures you studies in earlier chapters because they don't have corners, or vertices, and their sides are curved lines rather than straight ones. For this reason, the vocabulary used in studying circles is different from that used with polygons. In Figure 5.4, the various parts of a circle are labeled.

When measuring the distance around the outside of a circle, you are measuring its **circumference,** not its perimeter. The surface inside a circle is still its area and is measured in square units.

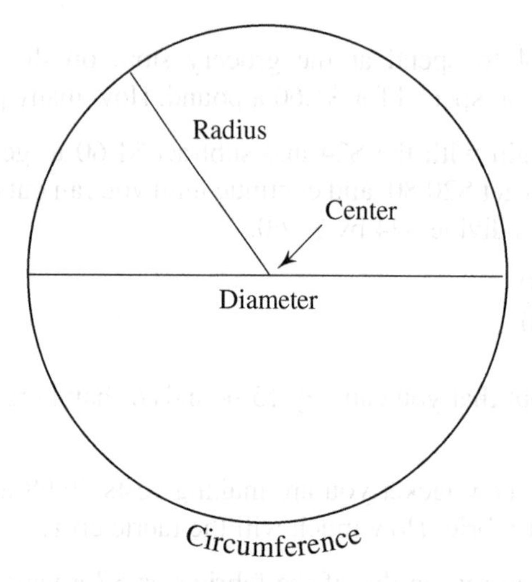

FIGURE 5.4

By definition, a **circle** is a set of points that lie the same distance from a point called its **center**. The distance from the center to a point on the circle is called its **radius**, and the length of a line segment from one side of the circle through its center to the other side is called its **diameter**. The radius, therefore, is half as long as the diameter.

Both the formula for circumference and the one for the area of a circle use a number denoted by the Greek letter π (read pi). Any time you measure the circumference of a circle and divide that quantity by the length of the diameter of the circle, you get the same result. The value is approximately equal to 3.14. Because this result, 3.14, is the same for all circles, it was given a special name, pi (π). Whenever you need to solve a formula using π, you can substitute 3.14, 3 1/7, or 22/7 for π; all are acceptable approximations. Remember, though, that your answers are approximations and not exact answers.

The formula for the circumference of a circle is $C = \pi d$, or $C = 2\pi r$, where d represents the diameter and r is the radius. (The formulas are equivalent because the radius goes from the center to the circle, and two radii lying in a straight line would form a diameter.) The formula for the area of a circle is $A = \pi r^2$. Sometimes you will be asked to substitute the decimal or fraction value for π, and in other problems you will be able to choose which to use.

Formulas for a Circle

The formula for the circumference of a circle is given by $C = 2\pi r$, or $C = \pi d$, where r is the radius, and d is the diameter, and $\pi \doteq 3.14$ or 22/7. (Recall that $\doteq$ means "is approximately equal to.")

The formula for the area of a circle is $A = \pi r^2$.

EXAMPLE 5 Find the circumference of a circle with a radius of 4 inches. Round your answer to the nearest tenth.

Solution First estimate an answer using 3 for π. $C = 2\pi r = 2(3)(4) = 24$ inches. Now find the actual answer.

$C = 2\pi r$

$C \doteq 2(3.14)(4)$

$C \doteq 6.28(4)$

$C \doteq 25.12$

The circumference is approximately 25.12 inches, which is close to the estimate.

EXAMPLE 6 Find the area of a circle with a radius of 3 inches. Round your answer to the nearest tenth.

Solution $A = \pi r^2$

$A \doteq (3.14)\ (3)^2$

$A \doteq (3.14)(9)$

$A \doteq 28.26$

The area is approximately 28.3 square inches.

EXAMPLE 7 Find the circumference of a circle that has a diameter of 5.6 feet. Give the answer in decimal form, rounded to the nearest tenth of a foot.

Solution $C = \pi d$

$C \doteq (3.14)(5.6)$

$C \doteq 17.584$

The circumference is approximately 17.6 feet.

EXAMPLE 8 Find the area of a circle that has a diameter of 7.5 feet. Round to the nearest tenth of a square foot.

Solution The formula calls for r, the radius, not the diameter that we are given. Since the radius is one-half of the diameter, we find 1/2 of 7.5 ($7.2 \div 2$) which is 3.75 feet.

$A = \pi r^2$

$A \doteq (3.14)(3.75)^2$

$A \doteq (3.14)(14.0625)$

$A \doteq 44.15625$

The area is approximately 44.2 square feet.

EXAMPLE 9 Find the area of triangle PQR if the base is 17.3 inches long and the height is 8.7 inches. Use $A = 1/2\ bh$.

Solution

$A = 1/2\ bh$	Write the formula for the area of a triangle.
$A = 1/2(17.3)(8.7)$	Substitute the values.
$A = 0.5(17.3)(8.7)$	Change 1/2 to 0.5.
$A = 75.255$	Perform the indicated multiplication.

The area is 75.255 square inches. (Remember that area is always measured in square units.)

EXAMPLE 10 Find the area of the composite shape in Figure 5.5. The height of the triangle is 1.2 inches and the rest of the dimensions are shown on the figure.

Solution A **composite shape** is a figure that is formed when two or more geometric figures are joined together. The composite shape in Figure 5.5 appears to be a rectangle with a triangle sitting on top of it. To find the area of the composite figure we need to calculate the area of each individual piece and then add the areas together.

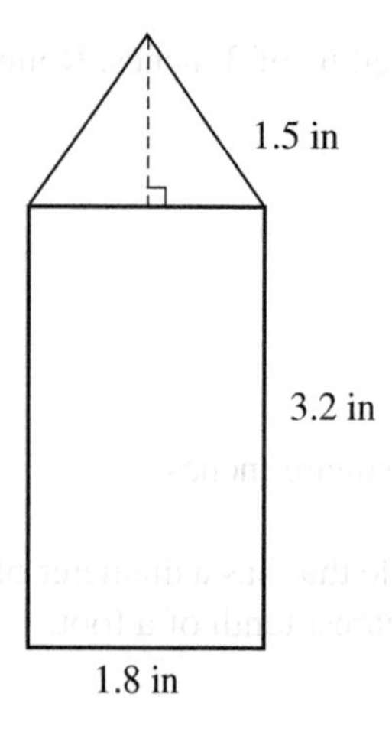

FIGURE 5.5

Area of the rectangle = length times width

$$A = LW$$

$$A = (3.2)(1.8)$$

$$A = 5.76 \text{ square inches}$$

Area of the triangle = 1/2 times base times height

$$A = 1/2\ bh$$

$$A = 1/2(1.8)(1.5)$$

$$A = (0.9)(1.5)$$

$$A = 1.35 \text{ square inches}$$

The area of the composite figure is the sum of the two areas:

5.76 + 1.35 = 7.11 square inches.

When finding the area of composite figures, you will sometimes need to add and sometimes to subtract. You might also be asked to find the perimeter of a composite figure. In that case, you need to be careful not to count a common side twice.

Now, let's consider some three-dimensional figures that have circular properties.

A **right circular cylinder** is shaped like a tin can. Its base and top are circles, and its sides are perpendicular to the bases (see Figure 5.6). You can think of the side of this figure as the label going around the can. What shape would the label be if you pulled it off the can? It would be a rectangle whose length is the distance around the can, the circumference of its top or bottom. The width of the label is equal to the height of the can. The total surface area (T) of a right circular cylinder is given by the formula $T = 2\pi r^2 + 2\pi rh$, where r is the length of a radius of a base, and h is the height of the cylinder. You are adding the area of the circular top and bottom, $\pi r^2 + \pi r^2$ or $2\pi r^2$, to the area of the rectangular label, $2\pi r$ times h. The volume of a circular cylinder is given by the formula $V = \pi r^2 h$, where r is the measure of the radius of a base and h is the measure of an altitude. Figure 5.6 shows the can and also the label and the top and bottom. The label of the can is a rectangle with height h and length the same as the circumference of the can, $2\pi r$. You can see by studying this figure how the formula for the surface area of a right circular cylinder was derived.

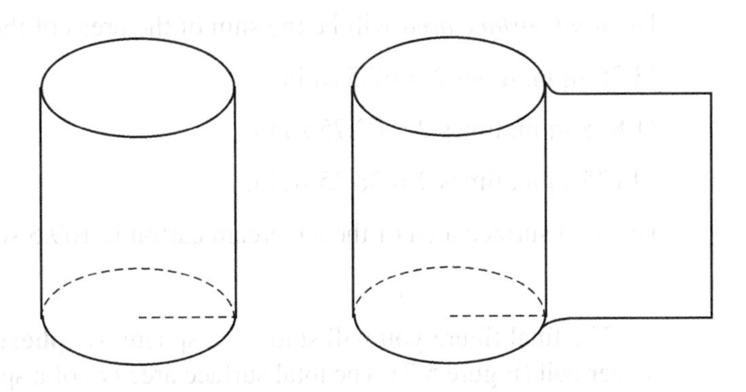

FIGURE 5.6

EXAMPLE 11 Find the surface area of a right circular cylinder with a radius of 3 inches and a height of 4.5 inches.

Solution The figure looks like a tin can. It has two circular ends, and its side can be thought of as a rectangular label wrapped around it. The formula for the surface area will be used.

Total surface area = area of 2 circular ends + area of rectangular side.

$T = 2\pi r^2 + 2\pi rh$

$T \doteq 2(3.14)(3)^2 + 2(3.14)(3)(4.5)$

$T \doteq 2(3.14)(9) + 2(3.14)(3)(4.5)$

$T \doteq 56.52 + 84.78$

$T \doteq 141.3$

The total surface area is approximately equal to 141.3 square inches.

◀

EXAMPLE 12 Find the total surface area of an ice cream carton with length of 7.5 inches, width of 4.5 inches, and height of 4.25 inches.

Solution To find the surface area we need to find the areas of the 6 surfaces—one top, one bottom, 2 sides and 2 ends. Each is a rectangle so we will use the formula $A = LW$.

The top and bottom are the same size and their dimensions are 7.5 inches by 4.5 inches.

$A = LW$

$A = (7.5)(4.5) = 33.75$ square inches *each.*

The two sides are the same and measure 7.5 inches by 4.25 inches

$A = LW$

$A = (7.5)(4.25) = 31.875$ square inches *each.*

The ends are the same and measure 4.25 inches by 4.5 inches.

$A = LW$

$A = (4.5)(4.25) = 19.125$ square inches *each.*

The *total surface area* will be the sum of the areas of the six rectangular sides.

33.75 sq in. times 2 = 67.5 sq in.

31.875 sq in. times 2 = 63.75 sq in.

19.125 sq in. times 2 = 38.25 sq in.

The total surface area of the ice cream carton is 169.5 square inches.

◀

The final figure you will study is a sphere. A **sphere** is shaped like a globe or a soccer ball (Figure 5.7). The total surface area (T) of a sphere with a radius of r units is given by the formula $T = 4\pi r^2$. The volume (V) of a sphere is given by the formula $V = 4/3\ \pi r^3$, where r is the measure of its radius.

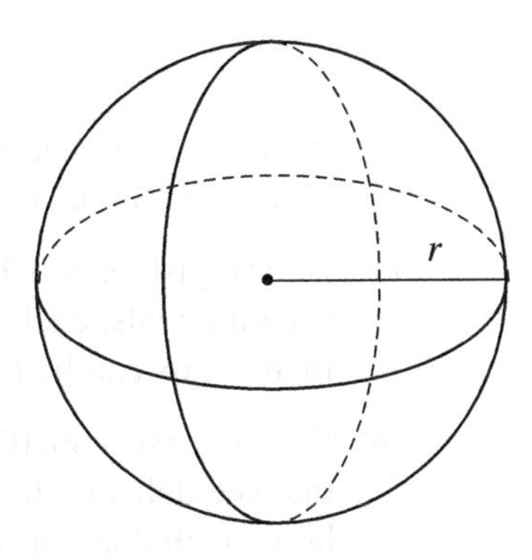

FIGURE 5.7

EXAMPLE 13 What is the volume of a basketball whose diameter is 12 inches? Use $\pi \doteq 3.14$.

Solution A basketball is a sphere, so you need to use the formula $V = 4/3\ \pi r^3$. Be sure to find the radius because that is what the formula calls for. If the diameter is 12 inches, the radius is 6 inches.

$V = 4/3\ \pi r^3$

$V \doteq 4/3(3.14)(6)^3$

$V \doteq 4/3(3.14)(216)$

$V \doteq 288(3.14)$

$V \doteq 904.32$

The volume of this basketball is approximately 904.32 cubic inches.

◀

Studying these three-dimensional figures should expand your ability to think in three dimensions. You need to be able to visualize what something looks like. Learn the correct names for these figures and keep a simple picture of each in your mind to help you work these problems. Chart 5.1 gives the formulas for the surface area and volume of each of these figures.

CHART 5.1

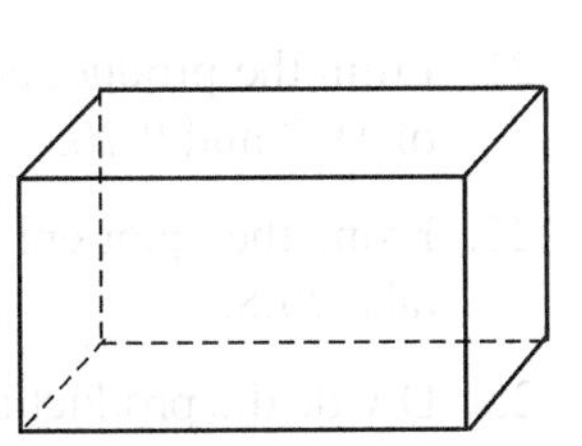

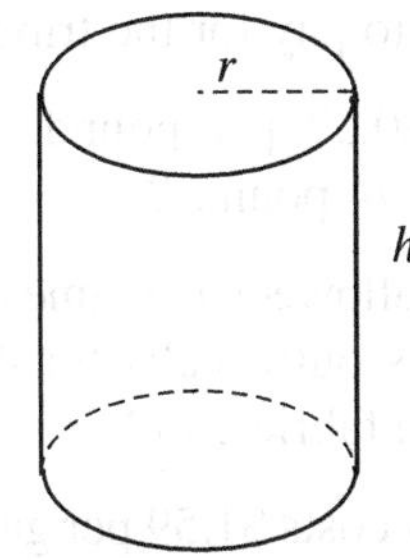

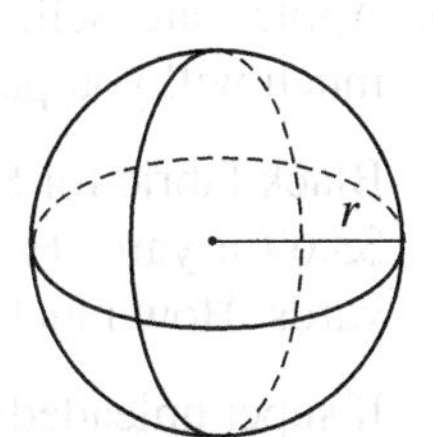

Figure:	Rectangular solid	Right circular cylinder	Sphere
Total surface area, *T*:	T = sum of areas of sides	$T = 2\pi r^2 + 2\pi rh$	$T = 4\pi r^2$
Volume, *V*:	$V = LWH$	$V = \pi r^2 h$	$V = 4/3\ \pi r^3$

5.6 EXERCISES

A. Estimate an answer to each of the following exercises. Remember that when you estimate, the numbers should be reasonable to work with so you can use mental math whenever possible.

1. Mark has $138.14 in his checking account. He writes checks for $8.32, $62.20, and $19.85. How much money is left in his account?
2. Four friends in a card club decide to go to dinner. The total bill, including the tip, is $46.60. How much is each person's share?
3. An auditor travels 218 miles. Her mileage payment for using her own car is 22 1/2 cents per mile. How much money will she receive for mileage on this trip?
4. Your son's birthday party guests are going bowling. How much will it cost for your son and his five guests if the cost of renting shoes and bowling three games is $3.75 for each child?
5. Find the circumference of a circle if its diameter is 54.01 centimeters.
6. Find the radius of the circle in Exercise 5.

B. Decide which mathematical operation to use, and then solve each of the following exercises. It will help to estimate your answers first. Be sure to label your answers.

7. If pork chops are selling for $2.89 per pound, how much will you pay for 3 1/2 pounds?
8. Bradbury Travel is planning a trip to Calgary for a group of 32 people. The airline tickets cost $289 per person, and each person's hotel bill will be $185.40. How much money will have to be collected from the group to pay for the trip?
9. Apples are selling for $0.37 per pound. How much will you pay for 3 3/4 pounds?
10. Black fabric for Sara's Halloween costume costs $3.69 a yard. Her witch's outfit calls for 4 2/3 yards. How much will the fabric cost?
11. If super unleaded gas now costs $1.59 per gallon, how many gallons can you buy with a $10 bill? (Round to the nearest tenth of a gallon.)
12. Candace worked 6.5 hours on Monday, 8 hours on Tuesday, 11 1/4 hours on Wednesday, 7 1/2 hours on Thursday, and 9.3 hours on Friday. How many hours did she work altogether?
13. A druggist puts 10.5 ounces of medicine into separate vials, each containing 0.75 ounce. How many vials can be filled?
14. Becca makes $0.40 per month for each paper that she delivers. If her route covers 103 houses, how much does she earn each month for delivering the papers?
15. How many pounds of ground beef can you buy with $8 if it is selling for $1.19 per pound? (Round to the nearest tenth of a pound.)
16. How much will your tuition be for next quarter if the current fees are $52.50 per credit hour and you are taking 11 hours?
17. A young college student's monthly utility bills were $18.26 for electricity, $12.53 for gas, $7.84 for water, and $31.18 for the telephone. Calculate his expected annual expenses for utilities.
18. Your car averages 22 miles per gallon. Calculate how much gas you will need for a trip to Washington, D.C., a distance of 407 miles.
19. Gold is currently selling for $415 an ounce on the trader's market. How much would Mr. Hutton pay for a gold bar that weights 3/4 ounce?
20. If the current balance in your checking account is $406.74, how much will be left in the account after you pay bills of $127.60, $34.55, $82.00 and $45.65?

C. You may use your calculator to help you find the answers to the following exercises. Be sure to label your answers where appropriate.

21. From the product of 82.4 and 0.65, take the sum of 18.7 and 9.26.
22. From the quotient of 533.875 divided by 6.5, take 79.8.
23. Divide the product of 583.9 and 218.6 by 32,506. Round your answer to the nearest thousandth.
24. Decrease the quotient of 546,823 divided by 879.42 by 495.9. Round your answer to the nearest thousandth.

25. Find the average of 458.9, 376.4, and 112.67.
26. Find the average of 872.4, 915.76, and 622.
27. The population of Columbus, Ohio, is approximately 30.4 times the population of Columbus, Nebraska. If the population of Columbus, Ohio, is about 566,100 people, what is the approximate population of Columbus, Nebraska?
28. The area of the state of New Mexico is approximately 121,660 square miles. The area of the country of Mexico is approximately 6.2 times as large. What is the approximate area of Mexico?

D. Answer each of the following questions. Use $\pi \doteq 3.14$, $d = 2r$, $C = \pi d$, $C = 2\pi r$, and $A = \pi r^2$. Round to the nearest hundredth.

29. What is the radius of a circle whose diameter is 9.3 feet?
30. What is the diameter of a circle whose radius is 16 inches?
31. Find the area of a circle whose radius is 10 meters.
32. Find the area of a circle whose radius is 4 feet.
33. Find the circumference of a circle whose diameter is 8 inches.
34. Find the circumference of a circle whose diameter is 12.2 meters.
35. Find the circumference of a circle whose radius is 3.5 feet.
36. Find the circumference of a circle whose radius is 10 feet.
37. Find the area of a circle whose diameter is 18 meters.
38. Find the area of a circle whose diameter is 12 inches.
39. Find the circumference of the semicircle (half-circle) in Figure 5.8. Use $C = 2\pi r$ and $\pi = 3.14$. (Remember that you only need half of the circumference.)
40. Find the dimensions of the rectangle in Figure 5.8.
41. Find the area of the semicircle in Figure 5.8. Use $A = \pi r^2$ and $\pi = 3.14$. (Remember that you only need to find half of the area of the circle.)
42. Find the area of the rectangle in Figure 5.8. Use $A = LW$.
43. Find the perimeter of the shape in Figure 5.8.

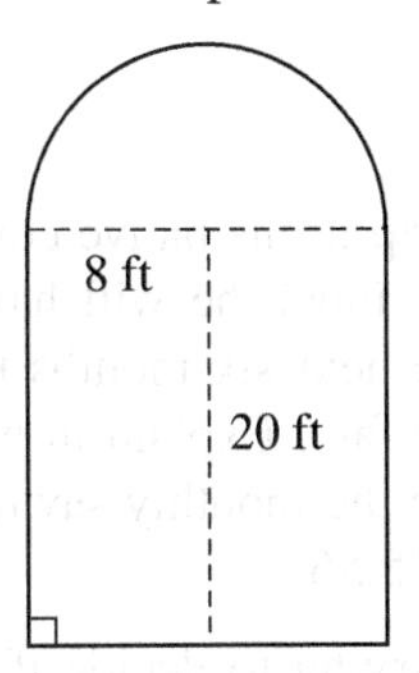

Figure 5.8

44. Find the area of the shape in Figure 5.8.
45. Find the area of the shape in Figure 5.9.

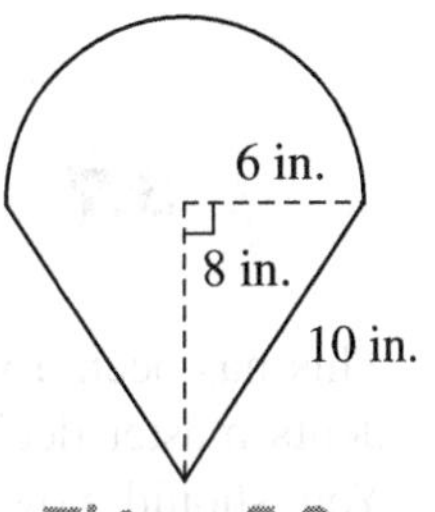

Figure 5.9

46. Find the perimeter of the shape in Figure 5.9.
47. Find the surface area of a juice can if its radius is 2 inches and it is 10 inches high.
48. Find the volume of the juice can in Exercise 47.
49. Find the volume of a basketball whose radius is 5 inches.
50. Find the surface area of a basketball whose radius is 6 inches.
51. Find the volume of a rectangular box if its length is 8 inches, its height is 5 inches, and its width is 3 inches.
52. Find the surface area of a juice can that is 10 inches high if its diameter is 8 inches.
53. Find the volume of the juice can in Exercise 52.
54. Find the surface area of the box described in Exercise 51.
55. The circumference of a dime is approximately 5.652 centimeters. Find its diameter.
56. The circumference of a quarter is approximately 7.85 centimeters. Find its area.

57. Find the area of the dime in Exercise 55.

58. Find the radius of the quarter in Exercise 56.

E. Read and respond to each of the following exercises.

59. Santiago is planning a trip to his native country and wants to know how much he will have to save each month for the next six months to be able to pay for his plane fare. Explain in words how he will approximate the monthly savings if his plane ticket costs $715.56.

60. Phillippe and Janis are trying to decide if they can afford to buy new living room furniture. The cost of the furniture is $1,126.95. Explain in words how they can approximate their monthly payments if they want to pay their bill off in full in 12 months.

61. Nancy is finishing a circular pillow by sewing a ruffle around the edge. Explain how she can calculate how much edging she will need to purchase if the pillow has a 22-inch diameter.

62. A gardener is ordering ivy plants to cover a circular planting area. He wants the plants to spread and fill in this circular space, which is 40 feet in diameter. Explain how you can help him to decide how many plants to buy if each plant should cover approximately 1 square foot.

63. If Nancy decides to make a second pillow with a 20-inch diameter, explain how she would calculate the amount of fabric needed to cover the front and back of the pillow.

64. Explain to the gardener in Exercise 62 how to figure out how much edging material he will need to buy to enclose his circular planting area.

5.7 CHAPTER REVIEW

This has been a very long chapter with many problems to be done. Normally, students master decimals readily because of their familiarity with prices and money. You should now be quite comfortable performing the standard operations and processes with decimals. In the following chapters you will be applying all these skills to whole numbers, fractions, and decimals.

When you have finished studying this chapter, you should be able to:

1. Discuss the following terms: place value, attaching zeros, rounding, sum, difference, product, quotient, dividend, divisor, powers of 10, fraction equivalent, and decimal equivalent.
2. Rank decimals using $>$ and $<$.
3. Calculate division results using these six steps:
   a. Decimal point in the dividend?
   b. Divisor a whole number?
   c. Rounding? Attach zeros as needed.
   d. Divide.
   e. Round the quotient as requested.
   f. Check by multiplication.
4. Change fractions to their decimal equivalents by dividing the denominator into the numerator and rounding as requested.
5. Change decimal values to their equivalent fraction values by using "Read, Write, and Reduce."
6. Simplify decimal expressions using the Order of Operations Agreement.
7. Estimate the solution to decimal application problems.
8. Solve application problems involving decimals, fractions, whole numbers, circles, cylinders, spheres, perimeter, area, surface area, and volume.

REVIEW EXERCISES

A. Write each of the following decimal numbers in words.

1. 24.063
2. 3.0058
3. 6.047
4. 35.98
5. 203.023
6. 607.204
7. 1,413.003
8. 3,548.058

B. Write each of the following verbal phrases as a decimal number.

9. four and sixteen thousandths
10. seventeen and five thousandths
11. four hundred nine and nine hundredths
12. six hundred two and twelve thousandths
13. four hundred eight ten-thousandths
14. seventy-seven ten-thousandths
15. twelve thousand and twelve thousandths
16. sixty thousand and sixteen thousandths

C. Round each of the following numbers to the place indicated.

17. 4.658 to the nearest tenth
18. 7.609 to the nearest hundredth
19. 45.675 to the nearest ten
20. 716.654 to the nearest hundred
21. 92.452 to the nearest tenth
22. 0.006 to the nearest hundredth
23. 0.1009 to the nearest thousandth
24. 37.863 to the nearest unit (one)

D. Arrange each of these groups of numbers from the smallest to the largest.

25. 0.61, 0.602, 0.609
26. 0.73, 0.729, 0.7
27. 2.2, 2.02, 2.002
28. 3.005, 3.5, 3.05
29. 0.801, 0.81, 0.08
30. 1.54, 1.405, 1.504

E. Arrange each of the following sets of numbers using $>$ or $<$ as requested.

31. 17.3, 17.03, 17.32, 17.2; $>$
32. 0.205, 0.255, 0.25, 0.22; $<$
33. 0.06, 0.006, 0.606, 0.066; $<$
34. 4.51, 4.503, 4.505, 4.05; $>$

F. Perform the indicated operations.

35. $45.2 + 31.567 + 14$
36. 26×9.42
37. $18.5 + 37 + 92.386$
38. 0.458×100
39. $145.8 - 87.982$
40. \$85 – \$29.81
41. $3,446.9 - 674.86$
42. 35×0.021
43. 24.8×76.5
44. 2.63×10
45. 18.95×2.8
46. $27.32 \times 1,000$
47. $18 + 92.7 + 24.78$
48. $1.49 + 38 + 16.5$
49. $14.9 + 27 + 33.86$
50. $16.8 \times 1,000$
51. \$84 – \$38.83
52. $140.08 - 87.596$
53. $91.5 + 37 + 86.43$
54. $205.4 - 97.65$

G. Divide in each of the following exercises, and round to the nearest hundredth if the answer does not come out even before that place.

55. $0.085 \div 1.2$
56. $0.096 \div 2.3$
57. $1.46 \div 0.7$
58. $3.45 \div 0.8$
59. $17 \div 0.19$
60. $23 \div 0.15$
61. $0.02 \div 10$
62. $4.56 \div 100$
63. $24 \div 0.3$
64. $35 \div 0.9$
65. $0.24 \div 7$
66. $0.68 \div 9$
67. $5.46 \div 100$
68. $8.72 \div 1,000$
69. $74.5 \div 15$
70. $86.3 \div 23$
71. $82 \div 17$
72. $56 \div 24$
73. $65.4 \div 1,000$
74. $0.159 \div 10$

H. Change each of the following decimals to fraction form. Be sure to reduce to lowest terms.

75. 0.28
76. 0.96
77. 22.45
78. 37.85
79. 0.008
80. 0.0045
81. 45.215
82. 93.148

I. Change each of the following fractions or mixed numbers to decimal forms. Round to the nearest thousandth, where necessary.

83. 5/8
84. 3/5
85. 7/9
86. 4/7
87. 3 5/6
88. 5 3/8
89. 4/1,000
90. 16/10,000

J. Solve each of the following application problems. Be sure to label your answers and round money answers to the nearest cent. You may want to estimate an answer first.

91. 12 1/2 pounds of birdseed costs $7.07. Find the cost per pound.

92. To train for the upcoming bicycle race, Mike rode 2.4 miles on Tuesday, 6.5 miles on Thursday, 9.8 miles on Saturday, and 5.9 miles on Sunday. How many miles did he ride that week?

93. Sara needs 11 yards of yarn for the garland she is making for holiday decorations. If the package that she bought has 7.2 yards of yarn in it, how many more yards does she need to buy?

94. Your car gets 19 miles to a gallon of gas. How many gallons of gas will you need to drive to Chicago, a distance of 376 miles? Round your answer to the nearest tenth of a gallon.

95. The paint mixture for a special color calls for 4 gallons of base to be mixed with 1.2 quarts of tint. If you have already put in 0.85 quart of tint, how much more must you add?

96. How many syringes of serum, each containing 2.3 cc (cubic centimeters), can be filled from a vial containing 19.55 cc of serum?

97. Mike's utility bills average $56.25 per month. How much must he budget per year to pay for his utilities?

98. Cinnamon is selling for $0.69 per ounce. How much will 2 1/4 ounces cost?

99. A pair of jeans that normally sells for $24.95 is on sale for $20.98. How much can you save by buying two pairs during this sale?

100. The new sheet metal product you are helping develop must be within 0.07 inch of the specifications required for a particular job. If the job calls for the metal to be 1.4 inches thick, what is the thinnest piece that can be used?

K. Complete each of the following exercises. Round answers to the nearest hundredth if necessary.

101. Find the area of a circle that has a diameter of 4.6 inches. (Use $\pi \doteq 3.14$.)

102. Find the volume of an oil drum with a diameter of 2 1/2 feet and a height of 3 1/2 feet.

103. Find the surface area to be covered when the entire oil drum in Exercise 102 needs to be painted.

104. Find the volume of air in a small kick ball if its diameter is 9.2 inches.

L. Read and respond to each of the following exercises.

105. Explain in words how to find the amount of leather needed to cover the kick ball in Exercise 104.

106. Explain in words how you would find the area of a label on a fruit juice can if the can is 9 inches high and has a radius of 1.4 inches.

107. Explain in words how you would calculate the number of pounds of grass seed needed to reseed a lawn that is 20 feet by 26 feet if 1 pound of grass seed will cover 25 square feet.

108. Explain in words how you could tell which was larger, a circular pizza with a diameter of 15 inches or a square pizza with a length of 14 inches.

M. Each solution that follows has at least one error in the steps performed. State what the errors are; then work the exercises correctly.

109. Find the sum of 1.207, 26.3, 18, and 14.37.

$$\begin{array}{r} 1.207 \\ 26.003 \\ 00.180 \\ +\,14.037 \\ \hline 41.427 \end{array}$$

110. Find the difference between 16.005 and 12.986.

$$\begin{array}{r} \scriptstyle 5\ 991 \\ 16.005 \\ -\,12.986 \\ \hline 13.019 \end{array}$$

111. Divide 143.261 by 100.

$143.261 \div 100 = 14{,}326.1$

112. Divide 17.36 by 8.

$$\begin{array}{r} 21.7 \\ 8\overline{)17.36} \\ 16 \\ \hline 1\,3 \\ -\,8 \\ \hline 5\,6 \\ -\,5\,6 \\ \hline 0 \end{array}$$

113. Find the quotient of 16 divided by 2.53. Round the answer to the nearest hundredth.

$$\begin{array}{r} 0.603 \\ 2.53\overline{)16\ 000} \\ -\,15\ 88 \\ \hline 820 \\ -\,759 \\ \hline 61 \end{array}$$

16 divided by 2.53 is approximately 0.60.

114. Find the volume of a cube whose depth is 8.3 inches.

$V = (8.3)(8.3)(8.3)$

$V = (68.89)(8.3)$

$V = 571.787$

The volume is 571.787 square inches

N. Answer each of the following questions. These exercises may require more than one step to complete.

115. Find the area of the shape in Figure 5.10.

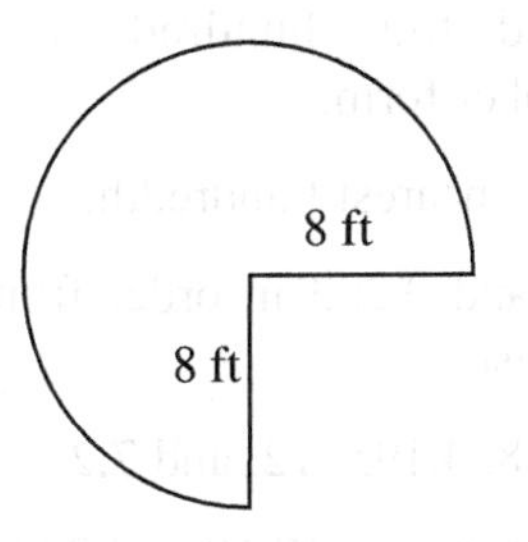

FIGURE 5.10

116. Find the area of the shaded portion in Figure 5.11.

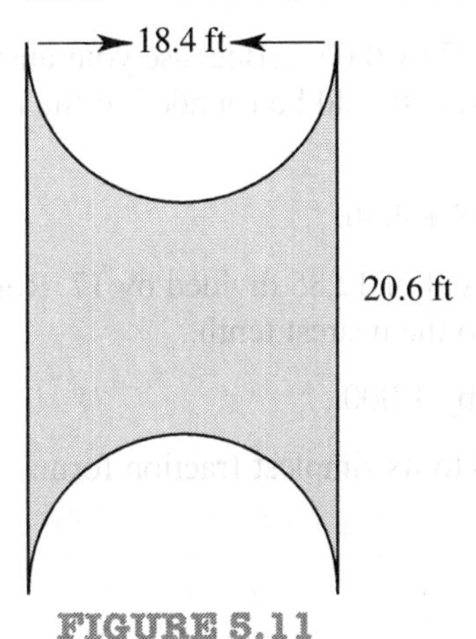

FIGURE 5.11

117. The Municipal Swimming Pool in Kokomo, Indiana, is circular, with the diving tower in the center. The distance from the center of the diving tower to the edge of the pool is 80 feet. The pool is surrounded by a 12 foot wide concrete apron. Calculate the area of this concrete apron. Draw a diagram to help you with this problem.

118. A circular spa has a depth of 3.5 feet and a diameter of 5 feet. Calculate the number of cubic feet of water the spa holds. If the weight of 1 cubic foot of water is 62 pounds, calculate the weight of the water in the spa.

119. If you check in an encyclopedia and find out that the circumference of the earth is approximately 6,400 kilometers, explain in words how you would estimate its diameter.

120. Explain in words how you would calculate the surface area of the earth given the information in Exercise 119.

CHAPTER TEST

1. Write 0.032 using words.
2. Write sixteen and four hundred six ten-thousandths in number form.
3. Round 27.885 to the nearest hundredth.
4. Rank 0.23, 0.023, and 0.203 in order from the smallest to the largest.
5. Find the sum of 0.38, 4.195, 12, and 7.2.
6. Find the difference between 17.68 and 9.895.
7. Subtract 14.834 from 52.
8. Find the product of 0.03 and 0.17.
9. Multiply $16.27 by 0.065. (Because your answer will be money, it should be rounded to the nearest cent.)
10. Simplify: 2 3/8 + 4.86.
11. Find the quotient of 12.85 divided by 17. Round your answer to the nearest tenth.
12. Divide 0.215 by 1,000.
13. Change 0.184 to its simplest fraction form.
14. Change 11/12 to its decimal form rounded to the nearest thousandth.
15. Find the area of a circle with a diameter of 6.4 inches. (Use $\pi \doteq 3.14$.)
16. Calculate the volume of a juice can with a radius of 1.5 inches and a height of 10 inches. (Use $\pi \doteq 3.14$.)
17. A large beach ball has a diameter of 24 inches. What is the volume of air in this beach ball? (Use $\pi \doteq 3.14$.)
18. How many gallons of gas can you buy with a $20 bill if the price per gallon is $1.26? Round your answer to the nearest tenth of a gallon.
19. What will a 12 3/4 pound turkey cost if turkey is selling for $0.65 a pound this week?
20. Explain in words how you would estimate the answer to this question: If the current balance in your checking account is $400.27, how much will remain in the account after you pay bills of $69.25 and $217.34?

STUDY SKILLS MODULE

6 Test Preparation

For centuries, college students have bragged about cramming for tests and pulling "all nighters." Actually, they should have been more embarrassed than proud of their need to learn all the material the night before the test. It meant that they had not studied and done homework after each class, nor had they read the textbook and reviewed their class notes before each class session. Their test grades probably reflected the "error of their ways."

The most successful students in any math course seldom have to do much studying the night before a test—not because they are "brains" and know-it-alls, but because they have been studying for the test since the first day the instructor began the material. They may even have begun before that, reading ahead so that the material was familiar when the instructor presented it to the class.

You can model the behavior of a successful student and become a better math student yourself. If you were not a particularly successful student, you may need to make major changes from the way you studied math when you were last in school. Remember that those practices were not very successful, so you definitely need to look for some new methods. Try to incorporate some of the following strategies as you begin the next chapter.

Test Preparation Strategies

1. Attend every class and take good notes. Review these notes and do the homework assignment after each class.
2. Highlight both in your notes and in the textbook any material that is unclear. Ask the instructor for help with this material during the next class.
3. Some students find it helpful to write vocabulary and example problems on index cards and to study these cards whenever they have a free moment.
4. Several days before the test, review your notes and homework, concentrating on weak areas. Try some problems out of each section, or work the example problems from the textbook and compare your procedure and solution to that in the book.

5. Get help from your instructor, a tutor, or a classmate if you still do not understand a concept.
6. List all the topics in the chapter and write down or recite all you know about each one. Spend extra time on the areas of which you are unsure.
7. The night before the test, you should only need to review the material and try a few practice problems. Then get a good night's sleep.

If you follow these suggestions, modifying them to match your needs and learning style, you should go into the test confident that you know the material and that you can do well on the test.

Your Point of View

1. List the advantages of studying the material throughout the chapter and the disadvantages of waiting until the last moment to read the material and do the homework.
2. Write one or two test preparation tips that were not listed.
3. Which three things from the list of strategies do you intend to do when you prepare for the next test?
4. How would your method of preparing for a math test differ from how you would prepare for a history test?
5. Look back at your results on the last test you took in this class. List the procedures you used to prepare for that test. Was the grade you received the best that you could achieve on that material? If not, what changes in test preparation could you make?

CHAPTER

6

Ratios and Proportions

LEARNING OBJECTIVES

When you have completed this chapter, you should be able to:

1. Write a ratio of two quantities in simplest terms.
2. Write ratios of measurement quantities in the same units.
3. Find rates and unit costs.
4. Determine if you have a proportion by using cross products.
5. Solve a proportion for the missing quantity.
6. Solve application problems that involve proportions.
7. Write and solve proportions for similar geometric figures.
8. Convert measurements using unit fractions.
9. Use ratios to compare statistical data.

Ratios and proportions have many applications outside the math classroom. You use them in science, business, nursing, and drafting classes, and they are very useful in your own kitchen or workshop. Proportions are used to adjust quantities in recipes, to find measurements in the course of enlarging or reducing a pattern, and to determine mileage between cities on a map.

Because of the practical nature of ratios and proportions, many of you are going to understand this material quickly and be very successful with it.

6.1 Ratios and Rates

Ratios

Ratios and rates are used to compare two quantities. If you have a bookcase that contains 23 westerns and 45 mysteries, you might express the comparison of westerns to mysteries as 23/45 or as 23 to 45 or as 23:45, but the fraction form is most commonly used. Each expresses the relationship between the number of westerns and the number of mysteries. This comparison is called a **ratio**.

When writing a ratio, be sure always to write the first quantity that is given as the numerator, or the first value in the statement. The second quantity given will be the denominator or the value after the symbol. If asked to write the ratio of the number of female students (5,400) to male students (4,800) in a college, you would write: 5,400/4,800. Is the relationship easier to understand as 5,400/4,800 or as 9/8? Anyone reading the information will have a better "feel" for the relative size of the female group compared to the male group if the fraction is in lowest terms. Simplify (reduce) all ratios to lowest terms. (It is easier to remember to do this when the fraction form is used.) Leave improper fractions as comparisons (common fraction form) rather than changing them to their mixed number form.

If the ratio of male to female students was requested, then 4,800/5,400, or 8/9, would be used. Because the total number of students is 5,400 + 4,800, or 10,200, the ratio of females to all students would be 5,400/10,200, or 9/17. The ratio of male students to the total student body would be 4,800/10,200, or 8/17.

Leaving fraction or decimal quantities in a ratio is unacceptable. By definition, a ratio is the comparison of *two whole numbers*. Fractions within a ratio are simplified by rewriting the expression as the division of two fractions and completing that operation.

The ratio 5/6 to 7/9 must be simplified so that both numerator and denominator are whole numbers. Rewrite the complex fraction as division of two fractions. (See Section 3.4.)

$$\frac{\frac{5}{6}}{\frac{7}{9}} = \frac{5}{6} \div \frac{7}{9}$$

$$= \frac{5}{6} \times \frac{9}{7}$$

$$= \frac{5 \times \overset{1}{\cancel{3}} \times 3}{2 \times \underset{1}{\cancel{3}} \times 7} \quad \text{or} \quad \frac{5}{\underset{2}{\cancel{6}}} \times \frac{\overset{3}{\cancel{9}}}{7}$$

$$= \frac{15}{14}$$

Leave the improper fraction as it is because a ratio compares two numbers.

Decimals within ratios are changed to whole numbers by multiplying by a value of 1 that is in the form of 10/10, 100/100, or the like. The fraction form of 1 to be used is determined by the maximum number of places that the decimal point needs to move to make all the decimals into whole numbers (see Section 5.3).

A ratio of 3.2 to 0.06 is simplified by multiplying by 100/100, because the maximum number of decimal places in either number is two, and multiplying by 100 will reposition the decimal point two places to the right.

$$\frac{3.2}{0.06} = \frac{3.2}{0.06} \times \frac{100}{100} = \frac{320}{6} = \frac{160}{3}$$

EXAMPLE 1 Write these ratios in fraction form. Remember to simplify all ratios to lowest terms. This means that in the simplified form, nothing but whole numbers (with no common factors) should remain.

a. 16:12

b. 3/4 to 5/8

c. 1.6 to 0.8

d. The Franklin Flying Disk factory can manufacture 12,600 disks each day. If 800 of those disks are defective, write the ratio of defective disks to the total number produced.

Solution

a. $\frac{16}{12} = \frac{4}{3}$

b. $\dfrac{\frac{3}{4}}{\frac{5}{8}}$

$= \frac{3}{4} \div \frac{5}{8}$ Rewrite the complex fraction as division.

$= \frac{3}{4} \times \frac{8}{5}$ Invert the second fraction and then multiply.

$= \frac{3}{\cancel{4}_1} \times \frac{\cancel{8}^2}{5}$ or $\frac{3 \times \cancel{2} \times \cancel{2} \times 2}{\cancel{2} \times \cancel{2} \times 5}$

$= \frac{6}{5}$ Because a ratio is the comparison of two quantities, leave the answer as an improper fraction.

c. $\frac{1.6}{0.8}$ Multiply both parts of the fraction by a value such as 10, 100, or 1000 that will make the decimal numbers whole numbers (Section 5.3).

$= \frac{1.6 \times 10}{0.8 \times 10}$

$= \frac{16}{8} = \frac{2}{1}$ Then reduce, but leave the answer as a fraction. Two would be an incorrect answer because a ratio must contain 2 whole numbers.

d. $\frac{800}{12{,}600}$ Write the number of defective disks over the total number of disks, and simplify to lowest terms.

$= \frac{4}{63}$

When writing the ratio of measurement numbers (length, volume, time, money, and so on), we must write the quantities in the same units. Otherwise the ratio is meaningless. In comparing 10 minutes to 1 hour, we cannot write 10 min/1 hr. Rather, we change 1 hour to 60 minutes.

$$\frac{10 \text{ min}}{1 \text{ hr}} = \frac{10 \text{ min}}{60 \text{ min}} = \frac{\overset{1}{\cancel{10 \text{ min}}}}{\underset{1}{\cancel{60 \text{ min}}}} = \frac{1}{6}$$

Note that units divide out just like numbers do.

Table 6.1 presents some common measurement conversion factors that you should already know. If you do not, now is a good time to memorize them. Ask your instructor whether you will be expected to know them for the test on this chapter.

TABLE 6.1 Units of Measure

Length	**Volume**
1 foot (ft) = 12 inches (in.)	1 pint (pt) = 2 cups (c)
1 yard (yd) = 3 feet (ft)	1 quart (qt) = 2 pints (pt)
1 mile (mi) = 5,280 feet (ft)	1 gallon (gal) = 4 quarts (qt)
Time	**Liquid volume**
1 minute (min) = 60 seconds (sec)	1 cup (c) = 8 fluid ounces (oz)
1 hour (hr) = 60 minutes (min)	1 pint (pt) = 16 fluid ounces (oz)
1 day = 24 hours (hr)	
1 week = 7 days	**Weight**
1 year (yr) = 52 weeks	1 pound (lb) = 16 ounces (oz)
1 year (yr) = 12 months	1 ton (T) = 2,000 pounds (lb)

EXAMPLE 2 Write the ratio of 4 yards to 8 feet.

Solution If 1 yard = 3 feet, then 4 yards = 4 × 3 or 12 feet.

$$\frac{4 \text{ yards}}{8 \text{ feet}} = \frac{12 \cancel{\text{feet}}}{8 \cancel{\text{feet}}} = \frac{3}{2}$$

EXAMPLE 3 Write the ratio of 4 nickels to 2 dollars.

Solution If 1 dollar = 20 nickels, then 2 dollars = 2 × 20 or 40 nickels.

$$\frac{4 \text{ nickels}}{2 \text{ dollars}} = \frac{4 \text{ nickels}}{40 \text{ nickels}} = \frac{1}{10}$$

If you had changed both measurements to cents, you would have had

4 nickels = 4 × 5 = 20 cents

2 dollars = 2 × 100 = 200 cents

$$\frac{4 \text{ nickels}}{2 \text{ dollars}} = \frac{20 \text{ cents}}{200 \text{ cents}} = \frac{1}{10}$$

Note that the ratios are the same whether you use nickels or cents.

Self-Check Simplify each comparison to a ratio in simplified form.

a. 27 to 15

b. $\frac{3}{4} \div \frac{5}{8}$

c. 7 nickels to 4 dimes

Answers: a. 9 to 5 b. 6:5 c. 7 to 8

Rates

A **rate** compares two different kinds of items. Comparisons such as miles per gallon, feet per second, and cost per ounce are common examples of rates. In miles per hour, the number of miles traveled in *one* hour is a **unit rate.**

EXAMPLE 4 On a recent vacation, Juan drove 350 miles in 6 hours.

a. Write his rate of travel in simplest form.

b. Find his average *unit* rate of travel in miles per hour.

Solution

a. $\frac{350 \text{ miles}}{6 \text{ hours}} = \frac{175 \text{ miles}}{3 \text{ hours}}$ = Rate of travel

b. To find the number of miles traveled in 1 hour (unit rate), divide the miles by the number of hours.

$$\frac{350 \text{ miles}}{6 \text{ hours}} \quad (350 \div 6)$$

$$= \frac{58\ 1/3 \text{ miles}}{1 \text{ hour}} = 58\ 1/3 \text{ or } 58.3 \text{ miles per hour (mph)}$$

EXAMPLE 5 On that 350-mile trip, Juan used 12 gallons of gasoline. How many miles per gallon (unit rate) did he average?

Solution To get the unit rate, the number of miles on 1 gallon, divide the miles by the gallons.

$$\frac{350 \text{ miles}}{12 \text{ gallons}} = \frac{29\ 1/6 \text{ miles}}{1 \text{ gallon}} = 29\ 1/6 \text{ or } 29.2 \text{ miles per gallon (mpg)}$$

Another common use of unit rates is in unit pricing. A **unit cost** is the cost of one item or one unit, such as the cost per ounce.

EXAMPLE 6 Oranges are priced at 6 for 99 cents. What is the cost of one orange (unit cost)?

Solution

$$\frac{99 \text{ cents}}{6 \text{ oranges}} \quad \text{Divide 99 by 6.}$$

$$= \frac{16.5 \text{ cents}}{1 \text{ orange}}$$

Unit price = 16.5 cents or $0.165 per orange

Notice that the second value given in the answer was expressed as a thousandth of a dollar. That is the form of the answer that we will request for unit cost in this textbook. Often the unit costs of items are so close that a difference is not seen in the hundredth of a dollar place.

EXAMPLE 7 A jar of peanut butter is selling for $1.29 for 18 ounces. The 24-ounce jar is priced at $1.89. Which is the better buy? Round these unit prices to the nearest thousandth of a dollar.

Solution The way to compare the prices is to look at the unit cost (cost per ounce) of each jar.

$$\frac{\$1.29}{18 \text{ oz}} = \frac{\$0.0716}{1 \text{ oz}} \text{ or } \$0.072 \text{ per ounce}$$

$$\frac{\$1.89}{24 \text{ oz}} = \frac{\$0.0787}{1 \text{ oz}} \text{ or } \$0.079 \text{ per ounce}$$

In this example, the smaller jar has the smaller unit cost so it is the better buy.

Self-Check Calculate each unit cost to the nearest thousandth of a dollar.

a. 16 ounces of jelly for $2.30 b. 6 toaster pastries for $1.87

Answers: a. $0.143 per ounce b. $0.312 per pastry

Definitions of Ratios and Rates

A *ratio* is the relationship of two whole numbers.

If a and b are whole numbers ($b \neq 0$), the ratio can be written a/b, $a{:}b$, or a to b.

A *rate* is the relationship of two whole numbers that represent different kinds of quantities.

A *unit rate* has a denominator of 1.

As you work with ratios and rates, remember to:

1. Write ratios and rates in the order given.
2. Simplify ratios to lowest terms containing two whole numbers.
3. Express unit rates with a denominator of 1.
4. Write ratios of measurement quantities with the same units of measure.

6.1 EXERCISES

A. Write these ratios in fraction form. Remember to simplify to lowest terms.

1. 3 to 9
2. 16 to 48
3. 5 : 9
4. 0.8 to 1.2
5. 3.2 to 0.24
6. 72 : 64
7. 6 : 3/4
8. 2/3 to 3/4
9. 1.2 : 4
10. 2/3 to 7
11. 1/2 : 5/8
12. 5 : 3.2
13. 1 5/9 to 6
14. 8.6 to 0.14
15. 11.6 : 5.04
16. 4/5 : 2
17. 9 to 3/8
18. 14 to 3 3/8

B. Write these ratios in lowest terms.

19. 10 feet to 80 inches
20. 3 weeks to 14 days
21. 15 seconds to 2 minutes
22. 11 nickels : 6 dimes
23. 3 dimes to 21 nickels
24. 14 inches to 3 feet
25. 6 pints to 5 quarts
26. 18 cups to 2 pints
27. 2 yards to 2 feet
28. 7 yards to 17 feet
29. 8 cents to 2 nickels
30. 1 hour to 19 minutes
31. 12 hours to 2 days
32. 3 pounds to 14 ounces
33. 2 pounds to 30 ounces
34. 5 gallons to 5 quarts
35. 3 cups to 2 pints
36. 400 pounds to 2 tons
37. 15 minutes to 3 hours
38. 3 quarts to 7 pints
39. 3.5 quarts to 14 pints
40. 5.5 feet to 9 inches

C. Answer each question with a ratio in lowest terms.

41. A vacation trip consisted of 640 miles traveled by airplane, 300 miles by ship, and 60 miles by car.

    a. What is the ratio of air travel to car travel?

    b. What is the ratio of ship travel to air travel?

    c. What is the ratio of car travel to air travel?

    d. What is the ratio of ship travel to the total trip?

    e. What is the ratio of air travel to the total trip?

42. A basket of fruit contains 3 bananas, 4 pears, 5 apples, and 10 plums.

    a. What is the ratio of pears to bananas?

    b. What is the ratio of apples to the total number of pieces of fruit?

    c. What is the ratio of apples to plums?

    d. What is the fraction relationship of bananas to plums?

    e. What is the ratio of total pieces of fruit to the number of pears?

43. The total enrollment of Metro Community College is 7,600. Of this number 4,200 are full-time students, and the rest are part-time. There are 150 full-time faculty members at the college.

    a. Write the ratio of full-time students to total students.

    b. What is the ratio of full-time to part-time students?

    c. What is the ratio of full-time faculty to full-time students?

    d. Write the ratio of full-time faculty to total students.

44. Yesterday the Easton Tool Company had a daily production total of 1,220 rakes and 1,400 shovels. Sixty of the rakes and 100 shovels were defective.

    a. What is the ratio of defective rakes to total rakes?

    b. What fraction of the shovels was defective?

    c. What is the ratio of the number of shovels to the total number of tools manufactured that day?

    d. What is the ratio of the total tools produced to the number of rakes?

D. For each question, write a ratio or rate in simplified form. If necessary, round answers to the nearest thousandth of a dollar.

45. A 16-ounce box of cereal sells for $2.40. What is the cost per ounce?

46. A recipe calls for 3 cups of flour and 1 1/2 cups of sugar. What is the ratio of sugar to flour?

47. A button manufacturer produces 24,000 buttons each day. If 600 each day are defective, what is the ratio of total buttons to defective ones?

48. A 22-ounce bottle of dish detergent costs $1.10. What is the cost per ounce?

49. A jogger ran 2 1/2 miles on Monday and 5 miles on Saturday. What is the ratio of Saturday's run to Monday's?

50. A grocery store receives 4,000 dozen eggs per week. If 200 dozen contain broken or cracked eggs, what fraction of the total number of dozens is damaged?

51. A motorist travels 240 miles on 10 gallons of gasoline. How many miles per gallon did he average?

52. An airplane can travel 750 miles in 2.5 hours. What is its average speed in miles per hour?

53. A race car can travel 165 miles in 1.5 hours. What is its average speed in miles per hour?

54. A speedboat can travel 50 miles on 4 gallons of gas. What is the boat's fuel consumption in miles per gallon?

55. An airplane can travel 560 miles on 16 gallons of fuel. What is its rate of fuel usage in miles per gallon?

56. A motorcyclist can travel 225 miles on 5 gallons of gas. Find the number of miles per gallon.

57. Socks are on sale this week at Kay's Mart. Six pairs cost $5.97. What is the price per pair?

58. Rory's Roast Beef has a special price on sandwiches. Four sandwiches cost $5.00. How much does each sandwich cost?

59. Ten bags of cypress mulch cost $25.00. What is the cost per bag?

60. Soup is priced at 6 cans for $4.00. What is the cost per can?

E. In Exercises 61 through 70, determine whether item A or item B is less expensive, and give its unit cost. If necessary, round answers to the nearest thousandth of a dollar.

	Item A	**Item B**
61.	6 for $0.98	15 for $2.49
62.	$1.49 for 12 oz	$0.98 for 8 oz
63.	$2.80 per dozen	$2.10 for 8
64.	8 for $6.00	3 for $2.25
65.	12 for $8.88	5 for $4.00
66.	$2.49 for 32 oz	$1.49 for 22 oz
67.	$0.72 per pint	$0.35 for 8 oz
68.	$1.56 per pint	$3.00 per quart
69.	10 oz for $1.79	16 oz for $2.49
70.	3 lb for $4.47	5 lb for $6.90

F. Round answers to the nearest hundredth.

71. A car traveled 213 miles on 13.2 gallons of gas. Find how many miles per gallon the car averaged. If gas costs $1.139 per gallon, how much was spent for gas for this trip?

72. How many gallons of gas were pumped into a car if gas costs $1.179 per gallon and the total charged was $13.69? Round the answer to the nearest hundredth.

73. A 10.75-ounce can of soup costs $0.89. What is the cost per ounce to the nearest thousandth of a dollar?

74. If 22.5 square yards of carpet cost $337.05, what is the cost per square yard?

G. Read and respond to the following exercises.

75. Describe in words the steps you would use to write the ratio of 16 inches to 6 yards in simplified form.

76. In the definition given in the box on page 231, why is ($b \neq 0$) included in the statement describing how to write a ratio?

77. If you know that a compact disk collection contains 14 disks that are jazz music and 11 that are rock music, describe in words the procedure you would use to write the ratio of jazz disks to the total number of these disks.

78. Describe in words how you can determine which is the better buy: a 23-ounce box of corn flakes for $2.69 or a 14-ounce box for $1.59.

6.2 Proportions

As we have said, **ratios** and **rates** compare two numbers and are usually written as fractions. A **proportion** is two equal ratios or rates.

$\frac{3}{12} = \frac{1}{4}$ and $\frac{6}{19} = \frac{24}{76}$ are proportions.

There are other ways of writing these proportions. One is

$3 : 12 :: 1 : 4$ and $6 : 19 :: 24 : 76$.

Another is to write "3 is to 12 as 1 is to 4" and "6 is to 19 as 24 is to 76."

Proportions

$\frac{a}{b} = \frac{c}{d}$ is a proportion because it expresses the equality of two ratios (where $b, d \neq 0$).

For all proportions, the cross products are equal.

$\frac{a}{b} \times \frac{c}{d}$ $\quad a \cdot d = b \cdot c$

When the $a : b :: c : d$ form is used, a and d are called the **extremes**, and b and c are the **means**. The cross-product rule is then stated as "The product of the means equals the product of the extremes."

You can test to see if a statement is a proportion by comparing cross products. This method is usually faster and easier than reducing both fractions to see if they simplify to identical fractions.

EXAMPLE 1 Is $\frac{3}{12} = \frac{1}{4}$ a proportion? Prove your answer.

Solution Cross-multiply $3 \times 4 = 12$ and $12 \times 1 = 12$.

$\frac{3}{12} \times \frac{1}{4}$

It is a proportion because 12 = 12. (Showing that the cross products are equal proves your answer.)

The process of computing cross products to show that a statement is a proportion is a convenient method, but it is not a true mathematical process. To show the equality of the two ratios, we should multiply both sides of the equation by a value that is a common denominator for the fractions. For the previous problem, a common denominator would be 48. Multiply both ratios by 48 and see if that produces a true statement.

$$48 \cdot \frac{3}{12} = \frac{1}{4} \cdot 48$$

$$\frac{\overset{4}{\cancel{48}}}{1} \cdot \frac{3}{\underset{1}{\cancel{12}}} = \frac{1}{\underset{1}{\cancel{4}}} \cdot \frac{\overset{12}{\cancel{48}}}{1} \quad \text{or} \quad \frac{\overset{1}{\cancel{2}} \cdot \overset{1}{\cancel{2}} \cdot 2 \cdot 2 \cdot \overset{1}{\cancel{3}} \cdot 3}{\underset{1}{\cancel{2}} \cdot \underset{1}{\cancel{2}} \cdot \underset{1}{\cancel{3}}} = \frac{1 \cdot \overset{1}{\cancel{2}} \cdot \overset{1}{\cancel{2}} \cdot 2 \cdot 2 \cdot 3}{\underset{1}{\cancel{2}} \cdot \underset{1}{\cancel{2}}}$$

$$12 = 12 \qquad \text{True.}$$

This is the same answer you got when using the shortcut of cross products. Either method can be used in the following problems, but we will do the cross-product method because it is more efficient.

EXAMPLE 2 Is $\frac{4}{64} = \frac{16}{192}$ a proportion? Why?

Solution Using cross products to find out yields

$4 \times 192 = 768 \quad 64 \times 16 = 1{,}024$

No, it is not a proportion because $768 \neq 1{,}024$.

EXAMPLE 3 Is $\frac{2\frac{1}{2}}{\frac{3}{4}} = \frac{15}{4\frac{1}{2}}$ a proportion? Why?

Solution Do not let the fractions change your approach. You still want to find out if the cross products are equal.

Multiply: $2\frac{1}{2} \times 4\frac{1}{2}$ and $\frac{3}{4} \times 15$

$= \frac{5}{2} \times \frac{9}{2} \qquad = \frac{3}{4} \times \frac{15}{1}$

$= \frac{45}{4} \qquad = \frac{45}{4}$

Yes, because $\frac{45}{4} = \frac{45}{4}$.

EXAMPLE 4 Is $\frac{7.2}{0.96} = \frac{35}{4}$ a proportion? Why?

Solution $7.2 \times 4 = 28.8 \qquad 0.96 \times 35 = 33.6$

No, because $28.8 \neq 33.6$.

Self-Check Are the following statements proportions? Why?

a. $\frac{10}{18} = \frac{4}{12}$

b. $\frac{4.32}{9} = \frac{1.2}{2.5}$

c. $\frac{4}{5\frac{3}{4}} = \frac{5}{8\frac{1}{2}}$

Answers: a. No, $72 \neq 120$ b. Yes, $10.8 = 10.8$ c. No, $34 \neq 28.75$

The property of equal cross products is also used to find missing parts of a proportion. If three of the four parts of a proportion are known, the fourth part can be found by computing the cross products and then dividing. The missing number is represented by a letter. Any letter can be used, and this letter is called a **variable.**

To solve for the value for N that makes

$$\frac{2}{3} = \frac{N}{9}$$

a proportion, find the cross products $2 \cdot 9$ and $3 \cdot N$. Set the products equal to each other.

$$3 \cdot N = 2 \cdot 9$$

Writing 3 and N side by side shows multiplication, so

$$3N = 18$$

To get N by itself, *undo the multiplication*, $3N$, by doing the *opposite* operation. *Divide* both sides by 3, then simplify.

$$\frac{3N}{3} = \frac{18}{3}$$

$1 \cdot N = 6$ or $N = 6$

Thus, 6 is the number missing from the original proportion.

Check Check your answer. Is $2/3 = 6/9$ a proportion?

$2 \times 9 = 18$ and $3 \times 6 = 18$

$18 = 18$ Yes.

EXAMPLE 5 Find the unknown value in $\frac{6}{p} = \frac{15}{8}$.

Solution $\frac{6}{p} = \frac{15}{8}$ Cross-multiply $15 \cdot p$ and $6 \cdot 8$. Set the products equal to each other.

$$15p = 48$$

$\frac{15p}{15} = \frac{48}{15}$ Divide both sides by 15 to isolate p.

$$p = \frac{16}{5} \text{ or } 3\frac{1}{5} \text{ or } 3.2$$

Check Is $\frac{6}{3.2} = \frac{15}{8}$ a proportion?

$6 \times 8 = 48$ and $3.2 \times 15 = 48$

$48 = 48.0$ Yes.

EXAMPLE 6 Find the missing number in $\frac{54}{85} = \frac{N}{26}$.

Solution $\frac{54}{85} = \frac{N}{26}$ Cross-multiply and set the products equal to each other.

$1404 = 85N$ Divide both sides by 85 to isolate N.

$$\frac{1404}{85} = \frac{85N}{85}$$

Rounding the answer to hundredths is a good choice if no rounding instructions are given.

$16.52 = N$

Check Does 54/85 = 16.52/26?

$54 \times 26 = 1404$ and $85 \times 16.52 = 1404.2$

Remember that 16.52 is a rounded value, not an exact number. Therefore, you can be pretty sure that you are right because 1404 = 1404.2.

◀

EXAMPLE 7 Find the value for C in $\frac{\frac{2}{3}}{\frac{1}{5}} = \frac{4}{C}$.

Solution $\frac{\frac{2}{3}}{\frac{1}{5}} = \frac{4}{C}$ Cross-multiply $2/3 \cdot C$ and $1/5 \cdot 4/1$.

$\frac{2}{3}C = \frac{4}{5}$ Set those products equal to each other.

$\frac{2}{3}C \div \frac{2}{3} = \frac{4}{5} \div \frac{2}{3}$ Divide by 2/3.

$$\frac{2}{3}C \cdot \frac{3}{2} = \frac{4}{5} \cdot \frac{3}{2}$$

$$C = \frac{6}{5}$$

Check Is $\dfrac{\frac{2}{3}}{\frac{1}{5}} = \dfrac{\frac{4}{6}}{5}$?

$\frac{2}{3} \times \frac{6}{5} = \frac{4}{5}$ and $\frac{1}{5} \times \frac{4}{1} = \frac{4}{5}$

$\frac{4}{5} = \frac{4}{5}$ Yes.

EXAMPLE 8 Solve $\frac{17.6}{24.8} = \frac{0.9}{B}$ for B. Round the answer to the nearest hundredth and check your answer.

Solution

$\frac{17.6}{24.8} = \frac{0.9}{B}$ Cross multiply 17.6(B) and 24.8(0.9)

$17.6B = 22.32$ Set the products equal to each other, then divide both sides by 17.6.

$\frac{17.6B}{17.6} = \frac{22.32}{17.6}$

$B \doteq 1.27$ (This answer is rounded, therefore the $\doteq$ symbol was used to indicate that the answer is approximately equal to 1.27.)

Check Is $\frac{17.6}{24.8} = \frac{0.9}{1.27}$ a proportion?

$17.6 \times 1.27 = 22.352$ and $24.8 \times 0.9 = 22.32$ Since one of the numbers was rounded, the products will not be equal, but will be very close to each other.

Self-Check Solve for the missing value in each proportion.

a. $\frac{A}{5} = \frac{18}{60}$ b. $\frac{1/2}{9} = \frac{C}{4\frac{1}{2}}$

Answers: a. $A = 1.5$ b. $C = \frac{1}{4}$

6.2 EXERCISES

A. Determine whether each of the following is a proportion. Prove your answer by computing the cross products.

1. $\frac{2}{3} = \frac{12}{18}$
2. $\frac{3}{4} = \frac{15}{20}$
3. $\frac{5}{9} = \frac{35}{45}$
4. $\frac{6}{7} = \frac{48}{56}$
5. $\frac{4}{49} = \frac{28}{343}$
6. $\frac{99}{55} = \frac{81}{45}$

7. $\frac{1.2}{3.6} = \frac{0.48}{1.44}$

8. $\frac{3.5}{0.49} = \frac{70}{9.8}$

9. $\frac{3.9}{5.85} = \frac{3}{4.5}$

10. $\frac{1/2}{3/4} = \frac{1/8}{3/16}$

11. $\frac{5/8}{15/2} = \frac{1/4}{3}$

12. $\frac{9/4}{8} = \frac{4/3}{2/3}$

B. Find the missing value in each proportion. If necessary, round your answers to the nearest hundredth.

13. $\frac{5}{9} = \frac{45}{N}$

14. $\frac{6}{W} = \frac{54}{36}$

15. $\frac{R}{7} = \frac{24}{56}$

16. $\frac{15}{16} = \frac{M}{80}$

17. $\frac{3}{8} = \frac{15}{A}$

18. $\frac{5}{6} = \frac{B}{36}$

19. $\frac{4}{c} = \frac{28}{35}$

20. $\frac{T}{9} = \frac{24}{36}$

21. $\frac{N}{3} = \frac{7}{5}$

22. $\frac{12}{9} = \frac{15}{F}$

23. $\frac{36}{5} = \frac{48}{R}$

24. $\frac{19}{6} = \frac{G}{4}$

25. $\frac{A}{19} = \frac{7}{8}$

26. $\frac{11}{M} = \frac{9}{14}$

27. $\frac{97}{B} = \frac{24}{47}$

28. $\frac{36}{49} = \frac{13}{W}$

29. $\frac{12}{21} = \frac{T}{28}$

30. $\frac{51}{81} = \frac{17}{M}$

31. $\frac{0.07}{40} = \frac{0.28}{A}$

32. $\frac{3\ 1/4}{2} = \frac{B}{2/3}$

33. $\frac{W}{1\ 1/8} = \frac{3\ 1/2}{1\ 1/2}$

34. $\frac{0.04}{0.006} = \frac{C}{5.4}$

35. $\frac{3/4}{5/8} = \frac{N}{1/2}$

36. $\frac{1.2}{P} = \frac{0.8}{3.9}$

37. $\frac{A}{0.05} = \frac{1.2}{0.006}$

38. $\frac{5\ 1/2}{B} = \frac{3\ 1/2}{4}$

39. $\frac{9.82}{4.1} = \frac{7.6}{R}$

40. $\frac{11.2}{M} = \frac{7.65}{18.8}$

C. Answer the following questions. You may use your calculator.

41. Is 7.59/5.22 = 25.3/17.4 a proportion? Why?

42. Is 0.0009/0.004 = 36,000/1,600,000 a proportion? Explain your answer.

Find the value of N in each of the following proportions. If necessary, round answers to the nearest hundredth.

43. $\frac{0.076}{1.34} = \frac{0.24}{N}$

44. $\frac{253}{891} = \frac{N}{567}$

45. $\frac{N}{965} = \frac{13{,}490}{2{,}340}$

46. $\frac{0.018}{N} = \frac{2.96}{6.45}$

D. Read and respond to the following exercises.

47. What does it mean that a proportion is true?

48. What are the "means" and the "extremes" in a proportion?

49. Explain in words how you would estimate the value for C in: C/12 = 19/62.

50. In the proportion: 4/7 = 26/*A*, estimate an answer for *A* and describe in words how you arrived at that estimated value.

6.3 Applications of Proportions

You will find proportions very useful in solving application problems that involve the relationship between two different types of quantities. Follow these steps to set up and solve this kind of application problem.

1. After reading the problem carefully, decide on a letter to represent the missing value.

2. Write a ratio or rate of two of the known values.

3. Set this ratio (rate) equal to one containing the missing value, represented by a variable. Be sure that the ratios (rates) are written in corresponding order. Writing a word fraction on each side might help keep the order the same.
4. Solve the proportion:
   a. Cross-multiply.
   b. Set the products equal to each other.
   c. Divide on both sides of the equals sign by the number that is multiplying the variable.
5. Label the answer with appropriate units or an identifying word.
6. It is a good idea to check your answer by testing to see if it makes sense in the problem.

General Applications

EXAMPLE 1 In a bag of "Melt in Your Mouth" candies, the ratio of yellow candies to red candies is 3 to 4. Find the number of yellow candies if there are 124 red ones.

Solution If Y is used to represent the number of yellow candies, then:

$$\frac{\text{yellow}}{\text{red}} \quad \frac{3}{4} = \frac{Y}{124} \quad \frac{\text{yellow}}{\text{red}}$$

If yellow/red is the first ratio, be sure that the order is the same on the other side. Now solve the proportion.

$4Y = 372$ Set cross products equal.

$\frac{4Y}{4} = \frac{372}{4}$ Divide both sides by 4.

$Y = 93$ yellow candies

Check Is 93 yellow candies reasonable? If the ratio of yellows to reds is 3 to 4, then there should be fewer yellows than reds. 93:124 is reasonable.

If you check your answer by substituting your solution in the proportion and finding the cross products, you are checking only the accuracy of your calculations. In application problems involving proportions, merely substituting will not catch errors of incorrectly written proportions. Estimating answers and testing to see if your answer is reasonable are more foolproof checks.

EXAMPLE 2 If 18 apples cost \$3.20, how much do 7 apples cost? Round your answer to the nearest cent.

Solution

$$\frac{\text{apples}}{\text{cost}} \quad \frac{18}{3.20} = \frac{7}{c} \quad \frac{\text{apples}}{\text{cost}}$$

In this proportion, the number of apples was written over the cost. This is just one of four correct proportions that could have been written. The others are

$$\frac{18}{7} = \frac{3.20}{c} \qquad \frac{7}{18} = \frac{c}{3.20} \qquad \frac{3.20}{18} = \frac{c}{7}$$

All would give the same cross products. Continuing with the solution, we find that

$$18c = 7 \times 3.20$$

$$18c = 22.4$$

$$\frac{18c}{18} = \frac{22.4}{18}$$

$$c = \$1.244 \quad \text{or} \quad \$1.24$$

Check Is $1.24 reasonable? Because 7 is less than half of 18, then the cost should be less than half of $3.20.

$\$3.20 \div 2 = \$1.60$ and $\$1.24 < \$1.60$

EXAMPLE 3 On a map, the scale says that a distance of 2 inches represents 15 miles. What is the actual distance between two cities if the map shows 4 1/2 inches between them?

Solution Let d = distance.

$$\frac{\text{inches}}{\text{miles}} \quad \frac{2}{15} = \frac{4\ 1/2}{d} \quad \frac{\text{inches}}{\text{miles}}$$

$$2d = 135/2$$

$$2d \div 2 = 135/2 \div 2$$

$$d = 135/4 \quad \text{or} \quad 33\ 3/4 \text{ miles}$$

Check Is 33.75 miles a reasonable answer? Because 4 1/2 is a little more than 2 times 2, the distance should be a little more than 2 times 15 miles.

$2(15) = 30$ and $33.75 > 30$

EXAMPLE 4 If a 4.6 pound roasting chicken costs $4.83, use a proportion to find the cost per pound of the chicken.

Solution You are looking for the cost of one pound of chicken. If the first ratio you write is $\frac{4.6}{4.83}$, then the other ratio must be $\frac{1}{C}$ (this is one pound of weight to the cost of one pound).

$$\frac{\textit{pounds}}{\text{cost}} \quad \frac{4.6}{4.83} = \frac{1}{C} \quad \frac{\textit{pounds}}{\text{cost}}$$

$$4.6\text{C} = 4.83$$

$$\frac{4.6C}{4.6} = \frac{4.83}{4.6}$$

$$\text{C} = 1.05$$

The cost of one pound is $1.05

Check Is $1.05 a reasonable answer? In the rate, 4.6 to 4.83, 4.6 is a little smaller than 4.83. In the rate, 1 to 1.05, 1 is a little smaller than 1.05.

EXAMPLE 5 Dosages for medication often are determined according to the body weight of the patient. If the label states that a 200-pound adult should be given 500 milligrams, how many milligrams should a 175-pound adult be given?

Solution Write a proportion and solve it.

$$\frac{\text{body weight}}{\text{dosage}} \quad \frac{200}{500} = \frac{175}{M} \quad \frac{\text{body weight}}{\text{dosage}}$$

$$200M = 87{,}500$$

$$\frac{200\,M}{200} = \frac{87{,}500}{200}$$

$$M = 437.5$$

The dose should be 437.5 milligrams.

Check We should expect an answer that is smaller than 500, but not a great deal smaller.

◀

Geometry Applications

Similar geometric figures are figures that have the same shape but are of different sizes. In similar figures, the corresponding angles have the same measure (are congruent), but the corresponding sides are not the same lengths—they are proportional. Two triangles are similar, $\sim$, if one is an enlargement or reduction of the other. Two triangles are congruent, $\cong$, if one is an exact duplicate of the other. In both similar and congruent triangles, the measures of the corresponding angles are the same; that is, the corresponding angles are congruent.

To better understand the ideas of similarity and congruence, think about your favorite photograph. Suppose that the photo is a 4 inches by 5 inches snapshot. A friend also really likes the picture, so you have another copy of the photo made in the 4×5 size. Your photo and your friend's are congruent—exact duplicates of each other. Later you decide that you want to enlarge the photo and put it in an 8×10 frame. The enlargement is similar to the original photo but is not congruent to it. The sides in the enlargement are twice as long as the sides in the original photo. But all the angles in the original photo, in your friend's copy, and in the enlargement, are right angles and are therefore congruent. Figure 6.1 shows how the copy and the enlargement are related to the original. Perhaps this explanation will help you to understand the two concepts and to remember what similarity and congruence mean.

Figure 6.2 shows two similar triangles. ΔABC is similar to ΔPQR; that is, $\Delta ABC \sim \Delta PQR$. We show the congruence of the corresponding angles by writing

$$\measuredangle A \cong \measuredangle P, \quad \measuredangle B \cong \measuredangle Q, \quad \text{and } \measuredangle C \cong \measuredangle R$$

You may want to identify the congruent angles with tick marks, as in Figure 6.3.

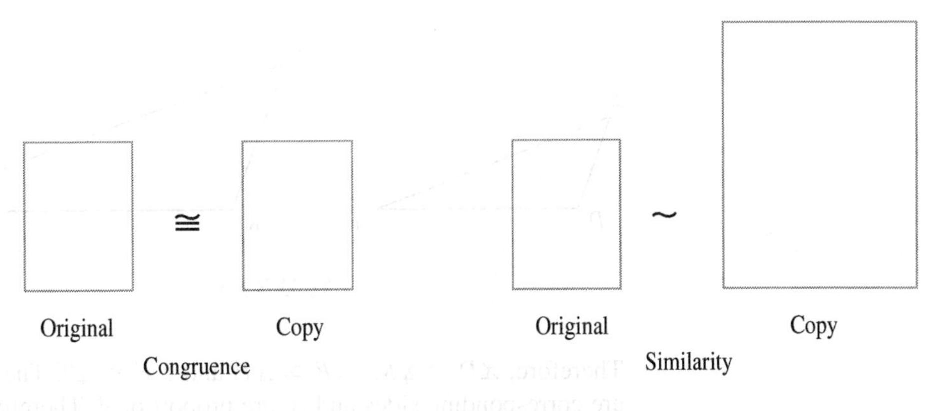

FIGURE 6.1

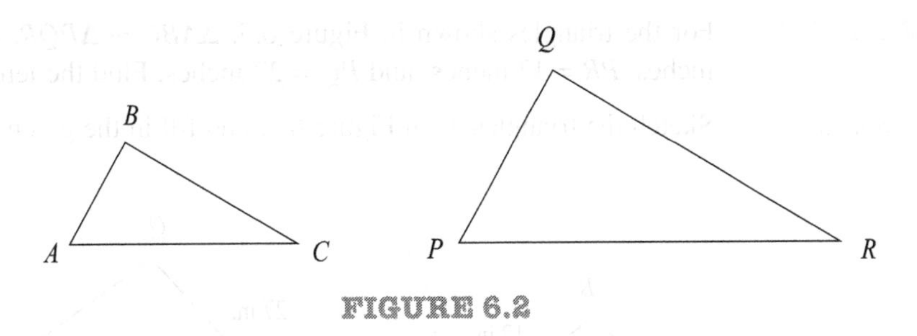

FIGURE 6.2

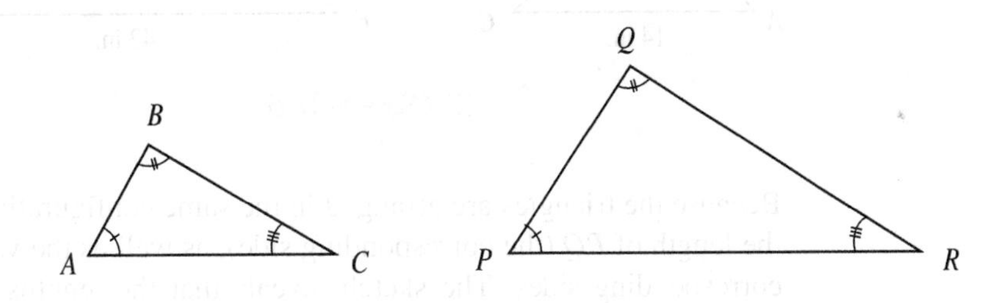

FIGURE 6.3

Because $\Delta ABC \sim \Delta PQR$, the corresponding sides are proportional. Therefore, the proportion showing the relationships between the lengths of the corresponding sides is written like this:

$$\frac{AB}{PQ} = \frac{BC}{QR} = \frac{AC}{PR}$$

EXAMPLE 6 Given that ΔDEF is similar to ΔKPT, identify the congruent angles and write the proportion relating the lengths of the sides.

Solution Making a sketch is always a good idea in solving a geometry problem. Although you do not know what the triangles actually look like, you do know how the corresponding angles need to be placed. Because $\Delta DEF \sim \Delta KPT$, angles D and K are in the same relative position, angles E and P are corresponding, and angles F and T need to match. The sketch might look like the one in Figure 6.4.

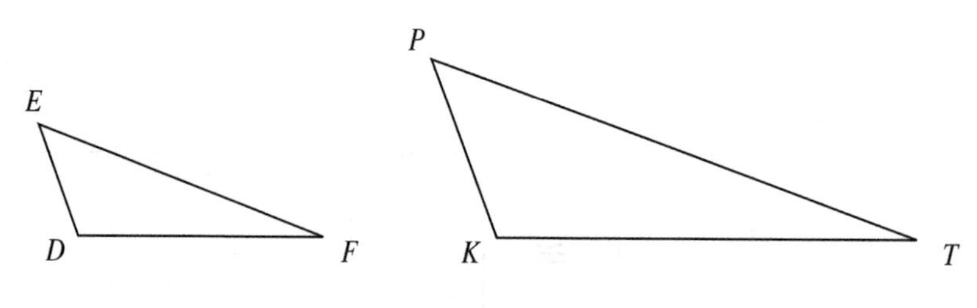

FIGURE 6.4

Therefore, $\measuredangle D \cong \measuredangle K$, $\measuredangle E \cong \measuredangle P$, and $\measuredangle F \cong \measuredangle T$. The sides opposite these angles are corresponding sides and so are proportional. Therefore,

$$\frac{EF}{PT} = \frac{DF}{KT} = \frac{DE}{KP}$$

EXAMPLE 7 For the triangles shown in Figure 6.5, $\Delta ABC \sim \Delta PQR$, $BC = 12$ inches, $AC = 14$ inches, $PR = 42$ inches, and $PQ = 27$ inches. Find the length of $\overline{AB}$.

Solution Sketch the triangles as in Figure 6.5, and fill in the given measurements.

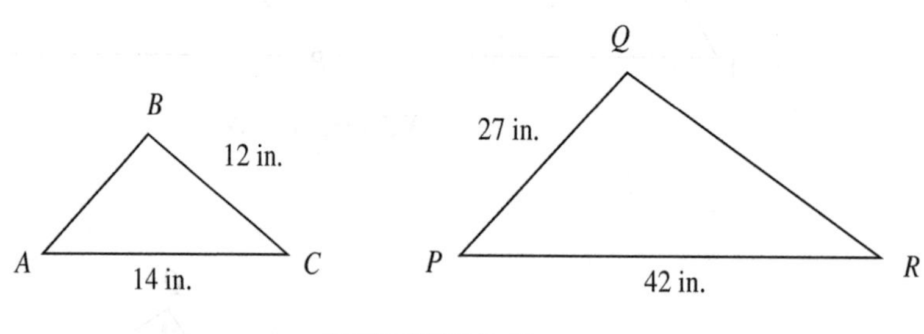

FIGURE 6.5

Because the triangles are arranged in the same configuration, you will need to know the length of PQ (the corresponding side), as well as the values for one other pair of corresponding sides. The sketch reveals that the lengths of sides $\overline{AC}$ and $\overline{PR}$ are given.

$$\frac{AB}{PQ} = \frac{AC}{PR} = \frac{BC}{QR}$$

Use the two parts of the proportion that relate these sides.

Fill in the known values: $AC = 14$, $PR = 42$, and $PQ = 27$.

$$\frac{AB}{27} = \frac{14}{42}$$

Solve the proportion. Compute the cross products.

$$42 \cdot AB = 27 \cdot 14$$
$$42\,AB = 378$$

Divide both sides by 42.

$$\frac{42 \cdot AB}{42} = \frac{378}{42}$$

$AB = 9$ inches

◀

EXAMPLE 8 Using the values given in Example 7, find the length of $\overline{QR}$ (see Figure 6.5).

Solution You will need the ratio of AC to PR and the ratio of BC to QR. Even though you now have a value for AB, it is safer to use the given information to do this second problem because you might have made a mistake in getting AB. Use the parts of the whole proportion that contain the unknown and the three given values.

$$\frac{BC}{QR} = \frac{AC}{PR}$$

Substitute the given values: $BC = 12$, $AC = 14$, and $PR = 42$.

$$\frac{12}{QR} = \frac{14}{42}$$

$14 \cdot QR = 504$

Solve by dividing both sides by 14.

$$\frac{14QR}{14} = \frac{504}{14}$$

$QR = 36$ inches

◀

EXAMPLE 9 ΔXYZ and ΔDCE, in Figure 6.6, are similar triangles in which $\measuredangle X \cong \measuredangle D$, $\measuredangle Y \cong \measuredangle C$, and $\measuredangle Z \cong \measuredangle E$. $YZ = 10$ feet, $XZ = 18$ feet, $CD = 6$ feet, and $CE = 4$ feet.

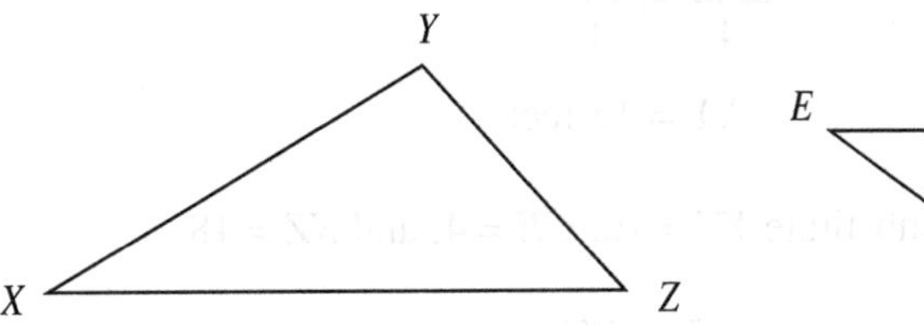

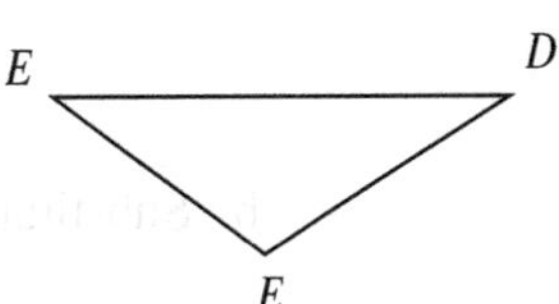

FIGURE 6.6

a. Find the length of $\overline{XY}$.

b. Find the length of $\overline{DE}$.

Solution Before labeling the given parts, rotate DDCE so that the equal angles are in the same position in both triangles.]C will need to be at the top,]D to the left, and]E to the right (Figure 6.7).

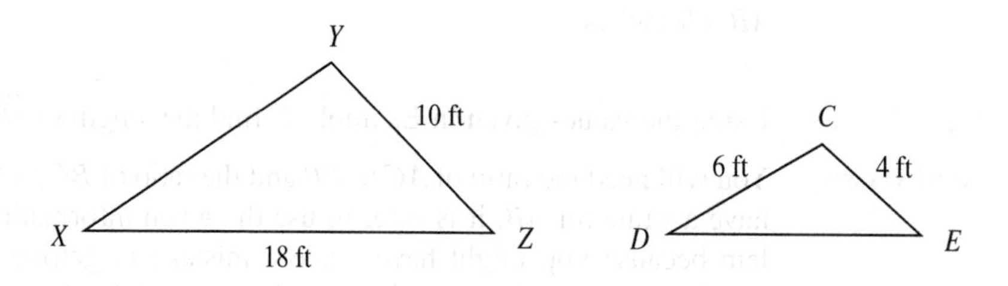

FIGURE 6.7

Now you can readily write the proportion for the corresponding sides.

$$\frac{XY}{DC} = \frac{YZ}{CE} = \frac{XZ}{DE}$$

a. *YZ* and *CE* are the pairs of corresponding sides that have both values given. Use that part of the proportion and the part with the side you are looking for. Then substitute $DC = 6$ feet, $YZ = 10$ feet, and $CE = 4$ feet.

$$\frac{XY}{DC} = \frac{YZ}{CE}$$

$$\frac{XY}{6} = \frac{10}{4}$$

$$4 \cdot XY = 60$$

$$\frac{\cancel{4}\,XY}{\cancel{4}} = \frac{60}{4}$$

$$XY = 15 \text{ feet}$$

b. Substitute $YZ = 10$, $CE = 4$, and $XZ = 18$.

$$\frac{YZ}{CE} = \frac{XZ}{DE}$$

$$\frac{10}{4} = \frac{18}{DE}$$

$$10 \cdot DE = 72$$

$$\frac{\cancel{10}\,DE}{\cancel{10}} = \frac{72}{10}$$

$$DE = 7.2 \text{ feet}$$

6.3 EXERCISES

A. Write a proportion for each application problem and solve it. If necessary, round answers to the nearest hundredth.

1. In a manufacturing plant, two VCRs are defective out of every 72 that are made. At this rate, how many would be defective out of 450 VCRs?
2. While evaluating its telephone system, a company finds that it receives 172 phone calls in 2 days. At that same rate, how many calls would it receive in 5 days?
3. If Carlos drove 159 miles in three hours, how far could he drive in five hours?
4. On a map, the scale says that 1 1/2 inches represent 12 miles. How many miles are there between two cities that are 5 1/4 inches apart on the map?
5. A doll house is built to scale so that 1 inch is equivalent to 1.5 feet. If a window in a real house measures 2.4 feet wide, how wide would it be in the doll house?
6. If a 6.4-pound roast costs $15.48, what is the cost per pound of that piece of meat? (Cost per pound means cost for 1 pound, or $C/1$.)
7. If a construction crew can frame a house in 2 1/4 days, how many houses can that crew frame in 6 days if each is framed at the same rate?
8. If a plane files 320 miles in 5 hours, how far can it fly in 3 hours?
9. How much per pound does chicken cost if 3 3/4 pounds cost $3.00?
10. If 5 out of every 250 parts manufactured by Camco Corporation are defective, how many are defective out of 1,200 parts?
11. If 3 pounds of onions cost 89 cents, how much would 2 pounds cost?
12. Dan received $42 interest on his savings account, which contained $700. How much interest would he receive on $1,750?
13. The hospitality committee plans to have five cans of soft drinks for every three guests attending the homecoming dance. How many cans will be needed if 150 guests attend?
14. A property tax rate is $1.09 for each $1,000 of property value. How much tax would be charged on a house valued at $58,500?
15. A concrete mixture calls for 1,000 pounds of gravel for every 6 bags of cement. How much gravel is needed for 33 bags of cement?
16. How long will an enlarged photograph be if its width will be 10 inches and the original picture is 2 1/2 inches wide and 3 1/4 inches long?
17. Monique can type 9 pages of text in two hours. How long will it take her to type a 54-page report?
18. If 2 pounds of ground beef cost $2.96, how much would 5 pounds cost?
19. Ishmial earned $15 in interest on the $120 he had in his savings account. How much interest would the account have earned if he had $600 in the account?
20. Don's cookbook says that 1 1/3 pounds of boneless roast should serve 4 people. How much meat will he need to serve 10 guests?
21. The nutritional information on the side of a box of cereal states that there are 170 milligrams of sodium in a 1.7-ounce serving. How many milligrams of sodium would be in a 1-ounce serving?
22. If 2 1/2 tablespoons of processed cheese spread contains 10 grams of fat, how many grams of fat are in 4 tablespoons of the spread?
23. To prevent accidents, safety regulations list the safe distance ratio for a ladder to be 4 to 1. This means that the top of the ladder should be placed no higher from the ground than four times the distance of the foot of the ladder from the wall. If the foot of a ladder is placed 3 1/2 feet from the wall, how high on the wall should the top of the ladder be?
24. A 36-pound bag of fertilizer and weed control mixture for lawns is supposed to cover 10,000 square feet. How many pounds of the mixture should be used on a lawn that is 3,500 square feet?
25. The dosage of a medicine is determined by the body weight of the patient. If 650 milligrams of an analgesic medication is recommended for a 200-pound adult, how much should be given to a 145-pound adult?

26. If 2 1/2 tablespoons of peanut butter contains 238 calories, how many calories from peanut butter are there in a recipe that calls for 1/2 cup (8 tablespoons) of peanut butter?

B. Complete the following sentences by filling in the blanks.

27. In ΔABC and ΔDEF are exact duplicates of each other. ΔABC is said to be ______ to ΔDEF.

28. If the sides of ΔXYZ are three times as long as the sides of ΔKLM, ΔXYZ is said to be ________ to ΔKLM.

29. The ratio of corresponding sides in two congruent triangles is ______.

30. Two equiangular triangles are always _____ and are sometimes _____.

31. In two similar triangles, the corresponding sides are _____.

32. In two similar triangles, the corresponding angles are _____.

33. Two equilateral triangles are always _____ and are also always _____.

34. In two congruent triangles, the corresponding sides are _____.

C. Solve for the missing parts of the pairs of similar triangles shown in Figures 6.10 through 6.15. If necessary, round answers to tenths.

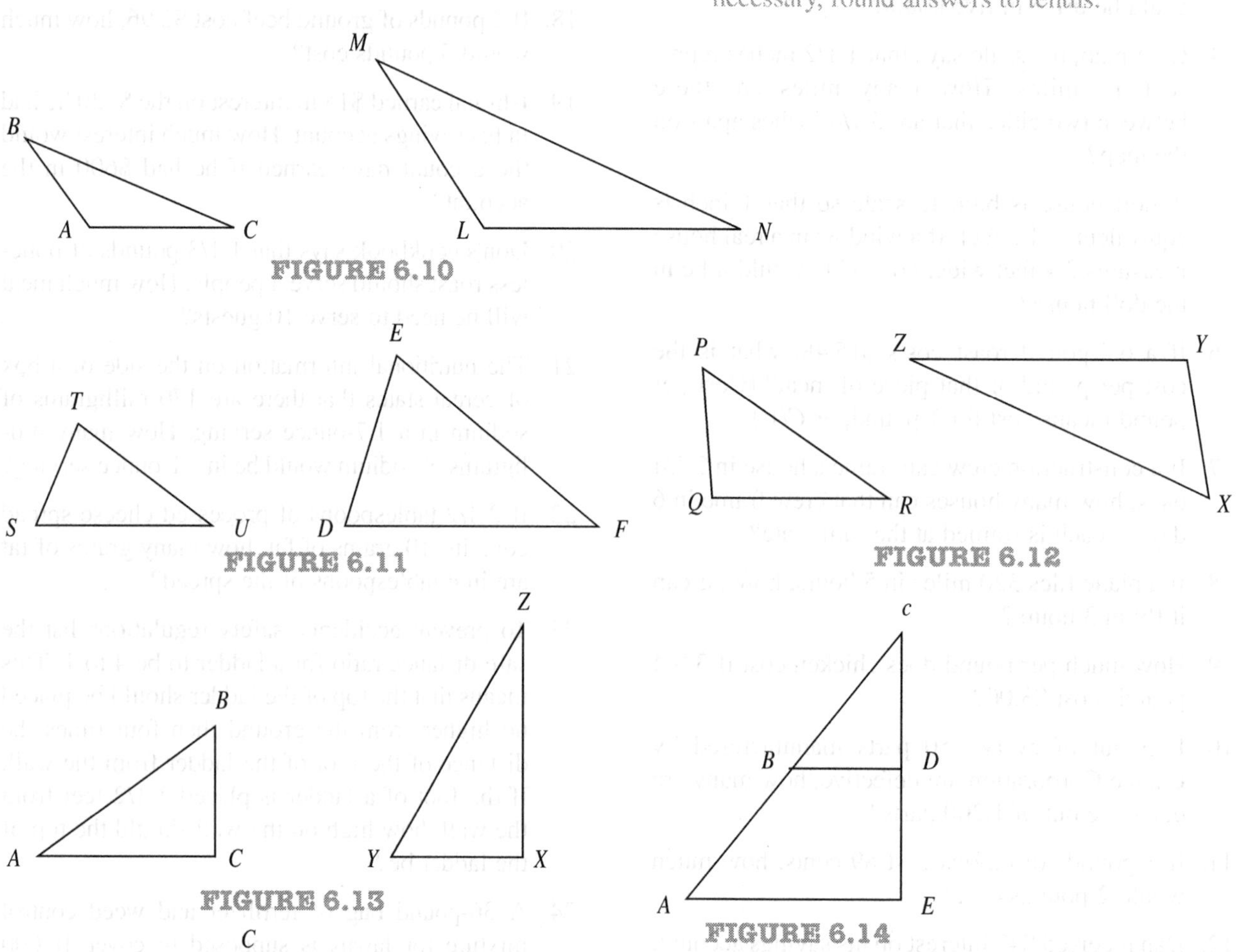

FIGURE 6.10

FIGURE 6.11

FIGURE 6.12

FIGURE 6.13

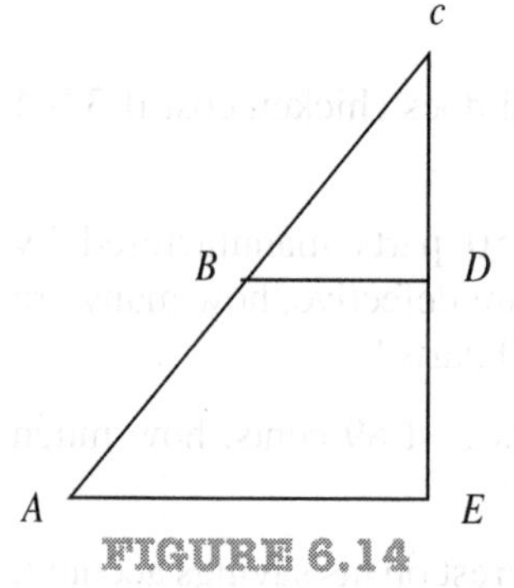

FIGURE 6.14

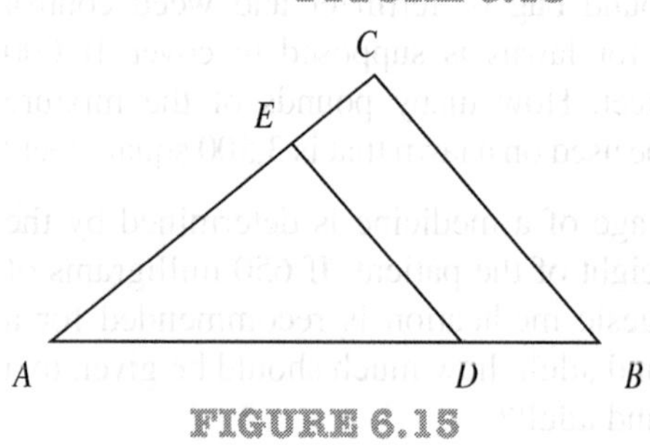

FIGURE 6.15

35. Use Figure 6.10, wherein ΔABC and ΔLMN are similar, $\measuredangle A \cong \measuredangle L$, $\measuredangle B \cong \measuredangle M$, $\measuredangle C \cong \measuredangle N$, $AB = 6$ yards, $AC = 8$ yards, $MN = 18$ yards, and $LN = 12$ yards. Find the length of $\overline{BC}$.

36. Use Figure 6.11, wherein ΔSTU and ΔDEF are similar, $\measuredangle S \cong \measuredangle D$, $\measuredangle T \cong \measuredangle E$, $\measuredangle U \cong \measuredangle F$, $ST = 6$ inches, $TU = 4$ inches, $SU = 8$ inches, and $DF = 12$ inches. Find DE.

37. Use Figure 6.10 and the information in Exercise 35 to find the length of $\overline{ML}$.

38. Use Figure 6.11 and the information in Exercise 36 to find the length of $\overline{EF}$.

39. In the similar triangles shown in Figure 6.12, $\measuredangle P \cong \measuredangle X$, $\measuredangle Q \cong \measuredangle Y$, $\measuredangle R \cong \measuredangle Z$, $PQ = 20$ meters, $QR = 40$ meters, $PR = 50$ meters, and $XZ = 45$ meters. Find the length of $\overline{XY}$.

40. Use this information and Figure 6.13 to find the length of $\overline{AC}$: ΔABC is similar to ΔXZY, $\measuredangle A \cong \measuredangle X$, $\measuredangle B \cong \measuredangle Z$, $\measuredangle C \cong \measuredangle Y$, $AB = 5$ feet, $BC = 3$ feet, $XY = 12$ feet, and $XZ = 15$ feet.

41. Use Figure 6.12 and the information in Exercise 39 to find the length of $\overline{YZ}$.

42. Use Figure 6.13 and the information in Exercise 40 to find the length of $\overline{YZ}$.

D. Read and respond to the following exercises by describing in words how you would find the answer to each problem.

43. Anja earned \$36.94 on the \$528 she had in her savings account. How much interest would the account have earned if \$1,350 was in the account?

44. Of every 100 widgets that are manufactured, nine are defective. How many defective widgets will be made during a 40-hour workweek if 1,750 widgets are produced per hour?

45. The amount of medication to be given is to be determined according to the body weight of the patient. If the label states that a 40-pound child should be given 50 milliliters, describe how you would find how many milliliters a 55-pound child should be given.

46. Proportions are often used to find the height of an object that cannot be measured directly. How would you find the height of a tree knowing that the shadow of the tree is 22 feet long and that a 6-foot-tall man standing near the tree casts a 4 1/2-foot shadow?

47. In Figure 6.14, ΔACE is similar to ΔBCD, $\measuredangle C \cong \measuredangle C$, $\measuredangle A \cong \measuredangle B$, $\measuredangle E \cong \measuredangle D$, $AB = 5$ yards, $BC = 5$ yards, $BD = 4$ yards, and $CD = 4$ yards. In words, describe how you would find the length of $\overline{CE}$.

48. ΔABC and ΔADE in Figure 6.15 are similar. Also, $\measuredangle A \cong \measuredangle A$, $\measuredangle E \cong \measuredangle C$, $\measuredangle B \cong \measuredangle D$, $AE = 12$ feet, $EC = 8$ feet, $ED = 6$ feet, and $AD = 15$ feet. Find the length of $\overline{BC}$.

6.4 Ratios and Conversion of Units

A very important application of ratios in scientific and technical courses is converting measurements to other units. The process involves writing conversion factors as fractions.

Because 1 foot = 12 inches, that fact could be written as the ratio 1 ft/12 in. or as the ratio 12 in./1ft. Both fractions have a value of 1 because the quantities are equal. For this reason, they are called **unit fractions**. Use this principle to change 6 feet to inches. Write the measurement in fraction form, and multiply it by the conversion factor, written as a unit fraction. The form of the fraction you should choose is the one that allows you to divide out the feet from the problem. This is easier than it sounds!

$$\frac{6 \text{ ft}}{1} \times ? = \text{inches}$$

Which conversion factor should we use, 1 ft/12 in. or 12 in./1 ft? We need to use 12 in./1 ft instead of 1 ft/12 in. so that the feet cancel.

$$\frac{6\ \text{ft}}{1} \times \frac{12\ \text{in.}}{1\ \text{ft}} = \frac{6 \times 12\ \text{in.}}{1} = 72\ \text{inches}$$

EXAMPLE 1 Change 30 miles per hour to miles per minute.

Solution The conversion factor needed is 1 hr = 60 min. Which unit fraction, 1 hr/60 min or 60 min/1 hr, will cause the hours to divide out and leave miles to minutes?

$$\frac{30\ \text{mi}}{1\ \text{hr}} \times ? = \frac{\text{mi}}{\text{min}}$$

$$\frac{30\ \text{mi}}{1\ \text{hr}} \times \frac{1\ \text{hr}}{60\ \text{min}} = \frac{30\ \text{miles}}{60\ \text{minutes}}$$

Now reduce the fraction or divide 30 by 60 to get a rate for 1 minute.

$$\frac{30\ \text{miles}}{60\ \text{minutes}} = 1/2 \text{ or } 0.5 \text{ mile per minute}$$

Look at the answer to see that only the desired units remain.

EXAMPLE 2 Change 45 miles per hour to miles per second.

Solution Two conversion factors will be needed: hours to minutes and minutes to seconds. Write the unit fractions so that all units divide out except miles in the numerator and seconds in the denominator.

$$\frac{45\ \text{mi}}{1\ \text{hr}} \times \frac{1\ \text{hr}}{60\ \text{min}} \times \frac{1\ \text{min}}{60\ \text{sec}} = \frac{45\ \text{mi}}{1\ \text{hr}} \times \frac{1\ \text{hr}}{60\ \text{min}} \times \frac{1\ \text{min}}{60\ \text{sec}} = \frac{45\ \text{mi}}{3600\ \text{sec}}$$

Divide 45 by 3600.

= 0.0125 mile per second

EXAMPLE 3 Using the conversion factors 1 cup = 8 ounces and 1 pint = 2 cups, change 6 pints to ounces.

Solution The conversion factors will take you from pints to cups to ounces. Note that you have to write 6 pints as 6 pints/1 so that it will be in fraction form when you multiply by the other unit fractions.

$$\frac{6\ \text{pt}}{1} \times \frac{2\ \text{c}}{1\ \text{pt}} \times \frac{8\ \text{oz}}{1\ \text{c}} = \frac{6\ \text{pt}}{1} \times \frac{2\ \text{c}}{1\ \text{pt}} \times \frac{8\ \text{oz}}{1\ \text{c}} = 96\ \text{ounces}$$

EXAMPLE 4 Using the conversion factors 1 inch = 2.54 centimeters and 1 meter = 100 centimeters, change 30 inches to meters. First write 30 inches in fraction form: 30 in./1.

Solution

$$\frac{30\ \text{in.}}{1} \times \frac{2.54\ \text{cm}}{1\ \text{in.}} \times \frac{1\ \text{m}}{100\ \text{cm}}$$

$$= \frac{30\ \text{in.}}{1} \times \frac{2.54\ \text{cm}}{1\ \text{in.}} \times \frac{1\ \text{m}}{100\ \text{cm}} = \frac{30 \times 2.54\ \text{m}}{100} = 0.762\ \text{meter}$$

Self-Check a. Change 130 centimeters to meters using 1 meter = 100 centimeters.

b. Change 18 quarts to cups using 1 quart = 2 pints and 1 pint = 2 cups.

Answers: a. 1.3 meters b. 72 cups

6.4 EXERCISES

A. Use unit fractions to convert the following measurements. You may use the chart of conversion factors, Table 6.1 in Section 6.1. If necessary, round answers to the nearest tenth.

1. Change 18 cups to quarts.
2. Change 68 inches to yards.
3. Change 3 days to minutes.
4. Change 33 yards to inches.
5. Change 8 hours to seconds.
6. Change 14 gallons to pints.
7. Change 40 miles per hour to miles per minute.
8. Change 20 miles per gallon to feet per gallon.
9. Change 100 miles per hour to feet per hour.
10. Change 50 miles per hour to miles per minute.

B. Change the following measurements using unit fractions. The conversion factors in Table 6.2 may be needed, as well as those in Table 6.1. If necessary, round answers to the nearest tenth.

TABLE 6.2 Conversion Factors for Length in the English and Metric Systems

Metric system

1 centimeter (cm) = 10 millimeters (mm)

1 decimeter (dm) = 10 centimeters (cm)

1 meter (m) = 100 centimeters (cm)

1 kilometer (km) = 1,000 meter (m)

English i metric conversion factors

1 inch (in.) = 2.54 centimeters (cm)

1 meter (m) = 39.37 inches (in.)

1 mile (mi) = 1.61 kilometers (km)

11. Change 3 meters to decimeters.
12. Change 0.4 kilometer to meters.
13. Change 150 millimeters to centimeters.
14. Change 44 decimeters to meters.
15. Change 28 inches to centimeters.
16. Change 3.2 miles to kilometers.
17. Change 4 miles per hour to kilometers per hour.
18. Change 14 yards to meters.
19. Change 65 centimeters to feet.
20. Change 101.6 centimeters to feet.

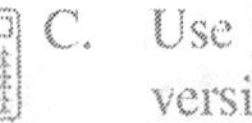

C. Use your calculator, unit fractions, and the conversion tables in this chapter to find the following measurements.

21. Change 0.165 meter to decimeters.
22. Change 421.5 kilometers to miles.
23. Change 42,000 millimeters to inches.
24. Change 5.6 meters to inches.
25. Change 67.2 yards to meters.
26. Change 49 miles per hour to kilometers per hour.
27. Change 32.9 feet per second to miles per hour.
28. Change 635 centimeters to feet.
29. Change 486 inches to centimeters.
30. Change 0.85 yard to meters.

D. Read and respond to the following exercises.

31. Explain, in words, how 12 inches/1 foot can be the same as 1 foot/12 inches when 2/3 is not the same as 3/2?

32. Marlo rented a golf cart for $36 an hour to tour Calabago Island. In three hours she had traveled 40 miles. Describe in words how you could use a "dividing out of units" setup to find the cost per mile of her excursion.

33. Describe in words the error made by a student trying to change 13 miles per gallon to meters per gallon.

$$\frac{13\text{ miles}}{1\text{ gallon}} \times \frac{1.61\text{ kilometers}}{1\text{ mile}} \times \frac{1\text{ kilometer}}{1{,}000\text{ meters}}$$

$= 0.02$ meter per gallon

6.5 Using Data

Information is often presented in chart or graph form. The ability to read these visual displays accurately is an important skill. You will be given practice in this section as you write ratios and proportions with data taken from a chart or graph.

EXAMPLE 1 Use the information in Figure 6.16 to answer the questions that follow.

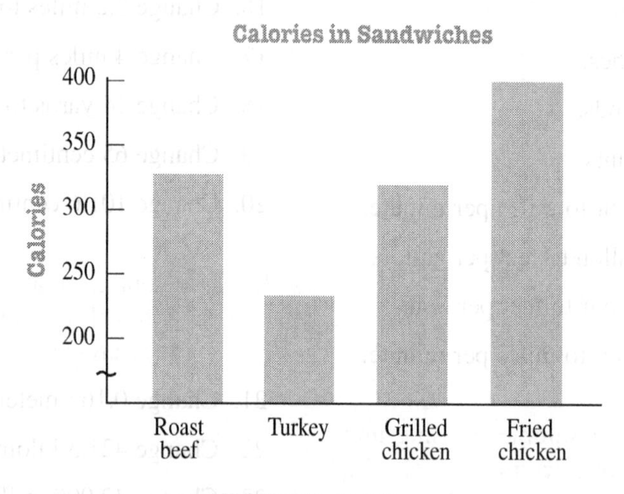

FIGURE 6.16

a. What is the ratio of the calories in the grilled chicken sandwich to the calories in the fried chicken sandwich?

b. What is the ratio of the calories in the fried chicken sandwich to the calories in the turkey sandwich?

c. What is the ratio of calories in the roast beef sandwich to the calories in the grilled chicken sandwich?

Solution a. Grilled chicken sandwich = 300 calories

Fried chicken sandwich = 400 calories

$$\frac{300\text{ calories}}{400\text{ calories}} = \frac{3}{4}$$

b. Fried chicken sandwich = 400 calories

Turkey sandwich = 250 calories

$$\frac{400\text{ calories}}{250\text{ calories}} = \frac{40}{25} = \frac{8}{5}$$

c. The calories for the roast beef sandwich must be estimated. The top of the bar appears to be halfway between 300 and 350, so 325 calories is a reasonable estimate.

$$\frac{325 \text{ calories}}{300 \text{ calories}} = \frac{13}{12}$$

EXAMPLE 2 The numbers of male and female students in five sections of Introductory Economics are displayed in Chart 6.1. Use that information to answer the questions that follow.

CHART 6.1 Number of Students by Gender

	Section 111	Section 112	Section 113	Section 001	Section 002
Males	13	15	7	20	17
Females	18	12	13	11	13

a. What is the ratio of males to females in section 112?

b. What is the ratio of the female students to the total number of students in section 001?

c. What is the mean (average) number of students in the five sections?

d. What is the median number of male students in the five sections?

e. What is the mode for the female students in the five sections?

Solution a. $\frac{15}{12} = \frac{5}{4}$

b. The total number of students in section 001 is 20 + 11 = 31. The ratio is 11/31.

c. For mean, median, and mode, review Section 1.5. To find the mean, add the total number of students in each section and divide by 5.

$$\text{Mean} = \frac{31 + 27 + 20 + 31 + 30}{5}$$

$$= \frac{139}{5} = 27.8 \text{ students per section}$$

d. The median is the middle number when the numbers are arranged in order of size: 20, 17, 15, 13, 7. The median number of male students is 15.

e. The mode for the female students is the most often repeated number of females.

Mode = 13 females

Self-Check Use Chart 6.2 to answer the questions that follow.

CHART 6.2 Cost of Shirts

Number of Shirts	1	2	4	6
Total Cost ($)	16.95	33.90	67.80	101.70

a. Use a proportion to find the cost of three shirts.

b. Use a proportion to find the cost of five shirts.

Answers: a. $50.85 b. $84.75

6.5 EXERCISES

All ratios should be written in simplest form. Answers should be rounded to the nearest hundredth, where necessary.

A. In Exercises 1 through 6, use Figure 6.17.

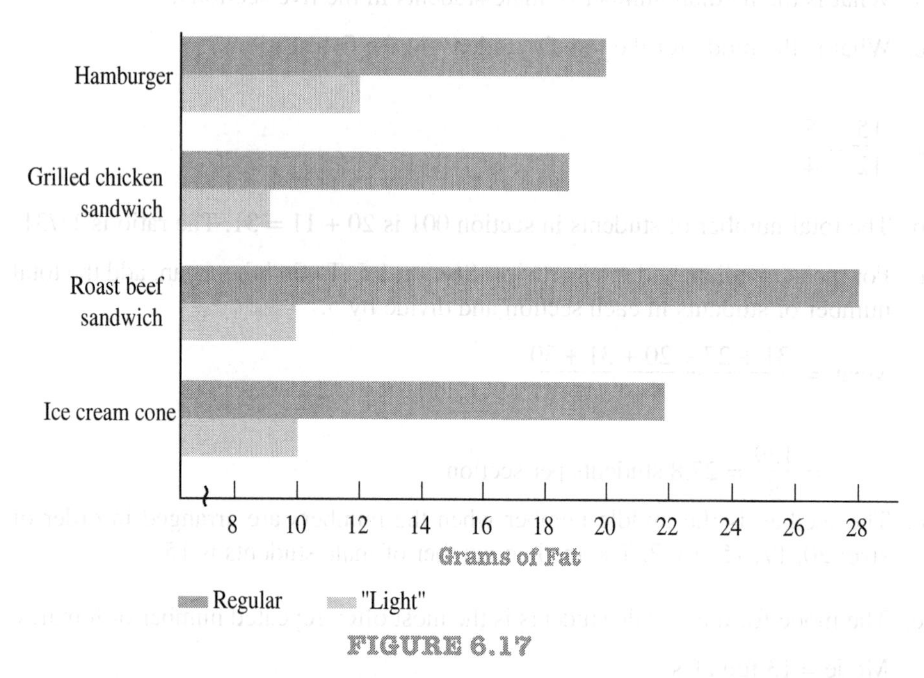

FIGURE 6.17

1. What is the ratio of grams of fat in "light" to grams of fat in regular hamburgers?

2. What is the ratio of grams of fat in regular to grams of fat in "light" ice cream cones?

3. What is the ratio of grams of fat in a regular hamburger to grams of fat in a regular roast beef sandwich?

4. What is the ratio of grams of fat in a "light" chicken sandwich to grams of fat in a regular roast beef sandwich?

5. What is the mean (average) number of grams of fat in the four regular food items displayed?

6. What is the mean (average) number of grams of fat in the four "light" food items displayed?

B. In Exercises 7 through 12, use Chart 6.3.

Chart 6.3 Cardinal Season Records

	1999	2000	2001	2002
Won	8	10	16	12
Lost	10	6	1	6
Tied	0	2	1	1

7. What was the ratio of wins to losses in 2000?
8. What was the ratio of losses to ties in 2000?
9. What was the ratio of losses to wins in 2001?
10. What was the ratio of wins to losses in 1999?
11. What was the mean number of wins for the four seasons?
12. What was the mean number of losses for the period?

C. In Exercises 13 through 18, use Figure 6.18.

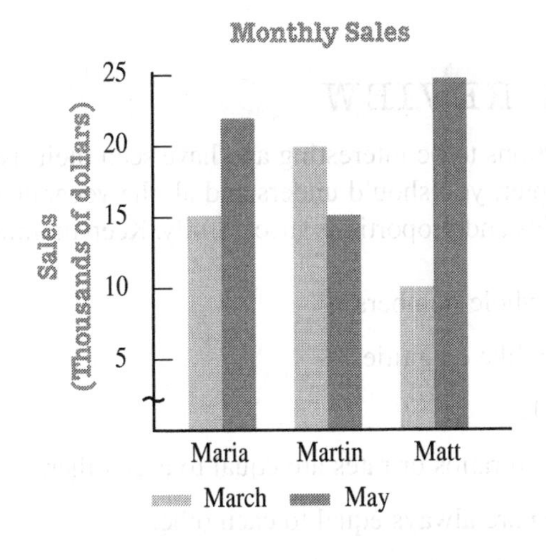

FIGURE 6.18

13. Write the ratio of Maria's March sales to her May sales.

14. What is the ratio of Martin's May sales to Matt's May sales?
15. What is the mean of the three sales associates' March sales?
16. What is the mean of the three sales associates' May sales?
17. If these are the only salespersons in the company, what were the total sales for the company for March?
18. If these are the only salespersons in the company, what were the total sales for the company for May?

D. In Exercises 19 through 24, use Figure 6.19.

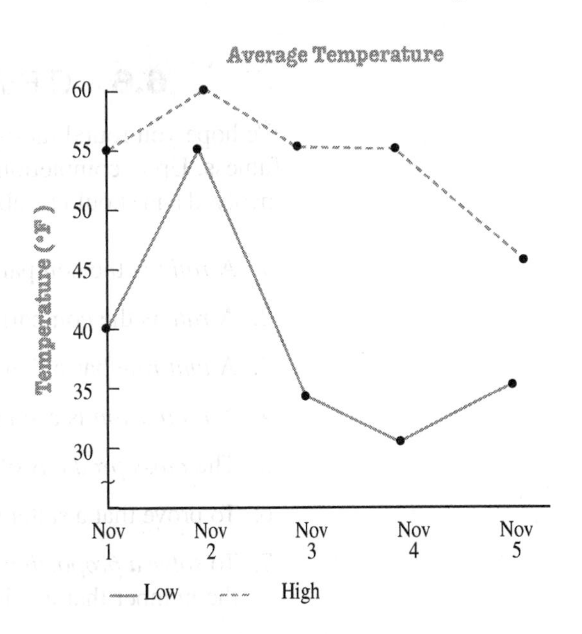

FIGURE 6.19

19. What is the ratio of the average low temperature to the average high temperature for November 4?

20. What is the ratio of the average high temperature to the average low temperature for November 5?

21. What is the ratio of the average high temperature on November 3 to the average high temperature on November 2?

22. Write the ratio of the average low temperature on November 1 to the average low temperature on November 4.

23. What is the mean average low temperature for the 5-day period?

24. What is the mean average high temperature for the 5-day period?

E. In Exercises 25 and 26, use Chart 6.4.

25. Write a proportion, and use it to find the number of miles traveled in three hours.

26. Write a proportion, and use it to find the number of miles traveled in seven hours.

CHART 6.4 Distance at 48 Miles per Hour

Hours	1	2	4	5	8
Miles	48	96	192	240	384

F. Read and respond, in words, to the following exercises.

27. Look at Figure 6.19 and describe, in words, how you would determine the greatest increase in the low temperatures (solid line) on two consecutive days.

28. Look at Figure 6.19 and describe, in words, how you would determine the greatest decrease in the high temperatures (dotted line) on two consecutive days.

29. Draw a line graph for the information displayed in Chart 6.4. Describe in words the graph.

30. Could the information displayed in Figure 6.18 be displayed by another type of graph? What kind would accurately display the information? Why?

6.6 CHAPTER REVIEW

We hope you found ratios and proportions to be interesting and have seen their usefulness. Upon completion of this chapter, you should understand all the vocabulary involved and should be able to use ratios and proportions successfully. Keep in mind:

1. A *ratio* is the comparison of two whole numbers.
2. A *rate* is the comparison of two unlike quantities.
3. A *unit rate* has a denominator of 1.
4. A *proportion* is a statement that two ratios or rates are equal to each other.
5. The *cross products* of a proportion are always equal to each other.
6. To prove that a statement is a proportion, show that the cross products are equal.
7. To *solve a proportion* of an unknown quantity, cross-multiply and then divide by the number that is with the variable.

8. To *check* the solution to a proportion, substitute the answer into the proportion and check the cross products.
9. For *application problems*, choose a letter for the unknown, and then write two ratios or rates in the same order and solve the proportion.
10. Corresponding sides of similar triangles are proportional.
11. To *convert measurement quantities* to other units, multiply by the unit fraction that will enable you to divide out the unwanted units.

REVIEW EXERCISES

A. Write each comparison as a ratio in lowest terms.

1. 14 to 63
2. 16 to 56
3. 28 to 8
4. 192 : 72
5. 6.4 to 0.12
6. 3/2 : 7/8
7. 2/5 to 9/15
8. 8 1/2 : 2/3
9. 0.8 to 7.2
10. 0.72 to 1.8
11. 6/5 to 3/8
12. 4 1/2 to 2
13. 6 minutes to 50 seconds
14. 11 feet to 29 inches
15. 20 cents to 5 dimes
16. 6 pints to 4 cups
17. 2 pounds to 14 ounces
18. 17 hours : 90 minutes
19. On a business trip, a sales representative traveled 120 miles by car and 460 miles by plane. Write the following ratios:
    a. Car mileage to plane mileage
    b. Car mileage to total mileage
    c. Plane mileage to total mileage
20. An appliance factory manufactures 120 washers and 200 dryers each day. Write the following rates:
    a. Number of washers to number of dryers
    b. Number of washers to the total number of both
    c. Number of dryers to the total number of both

B. Find the unit price for each of the following. If necessary, round your answer to the nearest thousandth of a dollar.

21. 6 apples for $0.78
22. $1.56 for 22 ounces
23. 5 cans for $4.90
24. 3 cans for 89 cents
25. 3 pounds for $1.29
26. $1.78 for 18 ounces
27. 3 shirts for $49.77
28. 4.2 pounds for $4.70
29. 6 donuts for $2.20
30. twelve for $2.28

C. Find out whether each of the following is a proportion. Prove each answer by showing your cross products.

31. $\frac{4}{5} = \frac{28}{35}$
32. $\frac{63}{72} = \frac{7}{8}$
33. $\frac{78}{16} = \frac{9}{2}$
34. $\frac{3}{2} = \frac{6.72}{4.48}$
35. $\frac{1.2}{0.36} = \frac{0.4}{0.12}$
36. $\frac{4/9}{3/2} = \frac{1/9}{3/8}$
37. $\frac{110}{212} = \frac{3}{8}$
38. $\frac{39}{48} = \frac{15}{16}$
39. $\frac{3/4}{5/8} = \frac{36}{25}$
40. $\frac{6.8}{9.5} = \frac{0.6}{1.2}$

D. Solve for the unknown in each proportion. If necessary, round your answers to the nearest hundredth.

41. $\frac{15}{40} = \frac{C}{8}$
42. $\frac{2}{3} = \frac{12}{A}$

43. $\frac{12}{27} = \frac{M}{6}$

44. $\frac{15}{P} = \frac{8}{21}$

45. $\frac{C}{5} = \frac{36}{25}$

46. $\frac{12}{39} = \frac{15}{R}$

47. $\frac{4}{F} = \frac{2.4}{0.28}$

48. $\frac{19}{57} = \frac{N}{9}$

49. $\frac{14}{96} = \frac{7}{K}$

50. $\frac{7.8}{5.6} = \frac{2.3}{M}$

51. $\frac{4/5}{3/8} = \frac{T}{1/2}$

52. $\frac{1\ 2/3}{2\ 1/4} = \frac{3/4}{H}$

E. Solve each application problem by using a proportion. If necessary, round to the nearest hundredth.

53. If six pens cost $1.98, how much would ten cost?

54. An airplane uses 20 gallons of fuel to fly 190 miles. How many miles could it fly on 25 gallons?

55. A map scale says that 1 1/2 inches represent 30 miles. How many miles do 7 inches represent?

56. If 6 dozen rolls cost $2.88, how much would 10 dozen cost?

57. The sales tax on a $360 television is $27. What would be the sales tax on the $98 table purchased at the same time?

58. A 6 1/2-pound roast costs $14.30. How much would an 8 1/4-pound roast cost?

59. At Mike's last dorm party, 8 students ate six pizzas. Only 5 students will be present this time. Now many pizzas will he need?

60. If a 12-ounce can of soda has 160 calories, how many calories will be in 8 ounces of the beverage?

61. On a map, 3 inches represent 175 miles. How far apart on the map are two cities that are really 220 miles apart?

62. If a 6-foot man casts a 4-foot shadow, how tall is a tree that cast a 15-foot shadow?

63. If 2 1/2 tablespoons of peanut butter contain 188 milligrams of sodium, how much sodium is in one-fourth cup (4 tablespoons) of peanut butter?

64. If Jaye's sewing machine stitch gauge is set at six stitches per inch, how many stitches will she need to rip out of a 14 1/2-inch seam that needs to be redone?

65. If a child weighing 25 pounds is to be given 1.5 milliliters of a cough medicine, how much should a child weighing 40 pounds be given?

66. If three-fourths of an ounce of cereal contains 25 milligrams of potassium, how much potassium is in 2 ounces of the cereal?

F. Use proportions to find the length of the sides of the following similar triangles. Round decimal answers to the nearest tenth.

67. $\Delta EDF \sim \Delta PKT$, $ED = 7$ yards, $EF = 18$ yards, $KT = 20$ yards, and $PT = 31$ yards. Find the length of $\overline{PK}$.

68. Using the values given in Exercise 67, find the length of $\overline{DF}$.

Using Figure 6.20, answer each of the following questions given that $\Delta ABC \sim \Delta DFE$.

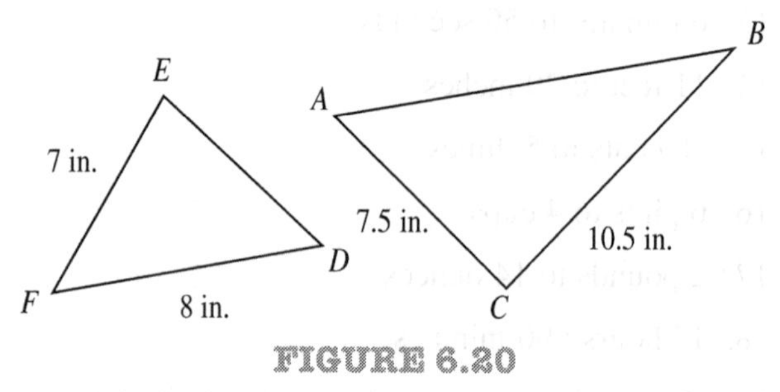

FIGURE 6.20

69. Find the length of $\overline{AB}$.

70. Find the length of $\overline{ED}$.

Exercises 71 through 74 refer to Figure 6.21, wherein ΔKJN is similar to ΔABC, $AB = 30$ yards, $AC = 37.5$ yards, $KJ = 20$ yards, and $JN = 15$ yards.

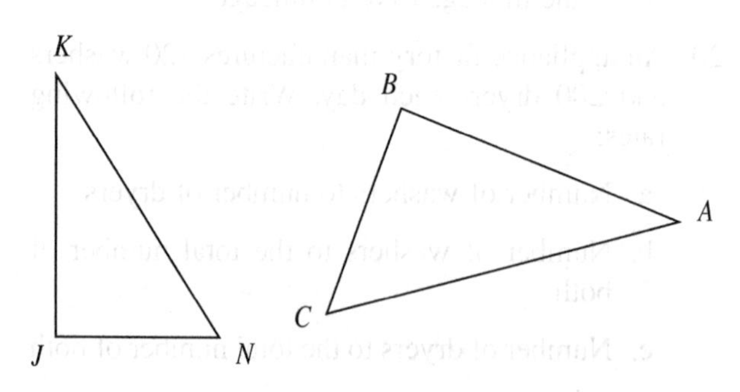

FIGURE 6.21

71. Find the length of $\overline{KN}$.

72. Find the length of $\overline{BC}$.

73. What is the ratio of AC to KN?

74. What is the ratio of the perimeter of ΔABC to the perimeter of ΔKJN?

G. Convert these measurements as indicated. You may need to refer to Table 6.1 on page 6 and Table 6.2 on page 29. If necessary, round answers to the nearest hundredth.

75. 16.25 inches to feet

76. 8 centimeters to inches.

77. 5 hours to seconds

78. 12 cups to pints

79. 4 gallons to cups

80. 15 miles per hour to feet per hour

81. 22 feet per second to miles per minute

82. 120 inches to meters

83. 5 feet to centimeters

H. Use the data represented in Figure 6.22 to do Exercises 84 through 91.

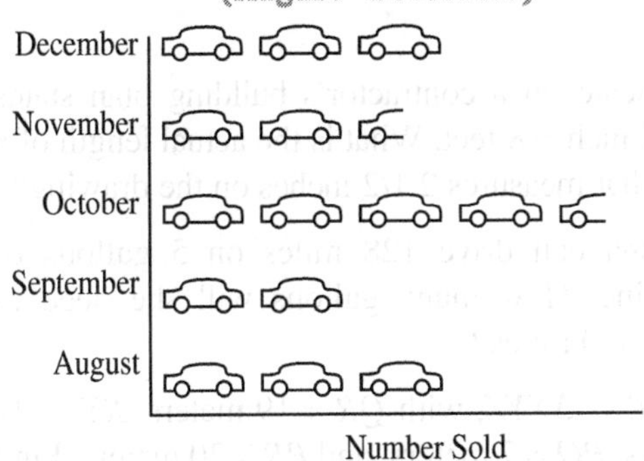

FIGURE 6.22

84. What is the ratio of August sales to October sales?

85. What is the ratio of September sales to August sales?

86. What is the mean of monthly sales for the period?

87. What is the median number of cars sold?

88. What is the mode of the sales for the period?

89. What is the ratio of October sales to the total sales for the 5-month period?

90. If the sale showed that each car pictured represented 10 cars sold, would the ratio you found in Exercise 88 change? Explain your answer.

91. If the scale showed that each car pictured represented 30 cars sold, would the ratio in Exercise 89 change? Explain your answer.

I. Read and respond to the following exercises.

92. Write an application problem that would require use of a proportion to get the answer. Use a real-life application from your work or home setting.

93. Find a graph representing data that is described in an article in a newspaper or magazine. Read both carefully. Does the graph accurately present the information given in the article? Is there enough information in the graph that you would not need to read the article? Explain those answers in detail. Include a copy of the article and the graph with your response.

94. Describe, in words, the procedure you would use to find the dimensions to use to build a doll house that is a copy of the White House. The scale you would use is that 1 foot in the White House is equal to 1 inch in the doll house.

95. If you were given two triangles that are congruent rather than similar, what would you know about the lengths of the corresponding sides of the triangles?

J. These exercises might require more than one step to reach the answer.

96. A can of soup costs $1.19. This 10.75-ounce can is mixed with a can of water; then the mixture is divided evenly into four servings. How would you determine the cost per ounce of the soup? How would you determine the cost per serving of the soup?

97. The cross products of a true proportion equal 75. If the two numerators of the ratios are 5 and 25, describe in words how you could find the denominators, and then write the proportion.

For Exercises 98 and 99, refer to Figure 6.23, where ΔKJN is similar to ΔKLM, $KL = 32$ yards, $4M = 40$ yards, $KJ = 20$ yards, and $JN = 15$ yards.

98. Find the length of $\overline{KN}$.

99. Using the information in Exercise 98, find the length of $\overline{LM}$.

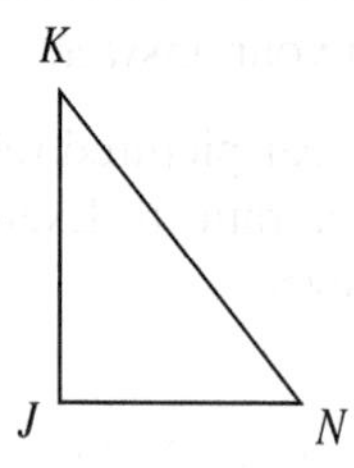

FIGURE 6.23

100. Change 50 miles per hour to feet per second using 1 mile = 5,280 feet, 1 hour = 60 minutes, and 1 minute = 60 seconds.

101. Change 25 miles per hour to kilometers per second using 1 mile = 1.61 kilometers, 1 hour = 60 minutes, and 1 minute = 60 seconds.

CHAPTER TEST

A. In Exercises 1 through 5, write each comparison as a ratio in simplest form.

1. 28 to 63
2. 7.2 to 0.12
3. 2/3 : 5/9
4. 2 dimes to 3 quarters
5. 5 feet to 17 inches
6. Find the unit price if 12 lemons cost $2.46.
7. Which is the better buy, a 12-ounce jar of coffee for $3.29 or an 8-ounce jar for $2.20? Tell which is the better buy, and state its unit price.
8. Change 4,680 seconds to hours using the conversion factors 1 hour = 60 minutes and 1 minute = 60 seconds.
9. Change 10 meters to inches using the conversion factor 1 meter = 39.37 inches.
10. Describe, in words, how to determine if $\frac{21}{35} = \frac{3}{5}$ is a proportion.
11. Is $\frac{3.2}{0.12} = \frac{0.12}{0.04}$ a proportion? Why or why not?

B. In Exercises 12 through 16, solve for the unknown. If necessary, round answers to the nearest hundredth.

12. $\frac{5}{9} = \frac{35}{W}$
13. $\frac{12}{C} = \frac{15}{32}$
14. $\frac{63}{X} = \frac{7}{8}$
15. $\frac{A}{14.5} = \frac{1.2}{0.5}$
16. $\frac{3/2}{3/4} = \frac{T}{5/8}$
17. The scale on a contractor's building plan states that 1 inch = 8 feet. What is the actual length of a wall that measures 2 1/2 inches on the drawing?
18. Lakisha can drive 128 miles on 5 gallons of gasoline. How many gallons will she need to drive 350 miles?
19. $\Delta PQR \sim \Delta XYZ$, with $QR = 19$ meters, $XY = 21$ meters, $PQ = 7$ meters, and $PR = 20$ meters. Find the lengths of $\overline{YZ}$ and $\overline{XZ}$.

20. Use Chart 6.5 to answer the following questions.

CHART 6.5 Grade Distribution for Math 108

	A	B	C	D	E
Fall Semester	12	38	49	17	21
Spring Semester	8	27	62	28	17

a. What is the ratio of A's in the fall to A's in the spring?

b. What is the ratio of E's in the spring to the total number of grades given during the spring semester?

STUDY SKILLS
MODULE

Test Taking and Test Anxiety

The announcement dreaded by most students in a math class has been made: "There will be a test on Monday." How you react to this news depends to a great extent on your past history. Math seems to be one of those things in life that carries with it a lot of history. Many students have horror stories about things that happened in their math classes. Let's see if we can overcome their negative impact once and for all. We need to begin, however, at the beginning and that means how you prepare for the test in the first place. The anxiety associated with it is only one piece of the picture.

As soon as you know that a test is coming, you need to begin your preparation. Putting it off until the night before is probably the worst thing you can do. Part of taking control of your feelings about test taking is believing that you really are in control in the situation. The best way for that to happen is to come to the test academically prepared for it. Reviewing and studying over several days is the best strategy. Taking notes and writing down sample test questions is a big help. Identifying the types of problems that you know you can do is very reassuring activity. The more time you spend becoming familiar with the subject, the less frightening it will be on test day.

When you come to test day, there are several things you can do to make things easier for yourself.

1. Try to come to class a few minutes early so you don't feel rushed or pressured and try not to have any other pressing appointments right afterward. That way you won't have to worry so much about time.
2. Practice some of the relaxation techniques that you learned about in Study Skills Module 3.
3. Listen carefully to any last-minute instructions that your teacher may give.
4. When you get the test in your hands, look it over carefully, read all the directions, and look for one or two problems that you know you can do.
5. Once you can begin the test, write down any facts or rules that you know you have studied but are afraid that you may forget. This is called a "memory dump."

6. Begin by working the problems that you know you can do. Don't work them too quickly or you may make careless errors. Just work steadily and accurately.

7. Go back to some of the more difficult problems. On second reading they often don't look so bad.

8. Don't worry about fellow students who finish the test quickly. That does not mean anything. They may know everything and do it very quickly, or they may have given up. There is no extra bonus for being the first person to turn in your paper.

9. Look over your test carefully before handing it in. Make sure you have answered each question. Look for little errors like nonreduced fractions or missing labels. If you can minimize these errors, your grade should be better.

10. Finally, when the test is returned to you, learn from your mistakes. Go back and correct each error and try to decide if there is a pattern to your mistakes. If there are still questions that you are unsure about, make an appointment to see your instructor and go over the mistakes. You are working to build a strong math foundation, so you don't want misunderstanding to remain to cause problems later.

Point of View

1. Describe how you feel when you know that a test is coming. What things contribute to that feeling?

2. What is your worst nightmare about taking a test? Many students have them: sleeping through a final exam, coming to take the test on the wrong day, and so on. If you can describe your greatest fear, then list some things you could do to make sure that it does not come to pass.

3. You have a friend in your class who is a real bundle of nerves when it comes to taking tests. What advice can you give him to help him overcome his fears? What things have worked for you?

4. Choose one suggestion from those made in this module and try it during preparation for your next test, or while taking your next test. Describe which activity you chose and comment on its benefit to you.

CHAPTER

Percent

LEARNING OBJECTIVES

When you have completed this chapter, you should be able to:

1. Understand the meaning of percent.
2. Interchange quantities among fraction form, decimal form, and percent form.
3. Identify the amount, rate, and base in a percent problem.
4. Use the percent formula, $A = RB$, to find the value of A, R, or B.
5. Solve a percent problem using the percent proportion, $A/B = P/100$.
6. Find the solutions to application problems involving percents.
7. Solve problems using the simple interest formula.
8. Interpret data from graphs.

We've all had experience with percents. Everyone who has taken a test understands that a 95 (95% correct) is a very good score. If your employer gives you an 8% raise, you are pleased that your good work has been recognized. These amounts, 95% and 8%, seem to be very different when you look at them as raw numbers. What makes them meaningful is the context in which they are discussed. Eight percent is a nice raise but a terrible grade on a test. Ninety-five percent is a grade to be proud of, but a 95% raise in salary is ridiculous—really nice, but unrealistic! In spite of the seemingly large difference between them, however, both 8% and 95% are very positive quantities in the situations we have mentioned. The point of all this is to help you realize that percents have no meaning by themselves.

Percent problems involve practical situations, such as sales tax, commission, percent increase or decrease, discount, and interest. We will solve these problems using either the percent formula, $A = RB$, or the percent proportion, $A/B = P/100$. Both approaches are algebraic in nature, but don't let that scare you. Most students are very successful with these problems once they understand the concepts involved.

7.1 The Meaning of Percent

The word *percent* comes from two Latin words, *per* and *centum*, which mean "by (or through) 100." The word *cent* comes from that same Latin word: A penny is a cent, one-hundredth of a dollar. If you can remember that **percent** means "by 100 or 1 per hundred or 1/100," much of what you will do in this chapter will make logical sense.

If there were 100 students on the roster in your math class, and 87% of them were in class today, that would mean 87 students were present out of the 100 students on the roster. If the sales tax rate in your city is 6%, that means you pay 6 cents in tax on a purchase that costs $1.00, and $6 in tax on a purchase that costs $100.

Suppose there were 100 students taking economics this term, and 17% of those students were doing A work. How many students would be earning an A? How many students would be doing less than A work? In order to answer these questions, you need to remember that percent means "per 100." If 17% of the 100 students are earning an A for the course, then 17 out of the 100 students will receive an A. If you need to figure out how many students are earning less than an A, you can approach the problem two different ways.

Method 1. If 17% are getting an A, then the rest are not. 100% represents the whole class. 100% – 17% = 83%, which represents the part of the class *not* earning an A. 83% of 100 students is 83 students.

Method 2. If 17% of a class of 100 students is earning an A, then 17 students are earning a grade of A. If 17 students are earning an A, then the rest are not. In a class of 100 students, 100 – 17 = 83 students are not earning an A.

It doesn't matter whether you begin with 100% and subtract the 17%, or figure the number of students first and subtract that number from the total of 100 students. What is important is to realize that you need to subtract *17%* from *100%*, or *17 students* from *100 students*. (Recall from previous chapters that you can add or subtract only the same kind of quantities.)

Because not all surveys ask 100 people to respond, and not all tests have exactly 100 questions, it is necessary to develop methods to find the number of people asked, or the percent of problems correct, in situations where the total quantity is not 100. That will be coming up soon.

7.1 EXERCISES

A. Answer each of the following questions based on percents.

1. One hundred people voted for their favorite flavor at Richard's Ice Cream Store. Of those 100, 62% voted for chocolate, 21% voted for strawberry, and 12% voted for peanut butter cup.
   a. How many people voted for chocolate?
   b. How many people voted for peanut butter cup?
   c. How many people voted for a flavor other than chocolate, strawberry, or peanut butter cup?
2. One hundred people were stopped on the street downtown and asked to name the mayor of the city. Of those 100, 43% gave the correct name, 29% gave the wrong name, and the rest admitted they didn't know.
   a. How many people knew the mayor's name?
   b. How many people gave the wrong name?
   c. How many people said they didn't know?
3. Loleetha was pleased to have earned an 89% on her vocabulary quiz. There were 100 questions on the quiz.
   a. How many questions did she answer correctly?
   b. How many did she answer incorrectly?
   c. What percent did she get wrong?
4. As part of a usage study, Sidney surveyed the first 100 students who walked into the library on a Wednesday morning. Eighteen of them were returning books, 36 were checking out books, and 38 were there to study. The rest of the students questioned were there for other reasons?
   a. What percent were returning books?
   b. What percent were there to study?
   c. What percent were there for activities other than returning a book, checking out a book, or studying?
5. If 72.6% of the people in a recent election voted in favor of the mass transportation levy, what percent voted against it?
6. When asked about a smoking ban for the school cafeteria, 63.8% of the students voted to have a "no smoking" policy in the cafeteria. What percent of the students thought smoking in the cafeteria should be allowed?
7. Marcus got a 92% on his last math exam. What percent of the questions did he miss?
8. Rose wants to make a least an 89% on her next French translation test to ensure that she maintains her B average. How many items can she afford to miss on this 100-question test?
9. When the state police set up a safety check on a well-traveled highway, they discovered that 18 of the first 100 cars stopped had one headlight that was not working. What percent of the cars stopped had a defective headlight?
10. The selling price of an item is a combination of the dealer's cost and the dealer's markup. (Markup covers profits, salaries, utilities, rent, insurance, and so on.) If the dealer's cost is 72% of the selling price, what percent represents the markup?

B. Read and respond to the following exercises.

11. Mrs. Hawkins told her students that 67% of them were doing very well in her class. Explain in words how you would find the percent of her 32 students that need to work harder to improve their grades.
12. The final exam makes up 28% of the grade for students in Mrs. Finnegan's math class. Describe in words how you would calculate what percent of the grade is dependent upon other factors such as quizzes, tests, and homework.
13. David wanted to calculate his home-to-business expenses for tax purposes. If he works from his home and can charge 24% of his expenses to his business, explain in words how he would calculate what percent of his bills is charged to normal household accounts.
14. If 400 students took the math placement test and 18% of those students placed into College Algebra, explain in words how you could decide what percent of the students placed into other math courses.

7.2 Converting among Fractions, Decimals, and Percents

Your ability to do percent problems will be much greater if you first learn how to change quantities from fractions and decimals to percents, and from percents to fractions and decimals. In fact, this ability to change percents to their fraction or decimal equivalents is crucial in many situations. Although you can add or subtract percents, as you did in the previous section, you cannot multiply or divide by any percents except 100%, which is a form of 1. Because you need to be able to use a fraction or a decimal equivalent in most problems involving percents, you will divide by 100%, as a form of 1, to change a percent number to a usable fraction or decimal form. You will also multiply by 100%, as a form of 1, to change from fractions or decimal numbers to their percent form when you are asked for a percent.

You already have a basic understanding of some of the relationships involving percents and fractions. If you score 100% on a test, you got one whole test completely right (100% = 1). If you were to get only 50% on a test, you would have done only one-half of the problems correctly (50% = 1/2). Getting 75% on a test means that you did 3/4 of the problems correctly (75% = 3/4). Students usually recognize these relationships from their previous experiences. Some of the other relationships or equivalences are not so common. You need to learn methods for changing from one form to another.

Changing Decimals to Percents

Let's first consider changing from a decimal to a percent.

Remember that *percent* means "by 100" and that 100% is another name for 1. When you multiply by 100%, you are multiplying by 1. This operation does not change the value of the quantity you are working with; it only changes the way the quantity looks. You used the same kind of process when you changed 3/8 to sixteenths. You multiplied 3/8 by 2/2 to get 6/16. You multiplied by the form of 1 that would produce the equivalent fraction you needed (Section 3.2).

Decimals to Percents

To change from a decimal to a percent, multiply the decimal quantity by 100%.

Consider the following examples to see how to change from decimal form to percent form.

EXAMPLE 1 Change 0.45 to a percent.

Solution When multiplying a decimal by 100%, move the decimal point two places to the right (Section 5.3), and attach the percent sign.

0.45 × 100% = .045.% = 45%

EXAMPLE 2 Change 0.384 to a percent.

Solution $0.384 \times 100\% = 0.38.4\% = 38.4\%$

EXAMPLE 3 Change 1.85 to a percent.

Solution $1.85 \times 100\% = 1.85.\% = 185\%$

This answer should not seem too large. Because 1.85 is more than 1, the equivalent percent should be more than 100%.

EXAMPLE 4 Change 0.006 to a percent.

Solution $0.006 \times 100\% = 0.00.6\% = 0.6\%$

0.006 is less than one-hundredth (0.01), so 0.006 in percent form should be less than 1%.

Changing Fractions to Percents

In order to change fractions to percents, you might change the fraction to its decimal form first and then change that decimal to a percent. In Section 5.5, you learned that to change a fraction to its decimal form, we divide the numerator by the denominator. Use that skill first, and then multiply the decimal answer by 100%.

> **Fractions to Percents**
>
> To change from a fraction to a percent, either multiply the fraction by 100%/1, or change the fraction to its decimal form and then multiply the decimal by 100%.

EXAMPLE 5 Change 3/4 to a percent.

Solution To multiply the fraction by 100%/1, the steps look like this:

$$\frac{3}{4} \times \frac{100\%}{1} = \frac{300\%}{4} = 75\%$$

To change the fraction to its decimal form and then multiply by 100% would be done in this manner:

$3/4 = 3 \div 4 = 0.75$ Change 3/4 to a decimal.

$0.75 \times 100\% = 0.75.\% = 75\%$ Change the decimal to a percent.

Therefore, 3/4 = 75%.

EXAMPLE 6 Change 5/8 to a percent.

Solution To multiply the fraction by 100%/1, the steps look like this:

$$\frac{5}{8} \times \frac{100\%}{1} = \frac{500\%}{8} = 62\ 1/2\%$$

To change the fraction to its decimal form and then multiply by 100% would be done in this manner:

5 ÷ 8 = 0.625

0.625 × 100% = 0.62.5% = 62.5%

Therefore, 5/8 = 62.5% or 62 1/2%.

EXAMPLE 7 Change 7/9 to a percent, rounded to the nearest tenth of a percent.

Solution As you well know, not all division comes out even, and some division answers need to be rounded. In the changing of *fractions to percents* in this book, the division answer is *carried out to four places, multiplied by 100%*, and then *rounded* to the nearest *tenth of a percent.* This is accurate enough for most situations.

7 ÷ 9 ≐ 0.7777

0.7777 × 100% = 0.77.77%

77.77% = 77.8%, rounded to the nearest tenth of a percent

Therefore, 7/9 = 77.8%. (This is read "seventy-seven and eight-tenths percent.")

EXAMPLE 8 Change 1/12 to a percent, rounded to the nearest *tenth of a percent.*

Solution 1 ÷ 12 ≐ 0.0833

0.0833 × 100% = 0.08.33% = 8.3%, rounded to the nearest tenth of a percent

Therefore, 1/12 = 8.3%.

EXAMPLE 9 Change 1 3/5 to a percent.

Solution First change 1 3/5 to an improper fraction, 8/5. Then change 8/5 to a decimal, 8/5 = 1.6.

1.6 × 100% = 1.60.% = 160%

Therefore, 1 3/5 = 160%. This answer makes sense; because 1 3/5 is more than 1, its percent form should be more than 100%.

Try these exercises and see if your answers match the ones given below. If they don't, look back over the examples carefully and see if you can find your mistake. If you are still having trouble, be sure to ask your instructor for help.

Self-Check Change the following decimals and fractions to their percent form. When necessary, round your answer to the nearest tenth of a percent.

1. 0.26	2. 0.789	3. 3.28	4. 0.0376
5. 4/5	6. 1/9	7. 2 3/8	8. 1/15

Answers:

1. 26%	**2.** 78.9%	**3.** 328%	**4.** 3.8%
5. 80%	**6.** 11.1%	**7.** 237.5%	**8.** 6.7%

Changing Percents to Decimals and Fractions

Changing from percent form to decimal or fraction form is the opposite of what you have just done. Because changing to a percent involves multiplying by 100%, it makes sense that *to change from a percent* involves *dividing by 100%*. Remember that to divide a number by 100, you move the decimal point two places to the left (Section 5.4).

> **Percents to Decimals**
>
> To change from a percent to a decimal, divide by 100%.

Consider the following examples to see how this can be done.

EXAMPLE 10 Change 64% to its decimal form.

Solution The decimal point in 64% currently follows the 4. Move it two places to the left and cross out the percent sign.

$64\% \div 100\% = .64.\% \div 100\% = 0.64$

Therefore, 64% = 0.64.

EXAMPLE 11 Change 0.7% to its decimal form.

Solution $0.7\% \div 100\% = 0.00.7\% \div 100\% = 0.007$

Therefore, 0.7% = 0.007.

EXAMPLE 12 Change 258% to its decimal form.

Solution $258\% \div 100\% = 2.58.\% \div 100\% = 2.58$

Therefore, 258% = 2.58. Because 258% is more than 100%, the decimal answer should be more than 1.

To change a percent to a fraction, you can either divide by 100%/1 or go through the decimal form, because you already know how to go from a decimal to an equivalent fraction.

Percents to Fractions

To change from a percent to a fraction, either divide by 100%/1, or change the percent to a decimal and then the decimal to its fraction form.

Consider these examples.

EXAMPLE 13 Change 48% to a fraction.

Solution To divide the fraction by 100%/1, the steps look like this:

$$48\% \div \frac{100\%}{1} = \frac{48\%}{1} \times \frac{1}{100\%} = \frac{48}{100} = \frac{12}{25}$$

To change the percent to its decimal form and then to the fraction would be done in this manner:

$48\% \div 100\% = 0.48.\% \div 100\% = 0.48$

0.48 is read "forty-eight hundredths," so it is written 48/100 in fraction form. You need to reduce 48/100 (by fours) to 12/25.

Therefore, 48% = 48/100 = 12/25.

EXAMPLE 14 Change 0.8% to a fraction.

Solution This problem has several steps. First change 0.8% from a decimal percent to a decimal number. Then change the decimal to its reduced fraction form.

$0.8\% \div 100\% = 0.00.8\% \div 100\% = 0.008$

0.008 is read "eight thousandths," so it is written 8/1,000. Reduce 8/1,000 (by eights) to 1/125.

Therefore, 0.8% = 1/125.

EXAMPLE 15 Change 145% to a fraction.

Solution To divide the fraction by 100%/1, the steps look like this:

$$145\% \div \frac{100\%}{1} = \frac{145\%}{1} \times \frac{1}{100\%} = \frac{145}{100} = \frac{29}{20} \text{ or } 1\frac{9}{20}$$

To change the percent to its decimal form and then to the fraction would be done in this manner:

$145\% \div 100\% = 1.45.\% \div 100\% = 1.45$

1.45 is read "one and forty-five hundredths," so it is written 1 45/100. The last step is to reduce 45/100 (by fives) to 9/20.

Therefore, 145% = 1 9/20.

EXAMPLE 16 Change 8 1/4% to a fraction.

Solution This problem also has several steps. First you must change 8 1/4 to 8.25.

Therefore, 8 1/4% = 8.25%.

8.25% ÷ 100% = 0.08.25% ÷ 100% = 0.0825

0.0825 is read "eight hundred twenty-five ten-thousandths," so it is written 825/10,000. Reduce 825/10,000 (by twenty-fives) to 33/400. This answer makes sense: 8 1/4% seems like a small quantity, and 33/400 seems equally small.

Therefore, 8 1/4% = 33/400.

◀

Try these exercises and check your answers. Make sure that you are comfortable changing percents to their fraction or decimal forms.

Self-Check Change each percent to its decimal or fraction form as indicated. Round all decimal answers to the nearest thousandth, and be sure to reduce all fraction answers.

1. Change 84% to a decimal.
2. Change 84% to a fraction.
3. Change 275% to a decimal.
4. Change 275% to a fraction.
5. Change 0.4% to a decimal.
6. Change 12.5% to a fraction.
7. Change 12 1/2% to a decimal.
8. Change 8 3/4% to a fraction.

Answers:

1. 0.84	2. 21/25	3. 2.75	4. 2 3/4 or 11/4
5. 0.004	6. 1/8	7. 0.125	8. 7/80

Sometimes when you are doing problems like these, it is easy to get lost in the steps. Remember that 100% is simply another name for 1, and keep these rules in mind:

Rules for Changing to and from Percents, Decimals, and Fractions

1. To change *from decimal form to the percent form, multiply by 100%.*
2. To change *from the percent form to decimal form, divide by 100%.*
3. If starting with a *fraction, change to its decimal* form first, and then *multiply by 100%* to get the percent form.
4. To change a *percent to a fraction, change to the decimal form first,* and then *read, write, and reduce.*
5. Always follow rounding instructions.
6. Always simplify fraction answers to their lowest terms.

7.2 EXERCISES

A. Change each of the following fractions or decimals to its equivalent percent form. When necessary, round your answers to the nearest tenth of a percent.

1. 0.32
2. 0.821
3. 3.47
4. 1.56
5. 0.4639
6. 0.2155
7. 12
8. 6
9. 3/5
10. 4/8
11. 5/7
12. 6/11
13. 9/4
14. 12/5
15. 3 1/2
16. 1 4/5

B. Change each of the following percents to its equivalent decimal form. When necessary, round your answer to the nearest thousandth.

17. 12%
18. 37%
19. 0.9%
20. 0.26%
21. 800%
22. 200%
23. 10 1/2%
24. 9 1/4%
25. 25 1/4%
26. 2 1/2%

C. Change each of the following percents to its equivalent fraction form. Remember always to reduce fractions to their lowest terms.

27. 19%
28. 5.2%
29. 24%
30. 800%
31. 650%
32. 0.12%

D. Complete the following table. Reduce all fraction answers to lowest terms. Round each decimal answer to three places. Round each percent answer to the nearest tenth of a percent.

	Fraction	Decimal	Percent
33.		0.286	
34.			85%
35.	4/5		
36.	5/9		
37.			2 1/2%
38.		1.4562	
39.			3/4%
40.		0.375	
41.	7/12		
42.			17.8%
43.		0.006	
44.	2 1/8		
45.			265%
46.	7/5		
47.		5	
48.			0.8%
49.	1/16		
50.		3.2	

E. Read and respond to the following exercises.

51. A student said that "3/5% is equal to 0.60, written in decimal form." Explain what his mistake was and what the correct decimal value would be.

52. A recent magazine article stated that 25% of the teachers in a large metropolitan area did not know their left hand from their right. The writer of the article was upset that this meant that 1 out of every 5 teachers was inept. Explain the error in the author's thinking. Also, give some reasons that teachers might have difficulty telling their left from their right.

53. Explain in words to a fellow student how you would change 62.4% to an equivalent fraction.

54. Explain in words how to arrange the following quantities in order from the smallest to the largest, and then arrange them:

a. 16.5% 1/8 0.17

b. 3 3/4 380% 3.4

7.3 Percent Formula

When you say "16 is one-half of 32" or "16 is 50% of 32," you are making a statement in what is called standard form. The 16 is a part of the 32. The 32 is the whole, original quantity that you started with. "One-half," or "50%," tells just how large a part of 32 the quantity 16 is. When you say that the amount of sales tax you pay is $5.00 on an item that costs $100.00 in a city where the sales tax is 5%, you could also be making a statement in standard form. You could restate it as $5.00 is 5% of $100.00." $5.00 is the part of the bill that you need to pay in sales tax based on the purchase price, or whole amount, of $100.00. The 5% is the *rate* of sales tax that is charged in your area.

When you talk about the relationship between two quantities in terms of "what percent one is of the other," you can restate that relationship using the percent formula. The **percent formula** says that *the amount (or part), A,* is equal to *the rate, R,* times *the base, B.* You can shorten all of that further and say $A = RB$. You need to remember that in formulas, two letters next to each other tell you to multiply those values. To work with these formulas, you need to determine which quantities in a problem are the amount (A), the rate (R), and the base (B).

The rate, R, is always the easiest to identify. It is the quantity with the percent sign, or it signifies the relationship between the two quantities. In "16 is one-half of 32," one-half is the rate. Another name for one-half, 50%, is the rate in "16 is 50% of 32." The rate in the sales tax example was 5%.

The base, B, is often the next easiest to identify. It usually follows the word *of* in one-sentence percent problems and stands for the original, or total, quantity in an application problem. In the first percent statement, "16 is 50% of 32," 32 follows the word *of* and is the base; in the tax example, 100.00 is the base because it is the amount before the tax was computed.

The amount or part, A, is the quantity that is yet to be identified. It represents the amount that is part of the base. In the first example, "16 is 50% of 32," we determined that R is 50% and B is 32. Therefore, the A value is 16. If you can describe percent or rate in an application problem (percent of increase, percent of commission, sales tax rate), you can describe the amount in a similar way. In the previous tax example, 5% is the *rate of sales tax*, so A would be *the amount of sales tax.*

Consider each of the following examples. First the values for R, B, and A are identified; and then the quantities are substituted into the percent formula.

EXAMPLE 1 Write "4 is 2% of 200" in percent formula form.

Solution 2% is the rate, R.

200 follows "of" and is the base, B.

4 is the part or amount, A.

Therefore, substituting the values into $A = RB$ yields

$4 = 2\% \times 200$.

Check Use 2% in its decimal form, 0.02, and multiply it by 200: $0.02 \times 200 = 4.00$, or 4.

4 *is* 2% of 200.

◀

EXAMPLE 2 Write "70% of 20 is 14" in percent formula form.

Solution 70% is the rate, R.

20 follows "of" and is the base, B.

14 is the amount or part, A.

Therefore, $A = RB$, or $14 = 70\% \times 20$.

Check Change 70% to its decimal form, 0.70, and multiply it by 20: $0.70 \times 20 = 14.00$, or 14.

70% of 20 *is* 14.

Now that you can write a percent statement in percent formula form, you can use the percent formula to solve for a missing value; the amount (A), the percent of rate (R), or the base (B).

Estimation

As in other calculating situations, developing an idea about the answer, or an estimate for the quantity sought, before you begin is often a good idea. You know that 1 is 100%, 1/2 is 50%, 1/4 is 25%, and 3/4 is 75%. You can use these figures as benchmarks when estimating in percent problems. We will learn that to find 10% of a number, we multiply that number by 0.1 or simply *move the decimal point one place to the left.* To find 5% of that same quantity, we can find one-half of the 10% answer. To find 20% of that same quantity, we multiply the 10% answer by 2. To find 30% of that same quantity, we multiply the 10% answer by 3, and so on. Remember that the estimating process is used to help you get a "ballpark" idea of what your actual answer should be. This takes some practice, but it is a good habit to develop. It can be used to help you know when the answer you have found makes reasonable sense. Keep estimating in mind as you calculate and then check the following example problems.

Finding A

The question "What number is 3.5% of 63?" could be looked at as a question in standard form. Because "is" translates to =, and "of" to multiply, you could rewrite the question as "What number = (3.5%)(63) or $3.5\% \cdot 63$"? If you were to use the percent formula, $A = RB$, R would be 3.5%, B would be 63, and you would be looking for A. Consider how this problem could be worked out using the percent formula.

After you have identified the values, follow the four steps for solving a problem by using a formula:

1. Write the formula.
2. Substitute the values you know.
3. Perform the necessary calculations.
4. Check the answer in the original problem.

$A = RB$	Write the formula.
$A = (3.5\%)(63)$	Substitute 3.5% for R and 63 for B.
$A = (0.035)(63)$	Because you cannot multiply by a percent number, change 3.5% to decimal form, 0.035, and then multiply.
$A = 2.205$	

Therefore, 2.205 is 3.5% of 63.

Check This answer seems to make sense because 3.5% of something is a very small part of that quantity and 2.205 is a very small part of 63.

Consider the following examples. You should first identify the *R, B,* and *A* values, then write the percent formula. Substitute these values into the formula, leaving the letter in place of the quantity you are trying to find. Finally, you multiply as implied by the variable *R* right next to the variable *B*.

EXAMPLE 3 What number is 42% of 87?

Solution *R* is 42%. 87 follows "of" and is the value for *B*. *A* must be the value you are to find.

$A = RB$	Write the formula.
$A = 42\% \cdot 87$	Substitute the known values.
$A = 0.42 \cdot 87$	Change 42% to its decimal form.
$A = 36.54$	Perform the needed operation. In this case, multiply.

Therefore, 36.54 is 42% of 87.

Check To try to use ballpark values to see if the answer seems reasonable, we might say that 42% of 88 is less than 50% (or 1/2) of 88 so our answer should be less than 50% of 88 which is 44. Since 36.54 is, in fact, less than 44, the answer seems reasonable.

◀

EXAMPLE 4 8 1/2% of 56 is what number?

Solution To estimate an answer we can realize that 10% of 56 would be 0.1 times 56 or 5.6 so 8 1/2% of 56 should be a little less than 5.6.

R is 8 1/2%, or 8.5% when you change 1/2 to its decimal form. The number following "of" is 56 and so 56 is the value for *B*. *A* must be the value you are looking for.

$A = RB$

$A = (8.5\%)(56)$

$A = (0.085)(56)$

$A = 4.760$

Therefore, 4.76 is 8 1/2% of 56.

Check 4.76 is, if fact, a little less than 5.6 and agrees with our estimate.

◀

Finding *B*

Suppose the question is "35 is 15% of what number?" In this case, you are given the rate, 15%, but the quantity following "of" is the one that is missing. In this type of problem, you are to find the value for B—the base—that will satisfy the question. You begin just as you did when finding the value for A. You follow the four steps to be used when solving a problem by applying a formula. To do that, remember to first pick out the values for R, B, and A, and then write the formula, substitute the values you have, and multiply or divide as needed. Study the following problem to see how this can be done.

30 is 15% of what number?

$R = 15\%$.

B is the quantity to be found because there is no quantity that follows "of."

A must then be 30.

Write the formula: $A = RB$

Substitute the given values: $30 = (15\%)(B)$ or $15\% \cdot B$

Change 15% to its decimal form: $30 = (0.15)(B)$

This is where you run into a problem. Before, to find the value for A, you had two quantities to multiply. This time the missing value, B, is not alone. It is multiplied by 0.15. To get the B value by itself, you need to divide by 0.15 to "undo" the multiplication. You used this same process in Chapter 6 when you solved proportions. When working with an equation (an equality statement), it is necessary to do the same operation to both sides of the equals sign to get the variable alone. This keeps the equation in balance. Here, then, you need to divide 30, as well as $0.15 \times B$, by 0.15. This is how the steps look:

$$30 = 0.15 \cdot B$$

$$\frac{30}{0.15} = \frac{\overset{1}{\cancel{0.15}} \cdot B}{\underset{1}{\cancel{0.15}}} \qquad 0.15.\overline{)30.00.} = 200.$$

$$200 = B$$

Therefore, 30 is 15% of 200. To check the answer, multiply 0.15 by 200, which is indeed equal to 30.00.

The following examples show how to find the value for B when you use the formula. Always follow the same steps and you won't get lost.

EXAMPLE 5 40% of what number is 25?

Solution R is 40%. B is the value you are to find because no quantity follows "of." A is 25.

A	$= RB$	Write the formula.
25	$= 40\% \cdot B$	Substitute the values.
25	$= 0.40 \cdot B$	Change percent to a decimal.
$\frac{25}{0.40}$	$= \frac{\overset{1}{\cancel{0.40}} \cdot B}{\underset{1}{\cancel{0.40}}}$	Divide both sides by 0.40 to "undo" the multiplication and isolate B. $25 \div 0.40 = 62.5$
62.5	$= B$	

Therefore, 25 is 40% of 62.5.

Check Multiply 0.40 by 62.5, which is equal to 25.00.

EXAMPLE 6 22.42 is 23.6% of what number?

Solution R is 23.6%. You must be looking for B because no quantity follows the word *of*. Therefore, 22.42 is A.

$$A = RB$$

$$22.42 = 23.6\% \cdot B$$

$$22.42 = 0.236 \cdot B$$

$$\frac{22.42}{0.236} = \frac{\overset{1}{\cancel{0.236}} \cdot B}{\underset{1}{\cancel{0.236}}}$$

$$B = 95$$

Therefore, 22.42 is 23.6% of 95.

Check Multiply 0.236×95, which is equal to 22.42.

◀

Finding R

When the percent, or rate, is the quantity you are asked to find, you follow a process very similar to the one for finding B. After you have identified the values, follow the four steps for solving a problem by using a formula:

1. Write the formula.
2. Substitute the values you know.
3. Perform the necessary calculations.
4. Check the answer to the original problem.

In this type of problem, you will not have a percent value. This is the value you are looking for. Remember that once you have found the decimal answer, you *must* write it in percent form. Consider the following problems to see how this will be done.

"25 is what percent of 200?"

Because no percent is given, you are looking for R. 200, which follows "of," is B. A must therefore be 25.

$$A = RB$$

$$25 = R \cdot 200$$

You are now at the same place you were when looking for the base, B. The letter B was not alone. To isolate it, you divided both sides of the equality by the number next to B to "undo" the multiplication. This time you will divide both sides by 200 to isolate the R.

$$\frac{25}{200} = \frac{R \cdot \overset{1}{\cancel{200}}}{\underset{1}{\cancel{200}}}$$ 200 over 200 cancels. $25 \div 200 = 0.125$

$$0.125 = R$$

Note that you found a decimal answer, but the question asked for a percent. You know from Section 7.2 that you need to multiply a decimal form by 100%, a form of the number 1, to change it to a percent.

$R = 0.12.5 \cdot 100\% = 12.5\%$

Therefore, 25 is 12.5% of 200.
Check the answer by multiplying 12.5% by 200:

$0.125 \cdot 200 = 25.000 = 25$

All problems that involve finding the rate, R, are done in this same manner. They are manageable when you use the percent formula and always follow the same steps. A formula is a real help. It shows what operation to perform with the numbers you are given in a problem. Consider each of the following examples and see how these problems are done.

EXAMPLE 7 14 is what percent of 700?

Solution No percent is given, so you are looking for R. 700, which follows "of," is B. 14 must then be A.

$A = RB$ Write the percent formula.

$14 = R \cdot 700$ Substitute the known values.

$\frac{14}{700} = \frac{R \times \cancel{700}^{1}}{\cancel{700}_{1}}$ To isolate R, you must divide by 700. $14 \div 700 = 0.02$

$0.02 = R$

$R = 0.02 \times 100\%$ Once you have found the decimal answer, it is necessary to change it to a percent, so you multiply by 100%.

$R = 2\%$

Therefore, 14 is 2% of 700.

Check Multiply 2% by 700: $0.02 \times 700 = 14.00 = 14$

EXAMPLE 8 What percent of 36 is 90?

Solution No percent is given, so you are looking for R. $B = 36$ because 36 follows "of." Therefore, A must be equal to 90.

$A = RB$ Write the percent formula.

$90 = R \times 36$ Substitute the given values.

$\frac{90}{36} = \frac{R \times \cancel{36}^{1}}{\cancel{36}_{1}}$ Divide both sides of the equation by 36 to "undo" the multiplication. 36 divided by 36 = 1.

$2.5 = R$ $90 \div 36 = 2.5$

$2.5 \times 100\% = 350\%$ To change from decimal to percent form, multiply by 100%.

Therefore, 90 is 250% of 36.

Check Multiply 250% by 36: $2.5 \cdot 36 = 90.0$

Note that when the amount is greater than the base, the percent is greater than 100%.

EXAMPLE 9 12 is what percent of 33? Round your answer to the nearest tenth of a percent.

Solution 12 is what percent of 33?

No percent value is given, so you are looking for R. B is 33 because 33 follows "of." A must therefore be 12.

$A = RB$ — Write the percent formula.

$12 = R \times 33$ — Substitute the given values.

$\frac{12}{33} = \frac{R \times \cancel{33}^{1}}{\cancel{33}_{1}}$ — Divide both sides of the equation by 33 to "undo" the multiplication. $33 \div 33 = 1$. $12 \div 33 = 0.3636$

$0.3636 = R$

$0.3636 \times 100\% = 36.36\%$ — Change from decimal form to percent form, and then round to the nearest tenth of a percent.

$R = 36.4\%$

Therefore, 12 is 36.4% of 33.

Check Multiply $36.4\% \times 33$: $0.364 \times 33 = 12.012$

The product will not be exactly 12 because 36.4% is a rounded answer.

Always remember to change the decimal answer to its percent form when asked to find the percent value. If percent answers are to be rounded to the nearest tenth of a percent, be sure to calculate your decimal answer to four places, multiply your answer by 100% to change it to a percent, and then round to the nearest tenth of a percent. If you were asked to round to the nearest hundredth of a percent, you would carry the division one place farther, multiply the decimal value by 100%, and then round to the nearest hundredth of a percent. This process could be carried out to the nearest thousandth of a percent (and beyond), but this is unlikely to be required in real-world situations. The nearest tenth of a percent is generally accurate enough.

7.3 EXERCISES

A. Estimate an answer to each of the following exercises. Remember that when you estimate, the numbers should be reasonable to work with so you can use mental math whenever possible.

1. What is 23% of 80?
2. Find 12 1/2% of 80.
3. What is 47% of 64?
4. Find 92% of 210.
5. Find 150% of 81.
6. What is 215% of 70?

B. For each of the following exercises, identify the values for R, B, and A. Then substitute the given values in the percent formula, and solve it to find the unknown value. Round all base and amount answers to the nearest hundredth and all percent answers to the nearest tenth of a percent. You may use your calculator to do the multiplication and division in this type of problem if your instructor agrees.

7. What is 7% of 95?
8. What is 143% of 60?
9. Find 124% of 70.

10. What is 0.5% of 93?
11. Find 12 1/2% of 100.
12. 35 is 7% of what number?
13. 30 is 125% of what number?
14. 2.8% of what number is 140?
15. 32% of what number is 8?
16. 74% of what number is 80?
17. 12% of what number is 43?
18. 45 is what percent of 50?
19. What percent of 40 is 4.8?
20. What percent of 120 is 50?
21. 16 is what percent of 30?
22. 125 is what percent of 200?
23. 7 is what percent of 8?
24. 24 is what percent of 30?
25. 16.5% of 42 is what number?
26. What percent of 18 is 11?
27. 24% of what number is 12?
28. What is 18.5% of 22?
29. 21 is what percent of 65?
30. 28 is 5 1/2% of what number?

C. Use your calculator to answer the following questions. Follow the rounding instructions given in Part B.

31. What is 77.3% of 92,476?
32. 17.5 is 21.8% of what number?
33. 91.4 is what percent of 146?
34. 11.9 is 43% of what number?
35. Find 29.4% of 37,645.
36. What percent of 8,372 is 43.5?
37. What percent of 8,276 is 189.4?
38. 81.2 is what percent of 375?
39. 127 is 38.4% of what number?
40. 83.6% of 42,756 is what number?
41. 67.4 is 82% of what number?
42. Find 17.9% of 46,384.
43. 35.4 is 46.5% of what number?
44. 238.5 is 21.6% of what number?
45. 1,276 is what percent of 8,375?

D. Read and respond to the following exercises.

46. How can you figure 10% of $36 in your head? How can you figure 15% of $36 in your head? Explain the process you are using.
47. You and three friends have gone out to dinner and the bill has been presented to your table. Because there are only four of you, the tip has not been figured into the bill. The service was good so you plan to pay the standard 15% gratuity. Explain how you would figure the amount of tip based on the $48 dinner check, and also what amount of the tip each of you should pay.
48. A fellow classmate is having trouble understanding why 250% of an amount is more than the original amount. What can you do to help him understand?
49. Explain, in words, how you would estimate the sale price of a suit at a "30% off" sale if the original price of the suit was $179.95.
50. Write out the steps you would use to estimate an answer to this question: "45 is 60% of what number?"

7.4 Percent Proportion

Another method for solving percent problems is to use the **percent proportion**. Because percent means "per 100," one side of the proportion can be expressed as a ratio of the percent number or rate (P), to 100. The other side of the proportion will be the ratio of the part or amount (A) to the total or base (B). The formula that expresses this relationship is

$$\frac{A}{B} = \frac{P}{100}$$

The most difficult part of solving percent problems by this method is identifying *A*, *B*, and *P*. Note that the percent quantity is denoted by *P* in the proportion. You will use the actual percent value, not its decimal equivalent as you did in the percent formula for *R*. Once you have identified the values and written the proportion, you will use the proportion solving skills that you learned in Chapter 6. Consider the following statement:

16 is 25% of 48.

Here 25 is *P*, 48 is *B* and 16 is *A*.

P and *B* are easy to identify, so do them first. *P* will be the number right before the percent symbol or the word *percent*. *B* always follows the word *of* in the one-sentence problems. Whatever is left will be *A*.

EXAMPLE 1 Identify *P*, *B*, and *A* in the following statements.

a. 65 is 50% of 130.

b. 12 is 75% of 16.

Solution

a. 50% means that 50 is *P*. The words *of 130* mean *B* is 130. Therefore, 65 must be *A*.

b. 75% means that 75 is *P*. The words *of 16* mean that 16 is *B*. Therefore, 12 is *A*.

◀

Once you have identified the parts of the problem, substitute the numbers for the letters in the percent proportion. Usually one of the parts is not known, and you will solve the proportion to get that value. But let's test the percent proportion with the two preceding examples. Substitute the parts you have identified, and see whether you have equal ratios when you compute the cross products. (See Section 6.2.)

$$\frac{65}{13} \times \frac{50}{100} \qquad \frac{12}{16} \times \frac{75}{100}$$

$$65 \cdot 100 = 50 \cdot 130 \qquad 120 \cdot 100 = 75 \cdot 16$$

$$6{,}500 = 6{,}500 \text{ True.} \qquad 1{,}200 = 1{,}200 \text{ True.}$$

EXAMPLE 2 Use the percent proportion to answer this question: What percent of 8 is 4?

Solution

$\frac{A}{B} = \frac{P}{100}$ Begin with the percent proportion. *B* is 8, you are looking for *P*, and *A* is 4.

$\frac{4}{8} = \frac{P}{100}$ Substitute the numbers for the letters.

$8 \cdot P = 4 \cdot 100$ Compute the cross products. Multiply $4 \cdot 100$ and $8 \cdot P$.

$\frac{\overset{1}{\cancel{8}}P}{\underset{1}{\cancel{8}}} = \frac{400}{8}$ Divide both sides by 8 to isolate the *P*.

$P = 50$

$P = 50\%$

Because P is a percent number, we do not move the decimal point but we simply attach the %. 50% of 8 is 4.

Check 50% of 8 is 4, so 4 is 50% of 8.

$4 = 0.50 \times 8$

$4 = 4.00$ True.

Note that when you finish solving the proportion for P, the decimal point is in the correct place and does not have to be moved. In fact, you won't have to worry about moving decimal points at all, except in decimal multiplication or division.

EXAMPLE 3 5% of 90 is what number?

Solution

$\frac{A}{B} = \frac{P}{100}$ — P is 5, B is 90, and A is unknown.

$\frac{A}{90} = \frac{5}{100}$ — Substitute. Be sure not to include the % with the 5 when you replace P.

$100A = 450$ — Cross-multiply: $100 \cdot A$ and $5 \cdot 90$

$\frac{\not{1}00A}{\not{1}00} = \frac{450}{100}$ — Divide by 100.

$A = 4.5$ — 4.5 is 5% of 90.

Check 4.5 is 5% of 90.

$4.5 = 0.05 \times 90$

$4.5 = 4.50$ True.

EXAMPLE 4 12% of what number is 80?

Solution

$\frac{A}{B} = \frac{P}{100}$ — 12 is P, B is unknown, and A is 80. Notice that the % is not written in the percent proportion.

$\frac{80}{B} = \frac{12}{100}$

$12B = 8000$

$\frac{\not{12}B}{\not{12}} = \frac{8000}{12}$

$B = 666\ 2/3$, or 666.7

80 is 12% of 666 2/3 or 80 is 12% of 666.7.

Check 80 is 12% of 666.7.

$80 = 0.12 \times 666.7$

$80 = 80.004$ — The answer is a rounded answer, so the check statement is very close but not exact.

7.4 EXERCISES

A. Solve these percent problems using the percent proportion, $A/B = P/100$. If necessary, round answers to the nearest hundredth, except for percents, which should be rounded to the nearest tenth of a percent. If you need more practice, do some of the practice exercises from Section 7.3, using the percent proportion this time. Again, you may want to use your calculator to help you do the actual calculation.

1. 24 is 15% of what number?
2. 22% of what number is 38?
3. What percent of 40 is 12?
4. 18 is what percent of 96?
5. 94% of 30 is what number?
6. What is 16% of 128?
7. 6 1/2 is 18% of what number?
8. 4.8 is what percent of 2.6?
9. 30 is what percent of 40?
10. 24.5 is 8% of what number?
11. What is 18% of 40?
12. What number is 12.5% of 200?
13. 48 is 8% of what number?
14. What percent of 70 is 4.5?
15. What percent of 80 is 45?
16. Find 210% of 65.
17. 1.8% of what number is 54?
18. 6 is what percent of 9?
19. What is 14.8% of 300?
20. 46% of what number is 70?
21. 36 is 8 1/2% of what number?
22. What is 3.4% of 96?
23. 52 is 120% of what number?
24. What percent of 12 is 7?
25. What percent of 32 is 60?
26. What is 25% of 84?
27. 45 is 150% of what number?
28. 41 is what percent of 65?
29. 21.5 is what percent of 38?
30. 48 is 110% of what number?

B. Use your calculator to solve the following exercises by applying the percent proportion. Follow the rounding instructions in Part A.

31. What percent of 2,006 is 18.4?
32. Find 176.2% of 37,658.
33. 21 is 43.2% of what number?
34. 70.2 is what percent of 9,840?
35. What is 67.5% of 13,465?
36. 82.3% of what number is 17,260?
37. What percent of 18,540 is 623?
38. 27.5% of 11,864 is what number?
39. 46.8% of what amount is 1,765?
40. What percent of 33,674 is 18,265?

C. Read and respond to the following exercises.

41. Explain how *P* in the percent proportion and *R* in the percent formula are similar and how they are different.
42. Which method, percent formula or percent proportion, do you prefer and why?
43. Explain in words how you would determine what percent 16 is of 40. Then explain how you would determine what fractional part 16 is of 40. Are the results equivalent?
44. Explain in words how you would estimate the answer to this percent question: "20 is 24% of what number?"

7.5 Applications Using Percent

General Percent Applications

Percent application problems are some of the most realistic problems you will study. Every day you might encounter sales tax, discounts, commission, and increases in the prices of almost everything. Solve these problems using the formula or proportion. For example, almost every state now has a sales tax based on the price of the item bought. Suppose you live in a state where the sales tax is 6%. To find the amount of tax on any item purchased, you need to find 6% of its selling price. **Commission**, based on the dollar amount an employee has sold; **discount**, the amount in dollars saved by buying an item on sale; and **percent of increase** or **percent of decrease** are other situations where percents are used. Percents are very much a part of everyday life.

When you are using the percent formula to solve application problems, the first step is to identify the values that will be substituted in place of *R, B,* and *A* in the formula. Follow the same steps that you have been using in the preceding sections. Identify *R* first; it is the percent (%). *B* stands for the base, the original quantity, the whole or total. *A* stands for the part, or amount. Consider the following situation and see how to pick out the values.

A sales tax is charged on the price of each item purchased in Franklin County. The sales tax rate is 5.5%. The tax is based on the purchase price of the item. The part that is charged in tax is added to the purchase price of the item to give the total bill. Find the amount of sales tax charged in Franklin County on an item costing $43.95.

$A = RB$

R is the sales tax rate, 5.5%. Sales tax is based on the purchase price, so *B* is $43.95. You are to find the amount paid in tax, *A*.

$A = RB$

$A = (5.5\%)(\$43.95)$

$A = (0.055)(43.95)$

$A = 2.41725$

The amount of tax charged is $2.42, rounded to the nearest cent.

The salesperson said that the tax on the new pair of slacks that Michael bought at the store would be $1.10. Can you figure out the price of the pants if he bought them at MAX's in Franklin County?

$A = RB$

R is the sales tax rate, 5.5%. Sales tax is based on the purchase price, *B*, which is what you are trying to find. $1.10 is the amount paid in tax, *A*.

$$A = RB$$

$$1.10 = 5.5\% \cdot B$$

$$1.10 = 0.055 \cdot B$$

$$\frac{1.10}{0.055} = \frac{\overset{1}{\cancel{0.055}} \cdot B}{\underset{1}{\cancel{0.055}}}$$

$$20.00 = B$$

Therefore, the slacks cost \$20.00.

The same pair of slacks in Chicago would carry a tax of \$1.40. What is the sales tax rate in Chicago?

$A = RB$

You are looking for the sales tax rate, R. This tax is based on the purchase price of \$20.00, which is B. The amount paid in tax is \$1.40, A.

$$A = RB$$

$$\$1.40 = R \times 20$$

$$\frac{1.40}{20} = \frac{R \times \overset{1}{\cancel{20}}}{\underset{1}{\cancel{20}}}$$

$$0.07 = R$$

Remember to change the decimal value for R to its percent form by multiplying by 100%.

$0.07 \times 100\% = 7\%$

Therefore, the sales tax rate in Chicago is 7%.

\$1.40 is 7% of \$20.

These next examples illustrate uses of the percent formula or percent proportion. As you can see from these problems, you proceed by identifying first the value for R or P (the percent), next the value for B (the base), and then the value for A (the part or amount). The following descriptions will help you to pick out these values in various application problems.

***A* (Amount or Part)**	***R* or *P* (Rate or Percent)**	***B* (Base)**
Amount of sales tax	Rate of sales tax	Original price
Amount of discount	Rate of discount	Original price
Amount of interest	Rate of interest	Original amount borrowed or saved
Amount of commission	Rate of commission	Amount of sales on which commission is based
Amount of decrease	Rate of decrease	Original quantity before the decrease
Amount of increase	Rate of increase	Original quantity before the increase

Study the following examples carefully to see if you can identify the values for R, B, and A. Once you have done that, you simply follow the steps outlined in the previous sections to solve for A, B, or R (when using the percent formula) or A, B, or P (when using the percent proportion). Even if your instructor wants you to practice only one method, you should at least read through all of the following examples to work on determining the values for A, B, P, or R. If you are working only on the percent formula, also study the method used in Examples 1, 2, 3, 4, and 5. If you are concentrating on the percent proportion, study the method used in Examples 6, 7, 9, 10, and 11.

EXAMPLE 1 Find the sales tax on an item costing \$42.85 if the sales tax rate is 6%.

Solution R is 6%. Sales tax is based on the price of an item, so $B = \$42.85$.

$A = RB$

$A = 6\% \text{ of } \$42.85 = 0.06 \cdot 42.85$

$A = \$2.5710$

Rounded to the nearest cent, a 6% tax on an item selling for \$42.85 is \$2.57.

EXAMPLE 2 Wendy wants to know how much she will save by buying a skirt at Sara's Shoppe while the store is having a "12% Off Sale." The skirt normally sells for \$22.95.

Solution To estimate an answer, we could find 10% of \$23 (0.10×23) and then realize that our answer should be a little more than \$2.30.

The rate of discount is 12%, and discount is based on the original price of the skirt, so B is \$22.95.

$A = RB$

$A = 12\% \cdot \$22.95 = 0.12 \cdot 22.95$

$A = 2.754$

Therefore, Wendy can save \$2.75 by buying the skirt on sale.

EXAMPLE 3 16 members on the board of directors voted to pass a no-smoking policy. This proposal passed with 80% of the board members voting for it. How many members are there on the board?

Solution $A = RB$

80% voted to pass the proposal, so R is 80%. You are looking for the total membership on the board, B. 16 represents A, the part of the board who voted to pass the policy.

$A = RB$	Write the percent formula.
$16 = 80\% \cdot B$	Substitute the given values.
$16 = 0.80 \cdot B$	Change 80% to its decimal form.
$\dfrac{16}{0.80} = \dfrac{\overset{1}{\cancel{0.80}} \cdot B}{\underset{1}{\cancel{0.80}}}$	Divide both sides by 0.80 to "undo" the multiplication and isolate B.
$20 = B$	$16 \div 0.80 = 20$
16 is 80% of 20.	There are 20 members on the board.

EXAMPLE 4 Maria bought a new car a year ago and paid $12,500 for it. Today the value of the car is $8,500. What is the percent of decrease in the value of the car?

Solution You are looking for R, the rate of decrease. Decrease is based on the original value before the decrease, so $B = 12{,}500$. The amount of decrease, A, is the difference between the original value and the current value. Therefore, $A = 12{,}500 - 8{,}500$, or 4,000.

$$A = RB$$

$$4{,}000 = R \cdot 12{,}500$$

$$\frac{4{,}000}{12{,}500} = \frac{R \cdot \overset{1}{\cancel{12{,}500}}}{\underset{1}{\cancel{12{,}500}}}$$

$$0.32 = R$$

$$0.32 \times 100\% = 32\%$$

Therefore, the percent of decrease was 32%.

EXAMPLE 5 Find the percent of increase in the value of a house if it recently sold for $98,672 but was originally purchased for $79,825. Round your answer to the nearest tenth of a percent.

Solution We first need to find the amount of the increase. This is a very important step.

Amount of increase: $98{,}672 - 79{,}825 = 18{,}847$

Base is the original price of the house: 79,825

You are looking for R.

$$A = RB$$

$$18{,}847 = R \cdot 79{,}825$$

$$\frac{18{,}847}{79{,}825} = \frac{R \cdot 79{,}825}{79{,}825}$$

$$0.236104 = R$$

$$0.236104 \times 100\% = 23.6104\%$$

We will follow our usual procedure and round to the nearest tenth of a percent.

Therefore, the percent increase is 23.6%.

The same type of approach is used when you are solving an application problem by applying the percent proportion, $A/B = P/100$. Begin by identifying the values for P, B, and A, and then substitute those values into the percent proportion. Solve this proportion as you would any other according to the steps you learned in Chapter 6.

EXAMPLE 6 Find the sales tax on an item costing \$42.85 if the sales tax rate is 6%.

Solution To estimate an answer, we could find 10% of \$42 (0.10×42) to get \$4.20 and then find half of that (5% of \$42) which is \$2.10 and realize that our answer (at 6%) should be a little more than \$2.10.

Sales tax is based on the price of the item, so $B = 42.85$. P is 6, and we are looking for the amount of tax.

$$\frac{A}{B} = \frac{P}{100}$$

$$\frac{A}{42.85} = \frac{6}{100}$$

$$100A = (6)(42.85)$$

$$\frac{100A}{100} = \frac{257.1}{100}$$

$$A = 2.571$$

Therefore, the sales tax is \$2.57, which is close to our estimate.

This is the same problem as Example 1 in this section. Compare the two methods and verify that the results are the same whichever method is used.

EXAMPLE 7 A \$7.80 discount is given on a dress that originally cost \$48.75. What is the percent or rate of discount?

Solution \$7.80 is what percent of \$48.75?

$$\frac{A}{B} = \frac{P}{100}$$

$$\frac{7.80}{48.75} = \frac{P}{100}$$

$$48.75P = (7.80)(100)$$

$$\frac{48.75P}{48.75} = \frac{780}{48.75}$$

$$P = 16$$

The rate of discount is 16%.

EXAMPLE 8 Jeremy, the store owner's son, earns a 15% commission on all of the furniture that he sells working part-time for Syburg's Furniture Store. Last week his sales totaled \$3,638.95. Estimate how much commission he earned.

Solution The rate of commission, R, is 15%. Commission is based on the amount sold, so B is \$3,638.95. Because we want an estimate, we can approximate the sales at \$4,000 and ask, "What is 15% of \$4,000?" We know that 10% of \$4,000 is \$400 and 5% is half of that, or \$200. Because 15% is 10% plus 5%, the commission would be \$400 + \$200, or approximately \$600.

EXAMPLE 9 Calculate the exact amount of Jeremy's commission from the information given in Example 8.

Solution Find 15% of $3,638.95.

$$\frac{A}{B} = \frac{P}{100}$$

$$\frac{A}{3,638.95} = \frac{15}{100}$$

$$100A = (15)(3,638.95)$$

$$\frac{100A}{100} = \frac{54,584.25}{100}$$

$$A = 545.8425$$

Therefore, Jeremy's commission was $545.85.

EXAMPLE 10 Enrollment for spring term is expected to be up 8 1/2% from last spring. If 8,150 students attended last spring, how many students can be expected on campus this spring term? Use the percent proportion.

Solution $\frac{A}{B} = \frac{P}{100}$

8 1/2% is the rate of increase, P. The increase is based on last year's figure, so B is 8,150. You need to find the amount of the increase, A.

$$\frac{A}{B} = \frac{P}{100}$$

$$\frac{A}{8,150} = \frac{8\frac{1}{2}}{100}$$

$$100 \cdot A = (8\ 1/2)(8,150)$$

$$100 \cdot A = 69,275$$

$$\frac{\cancel{100}^{1} \cdot A}{\cancel{100}_{1}} = \frac{69,275}{100}$$

$$A = 692.75$$

As so often happens in the real world, the answer is not a whole number, and an answer of 692.75 students does not make sense. The result should be rounded to suit the situation, and the answer will be reported as an increase of 693 students. The question asks how many students are expected on campus this spring, so the amount of increase needs to be added to last spring's enrollment.

8,150 + 693 = 8,843

Therefore, 8,843 students are expected to enroll for the spring term this year.

EXAMPLE 11 According to the 1990 census, the population of Glace Bay was 26,300, The new 2000 census reported a population of 28,141. What was the percent of increase over the 10-year period?

Solution You are not asked what percent one number is of the other; rather, you are asked to find the *percent increase.* The increase is based on the original population in the 1990 census. Therefore, B is 26,300. The amount of the increase, A, must be calculated.

$28{,}141 - 26{,}300 = 1{,}841$

Therefore, A is 1,841. You are looking for P, the percent of increase. Use the percent proportion.

$$\frac{A}{B} = \frac{P}{100}$$

$$\frac{1{,}841}{26{,}300} = \frac{P}{100}$$

$$26{,}300 \cdot P = 100 \cdot 1{,}841$$

$$26{,}300 \cdot P = 184{,}100$$

$$\frac{\overset{1}{\cancel{26{,}300}} \cdot P}{\underset{1}{\cancel{26{,}300}}} = \frac{184{,}100}{26{,}300}$$

$$P = 7$$

Therefore, the population increased by 7%.

Simple Interest Applications

One use for percent is to find interest. If you were to put some money into a savings account, this procedure would enable you to find the amount of interest that your deposit would earn in a year. To keep things uncomplicated, assume that this bank is paying only simple interest, not compound interest, and that you will earn interest only if the bank keeps your money for one year. Simple interest is figured once in a specific time period, and the interest is paid only on the principal. Compound interest is figured over and over again—compounded—so the principal increases each time the interest is added to it. With compound interest, an account earns interest on the principal and on the previously earned interest. The formula for compound interest is more complex than the simple interest formula, $I = PRT$. (Of course, earning only simple interest is not a very good deal financially, but it is good for our classroom purposes right now.)

EXAMPLE 12 Find the amount of interest that your account will earn in 1 year if you deposit $1,200 and the bank pays 4% interest per year.

Solution You can use the percent formula to answer this question because the time span is a single year. The question asked is "How much is 4% of $1,200?"

Change 4% to its decimal form, 0.04, and then multiply. 4% of $1,200 = 0.04 × 1,200 = 48.00, or $48.

Your account would earn \$48 in interest if you invested \$1,200 at 4% annual interest for 1 year.

Now expand the idea of finding interest by using the interest formula. It is a common formula, similar to $A = RB$, and is used when you need to compute simple interest for periods other than one year. The formula is an easy one:

$I = PRT$

where I is the amount of interest the account will earn.

P is the principal, the amount deposited or borrowed, the base quantity.

R is the rate or percent in its decimal form.

T represents the number of years involved.

EXAMPLE 13 Find the amount of interest you could earn on an account that pays 5.5% annually if you have invested \$3,200 for 4 years.

Solution Determine from the information given which letters in the formula you know the values of.

P = Money you invested = \$3,200

R = Rate of interest paid = 5.5% annually (per year)

T = Length of time that the money was invested = 4 years

$I = PRT$	Write the formula.
I = (\$3,200)(5.5%)(4 years)	Substitute given values.
I = (3,200)(0.055)(4)	Change the percent to a decimal.
I = 704.00 or \$704	Multiply

The account will earn \$704 over a 4-year period if you have deposited \$3,200 at 5.5% annual interest.

EXAMPLE 14 Find the amount of interest you would have to pay if you borrowed \$1,450 for your tuition and you were charged 7% annual interest for 3 years. Then find the total amount to be repaid.

Solution Determine which letters you know the values of.

P = Money that you borrowed = \$1,450

R = Rate of interest charged = 7% annual interest

T = Length of time for which the money was borrowed = 3 years

$I = PRT$	Write the formula
I = (\$1,450)(7%)(3 years)	Substitute given values.
I = (1,450) (0.07)(3)	Change the percent to a decimal.
I = 304.50 or \$304.50	Multiply.

You would be charged $304.50 in interest for the 3-year loan. To find out how much you would have to pay back at the end of the 3 years, you need to add the interest to the *principal,* the amount you borrowed.

$1,450 + $304.50 = $1,754.50 to be paid back.

7.5 EXERCISES

A. Estimate an answer to each of the following exercises. Remember that when you estimate, the numbers should be reasonable to work with so you can use mental math whenever possible.

1. Blue Bird Airlines is offering a 35% discount on regular round-trip airfares to Florida. If the ticket usually costs $178, estimate how much you will save by buying it during this sale.
2. National Car Company is offering a 12% discount on the sticker price of its new vans. If one of these vans normally sells for $17,895, estimate how much you save by buying it during this sale.
3. Allen County's sales tax rate is 6 1/2%. Estimate how much county sales tax you will pay for a new deluxe vacuum cleaner that costs $122.95.
4. Estimate how much you will save on a car priced at $8,285 if the dealer is advertising a "12% off sale."

B. Find the answers to the following application problems. Use the percent formula or the percent proportion. Round percent answers to the nearest tenth of a percent. Be sure to label the answers, because you are finding the solution to a real-world problem.

5. What will you pay for the discounted airplane ticket in Exercise 1?
6. How much will you pay for the van in Exercise 2 if you buy it during this sale?
7. A house originally purchased for $86,500 sold this month for $96,880. What was the percent of increase in the value of this house?
8. In a recent survey, 44% of the people who were asked said that they preferred candies with nuts to candies without. If 176 people preferred the nut candies, how many people were asked their preference?
9. In a recent taste test at Dwight Marketing Research Center, 24% of the people who participated in the tests liked cereal B better than cereal A. Cereal B was preferred by 120 people. How many people participated in the taste test between the two cereals?
10. A new washer to Mayrun Appliance regularly sells for $425.50. In order to compete with the new store down the street, Mayrun has recently reduced the price of this washer to $375.44. What is the percent of decrease for this sale? (See Examples 4 and 11.)
11. Marty's last credit statement showed an interest charge of $5.40. If the rate of interest on the card is 1.8%, what was his balance due to the account?
12. Amber earned $176 in interest on her certificate of deposit at the bank of Lewiston. This bank pays 5 1/4% simple interest per year. How much money did Amber have invested in her CD last year?
13. The population of Mayes Beach, Georgia, increased from 32,650 in 1990 to 35,262 in 2000. What was the percent growth? (See Examples 4 and 11.)
14. The average price of a new family car in 1982 was $8,250. The average price for the same type of car was $18,150 in 1996. What was the percent increase during this period? (See Examples 4 and 11.)
15. Marshall City's sales tax rate is 6 1/2%. How much will you pay for a new dining room set priced at $595?
16. A portable tool chest is on sale for $88.95. The sales tax rate in your area is 5 1/2%. How much will you pay for this deluxe tool chest?
17. José is selling furniture for Ortega and Sons. He earns a weekly salary of $225 plus a 6% commission of everything he sells. Last week his sales totaled $2,348. How much money did he earn for last week's work?

18. Salespeople working for Goodman Insurance Company earn a commission of 0.5% on the value of the policies they sell. They are also paid a monthly salary of $830. In April, Martina sold policies with a total worth of $320,000. What was her income for the month of April?

C. Use the simple interest formula, $I = PRT$, to find the answers to these questions.

19. Estimate how much interest you will be charged to borrow $800 for 3 years if the bank charges 9% annual interest.

20. Estimate how much interest your account will earn in 3 years if you deposit $1,450 and it earns 5% annual interest.

21. Martin borrowed $560 to pay his tuition for this term. He knows that he will pay it back in 1 year, and the bank is charging him 8% annual interest. Estimate how much money will have to be repaid. (*Note:* The amount to be repaid includes the amount borrowed as well as the interest. See Example 14.)

22. If you deposit $1,200 in a savings account that earns 4 1/4% annual interest for 2 years, estimate the balance in the account at the end of the 2 years.

23. How much interest will you earn on a deposit of $600 if you leave it in the bank for 2 years and it earns 6% per year?

24. How much interest will you have to pay on a loan of $850 if you borrow the money for 1 year at 7 1/2% interest per year?

25. Jacqueline received an inheritance from her grandfather. He left her $3,500 to help with her school expenses. She wants to leave the money in a savings account for 3 years. The bank will pay her 5 1/2% annual interest. How much interest will she earn in the 3 years?

26. How much interest will you pay on a $2,500 loan if you borrow the money for 2 years at 7 1/2% annual interest?

27. Amanda needs to apply for a car loan from her credit union. The credit union is charging 6.5% annual interest on used-car loans if the money is borrowed for 4 years. Amanda wants to borrow $3,250 for the car she is buying. How much interest will she have paid at the end of the 4 years? What is the total amount she will have to repay?

28. Paula deposited her holiday bonus check of $1,100 in a special savings account. This account earns 4.5% interest per year. How much money will she have in her account at the end of 2 years?

D. Read and respond to the following exercises.

29. Explain in words why you can change from a decimal to its percent equivalent by multiplying the value by 100%.

30. When using the percent formula to solve an application problem, which component of the formula do you identify first? Why?

31. Explain how you can figure in your head the amount of a 20% tip on the bill for a restaurant dinner costing $17.95.

32. Write an application problem involving percent of increase. Are your values reasonable? Is your wording clear and concise?

33. Read the problem below and explain in words what steps you would follow to find the answer to the question.

    Your part-time job pays $5.80 an hour. Your supervisor has just told you that because you have been doing such a good job, you have been given a raise and will be making $6.50. What is the percent of increase that you have earned? Round to the nearest whole percent.

34. Read the problem below and explain in words what steps you would follow to find the answer to the question.

    The average price for a gallon of gasoline increased by 225% in the 30-year period from 1958 to 1988. In 1958, the average price for a gallon of gas was approximately $0.52. Calculate the average price for a gallon of gasoline in 1988.

7.6 Interpreting Data

A **circle graph**, often called a **pie graph**, is circular and is divided into portions, or slices. The pie represents one whole body of data, and the slices represent different parts of that data. The sizes of the slices are proportional to the sizes of the different components. Most often, percents are used to represent the portions that make up the whole (100%) pie graph.

Consider the following example, which uses the pie chart shown in Figure 7.1 to convey information about the grading procedures in Dr. Hall's American Literature course.

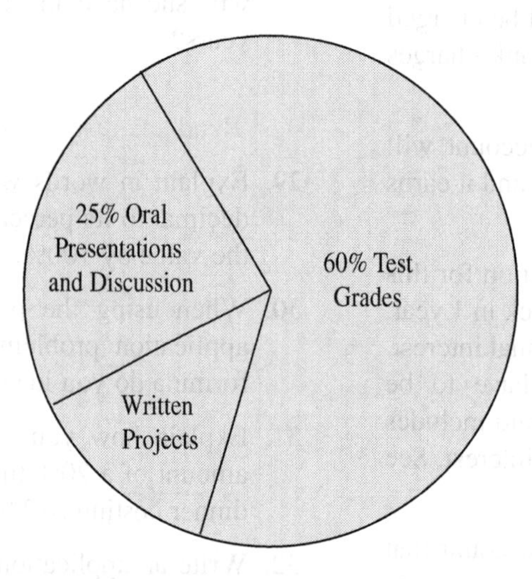

FIGURE 7.1

In Dr. Hall's American Literature course, a student's grade depends on three different components: test scores, oral presentations, and written projects. The three parts do not contribute equally to the final grade. The distribution is illustrated by the pie chart in Figure 7.1. It is quickly evident that the test grades have more impact on the final grade than do the other components. Use Figure 7.1 to answer the following questions.

a. What percent of the course grade comes from test scores?
b. What percent comes from written projects?
c. Is the sum of the oral and written components equal to, greater than, or less than the test score component? By what percent?
d. If the most it is possible to earn in the course is 800 points, how many points can come from test scores?

Answers:

a. The graph shows that 60% of the grade comes from test scores.
b. The percent that comes from written projects can be found by adding the other components together and subtracting that total from 100%.

   $60\% + 25\% = 85\%$

   $100\% - 85\% = 15\%$

   Therefore, 15% of the final grade comes from written projects.

c. Test scores account for 60% of the course grade. 15% + 25% = 40%, and 40% is less than 60%. Therefore, the sum of the written and oral presentations is less than the test score component. 60% – 40% = 20%, so the other components account for 20% less than the test scores.

d. 60% of the final grade comes from test scores.

60% of 800 points = 0.60 × 800 = 480 points

EXAMPLE 1 Use Figure 7.2 to answer the following questions.

a. What percent of the income is spent on utilities?

b. What percent is spent on miscellaneous items?

c. If the monthly income is $2,400, how much is spent each month for the mortgage payment?

d. From a monthly income of $2,400, how much is spent in paying taxes?

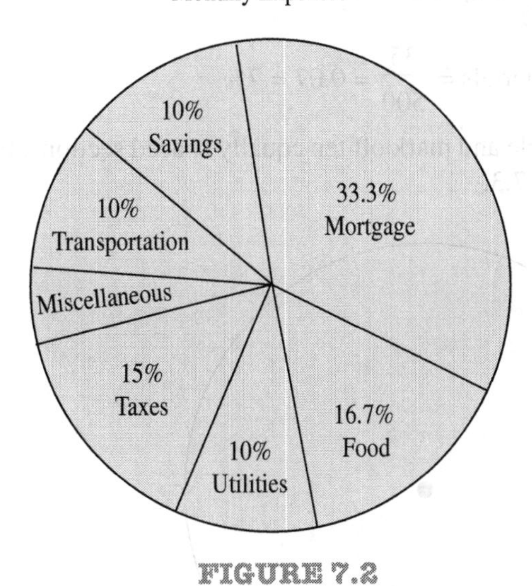

FIGURE 7.2

Solution From the graph, it can be determined that

a. 10% is spent on utilities.

b. 5% is spent on miscellaneous items because all the other portions added to 95%.

100% – 95% = 5% for miscellaneous items.

c. You need to find 33.3% of $2,400.

$A = 33.3\% \cdot \$2{,}400$ or $\dfrac{A}{2400} = \dfrac{33.3}{100}$

Therefore, the mortgage payment is $799.20.

d. Taxes take 15% of the monthly income.

$A = 15\% \cdot \$2{,}400$ or $\dfrac{A}{2400} = \dfrac{15}{100}$

Therefore, $360 is withheld for taxes each month.

EXAMPLE 2 Use the following data to construct a pie graph. Label the components (or slices) in percents.

College Student's Monthly Expenses

Rent = $250 Transportation = $50

Food = $165 School materials = $35

Solution To calculate the percent parts, first find the monthly total.

$250 + $165 + $50 + $35 = $500

Find each component as a part of the whole, and then write it in percent form.

$$\text{Rent} = \frac{250}{500} = 0.5 = 50\%$$

$$\text{Transportation} = \frac{50}{500} = 0.1 = 10\%$$

$$\text{Food} = \frac{165}{500} = 0.33 = 33\%$$

$$\text{School Materials} = \frac{35}{500} = 0.07 = 7\%$$

Draw a circle and mark off ten equally spaced sections. Each section represents 10%. See Figure 7.3.

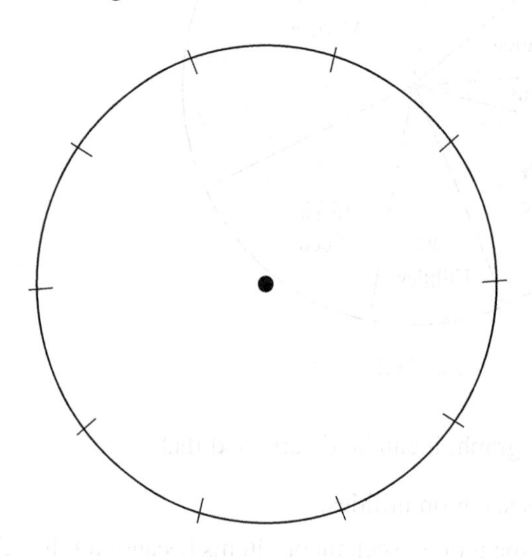

FIGURE 7.3

Mark off the expense percents, estimating where necessary and labeling each part. See Figure 7.4.

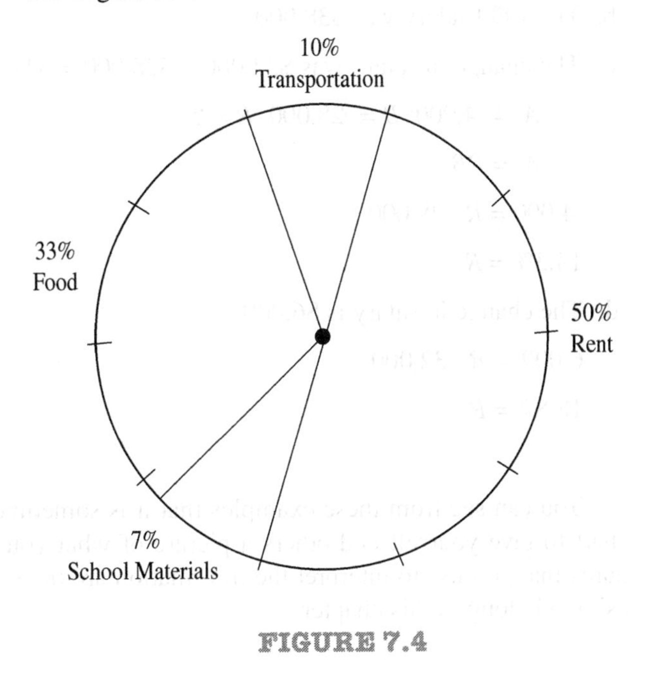

FIGURE 7.4

EXAMPLE 3 Use the bar graph in Figure 7.5 to answer these questions.

a. What was the computer operator's salary in 1995?

b. What was the annual salary in 1999?

c. What was the percent increase, to the nearest tenth of a percent, between 1993 and 1995?

d. What was the percent increase between 1995 and 1999?

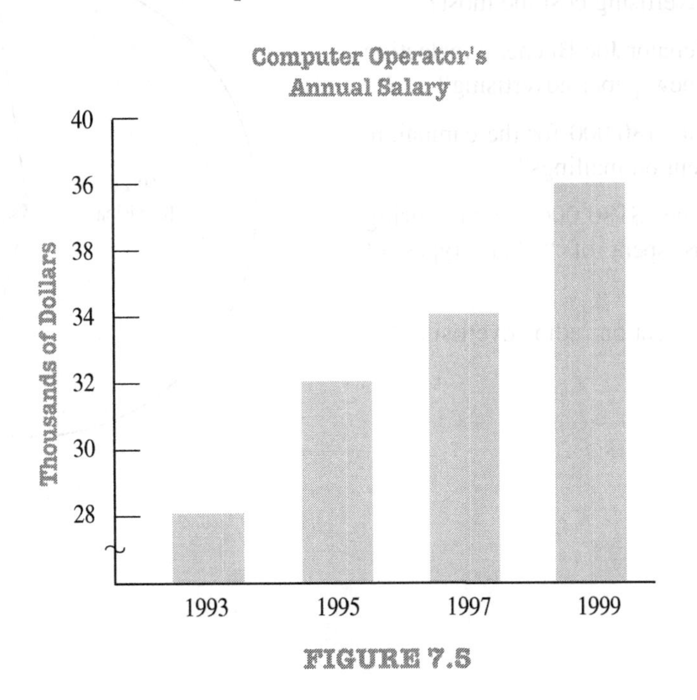

FIGURE 7.5

Solution a. The 1995 salary was $32,000.

b. The 1999 salary was $38,000.

c. The change in salary was $32,000 – $28,000 = $4,000

$A = 4{,}000;\ B = 28{,}000;\ R = ?$

$A = RB$

$4{,}000 = R \cdot 28{,}000$

$14.3\% = R$

d. The change in salary is $6,000.

$6{,}000 = R \cdot 32{,}000$

$18.8\% = R$

You can see from these examples that it is sometimes very helpful to use a pie chart to give yourself and others a picture of what you are discussing. The procedures that you use to interpret the information are the same as those you have been using all along in this chapter.

7.6 EXERCISES

A. Use Figure 7.6 for Exercises 1 through 5.

1. In Senator Joe Bucher's reelection campaign, what method of advertising cost the most?
2. What percent of Senator Joe Bucher's reelection expenses paid for newspaper advertising?
3. If the senator spent $180,000 for the campaign, how much was spent on mailings?
4. How much of the $180,000 spent during the campaign was spent of "other" types of advertising?
5. What amount was spent on radio advertising?

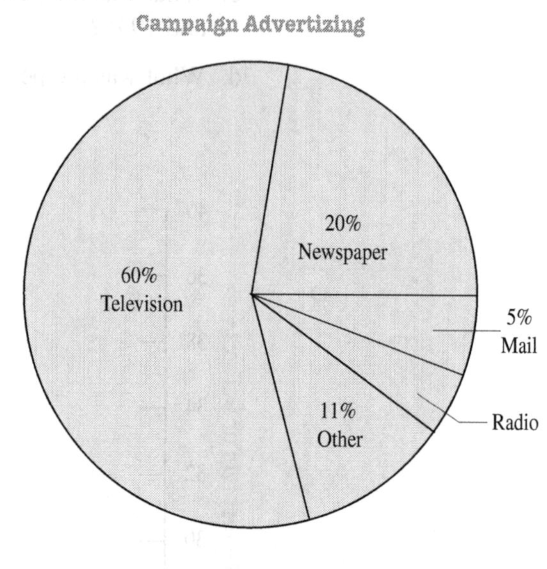

FIGURE 7.6

B. Use Figure 7.7 for Exercises 6 through 12.

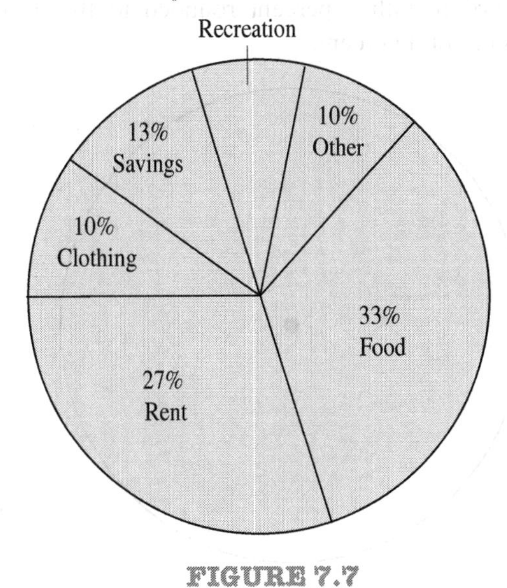

FIGURE 7.7

6. What percent of this family's income goes into savings?
7. What percent is used for food?
8. What percent is spent on recreation?
9. If the monthly income is $3,200, how much is spent for food?
10. How much is spent on rent if the monthly income is $3,200?
11. If the monthly income increases to $3,500, but the budget is maintained, how much will now be spent for clothing?
12. How much will be spent for recreation if the monthly income increases to $3,500?

C. Use Figure 7.8 for Exercises 13 through 18.

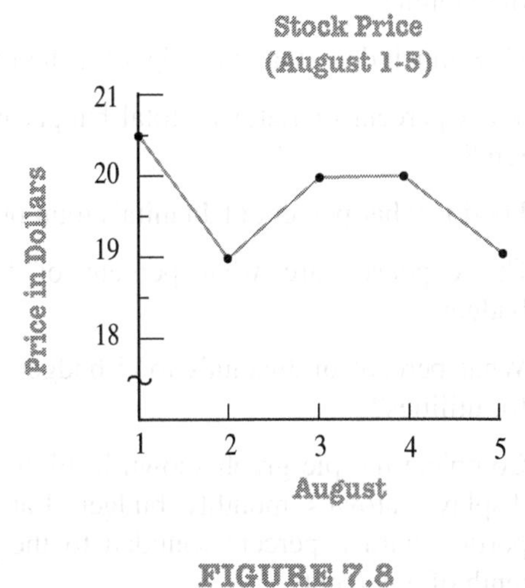

FIGURE 7.8

13. What was the stock's value on August 1?
14. What was the value of the stock on August 3?
15. What was the percent decrease in the value of the stock from August 4 to August 5?
16. What was the percent decrease from August 1 to August 2?
17. What was the average (mean) value of the stock for the 5-day period?
18. What were the median and the mode (see Section 1.6) for the stock in the 5-day period?

D. Use Table 7.1 for Exercises 19 through 26 and 29.

Table 7.1

Monthly budget		
	Latoya	Juanita
Rent	$325	$400
Utilities	$100	$150
Food	$200	$175
Car	$175	$260
Miscellaneous	$200	$215

19. What is the total amount Latoya has budgeted for the month?
20. How much does Juanita budget for the month?
21. What percent of Latoya's total budget goes for rent?
22. Food is what percent of Juanita's total budget?
23. Car expenses are what percent of Latoya's budget?
24. What percent of Juanita's total budget is spent for utilities?
25. Complete the pie graph shown in Figure 7.9 to display Latoya's monthly budget. Label each portion with a percent rounded to the nearest tenth of a percent.

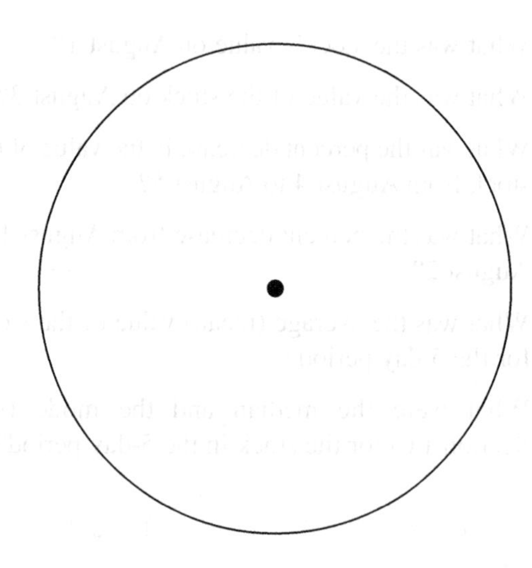

FIGURE 7.9

26. Complete the pie graph shown in Figure 7.10 to display Juanita's monthly budget. Label each portion with a percent rounded to the nearest tenth of a percent.

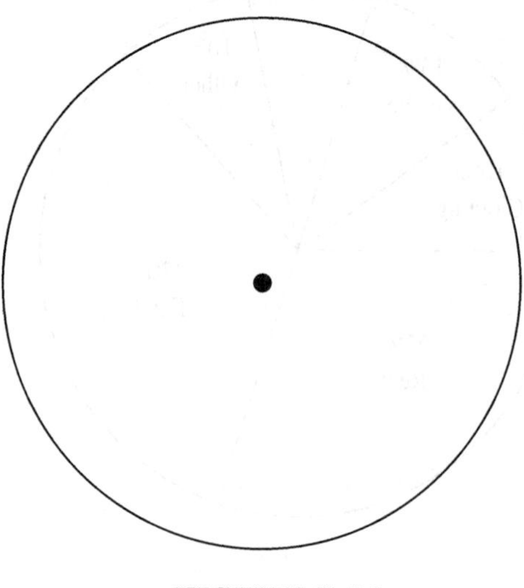

FIGURE 7.10

E. Read and respond to the following exercises.

27. Reread Exercise 9 and explain in words how you could estimate an answer to the question.
28. Study Figure 7.7 and then describe in words the distribution of monthly income in general terms, not in specific amounts. For example, you might notice that the amount for rent is a little more than one-fourth of the income. Make a statement about each of the categories.
29. Use Table 7.1 to make a bar graph comparing Latoya's and Juanita's monthly budgets. (See Section 1.5 if you need a refresher about how to make a bar graph.)
30. Bring in a copy of an article from a newspaper or magazine that includes percents. Make a graph or a pie chart that would help to explain the information given in the article.

7.7 CHAPTER REVIEW

In this chapter, you have studied percents. Because you cannot multiply or divide by a percent (other than 100% as a form of 1), you had to learn some techniques for changing percents to their decimal or fraction forms. In order to do this, you needed to realize that 100% is another form of 1. Because multiplying or dividing by 1 does not change the value of the quantity being used, multiplying or dividing by 100% does not change the *value* of the fraction, decimal, or percent. It changes only the way that figure looks.

The principles for changing among forms are:

1. Percent means *per hundred.*
2. To change *from a decimal form to a percent form, multiply by 100%.*
3. To change *from a fraction form to a percent form, either multiply the fraction by 100%/1, or change the fraction to its decimal form* by dividing, and then *multiply the decimal by 100%.*
4. To change *from a percent to a decimal or fraction, divide by 100%* and then *simplify* the answer.

It is important to remember that when you do problems using the percent formula $A = RB$, you must first identify the values for R, B, and A. These values must then by substituted into the percent formula. If a percent value, R, is given, it must be changed to its decimal equivalent before any calculating can be done. If both number values are on one side of the equals sign, the formula tells you to multiply. However, if one of the numbers is on the same side of the equals sign as a letter, it is necessary to "undo" the indicated multiplication to isolate the letter. You undo multiplication by dividing. If you are asked to find the percent, R, it is necessary to change the final decimal answer to its percent equivalent.

When you are doing problems by using the percent proportion, $A/B = P/100$, follow the same steps as for solving any formula. Identify the values you are given, substitute them into the formula, and solve for the missing value. The one thing you need to remember when using the percent proportion is that the P value is the percent value in its percent form. You do not need to change it to a decimal first before substituting into the proportion.

In the case of application problems, it is necessary to identify from the statement of the problem, the values for the rate, R or P; the base, B; and the part or amount, A. The rate, or percent, is the easiest to spot. It comes before the percent sign, %. The base, B, is usually the next easiest to pick out. It is the original value—the quantity before anything happens. The amount, A, is the part that you are finding. It might be the part paid in tax, the amount of decrease or discount, or the number of people who did a certain activity. It does not make any difference whether you choose to solve the problem using the percent formula or the percent proportion. The more application problems you do, the better you will become at picking out the appropriate values for R or P, B, and A.

When you have an interest problem to solve, you must use the simple interest formula, $I = PRT$. Identify the values you are given, and solve for the one you are asked to find.

To summarize, when you are solving a problem involving percent, follow these steps:

1. Identify the values for R or P, B, and A in that order.
2. Write the percent formula, $A = RB$, or the percent proportion, $A/B = P/100$.
3. Substitute the values you know.
4. Decide from the formula whether you are to multiply or divide to find the answer, and then solve for the unknown value.
5. Give the answer in the form requested, such as a percent (%), or money, or rounded to a particular place.
6. Reread the problem to see if your answer makes sense.
7. When you are asked to interpret percent data from a chart, you need to use all the skills that you learned before when you studied graphs.
8. When you are estimating an answer, use reasonable numbers and use mental math whenever possible.

This chapter has been a long but important one. If you are going on to study algebra, the procedures you have learned here will be very helpful. If you go on to business math instead, or to applied math courses, you will often use these "formula skills."

REVIEW EXERCISES

A. Answer each of the following questions.

1. Sixty-two percent of the students who were asked thought that their school should sponsor a food drive to support a shelter for the homeless near their campus. What percent of the students who were asked disagreed with this idea?
2. Miguel has decided to try to save 28% of his paycheck. He will use the rest to pay expenses. What percent will he use for paying his bills?
3. What percent of the problems could you afford to miss on your final exam and still keep your B average, 85%?
4. Eighteen percent of the people in Miss Blake's composition class failed to turn in their final drafts on time. What percent of the students met the deadline?

B. Fill in the following chart. Be sure to reduce fraction answers. Round decimal answers to three places, and round percent answers to the nearest tenth of a percent.

	Fraction	Decimal	Percent
5.		0.16	
6.			20%
7.	1/4		
8.		0.38	
9.	1/8		
10.			6.5%
11.		1.54	
12.	7/100		
13.			112%
14.		0.006	
15.	1/12		
16.			12 3/4%

C. Estimate an answer to each of the following exercises. Remember that when you estimate, the numbers should be reasonable to work with so you can use mental math whenever possible.

17. Find 19% of 77.

18. Find 16.5% of 102.
19. What is 0.8% of 53?
20. What is 1.2% of 48?
21. Find 210% of 80.
22. Find 148% of 205.

D. Answer the following exercises by using the percent formula, $A = RB$, or the percent proportion, $A/B = P/100$. Be sure to identify the values for R or P, B, and A before you begin to work the problem. Round all percent answers to the nearest tenth of a percent if necessary. All other answers can be rounded to the nearest hundredth.

23. Find 42% of 80.
24. What is 11.2% of 90?
25. 30 is 50% of what number?
26. 42 is 25% of what number?
27. 2.8% of what number is 70?
28. 32 is what percent of 80?
29. What percent of 35 is 15?
30. What percent of 67 is 30?
31. 60 is 8.5% of what number?
32. 4.5% of what number is 135?
33. 1.5 is what percent of 60?
34. What percent of 45 is 12?
35. Find 12.8% of 120.
36. 9 1/2% of 126 is what number?
37. 12 1/4% of what number is 88?
38. 4 1/4% of what number is 80?
39. 5 1/2% of what number is 50?
40. Sixteen percent of what number is twelve?
41. Nine is what percent of sixty?
42. Twenty-four percent of 96 is what number?
43. Eighty-two percent of 16 is what number?
44. Seventy-two is what percent of 20?

E. Estimate an answer to each of the following exercises. Remember that when you estimate, the numbers should be reasonable to work with so you can use mental math whenever possible.

45. Dalane is working on commission, selling service contracts on copiers. He makes 30% on what he sells. His most recent sale was for $235 for a year's service. Estimate how much commission he will make on that sale.
46. Franklin County is considering raising its sales tax to 6 3/4%. Estimate, if it does so, how much county sales tax will you have to pay on an item that costs $42.00?
47. Estimate the amount of money Quenten's savings account will earn this year if he deposits $2,300 and his bank is paying 4 1/2% annual interest.
48. If the usual error rate is 18%, estimate the number of faulty tax returns that could be expected in a group of 4,210 returns.
49. Williams Manufacturing Company is trying to improve its quality control. They produce approximately 1,850 widgets per week. If their error rate has been 6%, estimate how many defective widgets they should expect to find this week.
50. Of the 176 students taking math this term, estimate the number of students who will pass the class if the usual passing rate is 63%.

F. Solve the following application problems. Round percent answers to the nearest tenth of a percent, and round money answers to the nearest cent. Be sure to label your answers.

51. Sixteen people, stopped on a corner downtown, could not give the mayor's name. This represented 40% of the people who were asked. How many people were questioned?
52. A car purchased last year for $12,500 now has a market value of $9,500. What is the percent of decrease in the value of the car?
53. Sara has 110 math review problems to do. She has finished 95 of those problems. What percent of the problems has she completed? Round your answer to the nearest tenth of a percent.
54. How much money will you need to deposit in your savings account if you want to earn $450 in interest this year? Your account pays 5% simple interest per year.

55. Village Antiques is having a "12% off sale." How much will a quilt cost on sale if it is normally priced at $55?

56. A college is experiencing an 18% growth rate from one year to the next. How many students can be expected on campus next fall term if there are 10,500 students on campus this fall?

57. Of the students registered for biological sciences this term, 45 earned an A in the course. If there are 320 students taking that course, what percent earned an A? (Round to the nearest tenth of a percent.)

58. A dress is on sale for 30% off the original price. If a $24 discount is given, what was the original price of the dress?

59. Kawai is paid a 3% commission on his total sales at Class Act Furniture. His sales total last week was $3,219.95. How much commission did he earn?

60. The occupancy rate at the Holton Inn had been averaging 78% on weekend nights. Last Saturday, 38 of the 45 rooms were taken. Was this an average, above average, or below average weekend for the Inn?

G. Use the simple interest formula, $I = PRT$, to answer the following questions. You may use a calculator if you wish.

61. What is the interest earned in 3 years at 5 1/2% annual interest on a deposit of $1,200?

62. What is the amount of interest that you must pay on a 4-year loan if you borrow $800 at 8% annual interest?

63. How much interest will a $2,500 savings account earn in 2 years if it pays 6 1/4% annual interest?

64. How much will the same savings account earn if the bank pays 4 1/2% annual interest and the money is left on deposit for 3 years?

65. How much money must be paid back at the end of 5 years by a student who borrows $7,000 at 8.5% annual interest on a simple interest loan?

66. Hassan will transfer part of his savings to a local bank. How much interest will he earn in 2 years if the bank is paying 5 1/4% interest on a 2-year CD and he deposits $12,500?

H. Use Figure 7.11 for Exercises 67 through 72.

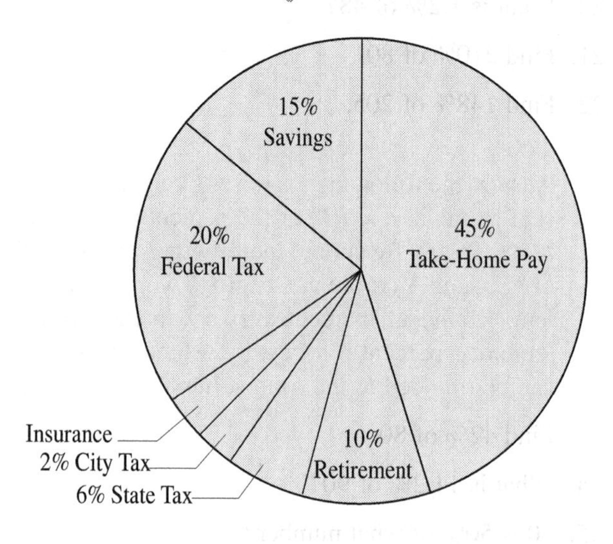

FIGURE 7.11

67. What percent of the salary goes to pay city taxes?

68. What percent of the pay is for federal taxes?

69. What percent is paid for insurance?

70. If a person's biweekly earnings are $1,750, how much is deducted for retirement?

71. If a person's biweekly earnings are $1,750, how much money does the person have to take home? (take-home pay)

72. Approximately how much money does this person take home in salary in a year's time?

I. Explain, in words, how you would estimate the answer to the following questions. Include the values you would use and describe the steps you would take.

73. The price on a midsize van in 1998 was $21,275. The price for a comparable van in 2002 was $24,495. What is the percent of increase?

74. 1,308 is what percent of 4,529?

75. How much simple interest will an account of $13,525 earn in 3 years at 6 3/4%?

76. How much should Zachary's gross paycheck (before taxes) be for the week when he earned 6% commission on the $2,950 website contract that he sold and also earned his base pay of $580 per week?

J. Answer each of the following questions. These problems may require multiple steps or approaches to reach a solution.

77. Using your calculator, find the amount of money in an account at the end of 4 years if $2,000 was deposited originally at 8.5% and the interest is compounded annually. (*Hint:* In annual compounding, the interest is computed at the end of each year, and the principal is increased by that amount. Thus, there is a new base amount each year.)

78. Is there a difference between investing $10,000 at 10% for 1 year and investing $10,000 at 5% for 6 months (1/2 year) and then reinvesting the new amount for another 6 months at 5%? Explain your answer. Isn't 10% for 1 year the same as 5% earned twice a year?

79. How much would you have paid at the end of 10 years on a simple interest loan if the agreement was for 8 3/4% and the amount of interest charged was $2,304.75?

80. At a recent meeting of the college president's cabinet, a discussion of projections for student enrollment was held. The vice-president for student services stated that the enrollment for next autumn was expected to increase by 9%. When asked about last autumn's enrollment, the vice-president reported that approximately 16,000 students enrolled. The president then immediately stated that an estimated projection for autumn term next year would be 17,600 students. Explain how he might have arrived at this figure.

CHAPTER TEST

1. Identify *R*, *B*, and *A* in this percent statement: 16% of 40 is 6.4.

2. In Shirley's division, 89% of the faculty were rated excellent by their students. What percent did not receive the excellent rating?

3. Change 7/1,000 to a percent.

4. Change 16 3/4% to a decimal.

5. Change 7 5/8 to a percent.

6. Change 14.5% to a fraction reduced to lowest terms.

7. Use $I = PRT$ to find what amount of interest must be paid on a 4-year, simple interest loan if $2,100 is borrowed at an annual rate of 5 1/2%.

8. Read the following problem and explain in words how you would estimate an answer.

   You must pay $17.60 in sales tax on a car stereo that sells for $320. What is the sales tax rate?

9. Use the percent formula to find the exact answer to Exercise 8. Show your work.

10. Use the percent proportion to find the exact answer to Exercise 8. Show your work.

11. What is 8 3/4% of 400?

12. 16 is what percent of 800?

13. What percent of 82 is 16.646?

14. 25.2 is 36% of what number?

15. Estimate how much commission Alaine will earn on sales of $2,740 if his commission rate is 5%.

16. A dress that regularly sells for $70 is on sale at 25% off. What is the sale price of the dress?

17. An airline ticket to Mexico City that usually costs $328 is on sale for $288.64. What is the percent of decrease in the cost?

18. How much money must you invest at 6% annual interest to earn $1,200 in 1 year?

19. In a pie chart used to show the breakdown of the college's annual budget, the money set aside for health insurance is 8.6%. How much is the cost of health insurance for the school if the annual budget is $13,560,000?

20. Explain in words how you would find the answer to the following question.

    What is the regular sticker price of a car that is selling for $14,025 at a "15% off sticker" sale?

Cumulative Review, Chapters 1–7

Solve each of the following exercises. Show all your work. Reduce all fraction answers to lowest terms. In Exercises 1 through 5, add.

1. 6.3 + 18 + 24.6
2. $\frac{1}{2} + \frac{3}{4}$
3. 1,214 + 2,765
4. 2 5/8 + 7 2/3
5. 0.065 + 1.2 + 25.62

In Exercises 6 through 10, subtract.

6. 14 7/9 – 5 2/3
7. 900 – 371
8. 64.36 – 8.97
9. 156 – 8.3
10. 49 – 18 4/5

In Exercises 11 through 15, multiply.

11. Find 8/15 of 90.
12. 0.08 × 13.75
13. \$491 × 23
14. 5 3/4 × 8 3/8
15. 0.076 × 100

In Exercises 16 through 20, divide. If necessary, round to the nearest hundredth.

16. 76.4 ÷ 0.06
17. 1,488 ÷ 4
18. 2 3/8 ÷ 3/4
19. $0.7\overline{)29}$
20. 6 3/16 ÷ 3

Find the answer to each of the following exercises.

21. Change 12 1/4 to an improper fraction.
22. What is the missing numerator in $\frac{3}{5} = \frac{\;}{20}$?
23. Reduce 65/210 to lowest terms.
24. Change 0.45 to a percent.
25. Change 12 1/2% to a fraction.
26. Round 65.187 to the nearest tenth.
27. Round 0.0053 to the nearest thousandth.
28. Change 2.375 to a fraction.
29. Change 5/12 to a decimal rounded to the nearest hundredth.
30. Write the ratio of 8 feet to 6 inches in simplest form.
31. Write the ratio of 3.2 to 2.4 in simplest form.
32. Find 62% of 13.8.
33. 12 is what percent of 90? Round the answer to the nearest tenth of a percent.
34. 87 is 60% of what number?

Simplify Exercises 35 through 39 using the Order of Operations Agreement.

35. 2 + 3[4 – (6 ÷ 3)]
36. $\left(\frac{2}{3}\right)^2 \div \sqrt{\frac{25}{36}}$
37. (11.8 – 2.9)(0.06 × 1.4)
38. 81.9 ÷ 0.9 × 2
39. 1/2 × 2/3 + 5/6 × 1/5

In Exercises 40 through 43, solve for N.

40. $\frac{7}{15} = \frac{56}{N}$
41. $\frac{11}{N} = \frac{27}{18}$
42. $\frac{5.6}{18.3} = \frac{N}{6.1}$
43. $\frac{N}{40} = \frac{36}{100}$

Find the answer to each of the following application problems.

44. If the sales tax rate is 6%, find the amount of tax on an item selling for \$158.95.

45. Two-thirds of the pieces of candy in an assorted box of chocolates have soft centers. If there are 36 pieces of candy in the box, how many have soft centers?

46. Max's checking account balance was $189.64 before he wrote checks for $12.96, $56.37, and $24.20. How much money remains in his account?

47. If 21 of the 35 points in the football game were scored on passing plays, what percent of the points was scored on passes?

48. Using the information from Exercise 47, find what fraction of the points was scored on passing plays.

49. If 15 bagels cost $3.75, what do 8 bagels cost?

50. The scale on a map shows that 2 inches equal 15 miles. How fat apart are cities that are 5 1/2 inches apart on the map?

STUDY SKILLS
MODULE

A Place to Study

Some students think that it doesn't matter where they study, whereas others think that studying in a lot of different places keeps them from getting bored. Actually, "experts" on study behaviors recommend that you choose one or two locations and always work in one of those places.

There are many reasons for this recommendation. To get the most benefit from the time you spend studying, you must be able to concentrate completely on what you are doing. That means that you must be someplace where you are able to control the amount of noise (or quiet) and are able to eliminate distractions, such as phones or people moving around you. Also, you can condition your mind and body to know that when you come to this place, it is time to work.

The place should be well lighted and the temperature should be comfortable. The area should not be one in which you do other things on a regular basis, like eat, sleep, watch television, or play cards. Many students say that they like to study while sitting on their bed. There are two unproductive things that can happen with this location. The most obvious is that you are so comfortable you fall asleep instead of study. The opposite of this is that students begin to associate the bed as a place to work and cannot relax and fall asleep when they should be doing that. Their minds are not able to shut down and allow them to get a good rest.

When you sit down to work, you want to get right to work and use the time wisely. By establishing a regular setting and routine, you are conditioning your mind and body to get on-task and get to work right away when you come to that place. Sitting up in a comfortable (but not too relaxing) chair at a table or desk is the best choice. You should be able to equip this area with proper lighting and all the supplies you will need. Keeping the area neat and orderly will also contribute to the efficient use of your time. You don't have time to waste looking for missing papers or a pencil or calculator.

You might find it beneficial also to have a regular place to study while on campus. Use that free hour between classes to do homework, study for a test, or write a paper. Stay a little longer instead of leaving immediately after class. It is easier to study without the distraction of home and family. There may be an empty classroom or an area in the Learning Center where you could work during those times. Try the library—there is usually a variety of study settings for you to choose from. Of course, you will need to have your supplies with you if you decide to study at school.

Think about it and see if you can't be creative in working out a good efficient place to study.

Your Point of View

1. Describe the place where you usually study. What are the benefits of this location? What are the disadvantages?
2. The level of noise that one person finds relaxing may be distracting to another person. Describe what your background noise tolerance is while studying. What can you do to control this level?
3. What are some things that you could have in your study area that might keep you on-task?
4. List three changes that you will make this week to improve your study location(s). Report on which of these changes were beneficial.

CHAPTER

The Why and What of Algebra

LEARNING OBJECTIVES

When you have completed this chapter, you should be able to:

1. Recognize situations where using algebra might be helpful.
2. Recognize variable, formulas, expressions, equations, and inequalities.
3. Simplify algebraic expressions by using the Order of Operations Agreement and by combining like terms.
4. Evaluate algebraic expressions and formulas.
5. Recognize the properties of real numbers.
6. Solve simple algebraic equations and inequalities by inspection or intuition.

In this chapter, we begin to tie arithmetic and algebra together. We will look at the differences between them and examine some applications that show why algebra is such a useful tool.

8.1 The "Why" of Algebra

You may think that you have never had any experience with algebra other than in a classroom. That, in fact, is probably not the case. Think about how you figure your grades in a class. If your grade for the term is based simply on the average of your scores on three tests, how do you calculate your grade? To find your average, you add the three scores together and divide the total by three. For instance, if your grades were 87, 95, and 94, you would figure the average this way:

$87 + 95 + 94 = 276 \qquad 276 \div 3 = 92$

Therefore, your average would be 92.

If your instructor wanted to find the averages for all the students in her class, she would have to figure each total separately and then divide by three. She might decide to use her computer to help her, instructing it to do the task over and over again with each student's grades. One way to do this would be to write a general statement that the computer could "understand" and would respond to by adding each set of three numbers and dividing the total by three to get the average. She would write a *formula*, using three different letters to represent the test grades. Using letters to represent the numbers allow her to make a general statement (one that could be used for many different student's scores) and the formula then gives the computer exactly those instructions. She might use this one:

$$A = \frac{a + b + c}{3}$$

where A stands for average and a, b, and c are the three scores. Any letter could be used to stand for the average, as long as what that letter represents is clearly defined, but A seems appropriate to stand for "average."

The use of letters and mathematical symbols to describe a process is **algebra**. Algebra and its formulas and symbols state in general terms what arithmetic numbers and symbols state in specific terms.

This isn't as complicated as it sounds. Think about the last vacation you planned. You knew approximately how fast you could drive in an hour (miles per hour) and about how many hours you could travel each day. Knowing that information, you could get an approximate idea of how many miles you could travel per day.

If you know you can drive about 50 miles in an hour, you know you can drive 100 miles in 2 hours and 150 miles in 3 hours, and so on. You can multiply the rate at which you travel by the number of hours you drive and find out how many miles you can travel. In general, the *distance* you can travel is equal to the *rate* at which you drive multiplied by your traveling *time*. In an algebraic formula,

Distance = Rate · Time or $D = rt$

Again, you do not have to use D, r, and t, but they are the most logical choices to help you remember what the parts of the formula represent.

You could write this information in table form as well. In this example, the average rate at which you drive never changes, so the only things that vary are the time and the distance. The following table displays the relationship between the time and the distance traveled.

t (hours)	1	2	3	4	5	6	7
D (miles)	50	100	150	200	250	300	350

In Chapter 15 you will learn how to display this type of information on a two-dimensional graph called a Cartesian coordinate system. You could indicate the hours (t) on the horizontal scale and the miles or distance (D) on the vertical scale. Figure 8.1 shows how this graph might be drawn.

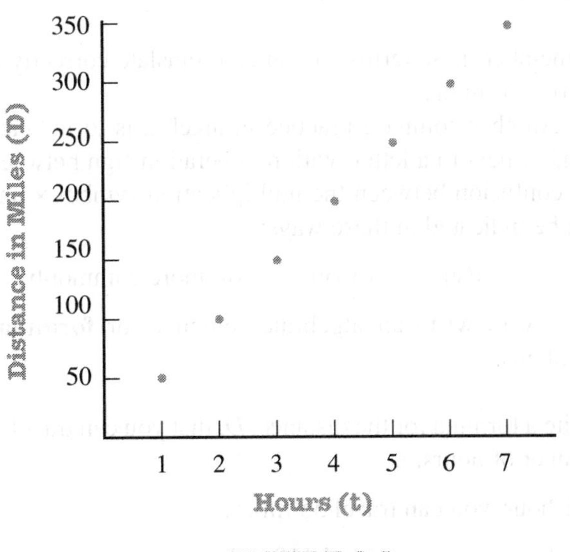

FIGURE 8.1

In any situation in which formulas are used to describe in a general manner something that is always true, algebraic techniques are used. Think about a sale at a store. The sale price is equal to the regular price minus the discount. You could also say $S = P - D$, where S stands for the sale price, P stands for the regular price, and D stands for the amount of the discount.

Computers were mentioned earlier. Any time a computer is used to keep records, figure prices, or do any other repetitive tasks, the person programming the computer must communicate with it in mathematical, algebraic formulas.

Think of other situations where you do a task over and over. Could you use a general statement and write a formula that would describe what you are doing?

These examples do not, of course, tell the whole story about algebra, but they do form a basis for understanding the difference between arithmetic and algebra. Arithmetic deals with very specific situations, and algebra generalizes those situations. Remember that the use of a letter symbol (called a **variable**) instead of a number symbol (called a **numeral**) generalizes what you are doing and says that the relationship can be true in more than one instance.

Algebra	**Arithmetic**
General statement of a relationship between quantities that is always true.	Specific problem involving calculations to arrive at an answer.
Sale price = Regular price – Discount $S = P - D$	Find the sale price of a \$120 coat if a \$30 discount is offered.
Distance = Rate · Time $D = rt$	How far can you travel in 6 hours if you average 55 miles per hour?

In order to translate from words to symbols, you should become familiar with certain terms commonly used to indicate arithmetic operations. You need to recognize these terms and to know which operation each implies.

Sum ⟶ addition
Difference ⟶ subtraction
Product ⟶ multiplication
Quotient ⟶ division

Remember these terms so you can translate correctly from words to symbols when writing formulas.

Another common practice in algebra is to indicate multiplication by putting a number next to a letter, with no operation sign between them. This is done to avoid the confusion between the multiplication symbol × and the letter X. Multiplication can be indicated in these ways:

$4 \cdot y$ $\quad$ $4(y)$ $\quad$ $(4)(y)$ $\quad$ or, more commonly, $\quad$ $4y$

Try to write an algebraic sentence, or **formula**, for each of the following problems.

EXAMPLE 1 Write a formula for the distance, D, that you can travel at 55 miles per hour in a given number of hours, t.

Solution In 1 hour, you can travel 55 miles.

In 2 hours, you can travel 110 miles.

In 3 hours, you can travel 165 miles.

From this pattern, it would seem you can multiply the number of hours by 55 to get the distance. Therefore, the formula is

$D = 55t$

(Remember that in algebra, writing a number next to a letter with no operation symbol in between means you are to multiply.)

EXAMPLE 2 Write a formula for the number of inches, I, in a given number of feet, represented by F.

Solution You know that 1 foot has 12 inches.

You know that 2 feet have 24 inches.

You know that 3 feet have 36 inches.

From this pattern, it would seem you could multiply the number of feet you are given by 12 to change the measurement to inches. Therefore, the formula is

$I = 12F$

EXAMPLE 3 Using the formula in Example 2, find the number of inches in 9 feet.

Solution

$I = 12F$ Write the formula.

$I = 12(9)$ Substitute the values and perform the indicated operations.

$I = 108$

There are 108 inches in 9 feet.

EXAMPLE 4 Write a formula for the volume of a box if the volume is equal to the product of its length, width, and height.

Solution Volume should be represented by V, length by L, width by W, and height by H. Because *product* means to multiply, the formula is

$V = LWH$

No symbol is necessary between the L and W or between W and H.

EXAMPLE 5 Write a formula for the cost (C) of buying t theater tickets if the tickets cost \$6.75 each. Then use your formula to find the cost of 8 tickets.

Solution You are looking for the cost of t tickets at \$6.75 each. You could add \$6.75 t times or multiply \$6.75 times t. Because you don't have a value for t, the formula to use is

$C = 6.75t$

To find the cost of 8 tickets, substitute 8 in place of t and do the calculation.

$C = 6.75t$

$C = 6.75(8)$

$C = 54.00$

Therefore, the cost for 8 tickets is \$54.00.

EXAMPLE 6 Look at the information given in the following table. See if you can write a formula to show the relationship between F and G.

F	5	6	7	8	9	10
G	10	11	12	13	14	15

Solution If you consider the first vertical pair of values, you could multiply 5 by 2 to get 10, or you could add 5 to 5 to get 10. You have two choices: Either multiply by 2 or add 5. Look at the second pair of values. Multiplying 6 by 2 does not give 11. Adding 5 to 6 does yield 11. Just to be sure, consider the third pair. If you add 5 to 7, you do get 12. Therefore, the relationship in the chart is such that if you add 5 to the top number, you get the bottom number .

$$\begin{array}{cccccc} 5 & 6 & 7 & 8 & 9 & 10 \\ \underline{+\ 5} & \underline{+\ 5} & \underline{+\ 5} & \underline{+\ 5} & \underline{+\ 5} & \underline{+\ 5} \\ 10 & 11 & 12 & 13 & 14 & 15 \end{array}$$

The formula that makes this same statement is

Top number	plus	five	equals	bottom number.
F	$+$	5	$=$	G

If you "began with" the bottom values, you would be subtracting 5, so you would also be safe in writing

Bottom number	minus	five	equals	top number.
G	$-$	5	$=$	F

EXAMPLE 7 Look at the information given in the following table, and write a formula that shows the relationship between the values for R and the values for S.

R	12	16	20	24	28
S	3	4	5	6	7

Solution

Top number	divided by	4	equals	bottom number.
R	$\div$	4	$=$	S

You could also write

4	times	bottom number	equals	top number.
4	$\cdot$	S	$=$	R

In this example, $R/4 = S$ or $4S = R$ would also be acceptable. Because an equality statement is true whether you read from left to right or from right to left, $S = R/4$ and $R = 4S$ are also acceptable.

Self-Check

1. Write a formula for the price, P, of an item if it is equal to the sum of the dealer's cost, C, and his markup, M.
2. Write a formula relating J and S from the relationship given in this table:

J	8	10	12	14	16
S	4	6	8	10	12

Answers:

1. $P = C + M$
2. $S = J - 4$ or $J = S + 4$

In this book, you will often be asked to write a formula that says in algebraic language what the specific numbers tell you to do. Don't be discouraged if this takes a lot of practice before you are comfortable doing it. It's not easy at first, but your skill will improve with effort and practice.

8.1 EXERCISES

A. Translate each of these statements into an algebraic formula. If you are not told what letters to use, choose letters that are appropriate to the quantities being represented.

1. The number of quarts (q) is equal to four times the number of gallons (g).
2. The number of yards (Y) is equal to the number of feet (f) divided by three.
3. The gasoline mileage for your car (M) is equal to the number of miles driven (m) divided by the number of gallons used (g). (*Hint*: Be careful when writing lowercase m and uppercase M. They stand for two different quantities.)
4. The average (A) of four grades is equal to the sum of the grades, a, b, c, and d, divided by 4. (*Hint*: Be careful to distinguish between A and a. They stand for two different quantities.)
5. The cost (C) of postage stamps is equal to 37 cents times the number of stamps (s).
6. The cost (C) of the movie tickets is $4 times the number of adult tickets (a) added to $3 times the number of student tickets (s).
7. The total number of students in the class is equal to the number of males added to the number of females.
8. The cost of the trip is the sum of the food costs and the hotel costs. (What operation does *sum* indicate?)
9. The number of feet is equal to the quotient of the number of inches and twelve. (What operation is signaled by *quotient*?)
10. The product of the rate at which you travel and the time you travel gives you the distance you can go. (What operation does *product* call for?)

B. Look at the information given in each of the following tables, and write a formula for each.

11.

p	3	4	5	6	7
q	6	7	8	9	10

12.

j	5	4	3	2	1
k	15	12	9	6	3

13.

a	4	6	8	10	12
b	8	12	16	20	24

14.

g	7	6	5	4	2
h	5	4	3	2	0

15.

w	1	3	5	7	9
x	6	8	10	12	14

16.

e	12	10	8	6	4
f	6	5	4	3	2

C. Write an algebraic formula for each of the following situations, and then answer the question.

17. Write a formula for A if it is equal to the difference between R and S. Then find A when $R = 17$ and $S = 9$.
18. Write a formula for K if it is equal to the product of A and B. Then find K when $A = 13$ and $B = 12$.
19. Write a formula for L if it is equal to the quotient of W and Y. Then find L when $W = 171$ and $Y = 19$.
20. Write a formula for M if it is equal to the sum of K and L. Then find M when $K = 32$ and $L = 4$.
21. Write a formula for the distance you can drive in h hours at 52 miles per hour. Find how many miles you can travel in 7 hours.
22. Write a formula for the total cost of an article if the price is c cents and you are to pay t cents in tax. Now find the total cost of a pair of shoes that cost $21.95 if the tax is $1.32.
23. Write a formula for the number of chairs in the theater if there are r rows of s seats each. How many people can be seated in a theater that has 54 rows of 36 seats each?
24. Write a formula for the cost of p pairs of socks at $1.19 per pair. How much will you pay for 5 pairs of socks?
25. Labor charges at the local garage include a flat fee of $32 and then a rate of $26 per hour. Write a formula for the labor charges, L, for a job that takes h hours to complete. What will the labor bill be for a repair that takes 3 hours to complete?

26. Car rental agencies run weekend specials with a flat fee of $59 for Saturday and Sunday and a mileage charge of $0.10 per mile. Write a formula for the weekend rental, R, when the car is driven m miles. Use the formula to find the actual charge if you drive 194 miles over the weekend to go home to see your family.

27. Measured phone service includes a flat monthly fee of $16.95 and a charge of $0.08 per call. Write a formula for the cost of phone service, P, if you make c calls per month. Use the formula to find the cost for phone service if you make 33 calls one month.

28. Write a formula for the cost, D, of day care if the cost for the first child in a family is $40 per week and each additional child in that family is $32 per week. (*Note*: If there are 3 children in the family, for how many children is the cost $32 per week? If there are 5 children in the family, for how many children is the cost $32 per week? If there are c children in the family, how many children pay $32 per week?) Use the formula to find the cost of day care for a family with 4 children.

D. Read and respond to the following exercises.

29. A fellow classmate is having trouble identifying which operation to use in writing a formula. Give him some suggestions for things that might help.

30. Write a table with five pairs of values for the following formula: $A = 2B$.

31. Write a table with five pairs of values for the following formula: $L = M + 3$.

32. Explain in three or four sentences some of the differences between arithmetic and algebra.

8.2 The "What" of Algebra

As we noted in the previous section, letters that stand for numerical values are called **variables**. That name makes sense because their values can change, or *vary*. The values of numbers do not change, so when they stand alone, they are called **constants**. When a number stands in front of a variable, indicating multiplication, it is called the **numerical coefficient**. When the variable is written alone—for example, Y—the numerical coefficient is understood to be 1. It can also be written as $1Y$. Any single variable or constant, or any grouping of variables or constants without an equals sign, is called an **expression**. Following are some examples of algebraic expressions:

$x + 3$	x is the variable; 1 is its coefficient. 3 is the constant term. $x + 3$ is the algebraic expression.
$3y - 2$	y is the variable; 2 is the constant. 3 is the numerical coefficient of y. $3y - 2$ is the expression.
$4 + 3c$	4 is the constant; c is the variable. 3 is the numerical coefficient of c. $4 + 3c$ is the expression.
$5xy$	5 is the numerical coefficient. x and y are the variables. $5xy$ is the expression.

An expression in algebra is like a phrase in English—it contains only a part of an idea. The phrase "gone fishing" expresses a part of the idea, but it does not tell you who has gone fishing or where! In order to get the complete idea, you need to write a complete sentence. "Mike has gone fishing at the pond" conveys a whole idea. An algebraic sentence that completes a thought or makes a statement and has values on both sides of an equals sign is called an equation or a formula. Whenever there is a statement that has an equals sign (=) in it, this statement is called an **equation**. Special equations that involve more than one variable are often called **formulas**. And, as we have said, an algebraic statement without an equals sign, such as $x - 4$, is called an **expression**.

Consider the following examples of equations or formulas:

$A = x + 3$

$14 = 3y - 2$

$4 - 3c = K$

$B = 5xy$

$P = a + b + c$

Note that each of these examples is a full "sentence" because it contains an equals sign. To distinguish between an equation and a formula, you need to consider the number of variables, or letters, in the sentence. Generally speaking, for our purposes here, an equation will involve only one variable, or letter, and a formula will involve more than one. Thus, $x + 3 = 7$ is an example of an equation, and $M = 3A - 2B + 4C$ is an example of a formula. (Later on, you will study equations that also have more than one variable.)

Let's look at the parts of an algebraic expression or equation that are called terms. A term can be made up of symbols, variables, or positive or negative constants. In an expression or equation, **terms** are separated by addition or subtraction signs.

Consider this expression with one term: $5a$.

The 5 is called the *numerical coefficient.*

The a is called the *variable.*

Consider the one-term expression $7ab^2$.

The 7 is called the *numerical coefficient.*

The a and b are *variables.*

The ^{2} is called the *power* or *exponent.*

Consider the expression $3y + 2$.

This expression has two terms, because $3y$ and 2 are separated by an addition sign.

The 3 is the *numerical coefficient* of the first term.

The y is the *variable* of the first term.

The 2 is the *constant* term in the expression.

Consider the formula $L = 4m + 3n$.

This formula has two terms because $4m$ and $3n$ are added together.

4 and 3 are the *numerical coefficients.*

m and n are the *variables*.

L is the *value* that results from the expression on the right.

Sometimes in an expression or an equation, there will be more than one term that has the same variable part. If you **combine** (add or subtract) those like-variable (same-variable) terms, you can write the equation or expression in simplified form. Consider the following expression, and decide for yourself how you might make it look simpler.

$4a + 6a - 2a$

Because all the terms have the same variable part—the letter a—you can combine those terms using the signs you are given. Work from left to right.

$4a + 6a - 2a$

$10a - 2a$

$8a$

Did you guess that answer correctly? If not, see where you made your error and try this one: $7y - 5y + 4y - 3y$. Remember to combine from left to right.

$7y - 5y + 4y - 3y$

$2y + 4y - 3y$

$6y - 3y$

$3y$

Like terms are terms that have exactly the same variable part but not necessarily the same numerical coefficients. Like terms in any algebraic expression should be combined. Use the operation sign before each term, combine the numerical coefficients, and write the variable part unchanged. Remember that a variable term with no numerical coefficient showing is understood to have a coefficient of 1. You might wish to write that 1 in the expression when you are combining like terms.

What do you suppose you would do with an expression like this?

$9W + 5Y - 3W + 2Y$

You can combine like terms, but only like terms. (Which are they?) You can rearrange the terms in any order as long as the sign in front of each term moves with it.

$9W + 5Y - 3W + 2Y$

$9W - 3W + 5Y + 2Y$

$6W + 7Y$

You have to stop there because $6W$ and $7Y$ are not like terms. What is unusual about the following problem?

$9P + 7R - 3P + R$

R is a variable by itself. Its numerical coefficient is 1 because $1 \cdot R = R$.

Rearrange and then combine the like terms.

$9P - 3P + 7R + 1R$

$6P + 8R$

Here is an example involving Y^2 instead of Y. Terms can be combined as long as their variable parts are exactly alike, but Y^2 is a different term from Y alone because of the exponent. The letter *and* any exponent parts must be *exactly* alike for the terms to be like terms.

$8Y^2 + 5Y - 3Y^2 + 4Y$

$8Y^2 - 3Y^2 + 5Y + 4Y$

$5Y^2 + 9Y$

Now try the following problem. Combine like terms.

$3R + 2S - 5T$

There aren't any like terms, so none can be combined. The expression is already in simplest form.

Self-Check Combine the like terms to simplify each of these expressions.

1. $13V - V + 4V - 8V$
2. $4X^2 - X^2 + 3X^2$
3. $9C^2 + 4C - C + 5C^2$
4. $6B - 3D - 2B + 7B$

Answers:

1. $8V$
2. $6X^2$
3. $14C^2 + 3C$
4. $11B - 3D$

8.2 EXERCISES

A. Fill in the blanks.

1. In the formula $K = 3W + 7Y$,
   a. 3 and 7 are _____.
   b. W and Y are _____.
2. In the formula $A = 2c + 3d + 5$,
   a. 2 and 3 are _____.
   b. c and d are _____.
   c. 5 is a(n) _____.
3. In the formula $4m + 3n + 6 = 8m$,
   a. the constant term is _____.
   b. the variables are _____.
   c. the numerical coefficients are _____.
4. In the formula $4p + 5q + 7 = 12$,
   a. the variables are _____.
   b. the numerical coefficients are _____.
   c. the constants are _____.
5. $a + b$ is called a(n) _____.
6. $W = p + q$ is called a(n) _____.
7. In $6a$, 6 is the _____.
8. In $9y$, y is the _____.
9. In $3a^2b$, ^{2} is called the _____.
10. In WXY^3, ^{3} is called the _____.

B. Combine like terms. Remember to combine from left to right.

11. $6w - 3w + 2w$
12. $9d - 5d + 4d$
13. $17x + 4x - 9x$
14. $6b^2 - 5b^2 + 2b^2$
15. $4a^3 - 3a^3 + a^3$
16. $9f^3 + 3f - 5f^3$
17. $11P^2 - P + 3P^2$
18. $14k^3 - 3k^3 - 5k^3$
19. $7c^2 + 4c^2 - 2c^2$
20. $16m - 9m - 5m$

C. Simplify each expression by combining like terms. Rearrange the order of the terms if it helps.

21. $6y + 4z + 5y + 3z$
22. $5W^2 + 2X^3 + 3W^2 + 4X^3$
23. $19f + 7g - 4f + 2g$
24. $9t - 6t + 8w - 4w$
25. $17a + 7b - 4a - 3b$
26. $23x + 7y - 8x + 3y$
27. $a^2 + 5b^2 + 6a^2 - b^2$
28. $24m - 7m + 2n + 8m$
29. $19b + 4c - 3c + 4c$
30. $11k - 4k + 6k - 3M$

D. Read and respond to the following exercises.

31. Explain in words the difference(s) between an expression and an equation.
32. Explain in words the difference(s) between and equation and a formula as they are used in this course.
33. A fellow student is having difficulty understanding why you can combine only *like* terms. Make up two real-world examples of things that have to be alike in order to be combined. For instance, you cannot add apples and oranges, but you can report your answer as the number of pieces of fruit you have.
34. Explain how adding 2/9 and 5/9 is similar to combining terms $2A$ and $5A$. What must be done if the problem asks for the sum of 3/5 and 7/10? Why?

8.3 Evaluating Expressions and Formulas

Review the Order of Operations Agreement, which was presented in Section 2.2. It is just as important in algebra as it was in arithmetic.

> **Order of Operations Agreement**
>
> In any problem containing more than one mathematical operation, always follow these steps:
>
> 1. Simplify within parentheses or grouping symbols, using steps 2, 3, and 4 below in order.
> 2. Simplify any expressions that involve exponents or roots.
> 3. Multiply or divide, **in order, from left to right.**
> 4. Add or subtract (combine), **in order, from left to right.**

In a moment we will try some arithmetic examples as a review. In order to do these, however, you need to know how to simplify expressions with exponents, or powers. In the expression 3^2, which is read "three to the second power" or "three squared," the 3 is called the **base**. The ^{2} is the **exponent** or **power**.

3^2 means $3 \cdot 3$, or 9

4^3 means $4 \cdot 4 \cdot 4$, or 64

5^2 means $5 \cdot 5$, or 25

2^5 means $2 \cdot 2 \cdot 2 \cdot 2 \cdot 2$, or 32

The last two examples make it clear that 5^2 and 2^5 are not the same.

Self-Check Try these problems and see if you can simplify them correctly. The answers follow.

1. 2^3 2. 7^2 3. 5^3 4. 1^5 5. 3^3

Answers:

1. 8 2. 49 3. 125 4. 1 5. 27

In terms that contain several factors, the exponent applies only to its base, the number or letter immediately to its left. For example, $3(2)^4 = 3(2 \cdot 2 \cdot 2 \cdot 2) = 3(16) = 48$. Only 2 is raised to the fourth power, not the product of 3 and 2. In another example, $2^3(5) = (2 \cdot 2 \cdot 2)(5) = 8(5) = 40$; only 2 is raised to the third power. In the expression $4a^2b$, only the a is affected by the exponent.

Now you should be ready to try some simplifying problems using the Order of Operations Agreement and the definition of exponent, or power. Be sure to look at the expression carefully. You will go through the list of steps in the Order of Operations Agreement each time to decide what needs to be done next. Do only one simplification at a time. We will review some arithmetic examples now.

EXAMPLE 1 Simplify: $26 - 8 \times 2 + 9 \div 3$

Solution

$26 - \underline{8 \times 2} + \underline{9 \div 3}$	Multiply or divide, left to right.
$\underline{26 - 16} + 3$	Add or subtract, left to right.
$\underline{10 + 3}$	Add or subtract, left to right.
13	

EXAMPLE 2 Simplify: $56 + 9 \times 2 - 4^3$

Solution

$56 + 9 \times 2 - \underline{4^3}$	Exponents first.
$56 + \underline{9 \times 2} - 64$	Multiply or divide, left to right.
$\underline{56 + 18} - 64$	Add or subtract, left to right.
$\underline{74 - 64}$	Add or subtract, left to right.
10	

EXAMPLE 3 Simplify: $45 - 4(8 + 3)$

Solution

$45 - 4\underline{(8 + 3)}$	Parentheses first.
$45 - \underline{4(11)}$	Multiply or divide, left to right.
$\underline{45 - 44}$	Add or subtract, left to right.
1	

Algebra and arithmetic come together when you are asked to find the value of a formula and are given specific values for the variables. A typical problem, for example, might be to find the value of K in $K = 5a + 3b$ when $a = 4$ and $b = 6$. First substitute the given values for a and b into the equation or formula. Then use the Order of Operations Agreement to simplify the resulting numerical expression. Replace a with 4 and b with 6 (remember that a number next to a letter means to multiply).

$K = 5a + 3b$	Substitute.
$K = 5(4) + 3(6)$	Multiply or divide.
$K = 20 + 18$	Add or subtract (combine).
$K = 38$	

Using the same formula, find the value of K if $a = 7$ and $b = 4$. Substitute for a and b.

$K = 5a + 3b$	Substitute.
$K = 5(7) + 3(4)$	Multiply or divide.
$K = 35 + 12$	Add or subtract (combine).
$K = 47$	

Any algebraic expression or equation can be evaluated by substituting the given numerical values and using the Order of Operations Agreement.

EXAMPLE 4 Find the value of $(P - S)(Q - R)$ when $P = 8$, $Q = 6$, $R = 2$, and $S = 1$.

Solution

$(P - S)(Q - R)$	Substitute for P, Q, R, and S.
$= (8 - 1)(6 - 2)$	Simplify within the parentheses first.
$= (7)(4)$	Multiply.
$= 28$	

EXAMPLE 5 Find the value of $6x^3y - 5x^2$ when $x = 2$ and $y = 5$.

Solution

$6x^3y - 5x^2$	Substitute for x and y.
$= 6(2)^3(5) - 5(2)^2$	Apply the rule for exponents.
$= 6(8)(5) - 5(4)$	Multiply.
$= 48(5) - 20$	Multiply.
$= 240 - 20$	Subtract.
$= 220$	

8.3 EXERCISES

A. Simplify each of the following expressions using the Order of Operations Agreement. Be sure to write out each step.

1. $16 + 3 \cdot 5$
2. $7 + 4 \cdot 3$
3. $4^2 - 5$
4. $3^4 + 2$
5. $12 + 3^2$
6. $2^3 - 3$
7. $14 - 3 \cdot 2$
8. $20 - 4 \cdot 3$
9. $4 + 5(3 - 1)$
10. $18 - 3(4 - 2)$
11. $7(8 + 3)$
12. $9(5 - 2)$
13. $19 - 5 + 4$
14. $26 - 9 - 4$
15. $(8 + 7)(3 + 2)$
16. $(7 - 3)(8 - 5)$
17. $12 + 3 \times 5 - 9$
18. $17 - 2 \times 4 + 5$
19. $30 - 6 \div 2 + 4$
20. $18 - 9 \div 3 + 4$

B. Find the value of each of the following expressions or formulas by substituting the given values for the variables and using the Order of Operations Agreement. Be sure to write out each step.

21. Find the value of $(a + b)(c + d)$ when $a = 4$, $b = 5$, $c = 2$, and $d = 6$.
22. Find the value of pqr when $p = 3$, $q = 4$, and $r = 6$.
23. Find the value of $ab - cd$ when $a = 7$, $b = 5$, $c = 3$, and $d = 4$.
24. Find the value of $2A^2B$ when $A = 3$ and $B = 4$.
25. Find the value of $3xy^2$ when $x = 4$ and $y = 3$.
26. Find the value of $K(P - R)$ when $K = 7$, $P = 8$, and $R = 3$.
27. Find the value of $(A - 3)(B - 9)$ when $A = 12$ and $B = 11$.
28. Find the value of $KL - M$ when $K = 5$, $L = 9$, and $M = 23$.
29. Find the value of $D + EF$ when $D = 7$, $E = 8$, and $F = 9$.
30. Find the value of AB^2C when $A = 3$, $B = 4$, and $C = 2$.
31. Find the value of K in $K = 3m - 4n$ when $m = 9$ and $n = 4$.
32. Find the value of P in $P = 5(R - S)$ when $R = 12$ and $S = 8$.
33. Find the value of A in $A = 4s^2t$ when $s = 2$ and $t = 3$.

34. Find the value of T in $T = 2(A + 4)(B - 5)$ when $A = 7$ and $B = 9$.

35. Find the value of R in $R = K - LM$ when $K = 23$, $L = 4$, and $M = 5$.

36. Find the value of f in $f = 8g - 4h$ when $g = 4$ and $h = 6$.

37. Find the value of g in $g = hk - 4j$ when $h = 9$, $k = 7$, and $j = 8$.

38. Find the value of m in $m = 3e^2 - 5f^2$ when $e = 5$ and $f = 3$.

39. Find the value of x in $x = 7w - 3y^2$ when $w = 8$ and $y = 2$.

40. Find the value of C in $C = 4f^3 - 2g$ when $f = 2$ and $g = 5$.

C. Describe, in words, the process used to go from one step to the next.

41. $46 - 12 \div 6 + 3 \times 9$

$46 - 2 + 3 \times 9$	1.
$46 - 2 + 27$	2.
$44 + 27$	3.
71	4.

42. $14 - 3(12 \div 3 - 2)$

$14 - 3(4 - 2)$	1.
$14 - 3(2)$	2.
$14 - 6$	3.
8	4.

43. $(17 - 8)(16 \div 8)$

$(9)(16 \div 8)$	1.
$(9)(2)$	2.
18	3.

44. $3^4 - (18 \div 2) + \sqrt{36}$

$3^4 - 9 + \sqrt{36}$	1.
$81 - 9 + 6$	2.
$72 + 6$	3.
78	4.

45. $5^3 \div \sqrt{25} + 2^5 \div 4 \times 3$

$125 \div 5 + 32 \div 4 \times 3$	1.
$25 + 8 \times 3$	2.
$25 + 24$	3.
49	4.

46. $36 \div 4 \times 2 \div 6 - 3$

$9 \times 2 \div 6 - 3$	1.
$18 \div 6 - 3$	2.
$3 - 3$	3.
0	4.

D. Read and respond to the following exercises.

47. In this expression, $17 - 3 + 9$, which operation is completed first, and why?

48. In this expression, $6 \div 2 \times 3$, which operation is completed first, and why?

49. Describe what strategies you used to learn the Order of Operations Agreement.

50. How is the expression $6x^2$ different from $(6x)^2$? Explain your answer in words.

8.4 The Number System and Its Properties

The Order of Operations Agreement is, of course, not the only property of whole numbers that applies to algebra. As a matter of fact, most whole number properties can also be stated algebraically.

One notable difference that you might have seen when comparing an arithmetic text to an algebra text is the use of many letters (variables). You now know that algebra is a general way of saying the same things that arithmetic has said. In order not to have to restate properties every time we use them with different values, we state them generally—algebraically—using variables rather than numbers.

Let's first look at the set of real numbers. This group includes almost every kind of number that you will encounter in beginning algebra. Once you have studied this system, you can be confident that, because of the way the properties are stated, they apply to all real numbers. Here, then, are all the kinds of numbers that make up the real number system.

Natural Numbers

These are also called counting numbers. From their names, you can guess that they include 1, 2, 3, 4, and 5 and go on forever. (Is there ever a largest number?) The set, or collection, of natural numbers is written this way:

{1, 2, 3, 4, 5, . . .}

The braces, { }, are used to show that a group or set of items has been formed and the ellipsis, . . . , indicates that the group continues on in the same pattern without ending.

Whole Numbers

This set of numbers includes the natural numbers, but it has one more member, zero. When the thinkers who developed our number system tried to show in nature what happened when they had five twigs and the wind came along and blew them all away, they realized there was a problem. There was no symbol for 5 take away 5. Because what remained was a void, "a hole," they wrote the symbol 0 to show that. The set, or collection, of whole numbers is written this way:

{0, 1, 2, 3, 4, 5, . . .}

Integers

The set of integers includes all the whole numbers and their opposites. You can have two pencils, +2, or be missing two pencils, –2. You can have $5, +5, or owe $5, –5. The set of integers is written this way:

{. . . , –3, –2, –1, 0, 1, 2, 3, . . .}

This set extends indefinitely in both directions.

Rational Numbers

What happens when you divide 8 by 2? You get 4. What happens when you divide 9 by 2? So far in the sets of numbers we have described, there is no way to show that result. This difficulty led to the development of the set of rational numbers, sometimes called fractions. Hence 9 ÷ 2 = 9/2. The rationals include all numbers that can be written in the form *a*/*b*, where *a* and *b* are integers and *b* does not equal 0. (Why can't we have *a*/0? Do you remember that division by zero is not possible? It is undefined.) This set of numbers includes the whole numbers because 6 can be written as 6/1, – 26 as – 26/1, and 0 as 0/7. The set of rationals also includes all repeating or terminating decimals; they too can be written in *a*/*b* form. This collection of numbers includes far too many numbers to show by example, so it is described as the set of all numbers that can be written in the form *a/b*, where *a* and *b* are integers, where $b \neq 0$. Using set notation, we express the set of all rational numbers like this:

$\{a/b \mid a \text{ and } b \text{ are integers}, b \neq 0\}$

This notation is read "the set of all numbers represented by *a* divided by *b*, such that *a* and *b* are integers and *b* is not equal to zero."

Irrational Numbers

This set of numbers is made up of numbers that cannot be written in *a/b* form. These are decimal numbers that cannot be written as fractions—for example, 2.010010001 .... There is a pattern, but the decimal is neither terminating nor repeating. The irrational numbers also include pi, π, the Greek letter that represents the quotient of the circumference of any circle divided by its diameter. π is a nonterminating, nonrepeating decimal. Numbers that are the square roots of quantities that are not perfect squares ($\sqrt{2}$, $\sqrt{5}$, and $\sqrt{17}$, for example) are also irrational numbers.

Real Numbers

When you put all of these numbers together, you have the set of real numbers. If you were to draw a line resembling a ruler but with *zero* in the middle of the scale, the real numbers would be represented by *all the points* that make up that line, not just the labeled points.

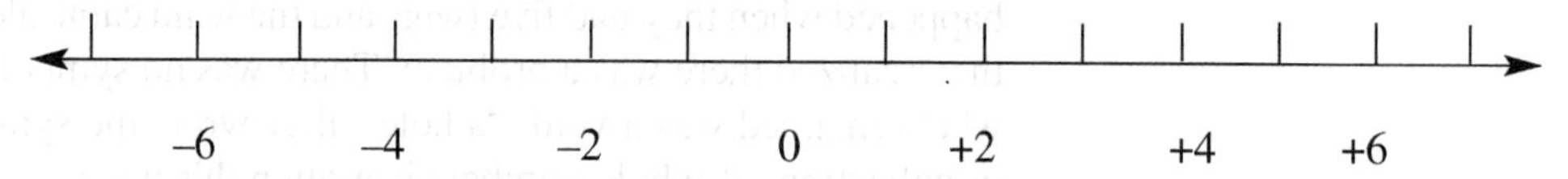

The line would also include values such as 3 1/2, –1.5, π, and $\sqrt{6}$, but for the sake of clarity, only the integer values are usually labeled on the scale. The numbers on the number line that fall to the left of the zero are called *negative numbers*. If you think in terms of temperature, those numbers would represent readings below zero degrees. If you lived in the Swiss Alps in the middle of winter, you might well see a temperature of –12°. The numbers that fall to the right of the zero on the number line are *positive numbers*. On a winter day in Pittsburgh, Pennsylvania, a temperature of +12° is to be expected. Negative 12 (–12) and positive 12 (+12) are **opposites**. They are the same distance from zero but in opposite directions. When we compare any two values on a number line, we find that the one to the left is the smaller (just as on a thermometer, the lower number on the scale is the colder temperature).

If you think about the logical need for different types of numbers, and consider what you know about their history, you can easily remember the real number system. When students begin to learn about signed numbers (so-called because such numbers have a numerical value and a positive or negative sign), they learn what these numbers are and how to do calculations with them. You will learn about these signed numbers in Chapter 11. Then you will become familiar with some of the processes for solving equations and finding solutions to application problems by using algebra. Once you are comfortable with these processes, you will be ready to expand your knowledge to include other topics in algebra.

Now, let's review the properties you studied in Chapter 2 and show that they apply to all real numbers.

Commutative Property of Addition

For all real numbers represented by a and b,

$a + b = b + a$

For example, $7 + 6 = 6 + 7$

because $13 = 13$

The Commutative Property of Addition says that the terms can be moved around, *commuted*, but the sum will be the same.

Commutative Property of Multiplication

For all real numbers represented by a and b,

$a \cdot b = b \cdot a$

For example, $9 \cdot 5 = 5 \cdot 9$

because $45 = 45$

The Commutative Property of Multiplication says that the factors can be moved around, *commuted,* but the product will be the same.

Associative Property of Addition

For all real numbers represented by a, b, and c,

$a + (b + c) = (a + b) + c$

Consider
$$2 + (3 + 4) = (2 + 3) + 4$$
$$2 + 7 = 5 + 4$$
$$9 = 9$$

The Associative Property of Addition says that you can regroup (or re-associate) the terms differently but the sum will not change.

Associative Property of Multiplication

For all real numbers represented by a, b, and c,

$a(bc) = (ab)c$

For example,

$$5(3 \cdot 6) = (5 \cdot 3)6$$
$$5(18) = (15)6$$
$$90 = 90$$

The Associative Property of Multiplication says that you can regroup (or re-associate) the factors differently but the product will not change.

Distributive Principle

For all real numbers represented by a, b, and c,

$$a(b + c) = ab + ac$$

and

$$a(b - c) = ab - ac$$

Consider

$$2(3 + 5) = 2 \cdot 3 + 2 \cdot 5$$
$$2(8) = 6 + 10$$
$$16 = 16$$

We have **proven** that the Distributive Principle is valid for this example since the resulting statement, $16 = 16$, is always true.

Addition Property of Zero

For all real numbers represented by a,
$a + 0 = a$

You know that

$$8 + 0 = 8$$

and

$$2.7 + 0 = 2.7$$

Multiplication Property of Zero

For all real numbers represented by a, $a \neq 0$,

$a \cdot 0 = 0$

You know that $12 \cdot 0 = 0$

and $5/8 \cdot 0 = 0$

Multiplication Property of One

For all real numbers represented by a,

$a \cdot 1 = a$

You know that $14 \cdot 1 = 14$

and $2\ 3/4 \cdot 1 = 2\ 3/4$

Division by One

For all real numbers represented by a,

$a \div 1 = a$

You know that $9 \div 1 = 9$

and $1/2 \div 1 = 1/2$

You may want to go back and review Section 1.3 if these next two properties are not clear to you.

Division by Zero

For all real numbers represented by a,

$a \div 0$ is undefined

It has no real number answer.

Division into Zero

For all real numbers represented by a, $a \neq 0$,

$0/a = 0$

8.4 EXERCISES

A. Match each of the statements in Column I with the property that it illustrates from Column II. You may use a property more than once.

Column I

1. $8 \times 1 = 1 \times 8$
2. $12 \div 0$ is undefined
3. $3 + (4 + 9) = (3 + 4) + 9$
4. $4(9 - 3) = 4(9) - 4(3)$
5. $15 \cdot 1 = 15$
6. $9 \div 1 = 9$
7. $14 + 0 = 0 + 14$
8. $(4 + 8) + 6 = 4 + (8 + 6)$
9. $0 \div 4 = 0$
10. $5(2 \cdot 3) = (5 \cdot 2)3$
11. $7(6 - 4) = 7(6) - 7(4)$
12. $0 = 4 \cdot 0$
13. $5 \cdot 9 = 9 \cdot 5$
14. $6 + 0 = 6$
15. $6(9 + 5) = 6(9) + 6(5)$
16. $3(7 \cdot 6) = (3 \cdot 7)6$
17. $7(3 + 5) = 7(3) + 7(5)$
18. $0 + 3 = 3 + 0$

Column II

a. Commutative Property of Addition
b. Commutative Property of Multiplication
c. Associative Property of Addition
d. Associative Property of Multiplication
e. Addition Property of Zero
f. Multiplication Property of Zero
g. Multiplication Property of One
h. Division by Zero
i. Division into Zero
j. Division by One
k. Distributive Principle

19. Write an arithmetic example of the Commutative Property of Addition and then simplify both sides of the equals sign to prove that it is true.
20. Write an arithmetic example of the Commutative Property of Multiplication and then simplify both sides of the equals sign to prove that it is true.
21. Write an arithmetic example of the Associative Property of Multiplication and then simplify both sides of the equals sign to prove that it is true.
22. Write an arithmetic example of the Associative Property of Addition and then simplify both sides of the equals sign to prove that it is true.
23. Write an arithmetic example of the Distributive Principle using subtraction and then simplify both sides of the equals sign to prove that it is true.
24. Write an arithmetic example of the Distributive Principle using addition and then simplify both sides of the equals sign to prove that it is true.

B. Read and respond to the exercises.

25. Describe two things that you do every day while getting ready for school or work that are commutative—the order in which you do them does not matter.
26. Describe two things that you do every day while getting ready for school or work that are not commutative. Explain why the order in which you do them makes a difference.
27. Explain in your own words why division by zero is not defined.
28. Give an example of a real-world situation where groups are associated in different ways throughout the day and this changing of groups does not adversely affect the outcome.

8.5 Solving Simple Equations and Inequalities

Suppose you are given a simple equation—for example, the statement $y + 5 = 12$. You know that y is a variable and can take on any value. When you talk about "solving the equation," you are trying to find the one value that, when it is substituted for y, will give a true statement. You could try a lot of different possibilities and, by *trial and error*, by *intuition*, or by *inspection*, find the solution. To solve $y + 5 = 12$, you could ask yourself intuitively, "What number must be added to 5 to get 12?" You might consider several possibilities. Because there are no restrictions placed on y, the value you try for y could be any number.

$y + 5 = 12$

Let $y = 4$.	Is $4 + 5 = 12$ a true statement? No.
Let $y = 9$.	Is $9 + 5 = 12$ a true statement? No.
Let $y = 0.4$.	Is $0.4 + 5 = 12$ a true statement? No.
Let $y = 2\ 1/2$.	Is $2\ 1/2 + 5 = 12$ a true statement? No.
Let $y = 7$.	Is $7 + 5 = 12$ a true statement? Yes.

Thus, 7 is the *solution* of the equation. It is the one value for y that makes the statement true. The value, 7, that makes the statement true is called the **solution**. It can be written in set notation, $\{7\}$, and $\{7\}$ can be called the **solution set**.

Try to solve these simple equations by inspection. Look at the equation and take a guess. Use your intuition. Then check your answer by substituting the number for the variable and simplifying. If you get a true statement, or an **identity** (that is, the number on the right equals the number on the left), then the solution you have found is the correct one.

Self-Check 1. $A - 5 = 3$ 2. $B + 4 = 9$ 3. $7C = 63$ 4. $D \div 5 = 3$

Answers:

1. $A = 8$; *Check:* $3 = 3$ 2. $B = 5$; *Check:* $9 = 9$
3. $C = 9$; *Check:* $63 = 63$ 4. $D = 15$; *Check:* $3 = 3$

It is not always possible or convenient to solve all equations by inspection or intuition. You will learn methods for solving more complex equations in Chapter 9. These methods will make solving equations more like solving a puzzle.

Suppose that instead of the equation $y + 5 = 12$, you were given the inequality $y + 5 > 12$. This is read "y plus five is greater than twelve." How is this statement different from the equation? Consider some simple statements first.

$A = 8$ means that A has a single value, 8, and the resulting equality statement will be true.

$B > 5$ means that B can assume any value larger than 5, and the resulting inequality statement will be true.

$C \leq 12$ means that C can assume any value less than or equal to 12, and the resulting inequality will be true.

In the equation, only *one* value for the variable will make a true statement. In the two inequalities, an *infinite number* of values will satisfy the inequality. (Notice that the third statement has an inequality and an equality symbol. It is read "less than or equal to.")

Because an inequality has a large number of solutions, we need a way to indicate that we are talking about a whole group of numbers. We do this by means of **set notation.** A set is a group of items that satisfy a particular condition. For example, we understand what is meant when someone talks about a set of dishes or a set of chess pieces. The solution to the second inequality is the group of numbers that are greater than 5. We write $\{B \mid B > 5\}$. It is read "the set of all numbers represented by the variable B such that B is greater than 5." The solution set for an inequality has too many members to list, so we describe them as members of a group. The solution set for the second inequality is written $\{C \mid C \leq 12\}$ and it is read "the set of an numbers represented by C such that C is less than or equal to 12."

Let's return now to the inequality $y + 5 > 12$. You begin by solving the *equation* $y + 5 = 12$. You know that $y = 7$ will make the equation true. Therefore, 7 is the **critical value** for the inequality. It separates the values that are solutions from those that are not. Consider two values less than 7 and two greater than 7 and see that it is the values greater than 7 that make the inequality true.

$y + 5 > 12$

Try 5.	$5 + 5 > 12$	⟶	$10 > 12$	False.
Try 6.	$6 + 5 > 12$	⟶	$11 > 12$	False.
Try 7.	$7 + 5 > 12$	⟶	$12 > 12$	False, but it's close.
Try 8.	$8 + 5 > 12$	⟶	$13 > 12$	True.
Try 9.	$9 + 5 > 12$	⟶	$14 > 12$	True.

You can see that the whole numbers greater than 7 will satisfy the inequality. In fact, you can correctly assume that every real number greater than 7 will make the inequality true.

Because there are many values for y that will satisfy the inequality $y + 5 > 12$, you show the solution by means of set notation. The variable used is y and you have determined that values greater than 7 make the inequality statement true. Write the solution set as $\{y \mid y > 7\}$; it is read "the set or group of all numbers represented by y such that y is greater than 7." This lets everyone know that 7.001, 8 3/4, 9.5, and 1,002 will all work as solutions for the inequality $y + 5 > 12$.

Consider $K - 2 < 6$. This is read "K decreased by two is less than six." You can see by inspection that $K - 2 = 6$ is true when $K = 8$. Therefore, 8 is the critical value. It separates the quantities that work from the ones that do not. You can try one value less than 8—for instance, 5—and note that $5 - 2 < 6$ is true because 3 is less than 6. Therefore, the solution set is $\{K \mid K < 8\}$.

Now try $d + 2 \geq 5$. This is read "the sum of d and two is greater than or equal to five." You know by inspection that $d + 2 = 5$ is true when $d = 3$. Therefore, 3 is the critical value. Because $7 + 2 \geq 5$, and 7 is greater than 3, all values greater than 3 and 3 itself will satisfy the inequality. Therefore, the solution set is $\{d \mid d \geq 3\}$.

Study these examples and see how some inequalities can be solved by inspection.

EXAMPLE 1 Solve $g + 3 < 8$ and report the answer in set notation.

Solution You can see by inspection that $g + 3 = 8$ when $g = 5$, so 5 is the critical value. If you substitute 6 in place of g, $6 + 3 < 8$ is not true. Therefore, the values that satisfy the inequality are those less than 5, not greater than 5. Therefore, the solution is $\{g \mid g < 5\}$.

EXAMPLE 2 Solve $6a \geq 24$ and report the answer in set notation.

Solution You know by inspection that $6a = 24$ when $a = 4$, so 4 is the critical value. Try $a = 11$. $6(11) \geq 24$ is true. Therefore, all values greater than the critical value 4, and 4 itself, are the solution. The solution set is $\{a | a \geq 4\}$.

In a later chapter, you will learn how to solve more complicated inequalities, but an intuitive idea of what they are and how they work is all you need for now.

8.5 EXERCISES

A. Fill in the blanks.

1. The statement $y - 9 = 4$ is called a(n) _____.
2. You can solve simple equations by inspection or by _____.
3. The value that makes the equation true is called the _____.
4. In $A + 6 = 7$, _____ is the solution.
5. In the equation $b - 5 = 4$, _____ is the value that makes the statement true when it is substituted in *place* of b.
6. 17 _______ (is or is not?) the solution of $25 = c + 8$.
7. 9 makes $Y - 8 = 17$ a(n) _____ (true or false?) statement.
8. If $D + 4 = 7$, 5 _____ (is or is not?) the solution.
9. An inequality has _____ solution(s).
10. The solutions to a(n) _____ are indicated by using set notation.
11. Is 1/2 a member of the solution set for $K + 3 < 4$?
12. Is 6 a member of the solution set for $Y + 3 > 9$?
13. Is 7 a member of the solution set for $3M \geq 21$?
14. Expressions can only be _____; equations and inequalities can be _______. (Use "solved" and "simplified".)

B. Solve each of the following equations by "looking," by inspection, or by intuition. Show a check for each equation.

15. $y + 8 = 17$
16. $2 + n = 11$
17. $M \div 3 = 5$
18. $4P = 28$
19. $12 = T - 5$
20. $12 = L - 4$
21. $17 = N + 5$
22. $8 = r \div 3$
23. $11 = 21 - K$
24. $6E = 42$
25. $12 = 8 + C$
26. $B \div 4 = 3$

C. Solve each of the following inequalities by inspection to find the critical value and then write the solution set.

27. $x + 3 < 9$
28. $c + 5 > 8$
29. $y - 2 > 10$
30. $3d \geq 12$
31. $4a \leq 20$
32. $k \div 4 < 24$
33. $30 \geq a + 10$
34. $14 \leq y + 5$
35. $m \div 2 > 5$
36. $n - 4 > 7$

D. Read and respond to the following exercises.

37. Explain in your own words the difference(s) between the solution to an equation and the solution to an inequality.
38. Explain the value of the checking process when you have solved an equation.
39. Is this statement true or false: "7 < 7"? Is this statement true or false: "3 ≤ 3"? Explain the difference between the two inequalities.
40. In your own words, explain why a solution set is an appropriate way to report the solution for an inequality.

8.6 CHAPTER REVIEW

Now that you have completed this chapter, you should be able to:

1. Write an algebraic formula from a given description.
2. Write an algebraic formula from information given in a table.
3. Recognize the following algebraic quantities: variable, constant, numerical coefficient, base, power or exponent, algebraic expression, equation, inequality, and solution set.
4. Combine like terms.
5. Use the Order of Operations Agreement to simplify numerical expressions.
6. Find the value of a given algebraic expression or formula.
7. Solve a simple algebraic equation by intuition or inspection.
8. Solve a simple inequality by inspection, finding the critical value and then writing the solution in set notation. (See Examples 1 and 2 in Section 8.5.)

REVIEW EXERCISES

A. Write a formula for each of the following situations.

1. The number of inches (i) in Y yards is equal to the product of 36 and Y.
2. The value (V) of s stamps at 45 cents each is equal to the product of 45 and s.
3. The cost (c) of 9 movie tickets at D dollars each is equal to 9 times D.
4. The tax (t) on an item if it is equal to 5% (0.05) times the price (p).
5. The number of students (s) in a class is equal to the sum of p, those who are present, and a, those who are absent.
6. The number of hours (h) is equal to the quotient of the number of minutes (m) and 60.
7. The rate (r) you can drive is equal to the quotient of the distance (d) and the travel time (t).
8. The sale price, S, of an item is equal to the difference between the regular price, R, and the discount, D.
9. The total cost of your vacation (V) is equal to the sum of the transportation costs (t) and the food costs (f).
10. The number of chocolates (c) in a box is equal to the product of r rows and p pieces in each row.

B. Write a formula from the information given in each of the following tables. There are two options for your answer, and either one is correct.

11.

K	9	8	7	6
L	3	2	1	0

12.

E	12	14	16	18
F	4	6	8	10

13.

A	9	12	15	18
B	3	4	5	6

14.

Y	8	12	18	22
A	4	6	9	11

15.

P	1	2	3	4
R	4	8	12	16

16.

M	3	7	11	15
N	7	11	15	19

17.

C	9	11	13	15
D	11	13	15	17

18.

E	2	5	8	11
F	6	15	24	33

19.

T	5	10	15	20
V	10	15	20	25

20.

S	12	20	28	36
W	3	5	7	9

C. Identify the parts asked for in each expression or equation.

21. the variables in $6x - 3y$
22. the constants in $4a - 5 + 3b + 7$
23. the numerical coefficient(s) in $7x^2 - 3x + 2$
24. the exponent(s) in $3a^4 - 6b^2$
25. the literal factor(s) in $7p^3r$
26. the base(s) in $- 8a^4b^5$
27. the solution set of $3 + y = 9$
28. the power(s) in $14 + 5x^3$
29. the numerical coefficient(s) in $5a + 3b + c$
30. the numerical coefficient of p

D. Combine like terms in each of the following exercises.

31. $5W + 7Y - 3W - 3Y$
32. $6a + 5a - 3a$
33. $9c - 6c + 8c$
34. $4a + 3b - 2c$
35. $3a^2 + 4a^2 - 2a^2$
36. $9d^2 - 4d^2 + 3d^2$
37. $4B - B + 2B$
38. $12Y - 3Y + Y$
39. $6w - 3x + 2y$
40. $9y^2 - 4y^2 + 6y$
41. $3y^2 + 6y^2 - y^2$
42. $8c^2 - 4c^2 + 3c^2$
43. $7a^2 - 5a^2 + 2a^2$
44. $6x^2 + 4x - 2x^2 + 5x$
45. $8t^3 + 3t^2 - 4t^3 + 8t^2$

E. Simplify the following numerical expressions using the Order of Operations Agreement.

46. $9 - 4 + 3$
47. $12 + 3(8 - 2)$
48. $14 - 2(6 - 2)$
49. $15 - 6 + 3$
50. $24 - 3 \times 2 + 8 \div 2$
51. $(7 - 4)(8 \times 2)$
52. $(11 + 3)(9 \div 3)$
53. $3^5 - 12 \div 3 + 9$
54. $9 \div 3 \times 2$
55. $6 \times 4 \div 3$
56. $2^5 + 3$
57. $4(2)^3$
58. $5^2 + 3 \times 2 - 4$
59. $8 + 3(9 - 5)$
60. $6 + 3^2 - 4$

F. Find the value of each expression or formula by substituting the given values and using the Order of Operations Agreement.

61. Evaluate $(P - T)(R + S)$ when $P = 7$, $T = 3$, $R = 5$, and $S = 2$.
62. Evaluate $5A^2B$ when $A = 4$ and $B = 2$.
63. Evaluate $3CD^2$ when $C = 5$ and $D = 2$.
64. Evaluate $K + 6(M - L)$ when $K = 5$, $M = 8$, and $L = 3$.
65. Evaluate M in $M = 4R - 3S$ when $R = 5$ and $S = 2$.
66. Evaluate L in $L = 5A + B - 2C$ when $A = 4$, $B = 6$, and $C = 5$.
67. Evaluate R in $R = 3S^2T$ when $S = 4$ and $T = 2$.
68. Evaluate K in $K = 3W - 4Y$ when $Y = 5$ and $W = 7$.
69. Evaluate r in $r = 5p^3 - 3q^2$ when $p = 2$ and $q = 3$.
70. Evaluate d in $d = 4a^2 - 3b^3$ when $a = 3$ and $b = 2$.

G. Find the solution for each equation by inspection or intuition.

71. $X - 9 = 4$
72. $16 - a = 9$
73. $f + 5 = 11$
74. $28 = 4Y$
75. $13 - B = 5$
76. $19 - K = 4$
77. $m/3 = 5$
78. $18 = 3Y$
79. $2y + 3 = 11$
80. $4a - 5 = 7$

H. Find the critical value for each inequality and then write the answer to the inequality in set notation. (See Examples 1 and 2 in Section 8.5.)

81. $g - 7 > 2$
82. $y + 2 < 8$
83. $y \div 6 < 3$
84. $3b \geq 15$
85. $2d \leq 12$
86. $m - 4 > 5$
87. $40 > m + 10$
88. $k \div 3 \leq 9$
89. $x + 7 \geq 11$
90. $18 < a + 4$

I. Each exercise that follows has an error. State what the error is; then correct the statement.

91. The formula for the data in this table is $L - 12 = K$.

K	16	20	22	26
L	4	8	10	14

92. The simplification for $9x^2$ is $9 \cdot 9 \cdot x \cdot x$ or $81x^2$.
93. The coefficient for the y-term in $3w - 4x + y - 5z$ is zero.
94. The simplification for $32 - 7(8 \div 2)$ is 100.
95. The solution for $a \div 3 = 12$ is $a = 4$.
96. The solution set for $K - 3 \geq 9$ is $\{K \mid K > 12\}$.

J. Read and respond to the following exercises.

97. In your own words, explain how algebra is different from arithmetic.
98. List five properties that are the same in algebra as they are in arithmetic.
99. Each of the algebraic expressions below could be derived from an English expression. Notice the operation symbol and then write an expression using words that might be represented by the given algebraic expression.
    a. $x + 5$
    b. $n - 4$
    c. $a \div 6$
    d. $8p$
    e. y^3
    f. $\sqrt{m}$
100. In what ways is solving an inequality like solving an equation? In what ways is it different?

CHAPTER TEST

In Exercises 1 through 6, write a formula for each situation.

1. The number of students present (p) is equal to the difference between the number of students enrolled (s) and the number absent (a).
2. The total price of an item (P) is equal to the sum of its cost (c) and the sales tax (t).
3. The number of seconds (s) in m minutes is equal to the product of 60 and the number of minutes (m).
4. The class average (A) on the last test is equal to the quotient of the total points earned (P) and the number of students (S) who took the test.
5.

X	8	10	12	16
Y	4	6	8	12

6.

G	3	5	7	9
H	18	30	42	54

7. Combine like terms: $7a - 3a + a$
8. Combine like terms: $16X + 4Y - 5X + 3Y$
9. Simplify: $(6 + 7)(18 \div 3)$
10. Simplify: $45 - 4(12 - 5)$
11. Simplify: $9 + 12 \div 3 - 2 \times 3$
12. Simplify: $26 - 5(16 \div 2^3)$
13. Evaluate $(K - P)(K + P)$ when $K = 7$ and $P = 4$.

14. Evaluate $5A^2B$ when $A = 3$ and $B = 2$.

15. Evaluate $7R - 3S - 4T$ when $R = 5$, $S = 6$, and $T = 3$.

16. Solve: $M - 5 = 6$

17. Solve: $4 = K \div 12$

In Exercises 18 and 19, solve the inequality for the critical value and then give the answer in set notation.

18. Solve: $Y + 3 < 11$

19. Solve $4p \geq 20$

20. Explain the error in the following simplification and find the correct value.

    Find the value of $3wy^2v$ when $w = 2$, $v = 3$, and $y = 4$.

$$3wy^2v$$
$$(3)(2)(3)^2(4)$$
$$(3)(2)(9)(4)$$
$$(6)(9)(4)$$
$$(54)(4)$$
$$216$$

STUDY SKILLS MODULE

Efficient Use of Study Time

Having a place to study was encouraged in Study Skills Module 8. This unit discusses studying and doing homework in that place efficiently and effectively. Many of today's students must find time to attend class, work full or part time, care for family, study, and do other essential things like eat and sleep. When a 24-hour day is over, often it is the studying component that was short changed. Yet, to be successful in college and to reach career and educational goals, studying is a necessity. Studying and learning is your job when you are a student.

It is very important for busy students to develop a study plan and schedule time for studying, doing homework, preparing for tests, and for doing special projects and writing papers. You might start by making a daily schedule, writing every task on your "to do" list. Make sure the schedule includes specific time to study for each course. You will also need a weekly and monthly layout, to ensure time has been allotted for large projects and for studying for tests.

As you begin to study during a scheduled period, you will want to use the time you have as efficiently as possible. Set a goal for that session of what you need to accomplish before you finish. You should start the session by working on your most difficult subject, so you face it when you are freshest. Save the subject you enjoy the most till the end of the session and use it as a reward. With each subject, work on learning the new material at the beginning of the session, and use the later part of the time to review material and do homework.

For every class, you should review the lecture notes, read the textbook material, and then complete any assigned work. Some classes require a lot of homework and you have a specific assignment to do. Other courses require mostly reading and thinking and studying, and you have very little product to show at the end of the study session. Whatever your homework is for a class, though, make sure you have something to show for each study period. It may be problems worked, note cards completed, or a short summary of what you have read. Producing something helps you feel a sense of accomplishment and gives you a way to measure how you have used your study time.

Don't treat yourself as a marathon runner who can't stop and rest—but, at the same time, avoid making excuses to take a break. Schedule your break times, just as you schedule the study time. You might try working 50 minutes, then taking a 10-minute break. If you find it hard to get back to work, think about your reasons for being in college and how you must be successful in your present courses to be able to reach your educational, career, or personal goals.

Your Point of View

1. Most colleges tell students they should expect to do 2 hours of work outside of class for every hour they spend in class. How many hours per week should you be studying for your current class schedule? How close are you to accomplishing this? What factors can you adjust to make more study time?
2. List four traits or behaviors that make a student efficient rather than inefficient.
3. For two days, write down exactly what you do with your study time. How long did you study for this course during those two days? What did you produce with this time? Was it efficient use of your time?
4. Some students study a subject for several hours every day and still don't fully understand the material. Other students spend a fraction of that time and do well in the course. If these students all have equal ability and background, describe the role that "efficient use of study time" might play in these varying successes.

CHAPTER

Equations

LEARNING OBJECTIVES

When you have completed this chapter, you should be able to:

1. Solve one-step equations.
2. Solve two-step equations.
3. Solve equations that require combining like terms.
4. Solve equations with unknown terms on both sides of the equals sign.
5. Solve equations that contain fraction or decimal numbers.
6. Understand the meaning of formulas, formula substitution, and rearrangement.

In Chapter 8, you found solutions to simple equations by inspection or intuition. These methods can be used only when you are trying to find the solution to very simple equations. In this chapter, we develop procedures to solve more complex equations. You will use these same procedures when you go on to other algebra courses and in science and technical courses where solving equations is required.

Later in the chapter you will be solving equations that contain decimals and fractions. In order to do this successfully, you must be very competent with these kinds of numbers. If you are unsure of these skills, do the Skills Checks in Appendix A. Check your answers, and for any errors you made, go back to the indicated sections in the text and review that material at this time.

9.1 Solving One-Step Equations

In this section, you will learn how to solve equations such as

$x + 5 = 7$ $y - 3 = 8$ $7t = 28$ $W/4 = 12$

To solve these equations, you rearrange the terms to get the variable alone on one side of the equals sign and the constant term on the other side. In each of these "one-step" equations, the variable has one operation performed on it. You will "undo" that operation by performing the opposite operation.

Addition ⟷ subtraction

Multiplication ⟷ division

To maintain equality, you must do the same operation on *both sides* of the equals sign. An equation is like a balance scale (Figure 9.1). If you add something to one pan of a scale that is in balance but not to the other, the scale is no longer in balance. If you add the same quantity to both sides, however, the balance is maintained. The same theory applies to solving equations.

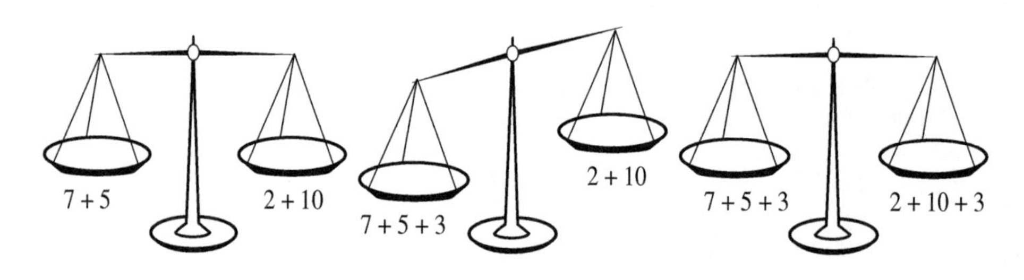

FIGURE 9.1

This process of doing addition or subtraction on both sides of the equation is an application of a property of real numbers called the Addition or Subtraction Property of Equality.

Addition or Subtraction Property of Equality

For all real numbers a, b, and c

$a = b$ if and only if $a + c = b + c$ and

$a = b$ if and only if $a - c = b - c$

In the first sample equation, $x + 5 = 7$, the variable x has 5 added to it. To "undo" that operation, we must do the opposite operation: subtraction. Five must be subtracted from both sides of the equation, as illustrated in Figure 9.2.

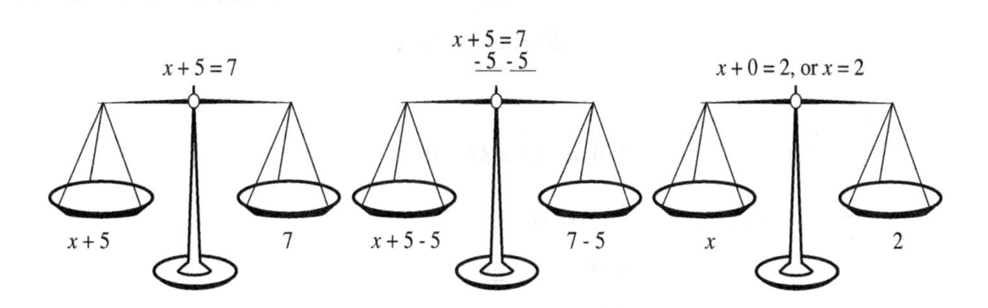

FIGURE 9.2

This means that the value for x that will give a true statement in $x + 5 = 7$ is 2. Try it.

$x + 5 = 7$

$2 + 5 = 7$

$7 = 7$ True.

This last process is the one used to check the answer to any equation. You should develop the habit of checking *every* equation that you solve. Substitute the answer back into the *original* equation, and simplify the arithmetic problem that has been created. If the answers on both sides come out to be the same value, that equation is called an **identity statement**, then you know your answer is correct. If you do not get the same value on both sides, then you need to go back and redo the problem. It is not often that there is a way you can check answers on homework or on a test and know for certain that they are correct! Use the opportunity to check whenever you can—and that should be every time you solve an equation.

The solution to the second sample problem is similar. In $y - 3 = 8$, the variable y has 3 subtracted from it. To solve and get y all alone, we must do the opposite operation: addition. Three must be added to both sides.

$$\begin{array}{rcr} y - 3 &=& 8 \\ \underline{+3} & & \underline{+3} \\ y + 0 &=& 11 \\ y &=& 11 \end{array}$$

Check: $y - 3 = 8$

$11 - 3 = 8$

$8 = 8$ True.

Here are more solutions to equations like these. Study every step and try to understand why each process (opposite operation, Addition or Subtraction Property of Equality, simplifying, and so forth) was performed.

EXAMPLE 1 Solve: $15 + T = 26$

Solution

$$15 + T = 26$$
$$\underline{-15} \quad \underline{-15}$$
$$T = 11$$

Check

$15 + T = 26$ Write the original equation.

$15 + 11 = 26$ Substitute 11 for T and simplify.

$26 = 26$ True.

EXAMPLE 2 Solve: $13 = M - 9$

Solution

$$13 = M - 9$$
$$\underline{+9} \quad \underline{+9}$$
$$22 = M$$

Check

$13 = M - 9$ Substitute 22 for M in the original equation.

$13 = 22 - 9$

$13 = 13$ True.

Self-Check Try these problems. Remember to check each answer.

1. $X + 7 = 13$
2. $B - 5 = 6$
3. $14 = J - 9$
4. $8 + G = 12$
5. $13 + P = 21$
6. $W + 4 = 9$

Answers:

1. $X = 6$; check: $13 = 13$
2. $B = 11$; check: $6 = 6$
3. $23 = J$; check: $14 = 14$
4. $G = 4$; check: $12 = 12$
5. $P = 8$; check: $21 = 21$
6. $W = 5$; check: $9 = 9$

In the third sample problem on page 346, $7t = 28$, the operation performed on t is multiplication (a number and variable written side by side indicate multiplication). Because division is the opposite operation, divide both sides by the numerical coefficient, 7. This will get t all alone with a coefficient of 1. Use the fractional form to indicate the division.

$$7t = 28$$

$$\frac{7t}{7} = \frac{28}{7} \quad \text{(Simplify: } 7/7 = 1,\ 28/7 = 4\text{)}$$

$$1t = 4 \quad \text{or} \quad t = 4$$

Check:

$$7t = 28$$
$$7(4) = 28$$
$$28 = 28 \quad \text{True.}$$

Remember, it is not necessary to write the coefficient 1 before a variable. Any literal term standing alone is understood to have a coefficient of 1.

In the last sample problem, $W/4 = 12$, the variable W is being divided by 4. Do the opposite operation to both sides of the equation: Multiply both sides by 4.

$$\frac{W}{4} = 12 \qquad \text{Check: } \frac{W}{4} = 12$$

$$(4)\frac{W}{4} = (4)12 \qquad \frac{48}{4} = 12$$

$$1W = 48 \quad \text{or} \quad W = 48 \qquad 12 = 12 \quad \text{True.}$$

The two previous sample problems utilize an important property of equations:

Multiplication or Division Property of Equality

For all real numbers a, b, and c,

$a = b$ if and only if $ac = bc$ and

$a = b$ if and only if $a/c = b/c$ where $c \neq 0$.

Here are two more solutions to one-step equations. Study every step and try to understand how each process (opposite operation, Multiplication or Division Property of Equality, do to both sides, simplify, and so forth) works.

EXAMPLE 3 Solve: $27 = 5N$

Solution $27 = 5N$ Divide by the coefficient of N.

$$\frac{27}{5} = \frac{5N}{5}$$

$$\frac{27}{5} \text{ or } 5\frac{2}{5} \text{ or } 5.4 = N$$

Don't be concerned that the answer is not a whole number—not all equation answers are whole numbers!

Check $27 = 5N$

$27 = 5(5.4)$

$27 = 27.0$ True.

EXAMPLE 4 Solve: $\frac{T}{4} = 24$

Solution

$$\frac{T}{4} = 24$$

$$(4)\frac{T}{4} = (4)24$$

$$T = 96$$

Check $\frac{T}{4} = 24$ Write the original equation.

$\frac{96}{4} = 24$ Substitute 96 for T and simplify.

$24 = 24$ True.

Self-Check Try these problems. Remember to check each answer.

1. $\frac{H}{5} = 35$
2. $32 = 8G$
3. $45 = 9R$
4. $\frac{M}{3} = 20$
5. $12P = 84$
6. $90 = \frac{F}{6}$

Answers:

1. $H = 175$, check: $35 = 35$
2. $4 = G$, check: $32 = 32$
3. $R = 5$, check: $45 = 45$
4. $M = 60$, check: $20 = 20$
5. $P = 7$, check: $84 = 84$
6. $F = 540$, check: $90 = 90$

Also included in this discussion of one-step equations should be equations such as 2/3 $Y = 12$ (this might also be written $2Y/3 = 12$). The variable Y is being multiplied by 2/3, so to "undo" the multiplication, you must divide both sides of the equation by 2/3. Review complex fractions (Section 3.4).

$$\frac{2}{3}Y = 12$$

$$\frac{\frac{2}{3}Y}{\frac{2}{3}} = \frac{12}{\frac{2}{3}}$$

Rewrite each side as division.

$$\frac{2}{3}Y \div \frac{2}{3} = 12 \div \frac{2}{3}$$

Rewrite the division as multiplication by the reciprocal; then divide out common factors.

$$\frac{\overset{1}{\cancel{2}}}{\underset{1}{\cancel{3}}}Y\frac{(\overset{1}{\cancel{3}})}{(\underset{1}{\cancel{2}})} = \frac{\overset{6}{\cancel{12}}}{1}\frac{(3)}{(\underset{1}{\cancel{2}})}$$

$$Y = 18$$

You can shorten the process and eliminate the need to write the equation as a division problem. *Multiply both sides by the reciprocal* of the fraction. The reciprocal of 2/3 is 3/2. This procedure is more efficient than working the division by a fraction.

$$\frac{2}{3}Y = 12$$

$$\frac{(3)}{(2)}\,\frac{2}{3}\,Y = \frac{(3)}{(2)}\,\frac{12}{1}$$

You could reduce by common factors.

$$\frac{(\overset{1}{\cancel{3}})}{(\underset{1}{\cancel{2}})}\,\frac{\overset{1}{\cancel{2}}}{\underset{1}{\cancel{3}}}\,Y = \frac{(3)}{(\underset{1}{\cancel{2}})}\,\frac{\overset{6}{\cancel{12}}}{1}$$

$$Y = 18$$

Check

$$\frac{2}{3}Y = 12$$

$$\frac{2}{3}(18) = 12$$

$$\frac{36}{3} = 12$$

$12 = 12$ True.

You can solve any one-step equation of the form 2/3 Y = 12 by using the shortcut. Multiply both sides of the equation by the reciprocal of the coefficient of the unknown term. Look at the steps in each of these examples before doing problems yourself.

EXAMPLE 5 Solve: $\frac{4}{5}t = \frac{3}{20}$

Solution

$$\frac{4}{5}t = \frac{3}{20}$$

$$\frac{(5)}{(4)}\,\frac{4}{5}\,t = \frac{(5)}{(4)}\,\frac{3}{20}$$

Multiply both sides by the reciprocal of 4/5 and simplify.

$$\frac{(\overset{1}{\cancel{5}})}{(\underset{1}{\cancel{4}})}\,\frac{\overset{1}{\cancel{4}}}{\underset{1}{\cancel{5}}}\,t = \frac{(\overset{1}{\cancel{5}})}{(4)}\,\frac{3}{\underset{4}{\cancel{20}}}$$

$$t = \frac{3}{16}$$

Check

$$\frac{4}{5}t = \frac{3}{20}$$

$$\frac{4}{5}\,\frac{(3)}{(16)} = \frac{3}{20}$$

Substitute 3/16 for t.

$$\frac{\overset{1}{\cancel{4}}}{5}\,\frac{(3)}{(\underset{4}{\cancel{16}})} = \frac{3}{20} \text{ or } \frac{\overset{1}{\cancel{2}} \cdot \overset{1}{\cancel{2}} \cdot 3}{5 \cdot \underset{1}{\cancel{2}} \cdot \underset{1}{\cancel{2}} \cdot 2 \cdot 2} = \frac{3}{20}$$

$$\frac{3}{20} = \frac{3}{20} \quad \text{True.}$$

EXAMPLE 6 Solve: $14 = \dfrac{3X}{10}$

Solution

$$14 = \frac{3X}{10}$$

$$\frac{(10)}{(3)}\,\frac{14}{1} = \frac{(10)}{(3)}\,\frac{3X}{10}$$ Multiply both sides by the reciprocal of 3/10.

$$\frac{140}{3} = \frac{30X}{30}$$

$$\frac{140}{3} \text{ or } 46\frac{2}{3} = X$$

Check

$$14 = \frac{3X}{10}$$ $\dfrac{3X}{10}$ is the same is $\dfrac{3}{10}X$.

$$14 = \frac{3}{10}\left(\frac{140}{3}\right)$$ Replace X with 140/3.

$$14 = \frac{420}{30}$$ Simplify to an identity statement.

$$14 = 14 \quad \text{True.}$$

EXAMPLE 7 Solve: $1\dfrac{2}{3}K = 30$

Solution

$$1\frac{2}{3}K = 30$$

$$\frac{5}{3}K = 30$$ Change the mixed number to an improper fraction.

$$\frac{(\overset{1}{\cancel{3}})}{(\underset{1}{\cancel{5}})}\,\frac{\overset{1}{\cancel{5}}}{\underset{1}{\cancel{3}}}K = \frac{(3)}{(\underset{1}{\cancel{5}})}\,\frac{\overset{6}{\cancel{30}}}{1}$$ Multiply by the reciprocal.

$$K = 18$$

Check $1\frac{2}{3}K = 30$

$\frac{5}{3}\frac{(18)}{(1)} = 30$ Substitute 18 for K in the original equation.

$\frac{90}{3} = 30$

$30 = 30$ True.

Self-Check Try these problems. Be sure to check your answers.

1. $\frac{4}{9}P = 60$ 2. $4\frac{2}{3}S = 56$ 3. $1\frac{4}{5}Y = \frac{6}{5}$ 4. $\frac{2}{3} = \frac{4}{9}M$

Answers:

1. $P = 135$; check: $60 = 60$
2. $S = 12$; check: $56 = 56$
3. $Y = 2/3$; check: $6/5 = 6/5$
4. $M = 3/2$; check: $2/3 = 2/3$

9.1 EXERCISES

A. Solve and check.

1. $B + 4 = 9$
2. $t + 3 = 15$
3. $8 + W = 12$
4. $11 + R = 13$
5. $m - 3 = 7$
6. $B - 9 = 7$
7. $6 = t - 3$
8. $15 = C - 4$
9. $12 = 3N$
10. $5q = 20$
11. $5s = 30$
12. $9T = 36$
13. $\frac{C}{3} = 6$
14. $\frac{d}{4} = 5$
15. $9 = \frac{P}{3}$
16. $7 = \frac{B}{4}$
17. $\frac{2y}{3} = \frac{5}{6}$
18. $\frac{3T}{4} = 12$
19. $7 = 1\frac{3}{4}X$
20. $10 = 3\frac{1}{3}R$
21. $3W = 21$
22. $9 = t - 2$
23. $\frac{1}{2}Y = 6$
24. $\frac{X}{8} = 3$
25. $S - 8 = 12$
26. $9A = 27$
27. $B - 14 = 3$
28. $\frac{3}{4}T = \frac{15}{24}$
29. $\frac{m}{7} = \frac{5}{9}$
30. $9 + A = 12$
31. $1\frac{5}{6}X = 22$
32. $4r = 15$
33. $10 = 3t$
34. $2\frac{3}{4}Y = \frac{33}{64}$
35. $\frac{W}{4} = 20$
36. $\frac{y}{5} = \frac{3}{8}$
37. $11 + A = 14$
38. $\frac{3Y}{4} = \frac{5}{8}$
39. $8 = 8 + R$
40. $56 = 8 + M$

B. Read and respond to the following exercises.

41. A friend solved the following problem as shown, but the answer would not check. Describe in words how the equation should have been solved and why.

$\frac{N}{8} = 16$

$\frac{N}{8} = \frac{16}{8}$

$N = 2$

42. Describe, in words, how to solve and check $18 = K + 8$.

43. If given a verbal statement—a number increased by 6 is equal to 17—describe, in words, the mental process you would use to determine the value of the number. Compare this process with the way you would solve the equation $N - 6 = 17$.

44. What is an identity statement in mathematics?

45. Is $M = 7$ the solution to the equation $M - 26 = 33$? Explain, in words, how you determined your answer.

9.2 Two-Step Equations

Solving equation such as $3X + 5 = 20$ and $11 = 2Y - 7$ requires two different operations. In the first equation, $3X + 5 = 20$, the variable X is multiplied by 3 *and* has 5 added to it. In $11 = 2Y - 7$, the variable Y is multiplied by 2 and has 7 subtracted from it. In such equations, we "undo" the addition or subtraction first and then do the multiplication or division. The thing to remember is that every step makes the problem simpler.

EXAMPLE 1 Solve: $3X + 5 = 20$

Solution

$$\begin{array}{rcl} 3X + 5 &=& 20 \\ \underline{-5} && \underline{-5} \\ 3X &=& 15 \end{array}$$

Subtract 5 from both sides.

Now a one-step equation remains.

$$\frac{3X}{3} = \frac{15}{3}$$

Divide by 3.

$$X = 5$$

Check

$3X + 5 = 20$ Substitute into the original equation.

$3(5) + 5 = 20$ Simplify using the Order of Operations Agreement.

$15 + 5 = 20$

$20 = 20$ True.

EXAMPLE 2 Solve: $11 = 2Y - 7$

Solution

$$\begin{array}{rcl} 11 &=& 2Y - 7 \\ \underline{+7} && \underline{+7} \\ 18 &=& 2Y \end{array}$$

Add 7 to both sides.

Divide by 2.

$$\frac{18}{2} = \frac{2Y}{2}$$

$$9 = Y$$

Check

$11 = 2Y - 7$

$11 = 2(9) - 7$

$11 = 18 - 7$

$11 = 11$ True.

EXAMPLE 3 Solve: $6A - 8 = 5$

Solution

$$
\begin{aligned}
6A - 8 &= 5 \\
+8 \quad & \quad +8 \\
6A &= 13
\end{aligned}
$$

Add 8 to both sides.

Because A must have a coefficient of 1, divide both sides by 6.

$$\frac{6A}{6} = \frac{13}{6}$$

$$A = \frac{13}{6} \text{ or } 2\frac{1}{6}$$

Check

$6A - 8 = 5$

$6\left(\frac{13}{6}\right) - 8 = 5$

$13 - 8 = 5$

$5 = 5$ True.

EXAMPLE 4 Solve: $7m + 4 = 18$

Solution

$$
\begin{aligned}
7m + 4 &= 18 \\
-4 \quad & \quad -4 \\
7m &= 14 \\
\frac{7m}{7} &= \frac{14}{7} \\
m &= 2
\end{aligned}
$$

Check

$7m + 4 = 18$

Substitute into the original equation and simplify.

$7(2) + 4 = 18$

$14 + 4 = 18$

$18 = 18$ True.

The presence of fraction or decimal numbers in the equation does not change the strategy for solving two-step equations. "Undo" addition or subtraction first; then "undo" multiplication or division. Use the rules of fraction and decimal arithmetic operations. (You may need to review parts of Chapters 3, 4, and 5.)

EXAMPLE 5 Solve: $\frac{H}{2} - 15 = 3$

Solution

$$\frac{H}{2} - 15 = 3$$

Add 15 to both sides because you need to undo addition or subtraction first.

$$\underline{+15} \quad \underline{+15}$$

$$\frac{H}{2} = 18$$

Division remains, so multiply both sides by 2.

$$(2)\frac{H}{2} = (2)18$$

$$H = 36$$

Check

$$\frac{H}{2} - 15 = 3$$

$$\frac{36}{2} - 15 = 3$$

$$18 - 15 = 3$$

$$3 = 3 \quad \text{True.}$$

EXAMPLE 6 Solve: $\frac{2}{3}R + 4 = 12$

Solution

$$\frac{2}{3}R + 4 = 12$$

Subtract 4 from both sides.

$$\underline{-4} \quad \underline{-4}$$

$$\frac{2}{3}R = 8$$

$$\frac{(3)}{(2)}\frac{2}{3}R = \frac{(3)}{(2)}\frac{(8)}{1}$$

Multiply both sides by the reciprocal of 2/3. You could divide out common factors.

$$\frac{\overset{1}{(\cancel{3})}}{\underset{1}{(\cancel{2})}}\frac{\overset{1}{\cancel{2}}}{\underset{1}{\cancel{3}}}R = \frac{(3)}{\underset{1}{(\cancel{2})}}\frac{\overset{4}{(\cancel{8})}}{1}$$

$$R = 12$$

Check

$$\frac{2}{3}R + 4 = 12$$

Substitute 12 for R into the original equation.

$$\frac{2}{\underset{1}{\cancel{3}}}\frac{\overset{4}{(\cancel{12})}}{1} + 4 = 12$$

Simplify using the Order of Operations Agreement.

$$8 + 4 = 12$$

$$12 = 12 \quad \text{True.}$$

EXAMPLE 7 Solve: $\frac{2}{3}Y + \frac{1}{2} = \frac{5}{6}$

Solution

$$\frac{2}{3}Y + \frac{1}{2} = \frac{5}{6}$$

$$\frac{2}{3}Y + \frac{1}{2} = \frac{5}{6}$$
$$\underline{\quad -\frac{1}{2} \quad -\frac{1}{2}}$$

Subtract 1/2 from both sides, using the common denominator form, 1/2 = 3/6.

$$\frac{2}{3}Y + \frac{1}{2} = \frac{5}{6}$$
$$\underline{\quad -\frac{1}{2} \quad -\frac{3}{6}}$$

$$\frac{2}{3}Y \quad = \frac{2}{6}$$

$$\frac{3}{2}\frac{(2)}{(3)}Y = \frac{3}{2}\frac{(2)}{(6)}$$

After subtraction is completed, multiply both sides by the reciprocal of 2/3. Then divide common factors or multiply and reduce.

$$\frac{\cancel{3}^{1}}{\cancel{2}_{1}}\frac{(\cancel{2})^{1}}{(\cancel{3})_{1}}Y = \frac{\cancel{3}^{1}}{\cancel{2}_{1}}\frac{(\cancel{2})^{1}}{(\cancel{6})_{2}}$$

$$Y = \frac{1}{2}$$

Check

$$\frac{2}{3}Y + \frac{1}{2} = \frac{5}{6}$$

$$\frac{\cancel{2}^{1}}{3}\frac{(1)}{(\cancel{2})_{1}} + \frac{1}{2} = \frac{5}{6}$$

$$\frac{1}{3} + \frac{1}{2} = \frac{5}{6}$$

$$\frac{2}{6} + \frac{3}{6} = \frac{5}{6}$$

$$\frac{5}{6} = \frac{5}{6} \quad \text{True.}$$

EXAMPLE 8 Solve: $14 = 0.8R + 1.2$

Solution

$$14.0 = 0.8R + 1.2$$
$$\underline{-\ 1.2 \qquad\quad -1.2}$$

Subtract 1.2 from both sides.

$$12.8 = 0.8R$$

Divide by 0.8.

$$\frac{12.8}{0.8} = \frac{0.8R}{0.8}$$

$$16 = R$$

Check

$14 = 0.8R + 1.2$

$14 = 0.8(16) + 1.2$

$14 = 12.8 + 1.2$

$14 = 14$ True.

EXAMPLE 9 Solve: $0.6X - 0.003 = 0.021$

Solution

$$\begin{aligned} 0.6X - 0.003 &= 0.021 && \text{Add 0.003 to both sides.} \\ +0.003 &+ 0.003 \\ 0.6X &= 0.024 \\ \frac{0.6X}{0.6} &= \frac{0.024}{0.6} && \text{Divide by 0.6.} \\ X &= 0.04 \end{aligned}$$

Check

$$\begin{aligned} 0.6X - 0.003 &= 0.021 \\ 0.6(0.04) - 0.003 &= 0.021 \\ 0.024 - 0.003 &= 0.021 \\ 0.021 &= 0.021 \quad \text{True.} \end{aligned}$$

Self-Check Try these problems. If necessary, use the examples as guides. Remember to check your answers.

1. $3H - 5 = 25$
2. $8M + 6 = 62$
3. $\frac{5}{6}X + 4 = 19$
4. $2.2 = 2X - 0.4$
5. $16 = 3Y - 2$
6. $4T + 3 = 23$

Answers:

1. $H = 10$; check: $25 = 25$
2. $M = 7$; check: $62 = 62$
3. $X = 18$; check: $19 = 19$
4. $X = 1.3$; check: $2.2 = 2.2$
5. $Y = 6$; check: $16 = 16$
6. $T = 5$; check: $23 = 23$

9.2 EXERCISES

A. Solve and check.

1. $2T - 7 = 11$
2. $2W - 8 = 10$
3. $4F + 1 = 25$
4. $3M + 5 = 17$
5. $8 + 5Y = 38$
6. $9 + 3T = 27$
7. $10 + 2M = 27$
8. $35 = 8 + 4T$
9. $47 = 11 + 12W$
10. $17 + 2C = 35$
11. $3.5M + 2.1 = 6.65$
12. $\frac{2}{3}Y - 12 = 48$
13. $\frac{4}{5}A + 17 = 57$
14. $2.8x - 1.5 = 0.18$
15. $18 = 5T + 18$
16. $4M + 17 = 17$
17. $1.2P - 8 = 1.6$
18. $11N - 4 = 7$
19. $5T + 6 = 41$
20. $3R - 1.66 = 0.17$
21. $9C + 12 = 48$
22. $0.06K + 1.1 = 7.7$
23. $7X + 3 = 3$
24. $6T + 7 = 7$
25. $0.2W - 1.6 = 2.4$
26. $3M - 21 = 15$

27. $56 + \frac{5F}{6} = 206$

28. $0.43 + 0.3F = 0.73$

29. $4 + \frac{5N}{2} = 44$

30. $9 + \frac{7Y}{3} = 30$

31. $2T - 6 = 0$

32. $7P - 16 = 0$

33. $8K + 4 = 13$

34. $9R + 5 = 15$

35. $\frac{1}{2}F - 7 = 17$

36. $\frac{1}{4}Y - 1 = 15$

37. $5A - \frac{2}{3} = 3$

38. $64 = 5C - 6$

39. $0.8 = 0.16 + 4W$

40. $\frac{5}{6} + 3R = \frac{7}{2}$

B. Read and respond to the following exercises.

41. Explain, in words, the relationship (similarity or difference) between the steps in the Order of Operations Agreement and the steps for solving equations.

42. Explain, in words, whether $3x - 2 = 7$ and $0.3x - 0.2 = 0.7$ are solved using different procedures because one has whole number coefficients and constants and the other has decimals.

43. Explain, in words, whether $3M + 6 = 18$ and $2/3M + 1/6 = 1\ 3/4$ are solved using different procedures because one has whole number coefficients and constants and the other has fractions.

44. Describe, in words, the steps you would use to solve and check each equation in Exercise 42.

45. Describe, in words, the steps you would use to solve and check each equation in Exercise 43.

46. Solve and check $5C + 6 = 6$, then describe your steps in words.

47. Solve and check $8Y + 16 = 16$, then describe your steps in words.

9.3 Equations That Require Combining Like Terms

An equation such as $6X - 3 + 2X = 29 - 8$ looks much more complex than the two-step equations that you just solved. But actually, after you simplify each side it will be a two-step equation. To simplify, combine like terms on each side of the equals sign. This concept was introduced in Chapter 8.

EXAMPLE 1 Solve: $6X - 3 + 2X = 29 - 8$

Solution

$$6X - 3 + 2X = 29 - 8$$

$6X$ and $2X$ are like terms. 29 and 8 are like terms. Combine the like terms, using the operation sign that appears in front of each term: $6X + 2X = 8X$ and $29 - 8 = 21$.

$$8X - 3 = 21$$
$$\underline{+3} = \underline{+3}$$
$$8X = 24$$
$$\frac{8X}{8} = \frac{24}{8}$$
$$X = 3$$

This should now look familiar. It's a two-step equation.

Check When checking an equation, *always start with the original* equation, not one of the simplified forms. Your may have made a mistake in getting to that step, and you will not find such an error if you don't take your answer into the original equation.

$$6X - 3 + 2X = 29 - 8$$
$$6(3) - 3 + 2(3) = 29 - 8$$
$$18 - 3 + 6 = 29 - 8$$
$$15 + 6 = 21$$
$$21 = 21 \quad \text{True.}$$

Note that for the first time the check answer, 21 = 21, is not a number that was carried down from the original equation. This happens in equations that have more than one term on each side of the equals sign.

EXAMPLE 2 Solve: $K + 4K + 3K = 180$

Solution

$$K + 4K + 3K = 180$$

Combine like terms: $1K + 4K + 3K = 8K$

$$8K = 180$$
$$\frac{8K}{8} = \frac{180}{8}$$
$$K = \frac{45}{2} \text{ or } 22\frac{1}{2} \text{ or } 22.5$$

Check

$$K + 4K + 3K = 180$$

Substitute 22.5 everywhere that K appears, then simplify.

$$22.5 + 4(22.5) + 3(22.5) = 180$$
$$22.5 + (90) + (67.5) = 180$$
$$180 = 180 \quad \text{True.}$$

EXAMPLE 3 Solve: $7m + 9 - 3m + 4 = 27 + 6$

Solution

$$7m + 9 - 3m + 4 = 27 + 6$$

$7m$ and $3m$, 9 and 4, and 27 and 6 are like terms. Combine like terms: $7m - 3m = 4m$, $9 + 4 = 13$, and $27 + 6 = 33$.

$$4m + 13 = 33$$
$$\underline{-13} = \underline{-13}$$
$$4m = 20$$
$$\frac{4m}{4} = \frac{20}{4}$$
$$m = 5$$

Check $7m + 9 - 3m + 4 = 27 + 6$ Substitute 5 for all the *m*s in the original equation, then simplify.

$$7(5) + 9 - 3(5) + 4 = 27 + 6$$

$$35 + 9 - 15 + 4 = 27 + 6$$

$$44 - 15 + 4 = 33$$

$$29 + 4 = 33$$

$$33 = 33 \quad \text{True.}$$

When solving equations, always look first to see if terms can be combined on either side of the equals sign. If so, do this simplifying first before "undoing" any operations.

Self-Check Try these equations before doing the exercises. Follow the steps shown in the examples if you need to. Be sure to check your answers.

1. $11X + 4X + X = 64$
2. $1.4 + 0.5R + 16 = 50$
3. $T + 7 + 2T - 2 = 22 + 4$
4. $M + 8 + 5M + 2 = 40$

Answers:

1. $X = 4$; check: $64 = 64$
2. $R = 65.2$; check: $50 = 50$
3. $T = 7$; check: $26 = 26$
4. $M = 5$; check: $40 = 40$

9.3 EXERCISES

A. Combine any like terms first; then solve and check.

1. $5T + 3T = 16$
2. $11M + 2M = 26$
3. $6F + 3 - 4F = 13$
4. $5X + 5 - 2X = 17$
5. $14 + 6R + 11 = 43$
6. $9 + 3Y + 4 = 28$
7. $240 = 4n + 2n$
8. $190 = 14R + 5R$
9. $12T + 6 - 4T = 22$
10. $17W + 4 - 4W = 43$
11. $3M - 9 = 24 - 6$
12. $5Y - 7 = 38 - 5$
13. $6r - 3r = 12 + 9$
14. $8t - 3t = 19 - 4$
15. $1.3m - 0.8m = 34 - 9$
16. $22 - 7 = 38W - 8W$
17. $4 + 28 = 10Y - 2Y$
18. $7.5 + 5 = 2.5K + 1.5K$
19. $14T - 4T = 30$
20. $80 = 55k - 15k$
21. $12M + M = 52$
22. $Y + 7Y = 8$
23. $7R + 2 - 3R + 6 = 21 - 13$
24. $6w + 13 - 2w + 15 = 28$
25. $240 = C + 19C$
26. $180 = X + 17X$
27. $12x - 8x + 12 = 28$
28. $5r + 2r + 2 = 23$
29. $6t + 6 + 4t = 16$

30. $9n + 3 - 6n = 15$

31. $4/5C + 12 - 9 = 15 - 4$

32. $7A + 4 - 3A = 9 + 3$

33. $9R - 6R = 36$

34. $54 - 18 = 3/5T - 3 + 7/10T$

35. $26T + 4 - 22T + 5 = 15$

36. $19M - 11M = 24$

37. $84 - 78 = 13M + 6 - 7M$

38. $8A - 7 + 11A = 12$

39. $8N + 15 + 9N - 6 = 62 - 53$

40. $47 - 8 = 24 + 8W + 15 - 3W$

B. Read and respond to the following exercises.

41. Your friend is having trouble with one of the problems in this section. His solution is shown here. Describe, in words, the error he is making, and then tell him what he should do to get the correct solution.

$$
\begin{aligned}
11C - 43 + 8C &= 14 \\
\underline{-8C} \qquad \underline{-8C}& \\
3C - 43 &= 14 \\
\underline{+43} \qquad &\underline{+43} \\
3C &= 57 \\
\frac{3C}{3} &= \frac{57}{3} \\
C &= 19
\end{aligned}
$$

Check:

$$
\begin{aligned}
11C - 43 + 8C &= 14 \\
11(19) - 43 + 8(19) &= 14 \\
209 - 43 + 152 &= 14 \\
166 + 152 &= 14 \\
318 &\neq 14
\end{aligned}
$$

42. Another student in your class had tried to help your friend with the problem in Exercise 41, but she thought his work was correct because her check showed $C = 19$ to be the correct solution. Why doesn't her check show that an error was made?

Check:

$$
\begin{aligned}
3C - 43 &= 14 \\
3(19) - 43 &= 14 \\
57 - 43 &= 14 \\
14 &= 14
\end{aligned}
$$

43. Write a list of general steps to follow to solve any equations that require combining of like terms.

44. Would your list from Exercise 43 need to change if the equation was $0.6x + 1.2 - 0.08 = 6.4$? Why?

9.4 Equations with Unknown Terms on Both Sides

The next equations you will solve in this chapter are those that have variable terms on both sides of the equals sign, such as:

$5Y + 4 = 3Y + 12$ and $16 - 7T = 3T - 4$

You are somewhat restricted in how you can proceed with these equations until you have studied Chapter 11 on signed numbers. For now you must work with positive numbers, so look at the numerical coefficients of the variable terms and picture their positions on a number line to determine which is *smaller*. The number that is farther to the left on the line is the smaller number.

In the first sample problem, if you compare the coefficients of the variable terms, 3 would be farther to the left than 5, so $3Y$ is the smaller variable term. You should start solving the equation by eliminating that smaller term. You will have to subtract $3Y$ from $3Y$ to eliminate it on the right side. Do the same to the other side of the equation, subtracting $3Y$ from $5Y$.

$$\begin{array}{rcl} 5Y + 4 & = & 3Y + 12 \\ -3Y & & -3Y \\ \hline 2Y + 4 & = & 0Y + 12 \end{array}$$

It's now a familiar two-step equation.

$$\begin{array}{rcr} 2Y + 4 & = & 12 \\ -4 & & -4 \\ \hline 2Y & = & 8 \end{array}$$

$$\frac{2Y}{2} = \frac{8}{2}$$

$$Y = 4$$

Check

$$5Y + 4 = 3Y + 12$$

$$5(4) + 4 = 3(4) + 12$$

$$20 + 4 = 12 + 12$$

$$24 = 24 \quad \text{True.}$$

As always, the check must be started in the original equation. Substitute 4 for Y on both sides.

In the second sample problem, look at the variable terms with the sign that is in front of them, $-7T$ and $3T$. 3 is assumed to be +3 and is to the right of the zero on the number line. The other variable coefficient, –7, is to the left of the zero and is therefore smaller than +3. You should add its opposite to both sides of the equation. (**Additive opposites** have the same number value but different signs. These opposites always add to zero.)

$$\begin{array}{rcl} 16 - 7T & = & 3T - 4 \\ +7T & & +7T \\ \hline 16 & = & 10T - 4 \end{array}$$

A two-step equation remains.

$$\begin{array}{rcl} 16 & = & 10T - 4 \\ +4 & & +4 \\ \hline 20 & = & 10T \end{array}$$

$$\frac{20}{10} = \frac{10T}{10}$$

$$2 = T$$

Check

$$16 - 7T = 3T - 4$$

$$16 - 7(2) = 3(2) - 4$$

$$16 - 14 = 6 - 4$$

$$2 = 2 \quad \text{True.}$$

EXAMPLE 1 Solve: $11N - 5 = 12 - 6N$

Solution

$$\begin{aligned} 11N - 5 &= 12 - 6N \\ \underline{+\,6N} &\quad \underline{+\,6N} \\ 17N - 5 &= 12 \\ \underline{+5} &= \underline{+5} \\ 17N &= 17 \\ \frac{17N}{17} &= \frac{17}{17} \\ N &= 1 \end{aligned}$$

$-6N$ is smaller than $+ 11N$. Add $+6N$ to both sides.

Check

$$\begin{aligned} 11N - 5 &= 12 - 6N \\ 11(1) - 5 &= 12 - 6(1) \\ 11 - 5 &= 12 - 6 \\ 6 &= 6 \quad \text{True.} \end{aligned}$$

In some problems, you might have to combine like terms first. Follow this example through carefully. Remember that no matter how complicated an equation looks at first, each step you do makes it less complicated.

EXAMPLE 2 Solve: $3W + 5 + W - 3 = 18 + 2W - 6$

Solution

$$\begin{aligned} 3W + 5 + W - 3 &= 18 + 2W - 6 \\ 4W + 2 &= 12 + 2W \\ \underline{-\,2W} &\quad \underline{-\,2W} \\ 2W + 2 &= 12 \\ \underline{-2} &\quad \underline{-2} \\ 2W &= 10 \\ \frac{2W}{2} &= \frac{10}{2} \\ W &= 5 \end{aligned}$$

Combine $3W + W$, $5 - 3$, and $18 - 6$. Which is smaller, $4W$ or $2W$?

Check

$$\begin{aligned} 3W + 5 + W - 3 &= 18 + 2W - 6 \\ 3(5) + 5 + 5 - 3 &= 18 + 2(5) - 6 \\ 15 + 5 + 5 - 3 &= 18 + 10 - 6 \\ 20 + 5 - 3 &= 28 - 6 \\ 25 - 3 &= 22 \\ 22 &= 22 \quad \text{True.} \end{aligned}$$

Use the original equation and substitute for all the Ws.

Self-Check Here are a few equations for you to try. Follow the steps of the examples as you solve these equations. Check your answers in the original equation first. Then look at the answers that follow if you need more proof that you are correct.

1. $3W - 7 = 8 - 2W$
2. $9M - 16 = 6 - 2M$
3. $6X + 4 = X + 14$
4. $5N + 3 = 2N + 30$

Answers:

1. $W = 3$; check: $2 = 2$
2. $M = 2$; check: $2 = 2$
3. $X = 2$; check: $16 = 16$
4. $N = 9$; check: $48 = 48$

In this chapter, you began by solving one-step equations and gradually learned to solve those that require three and four steps. The following list summarizes what you have learned.

Steps for Solving Equations

1. Combine like terms on each side of the equals sign.
2. If unknowns occur on both sides, start by eliminating the smaller variable term.
3. Add or subtract numerical terms.
4. Do multiplication or division.
5. Check your answer in the original equation.

9.4 EXERCISES

A. Solve and check.

1. $9R = 33 - 2R$
2. $7M = 15 + 2M$
3. $3 - t = 8t$
4. $5P = 2 + P$
5. $11W = 36 - 7W$
6. $3Y + 6 = 4Y$
7. $4k + 8 = 12k$
8. $5c = 21 = 2c$
9. $6m = 36 - 3m$
10. $8X + 5 = 4X + 13$
11. $6T + 2 = T + 17$
12. $7N + 4 = 6N + 7$
13. $5w - 4 = 2w + 5$
14. $9C - 10 = 3C + 2$
15. $12Y - 2 = 9Y + 7$
16. $13A - 1 = 4A + 17$
17. $1.5x - 0.22 = 0.4x + 1.1$
18. $7a - 5 = 2a + 20$
19. $3W + 1 = 11 - 2W$
20. $n - 2 = 6 - 3n$
21. $2x - 3 = 9 - 2x$
22. $4y - 5 = 16 - 3y$
23. $5a + 7 = 2a + 7$
24. $4y + 8 = y + 8$
25. $10x + 1 = 6x + 10 - 1$
26. $5r + 0.5 = 7 + 1.75r$
27. $11C - 4 = 7 + 4C + 10$
28. $5w + 6w + 15 = 3w + 15$
29. $4k + 8 = 20 + k$
30. $3Y + 6 = 55 - 4Y$
31. $22 + 10y = 6y + 26$
32. $7a - 15 = a + 21$

33. $2W + 11 = 3W + 1$
34. $3N - 6 = 2 - N$
35. $12X - 19 = 9 - 2X$
36. $14Y - 19 = 49 - 3Y$
37. $2 + 2a = 5a - 7$
38. $18 + y = 4y + 18$
39. $7c - 3 - 2c = 7 + 4c - 4$
40. $5M - 1 + 6M = 7M + 15 - 4M$

B. Solve and check. These equations include all types of problems found in Sections 9.1 through 9.4.

41. $b - 6 = 27$
42. $6W + 9 = 45$
43. $\frac{y}{3} - 8 = 2$
44. $3\frac{1}{2}P - 9 = 45$
45. $3a + 7 = 28$
46. $2 = y - 8$
47. $3T - 25 = 8$
48. $2X + 7 = 7$
49. $35 = 10y - 45$
50. $4M + 7 = 7M - 20$
51. $3c + 7 = 8c - 18$
52. $9w + 4 - 4w = 3w + 24$
53. $2a + 7 = 7$
54. $24 = 18c + 21$
55. $\frac{x}{5} = 6$
56. $\frac{m}{6} = 0$
57. $6Z - 17 = 8Z + 10 - 5Z$
58. $4 + 3X = 13$
59. $24R + 25 = 73$
60. $\frac{3}{5}B + 7 = 12$

C. Read and respond to the following exercises.

61. Describe, in words, how to decide which variable term to eliminate when beginning a problem with a variable on both sides of the equals sign.
62. Why must you eliminate the smaller variable term first in the equations in this section?
63. As you follow the Steps for Solving Equations listed in the box to solve any equation in this chapter, what are steps 1–4 accomplishing?
64. Describe in words how you would solve the equation $8M + 15 - M = 25 + 2M$. You do not actually have to solve it.
65. Describe in words how you would check an answer to the equation in Exercise 64.

9.5 Solving Equations with Fractions and Decimals

In this section, you will learn another method for solving equations that contain fraction or decimal numbers. The process for solving these equations includes the same steps as those you followed in Sections 9.1 through 9.4.

If you are to succeed with this section, it is important that you have good skills in fraction and decimal operations. If you have not reviewed or tested yourself on these skills, as was suggested at the beginning of the chapter, then do so now before going any farther (see the Skills Check, Appendix A).

In Section 9.2, you solved some equations that contained fractions and decimals by working with those numbers and following the same procedures you would have followed with whole numbers. In this section you will learn an alternative method that eliminates the fraction or decimal numbers and thus enables you to convert these equations to equations containing only whole numbers.

The new method requires that you multiply *every* term in the equation (on *both sides* of the equals sign) by the same value. Let's explore that idea first. Solve each of the following equations:

$X + 2 = 6 \qquad 2X + 4 = 12 \qquad 5X + 10 = 30$

Did you get $X = 4$ as the solution to each one? In fact, all three represent the same equation. See what happens when each term of $X + 2 = 6$ is multiplied by 2 or by 5.

$X + 2 = 6$ multiplied by 2 : $2(X) + 2(2) = 2(6)$

$$2X + 4 = 12$$

$X + 2 = 6$ multiplied by 5: $5(X) + 5(2) = 5(6)$

$$5X + 10 = 30$$

The fact that all three equations have a solution of $X = 4$ is not the only thing that makes them **equivalent equations.** They are also equivalent because each of the equations is just a *multiple* of the first equation. That is, each term of the first equation was multiplied by the same value to produce the second and third equations. Division can be used on each term as well, though you will not see such an example in this chapter.

Equivalent Equations

If every term in an equation is multiplied or divided by the same value, the resulting equation is equivalent to the original equation and has the same solution.

This process is similar to the balance illustration (Figure 9.1 in Section 9.1), which showed that adding or subtracting the same number on both sides of the equation maintains the balance. Multiplying or dividing every term on both sides of the equation by the same number also maintains the balance or equality. It produces an equivalent equation—one with a different appearance but the same solution. This is another application of the Multiplication or Division Property of Equality.

EXAMPLE 1 Are $Y - 2 = 5$ and $3Y - 6 = 15$ equivalent equations?

Solution Yes, because when each term in $Y - 2 = 5$ is multiplied by 3, $3(Y) - 3(2) = 3(5)$, it becomes $3Y - 6 = 15$.

or

$$\begin{aligned} Y - 2 &= 5 \\ +2 &\quad +2 \\ Y &= 7 \end{aligned}$$

$$\begin{aligned} 3Y - 6 &= 15 \\ +6 &\quad +6 \\ 3Y &= 21 \\ \frac{3Y}{3} &= \frac{21}{3} \\ Y &= 7 \end{aligned}$$

EXAMPLE 2 Are $6M - 15 = 21$ and $2M - 5 = 8$ equivalent equations?

Solution Inspection reveals that the second equation has a smaller coefficient with the variable. You could multiply $2M$ by 3 to get $6M$. Multiply every term in the second equation by 3 to see if you get an equation identical to the first.

$$(3)2M - (3)5 = (3)8$$
$$6M - 15 = 24$$

This is not exactly the same as $6M - 15 = 21$, so the equations are not equivalent.

or

$$\begin{aligned} 6M - 15 &= 21 \\ +15 \quad &\ \ +15 \\ 6M &= 36 \\ \frac{6M}{6} &= \frac{36}{6} \\ M &= 6 \end{aligned} \qquad \begin{aligned} 2M - 5 &= 8 \\ +5 \quad &\ \ +5 \\ 2M &= 13 \\ \frac{2M}{2} &= \frac{13}{2} \\ M &= 6\frac{1}{2} \end{aligned}$$

$6M - 15 = 21$ and $2M - 5 = 8$ are not equivalent because they do not have the same solution.

In solving a fraction or decimal equation by this new method, you must multiply through the equation by a value that eliminates the denominators or the decimal places. When all terms in a fraction or decimal equation are multiplied by the right value, the resulting equation contains *only* whole number terms and coefficients. In fraction equations, you multiply by the least common denominator (LCD) of all the denominators in the equation.

EXAMPLE 3 Solve for Y: $\frac{2}{3}Y + \frac{1}{2} = \frac{5}{6}$

Solution $\frac{2}{3}Y + \frac{1}{2} = \frac{5}{6}$

The least common denominator of the denominators 3, 2, and 6 is 6. Multiply every term by 6 to eliminate the fractions and thus produce an equation that you can easily solve using steps you already know.

$$\frac{6}{1}\left(\frac{2}{3}Y\right) + \frac{6}{1}\left(\frac{1}{2}\right) = \frac{6}{1}\left(\frac{5}{6}\right)$$

$$\frac{\overset{2}{\cancel{6}}}{1}\left(\frac{2}{\underset{1}{\cancel{3}}}Y\right) + \frac{\overset{3}{\cancel{6}}}{1}\left(\frac{1}{\underset{1}{\cancel{2}}}\right) = \frac{\overset{1}{\cancel{6}}}{1}\left(\frac{5}{\underset{1}{\cancel{6}}}\right)$$

Dividing out common factors could be used to simplify the terms.

$$\begin{aligned} 4Y + 3 &= 5 \\ -3 \quad &\ \ -3 \\ 4Y &= 2 \\ \frac{4Y}{4} &= \frac{2}{4} \\ Y &= \frac{1}{2} \end{aligned}$$

Now solve the whole number equation.

Check $\frac{2}{3}Y + \frac{1}{2} = \frac{5}{6}$ Always start with the original equation.

$$\frac{\overset{1}{\cancel{2}}}{3}\frac{(1)}{(\underset{1}{\cancel{2}})} + \frac{1}{2} = \frac{5}{6}$$

Substitute $\frac{1}{2}$ for Y and simplify.

$$\frac{1}{3} + \frac{1}{2} = \frac{5}{6}$$

$$\frac{2}{6} + \frac{3}{6} = \frac{5}{6}$$

$$\frac{5}{6} = \frac{5}{6} \quad \text{True.}$$

EXAMPLE 4 Solve: $A + \frac{2}{3}A = \frac{4}{9}$

Solution Multiply *every* term by the LCD, 9. Even the whole number term (1A) must be multiplied by 9.

$$A + \frac{2}{3}A = \frac{4}{9}$$

$$9(A) + \frac{\overset{3}{\cancel{9}}}{1}\frac{(\cancel{2})}{(\underset{1}{\cancel{3}})}A = \frac{\overset{1}{\cancel{9}}}{1}\frac{(4)}{(\underset{1}{\cancel{9}})}$$

$$9A + 6A = 4$$

$$15A = 4$$

$$\frac{15}{15}A = \frac{4}{15}$$

$$A = \frac{4}{15}$$

Check $A + \frac{2}{3}A = \frac{4}{9}$ Start with the original equation.

$$\frac{4}{15} + \frac{2(4)}{3(15)} = \frac{4}{9}$$

Substitute 4/15 for both of the As.

$$\frac{4}{15} + \frac{8}{45} = \frac{4}{9}$$

$$\frac{12}{45} + \frac{8}{45} = \frac{4}{9}$$

$$\frac{20}{45} = \frac{4}{9}$$

$$\frac{4}{9} = \frac{4}{9} \quad \text{True.}$$

In decimal equation, the method of eliminating decimal places involves multiplying by 10, 100, 1,000, and so forth, depending on the greatest number of decimal places in any term of the equation. Review the shortcut for doing this multiplication in Section 5.3.

EXAMPLE 5 Solve: $4.2B + 0.86 - 3.1B = 1.41$

Solution Multiply by 100 to eliminate the decimals.

$$100(4.2B) + 100(0.86) - 100(3.1B) = 100(1.41)$$

$$420B + 86 - 310B = 141$$

$$110B + 86 = 141$$

$$\underline{-86} \quad \underline{-86}$$

$$110B = 55$$

$$\frac{110B}{110} = \frac{55}{110}$$

$$B = 0.5$$

Check

$$4.2B + 0.86 - 3.1B = 1.41$$

$$4.2(0.5) + 0.86 - 3.1(0.5) = 1.41$$

$$2.1 + 0.86 - 1.55 = 1.41$$

$$2.96 - 1.55 = 1.41$$

$$1.41 = 1.41 \quad \text{True.}$$

EXAMPLE 6 Solve for M and, if necessary, round the answer to the nearest hundredth: $5.6M + 0.4 = 2.4M + 1.38$

Solution If you want to clear the decimals and have a whole number equation to solve, multiply every term in the equation by 100, because the greatest number of decimal places in those terms is two.

$5.6M + 0.4 = 2.4M + 1.38$ — Multiply through by 100.

$$(100)\,5.6M + (100)0.4 = (100)2.4M + (100)1.38$$

$560M + 40 = 240M + 138$ — Eliminate the smaller variable term, $240M$.

$$\underline{-240M} \quad \underline{-240M}$$

$$320M + 40 = 138$$

$$\underline{-40} \quad \underline{-40}$$

$$320M = 98$$

$$\frac{320M}{320} = \frac{98}{320}$$

$M = 0.30625$ — Round to hundredths.

$M \doteq 0.31$ — ($\doteq$ means approximately equal)

Check $5.6M + 0.4 = 2.4M + 1.38$ — Start with the original equation.

$5.6(0.31) + 0.4 = 2.4(0.31) + 1.38$

$2.136 \doteq 2.124$ — Because 0.31 is not an exact number (it was rounded), the check answers will be close, but not equal.

You have now experienced two methods for solving decimal and fraction equations: one where you work with those numbers given, and one where you eliminate them and work with whole numbers. As you solve the following problems, try using both methods and then decide which works better for you or which looks easier to use for that particular problem. Be sure to check your answers in the original equation.

Self-Check

1. $2W + \frac{1}{3} = \frac{7}{3}$
2. $7 = 0.4Y + 0.6$
3. $3.6 = 3.42 + 0.006X$
4. $\frac{7}{2} = \frac{2}{5}X + 3$

Answers:

1. $W = 1$; check: $7/3 = 7/3$
2. $Y = 16$; check: $7 = 7$
3. $X = 30$; check: $3.6 = 3.6$
4. $X = 5/4$; check: $7/2 = 7/2$

9.5 EXERCISES

A. Are these pairs of equations equivalent? Why or why not?

1. $3Y - 6 = 15$ and $Y - 2 = 5$
2. $7C + 14 = 35$ and $C + 2 = 5$
3. $2W - 9 = 12$ and $12W - 54 = 84$
4. $3P - 4 = 15$ and $21P - 28 = 90$
5. $T + 6 = 15$ and $75 = 5T + 30$
6. $18V + 12 = 24$ and $4 = 2 + 3V$

B. Write an equivalent equation containing only whole numbers by multipling every term by the value that will clear the denominators or decimal points. Check each answer.

7. $\frac{3}{4}K - \frac{1}{3} = \frac{1}{2}$
8. $12 + 0.4M = 14.32$
9. $2.4N - 0.08 = 7.12$
10. $\frac{3}{8}P + \frac{3}{2} = 6$
11. $0.2W - 1.6 = 2.4$
12. $\frac{3M}{4} - \frac{21}{5} = 15$
13. $\frac{3A}{5} + 39 = 144$
14. $0.3F + 0.43 = 0.73$

C. Solve the following equations. You don't have to use the same method for all exercises. Check your answers.

15. $\frac{1}{3}X + 1 = \frac{7}{2}$
16. $28 = 1.4Y + 2.1$
17. $1.4K - 3.5 = 1.12$
18. $\frac{1}{2}A - 3 = \frac{3}{4}$
19. $\frac{3}{4}N + 7 = \frac{15}{2}$
20. $0.735 + 2.4W = 12.9W$
21. $3.92 + 5.6X = 7.2X$

22. $\frac{3}{7}C + \frac{1}{4} = 1$

23. $2M - \frac{3}{4} = \frac{13}{4}$

24. $8.7P = 3.9P + 16.8$

25. $2.7X + 9 = 4.5X$

26. $3Y - \frac{5}{6} = \frac{13}{6}$

27. $0.2N + 1.3 = 1.5 - 0.6N$

28. $3\frac{3}{4}W + 1\frac{1}{2}W = 7$

29. $1\frac{5}{6}A + 2\frac{2}{3}A = 6$

30. $2.4X + 0.94 = 12.42 - 5.8X$

31. $1\frac{2}{5}x = x + \frac{2}{3}$

32. $7W - \frac{2}{3} = \frac{5}{2}W + 4$

33. $\frac{5}{9} + 11c = 12\frac{3}{8}c - \frac{2}{3}$

34. $2\frac{2}{3}y = y + 150$

35. $1.11X + 0.009 = 2.5X - 1.103$

36. $\frac{3}{8}A - \frac{1}{2} = \frac{4}{5}$

37. $0.008 + 1.6C = 3.64$

38. $56.8 + 1.2A = 48 + 1.6A$

39. $\frac{3}{5} + \frac{9}{4}C = \frac{3}{4}$

40. $0.4 + 3a + 1.6 = 4a - 0.48 + a$

D. Solve and check using a calculator. If necessary, round answers to the nearest hundredth.

41. $83.6y - 19.8 = 6.35$

42. $\frac{0.7R}{0.72} + 6.91 = 11.3$

43. $\frac{4.3}{18.46} = 0.94t$

44. $1.82m = 6.56 - 0.86m$

45. $2.026 = 1.48Y + 0.25$

46. $1.42K - 3.449 = 1.308$

47. $2.9X + 8.03 = 12.42 - 5.88X$

48. $\frac{0.445m}{0.89} - 6.72 = 1.775$

E. If you have a calculator with fraction capabilities, solve and check the following equations, working with the fraction numbers rather than eliminating the denominators.

49. $\frac{1}{5}x + 6 = 3x - 1\frac{3}{5}x$

50. $\frac{1}{3}y + \frac{3}{5} = \frac{3}{8}y + \frac{1}{8}$

51. $\frac{2}{3}a + 3a + 1\frac{1}{2}a = \frac{5}{12}$

52. $4 - \frac{5}{9}X = \frac{2}{3}X + 1\frac{1}{4}$

53. $\frac{15}{16}M + \frac{1}{3} = M - \frac{1}{12}$

54. $7\frac{1}{3} - 5\frac{3}{8} = 8\frac{1}{2}c - 4\frac{5}{6}c$

F. Read and respond to the following exercises.

55. Two equations can be described as being equivalent to each other because they have the same solution or because one equation is a multiple of the other. Which method do you think is "easier" to use when you have to prove that two equations are equivalent? Why?

56. Solve and check $0.03x + 7.5 = 8.3$; then describe your check's identity statement and explain what it means.

57. When asked to solve an equation such as Exercise 21 in this section, do you keep the decimals and work with them, or do you eliminate the decimals and work with whole numbers as you solve? What factors influence your decision to do the problem that way?

58. When asked to solve an equation such as Exercise 22 in this section, do you keep the fractions and work with them, or do you eliminate the fractions and work with whole numbers as you solve? What factors influence your decision to do the problem that way?

59. Your friend is having trouble with the problems in this section. One solution, shown here, does not check. State what his error is and then explain, in words, how to solve the equation correctly.

$$\frac{3}{4}x - 2 = \frac{5}{8}$$

$$\frac{8}{1}\frac{(3)}{(4)}x - 2 = \frac{8}{1}\frac{(5)}{(8)}$$

$$\frac{\overset{2}{\cancel{8}}}{1}\frac{(3)}{(\underset{1}{\cancel{4}})}x - 2 = \frac{\overset{1}{\cancel{8}}}{1}\frac{(5)}{(\underset{1}{\cancel{8}})}$$

$$\begin{aligned} 6x - 2 &= 5 \\ +2 &\quad +2 \\ 6x &= 7 \\ \frac{6x}{y} &= \frac{7}{6} \\ x &= \frac{7}{6} \end{aligned}$$

60. Your friend is having trouble again. His solution, shown here, does not check. State what his error is and then explain, in words, how to solve the equation correctly.

$$\begin{aligned} 0.65y + 0.58 &= 0.606 \\ -0.58 &\quad -0.58 \\ 0.65y &= 0.026 \\ \frac{0.65}{0.026}y &= \frac{0.026}{0.026} \\ y &= 25 \end{aligned}$$

61. Are $6x + 3 = 16$ and $\frac{3}{4}x + \frac{3}{8} = 2$ equivalent equations? In words, explain your answer.

62. Are $0.08y + 1.2 = 0.4y$ and $8y + 120 = 40y$ equivalent equations? In words, explain your answer.

9.6 Formulas

A **formula** is a general statement of how several quantities are related, so formulas usually contain more than one variable. The formula $P = 2L + 2W$, which you have seen before, is used to find the perimenter P of any rectangle of length L and width W (see Figure 9.3). The formula $A = LW$ is used to find the area A of any rectangle.

W

L

FIGURE 9.3

These formulas show the general relationships that apply to all rectangles of any size. If you are given dimensions (measurements) of a specific rectangle, then you can find the perimeter (distance around the figure) or area (size of the surface of the figure) by substituting those known values into the formula and solving the resulting equation.

Evaluating Formulas

EXAMPLE 1 Find the perimeter of the rectangle in Figure 9.4 using the formula $P = 2L + 2W$.

Solution

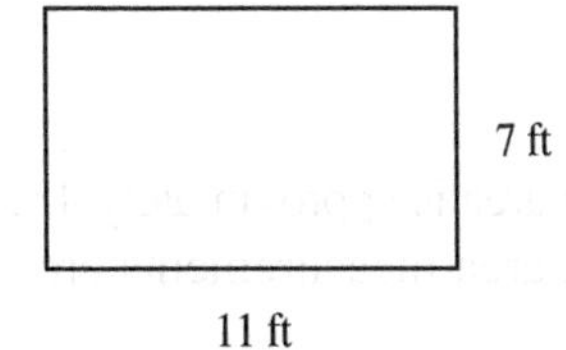

FIGURE 9.4

$P = 2L + 2W$	Write the formula.
$P = 2(11) + 2(7)$	Substitute for L and W.
$P = 22 + 14$	Simplify.
$P = 36$ feet	

Always label your answers with the correct units.

Check

$P = 2L + 2W$

$36 = 2(11) + 2(7)$

$36 = 22 + 14$

$36 = 36$ True.

EXAMPLE 2 Find the area of the rectangle in Figure 9.4 using the formula $A = LW$.

Solution $A = LW$

$A = (11)(7)$

$A = 77$ square feet

Note: When you multiply feet × feet, the resulting units are square feet. Area is *always* expressed in square units.

Check $A = LW$

$77 = (11)(7)$

$77 = 77$ True.

EXAMPLE 3 Find the circumference of (distance around) the circle in Figure 9.5 using $C = 2\pi r$ and $r = 8.1$ inches. π is a numerical constant that equals approximately 3.14. If necessary, round the answer to the nearest hundredth.

Solution

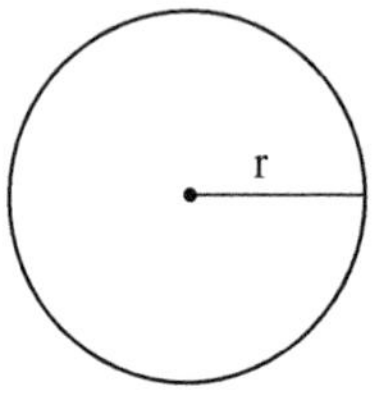

r

FIGURE 9.5

$C = 2\pi r$

$C \doteq 2(3.14)(8.1)$

$C \doteq 6.28(8.1)$

$C \doteq 50.868$

Rounded $C \doteq 50.87$

Since 3.14 is a rounded number, not an exact one, the symbol, $\doteq$, is used to show "approximately equal to." The circumference is approximately 50.87 inches.

Check $C = 2\pi r$

$50.87 \doteq 2(3.14)(8.1)$

$50.87 \doteq 6.28(8.1)$

$50.87 \doteq 50.868$ True.

EXAMPLE 4 Find the area of a circle using $A = \pi r^2$, $r = 12$ inches, and $\pi \doteq 3.14$.

Solution $A = \pi r^2$

$A \doteq (3.14)(12)^2$

$A \doteq (3.14)(144)$

$A \doteq 452.16$

The area is approximately 452.16 square inches. Remember that area measurements are always square units.

Check $A = \pi r^2$

$452.16 \doteq (3.14)(12)^2$

$452.16 \doteq (3.14)(144)$

$452.16 \doteq 452.16$ True.

EXAMPLE 5 Find the area of the trapezoid shown in Figure 9.6. The bases are 7 inches and 18 inches and the height is 9 inches. Use $A = 1/2\ h(b_1 + b_2)$.

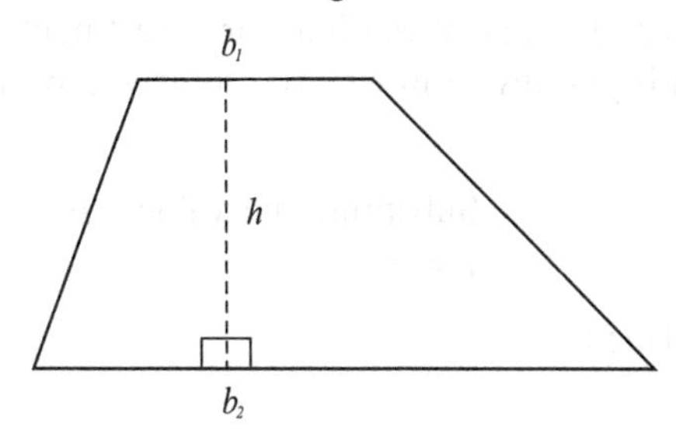

$A = 1/2\ h(b_1 + b_2)$

FIGURE 9.6

Solution In the formula, b_1 and b_2 mean that two sides are identified as bases in the trapezoid (these are the parallel sides), but they are not usually equal in length: $b_1 \neq b_2$. The subscripts (1 and 2) are used for identification but do not affect the calculations.

$A = 1/2h(b_1 + b_2)$

$A = \frac{1}{2}(9)(7 + 18)$

$A = \frac{1}{2}(9)(25)$

$A = 4.5(25)$

$A = 112.5$ The area is 112.5 square inches.

Check It is recommended that you check the answer to all formula evaluation problems by substituting and then simplifying (as was shown in Examples 1–4), but the checks will not be shown for the remaining examples in this section.

There are many formulas used in business, science, and mathematics with which you may be familiar. The formula for finding simple interest is $I = PRT$ (I is interest, P is principal, R is rate, and T is time). $S = C + M$ shows the relationship among the selling price (S) of an item, the cost (C) of the item, and the markup (M) needed to cover expenses and profit. You may need to review percents and the simple interest formula in Section 7.5.

EXAMPLE 6 What selling price should be placed on an item that cost \$19.36 if the manager requires a markup of \$2.17? Use $S = C + M$.

Solution

$S = C + M$

$S = 19.36 + 2.17$

$S = \$21.53$ The selling price should be \$21.53.

EXAMPLE 7 Find the amount of interest earned if \$1,500 is invested at 4% interest for 5 years. Use $I = PRT$.

Principal (P) is always the amount borrowed or invested. Rate (R) is the percent (you must change percents to decimals before you multiply or divide them). Time (T) is the number of years of the loan or investment. Interest (I) is the amount of money you earn if you invest money or pay if you borrow money.

Solution $I = PRT$ — Substitute the values: $P = \$1{,}500.$, $R = 4\% = 0.04$, and $T = 5$.

$I = (1{,}500)\ (0.04)\ (5)$

$I = 60(5)$

$I = 300$ — The interest earned is \$300.

In Examples 1–7, the variable whose value you were to determine was already isolated (all alone on one side of the equals sign). That will not be the case in Examples 8–16.

EXAMPLE 8 The perimeter of a rectangle is 98 centimeters and the width is 12 centimeters. Find the length, using $P = 2L + 2W$.

Solution $P = 2L + 2W$

$98 = 2L + 2(12)$ — Substitute the known values and simplify.

This is now an equation with one variable (L). Solve it as a two-step equation.

$$\begin{array}{rcl} 98 &=& 2L + 24 \\ \underline{-24} & & \underline{-24} \\ 74 &=& 2L \end{array}$$

$$\frac{74}{2} = \frac{2L}{2}$$

$37 = L$ — The length is 37 centimeters.

EXAMPLE 9 Find the amount of markup on an item that sells for \$114.50 and that costs \$69.79. Use $S = C + M$.

Solution

$$\begin{array}{rcl} S &=& C + M \\ 114.50 &=& 69.79 + M \\ \underline{-\ 69.79} & & \underline{-69.79} \\ 44.71 &=& M \end{array}$$

The markup is \$44.71.

EXAMPLE 10 Find the amount that was invested (principal) at 8% interest for 1 year if \$125.00 interest was earned. Use $I = PRT$.

Solution

$I = PRT$

$125 = P(8\%)\,(1)$ — Change 8% to its decimal equivalent.

$125 = P(0.08)\,(1)$

$125 = 0.08P$

$$\frac{125}{0.08} = \frac{0.08P}{0.08}$$

$1{,}562.50 = P$ — The amount invested is \$1,562.50.

EXAMPLE 11 Find h for a trapezoid for which the area is 1,000 square meters, and the bases are 60 meters and 40 meters in length. Use $A = 1/2\ h(b_1 + b_2)$.

Solution

$A = 1/2\ h(b_1 + b_2)$ — Substitute the values.

$1{,}000 = \frac{1}{2}h(60 + 40)$ — Simplify the addition.

$1{,}000 = \frac{1}{2}h(100)$ — Multiply 1/2 and 100.

$1{,}000 = 50h$ — Divide by 50.

$$\frac{1{,}000}{50} = \frac{50h}{50}$$

$20 = h$ — The height is 20 meters. The units for the answer are meters because $\frac{1000m^2}{50m} = \frac{1000m \cdot \cancel{m}}{50\cancel{m}}$. A pair of meters cancel out leaving 20 meters.

EXAMPLE 12 The circumference of a circle is 28.26 meters. Find the radius. Use $\pi \doteq 3.14$ and $C = 2\pi r$.

Solution You can substitute first and then solve for r:

$C = 2\pi r$ — Substitute.

$28.26 \doteq 2(3.14)r$ — Simplify.

$28.26 \doteq 6.28r$ — Solve for r.

$$\frac{28.26}{6.28} \doteq \frac{6.28}{6.28}r$$

$4.5 \doteq r$ — The radius is approximately 4.5 meters.

Solution Or you can solve for r first and then substitute:

$C = 2\pi r$ — Divide by 2π.

$$\frac{C}{2\pi} = \frac{2\pi r}{2\pi}$$

$$\frac{C}{2\pi} = r$$ Substitute the given values.

$$\frac{28.26}{2(3.14)} \doteq r$$

$$4.5 \doteq r$$ The radius is approximately 4.5 meters.

Rearranging Formulas

EXAMPLE 13 Another approach to Example 8 would be to rearrange or solve the formula for L before substituting the given values. The formula $P = 2L + 2W$ is solved for P because P is alone on one side of the equals sign. To solve that equation for L, add or subtract terms *on both sides* in such a way that $2L$ ends up alone on one side. Then divide both sides by 2 to leave L. After L is isolated, substitute the given values and simplify.

Solution

$$P = 2L + 2W$$ Solve for L.

$$\underline{-2W} \qquad \underline{-2W}$$

$$P - 2W = 2L$$

$$\frac{P-2W}{2} = \frac{2L}{2}$$

$$\frac{P-2W}{2} = L$$ Substitute given values.

$$\frac{98-2(12)}{2} = L$$

$$\frac{98-24}{2} = L$$

$$\frac{74}{2} = L$$

$$37 = L$$ The length is 37 centimeters.

EXAMPLE 14 Rearrange the formula first, and then substitute and simplify. The volume of a rectangular solid is found using the formula $V = LWH$. Find H if $V = 180$ cubic feet, $L = 9$ feet, and $W = 4$ feet.

Solution

$$V = LWH$$ Divide by LW to isolate H.

$$\frac{V}{LW} = \frac{LWH}{LW}$$

$$\frac{V}{LW} = H$$ Substitute the given values.

$$\frac{180}{(9)(4)} = H$$

$$\frac{180}{36} = H$$

$$5 = H$$ The height is 5 feet.

Here is what happens to the measurement units in this problem. When you have

$\frac{180}{(9)(4)}$ the numbers with units are: $\frac{180\ \textit{cubic feet}}{9\ \textit{feet}\ (4\ \textit{feet})} = \frac{180\ \textit{feet (feet) (feet)}}{9\ \textit{feet}\ (4\ \textit{feet})}$.

Two pairs of feet can cancel: $\frac{180\ \textit{feet}(\cancel{\textit{feet}})(\cancel{\textit{feet}})}{9\ \cancel{\textit{feet}}(4\ \cancel{\textit{feet}})} = \frac{180\ \textit{feet}}{36} = 5\ \textit{feet}$

In some instances, you may be asked only to rearrange a formula for a certain variable. Use the same procedures to solve the formula for the specific variable that you would use to solve any equation. Locate the variable that you need to isolate and determine what operations happened to it. Undo those operations by doing the opposite operations. Be sure to keep the equation in balance by doing the same operation to both sides of the equals sign.

EXAMPLE 15 Solve $S = C + M$ for C.

Solution Since C has M added to it, undo the addition by subtracting M from both sides of the equation.

$$S = C + M$$
$$\underline{-M} \quad \underline{-M}$$
$$S - M = C$$

The formula is solved for C.

EXAMPLE 16 Solve $W = AB$ for A.

Solution Because A is multiplied by B, divide by B to undo the operation.

$$W = AB$$
$$\frac{W}{B} = \frac{A\cancel{B}}{\cancel{B}}$$
$$\frac{W}{B} = A$$

EXAMPLE 17 Solve this formula for b: $P = a + b + c$

Solution The variable b has both a and c added to it. Undo those additions by subtracting a and c from both sides.

$$P = a + b + c$$
$$\underline{-a-c} \quad \underline{-a} \quad \underline{-c}$$
$$P - a - c = b$$

EXAMPLE 18 Solve $A = P + Prt$ for t.

Solution The variable t is being multiplied by Pr and has P added to it. Begin by undoing the addition (as you would do for any similar two-step equation).

$$A = P + Prt$$ Subtract P to isolate the term that contains t.

$$\underline{-P \quad -P}$$

$$A - P = Prt$$ Divide by Pr to isolate t.

$$\frac{A-P}{Pr} = \frac{\cancel{P}\cancel{r}t}{\cancel{P}\cancel{r}}$$ You can cancel P and r on the right side (there is only multiplication).

$$\frac{A-P}{Pr} = t$$ This is the correct answer. Do not cancel P into P on the left side because there is more than one term (subtraction) in the numerator.

EXAMPLE 19 Solve $V = \frac{1}{3}bh$ for b.

Solution

$$V = \frac{1}{3}bh$$ Multiply both sides by 3 to eliminate the fraction.

$$(3)V = \overset{1}{(\cancel{3})}\frac{1}{\underset{1}{\cancel{3}}}bh$$

$$3V = bh$$ Divide both sides by h to isolate b.

$$\frac{3V}{h} = \frac{bh}{h}$$

$$\frac{3V}{h} = b$$

9.6 EXERCISES

A. Substitute the given values into each formula and solve for the unknown. Label answers with correct units, and check your answers. If necessary, round decimal answers to the nearest hundredth.

1. The formula for the perimeter of a square is $P = 4s$. Find the perimenter if $s = 6$ feet.
2. The formula for the perimeter of a regular hexagon (six-sided figure) is $P = 6s$. Find the perimeter if the side (s) is 11 feet.
3. Find the perimeter of a rectangle ($P = 2L + 2W$) if the length is 5 centimeters and the width is 3 1/2 centimeters.
4. Find the perimeter of a rectangle ($P = 2L + 2W$) if the length is 11 1/2 yards and the width is 5 1/4 yards.
5. The formula for the perimeter of a rectangle is $P = 2L + 2W$. Find the width if the perimeter is 156 inches and the length is 28 inches.
6. The formula for the perimeter of a rectangle is $P = 2L + 2W$. Find the length if the perimeter is 96 meters and the width is 12 meters.
7. Find the amount of interest earned if $10,000 is invested at 10% annual interest for 3 years. Use $I = PRT$.
8. How much interest must be paid if $5,000 is borrowed at 8% annual interest for 2 years? Use $I = PRT$.
9. Find the circumference of a circle whose radius is 19 feet. Use $C = 2\pi r$ and $\pi \doteq 3.14$.
10. Use the circumference formula, $C = 2\pi r$ ($\pi \doteq 3.14$), to find the circumference of a circle that has a radius of 2.4 inches.
11. Find the area of the rectangle in Exercise 3. Use $A = LW$.
12. Find the area of the rectangle in Exercise 4. Use $A = LW$.

13. How long did it take to repay a loan if \$500 was borrowed at 13% interest and the interest charged was \$260? Use $I = PRT$.

14. How much money was borrowed at 15% interest if \$90 in interest was charged for the 3-year period? Use $I = PRT$.

15. Find the area of the square in Exercise 1. Use $A = s^2$.

16. Find the area of a square ($A = s^2$) if $s = 1.2$ centimeters.

17. How long does it take to earn \$240 when \$400 is invested at 12% annual interest? Use $I = PRT$.

18. How much money was invested at 11.5% annual interest if it earned \$172.50 in 3 years? Use $I = PRT$.

19. Find the area of the circle in Exercise 9. Use $A = \pi r^2$ and $\pi \doteq 3.14$.

20. Find the area of the circle in Exercise 10. Use $A = \pi r^2$ and $\pi \doteq 3.14$.

21. The formula $A = P + I$ is used to compute the total amount (A) of an investment or loan. Using this formula and the simple interest formula, $I = PRT$, find the amount of an investment at the end of 3 years if $P = \$12{,}500$ and $R = 12.5\%$.

22. Find the total amount to be repaid at the end of 2 years if \$10,000 is borrowed at 11.6%. The formula $A = P + I$ gives the total amount (A) of an investment or loan. Use this formula and the simple interest formula, $I = PRT$.

23. Formula: $S = C + M$. Find S if $C = \$117.30$ and $M = \$40.00$.

24. Formula: $S = C + M$. Find M if $S = \$19.99$ and $C = \$10.00$.

25. Formula: $A = 1/2\ bh$. Find the area (A) if $b = 13$ inches and $h = 10$ inches.

26. Formula: $P = a + b + c$. Find P if $a = 6$ feet, $b = 7.5$ feet, and $c = 11$ feet.

B. Using the given formula and information, solve for the unknown quantity. If necessary, round answers to the nearest hundredth. Be careful when labeling answers.

27. Formula: $P = a + b + c$. Find b if $P = 65$ centimeters, $c = 25$ centimeters, and $a = 28$ centimeters.

28. Formula: $P = 2L + 2W$. Find L if $P = 124$ yards and $W = 27$ yards.

29. Formula: $C = 2\pi r$. Find r if $C = 43.96$ inches and $\pi \doteq 3.14$.

30. Formula: $C = 2\pi r$. Find r if $C = 56.52$ centimeters and $\pi \doteq 3.14$.

31. Formula: $V = 1/3\ bh$. Find b if $V = 48$ cubic inches and $h = 6$ inches. (See Example 14 for help in labeling the answer.)

32. Formula: $S = 1/2\ gt^2$. Find g if $t = 6$ and $S = 324$.

33. Formula: $P = a + b + c + d$. Find c if $P = 82$ feet, $a = 17$ feet, $b = 12$ feet, and $d = 28$ feet.

34. Formula: $P = 4s$. Find s if $P = 62$ meters.

35. Formula: $S = 1/2\ gt^2$. Find g if $S = 64$ and $t = 2$.

36. Formula: $V = 1/3\ bh$. Find h if $V = 12$ cubic yards and $b = 8$ yards. (See Example 14 for help in labeling the answer.)

37. Formula: $P = 2a + b$. Find a if $P = 18$ centimeters and $b = 9$ centimeters.

38. Formula: $A = P + PRT$. Find T if $A = \$12{,}000$, $P = \$8{,}000$, and $R = 10\%$.

39. Formula: $a + b = c$. Find a if $b = 12$ kilometers and $c = 20$ kilometers.

40. Formula: $a + b = c$. Find b if $a = 22$ inches and $c = 43$ inches.

41. Formula: $I = PRT$. Find T if $P = \$2{,}400$, $R = 5\%$, and $I = \$360$.

42. Formula: $I = PRT$. Find P if $R = 4\%$, $T = 5$ years, and $I = \$480$.

C. Solve each formula or equation for the stated variable.

43. $S = C + M$ for M

44. $T = E + F$ for E

45. $E = AB$ for A

46. $E = AB$ for B

47. $V = \frac{1}{3}bh$ for h

48. $I = PRT$ for T

49. $I = PRT$ for P

50. $y = mx + b$ for m

51. $y = mx + b$ for x

52. $A = P + PRT$ for R

53. $A = P + PRT$ for T

54. $E = IR$ for I

55. $E = IR$ for R

56. $V = \frac{1}{3}\pi r^2 h$ for r^2

57. $V = \frac{1}{3}\pi r^2 h$ for h

58. $F = \frac{9}{5}C + 32$ for C

59. $V = \frac{4}{3}\pi r^3$ for r^3

60. $P = 2a + b$ for a

61. $P = a + b + c$ for a

62. $P = a + b + c$ for c

63. $S = 1/2\ gt^2$ for g

64. $S = 1/2\ gt^2$ for t^2

D. Read and respond to the following exercises.

65. Describe, in words, the difference(s) in procedure in finding the requested value for each of the following:
    a. Find P in $P = 2L + 2W$ if $L = 5$ feet and $W = 3$ feet.
    b. Find L in $P = 2L + 2W$ if $P = 19$ feet and $W = 4$ feet.

66. The formula $A = P + PRT$ is said to be solved for A. What does that mean?

67. Describe the similarities and differences between $2X + 8 = 26$ and $2a + b = P$. Then describe the similarities in solving the first for X and the second for A.

68. If you are helping your friend with the exercises in this section, how would you advise him to start the following exercise: Find W in $V = LWH$ if $V = 165$ cubic feet, $L = 11$ feet, and $H = 3$ feet.

69. Describe, in words, how you would determine the label for your answer in: Find W in $V = LWH$ when $V = 400$ cubic feet, $L = 20$ feet, and $H = 5$ feet.

70. In Example 12, the solution was found using two different methods. Describe which of the methods you prefer, now that you have done the homework for this section.

9.7 CHAPTER REVIEW

Solving an equation means finding the value for the variable that produces a true statement.

The steps for solving an equation are as follows:

1. If desired, eliminate fractions or decimal numbers by applying the Multiplication or Division Property of Equality to create an equivalent equation.
2. Combine like terms on each side of the equals sign.
3. If unknown terms are on both sides of the equals sign, start by eliminating the smaller variable term.
4. Add or subtract variable terms and numerical terms by using the Addition or Subtraction Property of Equality.
5. Do multiplication or division.
6. Check the answer in the *original* equation.

Formulas are equations that are general statements about how several quantities are related. You can substitute known values into the formula and then solve the resulting equation, or you can rearrange first and then substitute.

This chapter looks very long, but don't be too concerned about that as you prepare for the test. Each section in the chapter just added a step to the beginning of the process of solving equations. Once you learned the first sections of material, you used them in every other section. With that much practice, those earlier steps have become almost second nature to you!

Also, this material contains a confidence-building factor. Because you can check each answer, you know that you are doing things correctly. This is very reassuring especially at test time! Many people think of solving equations as "doing algebra." Thus, your success with this material should make you feel even more confident about your ability to learn algebra.

REVIEW EXERCISES

A. Solve and check.

1. $3X = 12$
2. $N + 5 = 7$
3. $F - 19 = 23$
4. $\frac{3}{4}C = 12$
5. $9 = H + 4$
6. $20 = 4Y$
7. $2A + 4 = 16$
8. $\frac{T}{9} = 4$
9. $11 = 5W - 9$
10. $J + 3.9 = 10.4$
11. $c - 1.37 = 9.44$
12. $R - 7 = 20$
13. $\frac{2}{3}Y = 24$
14. $\frac{F}{12} = 12$
15. $14 = 8M - 2$
16. $3N - 8 = 19$
17. $32.1 = 7S + 4.1$
18. $\frac{5}{6}X + 4 = 24$
19. $\frac{3}{4}X + 6 = 24$
20. $8.9 = 12W + 1.7$
21. $9P = 24$
22. $2X - 5 = 16$
23. $1.1Y + 0.3 = 3.6$
24. $\frac{4}{5}Y + 3 = 20$
25. $19C - 12 = 64$
26. $11W = 20$
27. $A + \frac{3}{4}A = 6$
28. $5X + 29 - X = 29$
29. $3S + 18 + S = 18$
30. $T + \frac{2}{3}T = 13$
31. $3F - 5 = F + 3$
32. $8R + 4 = 5R + 12$
33. $5Y - 4 = 22 - 6$
34. $180 = 4N + 2N$
35. $24 = 14R + 10R$
36. $4M - 8 = 20 - 6$
37. $\frac{5y}{6} = \frac{2y}{9} + 3$

38. $1.8K = 3 - 0.7K$

39. $1.1 + 0.45M = M$

40. $2 - \frac{2}{3}T = \frac{5}{6} + \frac{1}{2}T$

41. $\frac{5}{8}Y = \frac{2}{3}$

42. $3C + 16 = 6C - 2$

43. $8 - w + 6 = 7w - 2$

44. $11 = n - 1$

45. $8 = 5 + \frac{3}{4}R$

46. $6 - 2M = 8M - 14$

47. $17R - 5R + R = 45 - 6$

48. $3A - 21 - A - 1 = 0$

49. $3y - 16 = 4 + y$

50. $17 = \frac{5m}{2} - 1$

B. Substitute the given values into each formula and solve for the unknown. If necessary, round answers to the nearest hundredth. Label your answer.

51. If $r = 50$ miles per hour and $D = 375$ miles, find t in $D = rt$.

52. Find I in $I = PRT$ if \$1,500 is invested for 4 years at 7.5%.

53. $P = 2L + 2W$. Find P if $L = 22$ feet and $W = 13$ feet.

54. In $A = LW$, find W if $L = 10$ yards and $A = 95$ square yards.

55. If \$360 in interest is paid on a loan of \$2,000 borrowed at an annual interest rate of 6%, for how long was the money borrowed? Use $I = PRT$.

56. Find L in $P = 2L + 2W$ if P is 80 meters and W is 14 meters.

57. Find M in $S = C + M$ if S is \$6.49 and C is \$4.15.

58. If \$20.90 in interest is earned on a 2-year investment at 5.5% a year, what amount (principal) was originally deposited? Use $I = PRT$.

59. Use $A = 1/2\ bh$ to find b when A is 36 square inches and h is 9 inches.

60. Use $A = \pi r^2$ to find the area of a circle whose radius is 8 feet. (Use $\pi \doteq 3.14$.)

61. Use $C = 2\pi r$ to find the radius of a circle whose circumference is 150.72 centimeters. (Use $\pi \doteq 3.14$.)

62. Use $A = 1/2\ bh$ to find h when A is 24 square feet and b is 12 feet.

63. Find how many years it would take an investment of \$5,000 to earn \$1,000 in interest at an annual rate of 5%. Use $I = PRT$.

64. Use $C = 2\pi r$ to find the radius of a circle whose circumference is 15.7 inches. (Use $\pi \doteq 3.14$.)

C. Solve each equation for the variable indicated.

65. $C = A + B$ for B

66. $C = A + B$ for A

67. $W = MN$ for M

68. $W = MN$ for N

69. $A = 1/2\ bh$ for b

70. $A = 1/2\ bh$ for h

71. $P = 2L + 2W$ for W

72. $P = a + b + c$ for c

73. $V = LWH$ for H

74. $V = LWH$ for L

75. $\frac{2x}{3} + 3y = 12$ for x

76. $\frac{2x}{3} + 3y = 12$ for y

77. $a^2 + b^2 = c^2$ for a^2

78. $S = 1/2\ gt^2$ for g

79. $V = 1/3\ LWH$ for L

80. $V = \pi r^2 h$ for h

D. Describe what operation was performed between the previous line and the numbered line for each step of the solutions that follow. Study the example that follows:

$2x + 7 = 11$

$2x = 4$ 1. Subtracted 7 from both sides.

$x = 2$ 2. Divided both sides by 2.

Check:

$2x + 7 = 11$ 3. Wrote original equation.

$2(2) + 7 = 11$ 4. Substituted 2 for x.

$11 = 11$ 5. Simplified by Order of Operations Agreement.

81. $15M = 90$

$M = 6$ 1.

82. $4Y - 3 = 21$

$4Y = 24$ 1.

$Y = 6$ 2.

83. $3C - 5 + C = 17 - 6$

$4C - 5 = 17 - 6$ 1.

$4C - 5 = 11$ 2.

$4C = 16$ 3.

$C = 4$ 4.

Check: $3C - 5 + C = 17 - 6$ 5.

$3(4) - 5 + 4 = 17 - 6$ 6.

$11 = 11$ 7.

84. $4A - 3 = 7 - A$

$5A - 3 = 7$ 1.

$5A = 10$ 2.

$A = 2$ 3.

85. $\frac{7}{8}W + \frac{3}{4} = 6$

$7W + 6 = 48$ 1.

$7W = 42$ 2.

$W = 6$ 3.

86. Find L in $P = 2L + 2W$ if $P = 84$ yards and $W = 20$ yards.

$P = 2L + 2W$ 1.

$84 = 2L + 2(20)$ 2.

$84 = 2L + 40$ 3.

$44 = 2L$ 4.

$22 = L$ 5.

The length is 22 yards. 6.

E. Each solution below has an error made in one step. State, in words, what mistake was made and tell what should have been done.

87. $18A + 3 + 12A = 63$

$\underline{-12A} \quad \underline{-12A}$

$6A + 3 = 63$

$\underline{-3} \quad \underline{-3}$

$6A = 60$

$\frac{6A}{6} = \frac{60}{6}$

$A = 10$

88. $5.4M + 0.09 = 2.79$

$\underline{-0.09} \quad \underline{-0.09}$

$5.4M = 2.7$

$\frac{5.4M}{2.7} = \frac{2.7}{2.7}$

$M = 2$

89. $\frac{3}{4}C - 8 = \frac{5}{6}$

$(12)\frac{3}{4}C - 8 = (12)\frac{5}{6}$

$9C - 8 = 10$

$\underline{+8} \quad \underline{+8}$

$9C = 18$

$\frac{9C}{9} = \frac{18}{9}$

$C = 2$

90. $7.2 + 0.05R - 0.05 = 33$

$(10)7.2 + (100)0.05R - (100)0.05 = (100)3.3$

$72 + 5R - 5 = 330$

$67 + 5R = 330$

$\underline{-67} \quad \underline{-67}$

$5R = 263$

$\frac{5R}{5R} = \frac{263}{5}$

$R = 52.6$

91. Check for Exercise 89.

$$9C - 8 = 10$$
$$9(2) - 8 = 10$$
$$18 - 8 = 10$$
$$10 = 10$$

F. Read and respond to the following exercises.

92. Compare the process of solving an equation to the process used when solving a formula for one of the variables.

93. Describe, in words, how you would solve this equation:
$$\frac{3}{4}x + 0.68 = 1.2x - \frac{5}{8}$$

94. For each of the following equations, what property describes the step illustrated?

$$\frac{x}{4} = 7 \qquad\qquad Y - 5 = 8$$
$$(4)\frac{x}{4} = (4)7 \qquad\qquad \begin{array}{r} Y - 5 = 8 \\ \underline{+5} \quad \underline{+5} \end{array}$$

95. Describe in words, in detail, how you would solve $p = 3a + 2b - 5c$ for a.

96. Give an example of two equivalent equations and tell what makes them equivalent.

CHAPTER TEST

For Exercises 1–17, solve and check.

1. $3R = 81$
2. $5R + 3 = 38$
3. $2C - 7 + C = 35$
4. $\frac{w}{4} = 12$
5. $21 + 2A = 5A$
6. $0.4m + 12 = 14.32$
7. $\frac{1}{2}B - 3 = \frac{3}{4}$
8. $9R - 6R = 24 - 18$
9. $11w - 8 = 3$
10. $\frac{3}{5}T - \frac{1}{3} = 1$
11. $A + \frac{2}{3}A = 25$
12. $4Y - 1 = 11$
13. $c + 17 = 2 + 6c + 15$
14. $6N - 1 + 5N = 3N + 15$
15. $\frac{3T}{8} = 4$
16. $0.6M + 1.4 + 0.04M = 14.2$
17. $12P - 2 = 9P + 7$
18. Solve $PV = nRT$ for R.
19. The following solution has an error in it. Circle the error and describe, in words, what should have been done.

$$\begin{array}{rcl} 7Y + 2 - 2Y & = & 47 \\ \underline{+\ 2Y} \quad \underline{+\ 2Y} & & \\ 9Y + 2 & = & 47 \\ \underline{-2} & & \underline{-2} \\ 9Y & = & 45 \\ \frac{9Y}{9} & = & \frac{45}{9} \\ Y & = & 5 \end{array}$$

20. Describe, in words, the process used to go from one step to the next.

$12P - 2 = 9P + 7$	
$3P - 2 = 7$	1.
$3P = 9$	2.
$P = 3$	3.
Check: $12P - 2 = 9P + 7$	4.
$12(3) - 2 = 9(3) + 7$	5.
$34 = 34$	6.

STUDY SKILLS MODULE

10 Note Taking

Many students, when questioned, will tell you they don't take notes in class because it bothers their concentration, or they learn best by listening. Although it is true that some people are primarily auditory learners, everyone can benefit from a multiple approach to learning. The more senses and activity involved, the more successful the learning will be. Simply sitting and absorbing is hardly the best way to learn. Activity is crucial. Most students will attempt to work problems in class, but many find it difficult to take notes. Let's look at a few simple strategies that can make this process work better for you.

First of all, remember that you should take notes when you study as well as when you are in class. Developing the habit of taking notes makes the process easier for everyone.

1. Be sure to title and date your notes and keep them organized in a special notebook or folder.
2. Some students prefer a column approach, whereas others use a standard outline form (I. A. B. 1. 2. a. b.). Whichever you prefer, use some numbering system to help you organize the points.
3. Don't write in sentences. Develop your own abbreviations and use symbols whenever possible.
4. Don't try to take down everything that is said or every problem that is worked. Make references to textbook pages. Use short phrases to help you remember a particular idea.
5. Be sure to leave room for additional comments or questions that you need to include when you go back and review your notes. You will never get every detail the first time through.
6. Always go back, as soon as possible after class, to review your notes, fill in the missing steps, and rethink the processes that were discussed. This is a really good way to learn new material.
7. Put questions in the margins of your notes so you can easily see that you need to go back over that topic. These questions are a good resource when you go in to ask your instructor or a tutor for help.

8. Use note cards for formulas or rules or special problems that you need to learn. Include a sample for each formula or rule. Carry your note cards with you and review them often.

Taking notes helps you keep focused on the topic at hand. Writing keeps you actively involved in the class and in your studying, and the more you participate in the learning process, the better student you will become.

Your Point of View

1. Discuss your note-taking history. Do you take notes in all of your classes, some of your classes, none of your classes, only when you are studying, only when in class, and so on? Do you regularly review the notes you take? Share where you are now in using this learning tool.
2. List three things that you like about taking notes. List three things you find difficult.
3. Describe two specific things you will do this week to improve your note-taking skills. At the end of the week, report on your experience.

CHAPTER

10 Application Problems

LEARNING OBJECTIVES

When you have completed this chapter, you should be able to:

1. Translate words and expressions into algebraic symbols.
2. Translate word sentences into algebraic equations.
3. Write a correct equation from word problems by using one of the following methods of organizing information:
   a. by list
   b. by sketch
   c. by chart
4. Solve the equation you have written and check your answer.
5. State answers requested and label answers, when appropriate.

Students in math classes are often unenthusiastic when they hear that the next topic to be covered is application or word problems. Are you one of them? Perhaps if you understand the skills that are being developed in studying application problems, the topic will become more meaningful. Every time in real life that you need to make a choice, you look at all the factors that affect that decision. You consider the pros and the cons. If it is a major decision—to buy a house, to marry, to go to college, to change jobs, to move—you think long and hard about it. You look at all the facts and information. You weigh all sides of the decision and then make the best choice. Gathering information, outlining options, and logically considering all sides are the same procedures you follow in solving application problems. The situations may not be so complex, but the logical reasoning is the same.

This chapter will demonstrate several ways of organizing information found in word problems. You will need to decide which method is appropriate to each problem and to your style of organization—but organizing in some manner is essential for success in doing application problems.

10.1 Words to Symbols

You are now comfortable with solving equations in one variable. In this chapter, you will learn to write equations from word problems. This skill is important because there are many situations in algebra, business math, and science courses, and also at work, where writing and solving an equation will enable you to find the solution to a given problem.

The first skill you need is that of translating expressions given in words to expressions that consist of variables and symbols.

EXAMPLE 1 If the price of a movie ticket is \$5, then the price of two tickets could be expressed as \$5(2) and the cost of 10 tickets as \$5(10). The cost of an unknown number of tickets (n) would be expressed as \$5(n), or $5n$.

◀

EXAMPLE 2 If Uri is 23 years old now, you could express his age in 5 years as $23 + 5$ or 28 years. Eight years ago he would have been $23 - 8$ or 15 years old. If Maria won't tell you her age now, you can still express her age in 5 years as $Y + 5$ or her age 8 years ago as $Y - 8$, where Y = Maria's age now.

◀

EXAMPLE 3 If the temperature in Houston is D degrees, a 6-degree rise in temperature could be expressed as $D + 6$. A 10-degree fall in temperature would be $D - 10$.

◀

EXAMPLE 4 If one jar of peanut butter costs \$1.89, express the cost of J jars. The cost of J jars would be \$1.89J.

◀

When changing from words to symbols, look for key words that indicate which mathematical operation you should use. Many of these words were listed in Section 1.4.

Add	**Subtract**	**Multiply**	**Divide**
sum	difference	product	quotient
plus	less	times	divided
total	less than	total	split
increase	decrease	in all	each
more than	reduce	twice	shared
raise	remain	part of	average
combined	larger	altogether	ratio
altogether	fewer	area	per
in all	other comparison words	volume	equal parts

Use these key words to help you write the following word phrases as algebraic expressions. N will be used to represent the number in each example, but any letter, capital or lowercase, could have been chosen.

Word Expression	**Algebraic Expression**
1. six more than the number	$N + 6$
2. difference between a number and 8	$N - 8$
3. product of 12 and a number	$12N$ is the preferred form; but $12(N)$, $(12)(N)$, and $(12)N$ are also acceptable.
4. 15 divided by a number	$15/N$ or $15 \div N$
5. a number decreased by 11	$N - 11$
6. 11 *less than* a number	$N - 11$
7. 11 less a number	$11 - N$
8. a number divided into 3 equal parts	$N/3$ or $N \div 3$
9. 2/3 of a number	$2/3\ N$ or $2N/3$
10. 16% of a number	$0.16N$
11. three more than twice a number	$2N + 3$
12. six less than one-half of a number	$1/2\ N - 6$

A few of the algebraic expressions listed here need some explanation. In expression 1 in the list, $N + 6$ is the direct translation, though $6 + N$ would also be correct because of the Commutative Property of Addition. A variable represents any real number, so the properties of real numbers apply to variable expressions. In the multiplication examples (expressions 3, 9, and 10), although the Commutative Property does apply, the numerical coefficient should always be written before the literal factor ($12N$, not $N12$). This will avoid confusion later when exponents become involved: $N12$ could be misread as N^{12}.

The subtraction and division examples (expressions 2, 4, 5, 7, and 8) are not commutative and therefore *must* be written in the order given. Exceptions to writing in the order given occur only in expressions 1, 6, 11, and 12. In "six more than the number," the exact translation is $N + 6$. In "11 *less than* a number," the number (N) has to have 11 *subtracted from it*. Be aware that any time "less than" or "more than" is used, the order of the terms in the algebraic expression is reversed from the order that appears in the word expression. You may want to underline or circle "less than" and "more than" whenever they appear in a problem so that you will remember to write the expression correctly.

Self-Check Translate each expression to symbols using N to represent the number.

a. Seventeen less than a number

b. Twelve less a number

c. Four more than a number

Answers: a. $N - 17$ b. $12 - N$ c. $N + 4$

10.1 EXERCISES

A. Write the algebraic expression for each of the following. Use T to represent the unknown quantity.

1. eleven more than a number
2. the difference between a number and 69
3. a number decreased by ten
4. the sum of thirteen and a number
5. twice a number
6. one-fourth of a number
7. seventeen divided by a number
8. the quotient of a number and four
9. two-thirds of a number increased by 12
10. twice a number decreased by 14
11. eight less than a number
12. eight less a number
13. seventeen more than a number
14. three-eighths of a number
15. twenty less than a number
16. eighty more than a number
17. the difference between eight and a number
18. three times a number increased by twelve
19. four less than twice a number
20. fifteen less than four times a number
21. eighteen more than a number
22. the difference between six and twice a number

B. Match each expression in Column I with the correct phrase in Column II.

Column I	Column II
23. $3X + 6$	a. half of a number
24. $M - 12$	b. two-thirds of a number
25. $1/2F$	c. the product of 12 and a number
26. $12 - P$	d. the sum of three times a number and six
27. $5N + 13$	
28. $\frac{2}{3}T$	e. twelve less a number
29. $J + 6$	f. twelve less than a number
30. $4/B$	g. the sum of a number and six
31. $Y/4$	h. the quotient of a number and four
32. $12R$	i. the quotient of four and a number
	j. five times a number plus thirteen

C. Write the algebraic expression for each of the following.

33. One can of tuna costs 68 cents. Express the cost of x cans.
34. Pineapples cost $1.98 each. Express the cost of k pineapples.

35. Jerome is M years old now. Express his age 5 years ago.

36. Quick Key calculators cost $4 less than they did three years ago. Express the cost now if they cost L dollars three years ago.

37. R represents the total rainfall for the past four months. Represent the average rainfall for any month.

38. An estate of T dollars is equally divided among six heirs. Represent each person's inheritance.

39. Greg got P points on his term paper. Express Marie's grade if she got 11 points more than Greg.

40. In a company's benefits package, one health insurance plan costs $14 more per month than another. The less expensive policy costs P. Represent the monthly cost of the other policy.

41. There are S students enrolled in Dr. Huntley's Psychology 101 class. Represent the number of students in class today if 5/6 of the class is present.

42. LaTonya is Y years old. Represent her age 5 years ago.

43. If gasoline costs $1.09 per gallon, express the cost of G gallons.

44. Represent the amount of money you have if you have D dimes.

45. Represent Jamie's age in 4 years if he is Y years old now.

46. If a number is represented by N, how do you represent a number that is three times as big as that number? How do you represent one that is three more than the original number ? How do you represent the sum of all three of those quantities?

47. Represent the amount of money you have if you have Q quarters.

48. Each adult movie ticket costs $5. Express the total cost of T tickets.

49. Mr. Perez has D dollars. How do you represent an amount that is six dollars more than that? How do you represent an amount of money that is six times the original amount Mr. Perez has? How do you represent the sum of all three amounts?

50. Manuella has six books less than Maria. If Maria has N books, express the number of books that Manuella has.

D. For each expression that is given, write a translation to words. Be creative! For example: $x + 6$ can be "six more than x," or "x increased by six," or "the sum of x and six"; but more creatively it can be "Mary's age now is x, but her age in six years is $x + 6$".

51. $3y$

52. $M - 5$

53. $W/6$

54. $7 + R$

55. $5N + 2$

56. $2/3\ K$

E. Read and respond to the following exercises.

57. One of the paragraphs in this section cautioned you to very carefully translate expressions that contain "more than" or "less than" wording. Express that warning in your own words.

58. Are the expressions "six less a number" and "six less than a number" the same thing when translated to symbols? Explain your answer.

59. Are the expressions "the quotient of sixteen and a number" and "the quotient of a number and sixteen" the same thing when translated to symbols? Explain your answer.

60. Compare the translation of "four less than a number" and "four more than a number." Point out the similarities and the differences.

61. In the statement "the difference of a number and two," what is the key word that tells you what operation to perform? What is the role of the word *and*?

62. Explain, in words, how to express the value of D dimes.

10.2 Organizing by List

You can now take algebraic expressions and make them part of a full sentence. These sentences in algebra are equations. When forming algebraic sentences (equations) by converting from words to symbols, first state what variable you will use and what it represents. Key words such as *equals, is the same as, the result is, was,* and *are* will translate into the equals sign in the equation. In the following examples, N is used to represent the number in each equation.

Word Sentence	Algebraic Equation
The sum of a number and sixteen is twenty-seven.	$N + 16 = 27$
Thirty-six is equal to the product of nine and some number.	$36 = 9N$
The difference of a number and seventeen is equal to sixty-three.	$N - 17 = 63$
Eight is equal to the quotient of ninety-six and a number.	$8 = 96/N$
The product of six and some number is equal to the sum of that number and forty-five.	$6N = N + 45$
The product of five and what number equals thirty?	$5N = 30$

Of course, once you have written the equation from the words, you will need to solve it and check your answer.

Word Problem Procedures

1. Read the problem carefully to see what is given and what is to be found.
2. Use a variable to represent one of the items, usually the one you know the least about. Represent the other items using the same variable.
3. Organize the information.
4. Translate the words to an algebraic equation.
5. Solve the equation. Give all the answers requested.
6. Check your answer in the words of the problem.

After reading an application problem once, go back and carefully reread it; see what you know about that quantity (or quantities), and decide what you need to find. You will need to organize this information so you can write the correct equation for the problem, as well as find all the answers that are requested after you solve the equation.

In this chapter, we will demonstrate several methods for organizing information. The method shown for a problem will not be the only one that could be used; it is just the one the authors found to be most suited to that particular kind of problem. You are encouraged to try different methods of organizing information and to decide for yourself the one most appropriate to each kind of problem and to your style of organization.

You should use some means of organizing the information in a word problem and not just jump from reading the problem to writing the equation. The latter seldom works, especially as the problems become more complex or if they involve two or three different, but related, quantities.

The method of organization used in this section of application problems is making a list of your definitions of the quantities in the problem. If there is only one unknown quantity, we will define the variable, not make a list. Pay close attention to that step as you study the example problems that follow. Be sure to utilize this or some other method of organization as you do the homework for this section.

EXAMPLE 1 If a number is increased by 12, the result is 29. Find the number. We chose to use N to represent the number, so we list:

Define Variable Let N = The number

Equation

Number	increased by 12	result is	29
N	$+ 12$	$=$	29.

Solution

$$N + 12 = 29$$
$$\underline{-12} \quad \underline{-12}$$
$$N = 17$$

Check

$N + 12 = 29$

$17 + 12 = 29 \qquad 29 = 29$ True.

Note: This kind of check is not sufficient for word problems because it checks only the solution to the equation. The most common error in doing word problems is in not writing the correct equation. This check does not expose that kind of error. Instead, *check a word problem by substituting the answer into the wording of the problem.*

If a number is increased by 12, the result is 29.

If 17 is increased by 12, the result is 29.

$17 + 12 = 29$ True.

EXAMPLE 2 Six less than twice a number is 10. Find the number.

Define Variable Let N = The number

Equation

Twice a number	6 less than	is	10
$2N$	-6	$=$	10.

Solution

$$2N - 6 = 10$$
$$\underline{+6} \quad \underline{+6}$$
$$2N = 16$$
$$\frac{2N}{2} = \frac{16}{2}$$
$$N = 8$$

Check Six less than twice a number is 10.

Twice (8) less 6 is $2(8) - 6 = (16) - 6 = 10$ True.

EXAMPLE 3 Marina is three times as old as her son, Jason. The sum of their ages is 48 years. How old is each?

List This time there are two different quantities, but they must be represented by the same variable. Jason's age is the one you know the least about, so call it J. Now list the representation of the two ages.

Let J = Jason's age

Marina is three times as old as Jason (J), so her age is $3J$.

Let $3J$ = Marina's age

Now that the two ages are defined, write the equation.

Equation

The sum of their ages	is	48
$J + 3J$	$=$	48.

Solution

$$J + 3J = 48$$

$$4J = 48$$

$$\frac{4J}{4} = \frac{48}{4}$$

$$J = 12$$

Look at the list above the equation. Because J = Jason's age, he is 12 years old. Marina is $3J = 3(12) = 36$ years old. Be sure to give all the answers asked for! The problem said, "How old is each?" so you must give both answers.

Jason is 12 years old.

Marina is 36 years old.

Check Marina is three times as old as Jason: $3(12) = 36$. The sum of their ages is 48: $12 + 36 = 48$. Both parts of the problem are true when the answers are substituted.

EXAMPLE 4 Together, Marc and Mike purchased a hamburger franchise for $72,000. Mike's part was $18,000 more than Marc's. How much did each invest?

List Let D = Marc's amount

$D + 18{,}000$ = Mike's amount

Equation

Marc	and	Mike	is	72,000
D	$+$	$D + 18{,}000$	$=$	$72{,}000$.

Solution

$$
\begin{aligned}
D + D + 18{,}000 &= 72{,}000 \\
2D + 18{,}000 &= 72{,}000 \\
-18{,}000 &\quad -18{,}000 \\
2D &= 54{,}000 \\
\frac{2D}{2} &= \frac{54{,}000}{2} \\
D &= 27{,}000
\end{aligned}
$$

Marc's amount = \$27,000

Mike's amount = $D + 18{,}000 = \$45{,}000$

Check \$45,000 is \$18,000 more than \$27,000. \$45,000 + \$27,000 does equal \$72,000.

EXAMPLE 5 One number is 6 more than another number. The sum of the larger number and twice the smaller number is 27. What are the numbers?

List Let N = "Another" number = Smaller number

$N + 6$ = First number = Larger number

Equation

Larger number	sum	twice smaller number	is	27.
$N + 6$	$+$	$2N$	$=$	27

Solution

$$
\begin{aligned}
N + 6 + 2N &= 27 \\
3N + 6 &= 27 \\
-6 &\quad -6 \\
3N &= 21 \\
\frac{3N}{3} &= \frac{21}{3} \\
N &= 7
\end{aligned}
$$

The smaller number = 7, and the larger number is $N + 6 = 13$.

Check One number is 6 more than the other. $7 + 6 = 13$. The sum of the larger and twice the smaller is 27.

$13 + 2(7) = 27.$ $13 + 14 = 27.$ True.

EXAMPLE 6 Four-fifths of the students in a class received a grade of C or better. If 28 students received a C or better, how many students are in the class?

Define Variable Let S = Number of students in the class

"Part of" means to multiply.

Equation

Four-fifths of students got C or better	=	28 students received C or better
$\frac{4}{5}S$	$=$	28

Solution Multiply by the reciprocal and divide common factors.

$$\frac{(5)4}{(4)5}S = \frac{(5)28}{(4)(1)}$$

$$\frac{(\overset{1}{\cancel{5}})\overset{1}{\cancel{4}}}{(\underset{1}{\cancel{4}})\underset{1}{\cancel{5}}}S = \frac{(5)\overset{7}{\cancel{28}}}{(\underset{1}{\cancel{4}})(1)}$$

$S = 35$ students

Check Four-fifths of the students is 28 students. 4/5 of 35 = 4/5(35) = 28. True.

EXAMPLE 7 Misty's salary this week was eight dollars more than twice Jaye's salary. Together they earned $248. How much did Misty earn for the week?

List "Eight dollars more than twice Jaye's salary" describes Misty's salary. You know a lot about her earnings, but nothing about Jaye's, so:

Let x = Jaye's salary

$2x + 8$ = Misty's salary

Equation

Together	earned	$248
$x + 2x + 8$	$=$	248

Solution

$$3x + 8 = 248$$
$$-8 \quad -8$$
$$3x = 240$$
$$\frac{3x}{3} = \frac{240}{3}$$
$$x = 80$$

From the list above: $2x + 8$ = Misty's salary

$$2(80) + 8 = 168$$

Therefore, Misty earned $168.

Check Jaye's salary = x = $80. Misty's salary is eight more than twice that = 2(80) + 8 = $168. Together they earned $248: $80 + $168 = $248.

Self-Check Only make a list to represent the unknown quantities.

a. Malva has 4 more than three times the number of overtime hours that Allia has. Represent the number of overtime hours for each woman.

b. One number is eight less than another number. Represent the two numbers.

Answers: a. Allia's overtime = x, Malva's overtime = $3x + 4$

b. One number = N, Other number = $N - 8$

EXAMPLE 8 The sizes of two numbers are in the ratio of 7 to 3. Find the two numbers it their sum is 50.

List 7 to 3 means 7 parts to 3 parts. The parts are the same size.

Let x = one part

$7x$ = larger number

$3x$ = smaller number

Equation

The sum	is	50 .
$7x + 3x$	$=$	50

Solution

$$10x = 50$$

$$\frac{10x}{10} = \frac{50}{10}$$

$$x = 5$$

$7x = 7(5) = 35$ $3x = 3(5) = 15$ The numbers are 35 and 15.

Check $35 + 15 = 50$

EXAMPLE 9 Tran has been notified that he is a winner in a magazine sweepstakes, but he doesn't yet know the amount of his prize. He decides that he will divide the prize money into parts that have the ratio of 5:4:2. The largest portion will be for himself, the middle amount for his children, and the last amount for charity. What will be the amount of each portion if the prize is $93,500?

List 5:4:2 means if 5 parts to 4 parts to 2 parts. The "parts" are all the same size.

Let x = One part

$5x$ = Parts for Tran

$4x$ = Parts for the children

$2x$ = Part for charity

Altogether, the parts must equal $93,500.

Equation

Tran's + children's + charity	=	Prize money
$5x + 4x + 2x$	$=$	$93{,}500$

Solution

$$11x = 93{,}500$$

$$\frac{11x}{11} = \frac{93{,}500}{11}$$

$$x = 8{,}500$$

$5x = 5(8{,}500) = \$42{,}500$ = Tran's part

$4x = 4(8{,}500) = \$34{,}000$ = children's part

$2x = 2(8{,}500) = \$17{,}000$ = part for charity

Check \$42,500 + \$34,000 + \$17,000 does equal \$93,500.

Consecutive means that things follow one another successively without interruption. **Consecutive numbers or integers** are numbers that follow one another, such as 13, 14, 15 or 36, 37, 38, 39. In these sets of consecutive numbers, each number is 1 larger than the previous number. If you don't know the numbers but you do know that they are consecutive, then let

n = 1st number

$n + 1$ = 2nd consecutive number

$n + 2$ = 3rd consecutive number

$n + 3$ = 4th consecutive number

and so on

Use this notation to write equations for consecutive number problems.

EXAMPLE 10 The sum of three consecutive numbers is 69. What are the numbers?

List Let n = 1st number

$n + 1$ = 2nd number

$n + 2$ = 3rd number

Equation

1st number	+	2nd number	+	3rd number	is	69.
n	+	$n + 1$	+	$n + 2$	=	69

Solution

$$n + n + 1 + n + 2 = 69$$

$$3n + 3 = 69$$

$$\underline{-3} \quad \underline{-3}$$

$$3n = 66$$

$$\frac{3n}{3} = \frac{66}{3}$$

$$n = 22, n + 1 = 23, n + 2 = 24$$

The numbers are 22, 23, and 24.

Check The sum of three consecutive numbers is 69: 22, 23, and 24 are consecutive numbers. 22 + 23 + 24 = 69. The sum is 69. True.

EXAMPLE 11 The sum of the first and third of three consecutive numbers is 96. Find all three numbers.

List Let n = 1st number

$n + 1$ = 2nd number

$n + 2$ = 3rd number

Though all three numbers are asked for in the answer, only two are used in the equation.

Equation

First	sum	third	is	96.
n	$+$	$n+2$	$=$	96

Solution

$$n + n + 2 = 96$$
$$2n + 2 = 96$$
$$-2 \quad -2$$
$$2n = 94$$
$$\frac{2n}{2} = \frac{94}{2}$$
$$n = 47, n + 1 = 48, n + 2 = 49$$

The numbers are 47, 48, and 49. Make sure that you give all 3 numbers.

Check The numbers are consecutive: 47, 48, and 49.

The sum of the 1st and 3rd is 96. 47 + 49 = 96. True.

Problems may also be concerned with consecutive even numbers (for example, 18, 20, 22) or consecutive odd numbers (for example, 33, 35). Note that in either situation, the next number is 2 more than the previous number. Therefore, the notation that might be used in any problem involving *consecutive even or consecutive odd numbers* is:

Let n = 1st consecutive even/odd number

$n + 2$ = 2nd consecutive even/odd number

$n + 4$ = 3rd consecutive even/odd number

Don't be confused by the 2 and 4 notation. They are used so you will skip the number that comes between your consecutive even or odd numbers, not because they are even numbers.

EXAMPLE 12 The sum of four consecutive even numbers is 28. Find the numbers.

List Let n = 1st consecutive even number

$n + 2$ = 2nd consecutive even number

$n + 4$ = 3rd consecutive even number

$n + 6$ = 4th consecutive even number

Equation

1st	+	2nd	+	3rd	+	4th	is	28.
n	$+$	$n+2$	$+$	$n+4$	$+$	$n+6$	$=$	28

Solution

$$n + n + 2 + n + 4 + n + 6 = 28$$
$$4n + 12 = 28$$
$$-12 \quad -12$$
$$4n = 16$$
$$\frac{4n}{4} = \frac{16}{4}$$
$$n = 4; n + 2 = 6, n + 4 = 8, n + 6 = 10$$

The numbers are 4, 6, 8, and 10.

You need to list all four answers because the problem tells you to find the numbers.

Check The numbers 4, 6, 8, 10 are consecutive even numbers. $4 + 6 + 8 + 10 = 28$. The sum is 28. True.

EXAMPLE 13 The sum of the first and third of three consecutive odd numbers is 130. Find all of the numbers.

List Let n = 1st consecutive odd number

$n + 2$ = 2nd consecutive odd number

$n + 4$ = 3rd consecutive odd number

Equation

$$\frac{\text{1st}}{n} \; + \; \frac{\text{3rd}}{n+4} \; = \; \frac{130}{130}$$

Only add two of the numbers.

Solution

$$\begin{aligned} n + n + 4 &= 130 \\ 2n + 4 &= 130 \\ -4 \quad & \quad -4 \\ 2n \quad &= 126 \\ 2n \quad &= 126 \\ \frac{2n}{2} &= \frac{126}{2} \\ n &= 63 \end{aligned}$$

$n = 63$, $n + 2 = 63 + 2 = 65$, $n + 4 = 63 + 4 = 67$

The numbers are 63, 65, and 67.

Even though the equation used only two numbers, you must give all three answers.

Check The numbers 63, 65, 67 are consecutive odd numbers. The sum of the first and third is 130. $(63) + (67) = 130$. $130 = 130$ True.

10.2 EXERCISES

A. Organize the given information and write an algebraic equation for each of the following exercises. Then solve the equation and check your answer.

1. Seventeen more than a number is 39. Find the number.
2. A number increased by 21 is 54. What is the number?
3. Eleven times a number is 99. What is the number?
4. The quotient of a number and 4 is 9. Find the number.
5. When a number is divided by 12, the result is 4. What is the number?
6. The sum of three consecutive numbers is 153. Find the numbers.
7. Four less than a number is 13. Find the number.
8. Eight less than a number is 37. Find the number.

9. Find the four consecutive numbers whose sum is 178.
10. Charlene's children are saving their money to buy a Mother's Day present. Kalen has saved three times as much as Brittany, and together they have saved $12. How much has Kalen saved?
11. If the product of 3 and a number is increased by 12, the result is 48. What is the number?
12. Half a number decreased by 6 is equal to 48. Find the number.
13. Sixteen less than two-thirds of a number is equal to 56. Find the number.
14. The sum of two consecutive even numbers is 90. What are the numbers?
15. One numer is 4 less than another number, and their sum is 20. Find the numbers.
16. One number is 11 less than another number, and their sum is 55. Find the numbers.
17. The sum of two consecutive odd numbers is 152. Find the numbers.
18. Four more than twice a number is equal to the number increased by 16. Find the number.
19. Six times a number is equal to twice the number increased by 12. Find the number.
20. The sum of four consecutive numbers is 450. What are the numbers?
21. The sum of the first and third of three consecutive numbers is 146. Find all three numbers.
22. This week Latisha studied algebra 4 hours less than Sonia. Together they studied a total of 20 hours for the week. How long did each woman study?
23. Patty is three years younger than Betsy. If the sum of their ages is 113 years, how young is each woman?
24. The sum of two consecutive numbers is 93. What are the numbers?
25. The cost of a full-feature washing machine is four dollars more than twice the cost of a washer with only basic features. If the difference in cost between the full-feature and basic-feature machines is $184, how much does each model cost?
26. Two-thirds of the students in a class completed their assignment on schedule. If 16 students finished on time, how many students are in the class?
27. Three-fourths of the students in this school drive their own cars to campus each day. If 7,200 students drive, how many students come to campus each day?
28. One number is 4 times another number. The difference between the larger number and the smaller number is 54. Find the numbers.
29. The sum of the second and third of three consecutive numbers is 89. Find all three numbers.
30. If $1,800 is divided into two parts with the ratio of 3 to 5, what is the size of each part?
31. A 12-foot length of wire is to be divided into two pieces that are in the ratio of 1 to 5. What is the length of each piece?
32. One number is 5 times another. The difference between the larger and the smaller is 84. Find the numbers.
33. The Cardinals' star running back scored one-fourth of the team's total points in this week's game. He scored 12 points. What was the team's score in the game?
34. Ron deposits two-thirds of his paycheck in the bank each week. If he deposited $144 this week, what were his total earnings?
35. A woman earned $40 more than twice what her sister earned. Together they earned $580. Find each women's salary.
36. The sum of the largest and smallest of three consecutive even numbers is 68. What are the three numbers?
37. One number is four times another number. When twice the smaller number is subtracted from the larger number, the result is 28. What are the numbers?
38. The total cost of buying a used car is $9,816. If Alexandria pays $600 down and makes 48 monthly payments to purchase the car, how much will she pay each month?
39. The sum of the first two of three consecutive numbers is equal to the third number. What are the numbers?

40. Max received a car repair bill for $245. If the bill included $80 for parts, and if labor costs $45 per hour, how long did the mechanic work on the car?

41. This week Mrs. Chang worked 6 hours less than Mrs. Alvarez. Together they worked 70 hours. How many hours did Mrs. Chang work during this period?

42. The sum of the second and twice the first of two consecutive numbers is equal to 28. What are the two numbers?

43. The total cost of the VCR that Manuella purchased was $179. If she paid $47 down and made twelve monthly payments to finish paying for it, how much was each monthly payment?

44. The sum of two numbers is 52, and one of them is three times the other. What are the numbers?

45. While on a business trip, Jose paid a total of $64.50 to park his car at the airport. That total included a fee of $10.50 per day or part of a day for parking and a $12 charge for washing the car. How many days was the car parked?

46. Latonya is purchasing a new computer for $1,375 (including finance charges) by paying $350 as a down payment and making 20 monthly payments. What will be the amount of each monthly payment?

47. Together, a new chair and sofa cost $780. If the sofa costs three times as much as the chair, how much does each piece cost?

48. Jud has put aside three-fourths of the money for next month's rent. How much is his rent if he has $144 put aside? How much more does he need to save to have the full amount?

49. The ages of Mrs. Zink's three sons can be represented as consecutive odd numbers. If the sum of their ages is 57, what are the ages of the sons?

50. The lengths of the sides of a right triangle can be represented as three consecutive numbers. If the perimeter is 12 feet, what is the length of the longest side?

51. Two angles that are complementary (have a sum of 90°) are in the ratio of 4 to 5. Find the measure of each angle.

52. The measures of two angles have a ratio of 5 to 3. If the angles are supplementary (have a sum of 180°), what is the measure of each?

53. The sizes of two numbers are in the ratio of 6 : 5 and the sum of the numbers is 77. Find the numbers.

54. Find two numbers whose sum is 144 and whose sizes are in the ratio of 7 : 5.

55. The sizes of three numbers are in the ratio of 4 : 3 : 2 and their sum is 117. What is the largest number?

56. The sizes of three numbers are in the ratio of 9 : 5 : 2 and their sum is 64. What are the three numbers?

B. Read and respond to the following exercises.

57. In your own words, describe the steps that are recommended each time you do an application problem.

58. If you are to solve a word problem like Exercise 32, why couldn't you use two different variables to represent the larger and smaller numbers?

59. Describe, in words, the differences between an expression and an equation in algebra.

60. Why can consecutive odd and consecutive even numbers be represented by the same notation: n, $n + 2$, $n + 4$, . . . ?

10.3 Organizing by Sketch

Making a sketch and writing the relationship of the different values on the picture is another way of organizing information in an application problem. This method is very appropriate to use with geometry problems and will also be applied to motion problems in this section.

You have worked with some geometric relationships in previous chapters. As a review, the *perimeter* of a figure is the total distance around it—that is, the sum of all its sides. Figure 10.1 shows where the different perimeter formulas came from. Having a picture or diagram to work with is often very helpful.

You will not have to memorize any formulas to do this work. You can write the perimeter formula for any figure by adding all the sides together.

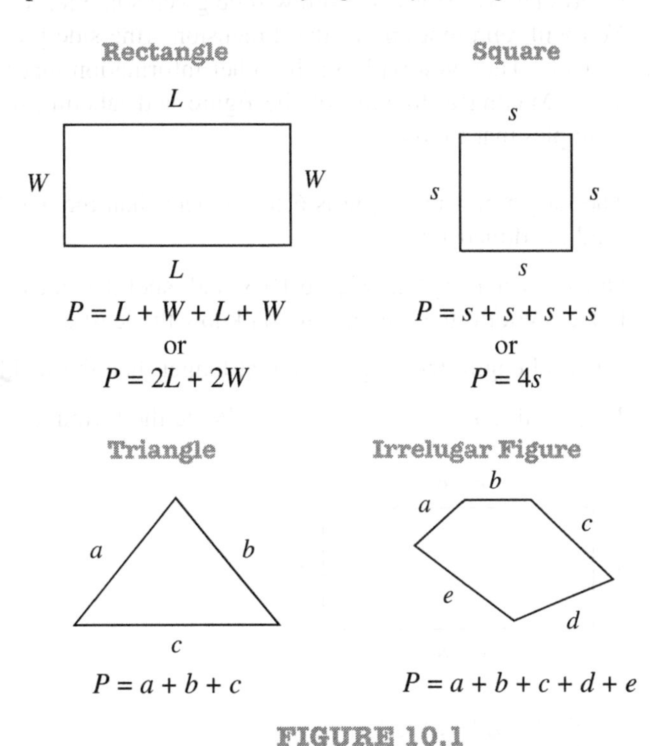

FIGURE 10.1

EXAMPLE 1 The perimeter of a quadrilateral is 65 feet. Three of the sides are 23 feet, 10 feet, and 17 feet in length. Find the length of the fourth side.

Sketch This quadrilateral is a four-sided figure with sides that could have different lengths. Sketch a four-sided figure and label the sides with the given lengths. Your sketch could be similar to Figure 10.2. Write a formula for the perimeter of a four-sided figure with different length sides:

$P = a + b + c + d$

Substitute the values in the drawing for *a, b, c,* and *d.*

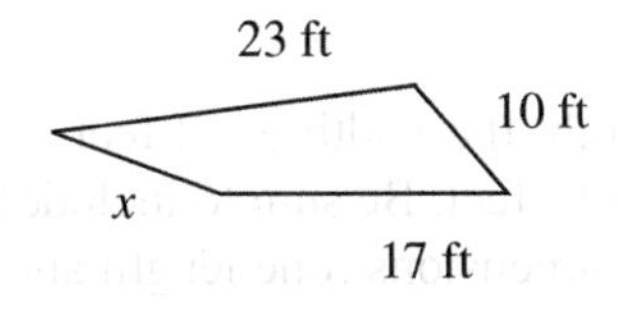

FIGURE 10.2

Equation $65 = 23 + 17 + 10 + x$ — Combine the numerical values and then isolate the variable.

Solution

$$\begin{aligned} 65 &= 50 + x \\ \underline{-50} &\quad \underline{-50} \\ 15 &= x \end{aligned}$$

$x = 15$ feet — The length of the fourth side is 15 feet.

Check 23 + 17 + 10 + 15 does equal 65.

In the problems you will do in this section, the actual lengths of the sides are not always known. However, you will be given some information about the dimensions. You will represent one of the dimensions (the side you know the least about) by a variable. Then you will use the other information that is given to express the other sides. Making a drawing of the figure and labeling its parts will help. Study the examples that follow.

EXAMPLE 2 The length of a rectangle is 6 feet longer than the width. The perimeter is 60 feet. Find the dimensions.

Sketch Draw a sketch such as Figure 10.3, and label the sides.
Let x = Width (the dimension you know the least about)

$x + 6$ = Length (the length is 6 feet longer than the width)

$P = L + W + L + W$ Write the formula.

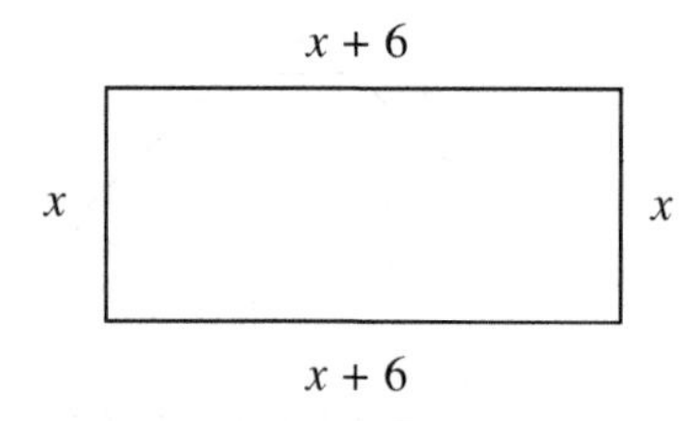

FIGURE 10.3

Equation $60 = \underline{x + 6} + \underline{x} + \underline{x + 6} + \underline{x}$ Substitute for P, L, and W.

Solution

$$60 = x + 6 + x + x + 6 + x$$

$$60 = 4x + 12 \quad \text{Combine like terms.}$$

$$\underline{-12} \quad \underline{-12}$$

$$48 = 4x$$

$$\frac{48}{4} = \frac{4x}{4}$$

$$12 = x$$

Because x represents the width, the width is 12 feet. Because $x + 6$ represents the length, the length is $12 + 6 = 18$ feet. Be sure to include the units of measurement in the answer and to give *both* dimensions (one length and one width).

Check $P = L + W + L + W$

$60 = 18 + 12 + 18 + 12$

$60 = 60$ True.

And the length, 18 feet, is 6 feet more than the width of 12 feet.

EXAMPLE 3 The length of a rectangle is twice the width. The perimeter is 66 yards. Find the length.

Sketch Let x = Width (you know the least about it)

$2x$ = Length (twice the width)

Draw a sketch similar to Figure 10.4.

$P = L + W + L + W$

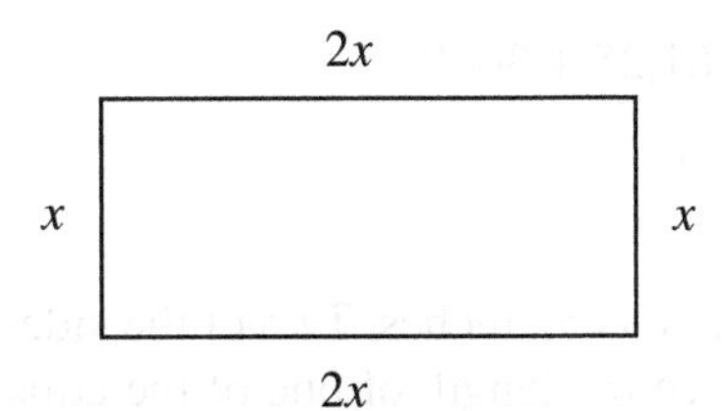

FIGURE 10.4

Equation $66 = \underline{2x} + \underline{x} + \underline{2x} + \underline{x}$

Solution $66 = 2x + x + 2x + x$

$66 = 6x$

$\frac{66}{6} = \frac{6x}{6}$

$11 = x$

Because length = $2x$, the length is 22 yards.

Check $P = L + W + L + W$

$66 = 22 + 11 + 22 + 11$

$66 = 66$ True. And 22 is twice 11.

EXAMPLE 4 The ratio of the length of a rectangle to the width is 5 to 3. Find the dimensions if the perimeter is 196 meters.

Sketch Let x = one part

$5x$ = length

$3x$ = width

Draw a sketch similar to Figure 10.5.

$P = L + W + L + W$

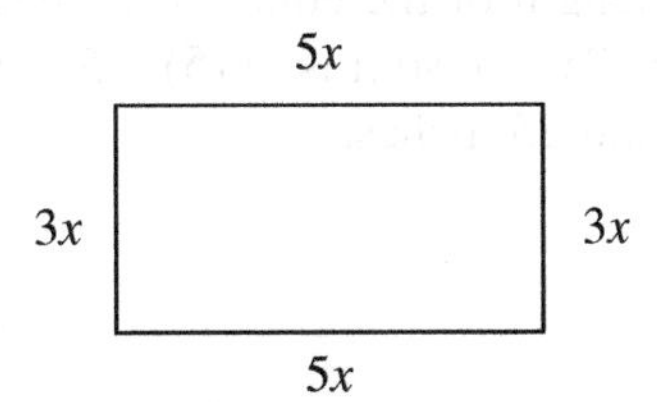

FIGURE 10.5

Equation $196 = 5x + 3x + 5x + 3x$

Solution $196 = 16x$

$$\frac{196}{16} = \frac{16x}{16}$$

$12.25 = x$

Length $= 5x = 5(12.25) = 61.25$ meters

Width $= 3x = 3(12.25) = 36.75$ meters

Check $196 = 61.25 + 36.75 + 61.25 + 36.75$

$196 = 196$ True

EXAMPLE 5 The perimeter of a triangle is 55 inches. Two of the sides are equal, and the third side is 5 inches less than twice the length of one of the equal sides. Find the dimensions of the triangle.

Sketch Let $x =$ Length of one equal side
$x =$ Length of other equal side

$2x - 5 =$ Length of third side (5 inches less than twice the length of one of the equal sides)

Is your sketch similar to Figure 10.6?

$P = a + b + c$

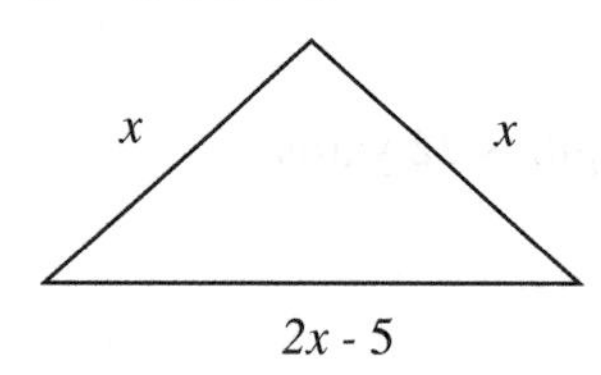

FIGURE 10.6

Equation $55 = \underline{x} + \underline{x} + \underline{2x - 5}$

Solution $55 = x + x + 2x - 5$

$55 = 4x - 5$

$\underline{+5} \quad \underline{+5}$

$60 = 4x$

$$\frac{60}{4} = \frac{4x}{4}$$

$15 = x$

Because x represents the length of the equal sides, each of them is 15 inches. The third side is represented by $2x - 5$, so it is $2(15) - 5 = 30 - 5 = 25$ inches. The sides are 15 inches, 15 inches, and 25 inches.

Check $P = a + b + c$

$55 = 15 + 15 + 25$

$55 = 55$ True.

Two of the sides are equal, and the third is 5 inches less than twice the length of one of the equal sides.

EXAMPLE 6 The first angle of a triangle is twice the size of the second angle. The third angle is 16 degrees more than the second angle. Find the size of the three angles.

Sketch The geometric fact you need to know in order to do this problem is that the sum of the three angles of any triangle is 180°. Make a sketch of a triangle and reread the problem. See Figure 10.7. Do you see that you know the least about the second angle? We called it x, and were then able to represent the other two angles in terms of that second angle, x.

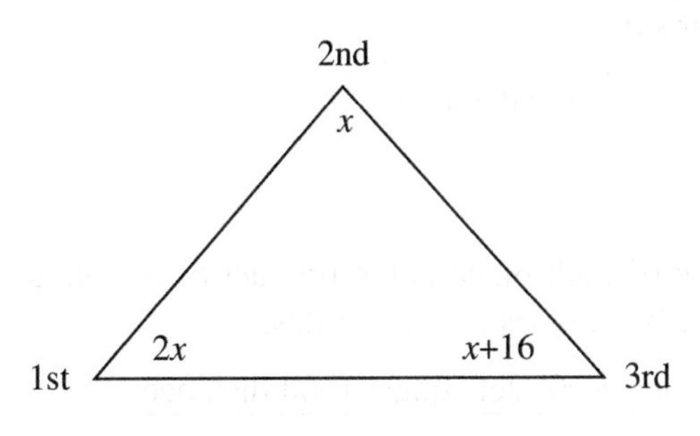

FIGURE 10.7

Equation The sum of the angles of any triangle equals 180°.

1st angle plus 2nd angle plus 3rd angle is 180°.

$$2x + x + x + 16 = 180$$

Solution

$$2x + x + x + 16 = 180$$
$$4x + 16 = 180$$
$$-16 \quad -16$$
$$4x = 164$$
$$\frac{4x}{4} = \frac{164}{4}$$
$$x = 41$$

First angle = $2x = 2(41) = 82°$

Second angle = $x = 41°$

Third angle = $x + 16 = 41 + 16 = 57°$

Therefore, the three angles are 82°, 41°, and 57°.

Check The sum of the angles must be 180: 82° + 41° + 57° = 180°.

Another formula that you worked with in Chapter 8 was the distance formula: Distance = Rate × Time, or D = rt.

A sketch can help you organize the information and can help you write the correct equation. Carefully study the process in the next examples.

EXAMPLE 7 A small private plane and a large commercial jet take off from Tri-Cities Airport and travel in opposite directions. The speed of the jet is three times the speed of the small plane. After two hours, the planes are 1,200 miles apart. How fast is each plane flying?

Sketch The time is known and you are looking for the speed (rate) of the planes. You know the least about the speed of the small plane and know that the jet is three times that amount. The product of the rate and the time for each gives the distance that each plane has traveled. See Figure 10.8.

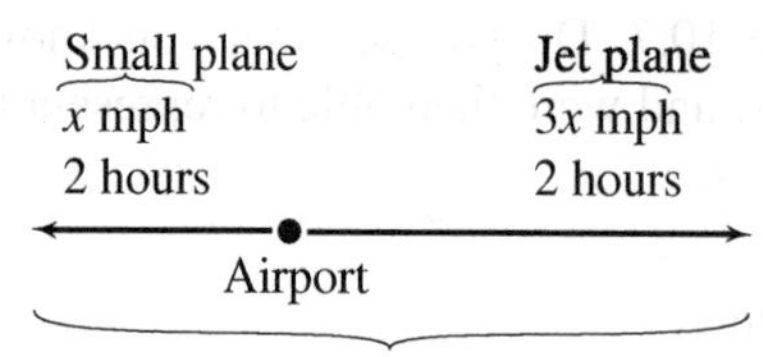

FIGURE 10.8

Equation The distance of each plane is the product of its rate and its time, and the total distance for the two planes is 1,200 miles.

$$\underline{r(t)\text{ Small plane}} + \underline{r(t)\text{ Jet}} \text{ equals } \underline{\text{Total distance}}$$

$$x(2) + 3x(2) = 1{,}200$$

Solution

$$2x + 6x = 1{,}200$$

$$8x = 1{,}200$$

$$\frac{8x}{8} = \frac{1{,}200}{8}$$

$$x = 150$$

Because x represents the rate of the small plane, its speed is 150 miles per hour. The rate of the jet is $3x = 3(150) = 450$ miles per hour.

Check The distance for the small plane $= 150(2) = 300$ miles. The distance for the jet $= 450(2) = 900$ miles. The total distance $= 300 + 900 = 1{,}200$ miles. Also, $450 =$ 3 times 150.

◀

EXAMPLE 8 The Municipal Park Department maintains a bike path that runs from the north edge of the city limits to the south edge, a distance of 32.5 miles. Darrel and Javene live at opposite ends of the bike path and decide to travel toward each other on the path and meet for lunch. Darrel will jog at an average rate of 5 miles per hour and Javene will ride her bike at an average rate of 8 miles per hour. If they start at the same time, how long will it take before they meet?

Sketch They leave at the same time and travel until they meet, so they are traveling the *same amount of time.* We have used t to represent the time for each. See Figure 10.9.

Darrel
5 mph
t hours

Javene
8 mph
t hours

Total distance 32.5 miles

FIGURE 10.9

Equation Because distance equals the product of rate and time: Darrel's distance = $5t$ and Javene's distance = $8t$. Together these distances total 32.5 miles.

$$\underline{\text{Darrel's distance}} + \underline{\text{Javene's distance}} = \underline{\text{Total distance}}$$
$$5t \quad + \quad 8t \quad = \quad 32.5$$

Solution

$$5t + 8t = 32.5$$
$$13t = 32.5$$
$$\frac{13t}{13} = \frac{32.5}{13}$$
$$t = 2.5$$

Therefore, they will each travel 2.5 hours before they meet.

Check Darrel's distance = $5t = 5(2.5) = 12.5$ miles

Javene's distance = $8t = 8(2.5) = 20.0$ miles

Total distance = 32.5 miles

Self-Check For each problem, make a sketch and write an equation. You do not need to solve it.

a. The width of a rectangle is 5 yards less than the length and the perimeter is 106 yards. Find the dimensions of the rectangle.

b. Two cars start at the same time and travel toward each other from cities that are 400 miles apart. One car averages 55 mph and the other 45 mph. In how many hours will they meet?

Answers: a. $106 = x + x - 5 + x + x - 5$

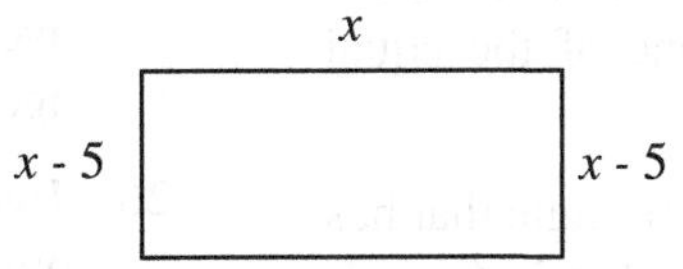

x

x - 5 x - 5

b. 55 mph 45 mph $55t + 45t = 400$
t hours t hours

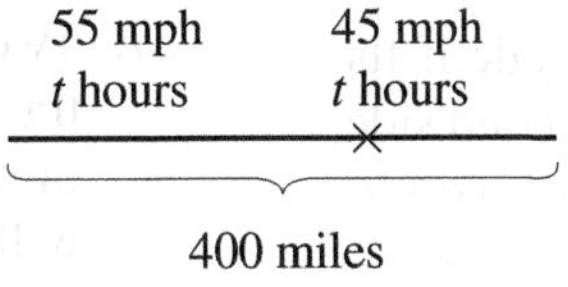

55 mph 45 mph
t hours t hours

400 miles

10.3 EXERCISES

A. For each exercise, make a sketch and write a formula and then an equation. Solve the equation and check your answer(s).

1. The perimeter of a triangle is 25 feet and the lengths of two of the sides are 7 feet and 11 feet. What is the length of the third side?

2. What is the length of the two equal sides of a triangle if the third side is 18 inches and the perimeter is 58 inches?

3. A triangle has two equal sides and a third side that is 90 meters in length. If the perimeter is 200 meters, what is the length of each of the equal sides?

4. The perimeter of a square is 96 meters. What is the length of each side?

5. A triangel with three equal sides (an *equilateral* triangle) has a perimeter of 135 feet. What is the length of each side?

6. If the perimeter of a regular octagon (a figure with eight equal sides) is 176 feet, what is the length of each side?

7. If the perimeter of a regular hexagon (a figure with six equal sides) is 144 centimeters, what is the length of each side?

8. The length of a rectangle is 6 inches more than the width. The perimeter is 132 inches. Find the dimensions.

9. The length of a rectangle is 20 centimeters more than the width. The perimeter is 96 centimeters. Find the dimensions.

10. The length of a rectangle is twice the width. The perimeter is 24 yards. Find the length.

11. The length of a rectangle is four times the width. The perimeter is 40 inches. Find the length.

12. The length of a rectangle is 6 feet more than twice the width. The perimeter is 84 feet. Find the dimensions.

13. The length of a rectangle is 4 inches more than three times the width, and the perimeter is 88 inches. Find the dimensions.

14. The perimeter of a triangle is 37 meters. Two of the sides are equal, and the third side is 7 meters less than twice the length of one of the equal sides. Find the dimensions.

15. Find the lengths of the sides of a triangle that has two equal sides and a third side that is 6 yards less than twice the length of one of the equal sides. Its perimeter is 66 yards.

16. Find the lengths of the sides of a triangle if the first side is 5 inches longer than the second side, and the third side is twice the length of the second side. Its perimeter is 41 inches.

17. The ratio of the length to the width of a rectangle is 4 : 3. Find the width if the perimeter is 252 meters.

18. The ratio of the lengths of the sides of a triangle is 4 to 3 to 2. Find each side if the perimeter is 126 feet.

19. The ratio of the length of a rectangle to the width is 9 to 4. Find the dimensions if the perimeter is 182 inches.

20. The ratio of the length of a rectangle to the width is 7 to 5. Find the dimensions if the perimeter is 168 feet.

21. The lengths of the three sides of a triangle are in the ratio of 3 : 4 : 5. Find the length of each side if the perimeter is 60 yards.

22. The lengths of the three sides of a triangle are in the ratio of 5 : 12 : 13. Find the length of each side if the perimeter is 90 meters.

B. For each exercise, make a sketch and write in the values. Then write an equation based on the formula $D = rt$.

23. A truck and a car leave the city at the same time and travel in opposite directions. The car travels at 62 mph and the truck at 56 mph. After how many hours will they be 531 miles apart?

24. At the same time, Kay and Lisa leave their homes, which are 24 miles apart, and walk toward each other on the city bike path. If Kay travels at 5 miles per hour and Lisa at 3 miles per hour, how long will it take before they meet?

25. Two cars start at the same point at the same time and travel in opposite directions until they are 472 miles apart. How long will they travel if one averages 60 miles per hour and the other averages 58 miles per hour?

26. Two cars leave Berlin at 8:30 A.M. and travel in opposite directions. One of the cars averages 55 mph and the other averages 65 mph. How long will it take for the cars to be 480 miles apart?

27. Two bicyclists leave Meridan at 10:00 A.M. and travel in opposite directions. One averages 22 mph and the other averages 28 mph. How long will it take for them to be 200 miles apart?

28. At 8:00 A.M., two cars start toward each other from cities that are 486 miles apart. One car travels on interstate highways and averages 60 miles per hour. The other car uses state routes and averages 48 miles per hour. In how many hours will they meet?

29. A truck traveling on state highways can travel at two-thirds the rate of a truck on the turnpike. If these trucks leave the same terminal at the same time and travel in opposite directions, they will be 500 miles apart in 5 hours. How fast does each truck travel?

30. Todd on his motorcycle can travel three times as fast as Jude on his motorbike. If they start at the same place at the same time and travel in opposite directions, they will be 280 miles apart in 3 1/2 hours. How fast is each man traveling?

31. Two cars start at the same time and drive toward each other from cities that are 360 miles apart. One car averages 64 mph and the other averages 56 mph. In how many hours will they meet?

C. Read and respond to the following exercises.

32. Describe, in words, the process you would use to find the solution to the following problem. You do not have to actually do the work. One side of a triangle is 7 feet longer than the second side. The third side is 6 feet longer than twice the length of the second side. Find the sides of the triangle if its perimeter is 37 feet.

33. Could you have used sketches to help you do the exercises in Section 10.2? Could you have used the list method for the exercises in this section? Explain your answer.

34. Is the answer to the equation always the answer to the question asked in the problem? Explain your answer.

10.4 Organizing by Chart

As we stated in the introduction to this chapter, when you need to sove an application problem (for a math class or in real life), you need to gather all the information you will need and decide what you must do with it to get to a solution. Two different techniques for organizing that information have been demonstrated so far. In general problems and consecutive number problems, we made a list and represented each quantity involved, using just one variable. For geometry and motion problems, we made a sketch and labeled the parts. Both of these techniques helped us to accurately translate words into an algebraic equation that we then solved. The equation answer was taken back up to the list or the sketch to find all the answers that were requested. In this section we will show you another strategy that works well with money-related problems.

In these problems, several like items (for example, coins, stamps, or tickets), each of different value or cost, are combined to give a total amount or a total cost. If you have six dimes, you know that you have 60 cents. What you probably don't realize—because you did it subconsciously—is that you must have multiplied six times ten cents to get that total value of the dimes: $6(0.10) = \$0.60$. It is important to remember that total value cannot be found by just combining the numbers of each item. The value or cost of each item must also be used. It is not just the fact that you have four coins that gives you $1.00. The fact that each of those coins is worth $0.25, or 4(0.25), is also essential.

As you work coin problems and mixture problems, you may find it easier to write the correct equation if you organize your information in a box like Chart 10.1. This should help prevent you from leaving important information out of the equation.

CHART 10.1

Items	Value/Cost of Each	Number of Each	Value/Cost of Each Kind
			Total Value/Cost

EXAMPLE 1 Denise bought stamps for her construction company while she was at the post office. She bought nine times as many 37-cent stamps as 30-cent stamps and spent a total of $36.30. How many 37-cent stamps did she buy?

Chart Fill in the information that is given: Items are 37-cent and 30-cent stamps, and the total amount is $36.30. Because the number of 30-cent stamps is the quantity you know the least about, call it x. That means the number of 37-cent stamps must be $9x$. Place all the information in the boxes shown in Chart 10.2.

CHART 10.2

Items	Value/Cost of Each	Number of Each	Value/Cost of Each Kind
30-cent stamps	0.30	x	
37-cent stamps	0.37	$9x$	
			Total Value/Cost
			36.30

The last box, *Value/Cost of Each Kind*, will be the *product* of the quantities in the *Value/Cost of Each* box and the *Number of Each* box (Chart 10.3).

CHART 10.3

Items	Value/Cost of Each	Number of Each	Value/Cost of Each Kind
30-cent stamps	0.30	x	0.30x
37-cent stamps	0.37	9x	0.37(9x)
			Total Value/Cost
			36.30

Equation

$$\underline{\text{Cost of \$0.30 stamps}} + \underline{\text{Cost of \$0.37 stamps}} = \underline{\text{Total cost}}$$

$$0.30(x) \quad + \quad 0.37(9x) \quad = \quad 36.30$$

Solution

$$0.30x + 3.33x = 36.30$$

$$3.63x = 36.30$$

$$\frac{3.63x}{3.63} = \frac{36.30}{3.63}$$

$$x = 10$$

Take this answer back into the *Number of Each* box. Because x is the number of 30-cent stamps and $9x$ is the number of 37-cent stamps, there are 10 30-cent and 90 37-cent stamps.

Check

10 30-cent stamps = 10(0.30) = $ 3.00

90 37-cent stamps = 90(0.37) = + $33.30

$36.30

And there are 9 times more 37-cent than 30-cent stamps.

EXAMPLE 2 The ticket booth for tonight's basketball game collected \$675. The Cougars had three times as many students as nonstudents attend the game. If student tickets cost \$2.00 and nonstudent tickets cost \$3.00, how many of each kind of ticket were sold?

Chart Because the number of nonstudent tickets is what you know the least about, call it N.

Draw a box like Chart 10.4, and fill in the information that is given.

CHART 10.4

Items	Value/Cost of Each	Number of Each	Value/Cost of Each Kind
Student tickets	2.00	$3N$	$2.00(3N)$
Nonstudent tickets	3.00	N	$3.00N$
			Total Value/Cost
			675.00

Equation The equation is the sum of the *Value/Cost of Each Kind* boxes, set equal to the *Total Value/Cost*, or the amount collected.

$$2.00(3N) + 3.00N = 675.00$$

Solution

$$2.00(3N) + 3.00N = 675.00$$
$$6.00N + 3.00N = 675.00$$
$$9.00N = 675.00 \text{ or}$$
$$9N = 675$$
$$\frac{9N}{9} = \frac{675}{9}$$
$$N = 75$$

Therefore, 75 nonstudent tickets and 3(75), or 225, student tickets were sold.

Check

$$75(3.00) = \$225$$
$$3(75)(2.00) = \underline{\$450}$$
$$\text{Total} = \$675$$

And there were three times as many student tickets as nonstudent tickets sold.

◀

EXAMPLE 3 While emptying his vending machine, Jerome noticed that he had five times as many dimes as quarters and twice as many nickels as dimes. How many of each coin did he have if there was a total of \$37.50 in the machine?

Chart Because the number of quarters is what you know the least about, call it X. Draw a box and fill in the rest of the information from the problem. Your box should look like Chart 10.5, but you can abbreviate the headings of the columns if you wish.

CHART 10.5

Items	Value/Cost of Each	Number of Each	Value/Cost of Each Kind
Dimes	0.10	$5X$	$0.10(5X)$
Quarters	0.25	X	$0.25(X)$
Nickels	0.05	$2(5X)$	$0.05(2)(5X)$
			Total Value/Cost
			37.50

Equation $0.10(5X) + 0.25(X) + 0.05(2)(5X) = 37.50$

Solution

$$0.10(5X) + 0.25(X) + 0.05(2)(5X) = 37.50$$

$$0.50X + 0.25X + 0.50X = 37.50$$

$$1.25X = 37.50$$

$$\frac{1.25X}{1.25} = \frac{37.50}{1.25}$$

$$X = 30$$

Therefore, there are 30 quarters; $5(x) = 5(30) = 150$ dimes; and $2(5X) = 2(5)(30) = 300$ nickels.

Check

30 quarters = 30(0.25) = $ 7.50

150 dimes = 150(0.10) = $15.00

300 nickels = 300(0.05) = $15.00

Total = $37.50

And there are five times as many dimes as quarters and twice as many nickels as dimes.

Another kind of application where a chart might be useful is problems where several investments are made so that a certain amount of interest will be earned by a specific time. A business might be looking to the future when it knows it will need to replace vehicles or equipment, and an individual might do it to be sure that he or she has money for a down payment on a house or car, for college tuition, or for a vacation. The amount of simple interest that is earned when money is invested is found using the formula $I = PRT$, where I is the amount of interest, P is the principal or amount invested, R is the rate of interest (percent), and T is the time the money is invested. Remember to change percent numbers to decimals before you multiply them (see Section 7.2).

EXAMPLE 4 Mega Mart plans to replace its delivery truck in three years and is making investments in two separate accounts now that will provide money for a down payment later. The company will invest $10,000 at 5% annual interest for the 3-year period. How much should they invest in an account paying 7 1/2% if they want to have earned a total of $4,200 interest at the end of three years?

Chart The chart will need columns for *P, R, T,* and *I*. See Chart 10.6.

CHART 10.6

Items	Principal	Rate	Time	Interest
5% Investment				
7 1/2% Investment				
				Total Interest

Fill in the given values (Chart 10.7). Remember to change percents to decimals by dividing by 100: 5% = 0.05, 7 1/2% = 7.5% = 0.075. The unknown is the amount to be invested (*P*) at 7 1/2%. Include a column for time, then multiply all three column values together for the interest.

CHART 10.7

Items	Principal	Rate	Time	Interest
5% Investment	$10,000	0.05	3	10,000(0.05)(3)
7 1/2% Investment	*P*	0.075	3	*P*(0.075)(3)
				Total Interest
				$4,200

Equation $10{,}000(0.05)(3) + P(0.075)(3) = 4{,}200$

Solution

$$\begin{aligned} 1{,}500 + 0.225P &= 4{,}200 \\ -1{,}500 \qquad\qquad &\quad -1{,}500 \\ 0.225P &= 2{,}700 \\ \frac{0.225P}{0.225} &= \frac{2{,}700}{0.255} \\ P &= 12{,}000 \end{aligned}$$

Therefore, $12,000 should be invested at 7 1/2%.

Check

$$\begin{aligned} (10{,}000)(0.05)(3) &= 1{,}500 \\ (12{,}000)(0.075)(3) &= \underline{2{,}700} \\ &\quad \$4{,}200 \end{aligned}$$

EXAMPLE 5 The Mendokovs want to take a trip to Europe for their 25th wedding anniversary in four years. They plan to invest money now so that it will earn $2,000 in interest to be used for plane fare for the trip. If they invest twice as much in a certificate of deposit as they invest in a money market account, how much must be put into each to earn that interest in four years? The certificate pays 7% and the money market pays 6% annual interest.

Chart The amount put in the money market account is the one you know the least about, and the certificate of deposit amount is twice that. See Chart 10.8.

CHART 10.8

Items	Principal	Rate	Time	Interest
Money Market Account	P	0.06	4	$P(0.06)(4)$
Certificate of Deposit	$2P$	0.07	4	$2P(0.07)(4)$
				Total Interest
				$2,000

Equation $P(0.06)(4) + 2P(0.07)(4) = 2{,}000$

Solution

$$0.24P + 0.56P = 2{,}000$$
$$0.80P = 2{,}000$$
$$\frac{0.8P}{0.8} = \frac{2{,}000}{0.8}$$
$$P = 2{,}500$$

Therefore, $5,000 should be invested in the certificate of deposit and $2,500 should be put into the money market account.

Check

$(2{,}500)(0.06)(4) = 600$
$(5{,}000)(0.07)(4) = 1{,}400$
$2,000

In Section 10.3 we organized motion problems by using a sketch. Those problems could also be organized by using a chart, because multiplication needs to be done.

EXAMPLE 6 A small private plane and a large commercial plane take off from Tri-Cities Airport at the same time and travel in opposite directions. The speed of the jet is three times the speed of the small plane. After two hours, the planes are 1,200 miles apart. How fast is each traveling?

Chart This is Example 6 from Section 10.3. This time we will organize the information in a chart using the formula $rt = D$. See Chart 10.9.

CHART 10.9

Items	Rate	Time	Distance
Private Plane	x	2	$x(2)$
Jet Plane	$3x$	2	$3x(2)$
			Total Distance
			1,200

Equation

$$x(2) + 3x(2) = 1{,}200$$
$$2x + 6x = 1{,}200$$
$$8x = 1{,}200$$
$$\frac{8x}{8} = \frac{1{,}200}{8}$$
$$x = 150$$

From the chart, we see that x represents the rate of the private plane, so its rate is 150 mph. The jet is $3x$ or $3(150) = 450$ mph.

Check The distance of the private plane = 150(2) = 300 miles. The distance of the jet = 450(2) = 900 miles. The total distance = 300 + 900 = 1 ,200 miles. Also, 450 mph = 3(150 mph).

Self-Check For each problem, organize the information in a chart and write an equation. You do not need to solve the equation.

a. A money bag contains only $10 and $20 bills. The number of $20 bills is three times the number of $10 bills and the total amount of money is $770. How many $20 bills are there?

b. Two cars start at the same time and travel toward each other from cities that are 440 miles apart. One car averages 60 mph and the other 50 mph. In how many hours will they meet?

Answers: **a. Chart 10.10**

Items	**Value of Each**	**Number of Each**	**Total Value of Each**
$10 Bills	10	x	$10(x)$
$20 Bills	20	$3x$	$20(3x)$
			Total Value
			770

$10(x) + 20(3x) = 770$

b. Chart 10.11

Items	**Rate**	**Time**	**Distance**
Car #1	60	t	$60t$
Car #2	50	t	$50t$
			Total Distance
			440

$60t + 50t = 440$

10.4 EXERCISES

A. Organize the information, write an equation, and then solve. Be sure to give all answers that are requested.

1. Niko sets off the alarm system when passing through an airport security gate. Upon emptying his pockets, he finds equal numbers of quarters and pennies that total $1.56 in value. How many of each coin does he have?
2. The toll collector at the Riverview exit notices that she has the same number of dimes as quarters in her box. If there is a total of $9.80 in the box, and she has only dimes and quarters, how many of each coin does she have?
3. When cleaning out her purse, Jan found $2.10 in loose change. There were eight times as many dimes as quarters. How many of each coin did she have?
4. Emptying his pants pockets one evening, Steve finds $1.00 in change. In the pile are five times as many nickels as quarters. How many nickels are there?
5. The afternoon ticket sales for the Plaza Movie Theater totaled $269.50. Twice as many children's tickets as adults' tickets were sold. How many of each kind of ticket were sold if children's tickets cost $1.50 and adults' tickets cost $2.50?
6. Elena collects the money from her co-workers and goes to the Sub House to get lunches for all of them. She collects $17.90 and has orders for twice as many sandwiches as drinks. If sub sandwiches cost $3.98 and large drinks cost $0.99, how many of each is she buying?
7. Tickets for the Marquee Playhouse are priced at $18.95 for adults, $12.95 for seniors, and $7.95 for children. Rochelle purchased tickets for a family outing and paid a total of $149.40. She purchased twice as many adult tickets as senior tickets, and three times as many children's tickets as senior tickets. How many of each did she buy?
8. A woman invests $1000 in a certificate of deposit that is paying 9% annual interest. She also invests some money in a high-risk stock that yields 12% annual interest. If after 1 year the investments earn $480 in interest, what amount did she invest in she stock?
9. A company retirement plan has $24,000 invested in a high-return venture paying 12% annual interest. How much should it invest in a project that is yielding a 10% return in order to earn $17,520 in four years from the combined investments?
10. Joanna had to buy twice as many textbooks as workbooks for her nursing classes next quarter. Workbooks cost $12.50 each, and textbooks cost $38.00 each. If her total bill is $177.00, how many textbooks did she buy?
11. The college bookstore charges $1.95 for notebooks and $0.85 for pens. When Cheryl bought twice as many notebooks as pens, her bill was $9.50. How many of each item did she purchase?
12. Nikki has decided to wait one year before going to college. She is going to invest the money she had saved for tuition for this year. She will put twice as much in a 1-year certificate of deposit that is paying 6 1/2% annual interest than she puts in municipal bonds that are paying 7% annual interest. Find the amount she needs to place in each investment so she will have $600 at the end of one year to use toward living expenses at college next year.
13. Nancy wants to invest money now so that she has money for her daughter's college tuition. She invests four times as much money in a certificate of deposit paying 6% as she invests in a money market account paying 5% annual interest. These investments will earn a total of $2900 in interest in 5 years. How much does she put into each investment?
14. If twice as much money is invested at 6 1/2% annual interest as is invested at 5 1/2 %, the amount of interest earned in 5 years is $3,237.50. What is the size of each investment?
15. If Roberto invests four times as much money at 7% as he invests at 6%, he will have $2,040 in interest in 3 years. What is the size of each investment?

16. Franz has half as many quarters as half-dollars. All together he has \$13.75. How many quarters does he have?

17. Juan noticed that his Salvation Army kettle had only dollar bills and quarters in it. There was \$14.00 in the kettle. If the number of dollar bills was one-fourth the number of quarters, how many dollar bills were there?

18. Franz invested 3 times as much money at 5% as he invested at 4% and earned a total of \$1,900 in interest in 2 years. How much money did he place in each investment?

19. Gate receipts for the tennis tournament's Friday night round were \$46,000. Adults' tickets cost \$25 and children's tickets cost \$15. If four times as many adults' as children's tickets were purchased, how many adults attended the matches on Friday evening?

20. Maurice is preparing the bank deposit for his restaurant. The "paper" money amounts to \$1,496 and is made up of \$1, \$5, \$10, and \$20 bills. There are sixteen times more \$1 than \$20 bills, twelve times more \$5 than \$20 bills, and four times more \$10 than \$20 bills. How many of each are there?

21. Four times as many adult than senior tickets were sold for the bargain matinee movie and the total collected for tickets was \$630. If adult tickets cost \$3 each and senior tickets cost \$2 each, how many adult tickets were sold?

22. Tom's coin sorting machine contains 5 times as many dimes as quarters and there is a total of \$4.50 from those two coins. How many dimes are there?

23. Sid invests 6 times more money at 7% as he invests at 5%. If at the end of 3 years he has earned \$1,410 in interest, how much did he put into each investment?

24. Karla invests 3 times as much money at 5% as she invests at 6% and she earns \$1,575 in interest in 5 years. How much did she put into each investment?

B. For each exercise, organize the information in a chart using $rt = D$. Then write an equation, solve, and give all answers that are requested.

25. Two cars that start at the same place at the same time traveling in opposite directions and are 270 miles apart after 3 hours. Because of road conditions, one car travels at twice the rate of the other car. How fast is each traveling?

26. Two sisters, who live in cities that are 400 miles apart, start at the same time and drive towards each other. In how many hours will they meet if one averages 55 mph and the other averages 45 mph?

27. Two trucks start at the same terminal at the same time and drive away in opposite directions. Because of heavy traffic, one truck averages 35 mph and the other truck averages 45 mph. In how many hours will they be 320 miles apart?

28. A car and a school bus start toward each other at the same time from cities that are 297 miles apart. If the car travels at twice the rate of the bus, they will meet in 3 hours. How fast is each traveling?

C. Read and respond to the following exercises.

29. Would a chart arrangement have been useful in organizing information for the exercises in Section 10.2? Explain your answer.

30. Using Example 7 in Section 10.3, draw a chart with columns headed Rate, Time, and Distance, and show how you could organize the information for that problem in the chart. Which method, sketch or chart, do you prefer? Why?

31. Describe, in words, how you find the value of nine quarters. Now describe how you express the value of x quarters. Is the same process involved?

32. Your friend is having difficulty with the problems in this section. When he goes back to study the example problems, he doesn't understand why a number like 6% is written as 0.06 in the chart and in the equation. Please write an explanation of that concept for him.

33. If you found the charts useful in these problems, explain why. If you didn't find them useful, explain why and describe, in words, how you organized your information for the exercises in this section.

10.5 CHAPTER REVIEW

In this chapter, you have learned and practiced how to change words to symbols and word sentences to equations. When solving application problems, remember these general and specific facts and suggestions:

1. Read the problem carefully to see what is given and what is to be found.
2. Use a variable to represent one of the items, often the one you know the least about. Then represent the other items in terms of this variable.
3. Organize information using a list, sketch, or chart.
4. Consecutive numbers can be represented as n, $n + 1$, $n + 2$, and so on.
5. Consecutive even or odd numbers can be represented as n, $n + 2$, $n + 4$, and so on.
6. Perimeter is the sum of all the sides of a figure.
7. In money problems you must consider both the *value/cost of each item* and the *number* of each item.
8. Write the equation and solve it.
9. Be sure to give all answers that are requested and to label answers with the correct units when appropriate.
10. Check your answers in the words of the problem, not just in the equation.

As you go on, both in school and at work, problems where you will need to use algebra will not just give you an equation to solve, but will present you with a situation (word problem) where there is something that needs to be found (unknown). You will need to determine the relationship among the items involved (define the variable), and make a statement involving these items (write an equation). Then you should be able to find the solution (solve the equation), and answer the question that was asked. These are the steps you have practiced in this chapter.

REVIEW EXERCISES

A. Match each phrase in Column I with the correct algebraic expression in Column II.

I	II
1. six more than x	a. $2X$
2. twice X	b. $6/X$
3. the difference between 6 and X	c. $6X$
4. six less than X	d. $1/2X$
5. the quotient of 6 and X	e. $X + 6$
6. the product of 6 and X	f. $X - 6$
7. half of X	g. $6 - X$
8. the quotient of X and 6	h. $X/6$

B. Match each sentence in Column I with the correct algebraic equation in Column II.

I	II
9. Eleven more than a number equals 19.	i. $n + 4 = 19$
10. When a number is decreased by 11, the result is 19.	j. $n - 4 = 19$
11. Two-thirds of a number is 12.	k. $2n = 19$
12. The quotient of a number and 12 is 3.	l. $n + 11 = 19$
13. When a number is increased by 4, the result is 19.	m. $1/2n = 3$
	n. $2/3n = 12$
	o. $n - 11 = 19$
	p. $2/3n - 4 = 12$
	q. $n/12 = 3$
	r. $2n + 6 = 12$
	s. $2/3n + 4 = 12$

14. Twice a number is 19.

15. The difference between a number and 4 is 19.

16. Half of a number is 3.

17. Six more than twice a number is 12.

18. Two-thirds of a number decreased by 4 is 12.

C. For each exercise, organize the information, write an equation, and then solve and check your answer(s).

19. Twice a number increased by 7 is 25. Find the number.

20. The difference between a number and 4 is 13. What is the number?

21. The sum of three consecutive numbers is 87. Find the numbers.

22. The perimeter of a rectangle is 64 feet. The length is 2 feet more than the width. Find the dimensions.

23. Four less than half a number if 16. Find the number.

24. Find three consecutive numbers whose sum is 174.

25. The length of a rectangle is 12 inches more than the width. Find the dimensions if the perimeter is 52 inches.

26. Two-thirds a number increased by 17 is 35. Find the number.

27. A pile of dimes and nickels contains equal numbers of the two coins. The value of the coins is $1.95. Find how many of each coin there are.

28. Equal numbers of adults' and children's tickets were sold for the movie *Cinderella*. Find the number of children's tickets sold if the total receipts were $225.00. Adults' tickets cost $3.00, and children's tickets cost $2.00.

29. One number is four times the other, and their sum is 35. Find the numbers.

30. Twice a number increased by 11 is 25. Find the number.

31. The sum of the first and third of three consecutive even numbers is 72. Find all the numbers.

32. A church collection basket contains as many dollar bills as quarters. The basket contains $21.25. How many quarters are there?

33. One number is 5 more than the other. The sum of the larger and three times the smaller is 37. Find the larger number.

34. The sum of the largest and smallest of three consecutive odd numbers is 58. Find all the numbers.

35. The length of a rectangle is 6 inches more than the width. The perimeter is 48 inches. Find the dimensions.

36. The ratio of the width to the length of a rectangle is 7:9. Find the dimensions of the rectangle if the perimeter is 192 feet.

37. The perimeter of a rectangle is 180 yards. The ratio of the width to the length is 2 to 3. Find the dimensions.

38. The ratio of the size of the three angles of a triangle is 22 to 9 to 5. What is the measure of each angle if their sum is 180°?

39. The perimeter of a triangle is 81 feet. All three sides have the same length (it is an equilateral triangle). Find the measure of each side.

40. The sum of two consecutive numbers is 19. Find the numbers.

41. If three times as much money is invested at 6 1/4% as at 5 1/2% annual interest, the amount of interest earned at the end of three years will be $2,910. How much must be invested at the higher rate?

42. Two cars leave an amusement park at the same time and travel in opposite directions to go home. One car encounters heavy traffic and can travel at half the rate of the other car. After one hour the cars are 90 miles apart. How fast is each traveling?

43. Eight less than three-fourths a number is 10. Find the number.

44. In the college cafeteria, pizza sells for $0.99 a slice and chicken sandwiches for $1.49 each. During a one-hour period, $31.23 was collected when twice as many pizza slices as sandwiches were sold. How many slices of pizza were purchased?

45. The perimeter of a triangle is 91 feet. If the length of the second side is twice the length of the first side, and the third side is 1 foot longer than the second side, what are the lengths of the three sides?

46. A jogger and walker start at the same point at the same time and travel in opposite directions around a lake. The jogger averages 6 miles per hour and the walker averages 2 1/2 miles per hour. If the distance around the lake is 4 1/4 miles, how long before the two pass each other?

47. Sara won a full scholarship to attend college, so her parents decide to put the money they had saved for tuition into investments and to give Sara the money earned in interest as a present at graduation. They invest twice as much at 7% as they invest at 6%, and the interest earned at the end of four years is $4,564. How much is invested at 7%?

48. The sum of the three angles of any triangle is always 180 degrees. If the first angle is half the measure of the second angle, and the third angle is 12 degrees more than the first, what are the measurements of the three angles?

49. The first angle in a triangle is twice as big as the second angle, and the third angle is 20° larger than the first angle. How big is each angle? The sum of the three angles of every triangle is 180°.

50. In a triangle, the first side is twice the length of the second side. The third side is 4 inches longer than the second side. Find the length of the longest side if the perimeter is 40 inches.

51. Six more than three-eighths of a number is 12. Find the number.

52. The coffee vending machine in the student lounge contains the same number of quarters as dimes. If they total $15.75, how many of each coin are there?

53. The ratio of the length to the width of a rectangle is 9 to 5. Find the dimensions of the rectangle if the perimeter is 112 meters.

54. The sum of the four angles in a quadrilateral is always 360°. In a certain quadrilateral, two of the angles are equal, the third angle is twice the size of one of the equal angles, and the fourth angle is 15° larger than one of the equal angles. What is the measure of each angle in this figure?

55. Monique invests 3 times as much money at 5% as she invests at 4% and earns total interest of $152 in two years. How much did she put in each investment?

56. Darryl is saving to buy a new car. In order to have a down payment when he is ready to buy, he decides to invest 4 times as much money at 6% as at 5% and can earn $870 in interest in 3 years. How much does he put into each investment?

57. The length of a rectangle is three times the width. The perimeter is 72 meters. Find the dimensions.

58. Lee Ann purchased twice as many letter stamps as postcard stamps when she went to the Post Office. Letter stamps cost $0.37, and postcard stamps cost $0.30. If she spent $18.72, how many of each did she purchase?

59. Find the lengths of the sides of a triangle if two sides are equal, and the third side is 2 feet shorter than one of the equal sides. The perimeter is 37 feet.

60. Two cars leave cities that are 480 miles apart at the same time and travel toward each other until they meet. One travels at 55 miles per hour and the other at 65 miles per hour. How long will it take before they meet?

D. Each of the following exercises has an error in the translation of the word problem to the equation. Describe the error and write the correct equation. You do not have to solve the problem.

61. A rectangle has a perimeter of 100 meters. The length is five times the width. Find the dimensions of the rectangle.

Equation: $5x + x = 100$

62. The number of children's books sold at the book fair was three times the number of adult books sold. Children's books sold for $1.50 each and adult books for $3.00 each. If a total of $180 was collected at the sale, how many children's books were sold?

Equation: $3x + x = 180$

63. Twice as much money is invested at 6% as at 5% annual interest. If at the end of 4 years, $1,020 in interest has been earned, what is the size of each investment?

    Equation: $2x(6)(4) + x(5)(4) = 1{,}020$

64. The length of a rectangle is 4 feet more than the width. If the perimeter is 96 feet, what is the length?

    Equation: $4x + x + 4x + x = 96$

65. The sum of the first and third of three consecutive odd numbers is 122. What are the three numbers?

    Equation: $x + x + 2 + x + 4 = 122$

E. Read and respond to the following exercises.

66. Describe, in words, the procedure you would use to find the solution to this exercise. You do not have to actually work the problem.

    Twice as much money is invested at 4 1/2% as at 6% annual interest. If at the end of 5 years, $3,750 in interest has been earned, what is the size of each investment?

67. After studying application problems that were solved using three different methods for organizing the information, do you consider that this step of organizing the information is important? Explain your answer.

68. List the kinds of information that you would need to consider if you were trying to decide whether to move to a new house or apartment in the same town or city. Where would you find each of these pieces of information?

69. It was stated at the beginning of the chapter that "the method shown for a problem will not be the only one that could be used; it is just the one that authors found to be most suited to that particular kind of problem." Give several reasons why you think the authors chose lists for general and consecutive number problems, sketches for geometry and motion problems, and charts for money problems.

70. Has your attitude toward solving word problems undergone any change as a result of your work in this chapter? Why or why not?

F. The following problems may require a little more thought than the previous ones. Organize the information, write an equation, solve, and give all the answers requested.

71. Mr. and Mrs. Yamato have just had a very happy month. During that time their first child was born and they won a sweepstakes prize of $100,000. The Yamatos decide to invest part of the winnings for the baby's college education. They invest the same amount in stocks paying 9% annual interest as they invest in municipal bonds paying 8% annual interest. Assuming that the interest rates don't change, how much should they invest in each program to have $76,500 in interest in 18 years? What is the total amount available for college at that time?

72. Shirley and her husband have decided to simplify their holiday shopping for gifts for their large family. They decide to buy gift certificates to a national department store. To be fair to everyone, they set the following guidelines. Their grandchildren's gift certificates are three times the value of those of the nieces and nephews, and their children's gift certificates are twice the value of those of the nieces and nephews. They have to buy for four children, eight grandchildren, and twenty-four nieces and nephews. If the total cost of the gift certificates was $560, what size gift certificates were purchased?

73. Ira and Norwin arrange to meet for lunch. They will each start at 11:00 A.M. from their homes, which are 7 miles apart, and will walk until they meet. If Ira walks at two-thirds the rate that Norwin walks, they will meet in 1 1/2 hours. How fast does Ira walk? How far does Norwin walk before they meet?

74. Joellen and Adrianne decide to start at the same place at the same time and walk around the lake in opposite directions until they meet, and then they can sit and rest before continuing their walk for exercise. Joellen walks at a rate of 3 1/2 miles per hour and Adrianne walks at a rate of 2 1/2 miles per hour. If the distance around the lake is 4 1/2 miles, how long will it be before they meet?

75. Leonard has 200 feet of chain-link fencing to enclose his rectangular back yard. If he uses his house for one of the lengths, what will be the dimensions of the fenced yard if the length is 8 feet more than twice the width?

76. The width of a rectangle is one-third the measure of the length. If each dimension is increased by 12 meters, the perimeter is doubled. What are the dimensions of the original rectangle?

77. Marvin invests in two different accounts. The sizes of the principals are in the ratio of 5 to 3. The larger investment pays 7% and the smaller pays 6% annual interest. If he earns $1060 in interest in 4 years, how much was in each investment?

CHAPTER TEST

In Exercises 1 through 4, write an algebraic expression.

1. Eight less than a number
2. Four more than three times a number
3. The average annual rainfall in Austin is R inches. The average annual rainfall in Tacoma is five inches more than four times the Austin average. Express Tacoma's annual rainfall.
4. Three-fourths a number increased by seven

In each of Exercises 5 through 18, write an equation and solve it.

Six less than twice a number is 20. What is the number?

5. Equation:
6. Solution:

The ages of the Montez children can be represented as consecutive odd numbers. If the sum of their ages is 27, how old is each of the three children?

7. Equation:
8. Solution:

Five-sixths of the students were in history class today. If 35 students were present, how many students are enrolled in the class?

9. Equation:
10. Solution:

The length of a rectangle is 8 meters more than the width. The perimeter is 180 meters. What are the dimensions?

11. Equation:
12. Solution:

The Whitestone Bridge toll collector noticed that she had twice as many quarters as dollar bills and had no other coins. If she had a total of $43.50, how many quarters did she have?

13. Equation:
14. Solution:

While completing a time management project, a student charts his daily activities for a week and sees that he spends a total of 37.5 hours studying, attending class, and commuting. If the ratio of the time for these activities is 8 : 4 : 3, how much time did the student study during the week?

15. Equation:
16. Solution:

A jet can travel twice as fast as a propeller plane. If the two planes leave the airport at the same time and travel in opposite directions, they will be 1,575 miles apart in 3 hours. What is the rate of each plane?

17. Equation:
18. Solution:

Twice as much money was invested at 5% as was invested at 8% annual interest. If after four years, $2,880 has been earned in interest, how much was invested at each rate?

19. Describe in words how to organize the information and write a correct equation for the problem just given.
20. After following the steps you described in Exercise 19 and solving the equation, how do you get the two pieces of information that are requested?

STUDY SKILLS MODULE

11 Time Management

Probably the single most difficult adjustment for students when they begin college is learning how to manage their time. If you are a traditional-age student, recently coming from high school, you are used to a structured day with five or more classes and parents and teachers who are constantly reminding you of what needs to be done and when. If you are an older student who has been away from high school for a few years or more, your life has become filled with lots of other things. You may have family responsibilities, job demands, social or community involvement, and a fairly set lifestyle. All of a sudden, in both scenarios, the routine is upset and major changes occur. Your friends don't understand why you aren't free to do the things you did before (such as talk on the phone for hours). Your family feels put out because you can no longer give as much to them as you did before you started school. As for yourself, you are wondering how you are ever going to get everything done and may even be questioning your decision to return to school in the first place.

It may help you to know that most students share your experiences and your frustrations. Feeling stressed and overwhelmed is a common experience for college students of all ages. The key to success is organization and focus. You need to take control of your life—all parts of it—and keep your mind on your goal of a better life through education.

The more organized you become, the better off you will be. You need to keep a daily, weekly, and monthly planner or calendar. If you get in the habit of writing everything down, you worry less about forgetting something or over- scheduling a day.

For most students, life and its demands can be put into 11 basic categories: Class, Studying, Job, Household, Food-Related, Personal Chores, Social/Recreation, Family, Sleeping, Travel, and Other. Your own situation may dictate some changes in the list. The rule of thumb for college work is: at least 2 hours of work and study outside of class for every hour in class. This means that a 4-hour course takes up at least 12 hours a week: 4 hours in class and 8 hours of study. Some classes, depending on your own skills and interest, may require more time than that. Therefore, when you plan your week, you need to make adjustments in your usual routine to allow for the extra demands on your time.

Try to make a list of all the things that you need to accomplish this week and make a note about any deadlines that you have. Just putting it all down on paper, where you can see it, is a helpful thing for most people. It will help you realize what is actually happening with your time and where adjustments may need to be made.

Remember these things:

1. Write everything down—become a list maker and keep track of your time.
2. Realize that you need to make changes in your life's routine and that those around you may resent those changes. Help them by explaining why your education is so important to you and why you need their support.
3. Keep focused on your long- and short-term goals. See yourself as successful.
4. Take care of yourself. Eat right. Exercise. Get enough rest. You cannot expect to do well in school, and enjoy the experience, if your body and mind are not well cared for. If you need help with any of this, see a counselor or advisor on your campus. They have worked with many students experiencing the same feelings and stresses that you are. Explain your situation and ask for their help.

An education is a precious thing, and it is well worth the struggles that it requires. Learn to make adjustments in your life, one at a time, that will help you reach your goal. Take control of your days and your time. Be good to yourself.

Your Point of View

1. How do you feel about the demands on your time now? Are you able to control your life or do you feel overwhelmed? Try to be specific about your areas of concern.
2. Make a *detailed* schedule of things you need to do for the next three days. Check things off as they happen or are completed, and evaluate your progress at the end of the three days.
3. Write down three changes, big or small, that you will make in your routine so you can manage your time more efficiently.

CHAPTER

11 Signed Numbers

LEARNING OBJECTIVES

When you have completed this chapter, you should be able to:

1. Translate verbal expressions into signed-number expressions.
2. Graph signed numbers on a number line.
3. Add, subtract, multiply, and divide signed numbers.
4. Make use of the Order of Operations Agreement in expressions with signed numbers.
5. Simplify signed-number expressions that include exponents.
6. Evaluate algebraic expressions using signed numbers.

The ability to do calculations with signed numbers is a fundamental math skill. Often such calculations are one of the first topics studied in an algebra course. What signed numbers are and what they mean are important concepts to grasp. One way to accomplish this is to begin by reading and rereading the rules, but be sure to keep reading through the explanations of those rules as well. The more problems you do, the more familiar the processes will become. Soon the reasoning behind the rules should become clearer too.

11.1 What Are Signed Numbers?

In the beginning, you might think of signed numbers as "temperature" numbers. Everyone knows the difference between 12 degrees above zero (+12) and 12 degrees below zero (–12). Money is also a practical application using signed numbers. Many people have had the experience of having money in a checking account (a positive balance) and overdrawing that account by writing a check that is larger than the amount of money in the account (leaving a negative balance).

Rather than a vertical scale like a thermometer, a horizontal number line is often used to illustrate signed numbers.

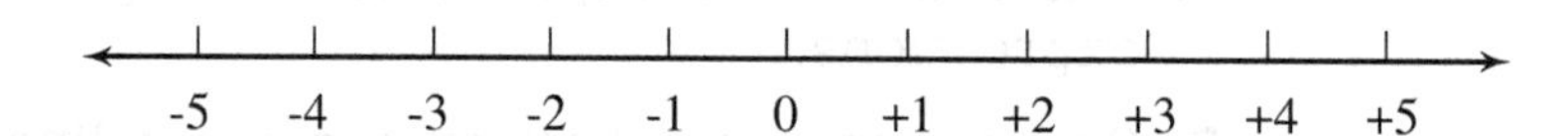

The zero is placed approximately in the middle, and it is thought of as the beginning point for all calculations. It separates the positive numbers on its right from the negative numbers on its left. Like a ruler, a number line is divided into units, and each unit is the same length. A number line extends without end in both directions, so arrows are used at each end of the line. Each signed number has distance and direction. For example,

+ 3 is 3 units to the right of zero.

– 5 is 5 units to the left of zero.

0 is in the middle and has no sign.

From the number line that was shown, it would seem that the signed numbers can be only integers. **Integers** are the positive and negative whole numbers and zero. Examples of integers that are also signed numbers are +5, – 6, 0, and –8. Fractions and decimals are not included in the group of numbers called integers. We could graph +2 3/4, –1.8, –2/3, and + 11.76, because they are signed numbers; but for the sake of clarity on the number line, we usually indicate only the integer values on the scale. On the following number line, – 4, – 1.5, +1, and + 2 1/4 are graphed.

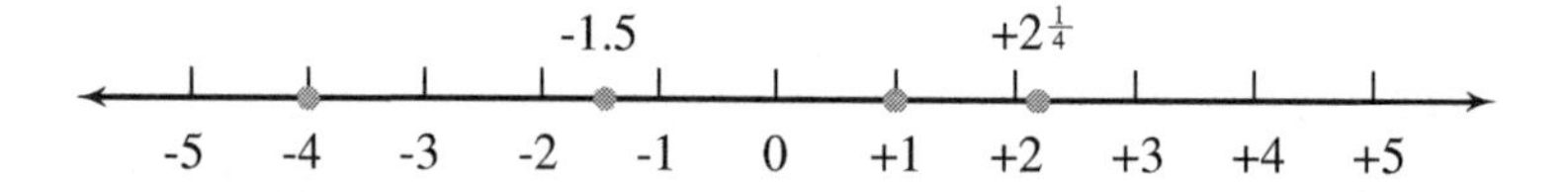

Because in many contexts (such as football and finances) it is customary to indicate a gain as a good thing, *a positive,* and to indicate a loss as a bad thing, *a negative;* it is easy to represent gains and losses as signed numbers. On a vertical graph, such as a thermometer, the positive values are above zero and the negative values are below zero. Applying these ideas to a map, where we could measure both north–south and east–west, north and east would be positive, and south and west would be negative.

Self-Check Represent each of the following expressions by a signed number.

1. four miles east
2. six feet below ground
3. eight degrees above zero
4. twenty-three miles north
5. a loss of twelve dollars
6. seventeen and one-half miles west
7. eighty-two miles south
8. ten points ahead in a basketball game

Answers:

1. +4 2. –6 3. +8 4. +23 5. –12

6. –17 1/2 7. –82 8. +10

Consider the relative positions of +5 and +3 on the number line.

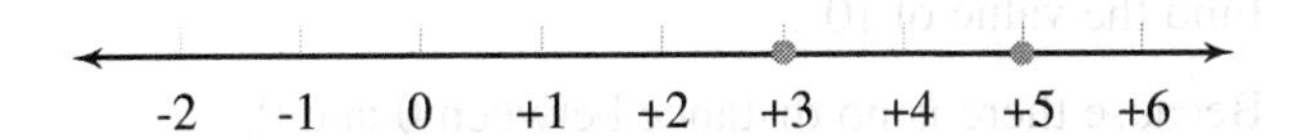

You know that +5 is larger than +3 and is to the right of +3 on the number line. Using inequality notation, you would write $+5 > +3$ or $+3 < +5$.

Consider the relative positions of +2 and –2 on the number line.

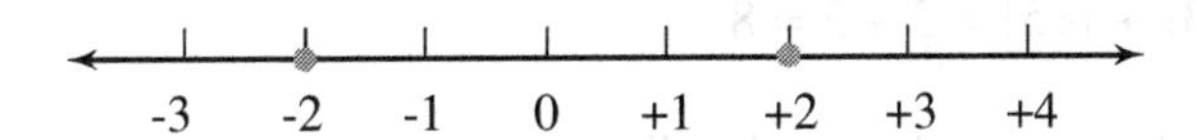

Because +2 is to the right of –2 on the number line, +2 must be the larger number. So $+2 > -2$ or $-2 < +2$.

Consider the relative positions of –4 and –10 on the number line.

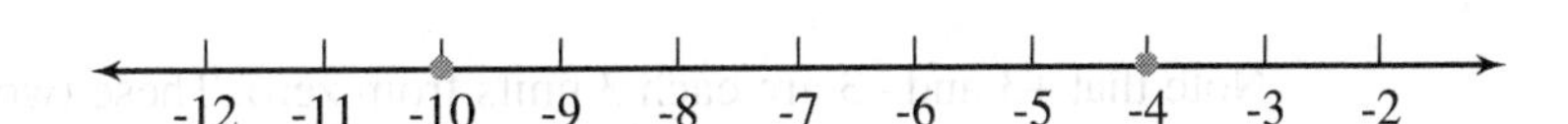

Because –4 is to the right of –10 on the number line, –4 is larger than –10. You would write $-4 > -10$ or $-10 < -4$.

Whenever you need to know which of two numbers is larger, picture their positions on the number line: the larger one is always to the right. You can also think of them as temperature numbers; the larger number is the "warmer" number. For example, +2 is to the right of –2, and 2 degrees above zero is warmer than 2 degrees below zero. Four degrees below zero, –4, is cold, but it is still warmer than 10 degrees below zero, –10. On the number line, –4 is to the right of –10.

Another concept that students usually study when they are learning about signed numbers is absolute value. Consider the relative positions of +3 and –3 on the number line.

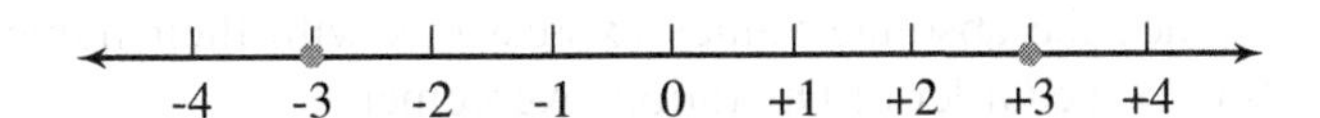

Each of these integers is three units from zero. We say that the absolute value of +3, which is written $|+3|$, is equal to 3. **Absolute value** is the distance that a given number is from zero. Distance is always positive (you can't walk a negative two miles), and the absolute value of a number is the positive value associated with it. The absolute value of –5, $|-5|$, is equal to +5. Similarly, $|-2.5| = 2.5$ and $|3\ 1/4| = 3\ 1/4$. Study these examples.

EXAMPLE 1 Find the value of $|-7|$.

Solution Because –7 is 7 units from zero,
$|-7| = 7$

EXAMPLE 2 Find the value of $|+6|$.

Solution Because +6 is 6 units from zero,
$|+6| = 6$

EXAMPLE 3 Find the value of $|-2.6|$.

Solution Because –2.6 is 2.6 units from zero,
$|-2.6| = 2.6$

EXAMPLE 4 Find the value of $|0|$.

Solution Because there is no distance between 0 and 0,
$|0| = 0$

EXAMPLE 5 Find the value of $|-3| + |+5|$.

Solution Because $|-3| = 3$ and $|+5| = 5$,
$|-3| + |+5| = 3 + 5 = 8$

Look again at the number line.

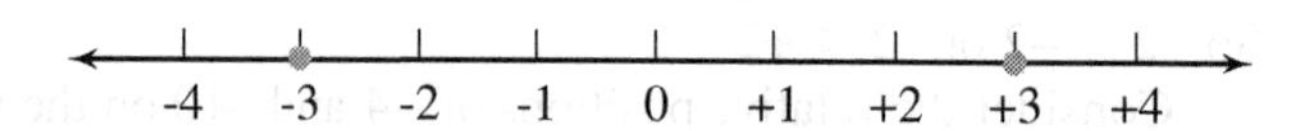

Note that +3 and –3 are each 3 units from zero. These two numbers are "opposites" of each other. They both have an absolute value of 3, but they lie on opposite sides of zero. Numbers that have the same absolute value but opposite signs are called **opposites.**

Now, you have three different ways of reading –3:

–3 can tell you to subtract three,

–3 can mean the signed integer, "negative three," and

–3 can mean "the opposite of three."

You will be able to tell from the problems themselves which meaning applies.

EXAMPLE 6 Find the value of $|-15| - |-3| \cdot |+4|$.

Solution Whenever you are simplifying an expression involving absolute values, you need to replace the absolute values expressions with their numerical equivalents and then follow the Order of Operations Agreement.

Because $|-15| = 15$ and $|-3| = 3$ and $|+4| = 4$,

$$|-15| - |-3| \cdot |+4|$$
$$= 15 - 3 \cdot 4$$
$$= 15 - 12$$
$$= 3$$

EXAMPLE 7 Simplify this expression $|-7| + 4\,|5|$.

Solution $|-7| + 4\,|5|$ First bring the values out of their absolute value notation.

$7 + 4 \cdot 5$ A number next to an absolute value symbol means to multiply and according to the Order of Operations Agreement, we need to multiply before we add.

$7 + 20$ Now we add.

27

Self-Check Simplify each of these expressions. Remember to remove the values from their absolute value notation first.

1. $|-3| + |-5| \cdot |2|$
2. $|7| - |-4|$
3. $2|-6| - |-8|$
4. $|-12| + (-3) - |-2|$

Answers:

1. 13 2. 3 3. 4 4. 7

11.1 EXERCISES

A. Represent each of the following descriptions as a signed number.

1. 4 steps down
2. 350 feet below sea level
3. winning $26
4. a drop of 18 degrees
5. 276 miles east
6. 143 miles north
7. 5,280 feet above sea level
8. a deposit of $125
9. 75 degrees above zero
10. losing by 14 points
11. a gain of 5 yards
12. writing a check for $37.50
13. a decrease in altitude of 2,200 feet
14. a loss of 7 yards
15. depositing $2,000 into a CD
16. a gain of 47 points in the stock market

B. Make a number line and graph each of the following signed numbers.

17. –5
18. +3 1/2
19. +2
20. –5.25
21. 0
22. –5/8
23. –3 1/3
24. +6
25. +4.2
26. +1.5

C. Write each of the following statements using inequality notation. Use $<$, $>$, or $=$ to fill in the blank.

27. –4 _____ –1
28. +8 _____ –8
29. –5 _____ –6
30. +9 _____ +12
31. +5 _____ +6
32. –8 _____ +8
33. 0 _____ –6
34. +7 _____ 0
35. –6 1/3 _____ –7
36. –8.5 _____ –2.4
37. $|-2|$ _____ –2
38. –16 _____ 16
39. $|5|$ _____ $|-6|$
40. $|18|$ _____ $|-18|$
41. $|0|$ _____ –12
42. $|-7|$ _____ $|-9|$
43. 8 _____ $-|-8|$
44. $-(-4)$ _____ $|-4|$

D. Simplify each of the following absolute value statements.

45. $|-6|$
46. $|0.4|$
47. $|+9|$
48. $|-4|$
49. $|0|$
50. $|+13|$

51. |–1.4|

52. |–1 1/4|

53. |5| + |–3|

54. |–2| + |–7|

55. |–12| – |+10|

56. |–8| – |+3|

57. |–3| · |–8| – 4|–5|

58. |–12| ÷ |–2| · |+3|

59. (|–2|)(|+8|) + (4)(|–2|)

60. |–42| – |–4|· |–1|

61. |–37| + |+3| · |–7|

62. (|+3|)(|–6|) – 3(|+3|)

E. Rank the members of each of the following groups in order from the least to the greatest.

63. |–7|, |+2|, |0|, |–3|, |+4|

64. |–2 1/2|, |+2 1/4|, |–5|, |+6|, |–3|

65. |+9|, |–10|, |+3|, |–7|, |–12|

66. |–2.4|, |5.3|, –3.5, |–0.2|, –|–3|

F. Read and respond to the following exercises.

67. Explain how you know which of two numbers is larger when they are graphed on a number line.

68. In your own words, explain the meaning of the term *absolute value.*

69. Write a practical problem that uses positive and negative numbers. You don't have to be able to find the solution yet.

70. Explain how you would read this expression, –|–5|, and then how you would simplify it.

11.2 Adding Signed Numbers

The addition of signed numbers can be shown easily by making use of the number line. The following graph shows the addition of (+3) and (+4). (Note: The parentheses are placed around the signed number to draw attention to the sign and do not normally indicate multiplication.) Begin at zero and travel 3 units in the positive direction. From that point, move 4 more units in the positive direction.

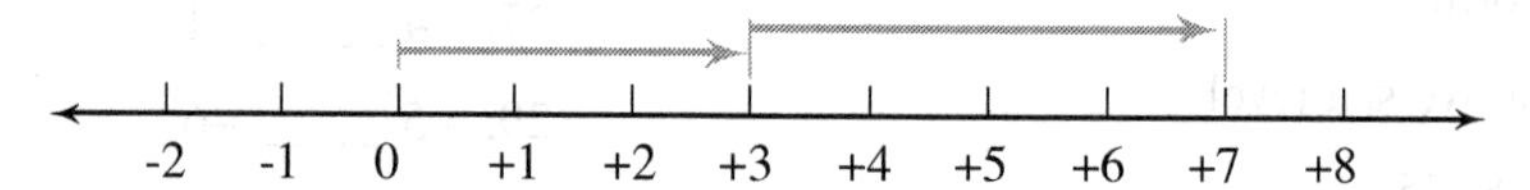

The number line shows what you already know, that

(+3) + (+4) = +7

Because the positive signs get in the way, you can simply write 3 + 4 = 7. But if you mean negative 3, you must write –3.

Consider (–5) + (–1).

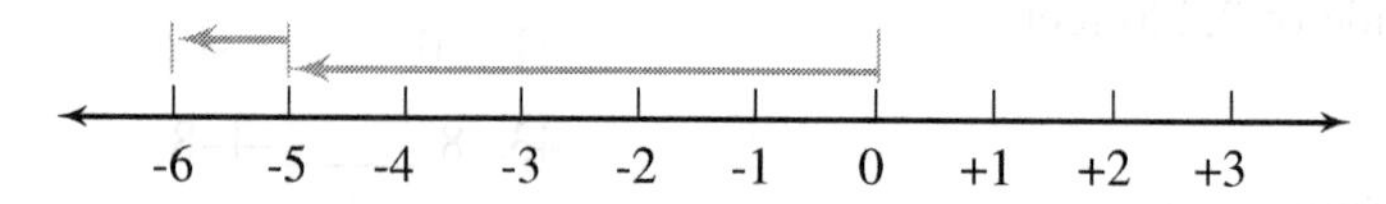

Therefore, (–5) + (–1) = –6

It makes sense that if you are combining positive numbers, you are moving farther in the positive direction, and your answer will be positive. Likewise, if you are combining negative values, you are moving farther in the negative direction, and your answer will be negative.

When combining numbers with like signs, we add the distances (absolute values), measured from starting point to ending point, and use the common sign. For example,

(–6) + (–8) = –14

(+7) + (+5) = +12

(–3) + (–8) + (–4) = –15

11 + 12 + 6 + 5 = 34

The next question is what happens when you are adding (combining) and the signs are different? Look at these examples, which were worked out using the number line, and see if you can figure out the rule.

Consider (–5) + (+3). Starting at zero, you will move 5 spaces to the left (in the negative direction), and then, from that point, you will move 3 spaces to the right (in the positive direction).

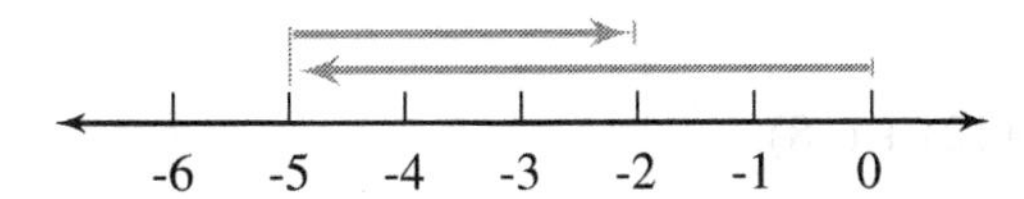

Therefore, (–5) + (+3) = –2.

Consider (+4) + (–7).

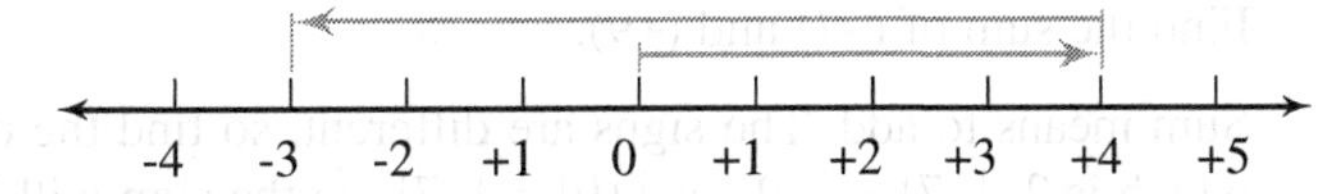

Therefore, (+4) + (–7) = –3

Consider (–3) + (+8).

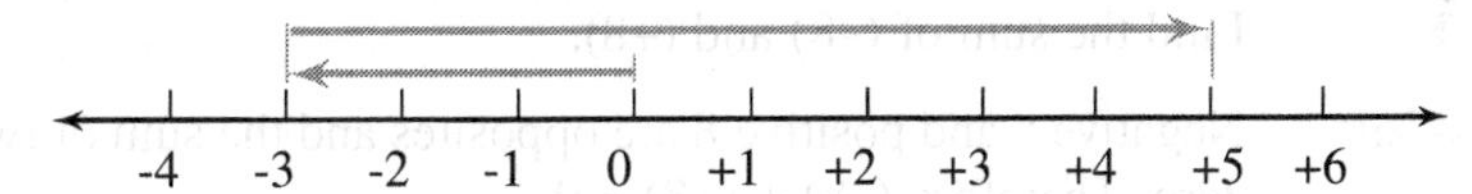

Therefore, (–3) + (+8) = +5.

Consider (–6) + (+6).

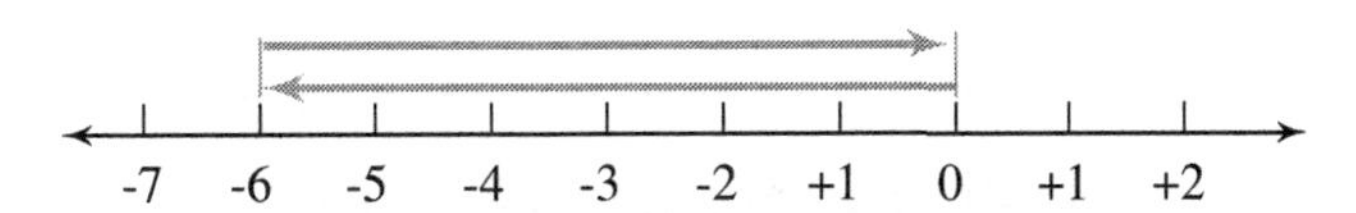

Therefore, (–6) + (+6) = 0.

These examples indicate that *when combining numbers of different signs, we find the answer by taking the difference between the two distances (absolute values) and using the sign of the longer distance.* The "longer distance" is the one with the larger absolute value. $|-5| = 5$ and $|+3| = 3$; we use the negative sign because 5 is greater than 3. Although the number line makes it easier to visualize what is going on when you combine signed numbers, you will not want to use it all the time. You need to learn the rules and be able to apply them.

Consider the following sample problems solved by using the rule. The answers are the same as in the previous graphs.

(–5) + (+3) = –2

The difference between 5 and 3 is 2, and 5 is the longer distance so you use its sign, –.

$(+4) + (-7) = -3$

The difference between 4 and 7 is 3, and $|-7| > |+4|$ so you use the sign from -7.

$(-3) + (+8) = +5$

The difference between 3 and 8 is 5, and $|+8| > |-3|$ so you use the sign from $+8$.

$(-6) + (+6) = 0$

The difference is 0 and it has no sign. *Note:* -6 and $+6$ are called *opposites* because they are the same number of units from zero but in opposite directions. *The sum of two opposites is always zero.*

Study the following examples and see how the rule applies.

EXAMPLE 1 Simplify $(-11) + (+4)$.

Solution The signs are different. The difference between 11 and 4 is 7 and $|-11| > |+4|$, so the sign is negative. Therefore $(-11) + (+4) = -7$.

◀

EXAMPLE 2 Simplify $(+12) + (-8)$.

Solution The signs are different, so find the difference between 12 and 8, which is 4. $|+12| > |-8|$, so the sign will be positive. Therefore, $(+12) + (-8) = +4$.

◀

EXAMPLE 3 Find the sum of (-7) and $(+9)$.

Solution Sum means to add. The signs are different, so find the difference between 7 and 9, which is 2. $|-7| < |-9|$ or $|+9| > |-7|$, so the sign will be positive. Therefore, $(-7) + (+9) = +2$.

◀

EXAMPLE 4 Find the sum of (-8) and $(+8)$.

Solution Negative 8 and positive 8 are opposites and the sum of two opposite values is always zero. Therefore $(-8) + (+8) = 0$.

You can now formally write the rules for the addition of signed numbers, which is also called *combining signed numbers*.

Combining Signed Numbers

If the numbers being added have the same sign, combine the absolute values and use that sign.

If the numbers being added have different signs, find the difference between the absolute values, and use the sign of the number with the larger absolute value.

Self-Check Try these problems to see if you understand how to combine (add) signed numbers.

1. $(-6) + (-9)$
2. $(+4) + (-7)$
3. $(-3) + (+8)$
4. $(-11) + (+11)$
5. $(+7) + (+4)$

Answers:

1. –15 2. –3 3. +5 4. 0 5. +11

The properties of real numbers that you studied in Chapter 8 also apply to signed numbers. The Commutative Property for Addition applies to addition with signed numbers.

Commutative Property for Addition

For all real numbers represented by a and b,

$a + b = b + a$

$$(-5) + (+3) = (+3) + (-5) \qquad (+4) + (-7) = (-7) + (+4)$$
$$-2 = -2 \qquad -3 = -3$$

The Associative Property for Addition also applies to calculations with signed numbers.

Associative Property for Addition

For all real numbers represented by a, b, and c,

$(a + b) + c = a + (b + c)$

$$[(+9) + (-12)] + (-4) = (+9) + [(-12) + (-4)]$$
$$[-3] + (-4) = (+9) + [-16]$$
$$-7 = -7$$
$$[(-5) + (-8)] + (+2) = (-5) + [(-8) + (+2)]$$
$$[-13] + (+2) = (-5) + [(-6)]$$
$$-11 = -11$$

You will find that by making use of these two properties, you can combine signed numbers in any order that seems most convenient to you, as long as the sign and the number are rearranged together. For instance, because it is easier to combine numbers with like signs, rearrange the following problem so that the like-signed terms are together.

$$(-8) + (+7) + (+3) + (-5)$$
$$= \underline{(-8) + (-5)} + \underline{(+7) + (+3)}$$
$$= \underline{(-13) + (+10)}$$
$$= -3$$

If you were to do the same problem combining from left to right as the order of operations suggests, it would look like this.

$$\begin{aligned} &\underline{(-8) + (+7)} + (+3) + (-5) \\ &= \underline{(-1) \quad + (+3)} + (-5) \\ &= \underline{(+2) \quad\quad + (-5)} \\ &= -3 \end{aligned}$$

The answers are the same no matter which approach you use.

EXAMPLE 5 Find the sum of (–6), (–4), (+11), (–3), (+7), and (+5).

Solution First regroup so that the like-signed values are together and then combine.

$$\begin{aligned} &\underline{(+11) + (+7)} + (+5) + \underline{(-6) + (-4)} + (-3) \\ &\underline{(+18) \quad + (+5)} + \quad \underline{(-10) \quad + (-3)} \\ &\underline{(+23) \quad + \quad (-13)} \\ &+10 \end{aligned}$$

Suppose you were to read this expression: –3 + 6 – 9 + 5 – 4. There is only one sign between each pair of numbers. This is another way of writing an addition or combine problem. It means the same as (–3) + (+6) + (–9) + (+5) + (–4), but the original expression is so much less complicated. When you have a series of numbers written as these are, simply rearrange the values with their signs to put the like-signed terms together and combine.

–3 + 6 –9 + 5 –4

(–3) + (+6) + (–9) + (+5) + (–4)

(–3) + (–9) + (–4) + (+6) + (+5) Rearrange like-signed values.

$\underline{(-3) + (-9) + (-4)} + \underline{(+6) + (+5)}$ Combine like-signed values.

$\underline{-16 \quad + \quad 11}$ Difference of the values and sign

–5 of the "larger"

EXAMPLE 6 Simplify 8 + (–2) + (–7) + 3 + (–6)

Solution You are to combine these five integers. First rewrite the expression, putting the like-signed values together. Then combine the like-signed values, find the difference of the two results, and use the sign of the "larger."

$$\begin{aligned} &\underline{8 + (-2)} + \underline{(-7) + 3 + (-6)} \\ &\underline{8 + 3} + \underline{(-2) + (-7) + (-6)} \\ &\underline{11 + \quad (-15)} \\ &-4 \end{aligned}$$

11.2 EXERCISES

A. Simplify each of the following expressions by combining the signed numbers.

1. (+4) + (–7)
2. (–6) + (+11)
3. (–8) + (–3)
4. (+7) + (+3)
5. (–11) + (+4)
6. (–2) + (–8)
7. (+6 1/2) + (+3 1/2)
8. (–7) + (+4 1/2)
9. 0 + (–5)
10. (–14) + (+14)
11. (+3) + (–3)
12. 0 + (–7)
13. (–8) + (–5) + (–3)
14. (–7) + (–6) + (–2)
15. (+6) + [(–9) + (–3)]
16. (+4) + [(–9) + (+3)]
17. (–4) + (0) + (–6)
18. (+8) + (–9) + 0
19. [(+4) + (–6)] + (+9)
20. [(–8) + (–4)] + (+9)
21. (–8) + (+6) + (–2) + (–3)
22. (+7) + (–9) + (+6) + (+4)
23. (–2) + (+3) + (+7) + (–8)
24. (–6) + (–5) + (+4) + (–3)
25. (–3) + (+7) + (–2) + (+5) + (–1)
26. 11 + (–16) + (+5) + (–8) + (+2)
27. (–8) + (–6) + (+9) + (+4) + (–3)
28. (–7) + (–2) + (–5) + (+3) + (+8) + (–1)
29. 6 + (–8) + 9 – 0 + 4 + (–3)
30. (–3) + (–5) + 0 + (–2) + 7 + (–4)

B. Write each of the following descriptions as a signed-number expression, and simplify the result. Label your answers where possible.

31. A loss of 14 points followed by a gain of 10 points
32. A gain of 7 pounds followed by a loss of 12 pounds
33. Driving up 250 feet from 150 feet below sea level
34. Putting $142 in your checking account and then writing a check for $87.35
35. You drive 23 miles east and then turn around and drive 40 miles west. How far and in what direction are you from your original position?
36. You fly 125 miles north and then turn around and fly 50 miles south. How far and in what direction are you from your original position?
37. Three days of 8-dollar losses in the price of a stock followed by 2 days of 10-dollar gains had what overall effect on the price of this stock?
38. A football team made a gain of 6 yards, had a loss of 8 yards, had another loss of 3 yards, and then a gain of 12 yards. What is the overall result of these four plays?
39. A pilot flies up from his cruising altitude of 20,000 feet to 26,000 feet and then comes back down to 24,000 feet to avoid turbulence. Where is the plane now in relationship to its original cruising altitude?
40. The student enrollment for the last four autumn terms showed an increase of 1,200, a decrease of 450, a decrease of 275, and an increase of 625. What is the overall change in autumn term's enrollment over the last four years?

C. Simplify each of the following expressions.

41. 12 + (–3 2/3)
42. (–8) + (+11 2/5)
43. (–9 3/4) + (–5 1/8) + (2 7/8)
45. (–1.4) + (+2.9) + (–3.8) + (+1.7) + (–2.4)
46. 7.2 + (–8.5) + (–11.4) + (+8.1) + (–3.2)
47. (–7 1/4) + (–6 2/3) + (+8 3/4) + (–5 1/2)
48. 6 5/8 + (–9 2/3) + (–8 1/4) + (+5 3/8)
49. 16.5 + (–18.7) + (–16.5) + (+18.7)
50. 7 2/5 + (–9 3/4) + (–17 6/15) + (+19 6/8)

D. Use your calculator to simplify each of the following addition problems.

51. (–2.673) + (+9.85)

52. (+14.892) + (–29.37)

53. (–31.4) + (+2.695) + (+17.43)

54. (–42.906) + (–18.479)

55. (+113.457) + (–206.384)

56. (+27.863) + (–19.095) + (–36.52)

E. Read and respond to the following exercises.

57. A fellow student is having trouble understanding why (+6) + (–11) equals (–5). Using a number line or a thermometer, explain why –5 is the correct answer.

58. Is this statement true or false? Why?

$|-8| < |+8|$

59. Is this statement true or false? Why?

$|-3| \geq |-1|$

60. Write a real-world addition problem that uses positive and negative numbers. Translate your problem into an addition expression and simplify it.

11.3 Subtracting Signed Numbers

Looking at the number line again may help you discover the rules used in subtracting signed numbers.

Consider (+3) – (+5)

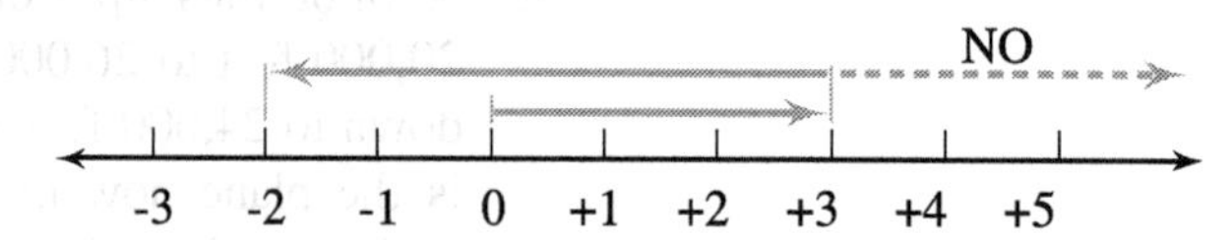

First go 3 spaces to the right of zero. From that point you cannot go 5 more spaces to the right, because that would be adding +5 to +3. You want to move 5 spaces, but you need to change direction because you are subtracting and doing the operation that is opposite to addition. You will go 5 spaces to the left.

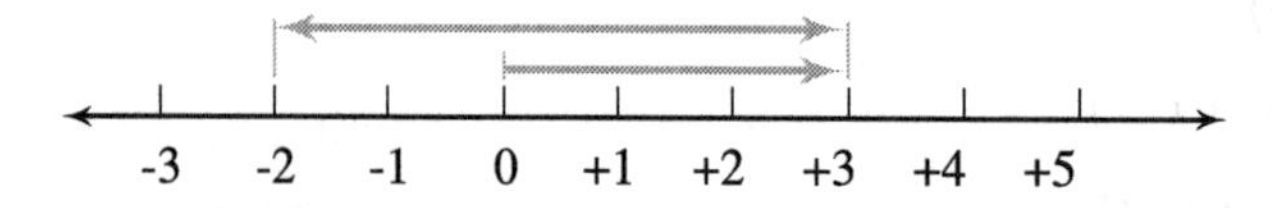

Therefore, (+3) – (+5) = –2

Consider (–4) – (–7).

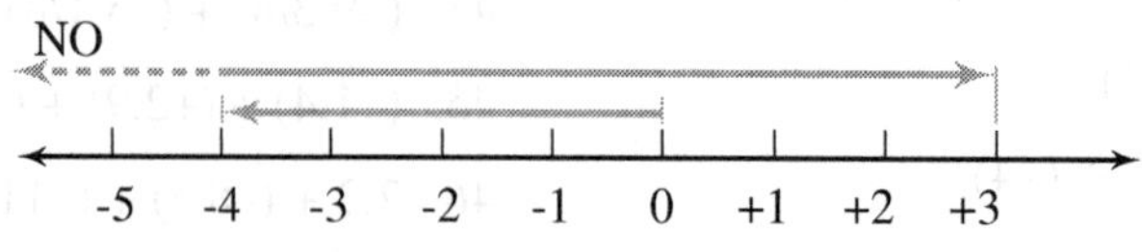

First go 4 spaces to the left of zero. From there you can't go 7 more spaces to the left, because that would be adding –7 to –4, so you need to change direction and go 7 spaces to the right.

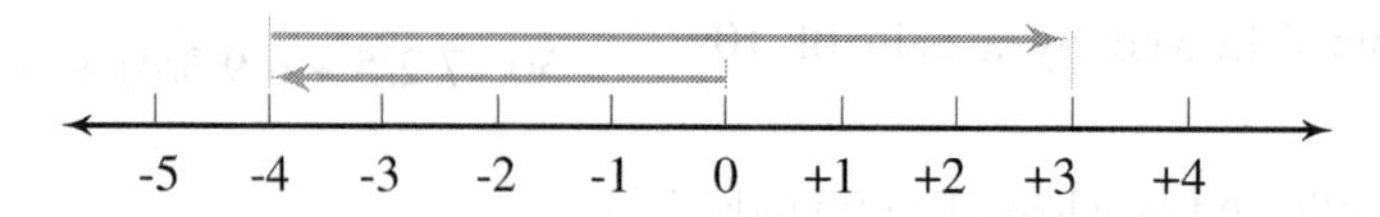

Therefore, (–4) – (–7) = +3

If you look *only* at the number line for (+3) – (+5), the arrows make it appear that you are adding (–5) to (+3).

S A
(+3) – (+5) looks like (+3) + (–5).

If you look *only* at the graph of (–4) – (–7), it would appear that you are adding +7 to –4.

S A
(–4) – (–7) looks like (–4) + (+7).

From these observations, you can formalize the rule for subtracting signed numbers.

Subtracting Signed Numbers

To subtract signed numbers, change the sign of the number being subtracted, and change the operation to addition. Symbolically, this may be written

$$a - b = a + (-b)$$

Consider the following true statements:

S A
1. (–6) – (+3) = (–6) + (–3)
 = –9

Now the signs are the same, so you add the distances and use the sign they share.

S A
2. (+9) – (+4) = (+9) + (–4)
 = +5

Now the signs are different, so you find the difference between the absolute values and use the sign of the number with the larger absolute value.

S A
3. (+14) – (+18) = (+14) + (–18) = –4

S A
4. (–8) – (–12) = (–8) + (+12) = +4

S A
5. (–9) – (–9) = (–9) + (+9) = 0

Self-Check Try these problems to see if you can apply the subtraction rule.

1. (+8) – (+3)
2. (+7) – (–12)
3. (–4) – (–9)
4. (–6) – (+6)
5. (–6) – (–6)
6. (+7) – (+11)

Answers:

1. +5 2. +19 3. +5 4. –12 5. 0 6. –4

You might also find it helpful to know that when you try to find the answer to the question "What is 7 – 4?" you are also asking "How far and in what direction is it from 4 to 7?" Beginning at 4 and moving to 7 is moving 3 spaces in a positive direction, and 7 –4 is equal to +3. Anytime you are finding the value of $a - b$, you can ask, "How far and in what direction is it from b to a?" Does this make a problem like (–9) – (–3) easier to understand? "How far and in what direction is it from (–3) to (–9)?" It is 6 spaces in a negative direction. Therefore, you know that (–9) – (–3) = –6. Of course, you get the same result when you work the problem by applying the rule for subtraction.

$$(-9) \overset{S}{-} (-3)$$

$$= (-9) \overset{A}{+} (+3)$$

$$= -6$$

If you are finding the subtraction rule difficult to understand, try thinking about it this way. When you subtract 8 from 11, your result is 3 because you need to find what number you add to 8 to reach 11.

(+11) – (+8) = +3 because +11 = (+8) + (+3)

(+6) – (+5) = ? What number do you add to +5 to reach +6?

(+6) – (+5) = +1 because +6 = (+5) + (+1)

(–6) – (+7) = ? What number do you add to +7 to reach –6?

(–6) – (+7) = –13 because –6 = (+7) + (–13)

(–3) – (–5) = ? What number do you add to (–5) to reach –3?

(–3) – (–5) = +2 because –3 = (–5) + (+2)

(+4) – (–6) = ? What number do you add to (–6) to reach +4?

(+4) – (–6) = +10 because +4 = (–6) + (+10)

When you are performing the operation of subtraction, you can do it by asking an addition question. Addition and subtraction are opposite, or inverse, operations. They are very closely related.

When you are faced with a subtraction problem, you will find the solution by adding the opposite of the number being subtracted. You have, in fact, done something similar before. When you divide fractions—for example, 7/8 ÷ 3/5—you keep the first fraction, 7/8, as it is, and multiply by the reciprocal of the divisor, 3/5:

$7/8 \div 3/5 = 7/8 \times 5/3 = 35/24$

When you subtract signed numbers, for example, (–7) – (–8), you keep the first quantity, –7, as it is, and add the opposite of the number being subtracted, –8. Therefore,

$$(-7) \overset{S}{-} (-8) \overset{A}{=} (-7) + (+8) = +1$$

There are two popular shortcuts for the rule for subtraction. One is "change the sign and add." The other is "add the opposite." It is fine to use these shortcuts as long as you understand what changes to make. Most students find it helpful to write S over the subtraction symbol in the original statement of the problem and to write A once they have made the changes. This technique will be useful in the next chapters as well.

11.3 EXERCISES

A. Subtract, following these examples:

S	S
(–8) – (+6)	(+11) – (–3)
A	A
= (–8) + (–6)	= (+11) + (+3)
= –14	= +14 or 14

1. (–4) – (–3)
2. (+8) – (+5)
3. (+6) – (+9)
4. (–11) – (–3)
5. (–12) – (+6)
6. (–4) – (+6)
7. (+7) – (–3)
8. (+12) – (–2)
9. 0 – (+6)
10. 0 – (–8)
11. (–6) – (–11)
12. (+8) – (–15)
13. (+7) – (+18)
14. (–3) – (–8)
15. (–2) – (+7)
16. (–9) – (+14)
17. (–8.3) – (+2.5)
18. (–7.1) – (–3.4)
19. (+9.5) – (–3.6)
20. (+8.4) – (+11.2)

B. Subtract, following these examples:

–8	–8	+11	+11
S –6	A +6	S – 4	A + 4
	–2		+15

21. + 6
S + 3
22. – 8
S – 2
23. – 9
S – 11
24. – 6
S + 9
25. + 12
S – 5
26. + 11
S + 15
27. – 14
S + 9
28. + 7
S – 9
29. – 5
S + 12
30. + 14
S + 21

C. Simplify each expression, following the examples given. First change all subtraction operations to addition operations, and then combine the resulting signed numbers.

S	S
(+6) – (+9) + (–8)	(–11 + (+6) – (–5)
A	A
= (+6) + (–9) + (–8)	= (–11) + (+6) + (+5)
= (–3) + (–8)	= (–5) + (+5)
= –11	= 0

31. (–11) – (–5) – (+6)
32. (+12) – (+4) – (–8)
33. (+20) + (–3) – (+8)
34. (+14) + (–11) – (+4)
35. (–15) – (–2) + (–6)
36. (–7) – (–9) + (–8)
37. (–9) – (+4) – (–8)
38. (–11) – (+6) – (–7)
39. (–5 3/4) – (–2 1/2)
40. (+8 6/7) – (+9 5/8)
41. (–3 1/5) + (–2 7/10) – (+9 3/5)
42. (–2 3/4) – (–3 5/12) – (+2 1/2)
43. (–6 1/2) + (+7 3/4) + (+8 1/4)
44. (+5 3/5) – (+7 2/5) + (–6 4/5)
45. (+7.3) – (–9.2) + (–12.4)
46. (–4.5) + (+8.6) – (–0.3)

D. Write a mathematical expression for each of the following exercises and then simplify the expression.

47. From the sum of (–8) and (+17) subtract (–3).
48. From the sum of (+11) and (–19) subtract (–5).
49. Find the difference between (–7) and (+12).
50. Find the difference between (+9) and (–6).
51. Uncle Milton's favorite stock lost 3/8 point on four days in a row last week. What was the total change in the stock's price?
52. A football team ran two plays with losses of 4 yards and 9 yards each and then completed a 22-yard pass. What was the net yardage change on these three plays?

E. Draw a number line showing the answer to each of these combinations.

53. (–4) + (+8) – (+7)

54. (+6) – (+3) + (–2)

55. (–5) – (+7) – (–4)

56. (0) – (+4) + (–3)

57. (+4) – (+6) + (–2)

58. (–3) + (–2) – (–1)

59. 0 – (–3) – (+7)

60. (+5) + (–3) – (+4)

61. (+3) + (–5) – (+2)

62. (–2) – (–6) + (–4)

F. Use your calculator to simplify each of the following expressions.

63. (–24.95) – (–13.268)

64. (+175.863) – (–205.93)

65. (–39.84) – (+42.765)

66. (+492.76) – (+508.375)

67. (–21.83) – (–19.276) + (–14.95)

68. (+123.9) – (+247.3) – (+142.8)

G. Read and respond to the following exercises.

69. Write an algebraic expression that describes the following situation, and then find the answer to the questions.

Your bank account has a balance of \$270. You write two checks, one for \$125 and another for \$82. Then you make a deposit of \$165 and write one more check for \$94. What is your ending balance after all of these transactions?

70. On Monday afternoon at 4PM when the TV weatherman was preparing for his evening broadcast, the temperature in Beam, Kentucky was 73º. During the next four days when he checked the temperature at 4 PM, he noticed the following changes: a drop of 8 degrees, then an increase of 12 degrees, an increase of 7 degrees and then a drop of 5 degrees. Write an algebraic expression that describes these temperature changes and then report the temperature reading on Friday of that week from the information you have been given.

71. Give your own definition of absolute value. Explain why |–6| = |+6|.

72. Explain to a fellow student how simplifying 3/5 ÷ 2/7 is similar to (–7) – (+4).

11.4 Multiplying Signed Numbers

The rules for multiplying signed numbers are fairly simple to develop. Consider the following pattern:

(+ 3)(+ 3) = +9

–1 ↓ ↓ –3

(+ 3)(+ 2) = +6

Note that a decrease of 1 in a factor produces a decrease of 3 in the answer (product).

(+ 3)(+ 1) = +3

A decrease of 1 in a factor produces a decrease of 3 in the answer.

(+ 3)(0) = 0

A decrease of 1 in a factor produces a decrease of 3 in the answer.

$(+3)(-1) = -3$ This must be correct because a decrease of 1 in a factor produces a decrease of 3 in the answer.

$-1 \downarrow \quad \downarrow -3$

$(+3)(-2) = -6$ A decrease of 1 in a factor again produces a decrease of 3 in the answer.

You already know that $3 \times 3 = 9$ and that $3 \times 0 = 0$, so you must have established a good working pattern to go by. From this pattern, you can see that the following rules hold true.

$+ \quad \times \quad +$

A positive × A positive = A positive

$+ \quad \times \quad -$

A positive × A negative = A negative

You know that multiplication is a commutative operation. For example, $6 \times 7 = 7 \times 6$. Therefore, if $(+3)(-2) = -6$, $(-2)(+3)$ must also equal -6. From this statement it would seem that

$- \quad \times \quad +$

A negative times a positive = A negative

You have only one case left to consider. What is a negative times a negative equal to? Let us again look at a pattern. You accept that $(-2)(+3) = -6$. Begin there.

$(-2)(+3) = -6$ A negative × A positive = A negative

$(-2)(+2) = -4$ Because you are multiplying by -2 this time, it would seem that a decrease of 1 in a factor produces an increase of 2 in the answer. You know that -4 is larger than -6.

$(-2)(+1) = -2$ A decrease of 1 in a factor produces an increase of 2 in the answer. (You know that -2 is larger than -4.)

$(-2)(0) = 0$ A decrease of 1 in a factor produces an increase of 2 in the answer. Also, you know that any number times 0 equals 0.

$-1 \downarrow \quad \downarrow +2$

(–2) (–1) = +2 — A decrease of 1 in the factor produces an increase of 2 in the answer.

(–2) (–2) = +4 — Given the pattern you have observed, this must be true.

It would seem from the pattern that

– × –

A negative times a negative = A positive

We can now summarize the rules for multiplication of signed numbers.

Multiplying Signed Numbers

When two factors with like signs are multiplied, the answer is positive

When two factors with unlike signs are multiplied, the answer is negative.

You can summarize these rules by using symbols as well.

+ × + = +

+ × – = –

– × + = –

– × – = +

Study the following sample problems.

(–8)(–2) = +16	(–7)(+4) = –28
(+6) (+5) = +30	(+4)(–3) = –12
(+7)(0) = 0	(0)(–8) = 0
(–1/4)(+24) = –6	(+2.3) (–1.4) = –3.22

Self-Check Try these problems to see if you understand the basic multiplication rules.

1. (–8)(+3) 2. (–2)(–5) 3. (+4) (+9)
4. (–7)(0) 5. (+9)(–5)

Answers:

1. –24 2. +10 3. +36 4. 0 5. –45

If a multiplication problem has more than two factors, you have two choices: (1) You can multiply from left to right, following the order of operations and dealing with the signs as you figure each individual product, or (2) you can determine the sign of the answer first and then multiply the quantities together.

Look at these problems, paying attention to the number of negative factors in each and the answer that you get.

$(-3)(+4) = -12$	1 negative factor = Negative answer
$(-3)(-5) = +15$	2 negative factors = Positive answer
$(-4)(-3)(-2) = -24$	3 negative factors = Negative answer
$(-2)(-2)(-2)(-2) = +16$	4 negative factors = Positive answer

It would seem that *an even number of negative factors produces a positive answer. An odd number of negative factors produces a negative answer.* This is true, and it enables you to determine the sign of the product (answer) first and then to multiply without worrying about the signs.

Odd-Even Rule for Signs

When multiplying or dividing, an even number of negative factors produces a positive result, and an odd number of negative factors produces a negative result.

Consider the following problems that are done both ways, and decide which you prefer. Either method is fine.

EXAMPLE 1 Simplify: $(-6)(+2)(-5)$

Solution

Method 1	Method 2	
$(-6)(+2)(-5)$	$(-6)(+2)(-5)$	Two negative factors, so the answer will be positive.
$= (-12)(-5)$	$= +(12)(5)$	
$= +60$	$= +60$	

EXAMPLE 2 Simplify: $(+6)(-5)(+3)$

Solution

Method 1	Method 2	
$(+6)(-5)(+3)$	$(+6)(-5)(+3)$	One negative factor, so the answer will be negative.
$= (-30)(+3)$	$= -(30)(3)$	
$= -90$	$= -90$	

EXAMPLE 3 Simplify: $(-2)(-3)(-4)$

Solution

Method 1	Method 2	
$(-2)(-3)(-4)$	$(-2)(-3)(-4)$	Odd number of negative factors, so the answer will be negative.
$= (+6)(-4)$	$= -(6)(4)$	
$= -24$	$= -24$	

EXAMPLE 4 Simplify: $(+3)(-4)(-2)(+1)$

Solution

Method 1	Method 2	
$(+3)(-4)(-2)(+1)$	$(+3)(-4)(-2)(+1)$	Even number of negative factors, so the answer will be positive.
$= (-12)(-2)(+1)$	$= +(3)(4)(2)(1)$	
$= (+24)(+1)$	$= +(12)(2)(+1)$	
$= +24$	$= +(24)(1)$	
	$= +24$	

Normally it is easier to determine the sign of the product first and then multiply.

When exponents are involved, the rules for counting negative signs are a real help. Remember that in multiplication, an even number of negative factors produces a positive answer, whereas an odd number of negative factors produces a negative answer. And exponents are just a shortcut notation for repeated multiplication. Study the following examples.

EXAMPLE 5 Simplify: $(-2)^5$

Solution

$$\begin{aligned} (-2)^5 &= (-2)(-2)(-2)(-2)(-2) \\ &= (+4)(-2)(-2)(-2) \\ &= (-8)(-2)(-2) \\ &= (+16)(-2) \\ &= (-32) \end{aligned}$$

Or $(-2)^5$ has an odd number of negative signs and will therefore give a negative answer.

$$\begin{aligned} (-2)^5 &= \text{negative } (2)^5 \\ &= -32 \end{aligned}$$

EXAMPLE 6 Simplify: $(-1)^4$

Solution

$$\begin{aligned} (-1)^4 &= (-1)(-1)(-1)(-1) \\ &= (+1)(-1)(-1) \\ &= (-1)(-1) \\ &= +1 \end{aligned}$$

Or $(-1)^4$ has an even number of negative factors and will therefore produce a positive answer.

$$\begin{aligned} (-1)^4 &= \text{positive } (1)^4 \\ &= +1 \end{aligned}$$

Of course, any number of positive factors will always produce a positive answer.

Study the next two examples to see the difference between $(-3)^2$, *negative three* raised to the second power, and -3^2, the opposite of 3 *raised to the second power.*

EXAMPLE 7 Simplify: $(-3)^2$

Solution

$$(-3)^2 = (-3)(-3)$$
$$= +9$$

EXAMPLE 8 Simplify: -3^2

Solution

-3^2 = The opposite of 3 squared = $-(3)(3)$

$$= -(9)$$
$$= -9$$

An exponent, or power, applies only to its base. In Example 8, only the 3 is the base. In Example 7 the base is negative 3, (–3). The parentheses make the difference in these two examples. Be very careful when working problems like these.

EXAMPLE 9 Simplify: -3^4

Solution -3^4 means to take the opposite of 3 raised to the fourth power. $3^4 = 81$, and the opposite of 81 is negative 81. Therefore, $-3^4 = -81$.

EXAMPLE 10 Simplify: $(-3)^4$

Solution In this problem the base is –3, so that whole quantity is raised to the fourth power.

$(-3)^4 = (-3)(-3)(-3)(-3) = 81.$

EXAMPLE 11 Simplify: $-22 - 3(-8)$

Solution In doing a simplification like Example 11, a shortcut can be applied. Notice that there is only one negative symbol *between* the 22 and the 3. This symbol can either be interpreted as the negative sign on the number 3, or be as the subtraction operation, but not both, because there is only one symbol. Look at both simplifications below. In the first, the – symbol is used to indicate subtraction.

$-22 - 3(-8)$

S $= -22 - (+3)(-8)$	If the – means to subtract, then the 3 is positive.
S $= -22 - (-24)$	Multiply before subtracting.
A $= -22 + (+24)$	Add the opposite.
$= +2$	Difference of the numbers, sign of the "larger"

If you consider the – symbol to mean a negative, you multiply (–3) by (–8) to get (+24) and then combine the result with the (–22).

$-22 - 3(-8)$	Multiply before combining.
$= -22 + 24$	Signs are different, so find the difference of the values and use the sign of the larger.
$= +2$	

EXAMPLE 12 Simplify: $(-6)(+7) - 4(-8)$

Solution

$(-6)(+7) - 4(-8)$	Multiply first, using the shortcut.
$= -42 + 32$	Now combine numbers with unlike signs.
$= -10$	

EXAMPLE 13 Simplify: $16 - 3(+4) - (-8)(-5)$

Solution

$16 - 3(+4) - (-8)(-5)$	Multiply using the first – symbol as a negative. The second – symbol must mean subtraction because there are two –'s between the 4 and the 8.
$= 16 - 12 \overset{S}{-} (+40)$	Use the subtraction rule.
$= 16 - 12 \overset{A}{+} (-40)$	Combine from left to right.
$= 4 \overset{A}{+} (-40)$	
$= -36$	

EXAMPLE 14 Simplify using the Order of Operations Agreement: $(-5)^2 - 2[7 - (-2)]$

Solution

$(-5)^2 - 2[7 - (-2)]$

$= (-5)^2 - 2[7 + (+2)]$

$= (-5)^2 - 2[9]$

$= 25 - 18$

$= 7$

11.4 EXERCISES

A. Find the product indicated in each of the following exercises.

1. $(+11)(+5)$
2. $(+7)(+9)$
3. $(-8)(+6)$
4. $(-7)(+4)$
5. $(-18)(-3)$
6. $(-22)(-6)$
7. $(+4)(-9)$
8. $(+8)(-3)$
9. $(-8)(0)(-4)$
10. $(-8)(-5)(0)$
11. $(-3)(-2)(-5)$
12. $(-6)(-2)(-3)$
13. $(+7)(-2)(+3)$
14. $(+9)(-3)(+2)$
15. $(-4)(-6)(0)$
16. $(-7)(0)(+6)$
17. $(-1)(-2)(-3)(-4)$
18. $(+2)(-3)(-5)(+4)$
19. $(+2)(-2)(-2)(+3)$
20. $(-1)(-1)(5)(+1)$

B. Simplify each of the following expressions using the Order of Operations Agreement (Section 8.3) and the rules for signed numbers. Remember to use S and A where they will help.

21. $23 - (+4)(-2)$
22. $37 - (-3)(+8)$
23. $17 + (-3)(+6)$
24. $12 + (-7)(+3)$
25. $(-6)(+2) + (-3)(+4)$
26. $(+7)(-2) + (-3)(+6)$
27. $(-6)(-2) - (-3)(-5)$
28. $(-8)(-4) - (-6)(-5)$
29. $12 - 6(3 - 8)$
30. $22 - 11(2 - 6)$

31. $-16 - 4(-3)$
32. $-19 + (-8)(+2)$
33. $+11 - 5(-8)$
34. $(+16)(-3) - 4(+5)$
35. $(+8)(-2) - 6(-3)$
36. $(-7)(-4) - 5(+6)$
37. $-2[16 - 3(+2)]$
38. $15 - 3[7 - 4(3)]$
39. $21 + 2(-3)(+4)$
40. $-16 - 4(-5)(-2)$

C. Simplify each of the following expressions involving exponents.

41. -3^2
42. -2^4
43. -4^3
44. -1^5
45. $(-2)^3$
46. $(+5)^2$
47. $(+2)^4$
48. $(-1)^6$
49. $(-1)^5(-2)^2$
50. $(-2)^3(-3)^2$
51. $(+3)^3(-2)^2$
52. $(-4)^3(+2)^3$
53. $(-1)^7 - 3(-2)^4$
54. $4(+2)^3 - (-3)^3$

D. Simplify each of the following expressions involving fractions by using the Order of Operations Agreement.

55. $(-1/2)(5/9) - (2/3)(-7/9)$
56. $(-1\ 3/4) - (2/3)(9/8)$
57. $(-7\ 1/4) - (-5/8)(7/2)$
58. $(2/5)^2 - 7\ 9/10$
59. $(-3/4)^2 + (-4\ 3/8)$
60. $(7/12)(-2/3) - (-5\ 5/6)$

E. Use your calculator to simplify each of the following expressions. Round answers to the nearest hundredth.

61. $(-2.45)^3$
62. $(-7.59)^4$
63. $(0.76)^2(-24.5)^3$
64. $(-16.4)^3(-2.45)^2$
65. $(-19.3)(-4.76)^3$
66. $(-9.1)^2(-17.65)$
67. $(17.3)(19.02) - (-16.3)(4.5)$
68. $(-21.4)(0.52) - (17.4)(-11.1)$
69. $(-12.4)(-2.35)(-17.8)$
70. $(-13.5)(+21.6)(-12.7)$
71. $27.5 - 16.2(8.75 - 11.3)$
72. $31.9 + 12.4(16.5 - 19.2)$
73. $14.27 + 3.9(-15.4 + 7.59)$
74. $42.7 - 24.3(18.4 - 27.1)$

F. Read and respond to the following exercises.

75. If 5(4) means five fours, or 4 + 4 + 4 + 4 + 4, explain in words why $3(-2) = -6$.
76. Explain in words the difference between the meaning of $(-6)(4)$ and $(-6)^4$.
77. Explain in your own words why $(-5)^4$ is $+625$ and -5^4 is -625.
78. Describe, in words, the process used to go from one step to the next.

$23 - 5[18 - 4(-3)]$	
$23 - 5[18 + 12]$	1.
$23 - 5[30]$	2.
$23 - 150$	3.
-127	4.

11.5 Dividing Signed Numbers

The division process for signed numbers is a simple one. Because dividing can be defined as "multiplying by the reciprocal," and because the sign of a number and its reciprocal are the same, the rules for division are exactly the same as those for multiplication. Looking at the process we use to check a division problem will help you understand the rules. You know that to check a division problem you multiply the quotient by the divisor and add on the remainder, if there is one. For example,

To check 28 ÷ 7 = 4, we see if 4 times 7 = 28.

The signs for 28, 7, and 4 are all positive, so

Positive ÷ Positive = Positive

Consider now what happens when you divide +28 by (–7).

(+ 28) ÷ (–7) = ?

? times (–7) must equal (+28), so ? must be (–4) because you know that (–4) times (–7) = +28. Therefore,

(+28) ÷ (–7) = –4 because (–4) times (–7) = +28.

Positive ÷ Negative = Negative

Consider now what happens when you divide (–28) by (+7).

(–28) ÷ (+7) = ?

? times (+ 7) must equal (–28), so ? must be (–4) because you know that (–4) times (+7) = –28. Therefore,

(–28) ÷ (+7) = –4 because (–4) times (+7) = –28.

Negative ÷ Positive = Negative

The only case left to look at is (–28) ÷ (–7).

(–28) ÷ (–7) = ?

The check tells you that ? times (–7) must equal (–28), so ? must be +4 because (+4) times (–7) = –28.

(–28) + (–7) = +4 because (+4) times (–7) = –28.

Negative ÷ Negative = Positive

These statements can be summarized in the rules that follow.

Dividing Signed Numbers

In division, two like signs produce a positive quotient (answer). Two unlike signs produce a negative quotient.

For example,

$(-16) \div (+4) = -4$ $\quad (+28) \div (+4) = +7$

$(-20) \div (-5) = +4$ $\quad (-42) \div (-7) = +6$

$(+12) \div (-2) = -6$ $\quad (+84) \div (-7) = -12$

$(-8/3) \div (-5/4) = (-8/3) \times (-4/5) = +32/15$

$\frac{-6}{+3} = -2 \qquad \frac{+12}{+4} = +3 \qquad \frac{-9}{-2} = \frac{9}{2}$ or 4 1/2

Note: You never leave a negative symbol in a denominator of a fraction answer. It is too easy for them to be lost or overlooked. Always calculate the final result based on like signs or unlike signs.

Self-Check Try these division problems to check your understanding of the division rules.

1. $(-12) \div (+6)$ 2. $(+18) \div (-2)$ 3. $(+40) \div (+4)$
4. $(-36) \div (-9)$ 5. $(0) \div (-3)$ 6. $(+11) \div (-5)$

Answers:

1. –2 2. –9 3. +10 4. +4 5. 0 6. –11/5 or –2 1/5

When multiplication and division are combined within the same problem, the Odd-Even Rule for Signs applies. *An even number of negative signs gives a positive answer, and an odd number of negative signs gives a negative answer.*

Consider these examples.

EXAMPLE 1 Simplify: $\frac{(-8)(+5)}{-10}$

Solution $\frac{(-8)(+5)}{-10} = \frac{-40}{-10} = +4$

Because this problem is *all* multiplication or division, you could first count the number of negative signs to determine the sign of the answer, and then calculate the numbers.

$\frac{(-8)(+5)}{-10} = 4$ $\quad$ 2 negatives ⟶ positive answer

EXAMPLE 2 Show the two ways of simplifying $\frac{(-6)(-4)(-8)}{(-3)(-2)}$.

Solution $\frac{(-6)(-4)(-8)}{(-3)(-2)} = \frac{(+24)(-8)}{+6} = \frac{-192}{+6} = -32$

Or, because there are 5 negative numbers in this problem and the operations are all either multiplication or division, the result will be negative (5 is an odd number).

$$-\frac{(\overset{2}{\cancel{6}})(\overset{2}{\cancel{4}})(8)}{(\underset{1}{\cancel{3}})(\underset{1}{\cancel{2}})} = -32$$

Therefore, the answer is –32.

When you have more than one operation in a problem, follow the Order of Operations Agreement (Section 8.3). Note that the fraction bar acts as a grouping symbol. You need to simplify the numerator and denominator separately and then perform the indicated division.

EXAMPLE 3 Simplify: $\frac{(-4)(-3)}{-2} - \frac{(-2)(+6)}{+3}$

Solution

$$\frac{(-4)(-3)}{-2} - \frac{(-2)(+6)}{+3}$$

$$= \frac{(+12)}{-2} - \frac{(-12)}{+3}$$

S

$$= (-6) - (-4)$$

A

$$= (-6) + (+4)$$

$$= -2$$

EXAMPLE 4 Simplify: $\frac{(-20) + (-5)}{(-20) - (-5)}$

Solution

$$\frac{(-20) + (-5)}{(-20) - (-5)}$$

Note: We cannot use the Odd-Even Rule for Signs when addition or subtraction are also present in the expression.

$$= \frac{(-20) + (-5)}{(-20) + (+5)}$$

$$= \frac{-25}{-15}$$

$$= \frac{+5}{3} \text{ or } +1\ 2/3$$

EXAMPLE 5 Find the quotient of (–8 1/3) divided by (+2 1/2).

Solution

(–8 1/3) ÷ (+2 1/2)	Rewrite as improper fractions.
= (–25/3) ÷ (+5/2)	Multiply by the reciprocal of the divisor.
= (–25/3) × (2/5)	Simplify by dividing out common factors. The answer will be negative.
$= -\frac{\overset{5}{\cancel{25}}}{3} \times \frac{2}{\underset{1}{\cancel{5}}}$	Simplify.
$= -\frac{10}{3}$ or – 3 1/3	

11.5 EXERCISES

A. Simplify each of the following expressions.

1. $(-10) \div (+5)$
2. $(+20) \div (-4)$
3. $(+8) \div (-2)$
4. $(-12) \div (+3)$
5. $-16/-4$
6. $-30/-6$
7. $\dfrac{+8}{-20}$
8. $\dfrac{+15}{-20}$
9. $\dfrac{-12}{+5}$
10. $\dfrac{-14}{+3}$
11. $\dfrac{(-3)(-12)}{+4}$
12. $\dfrac{(-5)(-8)}{+10}$
13. $(-6\ 2/3) \div (+1\ 1/3)$
14. $(-5\ 1/3) \div (+2\ 2/3)$
15. $(+2.4) \div (-0.06)$
16. $(+3.5) \div (-0.07)$
17. $\dfrac{-2+8}{-3}$
18. $\dfrac{-9+3}{+2}$
19. $\dfrac{(-7)-(-3)}{(-7)+(+3)}$
20. $\dfrac{(-8)-(-2)}{(-8)+(-2)}$

B. Simplify using the rules for signed numbers and the Order of Operations Agreement (Section 8.3).

21. $\dfrac{-8}{-2} \div \dfrac{+6}{+2}$
22. $\dfrac{-12}{-3} \div \dfrac{-20}{+2}$
23. $\dfrac{-12}{-4} \div \dfrac{-10}{+8}$
24. $\dfrac{(-8)(-3)(+4)}{(-6)(+10)}$
25. $-15 - \dfrac{+12}{-4}$
26. $18 - \dfrac{+10}{-5}$
27. $-14 - \dfrac{-8}{+4}$
28. $-10 - \dfrac{+12}{-3}$
29. $(-3)(-8) - (+6)(-2)$
30. $(-7)(+6) - (-5)(-4)$
31. $(-7)(+8) - (-3)(-2)$
32. $(-9)(+5) - (-4)(-5)$
33. $(-2)^4 \div (+2)^3$
34. $(-3)^5 \div (+3)^3$
35. $(-3)^6 \div (-3)^3$
36. $(-4)^3 \div (-2)^5$
37. $\dfrac{(-5)(-4)(-3)}{(-6)(-10)}$
38. $\dfrac{-18}{+6} \div \dfrac{+9}{-3}$
39. $\dfrac{17 + (-5)(+2)}{-6+4}$
40. $\dfrac{21 + (-3)(+7)}{-4+5}$
41. $\dfrac{(-3)(-8) - 2(+6)}{(-3)(-4)}$
42. $\dfrac{(-4)(-5) - 3(-8)}{(-2)(+6)}$
43. $\dfrac{12 - 5(-3)}{12 + 5(-3)}$
44. $\dfrac{20 - 6(-2)}{20 - 6(+2)}$
45. $\dfrac{-8 - 4 + 6}{-3 + 4}$
46. $\dfrac{-10 + 8 - 6}{7 - 3}$
47. $42 - 63 \div (-7) + 4(-5)$
48. $-18 - 12 \div (+3) - 5(-2)$
49. $\dfrac{-16}{+4} \div \dfrac{+12}{-3} + \dfrac{-10}{+2}$
50. $\dfrac{-36}{-9} \div \dfrac{-6}{+12} - \dfrac{(-8)}{-2}$

C. Write a mathematical expression for each of the following exercises and then simplify the expression.

51. From the product of (–7), (+6), and (–3) subtract (–9)
52. From the quotient of (–54) and (+6) subtract 11.
53. From the sum of (–6) and (+14) subtract the quotient of (–24) and (–3).
54. Find the difference between the product of (–12) and (+6) and (–5).
55. Find the quotient of the sum and difference of (–8) and (+12).

56. Find the sum of the product and quotient of (–36) and (+4).

D. Read and respond to the following exercises.

57. Explain how you determine the final sign in a problem involving only multiplication and division.

58. In your own words, describe why the sign rules for multiplication and division are the same.

59. Simplify $(-4)^3 \div -4^2$ and explain each step in words.

60. Simplify –3 3/8 ÷ 4 1/2 and explain each step in words.

11.6 Evaluating Expressions Using Signed Numbers

You substituted numbers into formulas in Chapter 9 and earlier chapters. You will now do the same thing to evaluate expressions using signed numbers. Review the Order of Operations Agreement first.

Order of Operations Agreement

When simplifying a mathematical expression involving more than one operation, follow these steps:

1. Simplify within parentheses or grouping symbols, using steps 2, 3, and 4 below in order.
2. Simplify any expressions that involve exponents or roots.
3. Multiply or divide **in order, from left to right.**
4. Add or subtract (combine), **in order, from left to right.**

Study the following examples. Note that when you substitute a value for a variable, it is a good idea to use parentheses to keep the sign with the number. This practice often eliminates a lot of confusion. You can remove the parentheses as you work through the simplification steps.

EXAMPLE 1 Evaluate abc when $a = -2$, $b = +3$, and $c = -4$.

Solution

abc

$= (-2)(+3)(-4)$

$= (-6)(-4)$

$= +24$

EXAMPLE 2 Evaluate $4b^2$ when $b = +3$.

Solution

$4b^2$

$= 4(+3)^2$

$= 4(+9)$

$= +36$

EXAMPLE 3 Evaluate $bd - ac$ when $a = -2$, $b = +3$, $c = -4$, and $d = -1$.

Solution $bd - ac$

S

$= (+3)(-1) - (-2)(-4)$ — Do not rewrite the subtraction until the multiplication has been done.

S

$= (-3) - (+8)$

A

$= (-3) + (-8)$

$= -11$

EXAMPLE 4 Evaluate $a - 3c$ when $a = -2$ and $c = -4$.

Solution $a - 3c$

S

$= (-2) - 3(-4)$

S

$= (-2) - (-12)$

A

$= (-2) + (+12)$

$= +10$

Look at Example 4 again. You can consider the –3 as negative 3 and multiply –4 by –3, writing the product's sign as the operation sign. Then combine the results.

$a - 3c$	Substitute –2 for a and –4 for c.
$(-2) - 3(-4)$	Multiply (–3)(–4).
$= -2 + 12$	Now combine.
$= +10$	

Treating this minus sign as a direction sign instead of as an operation sign is a shortcut. It usually saves steps.

EXAMPLE 5 Evaluate $a + 2b - 3c$ when $a = -2$, $b = +3$, and $c = -4$.

Solution

$a + 2b - 3c$

S

$= (-2) + 2(+3) - 3(-4)$

S

$= (-2) + (+6) - (-12)$

A

$= (-2) + (+6) + (+12)$

$= (+4) + (+12)$

$= +16$

or

$a + 2b - 3c$

$= (-2) + 2(+3) - 3(-4)$

$= (-2) + 6 + 12$

$= 4 + 12$

$= +16$

Follow through these next two examples to be sure you understand this alternative. It is usually shorter, and it always works.

EXAMPLE 6 Evaluate $cb - 5d$ when $b = +3$, $c = -4$, and $d = -1$.

Solution

$cb - 5d$

$= (-4)(+3) - 5(-1)$

$= -12 + 5$

$= -7$

EXAMPLE 7 Evaluate $-5c - 4b + 2a$ when $a = -2$, $b = +3$, and $c = -4$.

Solution

$-5c - 4b + 2a$

$= (-5)(-4) - 4(+3) + 2(-2)$

$= +20 - 12 - 4$

$= 8 - 4$

$= +4$

EXAMPLE 8 Evaluate $\frac{a^2 - b^2}{cd}$ when $a = -2$, $b = +3$, $c = -4$, and $d = -1$.

Solution

$\frac{a^2 - b^2}{cd}$

S

$= \frac{(-2)^2 - (+3)^2}{(-4)(-1)}$

S

$= \frac{4 - (+9)}{+4}$

A

$= \frac{4 + (-9)}{+4}$

$= \frac{-5}{+4} = \frac{-5}{4}$

A negative divided by a positive gives a negative answer, so the answer can be written as $-(5/4)$. Usual algebra practice is not to leave a negative value in the denominator, so if the calculations had yielded $+5/-4$, you would also have written $-(5/4)$.

Self-Check Evaluate these expressions when $x = 2$, $y = -3$ and $z = -1$.

1. xyz 2. $x - 2y$ 3. $-3x + 2y - z$ 4. $-xy^2 + 2z$

Answers:

1. +6 2. +8 3. −11 4. −20

11.6 EXERCISES

A. Find the value of each of the following expressions when a = –2, b = +4, c = –3, d = –1, and e = 0.

1. $a + b + c$
2. $d + bc$
3. $b + ad$
4. $a - b - c$
5. abc
6. bde
7. $2a - 3b$
8. $3c - 2d$
9. $a + 3b - 4c$
10. $a + 2b - 3c$
11. $4a - 5c$
12. $2b - 3d$
13. $3e - 4c$
14. $6e + 2d$
15. a^2c
16. b^2d
17. $5c^2$
18. $-3d^2$
19. $4a^3$
20. $2c^3$
21. $\frac{ab}{cd}$
22. $\frac{bc}{ad}$
23. $\frac{ab + cd}{c}$
24. $\frac{bd - ac}{d}$
25. $a^2 + 2b$
26. $c^2 - 2b$
27. $2b^2 - 3b + c$
28. $3a^2 + 2a - b$
29. $\frac{a^2 - b^2}{a^2 + b^2}$
30. $\frac{c^2 + d^2}{c^2 - d^2}$
31. $|a| - 5|c|$
32. $|b| \cdot |d| - a$
33. $|c| \cdot |a| - c$
34. $|e| - 4|d|$
35. $|a| \cdot |b| - |c| \cdot |d|$
36. $\frac{|a| + |d|}{|b| - |c|}$

B. Find the value of each of the following expressions when $w = 0.7$, $x = 1/4$, $y = -2/5$, and $z = -2.5$.

37. $wx - yz$
38. $x - 7z$
39. $3x + 10y$
40. $yz - 2wx$
41. xyz
42. $w + x + y$
43. $-2x + 3z$
44. $3w - 2z + y$
45. $3y - 2x^2 - 4z$
46. $4w - 20y^2$
47. $|x| \cdot |y|$
48. $|w| \div |z|$
49. $|x| + |y| - |w| \cdot |z|$
50. $x|z| - y|w|$

C. Each exercise that follows has an error in the solving process. State what the error is; then evaluate correctly. Use these values for the variables: $p = -6$, $r = +5$, $s = -3$, $t = +0.5$, and $u = 0$.

51. $pr - tu$

$(-6)(+5) - (+0.5)(0)$

$(-30) - (+0.05)$

-30.05

52. $rt - 3p$

$(+5)(-3) - 3(-6)$

$(-15) + 18$

$+3$

53. $r^2 - tp$

$(+5)^2 - (+0.5)(-6)$

$(+10) - (-3.0)$

$(+10) + (+3.0)$

$+13$

54. $\frac{pr}{st}$

$\frac{(-6)(+5)}{(-3)(+0.5)}$

$\frac{-30}{-0.15}$

$+200$

55. $p - r + s - t$

$(-6) - (+5) + (-3) - (+0.5)$

$(-6) + (-5) + (+3) + (-0.5)$

$(-11) + (+3) + (-0.5)$

$(-8) + (-0.5)$

-8.5

56. rst

$(+5)(-3)(-0.5)$

$(-15)(-0.5)$

$+7.5$

57. $p + s$

$(-6) + (-3)$

$+9$

58. $-s - 4p + 8t$

$(-3) - 4(-6) + 8(+0.5)$

$(-3) + 24 + 4.0$

$21 + 4$

25

11.7 CHAPTER REVIEW

Now that you have completed this chapter, you should be able to:

1. Describe or talk about these terms and expressions: integer, absolute value, negative number, positive number, the odd and even rules for signs in multiplication, and division, exponent, and division problems.
2. Plot values on a number line.
3. Translate verbal expressions into signed-number expressions and simplify them when possible.
4. Use the following rules for operations with signed numbers:

Combining Signed Numbers

If the numbers being added have the same sign, combine their absolute values and use the common sign.

If the numbers being added have different signs, find the difference between the absolute values and use the sign of the number with the larger absolute value.

Subtracting Signed Numbers

To subtract signed numbers, change the sign of the number being subtracted, and change the operation to addition.

Multiplying Signed Numbers

In the multiplication of two factors with like signs, the answer is positive.

In the multiplication of two factors with unlike signs, the answer is negative.

Dividing Signed Numbers

In division, two like signs produce a positive quotient (answer). Unlike signs produce a negative quotient.

Odd-Even Rule for Signs

When multiplying or dividing, an even number of negative factors produces a positive result, and an odd number of negative factors produces a negative result.

5. Substitute signed numbers into algebraic expressions.
6. Use the Order of Operations Agreement to simplify expressions involving signed numbers.

REVIEW EXERCISES

A. Complete the following sentences by filling in the blanks.

1. A signed number has both _____ and _____.
2. 0 has _____ sign.
3. –6 is _____ than –4. (less than or greater than?)
4. –3 is _____ than –8. (less than or greater than?)
5. A gain of 6 pounds is represented by _____.
6. A loss of $14 is represented by _____.
7. When adding or combining numbers with like signs, _____.
8. When adding or combining numbers with unlike signs, _____.
9. When multiplying signed numbers, you can determine that the answer is positive if there is an _____ number of negative signs in the problem.
10. When dividing signed numbers, you can determine that the answer is negative if there is an _____ number of negative signs in the problem.
11. $(-2)^4$ will have a _____ answer.
12. $(+3)^5$ will have a _____ answer.
13. $(-5)^3$ will have a _____ answer.
14. $(-1)^7 =$ _____.
15. -3^4 will have a _____ answer.
16. -2^5 will have a _____ answer.
17. What are the four steps in the Order of Operations Agreement?
18. In $-4p^3$, if p has the value –2, take –2 to the _____ power first, and then _____ by –4.
19. In evaluating $+3a - 2b$ when $a = +4$ and $b = +5$, it is easier to think of _____ times +5, rather than thinking of +2 times +5, before subtracting.
20. Zero times any quantity equals _____.

B. Graph the following numbers on a number line.

21. –3
22. +2 1/2
23. +1 1/4
24. 0
25. 3
26. –5
27. +1/3
28. –5/8
29. –1.5
30. +4.2

C. Perform the indicated operations.

31. $-6 + 9$
32. $4 - 8$
33. $(-3)(5)$
34. $(-2)(-4)$
35. $(-45) \div (+9)$
36. $(-18) \div (-3)$
37. $(-6)(0) + (-2)(+3)$
38. $(4)(-3) - (2)(-1)$
39. $-8 - 4 + 5 - 2$
40. $-11 + 6 - 5 - 4$
41. $+8 - (+11)$
42. $-9 - (-3)$
43. $\frac{-9}{+3}$
44. $\frac{+12}{-3}$
45. -2^4
46. -5^2
47. $(-2)^2(-1)^5$
48. $(-3)^3(+2)^3$
49. $-3(-2) - 2(4)$
50. $5(-1) - 3(-2)$
51. $\frac{-12}{-6} \div \frac{+3}{-2}$
52. $\frac{-30}{+6} \div \frac{+12}{-2}$
53. -3^2
54. -2^6
55. $(-4)^2 \div (-3)^3$
56. $(+5)^3 \div (-2)^4$
57. $-3/4(+12) - 1/2(-6)$
58. $2/3(-12) - 3/4(-16)$
59. $(1/2)^3(-2/5)^2$
60. $(3/4)^2(-2/3)^3$
61. $(-5/8)(-3/10)$
62. $(-4/5)(-1/2)(-5/8)$
63. $|-11| - 2|-7|$
64. $|-8| \cdot |+2| - |-5|$
65. $-6 + 4\ 2/5$
66. $-11 - 2\ 3/4$
67. $(-2.4)(+3.01)$
68. $(-1.7)(-2.05)$
69. $\frac{(-6) + (-3)(+2)}{(-4)(-3)}$
70. $\frac{8 + (-4)(+5)}{(2)(-6)}$

D. Evaluate each of the following expressions when $w = -2$, $x = +3$, $y = 0$, and $z = -4$.

71. xyz
72. wxz
73. $x + y - z$
74. $w + x - z$
75. $wx - wy$
76. $wy - wz$
77. $2w + 3x$
78. $-4w + 2z$
79. $-4x^2$
80. $-2z^3$
81. $-4xz$
82. $5wx$
83. $-3z - 2y$
84. $-6w + 5z$
85. $\frac{wx}{z}$
86. $\frac{wz}{x}$
87. $\frac{x + w}{x - w}$
88. $\frac{x + z}{x - z}$
89. $\frac{w^2 - x^2}{w^2 + x^2}$
90. $\frac{x^2 - z^2}{x^2 + z^2}$
91. $|x|$
92. $-|z|$
93. $|x| + 2|z|$
94. $|w| - |x| + |z|$
95. $w - |w|$
96. $x - 3|w|$
97. $|w| - z$
98. $(|w|)(|z|)$
99. $(|x|)(|y|)$
100. $|w| + |x| - |y| - |z|$

E. Using your calculator, simplify each of the following expressions.

101. $7(-8) - (-3)(+9)$
102. $(-3)^4 + (-5)^5$
103. $(0.26)^2(-1.4)^3$
104. $(-3)(+7)(-9)(-15)(+12)$
105. $|-25| \cdot |+17| \cdot |0|$
106. $|-4.2| + |-8|$
107. $|17.5| - |-12|$
108. $14.9 - 7.2(+5.3) - (-3.1)^2$

F. Each exercise that follows has an error in the solving process. State what the error is; then work the problem correctly.

109. $(-8) + (-7) = +15$
110. $(-12) - (+6) = -6$
111. $(-3)^5 = -15$
112. $(-1)^6 = +6$
113. $(+12) \div (-3) = +4$
114. $+18 - 6(-4) = -6$
115. $-30 - (-2)(+5) = -40$
116. $-6^4 = +1{,}296$

G. Read and respond to the following exercises.

117. A friend is just beginning to work on this chapter. Give her some hints about how best to learn this material.
118. Turn to page 364 in Section 9.4 and rework Example 1. This time, however, begin by adding $-11N$ to both sides of the equation. See if you can solve the equation and find again that $N = +1$ is the solution.

119. Turn to page 400 in Section 10.2 and rewrite consecutive number Example 10 so that that sum of the three numbers is –69. Write and solve the equation and see if you can find that the numbers are –24, –23, and –22. How is this solution different from the one you found when you worked the problem originally?

120. Describe, in words, the process used to go from one step to the next.

$(-16 \div (-8) - 7[12 - 4(9 - 4)]$	
$= (-16) \div (-8) - 7[12 - 4(5)]$	1.
$= (-16) \div (-8) - 7[12 - 20]$	2.
$= (-16) \div (-8) - 7[-8]$	3.
$= +2 - 7[-8]$	4.
$= +2 + 56$	5.
$= +58$	6.

CHAPTER TEST

A. Find the value of each of the following expressions.

1. $-8 + 5$
2. $(-3) - (-7)$
3. $(-3)(-4)(-2)$
4. $\frac{+21}{-3}$
5. $(-4)^3$
6. $2(-8) + 3(-4)$
7. $(+6) - (+11)$
8. $(-5) + (-2) - (+3)$
9. $(-24) \div (-3)$
10. $(-2)^3(+3)^2$
11. $21 - 4[(-2) + (-3)]$
12. $3(-6) - 2(+5)$

B. Evaluate each of the following expressions when $d = +3$, $e = -2$, $f = -1$, and $g = 0$.

13. def
14. $d - ef$
15. $fg + df$
16. $-2f^2 - 3f$
17. $\frac{d+e}{d-e}$
18. $|d| - |f| + |e|$

19. The example that follows has errors in the simplification process. State what the errors are, and then work the problem correctly.

$14 - 3[(-2) - (-4)]$

$14 - 3[(-2) + (-4)]$

$11[(-6)]$

-66

20. If $|M| = 3$, explain in words how you know what value(s) M could have.

Cumulative Review, Chapters 1–11

Find the answer to each of the following problems. Show all your work. Simplify all fraction answers to lowest terms.

In Exercises 1 through 5, add.

1. $11.56 + 3.8 + 18$
2. $11\ 4/9 + 7\ 2/3$
3. $5x + 8y + x + 3y$
4. $(-15) + (8) + (-3)$
5. $29 + (-17)$

In Exercises 6 through 10, subtract.

6. $(-15) - (-12)$
7. $0 - 8 - (-5)$
8. $4\ 1/4 - 2\ 5/6$
9. $18 - 3.56$
10. $12a - 3a - 18a$

In Exercises 11 through 15, multiply.

11. 0.176×100
12. $(-14)(-3)$
13. $5\ 2/3 \times 2\ 1/8$
14. Find 4/5 of 36.
15. $(-2)(3)(-1)(2)$

In Exercises 16 through 20, divide.

16. $(-32) \div (-8)$
17. $\dfrac{1.68}{0.04}$
18. $2\ 5/8 \div (-3)$
19. $\dfrac{-54}{9}$
20. $0.8\overline{)24}$

Simplify Exercises 21 and 22 using the Order of Operations Agreement.

21. $3 + 2[6 - 2(4 \div 2)]$
22. $\dfrac{25}{36} \div \left(\dfrac{2}{3}\right)^2 \div \dfrac{5}{9}$

Find the value for each of the expressions in Exercises 23 through 25 by substituting the given values.

23. $ab^2 - ac$ when $a = 2$, $b = -1$, and $c = -3$
24. $\dfrac{2bc}{b^2 - c^2}$ when $a = 2$, $b = -1$, and $c = -3$
25. $2ab - 4c + 3ac$ when $a = 2$, $b = -1$, and $c = -3$

In Exercises 26 through 28, write an algebraic expression for each of the English phrases.

26. The difference between a number and six
27. seven more than twice a number
28. the product of a number and eiht

In Exercises 29 and 30, write an algebraic equation for each of the English sentences.

29. The quotient of 15 and a number is –5.
30. The total cost (C) is equal to the product of the number of tickets (N) and $5.85.

Solve each of the following equations and inequalities.

31. $7Y = 56$
32. $\dfrac{15}{N} = \dfrac{75}{13}$
33. 16% of 90 is what number?
34. $3x + 5 = 20$
35. $\dfrac{3}{5}M + 6 = \dfrac{2}{3}$
36. $6A - 2 + 3A = 59 - 7$
37. $5.6W + 0.12 = 8W$
38. 18 is what percent of 90?
39. $2X > 14$
40. $4N - 5 = 2N + 9$

Solve Exercises 41 and 42 as requested.

41. Solve for W in $P = 2L + 2W$ when $P = 100$ inches and $L = 29$ inches.

42. Solve for M in $S = C + M$ when $S = \$129.95$ and $C = \$74.25$.

In Exercises 43 through 50, find the answer(s) to each of the questions asked. You may need to write an algebraic equation or a proportion to help you find the solution(s).

43. Sales tax on a $28 item is $1.54. What is the sales tax rate (percent)?

44. The sum of the three angles of any triangle is 180°. If one angle measures 39° and the second angle measures 78°, what is the measure of the third angle?

45. The sum of three consecutive odd integers is –75. What are the three integers?

46. Elena earns $2.25 per hour when she babysits for the O'Toole children. If she babysat for 6 1/2 hours, how much money did she earn?

47. If Tom can drive 341 miles in 5 1/2 hours, how far can he drive in 1 hour?

48. A10 3/4-ounce can of soup will be mixed with an equal amount of water, heated, and then divided into five equal servings. How many ounces will there be in each serving?

49. Using the formula $A = \pi r^2$, find the area of a circle whose radius is 10 inches. Use $\pi \doteq 3.14$.

50. Find the dimensions of a rectangle whose perimeter is 110 yards if the length is 4 times the width.

STUDY SKILLS MODULE

12 Writing in the Math Classroom

You have undoubtedly noticed by now that you are often asked to describe, in words, the approach you take to solve a problem, or to explain why you followed a particular procedure, or to discuss why an answer seems reasonable or not. There are several very good reasons for using writing in a math class. One reason is the belief that if you can explain something to someone else, you will better understand it yourself. This type of realization often happens with teachers. In trying to respond to a student's question, the teacher sees the topic in a new way. In writing a response, students have to formalize their own thinking about a topic and they often see a relationship that was previously unclear. When you write something down, you become precise and, by necessity, pick out the important points.

A second reason for writing in the math class is the carryover to problem solving in real life. At work and at home, problems rarely involve simply doing a calculation with a set of numbers. Instead, information is presented or gathered; it is examined and organized; then decisions about the situation are made; and, finally, a solution is reached. Although these problem-solving steps do not always involve writing, they do require the ability to communicate both with yourself and with others. Good communication is one of the skills you are practicing in written-response questions in mathematics. Writing down thoughts and procedures in problem solving helps you process what is there, and might help you recognize what steps to perform, or what facts or formulas may be needed, as well as how to use them.

A third benefit of writing in a math classroom involves comprehension of the special vocabulary of mathematics. Some terms that have very specific meanings in math have different meanings in everyday language (like *rational* or *product*.) In addition to knowing and understanding the vocabulary, you also have to understand the meaning of symbols. Being able to use the vocabulary and symbols in written responses helps you understand their meaning. If you can communicate clearly, the instructor can see any misunderstanding you have and can help you correct it.

Writing in the math classroom adds another dimension to the learning process. It makes you more actively involved in the process. You cannot let your mind wander as easily when you are writing. It is harder to fool yourself into thinking you understand something if you cannot write it down. Being able to write out a response, as when you are explaining it to others, lets you see that you do understand a concept.

Your Point of View

1. Describe a situation where you needed to make a decision and actually wrote down the pros and cons to help you make up your mind.
2. Write a note to another student describing a concept that you are having difficulty with (be very specific and clear) and ask for his or her suggestions.
3. Make a list of 15 math symbols or vocabulary terms and write a short explanation of each.
4. Write a detailed explanation of how to prepare a container of orange juice from a can of frozen concentrate. Then look at what you have written and decide if someone who has never done this task could complete the job following your directions. Did you include enough detail? Were your steps given in correct order?

CHAPTER

Expressions, Equations, and Inequalities

LEARNING OBJECTIVES

When you have completed this chapter, you should be able to:

1. Simplify expressions using the Distributive Principle.
2. Apply the Order of Operations Agreement to simplify expressions that include more than one set of grouping symbols.
3. Follow the Steps to Solve Equations procedures to solve and check first-degree equations that contain grouping symbols.
4. Solve first-degree inequalities, and graph the solution sets on number lines.
5. Organize information, write, and solve an equation or inequality for application problems.

To solve more complex equations, you will need to bring together many of the concepts you mastered in Chapters 8, 9, and 11, as well as several new ones. You will also use many of these concepts again to solve inequalities and application problems.

12.1 Distributive Principle

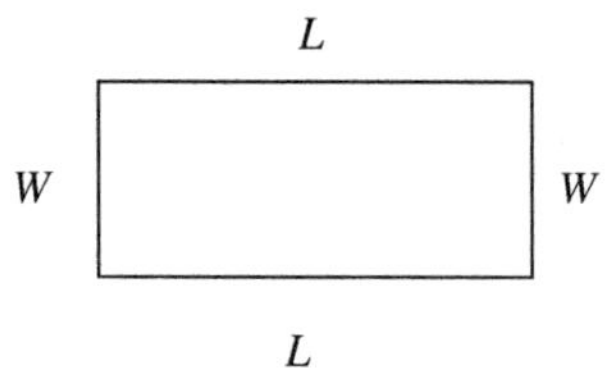

FIGURE 12.1

Do you remember the formula for finding the perimeter of a rectangle (Figure 12.1)? It was stated in several different ways:

$P = L + W + L + W$

or

$P = 2L + 2W$ Combined like terms

It could also have been written

$P = 2(L + W)$

Let us compare the last two forms of the formula and check to see that they are equivalent. If $L = 8$ feet and $W = 5$ feet, then

$P = 2(L + W)$	or	$P = 2L + 2W$
$P = 2(8 + 5)$		$P = 2(8) + 2(5)$
$P = 2(13)$		$P = 16 + 10$
$P = 26$ feet		$P = 26$ feet

You should conclude that $2(L + W)$ and $2L + 2W$ are the same; they are equivalent statements.

$2(L + W) = 2L + 2W$

This example illustrates a property of numbers called the **Distributive Principle**. You studied this property in both Sections 2.4 and 8.4. This rule concerns multiplication of a sum or difference. The Distributive Principle indicates that you get the same answer when you multiply first and then add (or subtract) as you get when you add (or subtract) first and then multiply. Read the examples that follow, remembering that when you multiply, you must multiply every term inside the parentheses by the factor that is outside.

EXAMPLE 1 Are $3(7 + 4)$ and $3(7) + 3(4)$ equal?

Solution

Add.	$3(7 + 4)$	$3(7) + 3(4)$	Multiply.
Multiply.	$3(11)$	$21 + 12$	Add.
	33	33	

Therefore, because $33 = 33$, $3(7 + 4) = 3(7) + 3(4)$.

EXAMPLE 2 Are $-5(4 + 9 - 3)$ and $(-5)(4) + (-5)(9) - (-5)(3)$ equal?

Solution

Add.	$-5(4 + 9 - 3)$	$(-5)(4) + (-5)(9) - (-5)(3)$	Multiply.
Subtract.	$-5(13 - 3)$	$-20 + (-45) - (-15)$	Add.
Multiply.	$-5(10)$	$(-65) - (-15)$	Subtract.
	-50	$(-65) + (+15)$	Combine.
		-50	

Therefore, the expressions are equivalent because $-50 = -50$.

Distributive Principle

For any numbers represented by a, b, and c,

$a(b + c) = ab + ac$

and

$a(b - c) = ab - ac$

The Distributive Property can be very useful to you in everyday situations. Suppose you need to multiply 3 times 26 and have neither pencil and paper nor calculator with you. Mentally picture the problem 3(26) as 3(20 + 6). You can multiply 3×20 and 3×6 in your head and add those quantities together.

$3(26) = 3(20 + 6) = 3(20) + 3(6) = 60 + 18 = 78$

This process will work with even larger numbers:

$8(357) = 8(300 + 50 + 7) = 8(300) + 8(50) + 8(7)$

$= 2{,}400 + 400 + 56 = 2{,}856$

Here are two other sample problems that make use of the Distributive Principle.

$3(7 - 5)$
$= 3(7) - 3(5)$
$= 21 - 15$
$= 6$

$-5(6 + 9)$
$= -5(6) + (-5)(9)$
$= -30 + (-45)$
$= -75$

In many algebraic expressions involving the Distributive Principle, you will not be able to do the addition or subtraction first because the terms will not be like terms. In $-3(x + 2)$, the x and the 2 cannot be added together; therefore, the only operation that can be done is multiplication: $-3(x + 2) = (-3)(x) + (-3)(2) = -3x + (-6)$. Using the definition of subtraction, you can write this expression in a simpler form as $-3x - 6$. You can use a shortcut when performing the distributive multiplication. When multiplying $-3(x + 2)$, treat the operation sign with the 2 as if it were a signed–number sign. You will be multiplying $(-3)(+x)$ and $(-3)(+2)$. You should get

$-3x - 6$, using the signed-number answer with 6 as if it were the operation sign. It sounds more confusing than it is! The following examples illustrate the shortcut.

EXAMPLE 3 Simplify: $6(5 - 2x)$

Solution $6(5 - 2x)$ $(+6)(+5)$ and $(+6)(-2x)$

$= 30 - 12x$

EXAMPLE 4 Simplify: $-11(-4y + 5)$

Solution $-11(-4y + 5)$ $-11(-4y)$ and $-11(+5)$ produces the preferred simplification without double signs such as $44y + (-55)$.

$= 44y - 55$

EXAMPLE 5 Simplify: $8(5P - 3R + 7T - 4)$

Solution $8(5P - 3R + 7T - 4)$ $8(5P)$, $8(-3R)$, $8(+7T)$, and $8(-4)$

$= 40P - 24R + 56T - 32$

EXAMPLE 6 Simplify: $-(x - 5)$

Solution $-(x - 5)$ Write a 1 before the parentheses and multiply through by -1:

$= -1(x - 5)$ $(-1)(+x)$ and $(-1)(-5)$

$= -1x + 5$ or $-x + 5$

In the examples and exercises in this section, you are simplifying algebraic expressions. Keep in mind that they are expressions, not equations. Equations can be solved; expressions (no equals sign) can only be simplified. We simplify by removing grouping symbols and combining like terms.

$3(y + 2)$ is not the same as $3(y + 2) = 15$.

$3(y + 2) = 15$ can be solved to find a value for y, whereas $3(y + 2)$ can only be simplified.

Algebraic Expression

Algebraic expressions contain terms and constants but have no equality symbol. Expressions cannot be solved, only simplified.

Examples of expressions:

$x + 2$ $3a - 2b + c$ $-2m$

Algebraic Equation

Algebraic equations contain terms and constants and have a symbol of equality. Equations can be solved.

Examples of equations:

$x + 3 = 7$ $\quad 2a - 7 = 5a + 4$

Self-Check Simplify each expression.

a. $4(a - 3)$ b. $-7(5 - 2k)$ c. $-(2y + 9)$

Answers:

a. $4a - 12$ b. $-35 + 14k$ c. $-2y - 9$

12.1 EXERCISES

A. Are these expressions equivalent? Prove your answer by simplifying each expression and showing whether the answers are the same. Use the following example:

$3(2 + 5)$ and $3(2) + 3(5)$

$3(7)$ $\quad 6 + 15$

21 $\quad 21$

Therefore, $3(2 + 5) = 3(2) + 3(5)$ because $21 = 21$.

1. $-11(1 + 3)$ and $-11(1) + (-11)(3)$
2. $5(8 - 5)$ and $5(8) - 5(-5)$
3. $4(10 - 6)$ and $4(10) - 4(6)$
4. $-9(7 + 5)$ and $(-9)(7) + (-9)(5)$
5. $\frac{2}{3}(9 - 6)$ and $\frac{2}{3}(9) - \frac{2}{3}(6)$
6. $\frac{3}{4}(16 - 12)$ and $\frac{3}{4}(16) - \frac{3}{4}(12)$
7. $-3(8 - 11)$ and $-3(8) - (-3)(8)$
8. $-2(15 - 24)$ and $(-2)(15) - (-2)(24)$
9. $0.4(1.7 - 1.2)$ and $0.4(1.7) - (0.4)(1.2)$
10. $1.5(2.1 - 3.6)$ and $1.5(2.1) - 1.5(3.6)$

B. Simplify the following expressions using the Distributive Principle.

11. $5(K + 2)$
12. $11(3 - 2y)$
13. $-2(5x + 1)$
14. $-7(6m + 5)$
15. $-(4 - 6c)$
16. $-(8T - 3)$
17. $-4(7A - 9)$
18. $3(5A + 6)$
19. $7(x + 6)$
20. $-5(9c - 13)$
21. $7(M + 5)$
22. $5(T + 12)$
23. $3(X + 2)$
24. $-2(y + 4)$
25. $-5(m + 4)$
26. $7(R + 5)$
27. $6(K - 2)$
28. $-5(X - 3)$
29. $-8(Y - 4)$
30. $15(B - 3)$
31. $17(2x - 3 + 4y)$
32. $8(9M - 5 + 7N)$
33. $3.2(5N + 2.1P)$
34. $-11(4S - 3R + 6)$
35. $-12(3A - 5B + 2C)$
36. $1.1(4.2G - 3.2H)$
37. $\frac{3}{8}(24A - 16C)$
38. $\frac{5}{6}(18x - 30y)$
39. $-\frac{1}{3}(12M + 9N)$
40. $-\frac{1}{5}(25A + 90B)$

C. Use your calculator to help you simplify the following expressions.

41. $0.65(3.8F - 0.06)$
42. $384(1{,}952X + 641)$
43. $-5.6(11.2 - 18.2W + 39.8X)$
44. $0.07(7.5C + 2.8 - 11.7E)$
45. $915(3{,}497M + 9{,}567N)$
46. $-52.7(16.7X - 10.9)$
47. $3\frac{5}{8}\left(\frac{7}{16}y - \frac{2}{3}\right)$
48. $\frac{4}{3}\left(15\frac{1}{4} + 7x\right)$
49. $13\frac{5}{9}\left(9\frac{5}{6} + 11\frac{2}{3}T\right)$
50. $\frac{6}{7}\left(49y + 3\frac{1}{3}\right)$

D. Read and respond to the following exercises.

51. Describe, in words, the process you could use to multiply large numbers together without using paper and pencil or calculator. Use an example such as 9×83.
52. Name two mathematical operations that are involved in the Distributive Principle. What do the parentheses in the expression tell you?

12.2 Expressions with Grouping Symbols

Grouping symbols such as parentheses, ( ), brackets, [ ], and braces, { }, are used in algebraic expressions or equations to show that terms are related to each other or that they have the same operation performed on them. For example, in $3(x + 2)$, the parentheses indicate that both x and 2 are to be multiplied by 3. In $6 - (Y + 4)$, the parentheses show that the entire quantity $Y + 4$ must be subtracted from 6. Without the parentheses in this example, $6 - Y + 4$, only the Y would be subtracted and the 4 would be added. Any time you wish to multiply, subtract, or divide a quantity that contains more than one term, you *must* enclose that quantity in parentheses.

EXAMPLE 1 Is $3(x + 2)$ equal to $3x + 6$?

Solution $3(x + 2)$ is equal to $3x + 6$ because the parentheses say to multiply everything inside by 3.

Without the parentheses, $3x + 2$, only the x would be multiplied by 3. The answers would not be the same:

$3(x + 2) = 3x + 6$ but $3x + 2 \neq 3x + 6$.

EXAMPLE 2 Is $6 - (y + 4)$ equal to $2 - y$?

Solution

$6 - (y + 4)$ — Simplify this expression by applying the rule of signed-number subtraction.

S

$= 6 - [(+y) + (+4)]$ — Rewrite the subtraction, changing the signs of *each* of the terms being subtracted.

A

$= 6 + [(-y) + (-4)]$

$= 6 + (-y) + (-4)$ — Combine like terms.

$= 2 + (-y)$ — Eliminate double signs by using the definition of subtraction.

$= 2 - y$

Therefore, $6 - (y + 4) = 2 - y$.

◀

In Example 2, the process of doing the subtraction was very awkward. You can get the same result by applying the Distributive Principle. Because multiplying by -1 and subtracting both result in changed signs, you can write a 1 in front of $(y - 4)$ and treat it as a distributive problem. Multiply each term in the parentheses by -1.

$6 - (y + 4)$	
$= 6 - 1(y + 4)$	Multiply by -1: $(-1)(y)$ and $(-1)(+4)$.
$= 6 - y - 4$	Combine like terms.
$= 2 - y$	

This process enables you to remove the parentheses and make the sign changes required in subtraction, but it is less complicated to do. Compare these processes:

Rewrite Subtraction Process	**Distributive Process**
S	
$7 - (3x - 4)$	$7 - (3x - 4)$
A	
$= 7 - [3x + (-4)]$	$= 7 - 1(3x - 4)$
S	
$= 7 - [3x + (-4)]$	$= 7 - 3x + 4$
A	
$= 7 + (-3x) + (+4)$	$= 7 + 4 - 3x$
$= 7 + (+4) + (-3x)$	$= 11 - 3x$
$= 11 + (-3x)$	
$= 11 - 3x$	

In the examples that follow, the Distributive Principle is used to remove parentheses, and then the like terms are combined. In Section 11.6, you learned that you can combine terms by using the operation signs as signed number signs. In the rest of the examples, this shortcut is used when like terms are combined.

EXAMPLE 3 Simplify: $5(A - B) + 4(A + B)$

Solution

$5(A - B) + 4(A + B)$	Multiply $5(A)$, $5(-B)$, $4(A)$, and $4(B)$.
$= 5A - 5B + 4A + 4B$	Combine like terms: $5A$ and $+4A$, $-5B$ and $+4B$.
$= 9A - 1B$ or $9A - B$	

◀

EXAMPLE 4 Simplify: $4(8x - 3) - 2(5 + 2x)$

Solution

$4(8x - 3) - 2(5 + 2x)$ Multiply $4(8x)$, $4(-3)$, $-2(5)$, and $-2(2x)$.

$= 32x - 12 - 10 - 4x$ Combine like terms.

$= 28x - 22$

EXAMPLE 5 Simplify: $4(y - 2) - (6y + 5)$

Solution

$4(y - 2) - (6y + 5)$ Write 1 before the second parentheses.

$= 4(y - 2) - 1(6y + 5)$ Multiply $4(y)$, $4(-2)$, $-1(6y)$, and $-1(+5)$.

$= 4y - 8 - 6y - 5$ Combine like terms.

$= -2y - 13$

EXAMPLE 6 Simplify: $3 - 2(5A - 4)$

Solution

$3 - 2(5A - 4)$ Multiply $(-2)(5A)$ and $(-2)(-4)$.

$= 3 - 10A + 8$ Combine like terms.

$= 11 - 10A$ or $-10A + 11$

EXAMPLE 7 Simplify: $3M + (5M - 4)$

Solution

$3M + (5M - 4)$ Write in 1 and multiply through by +1.

$= 3M + 1(5M - 4)$

$= 3M + 5M - 4$ Combine like terms.

$= 8M - 4$

EXAMPLE 8 Simplify: $9(2z - 6) - 4(3z - 8)$

Solution

$9(2z - 6) - 4(3z - 8)$

$= 18z - 54 - 12z + 32$

$= 6z - 22$

EXAMPLE 9 Simplify: $5(3r + 4 - 8r) - (11 + 2r)$

Solution

$5(3r + 4 - 8r) - (11 + 2r)$

$= 5(3r + 4 - 8r) - 1(11 + 2r)$

$= 15r + 20 - 40r - 11 - 2r$

$= -27r + 9$

Self-Check Simplify these problems and check your answers.

1. $15 - (m + 3)$ 2. $2(x - 3) + 5(2x + 4)$ 3. $-4(B - 8) - 2(5B + 4 - 7B)$

Answers:

1. $-m + 12$ 2. $12x + 14$ 3. 24

Expressions that contain grouping symbols within another set of grouping symbols require the use of different kinds of symbols so that you can tell where one group begins and the other ends. In addition to parentheses ( ), brackets [ ] and braces { } are often used.

When simplifying expressions that contain grouping symbols within other grouping symbols, always begin with the *innermost* set of symbols and work outward. As you simplify expressions like this, be sure to rewrite all parts of the expression that you have not worked with yet.

$7 - [4n - 5(n + 6)]$	First eliminate the parentheses by multiplying through by -5.
$= 7 - 1[4n - 5n - 30]$	Eliminate the brackets by multiplying through by -1.
$= 7 - 4n + 5n + 30$	Combine like terms.
$= 37 + n$ or $n + 37$	

In the previous problem, you could combine like terms after the first step and have a simpler expression to multiply by -1.

$7 - [4n - 5(n + 6)]$	Multiply by -5 to eliminate parentheses first.
$= 7 - [4n - 5n - 30]$	Combine inside the brackets.
$= 7 - [-n - 30]$	Now multiply by -1.
$= 7 - 1[-n - 30]$	
$= 7 + n + 30$	Combine like terms.
$= 37 + n$ or $n + 37$	

EXAMPLE 10 Simplify: $5[7x + 4(3x - 6)]$

Solution

$5[7x + 4(3x - 6)]$	Multiply by +4.
$= 5[7x + 12x - 24]$	$= 5[7x + 12x - 24]$

Combine inside next.	or	Multiply by 5 next.
$= 5[19x - 24]$		$= 35x + 60x - 120$
Then multiply by 5.		Then combine like terms.
$= 95x - 120$		$= 95x - 120$

The result is the same.

EXAMPLE 11 Simplify: $6 - \{12[y + 3(2y - 4)]\}$

Solution

$6 - \{12[y + 3(2y - 4)]\}$ — Multiply by +3 to remove parentheses first.

$= 6 - \{12[y + 6y - 12]\}$

	or	
$= 6 - \{12[7y - 12]\}$	or	$= 6 - \{12y + 72y - 144\}$
$= 6 - \{84y - 144\}$		$= 6 - 12y - 72y + 144$
$= 6 - 84y + 144$		$= 150 - 84y$ or $-84y + 150$
$= 150 - 84y$ or $-84y + 150$		

Skills Check Simplify each expression completely.

a. $3[2y - 6(5y - 4)]$ b. $-5[W - (4W + 1)]$

Answers:

a. $-84y + 72$ b. $15W + 5$

12.2 EXERCISES

A. Simplify the follwing expressions by removing the grouping symbols and combining like terms.

1. $11 - 6(x + 2)$
2. $3(2m - 5) + 4(m + 1)$
3. $7 - 5(k - 3)$
4. $5C - (C + 2)$
5. $2(3A - B) - (A - B)$
6. $-3(y + 2) - 2(y - 5)$
7. $6R - (2R - 4)$
8. $5(s + 2) - (s - 3)$
9. $-2(T - 5) - 4(2T + 3)$
10. $5 - 2(w + 2)$
11. $-2(m - 4) + 3(2m - 1)$
12. $2(x + 4) - (2x - 1)$
13. $8B - (B + 2)$
14. $15 - 2(y - 3)$
15. $-3(R + 1) - 5(2R - 3)$
16. $9 - (6c + 4)$
17. $18 - (7 + w)$
18. $3[4m - 2(m + 3)]$
19. $7 - 2[4N - (3 + 2N)]$
20. $9 - [6 - (S + 3)]$
21. $2[7Y - 3(2Y - 1)]$
22. $6[3(2A + 1)]$
23. $2[3(4x - 1)]$
24. $15 + [3N - 2(4N - 3)]$
25. $11 - [5(w + 4) - 6]$
26. $5[7(R - 4) + 9]$
27. $-2[3y - (5y - 2)]$
28. $2A - 3[A - 2(4 - A)]$
29. $12(y - 2) + 2(3y - 7)$
30. $4(a - 3b) - 3(a + b)$

31. $-3[2x + (x - 7)]$
32. $-5[3g + 2(5 - g)]$
33. $3(a - 2b) - 5(4a + 7b)$
34. $2(y - 3x) + 2(7x - y)$
35. $-2[3H - (5H - 4)] + 3H$
36. $2\{4[c - 2(c - 3)]\}$
37. $3\{2[p + 2(p + 7)]\}$
38. $-3\{6p - 2[2p - (p + 1)]\}$
39. $-\{2w - 3[w - 2(4 - w)]\}$
40. $-\{-5a + 2[3a - 7(3 - 2a)]\}$

B. Describe, in words, the process used to go from one step to the next.

41. $2(y + 4) + 3(2y - 5)$

$2y + 8 + 3(2y - 5)$	1.
$2y + 8 + 6y - 15$	2.
$8y + 8 - 15$	3.
$8y - 7$	4.

42. $3x - 2(x - 4)$

$3x - 2x + 8$	1.
$x + 8$	2.

43. $8A - 3(A - 5)$

$8A - 3A + 15$	1.
$5A + 15$	2.

44. $7(2c + 3) + 5(6 - c)$

$14c + 3 + 5(6 - c)$	1.
$14c + 3 + 30 - 5c$	2.
$9c + 3 + 30$	3.
$9c + 33$	4.

45. $6V - (2 - 5V)$

$6V - 1(2 - 5V)$	1.
$6V - 2 + 5V$	2.
$11V - 2$	3.

46. $14y - (8 - 5y)$

$14y - 1(8 - 5y)$	1.
$14y - 8 + 5y$	2.
$19y - 8$	3.

47. $-2(7x - 2[3x + 2(x + 4)])$

$-2(7x - 2[3x + 2x + 8])$	1.
$-2(7x - 6x - 4x - 16)$	2.
$-14x + 12x + 8x + 32$	3.
$6x + 32$	4.

48. $12 - [5F - 3(F + 2)]$

$12 - [5F - 3F - 6]$	1.
$12 - [2F - 6]$	2.
$12 - 2F + 6$	3.
$-2F + 18$	4.

C. Read and respond to the following exercises.

49. Describe, in words, how subtracting the quantity in parentheses in $4 - (7T + 5)$ is similar to multiplying the quantities in the parentheses by -1.
50. Describe, in words, the steps you would use to simplify $6(2k - 7) - 3(3 - 2k)$.
51. Explain, in words, why $14 - (2y - 4)$ and $14 - 1(2y - 4)$ are the same.
52. Why do some of the exercises in this section use three different kinds of grouping symbols?

12.3 Equations with Grouping Symbols

To solve more complex first-degree equations, you will use the concepts you learned in the two previous sections. You will also have to recall the procedures for solving equations. All of the following processes will be applied when solving the equations in this section.

1. When simplifying expressions that have more than one set of grouping symbols, begin by simplifying the innermost set and then work outward.

2. In expressions such as $-5(x - 2)$, apply the Distributive Principle and multiply each term inside the parentheses by -5.
3. After eliminating all grouping symbols, follow these Steps to Solve Equations:
   a. If desired, multiply every term in the equation by a number that will eliminate the fractions or decimals.
   b. Combine like terms on each side of the equals sign.
   c. Do steps that require addition or subtraction of variable and constant terms.
   d. Do steps that require multiplication or division.
   e. Always check answers by substituting the answer into the *original* equation.
4. Apply appropriate signed-number operation rules to all the foregoing steps.

Study the examples that follow to see where each step is used.

EXAMPLE 1 Solve and check: $3(x + 2) + 5 = 38$

Solution

$3(x + 2) + 5 = 38$ Distributive Principle

$3x + 6 + 5 = 38$ Combine like terms.

$3x + 11 = 38$ Do addition or subtraction.

$\underline{-11} \quad \underline{-11}$

$3x = 27$

$\frac{3x}{3} = \frac{27}{3}$ Do multiplication or division.

$x = 9$

Check

$3(x + 2) + 5 = 38$ Substitute into the *original* equation.

$3(9 + 2) + 5 = 38$

$3(11) + 5 = 38$

$33 + 5 = 38$

$38 = 38$ True.

EXAMPLE 2 Solve and check: $-24 = 3[2t - 4(t - 6)]$

Solution

$-24 = 3[2t - 4(t - 6)]$ Remove parentheses by multiplying by -4.

$-24 = 3[2t - 4t + 24]$ Remove brackets by multiplying by 3.

$-24 = 6t - 12t + 72$ Combine like terms.

$-24 = -6t + 72$ Subtract 72.

$\underline{-72} \quad \underline{-72}$

$-96 = -6t$ Divide by -6, since it is multiplying t and so that t ends up with a positive coefficient.

$\frac{-96}{-6} = \frac{-6t}{-6}$

$16 = t$

Check

$-24 = 3[2t - 4(t - 6)]$ Substitute into the *original* equation.

$-24 = 3[2(16) - 4(16 - 6)]$ Simplify using Order of Operations Agreement.

$-24 = 3[32 - 4(10)]$

$-24 = 3[32 - 40]$

$-24 = 3[-8]$

$-24 = -24$ True.

When solving an equation, be sure that the final step shows the variable with a positive one (+1) coefficient. In Example 2, both sides were divided by the coefficient and its sign: $-6t/-6$ and $-96/-6$. This resulted in $16 = (+1)t$, but with the +1 understood. Never leave the coefficient of the variable as anything but +1. You have not finished solving until the coefficient is +1. Follow through Example 3 to see this process.

EXAMPLE 3 Solve and check: $2(2k + 1) - 5(k - 2) = 36$

Solution

$$\begin{aligned} 2(2k + 1) - 5(k - 2) &= 36 \\ 4k + 2 - 5k + 10 &= 36 \\ -k + 12 &= 36 \\ \underline{-12} \quad & \quad \underline{-\ 12} \\ -k &= 24 \end{aligned}$$

You are not finished because the coefficient of k is -1.

$$\frac{-1k}{-1} = \frac{24}{-1}$$

$$k = -24$$

Check

$$\begin{aligned} 2(2k + 1) - 5(k - 2) &= 36 \\ 2[2(-24) + 1] - 5(-24 - 2) &= 36 \\ 2[-48 + 1] - 5(-26) &= 36 \\ 2[-47] - 5(-26) &= 36 \\ -94 + 130 &= 36 \\ 36 &= 36 \quad \text{True.} \end{aligned}$$

Substitute in the original equation and simplify.

EXAMPLE 4 Solve and check: $7x - (x + 2) = -2(x - 7)$

Solution

$$\begin{aligned} 7x - (x + 2) &= -2(x - 7) \\ 7x - 1(x + 2) &= -2(x - 7) \\ 7x - x - 2 &= -2x + 14 \end{aligned}$$

Remember to distribute the number in front of the parentheses. You can insert a 1. Then distribute -1 and -2.

$$\begin{aligned} 6x - 2 &= -2x + 14 \\ +2x \quad & \quad\ +2x \\ 8x - 2 &= \quad 14 \\ +2 \quad & \quad\ +2 \\ 8x &= 16 \\ \frac{8x}{8} &= \frac{16}{8} \\ x &= 2 \end{aligned}$$

You could eliminate any one of the terms now. This is just one choice.

Check

$$\begin{aligned} 7x - (x + 2) &= -2(x - 7) \\ 7(2) - (2 + 2) &= -2(2 - 7) \\ 14 - 4 &= -2(-5) \\ 10 &= 10 \quad \text{True.} \end{aligned}$$

EXAMPLE 5 Solve and check: $6[T - 3(T + 2)] = -36$

Solution

$$\begin{aligned} 6[T - 3(T + 2)] &= -36 \\ 6[T - 3T - 6] &= -36 \\ 6T - 18T - 36 &= -36 \\ -12T - 36 &= -36 \\ +36 \quad & \quad +36 \\ -12T &= 0 \\ \frac{-12T}{-12} &= \frac{0}{-12} \\ T &= 0 \end{aligned}$$

Check

$$\begin{aligned} 6[T - 3(T + 2)] &= -36 \\ 6[0 - 3(0 + 2)] &= -36 \\ 6[0 - 3(2)] &= -36 \\ 6[0 - 6] &= -36 \\ 6[-6] &= -36 \\ -36 &= -36 \quad \text{True.} \end{aligned}$$

EXAMPLE 6 Solve and check: $2(y - 4) = 7 + 2y$

Solution

$$\begin{aligned} 2(y - 4) &= 7 + 2y \\ 2y - 8 &= 7 + 2y \\ -2y \quad & \quad\ -2y \\ -8 &= 7 \quad \text{No solution} \end{aligned}$$

Because this can never be true, $-8 \neq 7$, there is no solution to this problem. When the variable disappears from the problem and a false statement (such as $-8 - 7$) is left, the equation has no solution. In other words, it has an empty solution set, which is represented as { } or ϕ (null set). No number substituted for y could give a true statement.

EXAMPLE 7 Solve and check: $2x + 4 = 2(2 + x)$

Solution

$$2x + 4 = 2(2 + x)$$
$$2x + 4 = 4 + 2x$$
$$\underline{-2x} \qquad \underline{-2x}$$
$$4 = 4$$

All real numbers

When the variable disappears and a true statement (an identity such as $4 = 4$) results, then any and all real numbers could be the solution to the equation. The solution is stated as "all real numbers."

Check In the original equation, test a few values such as $x = 1$ and $x = -8$.

$2x + 4 = 2(2 + x)$	$2x + 4 = 2(2 + x)$
$2(1) + 4 = 2(2 + 1)$	$2(-8) + 4 = 2[2 + (-8)]$
$2 + 4 = 2(3)$	$-16 + 4 = 2[-6]$
$6 = 6$	$-12 = -12$

Any real number would produce a true statement.

EXAMPLE 8 Solve and check: $3x + 4 = 2x + 5 + x$

Solution

$$3x + 4 = 2x + 5 + x$$
$$3x + 4 = 3x + 5$$
$$\underline{-3x} \qquad \underline{-3x}$$
$$4 = 5$$

No solution because $4 \neq 5$.

EXAMPLE 9 Solve and check: $2(Y - 6) + 4 = -8 + 2Y$

Solution

$$2(Y - 6) + 4 = -8 + 2Y$$
$$2Y - 12 + 4 = -8 + 2Y$$
$$2Y - 8 = -8 + 2Y$$
$$\underline{-2Y} = \underline{-2Y}$$
$$-8 = -8$$

All real numbers

Unusual Solutions

When solving an equation, if all the variable terms are eliminated and:

a. the resulting numerical statement is true, then the solution is stated as "all real numbers."

b. the resulting numerical statement is not true, then the solution is stated as "no solution."

12.3 EXERCISES

A. Solve each equation and check your answer by substituting in the *original* equation.

1. $3(x + 2) = 18$
2. $9 + 2(T + 1) = 27$
3. $5 - (C - 2) = -4$
4. $5(y - 1) = 40$
5. $24 = -5(P - 5)$
6. $14 - (2A - 1) = 7$
7. $8 + 3(N + 1) = 38$
8. $6 + 2/3(p + 8) = 6$
9. $2(x - 1) = 3(2x + 3)$
10. $-3(M + 2) = -12$
11. $1 - (2c + 5) = 10$
12. $8(w - 2) = w + 5$
13. $-\frac{1}{6} = \frac{1}{3}(B - 4)$
14. $1 + 3(c - 1) = 10$
15. $4 - \frac{1}{2}(K + 1) = 16$
16. $6(N - 3) = 3(N + 1)$
17. $-5(r + 2) = 25$
18. $1/2 + 3/4(a + 6) = 29$
19. $7(5m + 3) = -49$
20. $18 = 3[3R - 2(R + 5)]$
21. $4(x + 2) + 5 = -35$
22. $3(y + 5) + 8 = 44$
23. $0.8(R - 2) = 0.4(3 + 2R)$
24. $-12 = 2[K - 3(K + 2)]$
25. $6T - 1 = 3(2T + 5)$
26. $-2[5M + 2(3M - 1)] = 70$
27. $21 = -3[2A + 3(2A + 3)]$
28. $4(Y - 2) - 3(2Y + 5) = -1$
29. $12x - 1/2(4x - 6) = 28$
30. $7a - (3a - 4) = 12 - 4a$
31. $7m - 4(2m - 3) = 3(11 + 2m)$
32. $3/4(8y - 4) = 3(2y - 1)$
33. $6R - 2(R - 2) = 4(R - 1)$
34. $5 - 1/3(9 - 6Y) = 16 - 5Y$
35. $2T - \frac{1}{2} = \frac{1}{2}(4T - 1)$
36. $-36 = -2(8T - 1) + 3(T - 4)$
37. $1.6 = 0.5(3.2R + 4.8)$
38. $6Y - 2(5 + 3Y) = 15$
39. $11m = 5(m + 2) - (1 - 6m)$
40. $0.3(4.5x + 0.6) = 5.85(x - 2)$
41. $1/3(6x - 15) = 2x - 5$
42. $2m + 3 = 8m - 3(2m - 1)$

B. Describe, in words, the process used to go from one step to the next.

43. $-12 = 2(A - 3)$

$-12 = 2A - 6$	1.
$-6 = 2A$	2.
$-3 = A$	3.
Check: $-12 = 2(A - 3)$	
$-12 = 2(-3 -3)$	4.
$-12 = 2(-6)$	5.
$-12 = -12$	6.

44. $7(W - 2) = 3(2W + 1)$

$7W - 14 = 3(2W + 1)$	1.
$7W - 14 = 6W + 3$	2.
$W - 14 = 3$	3.
$W = 17$	4.
$7(W - 2) = 3(2W + 1)$	
$7(17 - 2) = 3(2 \cdot 17 + 1)$	5.
$105 = 105$	6.

45. $8(c + 2) = 4(2c - 3)$

$8c + 16 = 4(2c - 3)$	1.
$8c + 16 = 8c - 12$	2.
$16 = -12$	3.
No solution	4.

46. $-8(k - 2) - (2k - 10) = -24$

$-8k + 16 - (2k - 10) = -24$	1.
$-8k + 16 - 2k + 10 = -24$	2.
$-10k + 26 = -24$	3.
$-10k = -50$	4.
$k = 5$	5.

C. Read and respond the the following exercises.

47. Describe, in words, all the steps you would use to find the solution to: $2/3(6Y - 9) = 1/2(12 - 2Y)$.

48. Describe, in words, the steps you would use to check the solution to the equation in Exercise 45.

49. Study Example 6 in Section 12.3 and describe, in words, what happened when solving that equation that hasn't happened before. How do you express the answer when this happens?

50. Study Example 7 in Section 12.3 and describe, in words, what happened when solving that equation that hasn't happened before. How do you express the answer when this happens?

51. Describe, in words, the similarities in the situations when an equation has a solution described as "all real numbers" to one that has a solution of "no solution."

52. Why is it constantly stated that you must use the original equation when checking your solution?

12.4 Application Problems

In this section, you will solve application problems whose equation may contain grouping symbols and signed numbers, and you will use the Distributive Principle. Except for the added complexity of the equations, the types of application problems will be familiar because you worked with them in Chapter 10. They include general, number, perimeter, motion, and money problems.

The emphasis will continue to be on demonstrating ways of organizing the information when solving application problems. You are encouraged to experiment and develop methods that work best for you. Where we show a sketch with the parts labeled, you might prefer to use a list or a chart. The important point is that you recognize and use some means of organizing and presenting the information that is given. This will help you in writing the correct equation and in giving the answers for the problem.

For all application problems you should:

1. Read the problem carefully to see what is given and what is needed.
2. State what variable you will use and use the variable to represent the item you know the least about; then define the other items in the problem in terms of that same variable.
3. Organize the information in a chart, in a list, or on a sketch.
4. Translate the words of the problem into an equation.
5. Solve the equation.
6. State all the answers requested and include labels if they are needed.
7. Check the answer(s) in the words of the problem.

Read the following examples carefully and look for the procedures that were just listed. Two new skills will be needed for the application problems in this section that were not needed for the problems in Chapter 10. If you multiply, divide, or subtract an expression that has more than one term, be sure to enclose the expression in parentheses so that the entire quantity is included in the operation. Also, you will need to work with negative numbers in some of the equations.

Organizing by List

EXAMPLE 1 Four times the sum of a number and 5 is equal to eight times the difference between 7 and the number.

List Let n = number

Equation 4 times the sum of no. and 5 equals 8 times difference of 7 and no.

$$4(n + 5) = 8(7 - n)$$

When a sum or difference is being multiplied, it must be enclosed in parentheses so that all the terms get multiplied.

Solution

$$4(n + 5) = 8(7 - n)$$
$$4n + 20 = 56 - 8n$$
$$+8n \qquad +8n$$
$$12n + 20 = 56$$
$$-20 \quad -20$$
$$\frac{12n}{12} = \frac{36}{12}$$
$$n = 3$$

The number is 3.

Check 4 times the sum of a number and 5 = $4(n + 5) = 4(3 + 5) = 4(8) = 32$.

8 times the difference of 7 and the number = $8(7 - n) = 8(7 - 3) = 8(4) = 32$. $32 = 32$ True.

EXAMPLE 2 Sarah, Laura, and Kristen work in the learning center. Last week Sarah worked 4 hours less than Kristen, and Laura worked twice as many hours as Sarah. If the three women worked a total of 52 hours, how many hours did Sarah work?

List Look at two people at a time. Because Sarah worked 4 hours less than Kristen, you know the least about Kristen's time. Let Kristen's time be x. Sarah's time would then be $x - 4$. Now that you have represented these two items, you can multiply your statement of Sarah's time by 2 to represent Laura's time because Laura worked twice as many hours as Sarah.

Let x = Kristen's hours

$x - 4$ = Sarah's hours

$2(x - 4)$ = Laura's hours

52 = Total hours worked

Equation $x + (x - 4) + 2(x - 4) = 52$

Solution

$$x + x - 4 + 2x - 8 = 52$$
$$4x - 12 = 52$$
$$\underline{+12} \quad \underline{+12}$$
$$4x = 64$$
$$\frac{4x}{4} = \frac{64}{4}$$
$$x = 16$$

The only answer requested is Sarah's time:

$x - 4 = 16 - 4 = 12$ hours

Check See if the times add up to 52:

Kristen: $x = 16$

Sarah: $x - 4 = 16 - 4 = 12$

Laura: $2(x - 4) = 2(16 - 4) = 2(12) = 24$

Total time: $16 + 12 + 24 = 52$ hours. True.

Sarah's time, 12 hours, is 4 less than Kristen's 16 hours. Laura's 24 hours is twice Sarah's time: $2(12) = 24$. True.

◀

EXAMPLE 3 The average daily temperature in Indianapolis for March is 4 degrees more than twice the average daily temperature in January. The difference between the January and the March averages is –23°. What are the average temperatures for the two months?

List Let n = Average temperature for January
$2n + 4$ = Average temperature for March
Do not split the information, "Four degrees more than twice the other number" is all describing one quantity, the March average temperature.

Equation January – March = –23 Difference (subtraction) is written in the given order.

$$n - (2n + 4) = -23$$

Solution

$$n - (2n + 4) = -23$$
$$n - 1(2n + 4) = -23$$
$$n - 2n - 4 = -23 \qquad \text{Insert a 1 and distribute } -1.$$
$$-n - 4 = -23$$
$$\underline{+4} \quad \underline{+4}$$
$$-n = -19$$
$$\frac{n}{-1} = \frac{-19}{-1} \qquad \text{You must divide by } -1 \text{ to isolate } n.$$
$$n = 19$$

n = January average = 19°

$2n + 4$ = March average = 42°

Check March is 4 degrees more than twice January.

$2(19) + 4 = 38 + 4 = 42$

January – March = $19 - 42 = -23$ True.

EXAMPLE 4 The sum of three consecutive integers is –75. Find the integers.

List Let n = 1st consecutive integer

$n + 1$ = 2nd consecutive integer

$n + 2$ = 3rd consecutive integer

Equation Sum of three consecutive integers is –75.

$$n + n + 1 + n + 2 = -75$$

Solution

$$3n + 3 = -75$$
$$-3 \quad -3$$
$$3n = -78$$
$$\frac{3n}{3} = \frac{-78}{3}$$
$$n = -26$$
$$n + 1 = -25$$
$$n + 2 = -24$$

Check –26, –25, and –24 are consecutive integers. Note that –26 is smaller than –24 because it is farther to the left on the number line.

$(-26) + (-25) + (-24) = -75$ True.

EXAMPLE 5 The sum of the smallest and twice the largest of three consecutive even integers is 56. Find the three integers.

List Let n = 1st consecutive even integer

$n + 2$ = 2nd consecutive even integer

$n + 4$ = 3rd consecutive even integer

Equation Smallest sum twice the largest is 56.

$$n + 2(n + 4) = 56.$$

Solution

$$n + 2(n + 4) = 56$$
$$n + 2n + 8 = 56$$
$$3n + 8 = 56$$
$$-8 \quad -8$$
$$3n = 48$$
$$\frac{3n}{3} = \frac{48}{3}$$
$$n = 16, n + 2 = 18, n + 4 = 20$$

Check 16, 18, and 20 are consecutive even integers. $16 + 2(20) = 16 + 40 = 56$

Organizing by Sketch

EXAMPLE 6 Find the width of a rectangle if the perimeter is 62 feet and the width is 4 feet more than one-half of the length. Draw a sketch like Figure 12.2.

Sketch Let x = Length

$\frac{1}{2}x + 4$ = Width

x

$\frac{1}{2}x + 4$ $\frac{1}{2}x + 4$

x

FIGURE 12.2

If $P = 2L + 2W$, then

Equation $62 = 2x + 2\left(\frac{1}{2}x + 4\right)$

Solution

$$62 = 2x + x + 8$$

$$62 = 3x + 8$$

$$\underline{-8} \qquad \underline{-8}$$

$$54 = 3x$$

$$\frac{54}{3} = \frac{3x}{3}$$

$$18 = x$$

Width is $\frac{1}{2}x + 4 = \frac{1}{2}(18) + 4 = 9 + 4 = 13$ feet.

Check $18 + 13 + 18 + 13 = 62$ True.

EXAMPLE 7 Jo needs to change the dimensions of her rectangular garden. The original garden had a length that was twice the width. If she increases the width by 6 feet and decreases the length by 3 feet, the new perimeter will be 54 feet. What were the dimensions of the original garden?

Sketch Draw a sketch similar to Figure 12.3.

Let x = Original width

$2x$ = Original length

$x + 6$ = New width

$2x - 3$ = New length

If $P = 2L + 2W$, then

$54 = 2(2x - 3) + 2(x + 6)$

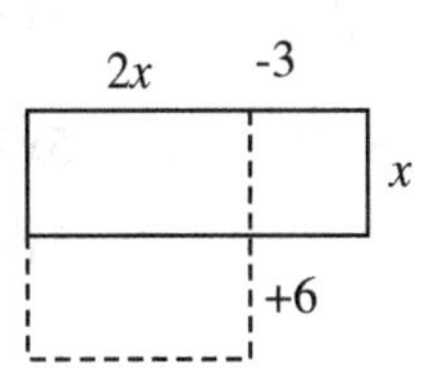

FIGURE 12.3

Solution

$$54 = 2(2x - 3) + 2(x + 6)$$
$$54 = 4x - 6 + 2x + 12$$
$$54 = 6x + 6$$
$$\underline{-6} \qquad \underline{-6}$$
$$48 = 6x$$
$$\frac{48}{6} = \frac{6x}{6}$$
$$8 = x$$

The original width is $x = 8$ feet, and the original length is $2x = 2(8) = 16$ feet.

Check Original length is twice the width: $16 = 2(8)$. New length $= 16 - 3 = 13$ feet; New width $= 8 + 6 = 14$ feet. $13 + 14 + 13 + 14 = 54$ feet. True.

Organizing by Chart

EXAMPLE 8 The hot dog stand at the football game last Friday took in $79.00. The money consisted of $5 bills, $1 bills, and quarters. There were twice as many $1 bills as $5 bills, and the number of quarters was six more than three times the number of $5 bills. How many of each were there?

Chart You could draw a box and fill in the given information. Compare your box with Chart 12.1.

CHART 12.1

Item	Value/Cost of Each	Number of Each	Value/Cost of Each Kind
$5 Bills	5	x	$5x$
$1 Bills	1	$2x$	$1(2x)$
Quarters	0.25	$3x + 6$	$0.25(3x + 6)$
			Total Value/Cost
			79.00

Equation $5x + 1(2x) + 0.25(3x + 6) = 79.00$

Solution

$$5x + 2x + 0.25(3x + 6) = 79.00 \qquad \text{Distribute 0.25.}$$
$$5x + 2x + 0.75x + 1.50 = 79.00 \qquad \text{Combine like terms.}$$
$$7.75x + 1.50 = 79.00$$
$$\underline{-1.50} \qquad \underline{-1.50}$$
$$7.75x = 77.50$$
$$\frac{7.75x}{7.75} = \frac{77.50}{7.75}$$

$$x = 10$$

$$2x = 2(10) = 20$$

$$3x + 6 = 3(10) + 6 = 36$$

There were ten $5 bills, twenty $1 bills, and thirty-six quarters.

Check 20 is 2 times 10, and 36 is 6 more than three times 10.

Ten $5 bills = $50.00

Twenty $1 bills = $20.00

Thirty-six $0.25 = $ 9.00

Total = $79.00 True.

EXAMPLE 9 A bank contains only dimes and half-dollars and there is a total of 29 coins. If this money has a value of $4.90, how many of each are there?

Chart You could draw a chart and fill in the given information. Compare your chart to Chart 12.2.

CHART 12.2

Items	Value/Cost of Each	Number of Each	Total Value/Cost of Each
Dimes	0.10		
Half-Dollars	0.50		
			Total Value/Cost
			$4.90

How do we represent the number of each kind of coin when all we know is that there are 29 altogether? When all you know is the total number of two different items, assign a variable such as x to represent the number of one of the items; it doesn't matter which one. You will represent the number of the other item as the total minus x. Fill those values into your chart. Notice in our chart, Chart 12.3, we decided to let x represent the number of dimes, so the number of half-dollars is $29 - x$.

CHART 12.3

Items	Value/Cost of Each	Number of Each	Total Value/Cost of Each
Dimes	0.10	x	$0.10(x)$
Half-Dollars	0.50	$29 - x$	$0.50(29 - x)$
			Total Value/Cost
			$4.90

Equation

$0.10(x) + 0.50(29 - x) = 4.90$ — Distribute 0.50.

$0.10x + 14.5 - 0.50x = 4.90$ — Combine like terms and be careful with the signs.

$-0.4x + 14.5 = 4.90$

$\underline{-14.5} \quad \underline{-14.50}$

$-0.4x = -9.6$

$\frac{-0.4x}{-0.4} = \frac{-9.6}{-0.4}$

$x = 24$ — Since x = number of dimes = 24 dimes. The half-dollars = 29 – x = 29 – 24 = 5. There are 5 half-dollars.

Check 24 + 5 = 29 coins and 0.10(24) + 0.50(5) = 2.40 + 2.50 = $4.90

Similar to the money problems you have studied are mixture problems, where variously priced items are put together to produce a product that might be more appealing or useful to the consumer.

A person who really likes to drink top-quality imported coffee might not be able to afford to spend $10 per pound for those coffee beans. But if a store can offer a blend that incorporates those costly beans with another kind of bean that is not so expensive, then the mixture can be sold at a more moderate price, and the customer might be able to indulge in this "luxury." Look at the example that follows.

EXAMPLE 10 The Gourmet Coffee Shoppe blends two grades of coffee beans to produce its Deluxe Mix. How many pounds of Venetian beans, costing $10 per pound, should be blended with 40 pounds of Colombian beans, costing $4 per pound, to produce a mixture costing $6 per pound?

Chart You could draw a chart similar to the money charts. To represent the number of pounds of mix, you will need to think about the process that is happening in the store as this mixture is being made. The x pounds of Venetian beans are being *added* to the 40 pounds of Colombian beans to get the Deluxe Mix, so the number of pounds of the Mix is $x + 40$, and the total cost of the Mix is the product of $6 and $(x + 40)$ pounds. Note that the third row contains information about the final mixture (Chart 12.4).

CHART 12.4

Item	Cost per Pound	Number of Pounds of Each	Total Cost of Each
Venetian	10	x	$10x$
Colombian	4	40	4(40)
Deluxe	6	$x + 40$	$6(x + 40)$

Equation Cost of Venetian beans plus Cost of Colombian beans equals Cost of Deluxe Mix.

$$10x + 4(40) = 6(x + 40)$$

Solution

$$10x + 4(40) = 6(x + 40)$$

$$10x + 160 = 6x + 240$$

$$\underline{-6x} \qquad \underline{-6x}$$

$$4x + 160 = 240$$

$$\underline{-160} \qquad \underline{-160}$$

$$4x = 80$$

$$\frac{4x}{4} = \frac{80}{4}$$

$x = 20$ pounds of Venetian beans

Check

$$10(20) + 4(40) = 6(20 + 40)$$

$$200 + 160 = 6(60)$$

$360 = 360$ True.

EXAMPLE 11 Craft and Company is preparing a new shredded cheese mixture. How many pounds of $4.30-per-pound romano cheese should be combined with 20 pounds of $2.25-per-pound mozzarella to produce a pizza cheese mix costing $3.80 per pound?

Sketch An organizational technique that you might prefer to use with mixture problems is to make a sketch. (see Figure 12.4).

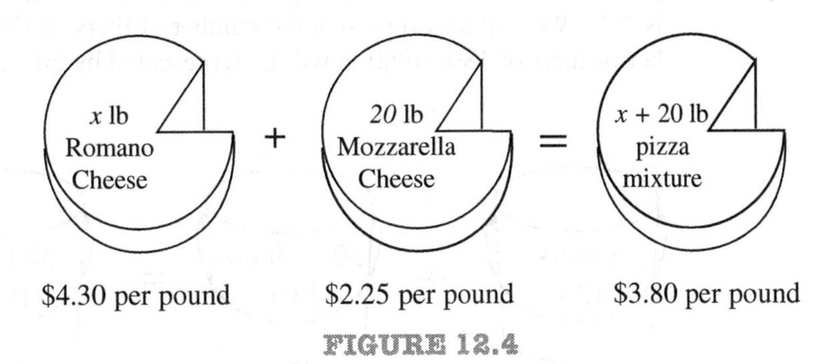

FIGURE 12.4

Chart Or you could record the information that is given in a box, and compare your box with Chart 12.5.

CHART 12.5

Item	Cost per Pound	Number of Pounds of Each	Total Cost of Each
Romano	4.30	x	$4.30x$
Mozzarella	2.25	20	2.25(20)
Mixture	3.80	$x + 20$	$3.80(x + 20)$

Equation Cost of Romano plus Cost of Mozzarella equals Cost of Mixture

$$4.30x + 2.25(20) = 3.80(x + 20)$$

Solution

$$4.30x + 45 = 3.80x + 76$$
$$-3.80x \qquad -3.80x$$
$$0.50x + 45 = 76$$
$$-45 \qquad -45$$
$$0.50x = 31$$
$$\frac{0.50x}{0.50} = \frac{31}{0.50}$$

$x = 62$ pounds of romano cheese

Check

$$4.30(62) + 2.25(20) = 3.80(62 + 20)$$
$$266.6 + 45.00 = 3.80(82)$$
$$\$311.60 = \$311.60 \quad \text{True.}$$

EXAMPLE 12 How much of a 13% salt solution and how much of an 18% salt solution should be combined to produce 50 liters of a 16% salt solution? (Two methods of organization will be shown for this example, where percent concentration is used instead of cost per pound.)

Sketch You could organize information with a sketch like Figure 12.5. If you have a total amount that is to be divided into two unequal parts, you will represent one of the parts by a variable and the other part as the total minus this variable. In this example, there is a total of 50 liters, part of which is 13% salt solution and part of which is 18%. We will let x represent the number of liters of 13% solution. Then the number of liters of 18% solution will be represented by $50 - x$, the total minus x.

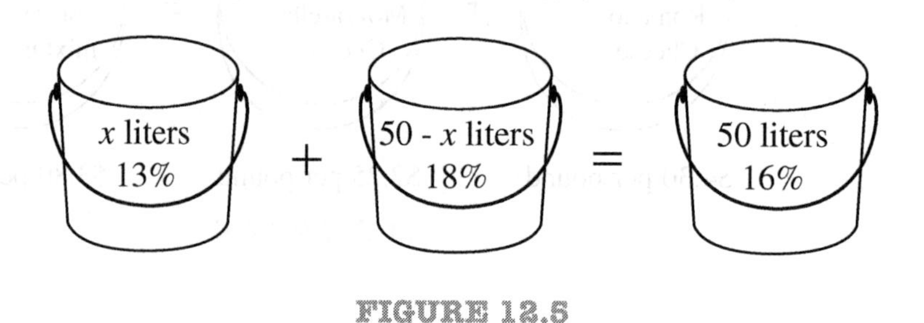

FIGURE 12.5

Chart Or you could set up a chart, but change the Cost per Pound column to *Percent of Each* (see Chart 12.6). When placing percent values in that column, put them in as decimals so you will remember to multiply by the decimal rather than by a percent.

CHART 12.6

Item	Percent of Each	Number of Liters of Each	Each Quantity
13% Salt	0.13	x	$0.13x$
18% Salt	0.18	$50 - x$	$0.18(50 - x)$
16% Salt	0.16	50	$0.16(50)$

Equation $0.13x + 0.18(50 - x) = 0.16(50)$

Solution

$$0.13x + 9 - 0.18x = 8$$ Carefully combine like terms.

$$-0.05x + 9 = 8$$

$$\underline{-9} \quad \underline{-9}$$

$$-0.05x = -1$$

$$\frac{-0.05x}{-0.05} = \frac{-1}{-0.05}$$

$$x = 20$$

Therefore, 20 liters of 13% solution and 50 – 20, or 30 liters of 18% solution must be combined.

Check Substitute the answer into the expressions given in the last column to see if the sum of the two components does equal the final mixture.

$$0.13x = 0.13(20) = 2.6$$

$$0.18(50 - x) = 0.18(50 - 20) = \underline{+5.4}$$

$$8.0$$

$$\text{and } 0.16(50) = 8.0 \quad \text{True.}$$

◀

Another kind of percent mixture problem was introduced in Chapter 10 and involved investments that were made at different interest rates. The simple interest formula, $I = PRT$, was used to set up the chart that helped to organize the information. Study the example problem that follows.

EXAMPLE 13 If $4,000 more is invested at 5% than is invested at 4% annual interest, the combined investments will earn $1,950 in interest over 3 years. How much is each investment?

Chart You could draw a chart and fill in the given information. Compare your chart to Chart 12.7.

CHART 12.7

Items	Principal	Rate	Time	Interest for Each
5% Investment	$x + 4000$	0.05	3	$(x + 4000)(0.05)(3)$
4% Investment	x	0.04	3	$(x)(0.04)(3)$
				Total Interest
				1950

Equation $(x + 4000)(0.05)(3) + (x)(0.04)(3) = 1950$

The equation is more complex looking than most you have dealt with, but it just needs some simplifying before you begin the solving steps. In the first term, $(x + 4000)(0.05)(3)$, multiply $(0.05)(3)$ and put that product in front of the parentheses: $0.15(x + 4000)$. In the second term, multiply $(0.04)(3)$ and write that product in front of x: $0.12x$. The equation now reads:

$0.15(x + 4000) + 0.12x = 1950$ Distribute 0.15

$0.15x + 600 + 0.12x = 1950$ Combine like terms.

$0.27x + 600 = 1950$ Subtract 600.

$\underline{-600} \quad \underline{-600}$

$0.27x = 1350$ Divide by 0.27.

$$\frac{0.27x}{0.27} = \frac{1350}{0.27}$$

$x = 5000$ \$5,000 was invested at 4% and $x + 4000 = 5000 + 4000 =$ \$9,000 at 5%.

Check Does $9000(0.05)(3) + 5000(0.04)(3) = 1950$?

$1350 \quad + \quad 600 \quad = 1950$

$1950 = 1950$ True.

In the motion problems that were worked with in Chapter 10, the information was organized by using either a chart or a sketch. In the examples that follow, we will use a sketch and a chart to show two different ways that you might organize information for motion problems.

In Example 14, the two planes will be traveling in opposite directions and a total distance will be given. In Examples 14 and 15, the sketches will show you that the two persons or vehicles are traveling the same distance. The difference in these two situations will show up when you write the equation. Study the next example problems very carefully.

EXAMPLE 14 Two planes take off from the same airport and travel in opposite directions. After three hours they are 2,250 miles apart. How fast is each plane flying if one travels 150 miles per hour faster than the other? See Figure 12.6 and Chart 12.8.

Sketch

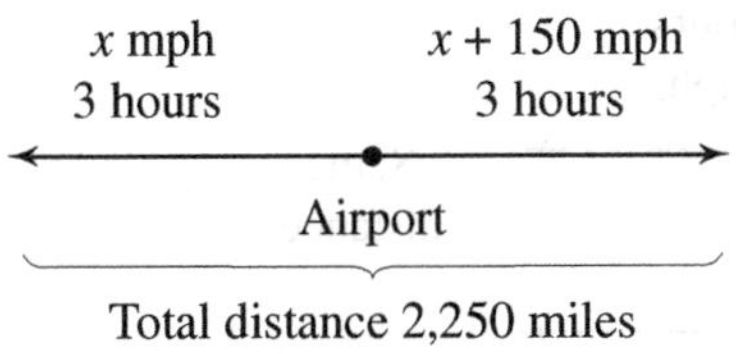

x mph, 3 hours — $x + 150$ mph, 3 hours

Airport

Total distance 2,250 miles

FIGURE 12.6

Chart

CHART 12.8

	Rate	Time	Distance
Plane 1	x	3	$3x$
Plane 2	$x + 150$	3	$3(x + 150)$
			Total Distance 2,250

Equation The sum of the distances equals 2,250 miles

$$3x + 3(x + 150) = 2{,}250$$
$$3x + 3x + 450 = 2{,}250$$
$$6x + 450 = 2{,}250$$
$$\underline{-450} \quad \underline{-450}$$
$$6x = 1{,}800$$
$$\frac{6x}{6} = \frac{1{,}800}{6}$$
$$x = 300$$

Therefore, Plane 1 is traveling at 300 miles per hour and Plane 2 is traveling 150 miles per hour faster, or 450 miles per hour.

Check 300(3)= 900 miles and 450(3) = 1,350 miles. The sum of 900 and 1,350 is 2,250 miles. 450 mph is 150 mph faster than 300 mph.

EXAMPLE 15 Maria leaves for her evening walk traveling at 3 mph. One hour later, her husband, Miguel, leaves the house, following the same route and jogging at 5 mph. How long will it take him to overtake Maria?

Sketch You could make a sketch representing the problem (Figure 12.7). You don't know anything about Maria's time, so call it x. Because Miguel leaves the house 1 hour later, his traveling time is $x - 1$ (he traveled 1 hour *less than* Maria). No total distance is given; in fact, they travel the same distance from their home to the point where Miguel catches up with Maria. The equation then will show that Maria's distance is equal to Miguel's distance.

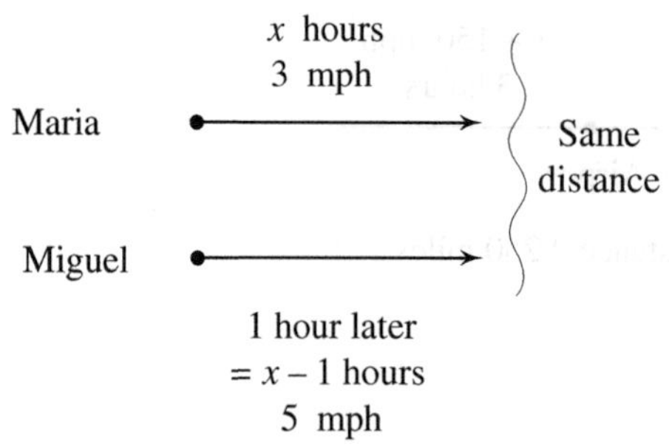

x hours
3 mph

Maria •———→ Same distance

Miguel •———→

1 hour later
$= x - 1$ hours
5 mph

FIGURE 12.7

Chart Or you could draw a box and fill in the information that is given. Does your box look like Chart 12.9?

CHART 12.9

	Rate	Time	Distance of Each
Maria	3	x	$3x$
Miguel	5	$x - 1$	$5(x - 1)$
			Total Distance
			Same

Equation Because they travel the same distance, the equation is

Maria's distance = Miguel's distance.

$$3x = 5(x - 1)$$

Solution

$$3x = 5x - 5$$
$$\underline{-5x \quad -5x}$$
$$-2x = -5$$
$$\frac{-2x}{-2} = \frac{-5}{-2}$$
$$x = 2\frac{1}{2} \text{ or } 2.5$$

Because you are looking for Miguel's time,

$x - 1 = 2.5 - 1 = 1.5$ hours

Check See if Maria's distance does equal Miguel's distance:

Maria's distance $= 3x = 3(2.5) = 7.5$ miles

Miguel's distance $= 5(x - 1) = 5(2.5 - 1)$

$= 5(1.5) = 7.5$ miles True.

EXAMPLE 16 A motorboat, traveling at 18 miles per hour, sets off from a marina 2 hours after a sailboat left, traveling at 6 miles per hour. If they follow the same route, how long will it take the motorboat to catch up with the sailboat?

Sketch

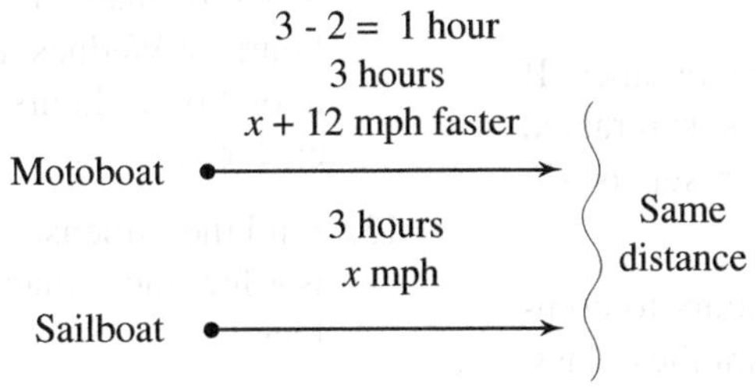

3 - 2 = 1 hour
3 hours
$x + 12$ mph faster
Motoboat
3 hours
x mph
Sailboat
Same distance

FIGURE 12.8

Chart **CHART 12.10**

	Rate	Time	Distance of Each
Sailboat	6	x	$6x$
Motorboat	18	$x - 2$	$18(x - 2)$
			Total Distance
			Same

Equation They travel the same distance.

$$6x = 18(x-2)$$

Solution

$$6x = 18x - 36$$

$$\underline{-18x \quad -18x}$$

$$\frac{-12x}{-12} = \frac{-36}{-12}$$

$$x = 3$$

Sailboat = 3 hours; Motorboat = $x - 2 = 3 - 2 = 1$ hour

The motorboat catches the sailboat in 1 hour.

Check See if the sailboat's distance is equal to the motorboat's distance.

Sailboat: $6x = 6(3) = 18$ miles

Motorboat: $18(x - 2) = 18(3 - 2) = 18(1) - 18$ miles

Sailboat's distance = motorboat's distance True.

As you go on to the exercises in this section, to the test of this chapter, and even in later math courses, keep in mind the importance of organizing the information before writing the equation or solving the problem. We demonstrated charts, lists, and sketches, but you may develop a method of your own or use a combination of the ones shown.

12.4 EXERCISES

A. Organize the information, write and equation, solve it, and give the answers requested.

1. One number is 5 less than another number. If three times the smaller number is subtracted from twice the larger number, the result is –7. Find the numbers.

2. Three students took a total of 39 hours to complete a group project for their hydraulics class. Tom worked 4 hours less than Ted. Terry worked three times as much as Tom. How much time did each student contribute to the project?

3. Mary, Diane, and Jean are servers at Pop's Pizza Parlor. Last week Mary earned $19 less than Jean, and Diane earned twice as much as Mary. Altogether they earned $163. How much was each woman's salary? (See Example 2.)

4. The sum of the smallest and twice the largest of three consecutive odd integers is –13. Find the three integers. (See Example 5)

5. When you subtract the third of three consecutive even integers from 4 times the first, the difference is –40. Find the largest integer.

6. One number is 8 less than another number. When twice the smaller number is subtracted from the larger number, the result is –10. What are the numbers?

7. Six times the difference between a number and 3 is equal to the difference between the number and 3. What is the number? (See Example 1)

8. Five times the difference between twice a number and 1 is equal to the sum of twice the number and 3. What is the number? (See Example 1)

9. When three times the first of three consecutive even integers is subtracted from twice the third, the result is –4. What are the three even integers?

10. Find two consecutive even integers such that twice the second equals three times the first.

11. Bob, Connie, and Lorraine all work as receptionists in the learning center. During the last pay period they earned a total of $464. Find out how much Lorraine earned if Bob earned $16 more than Connie, and Lorraine earned twice as much as Bob.

12. A nurse's aid worked two more hours on Monday than on Tuesday, and twice as many hours on Wednesday as on Monday. If he worked a total of 42 hours, how many hours did he work each day?

13. Find the dimensions of a rectangle whose length is 8 feet more than its width. The perimeter is 76 feet.

14. Find the width of a rectangle if the perimeter is 138 inches, and the length is 3 inches more than the width. (See Example 6)

15. A rectangular field is enclosed by 440 feet of fencing. Find the dimensions of the field if the length is 20 feet less than 3 times the width.

16. The second side of a triangle is 4 feet shorter than the first side. The third side is 2 times as long as the second side. Find the lengths of the sides if the perimeter is 28 feet.

17. A college student saves dimes and quarters so that he can go to the laundromat. The number of quarters he has is four more than twice the number of dimes. Altogether he has $3.40. How many of each coin does he have? (See Example 8)

18. A college student working in the mailroom of an engineering company is sent to the post office with $20. She is to purchase forty-five 32-cent stamps and to use the rest of the money to buy prestamped envelopes that cost 40 cents each. How many envelopes will she be able to buy?

19. Tickets for the Cougar basketball games cost $2.00 for students and $3.50 for nonstudents. On Saturday, 120 tickets were sold and $360 was collected. How many of each kind of ticket was sold?

20. The Nut House blends $3.00-per-pound peanuts with $7.00-per-pound mixed nuts to produce an economy mix that sells for $6.00 per pound. How many pounds of peanuts and mixed nuts need to be blended to produce 20 pounds of economy mix?

21. The Gamma Gamma Sorority is selling cakes to raise money for Children's Hospital. Pound cakes sell for $5 each, chocolate layer cakes for $7, and hazelnut tortes for $8. Group members sold six more tortes than pound cakes and twice

as many chocolate cakes as tortes. They raised \$321. How many chocolate cakes were sold?

22. Two motorcycles, leaving at the same time, are traveling in opposite directions from the same city. They are 658 miles apart after 7 hours. Find the rate of each motorcycle if one travels 10 miles per hour faster than the other.

23. Two trains start from the same station at the same time and travel in opposite directions for 6 hours, at which time they are 732 miles apart. One train travels 12 miles per hour faster than the other. How fast is each traveling?

24. Lil's Card Shop sells two different styles of wrapping paper. One retails for \$1.20 a roll and the other for \$2.00 a roll. Last month the store sold a total of 40 rolls for \$72.00. How many rolls of each kind were sold? (See Example 9)

25. If three times as much money is invested at 4 1/2% as at 3 1/2% annual interest, the total amount of interest earned in four years will be \$1,020. How much money was put into each investment? Use $I = PRT$. (See Example 13)

26. Pete's Produce Stand currently is selling only corn and cantaloupes. Corn sells for \$0.20 per ear and cantaloupes for \$1.39. The number of ears of corn sold yesterday was six more than twice the number of cantaloupes sold. If \$96.07 worth of produce was sold, how many cantaloupes were sold?

27. How much of an 8% alcohol solution and how much of a 5% alcohol solution must be combined to produce 300 milliliters of a 6% alcohol solution? (See Example 12)

28. If \$10,000 more is invested in a certificate of deposit paying 5 1/2% than in a mutual fund paying 6% annual interest they will yield a total of \$2,850 in interest in 1 year. How much should be placed in each investment? Use $I = PRT$.

29. The petty cash box contains \$50.00. There are 30 bills, all either \$1 or \$5 bills. How many \$5 bills are there? (See Example 9)

30. Sara left home on her bicycle traveling at a rate of 6 mph. One hour later Becca left to catch up with her sister, traveling at 10 mph. How long will it take Becca to overtake Sara? (See Example 15)

31. How many quarts of 20% salt solution must be added to 9 quarts of a 15% salt solution to get a solution that is 17% salt?

32. How many liters of a 30% sulfuric acid solution must be added to a 50% sulfuric acid solution to produce 10 liters of a 35% sulfuric acid solution? (See Example 12)

33. How many liters of a 25% alcohol solution should be added to 20 liters of a 20% alcohol solution, to produce a 23% alcohol solution?

34. A freight train left Cleveland traveling at 40 mph. Two hours later a passenger train left the same station, following the same route and traveling at 60 mph. How long will it take the passenger train to overtake the freight train?

35. At the same time, John and Nan start from two different cities that are 330 miles apart. They travel toward each other and meet for lunch 3 hours later. Find their rates if Nan travels 10 mph faster than John.

36. The length of Tony's family room is 6 feet less than twice the width. Find the dimensions if the perimeter is 54 feet.

37. Sarah invests \$5,000 more in a certificate of deposit paying 5% than she invests in municipal bonds paying 4% annual interest. If these investments earn \$1,290 in interest in three years, how much did she put into each investment? Use $I = PRT$.

38. If \$2,000 more is invested at 6 1/2% than is invested at 6% annual interest, the amount of interest at the end of 3 years will be \$1,515. How much money is put into each investment? Use $I = PRT$.

39. The width of a rectangle is seven yards more than half of the length. Find the width if the perimeter is 86 yards.

40. At the same time, Kay and Lisa leave their homes, which are 24 miles apart, and walk toward each other on the city bike path. If Kay travels at 5 miles per hour and Lisa at 3 miles per hour, how long will it take before they meet?

B. Read and respond to the following exercises.

41. For the following problem, without actually writing an equation and solving it, decide which woman will travel the greater distance. How do you know this?

Manuella and Marisol live in cities that are 440 miles apart. They leave their homes at the same time and drive toward each other until they meet 4 hours later. What is Marisol's average speed if she drives 20 miles per hour faster than Manuella?

42. For the problem given in Exercise 41, describe in words how you would proceed if you were asked for the exact answer.

43. For the following problem, without actually writing an equation and solving it, decide which of the meats the mixture would use more of. In words explain your answer.

How many pounds of ground beef that is 10% fat and how many pounds of ground beef that is 15% fat must be combined to produce 50 pounds of ground beef that is 12% fat?

44. For the problem given in Exercise 43, describe, in words, how would you proceed if asked for the exact answer?

12.5 Solving Inequalities

Simple Inequalities

Let's review the definition of inequality that was introduced in Section 8.5. On the number line, larger numbers are farther to the right than smaller numbers. Therefore, $8 > 3$ (8 is greater than 3) because 8 is farther to the right than 3, and $-6 < 3$ because 3 is farther to the right than -6.

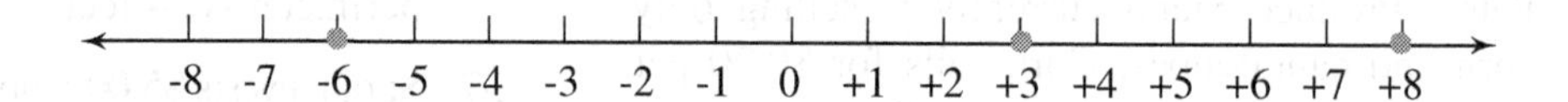

The points on the number line correspond to all the *real* numbers, not just to the integers shown. The set of real numbers is made up of the *rational numbers* (integers, and fractions with integer numerators and denominators) and the *irrational numbers* (numbers such as $\sqrt{2}$, π and $\sqrt{5}$ that are non-repeating and non-terminating in decimal form). Examples of rational numbers include -6, 4/5, and 2.333 . . . , and examples of irrational numbers include $\sqrt{2} = 1.4142135\ldots$ and $\pi = 3.14159265\ldots$.

An inequality such as $x > 3$ includes *all* real numbers (all points on the number line) greater than 3. This is illustrated by:

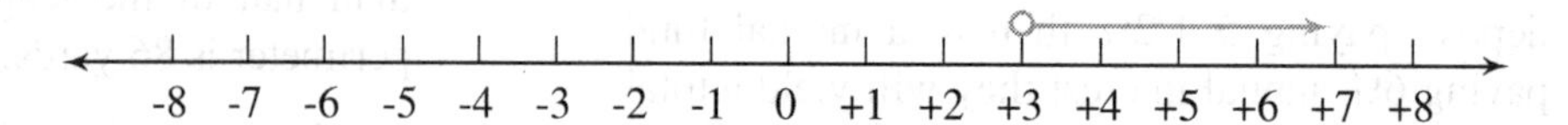

The open circle shows that 3 is not included in the answer (solution set), but all numbers to the right of that point are included.

If the statement had been $x \geq 3$, the graph would have had a filled-in circle around 3, indicating that 3 was included because of the equality symbol.

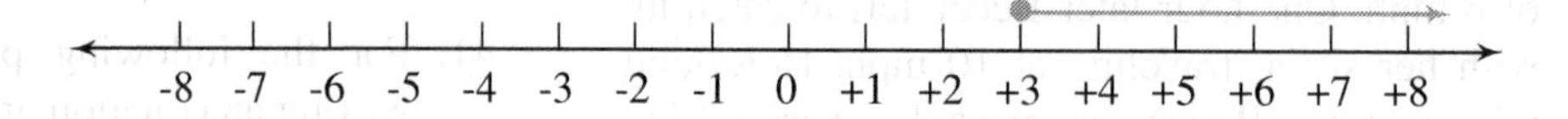

The line with an arrowhead extending to the right means that it continues in that direction forever. There are an infinite number of answers to the problem. Unlike **first-degree equations** (the variable has an exponent of 1) which have only one answer, first-degree inequalities have many answers. Graphing on the number line is one way to show the solution set. Another way to show it is to use set notation. Let's go back to the statement $x > 3$. In set notation, the solution is $\{x \mid x < 3\}$ .This is read "the set of all real numbers represented by x such that x is greater than 3."

Examine the following examples of algebraic inequalities, solution set notation, and graphs.

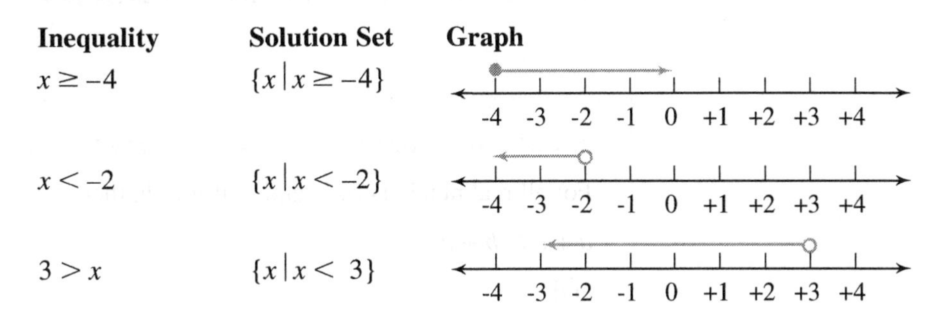

Inequality	Solution Set	Graph
$x \geq -4$	$\{x \mid x \geq -4\}$	
$x < -2$	$\{x \mid x < -2\}$	
$3 > x$	$\{x \mid x < 3\}$	

In the last example, did you notice that $3 > x$ and $x < 3$ express the same relationship? That is, "3 is greater than x" and "x is less than 3" are graphed the same way. You may find it easier to write the set notation and graph correctly if you read the inequality from the variable side of the symbol. Read $3 > x$ from the variable side as "x is less than 3"; write the set notation, $\{x \mid x < 3\}$; and draw the graph from that statement. The inequality $5 \leq y$ might be easier to work with if it is stated as $y \geq 5$, "y is greater than or equal to 5."

Both set notation and graphs will be required when you solve the algebraic inequalities in this section. To understand how to solve inequalities, look at the solutions to the equation and the inequality that follow. Do you see any difference in procedure?

Equation	Inequality
$2x + 5 = 17$	$2x + 5 > 17$
$\underline{-5} \quad \underline{-5}$	$\underline{-5} \quad \underline{-5}$
$2x = 12$	$2x > 12$
$\frac{2x}{2} = \frac{12}{2}$	$\frac{2x}{2} > \frac{12}{2}$
$x = 6$	$x > 6$

There was no difference in the steps for solving the equation and the inequality. The same number, 5, was subtracted from both sides of the symbol, and then both sides were divided by 2, the coefficient of the variable term. With the inequality, you would continue and state the solution in set notation and perhaps draw the graph for that solution.

The addition and subtraction properties are the same for equations and inequalities.

Addition–Subtraction Property of Equality

For all real numbers a, b, and c, if $a = b$, then

$a + c = b + c$

and

$a - c = b - c$

Addition–Subtraction Property of Inequality

For all real numbers a, b, and c, if $a > b$, then

$a + c > b + c$

and

$a - c > b - c$

The following arithmetic (no variables) examples show that these addition and subtraction properties for inequalities are true.

$6 > 3$	$14 < 20$
$6 + 2 > 3 + 2$	$14 - 5 < 20 - 5$
$8 > 5$ True.	$9 < 15$ True.

EXAMPLE 1 Solve $7 > x + 5$, state the answer in set notation, and graph the solution.

Solution $7 > x + 5$ Use the same procedure you would use to solve an equation.

$\underline{-5} \quad \underline{-5}$

$2 > x$

Reading from the x side, "x is less than 2," so

$\{x \mid x < 2\}$

Remember that if 2 is greater than x, then x is also less than 2; $2 > x$ and $x < 2$ are the same.

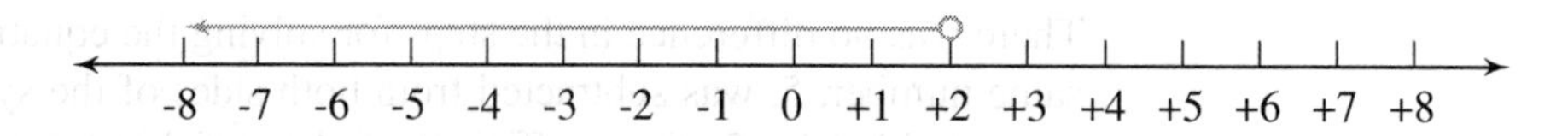

Check You would not check the solution by substituting 2 into the original problem because x does not equal 2. Instead, pick any value that is *less than* 2, substitute it into the original inequality, and see if you get a true statement. Because you can choose any value in the solution set, look for one that is easy to use.

$7 > x + 5$ The solution was $x < 2$, so you might try 0. The choice is yours.

$7 > 0 + 5$

$7 > 5$ True.

EXAMPLE 2 Solve $x + 2 \geq -3$, state the answer in set notation, and graph the solution.

Solution

$$\begin{aligned} x + 2 &\geq -3 \\ \underline{-2} \quad &\underline{-2} \\ x \quad &\geq -5 \end{aligned}$$

$\{x \mid x \geq -5\}$

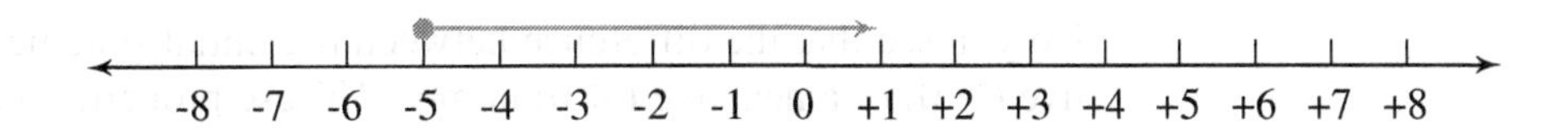

Remember to use the closed circle because x also equals -5.

Check

$x + 2 \geq -3$

$0 + 2 \geq -3$

$2 \geq -3$ True.

We will check with 0 because $0 > -5$, and it is an easy value to use.

Now let's look at some more arithmetic examples that involve multiplication and division to see if there is a pattern.

$4 > -2$

Multiply both sides by 3.

$3(4) > 3(-2)$

$12 > -6$ True.

$15 < 35$

Divide both sides by 5.

$15/5 < 35/5$

$3 < 7$ True.

$4 > -2$

Multiply both sides by -3.

$-3(4) > -3(-2)$

$-12 > +6$ False.

$15 < 35$

Divide both sides by -5.

$15/(-5) < 35/(-5)$

$-3 < -7$ False.

It appears that multiplying or dividing both sides of an inequality by a *positive* number does not affect the sense, or direction, of the inequality. However, multiplying or dividing by a *negative* number produces a false statement. In fact, the symbols in those answers are just the opposite of what they should be to give a true statement.

$-12 > +6$ False. $-3 < -7$ False.

But $-12 < +6$ True. But $-3 > -7$ True.

This step, reversing the inequality symbol, must be done each time you multiply or divide both sides of an inequality by a negative number.

Multiplication and Division Properties of Inequalities

1. For all real numbers a, b, and c, with $c > 0$:

   If $a > b$, then $ac > bc$.

   If $a > b$, then $a/c > b/c$.

2. For all real numbers a, b, and c, with $c < 0$:

   If $a > b$, then $ac < bc$.

   If $a > b$, then $a/c < b/c$.

Do you see that the difference between the initial statements is whether c is a positive (> 0) or a negative (< 0) number? These properties could have been expressed with $\geq$, $<$, or $\leq$ in the $a > b$ statement. For example,

For all real numbers a, b, and c, with $c < 0$:

If $a \leq b$, then $ac \geq bc$.

EXAMPLE 3 Solve, state the set notation, and graph: $4x - 3 \leq 15$

Solution

$$4x - 3 \leq 15$$
$$\underline{+3 \quad +3}$$
$$4x \leq 18$$

You are dividing by a positive value, so do not reverse the symbol.

$$\frac{4x}{4} \leq \frac{18}{4}$$
$$x \leq 4.5$$

$\{x \mid x \leq 4.5\}$

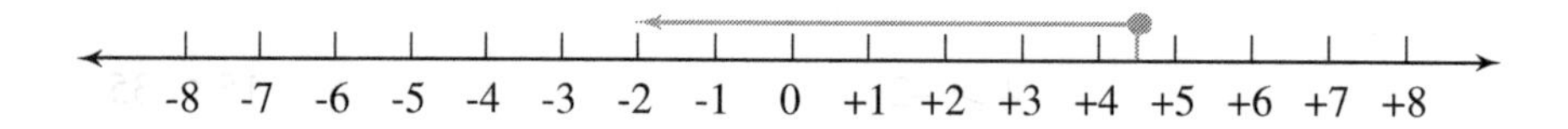

Check

$4x - 3 \leq 15$ Use $x = 0$ to check because 0 is less than 4.5.

$4(0) - 3 \leq 15$

$0 - 3 \leq 15$

$-3 \leq 15$ True.

EXAMPLE 4 Solve, state the answer in set notation, and graph: $5 - 3x > 11$

Solution

$$5 - 3x > 11$$
$$\underline{-5} \qquad \underline{-5}$$
$$-3x > 6$$
$$\frac{-3x}{-3} < \frac{6}{-3}$$ To isolate x, divide by -3, which reverses the symbol.

$$x < -2$$ $\{x \mid x < -2\}$

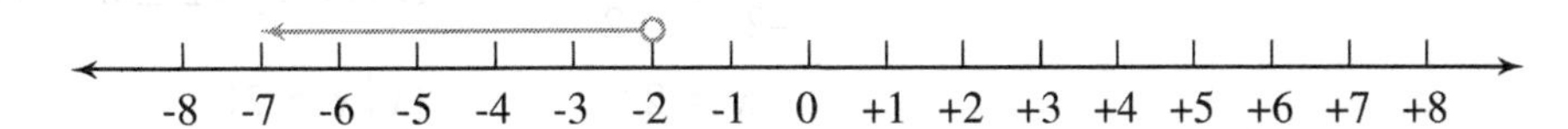

Check

$$5 - 3x > 11$$
$$5 - 3(-4) > 11$$
$$5 + 12 > 11$$
$$17 > 11$$ True.

We will use $x = -4$ to check, though any value less than -2 could have been chosen.

To solve more complex-looking inequalities, we use the same procedures we apply to equations that require more steps. Be especially careful with steps that require multiplication or division of *both* sides of the inequality by a negative number.

EXAMPLE 5 Solve, state the answer in set notation, and graph: $3x + 6 > x - 4$

Solution

$$3x + 6 > x - 4$$ Subtract x from both parts.
$$\underline{-x} \qquad \underline{-x}$$
$$2x + 6 > -4$$
$$\underline{-6} \qquad \underline{-6}$$
$$2x > -10$$
$$\frac{2x}{2} > \frac{-10}{2}$$
$$x > -5$$ $\{x \mid x > -5\}$

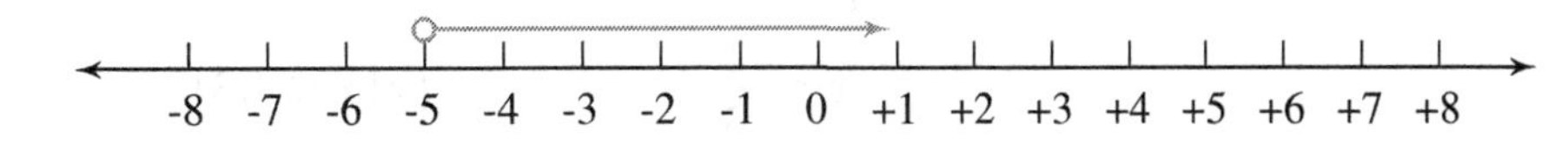

Check

$$3x + 6 > x - 4$$ Try $x = 0$.
$$3(0) + 6 > 0 - 4$$
$$0 + 6 > 0 - 4$$
$$6 > -4$$ True.

EXAMPLE 6 Solve, and state the answer in set notation, and graph: $-5(x - 3) + 3(x + 1) < 14$

Solution

$$-5(x - 3) + 3(x + 1) < 14$$
$$-5x + 15 + 3x + 3 < 14$$
$$-2x + 18 < 14$$
$$\underline{-18} \quad \underline{-18}$$
$$-2x \quad < -4$$
$$\frac{-2x}{-2} > \frac{-4}{-2} \qquad \text{Divide by } -2 \text{ and reverse the symbol.}$$
$$x > 2 \qquad \{x \mid x > 2\}$$

-8 -7 -6 -5 -4 -3 -2 -1 0 +1 +2 +3 +4 +5 +6 +7 +8

Check

$$-5(x - 3) + 3(x + 1) < 14 \qquad \text{Try } x = 4.$$
$$-5(4 - 3) + 3(4 + 1) < 14$$
$$-5(1) + 3(5) < 14$$
$$-5 + 15 < 14$$
$$10 < 14 \quad \text{True.}$$

EXAMPLE 7 Solve, and state the answer in set notation, and graph: $\frac{1}{2}x + \frac{2}{3} < \frac{5}{6}$

Solution

$$\frac{1}{2}x + \frac{2}{3} < \frac{5}{6} \qquad \text{Multiply every term by the LCD of 6.}$$
$$6\left(\frac{1}{2}x\right) + 6\left(\frac{2}{3}\right) < 6\left(\frac{5}{6}\right)$$
$$3x + 4 < 5$$
$$\underline{-4} \quad \underline{-4}$$
$$3x < 1$$
$$\frac{3x}{3} < \frac{1}{3}$$
$$x < \frac{1}{3} \qquad \{x \mid x < 1/3\}$$

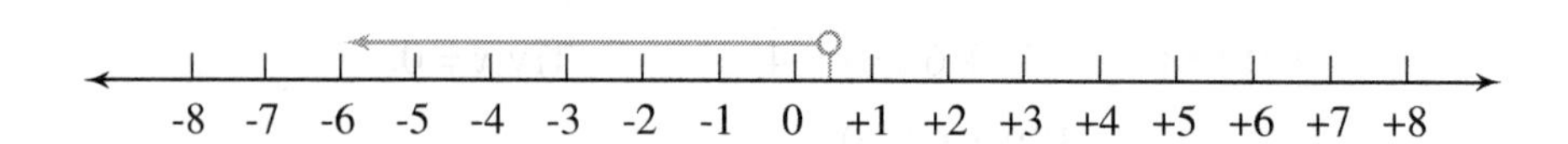

Check $\frac{1}{2}x + \frac{2}{3} < \frac{5}{6}$ Use $x = 0$, and substitute into the original inequality.

$\frac{1}{2}(0) + \frac{2}{3} < \frac{5}{6}$

$0 + \frac{2}{3} < \frac{5}{6}$ True, because $2/3 = 4/6$ and $4/6 < 5/6$.

Compound Inequalities

EXAMPLE 8 State the inequality in set notation and graph on the number line: $-1 < x \leq 5$.

Solution x has two different statements of inequality. If you read from the variable part of the statement (middle), you read "x is greater than -1" *and* "x is less than or equal to 5." Both conditions are true at the same time. This inequality doesn't have to be solved because x is already isolated, but you do need to express set notation and show the graph.

$\{x \mid -1 < x \leq 5\}$ The set of values for x such that -1 is less than x and x is less than or equal to 5.

The graph for $-1 < x$ (x is greater than -1) would include an open circle at -1 and an arrow to the right. The graph for $x \leq 5$ would be a closed circle at 5 and an arrow to the left. Neither of these arrows could go past the other point because, although x is greater than -1, it also must be less than or equal to 5.

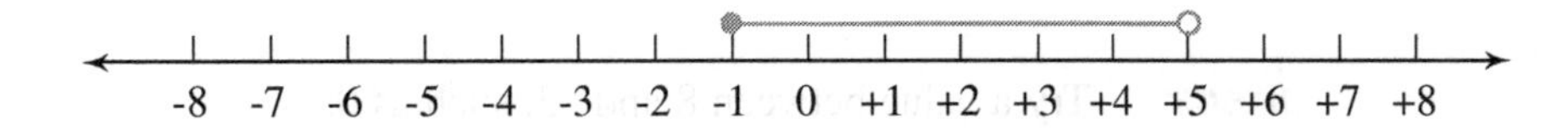

Check Try a point included in the graph such as 0 because it is between -1 and 5.

$-1 < x \leq 5$ $-1 < 0 \leq 5$ Both conditions are true. 0 is greater than -1 and 0 is less than or equal to 5.

EXAMPLE 9 Solve, and state in set notation, and graph: $0 < x + 3 < 6$

Solution Work with all parts (left, center and right) at the same time. focus on the center to decide what step to take since that is where the variable is.

$0 < x + 3 < 6$ Subtract 3 from all parts to isolate x.

$\underline{-3} \quad \underline{-3} \quad \underline{-3}$

$-3 < x < 3$ $\{x \mid -3 < x < 3\}$

The graph must show all points that are greater than -3 *and are also* less than 3.

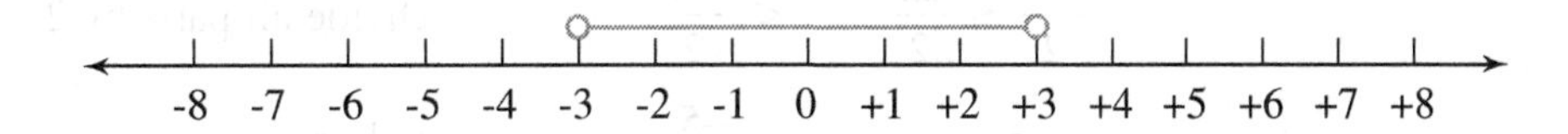

Check $0 < x + 3 < 6$ Try $x = 0$ because it is between –3 and 3.

$0 < 0 + 3 < 6$

$0 < 3 < 6$ True, because 3 is greater than 0 *and* is also less than 6.

EXAMPLE 10 Solve and graph: $-9 \leq 2x + 7 < 13$

Solution Work with all parts at the same time to isolate x, focusing on the middle section to decide what steps to take.

$$-9 \leq -2x + 7 < 13$$

$$\underline{-7} \quad \underline{-7} \quad \underline{-7}$$

$$-16 \leq -2x < 6$$

$$\frac{-16}{-2} \geq \frac{-2x}{-2} > \frac{6}{-2}$$

Remember to *reverse* both inequality symbols because you are dividing by a *negative* number.

$8 \geq x > -3$ $\{x \mid 8 \geq x > -3\}$

The graph must show all values that are less than or equal to 8 *and are also* greater than –3.

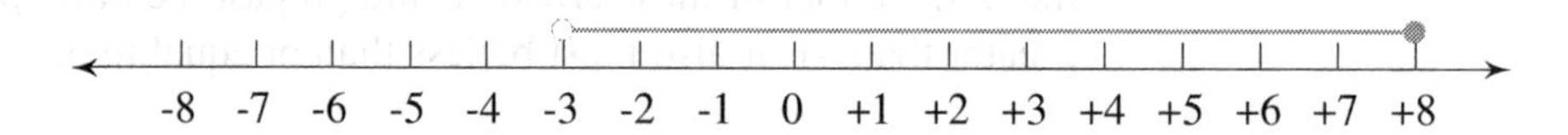

Check Try a value between 8 and –3, such as 0.

$$-9 \leq -2x + 7 < 13$$

$$-9 \leq -2(0) + 7 < 13$$

$$-9 \leq 0 + 7 < 13$$

$$-9 \leq 7 < 13$$ Both parts are true.

EXAMPLE 11 Solve, and state in set notation, and graph: $-4 \leq 2(x + 3) < 1$

Solution Work with all parts at the same time.

$-4 \leq 2(x + 3) < 1$ Distribute 2.

$-4 \leq 2x + 6 < 1$ Subtract 6 from all parts.

$$\underline{-6} \quad \underline{-6} \quad \underline{-6}$$

$$-10 \leq 2x < -5$$

$\dfrac{-10}{2} \leq \dfrac{2x}{2} < \dfrac{-5}{2}$ Divide all parts by 2.

$-5 \leq x < \dfrac{-5}{2}$ $\{x \mid -5 \leq x < -5/2\}$

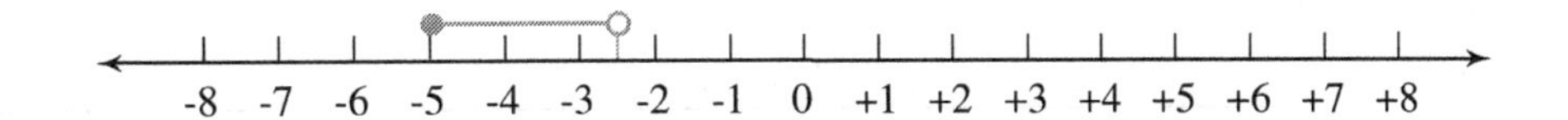

The graph shows all the points that are greater than or equal to –5 *and are also* less then –5/2.

Check $-4 \leq 2(x + 3) < 1$ Let $x = -4$ because it is between –5 and –5/2.

$-4 \leq 2[(-4) + 3] < 1$

$-4 \leq 2[-1] < 1$

$-4 \leq -2 < 1$ True, because –2 is greater than or equal to –4 *and* is also less than 1.

Another way of displaying the graph for Example 11 would be to use a bracket, [, for the closed circle and a parenthesis,), for the open circle.

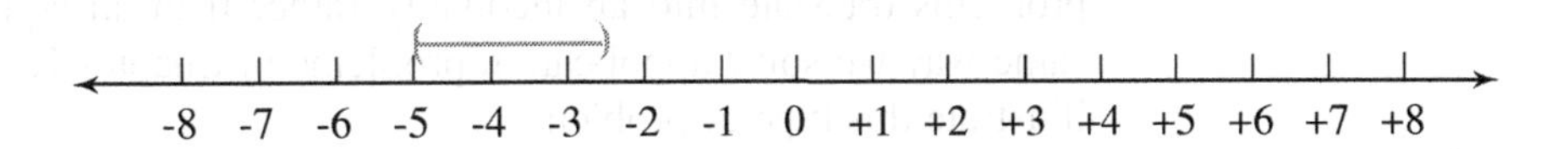

Also, the solution set might be described in terms of an interval, rather than in set notation. The interval states the minimum and maximum values for the solution set. For instance, the interval for Example 11 would be [–5, –5/2). The first number tells the lower limit of the set of numbers and the bracket says to include –5, so it is greater than or equal to –5. The second value, separated by a comma in the interval, tells the upper limit of the solution set and the parenthesis says not to include that value, so it is less than –5/2.

The solution set for Example 1 in this section was $\{x \mid x < 2)$, and the graph showed an open circle at 2 and an arrow going to the left. The arrow would go on forever to the left, or, mathematically, it goes to *negative infinity* $(-\infty)$. Using the new notation:

-8 -7 -6 -5 -4 -3 -2 -1 0 +1 +2 +3 +4 +5 +6 +7 +8 $(-\infty, 2)$

Let's compare some of the earlier solutions and graphs to this new notation.

EXAMPLE 2

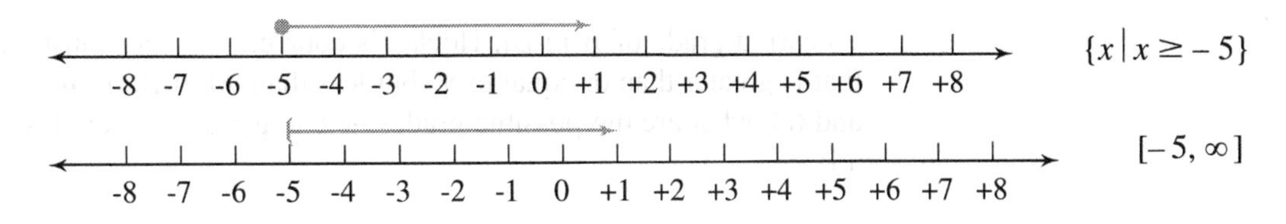

EXAMPLE 7

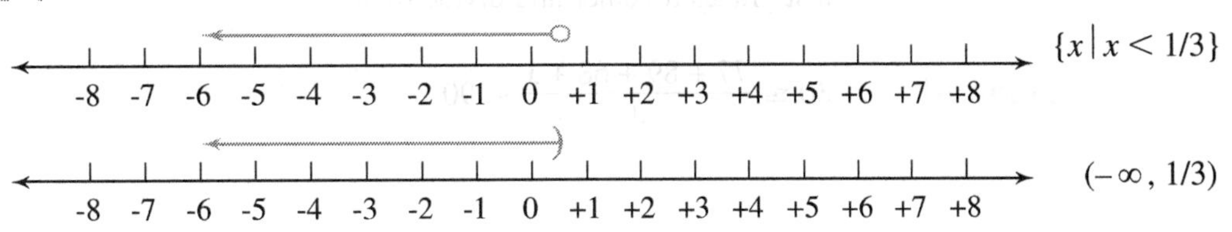

EXAMPLE 9

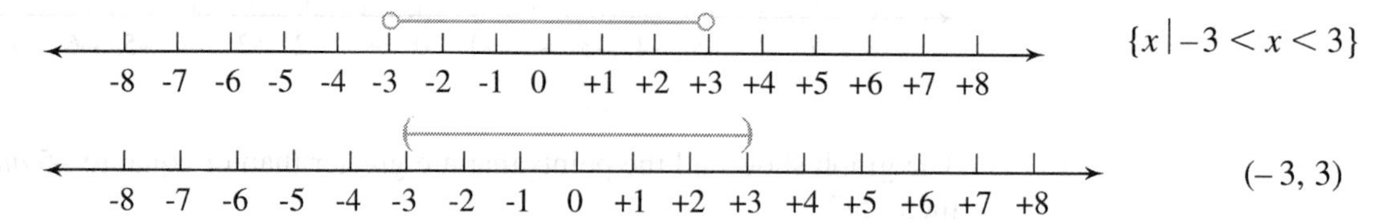

$\{x \mid -3 < x < 3\}$

$(-3, 3)$

Be careful with this last interval. It looks very similar to coordinates for a point on the Cartesian coordinate system (see Chapter 15). You will know whether it is intended as an interval or as coordinates by the context of the problem.

Applications with Inequalities

Some application problems require an answer that is a ragne of numbers rather than a single number. Others might be looking for a minimum or maximum value. Such problems translate into an inequality rather than an equation. We solve them the same way we solve appplication problems in Chapter 10. The examples that follow illustrate this type of problem.

EXAMPLE 12 Find the numbers that satisfy the following relationship: Six more than twice a number is greater than 32.

List Let n = Number

Inequality Six more than twice the number is greater than 32.

$$6 + 2n > 32$$

Solution

$$6 + 2n > 32$$
$$-6 \qquad -6$$
$$2n > 26$$
$$n > 13 \qquad \{n \mid n > 13\}$$

All numbers greater than 13

Check Check in the words of the problem, using some value that is greater than 13, such as 15. Six more than twice fifteen is $2(15) + 6 = 36$, and 36 is greater than 32.

EXAMPLE 13 To earn a grade of B in Dr. Hocken's course, a student must have a grade average that is greater than or equal to 80 but less than 90. If Barry has test grades of 77, 89, and 68, what are the possible grades he can get on the next test so that he will earn a B?

List The grade on the next test is unknown, so call it x. To find an *average*, add the four test grades together and divide by 4.

Inequality

$$80 \leq \frac{77 + 89 + 68 + x}{4} < 90$$

Solution Combine the like terms in the numerator, and then multiply all parts of the inequality by 4 to eliminate the denominator.

$$80 \leq \frac{234 + x}{4} < 90$$

$$4(80) \leq 4\left(\frac{234 + x}{4}\right) < 4(90)$$

$$320 \leq 234 + x < 360$$

$$\underline{-234} \quad \underline{-234} \qquad \underline{-234}$$

$$86 \leq x < 126 \qquad \{x \mid 86 \leq x < 126\}$$

Assuming that a grade greater than 100 is not possible, Barry must get a test grade greater than or equal to 86 and less than or equal to 100.

Check Use a value between 86 and 100, such as 88.

$$80 \leq \frac{77 + 89 + 68 + 88}{4} < 90$$

$$80 \leq \frac{322}{4} < 90$$

$$80 \leq 80.5 < 90 \qquad \text{Both parts are true.}$$

EXAMPLE 14 The Performing Arts Center has 500 seats. If admission for adults is \$5.00 and admission for children is \$3.00, what is the least number of adult tickets that can be sold and still cover the \$1,800 in production costs for a show?

Chart You could use a chart like Chart 12.9 to organize the information.

CHART 12.9

Item	Cost of Each	Number of Each	Total Cost of Each
Adults' tickets	5.00	x	$5.00x$
Children's tickets	3.00	$500 - x$	$3.00(500 - x)$
			Total Needed
			1,800

Inequality

$$5.00x + 3.00(500 - x) \geq 1{,}800$$

$$\text{or } 5x + 3(500 - x) \geq 1{,}800$$

Solution

$$5x + 1{,}500 - 3x \geq 1{,}800$$

$$2x + 1{,}500 \geq 1{,}800$$

$$\underline{-1{,}500} \quad \underline{-1{,}500}$$

$$2x \geq 300$$

$$x \geq 150 \qquad \{x \mid x \geq 150\}$$

At least 150 adults' tickets must be sold.

Check See that at least $1,800 is collected if 150 adults' tickets and 350(500 – 150) children's tickets are sold.

$5(150) = \$750$

$3(350) = \underline{\$1,050}$

$\$1,800$ True.

12.5 EXERCISES

A. Graph each inequality on a number line.

1. $x < 0$
2. $x \leq -1$
3. $x \leq -2$
4. $x > 0$
5. $3 < x$
6. $4 < x$
7. $-4 < x \leq 4$
8. $-3 < x < 3$
9. $0 < x \leq 5$
10. $-1 \leq x < 4$
11. $-3 \leq x \leq 1$
12. $-4 \leq x \leq 0$

B. Solve each inequality, state the answer in set notation (such as $\{x \mid x > 2\}$) or interval notation (such as $(2,\infty)$). Graph the solution on a number line.

13. $x - 3 \leq 4$
14. $5 - x > 4$
15. $2x + 5 > 17$
16. $x + 7 \leq 2$
17. $7 - x > 3$
18. $4x - 3 \geq 17$
19. $11x - 3 \leq 30$
20. $40 \geq 6x - 2$
21. $3 - x < 3x - 13$
22. $x + 2 < 3x - 10$
23. $6x - 5 \leq 11 - 2x$
24. $21 - 3x > 12x - 9$
25. $3(x + 2) > 72$
26. $44 < 2(x - 1)$
27. $5(2x - 1) \leq 6(2x - 1)$
28. $x - (4x - 9) \geq 1 - 2(x + 4)$
29. $7 - (-6 - x) \geq x - (2x + 5)$
30. $-2(3x + 1) < 4(x - 6)$
31. $-8 \leq 2x < -4$
32. $-12 \leq 3x \leq -6$
33. $8 \leq x + 2 < 10$
34. $-12 \leq x - 2 \leq -8$
35. $-7 \leq 2x + 3 < 9$
36. $-5 < 3x + 1 < 7$
37. $2(x + 2) \geq -(x + 5)$
38. $x + 1 \leq 5(1 - x)$
39. $5(x + 1) > x - 1$
40. $2(2x - 1) < 6(x + 1)$
41. $-2 < 3x + 5 < 8$
42. $-5 \leq 2x - 1 < 3$
43. $3 \leq 4 - x < 5$
44. $-4 \leq 8 - x \leq 0$
45. $-4 \leq \frac{2}{3}x - 2 \leq 4$
46. $-5 < \frac{3}{4}x + 1 < 3$

C. For each exercise, organize the information; then write an inequality and solve it.

47. What is the smallest integer such that four less than three times the integer is greater than five plus twice the integer?
48. For what numbers is seven more than twice the number greater than three?
49. Jillian has received grades of 68 and 82 on her first two writing assignments. What grade must she get on the next assignment to have at least an 80 average?
50. Jack scored 69 and 73 during the first two rounds of a golf tournament. What is the minimum he can score in the third round to maintain his average of 70 on this course?
51. The length of a rectangle is three feet more than the width. What is the largest possible length if the perimeter cannot exceed 72 feet?
52. Sarah is going to buy 15 stamps. Some will be 37-cent stamps and the others will be 3-cent stamps. What is the maximum number of 37-cent stamps she can buy if she spends no more than $3.85?

D. Describe the operation performed when going from the previous line to the next line in each solution.

53. $3x + 1 \le -2$

$3x \le -3$	1.
$x \le -1$	2.

54. $-5 \ge 11 - 2x$

$-16 \ge -2x$	1.
$8 \le x$	2.

55. $14 \le 8 - 3x$

$6 \le -3x$	1.
$-2 \ge x$	2.

56. $4x + 5 \ge -3$

$4x \ge -8$	1.
$x \ge -2$	2.

57. $-2 \le 2(x-4) < 6$

$-2 \le 2x - 8 < 6$	1.
$6 \le 2x < 14$	2.
$3 \le x < 7$	3.

58. $-9 \le 3(x+4) < 12$

$-9 \le 3x + 12 < 12$	1.
$-21 \le 3x < 0$	2.
$-7 \le x < 0$	3.
$\{x \mid -7 \le x < 0\}$	4.

59. Select the letter of the correct graph for the inequalities that were solved in Exercises 53, 54, and 57.

a. 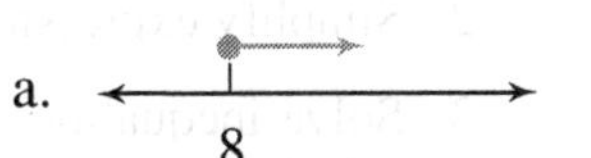

b. 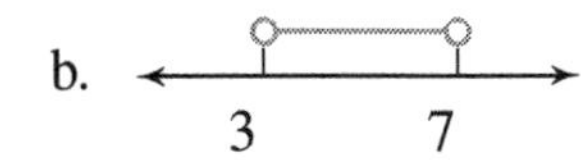

c. 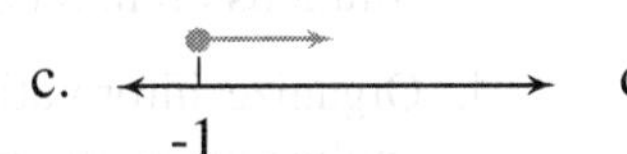

d. 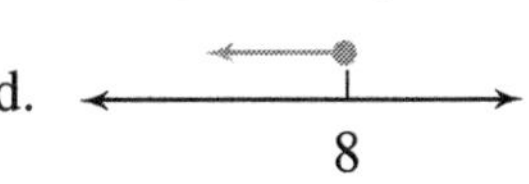

e. 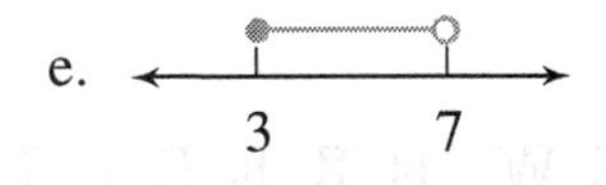

f. 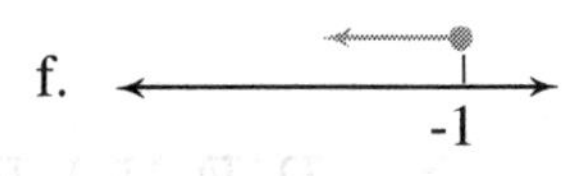

E. Write a set notation or interval notation statement for these descriptions of inequality graphs.

60. An arrow to the left that has a closed circle at –3

61. An arrow to the right that has an open circle at 7

62. A line between –3 and 2 that has an open circle at –3 and a closed circle at 2

63. An arrow to the left that has an open circle at 5

64. A line between –4 and 0 that has parentheses at both ends

65. A line between –7 and 3 that has brackets at both ends

F. Read and respond to the following exercises.

66. Describe, in words, the differences between the solution $x = 5$ and $x > 5$.

67. Compare, in words, the steps in solving $3(x - 2) = 15$ with those used to solve $3(x - 2) \le 15$.

68. Describe, in words, any differences in the steps to solve $-3x > -63$ and $3x > 63$.

12.6 CHAPTER REVIEW

There have been few new concepts to master in this chapter, but many of the concepts from previous chapters come together here. You have continued to work with simplifying expressions and solving equations, which you began in Chapters 8 and 9, and have dealt with signed numbers and the Distributive Principle at the same time. You should recognize the differences between simplifying an expression and solving an equation or inequality.

Now that you have completed this chapter, you should be able to:

1. Simplify expressions or solve equations using the Distributive Principle.
2. Simplify expressions or solve equations that contain grouping symbols.
3. Solve inequalities, state the answers in set or interval notation, and graph the solutions on number lines.
4. Organize information, and then write and solve an equation or inequality for application problems.

REVIEW EXERCISES

A. Simplify these expressions.

1. $7(8 - 3y)$
2. $-2(x + 4)$
3. $-3(b + 6)$
4. $4(9 - 2a)$
5. $1/4(8x + 12)$
6. $2(b + 3) - (b - 4)$
7. $5(a + 9) - (2a + 3)$
8. $2/3(9x - 24)$
9. $6 - [3r - 4(r + 5)]$
10. $7(6p - 4) + 3(5p + 1)$
11. $2(n - 5) + 5(3n + 1)$
12. $10c - 4[c + 2(5c + 1)]$
13. $3[2(5p + 3)]$
14. $5(3x - 2y + 4z - 1)$
15. $3(2x - 4y + 5)$
16. $4[3(5a - 2)]$
17. $4[t - (t - 3)]$
18. $2\{5[2 + 3(5d - 2)]\}$
19. $10 - \{2[3c + 5(2c - 1)]\}$
20. $-2[3x - (x - 5)]$
21. $2y - \{3[y + 2(2y + 3)]\}$
22. $-3\{2[m - (m - 5)]\}$
23. $-2\{5[3d - 2(d + 4)]\}$
24. $2\{-3[2b - (b + 4)]\}$

B. Solve and check each of these equations.

25. $6x - 2 = 7x + 5$
26. $10 - 4y = 16 - y$
27. $2m - 4 = 6m$
28. $2(5n - 4) = 22$
29. $3(2p - 5) = 45$
30. $2c - 10 = 7c$
31. $5 - (-2a + 9) = 2(a - 1)$
32. $3x - 7 = 5(2x + 7)$
33. $2y - 5 = 3(4y + 5)$
34. $4p + 3 = 7 - (-8p + 5)$
35. $7 - 2(r - 2) = 5r - 3$
36. $7 - 2(2m + 5) = -3m + 4$
37. $10[4 - (4x - 8)] = 4(10 - 6x)$
38. $2(3y + 5) - 1 = 6y + 1$
39. $-8[2y - 4(4y - 6) + 1] = 4(y + 4)$
40. $3[2 - 4(a - 1)] = 2(a - 5)$
41. $-2[4 - (3b + 2)] = 3b + 5$
42. $5 + 3[1 + 2(2x - 3)] = 2(x + 5)$

C. Solve each inequality, express the solution in set or interval notation, and graph the solution on a number line.

43. $-x \geq -2$
44. $4 < x - 1$
45. $7 < x + 5$
46. $-x \geq 0$
47. $6(x + 1) > 7 - x$
48. $-4 < x - 3 \leq 1$
49. $0 < x + 2 < 3$
50. $4 - 3x \leq 2(x + 1)$

D. Organize the information, write an equation, solve it, and give the answers requested.

51. Three co-workers contributed toward a wedding gift for their secretary. The first gave $10 more than the second. The third worker gave twice as much as the first. Altogether they donated $90. How much did each contribute?

52. Three numbers have a sum of 39. The second number is four less than the first, and the third number is equal to three times the second. What are the three numbers?

53. Jillian goes to the drugstore to buy some snacks. Candy bars cost 45 cents each, and gum is 30 cents a pack. The number of candy bars that she buys is twice as large as the number of packs of gum, and she spends $2.40. How many candy bars does she buy?

54. The larger of two numbers is three more than twice the smaller. And when the larger number is subtracted from the smaller, the result is –15. What are the numbers?

55. Find three consecutive odd numbers such that the sum of three times the largest and twice the smallest is 97.

56. The total cost for padding and upholstery fabric to re-cover a small living room chair will be $73.50. The amount of fabric to be used is two-and-a-half times the amount of padding needed. Padding costs $3.25 per yard, and fabric costs $8.50 per yard. How much fabric will be needed?

57. The Birch and Maple Tree Farm needs to run fencing around its newly planted trees. The length of this new rectangular plot is 4.5 times its width. If the perimeter of the lot is 440 yards, what is its length?

58. Find three consecutive even integers such that the difference between three times the smallest and four times the largest is zero.

59. To mix a perfect blend of chocolate almond coffee, the clerk at the Roasted Bean needs to grind 1 ounce more of chocolate beans than the amount he uses of the almond-flavored beans. How many ounces of each will he mix if Mr. Harris asks for $4.00 worth of the blend? (The almond-flavored beans cost $0.95 an ounce, and the chocolate beans cost $0.70 an ounce.)

60. A propeller-driven plane leaves the Municipal Airport traveling at 200 mph. Two hours later, a jet leaves the same airport and follows the same flight path. It travels at 600 mph.
    a. In how many hours will the jet overtake the propeller plane?
    b. How far will the propeller plane have traveled in that time?

61. The width of a rectangular yard is one-fourth its length. Find the width if the perimeter is 250 feet.

62. A car and a bus set out from Rochester, New York, traveling in opposite directions. The car averaged 40 miles per hour less than twice the speed of the bus. After 2 hours, they were 220 miles apart. How fast was the car traveling?

63. Becky is putting a wallpaper border around her baby's nursery walls. She uses 46 feet of border paper. If the width of the room is 7 feet less than the length, what are the room's dimensions?

64. Four less than twice a number is the same as five times the sum of the number and 1. What is the number?

65. If twice as much money is invested at 4 1/2% as at 3 1/2 % annual interest, the total amount of interest in 5 years would be $6,125. What is the size of each investment? Use $I = PRT$.

66. $5,000 more is invested in a 2-year certificate of deposit paying 6 1/2% per year than in a money market account paying 6% annual interest. Together they earn simple interest of $1,650 in the 2-year period. How much was put into each investment?

67. How many liters of 22% and 12% acetic acid should be mixed to produce 48 liters of 15% acetic acid?

68. How many gallons of 1/2% milk (1/2% fat) should be mixed with 24 gallons of 2% milk (2% fat) to produce a mixture that is 1% fat?

E. For each exercise, write an inequality and then solve it.

69. Three times a number is greater than five times the sum of the number and four. Find the possible values for the number.

70. One-half a number is less than twice the difference between the number and 6. Find the possible values for the number.

71. Jo is preparing a new garden plot. She wants to use no more than one 72-foot package of edging to outline this rectangular plot. The width must be 20 feet. What is the maximum length she can make this garden?

72. Marian has bowled a 126 and a 147 in her league this evening. What score can she get in her third game to exceed her three-game average of 145? (A perfect game in bowling is 300.)

F. State what error has been made in each simplification or solution that follows and correctly complete the exercise.

73. $-6(5x-3)$

$= -30x - 18$

74. $3(2x + 2y - 7z)$

$= 6x + 6y - 7z$

75. $7(a-1) - (5-a)$

$= 7a - 7 - 5 - a$

$= 6a - 12$

76. $14 = -2(3x - 4)$

$14 = -6x - 4$

$\underline{+4} \quad \underline{+4}$

$18 = -6x$

$\frac{18}{-6} = \frac{-6x}{-6}$

$-3 = x$

77. $14 - c = 11$

$\underline{-14} \quad \underline{-14}$

$-c = -3$

78. $5(2y-3) = -3(4y-1)$

$10y - 15 = -12y + 3$

$\underline{-12y} \quad \underline{12y}$

$-2y - 15 = +3$

$\underline{+15} \quad \underline{+15}$

$-2y = 18$

$\frac{-2y}{-2} = \frac{18}{-2}$

$y = -9$

79. $3 - 2a > 7$

$\underline{-3} \quad \underline{-3}$

$-2a > 4$

$\frac{-2a}{-2} > \frac{4}{-2}$

$a > -2$

$\{a \mid a > -2\}$

80. $12(2m-7) = 5(3-7m) + 9$

$24m - 84 = 15 - 35m + 9$

$24m - 84 = 24 - 35m$

$\underline{-24m} \quad \underline{-24m}$

$-84 = 24 - 59m$

$\underline{+84} \quad \underline{+84}$

$108 = -59m$

$\frac{108}{-59} = \frac{-59m}{-59}$

$-2 = m$

81. $-6 < x + 2 < 3$

$-6 < x + 2 < 3$

$\underline{-2} \quad \underline{-2}$

$-8 < x \quad < 3$

$\{x \mid -8 < x < 3\}$

82. $5 \leq 3T + 2$

$5 \leq 3T + 2$

$\underline{-2} \quad \underline{-2}$

$3 \leq 3T$

$\frac{3}{3} < \frac{3T}{3}$

$1 \leq T$

$\{T \mid T \leq 1\}$

83. $-8R > 56$

$\frac{-8R}{-8} > \frac{56}{-8}$

$R > -7$

$\{R \mid R > -7\}$

84. A food company wants to improve its fruit punch product by adding more pure orange juice. How much orange juice (100% orange juice) should be added to 50 gallons of the fruit punch that is 20% orange juice to create a more healthful mixture that is 50% orange juice?

$$1.00x + 0.2(50 - x) = 0.5(50)$$
$$1x + 10 - 0.2x = 25$$
$$0.8x + 10 = 25$$
$$\underline{-10} \quad \underline{-10}$$
$$0.8x = 15$$
$$\frac{0.8x}{0.8} = \frac{15}{0.8}$$

$x = 18.75$ gallons of orange juice

G. Read and respond to the following exercises.

85. Graph the solutions $x - 3$ and $x \geq 3$ and describe, in words, the similarities and differences between the graphs.

86. Describe, in words, any differences in procedure between solving $2x - 4 = 10$ and $2x - 4 \leq 10$.

87. Describe, in words, the differences among an equation, an expression, and an inequality.

88. Do you organize information in some manner before trying to write an equation for an application problem? What importance do you see in doing (or not doing) this step?

89. Describe a real-life situation (not necessarily related to math) wherein the problem-solving skills you have used in Chapters 10 and 12 could be applied. Explain each step of the process.

90. Write a mixture application problem that would have this equation: $8.95(x) + 6.50(30 - x) = 7.25(30)$.

H. The following problems may require a little more thought than the previous ones. Organize the information, write an equation, solve, and give all the answers requested.

91. Mrs. Zink is going to invest $5,000 in a certificate of deposit paying 7% annual interest. How much *more than $5,000* should she invest in municipal bonds paying 6 1/2% if she wants to have $2,415 in interest from the two investments at the end of 3 years? Use $I = PRT$.

92. Tom is having his family room enlarged. The original room had a width that was 6 feet less than the length. With remodeling, the length will not change but the width will double. The perimeter of the new room will be 66 feet. What were the dimensions of the original room?

93. The width of a rectangle is one-half the length. If the width is increased by 4 feet and the length is decreased by 3 feet, the perimeter of the new rectangle will be 62 feet. What are the dimensions of the original rectangle?

94. $5,000 more is invested in a 2-year certificate of deposit paying 6 1/2% per year than in a money market account paying 6% annual interest. Together they earn simple interest of $1,650 in the 2-year period. How much was put into each investment?

CHAPTER TEST

Simplify each expression completely.

1. $6(x + 2)$
2. $5 - (2n - 1)$
3. $2[8y - 2(3y + 2)]$

Solve and check each equation.

4. $3(x + 1) = 33$
5. $6 - (t + 1) = 13$
6. $24 = -2(p - 5)$
7. $28 - (2A - 1) = 7 - 2(A + 1)$
8. $2[K - 3(K + 2)] = -24$
9. $-40 = 5(T + 1)$
10. $-4 = 5 - (c - 2)$

11. $3(2y + 5) - 4(y - 2) = -1$
12. $5[2 - (2w - 4)] = 2(5 - 3w)$

Solve each inequality, express the solution in set or interval notation, and graph the solution on a number line.

13. $2x + 3 > 5$
14. $3 - 4x \leq -13$
15. $-14 \leq 4x + 6 < 18$

For each application problem, organize the information, write an equation, and solve it.

When Pat drove to Jamestown on Friday evening to visit her parents, it took her 6 hours. When she made the return trip on Sunday, with lighter traffic, the trip took only 5 hours. What was her average speed on the trip on Sunday if she averaged 10 miles per hour faster than on Friday?

16. Equation ____________________
17. Solution ____________________

How many quarts of 8% fat buttermilk and 3% fat buttermilk should be blended to produce 25 quarts of 5% fat buttermilk?

18. Equation ____________________
19. Solution ____________________
20. In words, describe the differences among the following:

$5x + 3$

$5x + 3 = 0$

$5x + 3 > 0$

STUDY SKILLS MODULE

13 Working in Groups

More and more students in classrooms on college campuses are being given assignments where they must work on a group project. With the variety of class, study, and work schedules that students have, the difficulty of getting even a small group of students together outside of class seems almost insurmountable. You are fortunate that there are very few outside group projects assigned in math classes. However, instructors do ask students to work in small groups to solve problems in math classes. Why this push for group work? Why can't individual students be left alone to study and work on their own? Why do you need to learn to work together? The answer is simple. Most work situations in our society today are too complex for an individual to handle on his or her own. Employers are looking for workers who can work together to solve problems, make presentations, and complete projects. Small-group work in a math class is a relatively safe environment in which to develop good collaborative skills.

Several qualities characterize well-functioning groups:

1. A leader surfaces who can help to guide the group.
2. Someone acts as record-keeper who can help keep the group on-task.
3. Each member of the group makes a contribution to the discussion or project.
4. Members of the group discuss the problem in a supportive manner, listening to each person's opinion.
5. The group produces a final product in which each group member shares ownership.

In the case of group work in a math class, the instructor may assign a "part" to each student or allow the group itself to divide up the work. When students work in groups to solve math problems, it is very important that each person in the group understand how the solution is reached. It is the responsibility of each group member to try to help all the other group members. Personality differences need to be addressed on a respectful and constructive level. Learning how to resolve differences on this level can be an invaluable tool in an employment situation. If you work in an ongoing math group, try each of the different "jobs" in the group. Even if you don't see yourself as a leader, try it once or twice when you feel confident about the topic at hand.

Good people skills are becoming very important in our society, and learning how to be a contributing member of a productive group is one of those skills.

1. Do you like to work better by yourself or as part of a group? Explain your preference.

2. Discuss a situation where you were a member of a group and were very pleased about the outcome from that situation. What things contributed to the success?

3. Describe a group situation that was not productive, and analyze what went wrong and how it could have been fixed.

4. List three personality traits that you possess that make you a valuable member of a group. Explain why.

CHAPTER

13 Polynomial Operations

LEARNING OBJECTIVES

When you have completed this chapter, you should be able to:

1. Add and subtract monomials.

2. Multiply and divide monomials using the laws of exponents.

3. Simplify expressions that contain negative exponents.

4. Add and subtract polynomials.

5. Multiply polynomials:

- **a.** Multiply a monomial times a polynomial using the Distributive Principle.
- **b.** Multiply a polynomial times a polynomial using the Distributive Principle.
- **c.** Multiply a binomial times a binomial using FOIL.

6. Divide polynomials:

- **a.** Divide a polynomial by a monomial.
- **b.** Divide a polynomial by a polynomial.

7. Work with scientific notation.

In the work you have been doing in Chapters 8 through 12, you have learned how to add and subtract like terms and how to multiply variable terms by numerical constants. You will extend those operations in this chapter.

13.1 Adding and Subtracting Monomials

Let's review some definitions and concepts that you have been working with since Chapter 8. An algebraic expression contains one or more terms and has no equality symbol. Algebraic terms are made up of variables and/or constants. In an algebraic expression, terms are connected by addition or subtraction.

In this chapter, algebraic expressions will be referred to as polynomials. A **polynomial** refers to an algebraic expression of one or more terms that has positive integer exponents. A polynomial can have an infinite number of terms.

Polynomials

$8Y - 4$

$7b^2 + 4b - 15$

$18k$

$2c + 9d - 5e + 14f$

> **Definition of Polynomial**
>
> An algebraic expression of one or more terms in which the terms are of the form ax^n where n is a positive integer or zero.

This definition will become clearer after you have studied zero exponents in Section 13.3.

Polynomials of one, two, or three terms have specific names. One-term polynomials are **monomials**, two-term polynomials are **binomials**, and polynomials of three terms are **trinomials**.

Monomials	**Binomials**	**Trinomials**
$3r^2s^3t$	$2x + 3$	$2m^2 - 3m + 7$
$4x$	$a^2 - b^2$	$c^2 - 7cd + 5d^2$
-19	$8f - g$	$-11 + 5t + 8s$

The chapter will begin by looking at operations with one-term expressions.

When you add or subtract monomials, you combine terms such as $7s$ and $-15s$. Such quantities can be added or subtracted only if they are *like terms*; that is, their variable and exponent parts are exactly the same. You have been combining like terms since Chapter 8. Combining these terms can be viewed as an application of the Distributive Principle, though it is seldom written this way when actually being calculated.

$7s + (-15s) = [7 + (-15)]s = -8s$

You can line up the terms vertically or horizontally. Remember to apply signed-number rules.

$$\begin{array}{r} 7s \\ + \underline{-15s} \\ -8s \end{array} \qquad \text{or} \qquad 7s + (-15s) = -8s$$

If you were to subtract $-13x$ from $9x$, your work would look like this:

$$\begin{array}{l} \quad 9x \\ \text{S} \\ \underline{-(-13x)} \end{array} \qquad \text{becomes} \qquad \begin{array}{r} 9x \\ \text{A} \quad \\ \underline{+(+13x)} \\ 22x \end{array}$$

$$\text{or } 9x \overset{\text{S}}{-} (-13x) = 9x \overset{\text{A}}{+} (+13x) = 22x$$

EXAMPLE 1 Find the difference between $-4a^2b^2c$ and $3a^2b^2c$.

Solution $-4a^2b^2c$ and $+3a^2b^2c$ are like terms because the variable part, a^2b^2c, is the same in both terms.

$$\begin{array}{l} -4a^2b^2c \\ \text{S} \\ \underline{-+3a^2b^2c} \end{array} \qquad \text{becomes} \qquad \begin{array}{r} -4a^2b^2c \\ \text{A} \quad \\ \underline{+-3a^2b^2c} \\ -7a^2b^2c \end{array}$$

or

$$-4a^2b^2c \overset{\text{S}}{-} (+3a^2b^2c)$$

$$= -4a^2b^2c \overset{\text{A}}{+} (-3a^2b^2c)$$

$$= -7a^2b^2c$$

EXAMPLE 2 Simplify: $11x^2 + (-2x^2) + (-7x^2) + 5x^2$

Solution There are two ways to simplify this problem. You can do it in the order given or do it by rearranging the terms (Associative and Commutative Properties of Addition). Both will be shown.

To do the problem in the order given, write

$$\underline{11x^2 + (-2x^2)} + (-7x^2) + 5x^2$$

$$= \underline{9x^2 + (-7x^2)} + 5x^2$$

$$= \underline{2x^2 + 5x^2}$$

$$= 7x^2$$

To do the problem by rearranging the terms, combine all the positive terms and then all the negative terms.

$$11x^2 + (-2x^2) + (-7x^2) + 5x^2$$
$$= \underline{11x^2 + 5x^2} + \underline{(-2x^2) + (-7x^2)}$$
$$= 16x^2 \quad + \quad (-9x^2)$$
$$= 7x^2$$

EXAMPLE 3 Simplify: $2a^2b - 3ab + 5ab^2$.

Solution There are no like terms, so there is nothing to add or subtract, and the expression is already in simplest form.

EXAMPLE 4 From the sum of $8c^2d^3$ and $(-11c^2d^3)$, subtract $(-5c^2d^3)$.

Solution Be sure to write a subtraction indicator and then $(-5c^2d^3)$.

Sum: $\begin{array}{r} 8c^2d^3 \\ \underline{+ -11c^2d^3} \\ -3c^2d^3 \end{array}$ Difference: $\begin{array}{r} -3c^2d^3 \\ \text{S} \\ \underline{- -5c^2d^3} \end{array}$ becomes $\begin{array}{r} -3c^2d^3 \\ \text{A} \\ \underline{+ +5c^2d^3} \\ +2c^2d^3 \end{array}$

or

$$\underbrace{8c^2d^3 + (-11c^2d^3)}_{\text{A}} \overset{\text{S}}{-} (-5c^2d^3)$$
$$= -3c^2d^3 + (+5c^2d^3)$$
$$= 2c^2d^3$$

Self-Check Simplify each expression.

a. $7a^2bc - a^2bc$ b. $4x^2y + 3xy - 5x^2y + xy$

Answers a. $6a^2bc$ b. $-x^2y + 4xy$

13.1 EXERCISES

A. Add the monomial terms where possible.

1. $\begin{array}{r} -3a \\ \underline{+ +4a} \end{array}$
2. $\begin{array}{r} 6x^2y \\ \underline{+ -2x^2y} \end{array}$
3. $\begin{array}{r} -15g \\ \underline{+ -12g} \end{array}$
4. $\begin{array}{r} +9abc^3 \\ \underline{+ -9abc^3} \end{array}$
5. $12r^2s + (-7r^2s)$
6. $(-150x^2) + (-70x^2)$
7. $4b^2d^2 + 3b^2d$
8. $(-7yz) + (-3yz)$
9. $8m^2n + (-3m^2n)$
10. $5g^2h + 4gh^2$

B. Subtract the polynomials where possible.

11. $\begin{array}{r} -14k \\ \underline{- +11k} \end{array}$
12. $\begin{array}{r} -5a^2b \\ \underline{- -7a^2b} \end{array}$
13. $\begin{array}{r} 6w^3 \\ \underline{- -w^3} \end{array}$
14. $\begin{array}{r} +8rs^2 \\ \underline{- +12rs^2} \end{array}$
15. $6wxy^2 - (-5wxy^2)$
16. $(-3f^2) - (-5f^2)$
17. $8b^3c - 17b^3c$
18. $a^2b^2c^2 - 7a^2b^2c^2$
19. $44xyz - (-xyz)$
20. $-9m^2n - (-12m^2n)$

C. Perform the indicated operation(s) in each of these exercises.

21. $16gh + (-5gh) + 3gh$
22. $7q^3 + (-4q^3) + (-5q^3)$
23. $-7x^2 + 9x^2 - (-4x^2)$
24. $8m - 11m + 3m + (-m)$
25. $11a^2b^3 - 8a^2b^3 - 6a^2b^3 + a^2b^3$
26. $49x^2y - 11x^2y + 33x^2y - 16x^2y$
27. From the sum of $(-9mn)$ and $(17mn)$, subtract $(-3mn)$.
28. Subtract $(14fgh)$ from the sum of $(-6fgh)$ and $(11fgh)$.
29. Find the sum of $16r^2s$, $-3r^2s$, and $-6r^2s$.
30. Find the sum of $(44z^2)$ and $(-56z^2)$, subtract $(20z^2)$.
31. Find the difference between $153p^3q^2$ and $(-119p^3q^2)$.
32. Find the sum of $75r^2s^3t^2$ and $(-87r^2s^3t^2)$.
33. Find the sum of $-86a^2b^2c^3$, $14a^2b^2c^3$, and $44a^2b^2c^3$.
34. Find the difference: $-128m^2n^3 - 116m^2n^2$.
35. From the sum of $(-8c^2d^3)$ and $(14c^2d^3)$, subtract $(-3c^2d^3)$.

D. Read and respond to the following exercises.

36. Compare a trinomial to binomial. State some similarities and some differences.
37. What process that was presented in Chapter 8 is the same thing as adding or subtracting monomials?
38. Do you prefer to simplify the addition or subtraction exercises horizontally or vertically? Explain why.
39. Find the sum of $-3p^2q^3$ and $5p^2q^2$. Explain your answer.
40. In simplifying Exercise 24, how do you know which terms to combine first?

13.2 Multiplying Monomials

When multiplying monomials, you do not need to work with only like terms as you do when adding and subtracting. Look at these monomial multiplications and see if you can discover a pattern and write a rule to describe that pattern.

In $2^3 \cdot 2^2$, the exponent (or power) tells you how many times the base is used as a factor. This means that

$$2^3 \cdot 2^2 = \underline{2 \cdot 2 \cdot 2} \cdot \underline{2 \cdot 2}$$

But, by the definition of exponents, $2 \cdot 2 \cdot 2 \cdot 2 \cdot 2$ can be written 2^5. Therefore,

$$2^3 \cdot 2^2 = 2^5$$

Note: Variables and numbers with no exponent showing are understood to have an exponent of 1.

$k = k^1 \qquad 2 = 2^1$

EXAMPLE 1 Multiply: $k^2 \cdot k$

Solution

$k^2 \cdot k$

$= \underline{k \cdot k} \cdot \underline{k}$

$= k^3$

EXAMPLE 2 Multiply: $a^4 \cdot a^3$

Solution $a^4 \cdot a^3$

$= \underline{a \cdot a \cdot a \cdot a} \cdot \underline{a \cdot a \cdot a}$

$= a^7$

Can you see a pattern in the expressions and answers? What relationship is there between the exponents in the original factors and the exponent in the answer? What happens to the base quantities?

Did you notice that the exponent in the answer is the *sum* of the exponents of the factors? And did you notice that the base quantity did *not* change?

$2^3 \cdot 2^2 = 2^{3+2} = 2^5$ $k^2 \cdot k = k^{2+1} = k^3$ $a^4 \cdot a^3 = a^{4+3} = a^7$

Multiplying Monomials with the Same Base

To multiply exponential quantities that have the same base, keep the base the same and add the exponents.

$a^b \cdot a^c = a^{b+c}$

If the monomials have numerical coefficients, then the Associative and Commutative Properties for Multiplication enable you to rearrange the factors, as in Examples 3 through 6.

EXAMPLE 3 Multiply: $2b^3 \cdot 3b^2$

Solution $2b^3 \cdot 3b^2$

$= \underline{2 \cdot b \cdot b \cdot b} \cdot \underline{3 \cdot b \cdot b}$

$= 2 \cdot 3 \cdot b \cdot b \cdot b \cdot b \cdot b$

$= 6b^5$

EXAMPLE 4 Multiply: $(2x^3)(3x^4)$

Solution $(2x^3)(3x^4)$

$= 2 \cdot 3 \cdot x^3 \cdot x^4$

$= 6x^7$

EXAMPLE 5 Multiply: $11a(-2a^2b)$

Solution $11a(-2a^2b)$

$= 11 \cdot -2 \cdot a \cdot a^2 \cdot b$

$= -22a^3b$

EXAMPLE 6 Multiply: $(3r^2s)(2rs)(-5s^2)$

Solution $(3r^2s)(2rs)(-5s^2)$

$= 3 \cdot 2 \cdot -5 \cdot r^2 \cdot r \cdot s \cdot s \cdot s^2$

$= -30r^3s^4$

You do not need to write the intermediate steps in such problems. First multiply the numerical coefficients; then multiply the variables using the rule of exponents.

Self-Check Try these exercises. If necessary, look at the examples.

1. $3ab(2a^2b)$ 2. $-2(-3x^2y)(2y^2)$ 3. $c^2d^2(cd^2)(c^2)$

Answers:

1. $6a^3b^2$ 2. $12x^2y^3$ 3. c^5d^4

There is another type of monomial multiplication. See if you can write a rule to describe the process you observe in Examples 7, 8, and 9.

EXAMPLE 7 Simplify: $(x^2)^3$

Solution $(x^2)^3$

$= (x \cdot x)^3$

$= \underline{x \cdot x} \cdot \underline{x \cdot x} \cdot \underline{x \cdot x}$

$= x^6$

EXAMPLE 8 Simplify: $(y^3)^3$

Solution $(y^3)^3$

$= (y \cdot y \cdot y)^3$

$= \underline{y \cdot y \cdot y} \cdot \underline{y \cdot y \cdot y} \cdot \underline{y \cdot y \cdot y}$

$= y^9$

EXAMPLE 9 Simplify: $(a^2)^4$

Solution $(a^2)^4$

$= (a \cdot a)^4$

$= \underline{a \cdot a} \cdot \underline{a \cdot a} \cdot \underline{a \cdot a} \cdot \underline{a \cdot a}$

$= a^8$

Can you see a pattern in $(x^2)^3 = x^6$, $(y^3)^3 = y^9$, and $(a^2)^4 = a^8$? In each case, when you raised an exponent form of the base to the power represented by an exponent, the base remained *the same* and the exponents were *multiplied* together to get the answer.

Raising a Base in Exponent Form to a Power

To raise a base in exponent form to a power, keep the base the same and multiply the exponents.

$(a^b)^c = a^{bc}$

Self-Check Simplify each of these expressions, following the preceding examples if necessary.

1. $(m^3)^2$ 2. $(q^3)^4$ 3. $(y^6)^3$

Answers:

1. m^6 2. q^{12} 3. y^{18}

In exercises where you are told to raise a product to the power represented by an exponent, raise each factor in the product to that power and apply the rules of exponents. For example, $(2b)^4$ is the same as $2b \cdot 2b \cdot 2b \cdot 2b = 2^4b^4 = 16b^4$, but it is easier to apply the rule and get $(2b)^4 = 2^4b^4 = 16b^4$.

Raising a Product to a Power

To raise a product to a power, raise each factor to the power represented by that exponent.

$(ab)^c = a^cb^c$

EXAMPLE 10 Simplify: $(a^2b)^3$

Solution

$(a^2b)^3$

$= (a^2)^3(b)^3$

$= a^6b^3$

EXAMPLE 11 Simplify: $(3x^2)^4$

Solution

$(3x^2)^4$

$= (3)^4(x^2)^4$

$= 81x^8$

Raising a Fraction to a Power

To raise a fraction to a power, raise both the numerator and the denominator to that power.

$$\left(\frac{a}{b}\right)^c = \frac{a^c}{b^c}$$

A similar rule applies to raising a fraction (quotient) to a power. You will raise each part (numerator and denominator) of the fraction to that power.

EXAMPLE 12 Simplify: $\left(\frac{a^2}{4}\right)^3$

Solution

$$\left(\frac{a^2}{4}\right)^3$$

$$= \frac{(a^2)^3}{4^3}$$

$$= \frac{a^6}{64}$$

EXAMPLE 13 Simplify: $\left(\frac{2x}{3y^3}\right)^2$

Solution

$$\left(\frac{2x}{3y^3}\right)^2$$

$$= \frac{(2x)^2}{(3y^3)^2}$$

$$= \frac{2^2x^2}{3^2(y^3)^2}$$

$$= \frac{4x^2}{9y^6}$$

Self-Check Simplify each expression.

1. $(-2b^2)^2$ 2. $(5ax^2)^3$ 3. $\left(\frac{m^2}{n}\right)^3$

4. $\left(\frac{0.4T^2}{R^3}\right)^2$ 5. $\left(\frac{2f^3}{3}\right)^2$

Answers:

1. $4b^4$ 2. $125a^3x^6$ 3. $\frac{m^6}{n^3}$ 4. $\frac{0.16T^4}{R^6}$ 5. $\frac{4f^6}{9}$

Pay close attention to how the problem is written! $(4x)^2$ and $4x^2$ are not equivalent because $(4x)^2 = 4^2x^2 = 16x^2$. In $4x^2$, only the x is raised to the second power. In the expression $4(x^2y^3)^5$, the exponent ^{5} is applied only to the factors inside the parentheses, not to the factor 4, which is outside. Thus, $4(x^2y^3)^5 = 4(x^2)^5(y^3)^5 = 4x^{10}y^{15}$.

Because an exponent applies only to the quantity that is immediately to its left, the expressions $(-2)^4$ and -2^4 are not equivalent. In $(-2)^4$, the quantity inside the parentheses is raised to the fourth power: $(-2)(-2)(-2)(-2) = 16$. In -2^4, only 2 is raised to the fourth power, and the negative sign means "the opposite of." Therefore, $-2^4 = -(2 \cdot 2 \cdot 2 \cdot 2) = -(16) = -16$, because the opposite of $+16$ is -16.

EXAMPLE 14 Simplify: -5^2

Solution $-5^2 =$ The opposite of 5 squared $= -(5 \cdot 5) = -(25) = -25$

EXAMPLE 15 Simplify: $(-5)^2$

Solution $(-5)^2$ = The quantity, – 5, squared = $(-5)(-5) = 25$

13.2 EXERCISES

A. Simplify each expression using the rules of exponents introduced in this section.

1. $y^2 \cdot y^3$
2. $a^4 \cdot a^2$
3. $b^3 \cdot b$
4. $m^2 \cdot m^2$
5. $a^3 \cdot a^2$
6. $m^3 \cdot m$
7. $z^3 \cdot z^4$
8. $d \cdot d^2$
9. $2^2 \cdot 2^3$
10. $3^2 \cdot 3^2$
11. $2ab(3a^2b)$
12. $3z(-2z^2)$
13. $A^2 \cdot A^3 \cdot A$
14. $x \cdot x^2 \cdot x^4$
15. $(-y)^3$
16. $(-x)^4$
17. $y^2 \cdot y^4 \cdot y$
18. $b \cdot b^2 \cdot b$
19. $(2x)^3$
20. $(3y)^2$
21. $(2m^2)^3$
22. $(6b^2)^2$
23. $\left(\frac{-2g^2}{3}\right)^3$
24. $\left(\frac{-x^3}{2y}\right)^2$
25. -2^6
26. -3^2
27. $(-2)^6$
28. $(-3)^2$
29. -1^4
30. -1^6

B. Simplify each exercise.

31. $(-3a^2)(2ab)(a^2b^2)$
32. $x^2y(3xy^3)$
33. $(-4x^2)(-8x^3z^2)$
34. $(7m^2n)(-2mn)(n^2)$
35. $(-a^3b^2)^5$
36. $(-4x^3y^2)(-9x^5z^3)$
37. $y(x^2y^2)^3$
38. $(-5m^2)^3$
39. $(-4abc)^3$
40. $8a(a^2b)^3$
41. $-7xy^2(2xy)$
42. $(3ab^2)^2(5a^2b)$
43. $(2mn^2)^3(6m^2n^2)$
44. $6a^2(-3a^2b)$
45. $(3/4\ PR^2)^2$
46. $(5/6\ A^3B^3)^2$
47. $\left(\frac{2m^2n}{3}\right)^2$
48. $\left(\frac{3a^2b^2}{4}\right)^2$
49. $(3a^2)^2(2a^3)^3$
50. $(4x^3)^2(3x^2)^3$
51. $\left(\frac{a^3b^2}{2}\right)^3$
52. $\left(\frac{-x^2y}{2}\right)^3$
53. $(2m^2)(3mn^2)(-mn)$
54. $\left(\frac{3}{4}c^3\right)^2$
55. $(2/5\ R^2S)^3$
56. $6m(3m^4)^2$
57. $(5x^4)^2(-2x^5)^3$
58. $(2y^2)^3(y^4)^2$
59. $(8a^2b^3)^2$
60. $(-7m^2n^2)(-2mn)(-3m^2n)$

C. Read and respond to the following exercises.

61. Are $R^2 \cdot R^3$ and $(R^2)^3$ equivalent? Explain your answer.
62. Is $2^3 \cdot 2^2 = 4^5$? Explain your answer.
63. Explain, in words, the process you would use to simplify $\{[(-2a^2b)^2]^2\}^2$
64. Why is $(y/3)^2 = y^2/9$?
65. Could you use the Odd-Even Rule of Signs to determine the sign of the answer for Exercise 63? Why or why not?

13.3 Dividing Monomials

The rules for division will be developed as the rules for multiplication were developed. You will examine some step-by-step solutions to division problems and see if you can spot a pattern.

EXAMPLE 1 Divide: $\frac{x^3}{x}$

Solution $\frac{x^3}{x}$

$= \frac{x \cdot x \cdot x}{x}$

$= \frac{x \cdot x \cdot \overset{1}{\cancel{x}}}{\underset{1}{\cancel{x}}}$

$= x^2$

◀

Because division by zero is undefined, a quotient such as $x^3 \div x$ has no value if $x = 0$. In this problem and in the rest of this chapter, assume that no denominator is zero.

EXAMPLE 2 Divide: $\frac{2^4}{2^3}$

Solution $\frac{2^4}{2^3}$

$= \frac{2 \cdot 2 \cdot 2 \cdot 2}{2 \cdot 2 \cdot 2}$

$= \frac{2 \cdot \overset{1}{\cancel{2}} \cdot \overset{1}{\cancel{2}} \cdot \overset{1}{\cancel{2}}}{\underset{1}{\cancel{2}} \cdot \underset{1}{\cancel{2}} \cdot \underset{1}{\cancel{2}}}$

$= 2^1$ or 2

◀

EXAMPLE 3 Divide: $\frac{m^5}{m^2}$

Solution $\frac{m^5}{m^2}$

$= \frac{m \cdot m \cdot m \cdot m \cdot m}{m \cdot m}$

$= \frac{m \cdot m \cdot m \cdot \overset{1}{\cancel{m}} \cdot \overset{1}{\cancel{m}}}{\underset{1}{\cancel{m}} \cdot \underset{1}{\cancel{m}}}$

$= m^3$

◀

Look at the examples and the answers again. What has happened to the base quantity? What about the exponents?

$\frac{x^3}{x} = x^2 \qquad \frac{2^4}{2^3} = 2 \qquad \frac{m^5}{m^2} = m^3$

In each example, the exponent of the answer was the *difference between* the exponents in the original fraction, and the base quantity was *unchanged.*

$\frac{x^3}{x} = x^{3-1} = x^2 \qquad \frac{2^4}{2^3} = 2^{4-3} = 2^1 = 2 \qquad \frac{m^5}{m^2} = m^{5-2} = m^3$

Dividing Monomials with the Same Base

To divide monomials with the same base, keep the base the same and subtract the exponent in the divisor from the exponent in the dividend.

$$\frac{a^x}{a^y} = a^{x-y}$$

$$\frac{3^5}{3^2} = 3^{5-2} = 3^3 = 27 \qquad \frac{p^4}{p} = p^{4-1} = p^3$$

EXAMPLE 4 Divide: $\frac{x^{20}}{x^{11}}$

Solution $\frac{x^{20}}{x^{11}} = x^9$

EXAMPLE 5 Divide: $\frac{b^3}{b}$

Solution $\frac{b^3}{b} = b^2$

(Remember, the b in the denominator is understood to have an exponent of 1.)

Self-Check Simplify these expressions and then check your answers with those given.

1. $\frac{z^3}{z}$ 2. $\frac{h^5}{h^4}$ 3. $\frac{a^6}{a^3}$ 4. $\frac{y^{12}}{y^4}$ 5. $\frac{m^9}{m^7}$

Answers:

1. z^2 2. h 3. a^3 4. y^8 5. m^2

In all the division problems we have done so far, the difference between the exponents has been positive. That is, the exponent in the numerator was larger than the exponent in the denominator. In the next section you will learn to work with negative exponents, and later in this section you will learn about zero exponents.

When the monomials contain numerical coefficients as well as variables, you will divide or reduce those factors and then apply the exponent rule to the variables.

EXAMPLE 6 Divide: $\frac{10x^6}{2x^4}$

Solution $\frac{10x^6}{2x^4}$ $\qquad 10 \div 2 = 5$ and $x^6 \div x^4 = x^{6-4} = x^2$

$= 5x^2$

EXAMPLE 7 Divide: $\frac{-27a^8}{9a^3}$

Solution $\frac{-27a^8}{9a^3}$

$= -3a^5$

EXAMPLE 8 Divide: $\frac{12m^4}{16m}$

Solution $\frac{12m^4}{16m}$

$= \frac{3m^3}{4}$ or $\frac{3}{4}m^3$

When monomials have more than one variable, you will repeat the procedures for each variable after reducing the numerical factors.

EXAMPLE 9 Divide: $\frac{36x^2y^4z^3}{12xy^2z^2}$

Solution $\frac{36x^2y^4z^3}{12xy^2z^2}$

$= \frac{36}{12} \cdot \frac{x^2}{x} \cdot \frac{y^4}{y^2} \cdot \frac{z^3}{z^2}$

$= 3xy^2z$

EXAMPLE 10 Divide: $\frac{-2a^5b^6}{8a^3b}$

Solution $\frac{-2a^5b^6}{8a^3b}$

$= \frac{-2}{8} \cdot \frac{a^5}{a^3} \cdot \frac{b^6}{b}$

$= -\frac{a^2b^5}{4}$ or $-\frac{1}{4}a^2b^5$

If either the numerator or the denominator is negative and the other part of the fraction is positive, the negative sign is written in front of the fraction (in dividing signed numbers, unlike signs give a negative answer). Never leave a negative symbol in the denominator.

Now let's look at the meaning of an exponent of zero. This could occur in a division problem such as

$$\frac{y^5}{y^5} = y^{5-5} = y^0$$

Or you might see it in multiplication: $y^0 \cdot y^4 = y^{0+4} = y^4$. Both problems should suggest to you that $y^0 = 1$. In $y^5 \div y^5$, you divided a quantity by itself, and that always equals 1.

$$\frac{y^5}{y^5} = \frac{\overset{1}{\cancel{y}} \cdot \overset{1}{\cancel{y}} \cdot \overset{1}{\cancel{y}} \cdot \overset{1}{\cancel{y}} \cdot \overset{1}{\cancel{y}}}{\underset{1}{\cancel{y}} \cdot \underset{1}{\cancel{y}} \cdot \underset{1}{\cancel{y}} \cdot \underset{1}{\cancel{y}} \cdot \underset{1}{\cancel{y}}} = \frac{1}{1} = 1 \quad \text{or} \quad \frac{y^5}{y^5} = y^{5-5} = y^0 = 1$$

In the multiplication problem, $y^0 \cdot y^4 = y^4$. The expression y^4 is multiplied by y^0, but the answer is the same as the original quantity. The only number that can do this is 1.

$y^0 \cdot y^4 = y^4$ and $1 \cdot y^4 = y^4$, so $y^0 = 1$.

Zero as an Exponent

Any nonzero quantity raised to a zero exponent is equal to 1.

$a^0 = 1$

$2^0 = 1 \qquad (10{,}000)^0 = 1 \qquad \left(-24\frac{2}{3}\right)^0 = 1 \qquad \left(a^2b^3c\right)^0 = 1$

You should always replace a quantity raised to a zero exponent with 1. Having a zero exponent in an answer is similar to not reducing a fraction to lowest terms. You have not finished simplifying the answer in either case.

EXAMPLE 11 Divide: $\frac{3x^5}{x^5}$

Solution $\frac{3x^5}{x^5}$

$= 3x^0$

$= 3 \cdot 1$

$= 3$

EXAMPLE 12 Divide: $\frac{16a^2b^3c}{24abc}$

Solution $\frac{16a^2b^3c}{24abc}$

$\frac{16}{24} \cdot \frac{a^2}{a} \cdot \frac{b^3}{b} \cdot \frac{c}{c}$

$= \frac{2}{3}ab^2c^0$

$= \frac{2}{3}ab^2 \cdot 1$

$= \frac{2}{3}ab^2$

Self-Check Simplify these expressions and check your answers.

1. $\dfrac{x^7y^6}{x^7y^2}$ 2. $\dfrac{12a^2b^2c^2}{4ab^2c}$ 3. $\dfrac{144m^2n^3}{24m^2n}$ 4. $\dfrac{-60a^4b^3}{15a^2b^3}$ 5. $\dfrac{28x^5y^2}{-12x^4y}$

Answers:

1. y^4 2. $3ac$ 3. $6n^2$ 4. $-4a^2$ 5. $-\frac{7}{3}xy$

In division problems such as Examples 13 and 14, you must simplify the numerator and denominator separately before doing the division.

EXAMPLE 13 Divide: $\dfrac{(2a^2)(-3ab^3)}{4a^2b}$

Solution

$$\frac{(2a^2)(-3ab^3)}{4a^2b}$$

$$= \frac{-6a^3b^3}{4a^2b}$$

$$= -\frac{3ab^2}{2} \quad \text{or} \quad -\frac{3}{2}ab^2$$

EXAMPLE 14 Divide: $\dfrac{(3m^2n)^2}{m^2(mn^2)}$

Solution

$$\frac{(3m^2n)^2}{m^2(mn^2)}$$

$$= \frac{(3)^2(m^2)^2(n)^2}{m^2(m)(n^2)}$$

$$= \frac{9m^4n^2}{m^3n^2}$$

$$= 9m$$

EXAMPLE 15 Simplify completely: $\dfrac{(2x^3y)^2(3xy^4)}{(5x^2y^2)^2}$

Solution Simplify the numerator and the denominator separately, then divide.

$$\frac{(2x^3y)^2(3xy^4)}{(5x^2y^2)^2}$$

$$= \frac{(4x^6y^2)(3xy^4)}{25x^4y^4}$$

$$= \frac{12x^7y^6}{25x^4y^4}$$

$$= \frac{12x^3y^2}{25} \quad \text{or} \quad \frac{12}{25}x^3y^2 \quad \text{or} \quad 0.48x^3y^2$$

13.3 EXERCISES

A. Simplify each expression completely.

1. $\frac{x^7}{x^4}$
2. $\frac{a^2}{a}$
3. $\frac{c^5}{c^4}$
4. $\frac{w^7}{w^3}$
5. $\frac{8d^4}{2d^2}$
6. $\frac{2^8}{2^6}$
7. $\frac{d^4}{d^4}$
8. $\frac{64b^5}{72b^3}$
9. $\frac{-28g^3}{-7g}$
10. $\frac{14c^2}{7c}$
11. $\frac{3^4}{3^2}$
12. $\frac{-81c^3}{-27c^2}$
13. $\frac{2^8}{2^8}$
14. $\frac{-15z^4}{3z^2}$
15. $\frac{12x^3}{27x}$
16. $\frac{4^4}{4^4}$
17. $(15x^2) \div (24x^2)$
18. $(-11a^7) \div (44a^4)$
19. $(16x^2y^3) \div (4xy)$
20. $(72a^2b^3) \div (12ab^2)$
21. $(-90b^2c^2d^2) \div (15cd)$
22. $(-144m^4n^3) \div (-24m^2n^2)$
23. $(4g^2h^4) \div (12g^2h^2)$
24. $(15r^2s^2t^3) \div (3rst)$
25. $(-9a^4b^4) \div (3ab^4)$
26. $(21x^3yz^2) \div (-6x^2z)$
27. $(-8m^2n) \div (16\ mn)$
28. $(17c^3d^2) \div (51c^2d^2)$
29. $(72b^2c^2) \div (9b^2c)$
30. $(14x^2y^5) \div (-21x^2y^2)$

B. Simplify completely.

31. $\frac{2^8a^5b^4}{2^5a^3b^3}$
32. $\frac{(3x^2)(-4y^3)}{2x^2y^2}$
33. $\frac{(-2m^2n)^2}{m^3n^2}$
34. $\frac{3^3r^3s^3}{-3rs}$
35. $\frac{(3a^3)(3a^4)(-a)}{(9a)(-3a^2)}$
36. $\frac{(-6x^4)(7x^6)}{7x}$
37. $\frac{(5m^2)(-4n^3)}{2m^2n^2}$
38. $\frac{3^3a^4b^5}{3^2ab^2}$
39. $\frac{3^4a^2c^3}{(-3)^2ac^2}$
40. $\frac{(-2m^2n^3)^3}{m^3n^2}$
41. $\frac{(-5y^2)(y^3)(2y)}{(2y)(-3y^3)}$
42. $\frac{(-3x^2)(2x)(-x^4)}{(2x^2)}$
43. $\frac{(-2x^3y^3)(x^2y^2)}{(-2x^2y^3)(-2x^2y^2)}$
44. $\frac{(9a^2b^3)(-7b^5)}{(7ab^2)(-11ab^3)}$
45. $\frac{(2ab^2)^3(3a^2b)}{(2a^2b^2)^2}$
46. $\frac{(-3x^2y^3)^2}{(-4xy)(9x^2y)}$
47. $\frac{-(5m^2n)^2(3mn^2)}{(-5m)^2(mn^2)}$
48. $\frac{-(a^2b)^2(-ab^3)^2}{(-a^2b)^3}$
49. $\frac{(-5x^2y)(2x^3)(-xy^2)}{3xy(-5x^2y)^2}$
50. $\frac{-(pqr)^2(-p^2q^2r^3)}{(4pq^2)(-pr)^3}$

C. Read and respond to the following exercises.

51. Is $2^5/2^4 = 1^1 = 1$? Explain your answer.
52. Explain in words why $x^0 = 1$.
53. In Exercise 49, what simplification step was done first? Why?
54. In $(y^5/y^3)^2$, would it matter whether you divided first or raised to the power first? Explain your answer.
55. In Exercise 43, how could you use the Odd–Even Rule for Signs to determine the sign for the answer without actually simplifying the problem?
56. Compare how you would use the Odd-Even Rule for Signs to determine the sign of the answer for Exercises 43 and 49.

13.4 Negative Exponents

As we noted in Section 13.3, negative exponents can occur in monomial division. In $y^3 \div y^5$, the answer that results from subtracting exponents, y^{3-5}, is y^{-2}. The meaning of y^{-2} is evident when we look at the problem in expanded form.

$$\frac{y^3}{y^5} = \frac{\overset{1}{\cancel{y}} \cdot \overset{1}{\cancel{y}} \cdot \overset{1}{\cancel{y}}}{\underset{1}{\cancel{y}} \cdot \underset{1}{\cancel{y}} \cdot \underset{1}{\cancel{y}} \cdot y \cdot y} = \frac{1}{y^2}$$

Therefore,

$$y^{-2} = \frac{1}{y^2}$$

> **Negative Exponents**
>
> If n is an integer,
>
> $$a^{-n} = \frac{1}{a^n}$$

In simplifying the problems in this section, do not leave negative exponents in an answer. Replace them with the fraction form of the expression. This is usually the last step, after you have done the rest of the problem using the rules for exponents.

EXAMPLE 1 Simplify: $a^3 \div a^7$

Solution

$a^3 \div a^7$ Keep the base and subtract the exponents.

$= a^{3-7}$

$= a^{-4}$ Replace the negative exponent with its fractional equivalent.

$= \frac{1}{a^4}$

◀

EXAMPLE 2 Divide m by m^4.

Solution

$m \div m^4$

$= m^{1-4}$ Subtract exponents in division.

$= m^{-3}$ Replace the negative exponent with its equivalent fraction form.

$= \frac{1}{m^3}$

◀

EXAMPLE 3 Find the product of $-2a^{-2}b^3$ and $4ab^{-2}$.

Solution Multiply the numerical coefficients, -2 and 4.

$(-2)(4) = -8$

Multiply the same bases, adding exponents.

$(a^{-2})(a) = a^{-2+1} = a^{-1}$

$(b^{3})(b^{-2}) = b^{3+(-2)} = b^{1} = b$

Therefore, $(-2a^{-2}b^{3})(4ab^{-2})$

$= -8a^{-1}b$ Replace a^{-1} with $\frac{1}{a}$.

$= -8\left(\frac{1}{a}\right)b$

$= -\frac{8b}{a}$

EXAMPLE 4 Simplify: $(2m^{-3}n^{2})^{-2}$

Solution $(2m^{-3}n^{2})^{-2}$

Raise each factor to the –2 power using the rules for exponents that appear in Section 13.2. Remember that the coefficient 2 has an exponent of 1. You may want to write it in.

$= (2^{1})^{-2}(m^{-3})^{-2}(n^{2})^{-2}$

$= 2^{-2}m^{+6}n^{-4}$

Replace the negative exponents.

$= \frac{1}{2^{2}} \cdot \frac{m^{6}}{1} \cdot \frac{1}{n^{4}}$

$= \frac{m^{6}}{4n^{4}}$

EXAMPLE 5 Simplify: $\frac{X^{-8}}{X^{-5}}$

Solution $\frac{X^{-8}}{X^{-5}}$

$= X^{(-8)-(-5)}$

$= X^{(-8)+(+5)}$

$= X^{-3}$

$= \frac{1}{X^{3}}$

EXAMPLE 6 Simplify: $\frac{y^{3} \cdot y^{-5}}{y^{-2}}$

Solution $\frac{y^{3} \cdot y^{-5}}{y^{-2}}$ Simplify the numerator.

$= \frac{y^{3+(-5)}}{y^{-2}}$

$= \frac{y^{-2}}{y^{-2}}$

$= y^{(-2)-(-2)}$

$= y^{(-2)+(+2)}$

$= y^0 = 1$

EXAMPLE 7 Simplify: $\dfrac{1}{c^{-3}}$

Solution $\dfrac{1}{c^{-3}}$ Replace c^{-3} with $\dfrac{1}{c^3}$.

$= \dfrac{1}{\frac{1}{c^3}}$ Rewrite the complex fraction as division (Section 3.4)

$= 1 \div \dfrac{1}{c^3}$ Multiply by the reciprocal.

$= \dfrac{1}{1} \cdot \dfrac{c^3}{1}$

$= c^3$

From Example 7 can you see that $1/c^{-3} = c^3$?

This relationship will be true all the time (if $c \neq 0$). It is stated in the following rule.

> **Negative Exponents**
>
> Because $a^{-n} = \dfrac{1}{a^n}$, then $\dfrac{1}{a^{-n}} = a^n$ (where $a \neq 0$).

Factors that have negative exponents and are in the numerator will move to the denominator and the exponent will become positive. Factors that have negative exponents and are in the denominator will move to the numerator in the same way.

Example 5 could have been simplified this way:

$$\frac{X^{-8}}{X^{-5}} = \frac{X^5}{X^8} = X^{-3} = \frac{1}{X^3}$$

And Example 6 could be done as follows:

$$\frac{y^3 \cdot y^{-5}}{y^{-2}} = \frac{y^3y^2}{y^5} = \frac{y^5}{y^5}$$

$$= y^0 = 1$$

And Example 7 would very quickly simplify as:

$$\frac{1}{c^{-3}} = \frac{c^3}{1} = c^3$$

EXAMPLE 8 Simplify: $\left(\dfrac{a^4}{b^5}\right)^{-3}$

Solution

$\left(\dfrac{a^4}{b^5}\right)^{-3}$ Raise the numerator and denominator to the – 3 power.

$= \dfrac{a^{-12}}{b^{-15}}$ Move terms with negative exponents to the opposite part of the fraction.

$= \dfrac{b^{15}}{a^{12}}$

EXAMPLE 9 Simplify: $\left(\dfrac{4x^2}{3^{-1}y^3}\right)^{-2}$

Solution

$\left(\dfrac{4x^2}{3^{-1}y^3}\right)^{-2}$ Apply the – 2 exponent to both numerator and denominator.

$= \dfrac{(4x^2)^{-2}}{(3^{-1}y^3)^{-2}}$ Raise each factor to the power.

$= \dfrac{4^{-2}x^{-4}}{3^2y^{-6}}$ Move factors with negative powers and change to positive exponents. (Do not move factors with positive exponents.)

$= \dfrac{y^6}{3^2 \cdot 4^2x^4}$ Simplify.

$= \dfrac{y^6}{9 \cdot 16x^4}$

$= \dfrac{y^6}{144x^4}$

EXAMPLE 10 Simplify: $\left(\dfrac{3a^2}{b^3}\right)^{-2}$

Solution

$$\left(\frac{3a^2}{b^3}\right)^{-2} = \frac{(3^1a^2)^{-2}}{(b^3)^{-2}}$$

$$= \frac{(3^1)^{-2}(a^2)^{-2}}{(b^3)^{-2}}$$

$$= \frac{3^{-2}a^{-4}}{b^{-6}}$$

$$= \frac{b^6}{3^2a^4}$$

$$= \frac{b^6}{9a^4}$$

EXAMPLE 11 Simplify: $\dfrac{(-3x^2y^{-1})^3}{x^{-3}y^2}$

Solution

$\dfrac{(-3x^2y^{-1})^3}{x^{-3}y^2}$ Raise each factor in the numerator to the 3rd power.

$= \dfrac{-27x^6y^{-3}}{x^{-3}y^2}$ Move factors with negative exponents. Note: – 27 is a negative number, not a number with a negative exponent, so it does not move.

$= \dfrac{-27x^6x^3}{y^2y^3}$

$= \dfrac{-27x^9}{y^5}$

Self-Check Simplify applying exponent rules and eliminating negative exponents.

a. $x^4 \div x^9$ b. $(-3a^2b^{-3})^2$ c. $\dfrac{m^2n}{(-2m^{-1}n^2)^3}$

Answers:

a. $\dfrac{1}{x^5}$ b. $\dfrac{9a^4}{b^6}$ c. $-\dfrac{m^5}{8n^5}$

13.4 EXERCISES

A. Simplify each expression, applying exponent rules and eliminating negative exponents.

1. $a^2 \div a^5$
2. $x \div x^3$
3. $4R^2 \div R^3$
4. $b^5 \div 5b^7$
5. $9M^2 \div 6M^3$
6. $8n^4 \div 12n^7$
7. $(x^2)^{-2}$
8. $(a^3)^{-2}$
9. $(b^{-2})^{-2}$
10. $(c^{-1})^{-2}$
11. $(2a^{-1})^{-2}$
12. $(2c^2)^{-1}$
13. $(2x^2y^{-1}z^3)^{-2}$
14. $(3a^{-2}b^2c)^{-2}$
15. $\dfrac{3}{m^{-2}}$
16. $\dfrac{b^{-2}}{4}$
17. $\dfrac{x^{-2}}{x^{-5}}$
18. $\dfrac{n^{-4}}{n}$
19. $\left(\dfrac{a^2}{b^3}\right)^{-2}$
20. $\left(\dfrac{c^3}{d^{-1}}\right)^{-2}$
21. $\dfrac{3x^2y^{-2}z}{2^{-1}xy^2z^3}$
22. $\dfrac{5^{-1}m^{-3}n^2}{3m^2n^{-3}}$
23. $\dfrac{(-2a^2b^{-1}c^{-2})^2}{a^{-3}bc^{-1}}$
24. $\dfrac{(-5x^2y^{-3}z)^{-2}}{3x^{-1}y^2z}$
25. $\dfrac{(2^{-1}x^2y^{-2})^{-2}}{(3x^{-1}y^2)(-2x^2y^{-1})}$
26. $\left(\dfrac{-a^2b^{-2}}{c^{-3}}\right)^{-2}$
27. $\left(\dfrac{2m^{-2}n}{-m^3n^2}\right)^{-3}$
28. $\dfrac{4m^3n^{-2}}{2^{-3}m^2n^{-3}}$
29. $\dfrac{5a^4b^{-2}}{5^{-2}a^2b^{-3}}$
30. $\dfrac{(-3x^2y^{-3})^{-1}}{3^2x^{-4}y^3}$
31. $\dfrac{(-2m^3n^{-2})^{-1}}{2^3m^{-4}n^2}$
32. $(-3a^3b^{-2})(-2a^{-1}b)$
33. $[(5m^2n^{-2})(-2m^{-3}n)]^{-2}$
34. $\dfrac{(2a^2b^{-3}c)^{-1}}{(5a^{-3}bc^2)^{-2}}$
35. $[(-2a^2b^{-2})(3a^{-1}b)]^{-2}$
36. $[(6c^3d^{-2})(-b^{-2}cd^{-1})]^{-1}$
37. $(2ab^{-2})(4a^{-1}b)(a^2b^3)$
38. $(-5xy^{-2})(2x^{-1}y)(x^2y^{-2})$

39. $\dfrac{(2m^2n^{-1})^{-2}}{(m^{-1}n^{-2})(m^3n^{-1})}$

40. $\dfrac{(3a^{-1}b^{-2})^{-2}}{(a^2b^{-1})(a^{-3}b^{-2})}$

41. $\dfrac{(-2a^2b^{-3}c^{-1})^{-2}}{(3a^{-1}b^{-2}c^{-3})^3}$

42. $[(x^{-3}y^{-2})^2]^{-1}$

43. $[(x^{-3}y^{-2})^{-2}]^{-1}$

44. $\left(\dfrac{(m^2n^{-2})^{-1}}{(m^{-1}n^3)^{-1}}\right)^{-2}$

B. Read and respond to the following exercises.

45. Would you describe your feelings about negative exponents as positive or negative? Explain what you mean.

46. In Exercise 41, what step in the simplification should be done first? Why?

47. What is the simplification of: $\{[(y^2/3)^{-1}]^{-1}\}^{-1}$? Explain in words now you reached that answer?

48. Is $(-2)^{-3} = 8$? Explain your answer, in words.

13.5 Adding and Subtracting Polynomials

The process of adding or subtracting polynomials simply involves combining like terms.

EXAMPLE 1 Simplify:

$$\begin{array}{r} -3x^2 + 4x - 6 \\ +\quad 6x^2 - 3x + 5 \\ \hline \end{array}$$

Solution

$$\begin{array}{r} -3x^2 + 4x - 6 \\ +\quad 6x^2 - 3x + 5 \\ \hline 3x^2 + x - 1 \end{array}$$

EXAMPLE 2 Combine: $(13a^2b - 5ab + 4) + (7a^2b - 2ab + 5)$

Solution

$(13a^2b - 5ab + 4) + (7a^2b - 2ab + 5)$ Remove the parentheses.

$= 13a^2b - 5ab + 4 + 7a^2b - 2ab + 5$ Rearrange like terms.

$= 13a^2b + 7a^2b - 5ab - 2ab + 4 + 5$

$= 20a^2b - 7ab + 9$

Note that the shortcut presented in Chapters 11 and 12 was used. Terms were combined by using the operation signs as if they were signed–number signs: $(-3x^2) + (+6x^2)$, $(+4x) + (-3x)$, $(-6) + (+5)$, $(-5ab) + (-2ab)$, and so on. You can write the problem either vertically or horizontally, whichever you like better.

EXAMPLE 3 Simplify:

$(-7a^2b^2 + 4ab - 8) + (15a^2b^2 + 3 - 4ab)$

Solution You could line up the terms vertically:

$$\begin{array}{r} -7a^2b^2 + 4ab - 8 \\ +\ 15a^2b^2 - 4ab + 3 \\ \hline 8a^2b^2 + 0ab - 5 = 8a^2b^2 - 5 \end{array}$$

Line up like terms.

or you could work horizontally:

$$(-7a^2b^2 + 4ab - 8) + (15a^2b^2 + 3 - 4ab)$$
$$= -7a^2b^2 + 4ab - 8 + 15a^2b^2 + 3 - 4ab$$
$$= -7a^2b^2 + 15a^2b^2 + 4ab - 4ab - 8 + 3$$
$$= 8a^2b^2 - 5$$

EXAMPLE 4 Simplify completely:

$$(9m^2 + 14m^2n - 5mn^2 + 3) + (2m^2 - 5m^2n + mn^2 - 1)$$

Solution

$$(9m^2 + 14m^2n - 5mn^2 + 3) + (2m^2 - 5m^2n + mn^2 - 1)$$
$$9m^2 + 14m^2n - 5mn^2 + 3 + 2m^2 - 5m^2n + mn^2 - 1$$
$$= 11m^2 + 9m^2n - 4mn^2 + 2$$

or

$$\begin{array}{r} 9m^2 + 14m^2n - 5mn^2 + 3 \\ + \quad 2m^2 - 5m^2n + mn^2 - 1 \\ \hline 11m^2 + 9m^2n - 4mn^2 + 2 \end{array}$$

When subtracting polynomials, remember the signed-number rule (Chapter 11) about rewriting subtraction as addition of the opposite. You change the operation sign to addition and change the sign(s) of the number(s) you are taking away; then you proceed as in addition. In the problem $(11x^2 - 8x - 5) - (3x^2 + 9x + 1)$, the entire second polynomial is being subtracted. Therefore, the operation sign will be changed to addition, and the sign of *every* term in the second polynomial must be changed before you do the addition.

Vertically:

$$\begin{array}{r} 11x^2 - 8x - 5 \\ \text{S} \qquad\qquad \\ - \quad 3x^2 + 9x + 1 \\ \hline \end{array}$$

Change all the signs in the subtrahend (the number you are subtracting). Then add.

$$\begin{array}{r} 11x^2 - 8x - 5 \\ \text{A} \qquad\qquad\qquad \\ + \quad -3x^2 - 9x - 1 \\ \hline 8x^2 - 17x - 6 \end{array}$$

Horizontally: Multiply by – 1 to do the sign changes, and then combine like terms.

$$(11x^2 - 8x - 5) - (3x^2 + 9x + 1)$$
$$= (11x^2 - 8x - 5) - 1(3x^2 + 9x + 1)$$
$$= 11x^2 - 8x - 5 - 3x^2 - 9x - 1$$
$$= 8x^2 - 17x - 6$$

Remember that multiplying the second polynomial by – 1 accomplishes the sign changes required in subtraction.

EXAMPLE 5 Simplify completely: $(3x + 6) - (2x - 3)$

Solution

$(3x + 6) - (2x - 3)$

$= (3x + 6) - 1(2x - 3)$

$= 3x + 6 - 2x + 3$

$= x + 9$

EXAMPLE 6 Simplify completely:

$$\begin{array}{r} 17a^2 - 4a + 6 \\ \underline{- -5a^2 - 4a - 3} \end{array}$$

Solution

$$\begin{array}{rr} 17a^2 - 4a + 6 & 17a^2 - 4a + 6 \\ \text{S} & \text{A} \\ \underline{- -5a^2 - 4a - 3} & \underline{+ +5a^2 + 4a + 3} \\ & 22a^2 \quad + 9 \end{array}$$

EXAMPLE 7 From the sum of $(4a^2 - 3)$ and $(5a^2 + 7)$, subtract $(-3a^2 + 4)$.

Solution

$(4a^2 - 3) + (5a^2 + 7) - (-3a^2 + 4)$

Note: Be sure to subtract the whole second polynomial

$= (4a^2 - 3) + (5a^2 + 7) - 1(-3a^2 + 4)$

$= 4a^2 - 3 + 5a^2 + 7 + 3a^2 - 4$

$= 12a^2$

or

$$\begin{array}{rrr} 4a^2 - 3 & 9a^2 + 4 & 9a^2 + 4 \\ & \text{S} & \text{A} \\ \underline{+ 5a^2 + 7} & \underline{- -3a^2 + 4} & \underline{+ +3a^2 - 4} \\ 9a^2 + 4 & & 12a^2 \end{array}$$

Self-Check Simplify these expressions. If you have difficulty, review the examples.

1. $(2x^2 - 3x + 5) + (x^2 - 2x - 4)$
2. $(3m^4 - 4m + 1) - (8m^4 - m - 2)$
3. From the sum of $(-5a^2b - 13ab^2)$ and $(7a^2b - 3ab^2)$, subtract $(-4ab^2 + 12a^2b)$

Answers:

1. $3x^2 - 5x + 1$
2. $-5m^4 - 3m + 3$
3. $-10a^2b - 12ab^2$

13.5 EXERCISES

A. In each of the following exercises, add.

1. $\begin{array}{r} 23x + 14y - 27 \\ \underline{+\,15x - 8y - 12} \end{array}$

2. $\begin{array}{r} 7q + 2r + 9s \\ \underline{+\,5q + 6rs - 3s} \end{array}$

3. $\begin{array}{r} 17m^2 + 5m - 14 \\ \underline{+\,11m^2 - 7m + 5} \end{array}$

4. $\begin{array}{r} -4a + 12b - 6c \\ \underline{+\,5a + 9b + 2c} \end{array}$

5. $\begin{array}{r} 6R - 3S + 5T \\ \underline{+\,R + 4S - 2T} \end{array}$

6. $\begin{array}{r} 6y^2 + 13y - 7 \\ \underline{+\,-3y^2 + 6y - 1} \end{array}$

B. In each of the following exercises, subtract. (Remember to use "S" and "A" if it helps.)

7. $\begin{array}{r} 11a + 4ab - 5b \\ \underline{-\,-7a + ab + 12b} \end{array}$

8. $\begin{array}{r} 9mn + 4m - n \\ \underline{-\,mn + 2m - 3n} \end{array}$

9. $\begin{array}{r} 15v + 2w - 7x \\ \underline{-\,23v - w + 5x} \end{array}$

10. $\begin{array}{r} -3c^2 - 4c + 8 \\ \underline{-\,c^2 + 5c - 1} \end{array}$

11. $\begin{array}{r} 15f^2 - 12f + h \\ \underline{-\,-3f^2 \qquad - 7h} \end{array}$

12. $\begin{array}{r} 17xy - 2x + 9y \\ \underline{-\,-5xy - 4x \qquad} \end{array}$

C. Add or subtract as indicated.

13. $(3a + 4b) + (-7a + 6b)$
14. $(18 - 6x) + (15x - 3)$
15. $(-5m + 3) - (2m - 4)$
16. $(-3k + 2m) - (6m - k)$
17. $(9x^2 + 2x + 12) - (-7x + 9)$
18. $(-18y + 2z + 9) + (16 + 4z - 11y)$
19. $(2R - 4S + T) + (3S - 7T)$
20. $(5L - 2M + N) - (4N + 3L - 2M)$
21. $(6x^2y - 4xy + 2y^2) + (3y^2 - 5x^2y)$
22. $(9mn^2 - 2mn) + (3 + 4mn^2 - 2mn)$
23. $(3b^2 - 6b - 7) - (4b^2 + 5b + 6)$
24. $(5x^3 + 3x^2 - 7x) + (-x^2 + 5x - 7)$
25. $(4y^2 - 4y + 3) + (7y^3 + 3y^2 - 1)$
26. $(17c^3 - 4c^2) - (8c^2 + 2c + 3)$
27. $(m^2 - m^3 + 2 - 4m) + (3m - 2m^2 - 6)$
28. $(c^3 + 2c^2 - c + 1) + (5c^2 - 2c + 9) + (2c - 3c^3 + 2c^2 - 4)$
29. $(3x^2 + 2x^3 - 5) + (7x - 2x^2 + 3x^3 - 1) + (x^3 - 2x^2 - 3x)$
30. $(4w^3 - w - 1) - (2w^2 + 3w + 5)$
31. Subtract $(-p^2q - 2pq^2)$ from $(6p^2q + 2pq - pq^2)$.
32. Find the sum of $(2a - 9)$, $(3a + 4)$, and $(5a - 2)$.
33. Find the sum of $(4x^2 - 2x + 5)$, $(2x^2 - 3)$, and $(5x - 9)$.
34. From $(6c^2 - 7cd + d^3)$ subtract $(3d^3 - c^2 + 2cd)$.
35. From the sum of $(2x + 3y)$ and $(-x - 4y)$, subtract $(5x + 2y)$.
36. From the sum of $(4a - b)$ and $(-3a + 6b)$, subtract $(a - b)$.
37. From $(7y^2 + 5y - 3)$ subtract $(-3y^2 - 4y + 2)$.
38. Subtract $(-2R^2S^2 - 3R^2S)$ from $(-R^2S^2 + 2R^2S)$.
39. Find the difference between $(3m^2n^2 - 2m^2n + 5)$ and $(2m^2n^2 + mn^2 - 2)$.
40. Find the difference between $(5a^2b^2c^3 - 2ab^2c^2 + 7a^2b^2c^2)$ and $(4ab^2c^2 - 2a^2b^2c)^2 + a^2b^2c)$.

D. Read and respond to the following exercises.

41. How does adding and subtracting polynomials differ from adding and subtracting monomials?
42. Your friend is having difficulty with the exercises in this section. Look at the work he did below and explain to him his error and the correct way to do the exercise.

$$\begin{array}{r} -8y^2 + 2y - 1 \\ \underline{-\quad 6y^2 - 3y + 7} \end{array} \qquad \begin{array}{r} -8y^2 + 2y - 1 \\ \underline{+\,-6y^2 - 3y + 7} \\ -14y^2 - y + 6 \end{array}$$

43. Do you prefer to do addition exercises in this section by setting them up vertically or horizontally? Do you do the same thing with the subtraction exercises? Why?
44. Why can you distribute -1 in a subtraction exercise?

13.6 Multiplying Polynomials

Multiplying Polynomials by Monomials

Multiplying polynomials by monomials combines the Distributive Principle and the rules for exponents. You will multiply every term inside the parentheses by the factor outside and, in doing so, will add exponents when the bases are the same.

EXAMPLE 1 Multiply: $3ab(2a^2b - b^2)$

Solution $(3ab)(2a^2b - b^2)$ Multiply $(3ab)(2a^2b)$ and $(3ab)(-b^2)$.

$= 6a^3b^2 - 3ab^3$

EXAMPLE 2 Simplify: $5x^2y(x^2y^2 + 2xy - y^2)$

Solution $5x^2y(x^2y^2 + 2xy - y^2)$ Multiply $(5x^2y)(x^2y^2)$, $(5x^2y)(+2xy)$, and $(5x^2y)(-y^2)$.

$= 5x^4y^3 + 10x^3y^2 - 5x^2y^3$

EXAMPLE 3 Multiply: $(3mn^2)^2(4m^2n^2 - 5mn + 6)$

Solution By the Order of Operations Agreement, simplify the first factor with the exponent before doing the multiplication.

$(3mn^2)^2 = (3)^2(m)^2(n^2)^2 = 9m^2n^4$

Then

$(3mn^2)^2(4m^2n^2 - 5mn + 6)$

$= 9m^2n^4(4m^2n^2 - 5mn + 6)$

$= 36m^4n^6 - 45m^3n^5 + 54m^2n^4$

Self-Check Simplify Completely:

1. $-3ab(2a^2b^3 - 5a^3)$ 2. $m^2n^2(4m^3n - 5m^2n^4)$

Answers:

1. $-6a^3b^4 + 15a^4b$ 2. $4m^5n^3 - 5m^4n^6$

Multiplying Polynomials by Polynomials

When multiplying a polynomial by a polynomial, you will use a process similar to the one used in the previous examples. To multiply $(y + 3)$ by $(4y - 1)$, distribute each term of the first polynomial over every other term of the second polynomial.

$(y + 3)(4y - 1)$

$y(4y - 1) + 3(4y - 1)$

Then multiply through using the Distributive Principle. Be very careful with the signed numbers and exponents as you multiply and combine like terms.

$= 4y^2 - y + 12y - 3$

$= 4y^2 + 11y - 3$

This multiplication can also be arranged vertically, much like long multiplication in arithmetic. Multiply all the terms in the top binomial by – 1; then multiply each term by 4y and line up the like terms.

$$
\begin{array}{r}
y + 3 \\
\underline{4y - 1} \\
-1y - 3 \\
\underline{4y^2 + 12y \quad\;\;} \\
4y^2 + 11y - 3
\end{array}
$$

Add the partial products.

Let's look at another example of polynomial multiplication:

$(2a + 3b)(a - 2b - 1)$

$2a(a - 2b - 1) + 3b(a - 2b - 1)$ Distribute.

$= 2a^2 - 4ab - 2a + 3ab - 6b^2 - 3b$ Combine like terms.

$= 2a^2 - 2a - ab - 3b - 6b^2$

Doing this distributive multiplication vertically would look like this:

$$
\begin{array}{r}
a - 2b - 1 \\
\underline{2a + 3b} \\
3ab - 6b^2 - 3b \\
\underline{2a^2 - 2a - 4ab \qquad\qquad\quad} \\
2a^2 - 2a - ab - 6b^2 - 3b
\end{array}
$$

Are the answers the two methods give the same? If you arranged the terms in the same order, they would be the same. Terms should be arranged in descending order of the exponents of the first(alphabetically first) variable and then in ascending order of the exponents of the next variable.

$2a^2 - 2a - ab - 6b^2 - 3b = 2a^2 - 2a - ab - 3b - 6b^2$

Watch the arrangement of answers in the next examples and see if you can follow the pattern as you do problems. If you happen to write the terms out of order, the answer will not be incorrect, but it is much easier for you—and everyone else—to check answers if they are all written in the same way.

EXAMPLE 4 Multiply: $(4x + 8)(x^2) + 2x - 9$

Solution By the distributive method:

$(4x + 8)(x^2 + 2x - 9)$	Write each term from the first factor in front of the second polynomial.
$= 4x(x^2 + 2x - 9) + 8(x^2 + 2x - 9)$	Multiply using Distributive Principle.
$= 4x^3 + 8x^2 - 36x + 8x^2 + 16x - 72$	Combine like terms.
$= 4x^3 + 16x^2 - 20x - 72$	Notice the decreasing order of powers of x.

By long multiplication (which also uses the Distributive Principle):

$x^2 + 2x - 9$	
$\underline{4x + 8}$	
$8x^2 + 16x - 72$	Multiply by + 8.
$\underline{4x^3 + 8x^2 - 36x}$	Multiply by + 4x, lining up like terms.
$4x^3 + 16x^2 - 20x - 72$	Add the partial products.

EXAMPLE 5 Multiply: $(y^2 + 2y - 3)(y^2 - 4y - 8)$

Solution By the distributive method:

$$(y^2 + 2y - 3)(y^2 - 4y - 8)$$
$$= y^2(y^2 - 4y - 8) + 2y(y^2 - 4y - 8) - 3(y^2 - 4y - 8)$$
$$= y^4 - 4y^2 - 8y^2 + 2y^3 - 8y^2 - 16y - 3y^2 + 12y + 24$$
$$= y^4 - 2y^3 - 19y^2 - 4y + 24$$

By long multiplication:

$$y^2 + 2y - 3$$
$$\underline{y^2 - 4y - 8}$$
$$-8y^2 - 16y + 24$$
$$-4y^3 - 8y^2 + 12y$$
$$\underline{y^4 + 2y^3 - 3y^2}$$
$$y^4 - 2y^3 - 19y^2 - 4y + 24$$

EXAMPLE 6 Find the product: $(k + 3)(2k - 7)$

Solution

$$(k + 3)(2k - 7)$$
$$= k(2k - 7) + 3(2k - 7)$$
$$= 2k^2 - 7k + 6k - 21$$
$$= 2k^2 - k - 21$$

EXAMPLE 7 Multiply: $(3m^2 - 2)(2m + 5)$

Solution

$(3m^2 - 2)(2m + 5)$

$= 3m^2(2m + 5) - 2(2m + 5)$

$= 6m^3 + 15m^2 - 4m - 10$

Self-Check Simplify completely:

1. $(x - 3)(x + 7)$ 2. $(3y - 5)(2y + 1)$ 3. $(a + 2)(3a^2 - 2a + 4)$

Answers:

1. $x^2 + 4x - 21$ 2. $6y^2 - 7y - 5$ 3. $3a^3 + 4a^2 + 8$

◀

EXAMPLE 8 Simplify: $(3a - 4)^2$

Solution

$(3a - 4)^2$ Rewrite as the product of the two binomials.

$= (3a - 4)(3a - 4)$

$= 3a(3a - 4) - 4(3a - 4)$

$= 9a^2 - 12a - 12a + 16$

$= 9a^2 - 24a + 16$

◀

Students often want to do this problem by squaring the terms inside, $(3a)^2$ and $(-4)^2$. The rule that they are trying to apply to this difference of terms is the rule for raising a *product* to a power (Section 13.2). Note the difference between the expressions $(3ab^2)^2$ and $(3a - 4)^2$. In the product, all factors are to be raised to the second power: $3^2a^2(b^2)^2 = 9a^2b^4$. This is not the correct procedure for a binomial. The factor $(3a - 4)$ *as a whole* must be multiplied by itself *as a whole*: $(3a - 4)(3a - 4) = 9a^2 - 24a + 16$. Note that there is a middle term that would not have occurred if you had squared only $(3a)$ and (-4). When you see a binomial in parentheses and the exponent outside, rewrite the problem as the product of the binomials.

$(3a - 4)^2 = (3a - 4)(3a - 4)$

Doing so should remind you to multiply it completely.

EXAMPLE 9 Multiply: $(R - 5)^2$

Solution

$(R - 5)^2$

$= (R - 5)(R - 5)$

$= R(R - 5) - 5(R - 5)$

$= R^2 - 5R - 5R + 25$

$= R^2 - 10R + 25$

◀

EXAMPLE 10 Simplify: $(x + 2)^3$

Solution

$(x + 2)^3$

$= (x + 2)(x + 2)(x + 2)$ Multiply the first two binomials together.

$= [x(x + 2) + 2(x + 2)](x + 2)$

$= [x^2 + 2x + 2x + 4](x + 2)$

$= (x^2 + 4x + 4)(x + 2)$ Then multiply that product by the remaining binomial.

$= x(x^2 + 4x + 4) + 2(x^2 + 4x + 4)$

$= x^3 + 4x^2 + 4x + 2x^2 + 8x + 8$ Combine like terms.

$= x^3 + 6x^2 + 12x + 8$

EXAMPLE 11 Find the product: $(a^3 + b^3)^2$

Solution

$(a^3 + b^3)(a^3 + b^3)$

$= a^3(a^3 + b^3) + b^3(a^3 + b^3)$

$= a^6 + a^3b^3 + b^3a^3 + b^6$

$= a^6 + 2a^3b^3 + b^6$

Multiplying Binomials by Binomials Using FOIL

When multiplying two binomials together, it is sometimes easier to use a memory aid to remember the order in which you do the multiplication. This aid, FOIL, lets you use the Distributive Principle without actually writing out the steps. The letters in FOIL stand for First, Outside, Inside, and Last. In $(2x + 1)(3x - 5)$:

First terms in each set of parentheses are $2x$ and $3x$.

Outside terms are the ones that are farthest apart in the expression and are $2x$ and -5.

Inside terms are closest together and are $+1$ and $3x$.

Last terms are at the end of each set of parentheses and are $+1$ and -5.

Multiply the first, outside, inside, and last terms and then combine like terms, if there are any. Remember to include a sign (+ or –) with each product.

F	O	I	L
$2x(3x)$	$2x(-5)$	$+1(3x)$	$+1(-5)$
$= 6x^2$	$-10x$	$+3x$	-5

$= 6x^2 - 7x - 5$

EXAMPLE 12 Use FOIL to find the product of $(w - 1)(w - 4)$.

Solution

$(w - 1)(w - 4)$

First $(w)(w) = w^2$

Outside $(w)(-4) = -4w$

Inside $(-1)(w) = -1w$

Last $(-1)(-4) = +4$

Combine like terms.

$(w-1)(w-4) = w^2 - 5w + 4$

EXAMPLE 13 Find the product: $(a^2 + b^2)(2a^2 + 3a)$

Solution $(a^2 + b^2)(2a^2 + 3a)$

F	O	I	L
$(a^2)(2a^2)$	$(a^2)(3a)$	$(b^2)(2a^2)$	$(b^2)(3a)$

$= 2a^4 + 3a^3 + 2a^2b^2 + 3ab^2$

EXAMPLE 14 Simplify: $(7x-2)^2$

Solution $(7x-2)^2 = (7x-2)(7x-2)$

F	O	I	L
$(7x)(7x)$	$(7x)(-2)$	$(-2)(7x)$	$(-2)(-2)$

$= 49x^2 - 14x - 14x + 4$

$= 49x^2 - 28x + 4$

Remember that FOIL can only be used when you are multiplying two binomials. If one of the polynomials contains more than two terms, then distributive multiplication must be done.

Self-Check Try these problems and check your answers before doing the Exercises. Review the examples if you need help.

1. $(2y-6)^2$ 2. $(W+4)^2$

Answers:

1. $4y^2 - 24y + 36$ 2. $W^2 + 8W + 16$

13.6 EXERCISES

A. Multiply, and combine like terms where possible.

1. $3a(a^2 - 4)$
2. $5cd(2c^2d + 3cd + d^2)$
3. $(-8mn)(7m^2 + 4mn)$
4. $(-2x^2y)(3xy + 2y)$
5. $(5x+2)(3x-1)$
6. $(8a^2 + a + 3)(a^2 - 2a - 4)$
7. $(p^2 + 9p - 2)(-4p + 1)$
8. $(2g-1)(g-4)$
9. $(-x^2y)(3x^2y^2 + 2xy)$
10. $(m^2 - 2m)(2m - 3)$
11. $(c^2d - 4)(cd + 2d)$
12. $(-a^2)(2a^2 + 4a - 1)$
13. $-5x(x^2 + 2xy + 2y^2)$
14. $(4a - 5b)(4a - 5b)$
15. $(y+2)(y^2 + 2y + 4)$
16. $(3a-8)(2a-1)$

B. Simplify completely.

17. $(a^2 - 7)(a^2 + 3)$
18. $(3x - 5y)(x + 2y)$
19. $(0.3x - 0.4y)^2$
20. $(4a - 5b)^2$
21. $(x - 3y)^2$
22. $(h^2 + 0.4)^2$
23. $(g - 4)(g + 4)$
24. $(h + 2)(h - 2)$
25. $(a - 2b)(3a^2 - 4ab + b)$
26. $(x + 3y)(x^2 - 2xy)$
27. $(-7x^2y)(4x^2y^2 - 2xy + 3y^2)$
28. $(5 + 4a)^2$
29. $(-2mn)(5m^3 - 4m^2n + 2n)$
30. $-5ab(a - b) + 2ab(3a - 2b)$
31. $(3 + 2w)^2$
32. $4x^2y(x + 2y) - 2xy(3x - y)$
33. $(y + 1)^3$
34. $(k - 3)^3$
35. $(3y - 1)(3y + 1)$
36. $(5w + 2)(5w - 2)$
37. $(x^2 - 3)(x^2 + 3)$
38. $(m^3 - 2)(m^3 + 2)$
39. $(x^2y^2 - 3xy)(xy - 2)$
40. $(ab - 5)(2a^2b^2 + 1)$

C. Read and respond to the following exercises.

41. Compare the problems and your answers in Exercises 21 and 31 with those in 35 and 37. What is the same in these problems and answers and what is different?
42. Compare the problems and your answers in Exercises 20 and 28 with those in 36 and 38. What is the same in these problems and answers and what is different?
43. Is $(5 + 6)^2 = 5^2 + 6^2$? Explain your answer, in words.
44. Is $(x + 3)^2 = x^2 + 3^2$? Explain your answer, in words.
45. Can FOIL be used to simplify $(m^2 + 2m - 1)^2$? Tell why or why not.

13.7 Dividing Polynomials

Dividing Polynomials by Monomials

Dividing a polynomial by a monomial will be a very familiar operation if you first rewrite the problem as separate terms. This could be viewed as an application of the Distributive Principle:

$$\frac{9x^2y^2 + 24x^2y - 12xy^2}{3xy}$$

$$= \frac{1}{3xy} \cdot (9x^2y^2 + 24x^2y - 12xy^2)$$

$$= \frac{9x^2y^2}{3xy} + \frac{24x^2y}{3xy} - \frac{12xy^2}{3xy}$$

Now each term represents a monomial division like those you did earlier in this chapter. Divide the coefficients and subtract the exponents when the bases are the same.

$= 3xy + 8x - 4y$

In rewriting the problem, you are just putting each term of the numerator over the common denominator.

EXAMPLE 1 Divide: $\dfrac{a^2b^2 + 4a^2b - 8ab}{2ab}$

Solution

$$\frac{a^2b^2 + 4a^2b - 8ab}{2ab}$$ Rewrite as separate terms.

$$= \frac{a^2b^2}{2ab} + \frac{4a^2b}{2ab} - \frac{8ab}{2ab}$$ Divide each term.

$$= \frac{1}{2}ab + 2a - 4$$

EXAMPLE 2 Divide: $\dfrac{36x^2y^2 - 18x^2y}{-6xy}$

Solution

$$\frac{36x^2y^2 - 18x^2y}{-6xy}$$

$$= \frac{36x^2y^2}{-6xy} - \frac{18x^2y}{-6xy}$$ Be careful with the signs.

$$= -6xy - (-3x)$$

$$= -6xy + 3x$$

EXAMPLE 3 Simplify: $\dfrac{14m^2 - 35m + 7}{7m}$

Solution

$$\frac{14m^2 - 35m + 7}{7m}$$

$$= \frac{14m^2}{7m} - \frac{35m}{7m} + \frac{7}{7m}$$

$$= 2m - 5m^0 + m^{-1}$$

Remember to simplify all negative and zero exponents.

$$= 2m - 5 + \frac{1}{m}$$

Self-Check Simplify completely. All negative and zero exponents should be simplified, and fractions should be reduced to lowest terms.

1. $\dfrac{20a^2b^2 - 16a^2b + 12ab^2}{-4ab}$ 2. $\dfrac{4x^2yz^2 - 18xy^2z}{6x^2yz}$ 3. $\dfrac{-20x^3y^4 + 10xy^2}{-5x^2y^3}$

Answers:

1. $-5ab + 4a - 3b$ 2. $\dfrac{2z}{3} - \dfrac{3y}{x}$ 3. $4xy - \dfrac{2}{xy}$

Dividing Polynomials by Polynomials

Dividing a polynomial by a polynomial is very similar to arithmetic long division. First arrange the terms of the polynomials in descending order of the exponents. In $(x^2 - x - 2) \div (x + 1)$, the terms are already in descending order, so just write them in the long division format.

$$x + 1\overline{)x^2 - x - 2}$$

Determine how many times x will divide into x^2, and write that answer above the x^2 term.

$$\begin{array}{l} \quad\quad\; x \\ x + 1\overline{)x^2 - x - 2} \end{array}$$

Multiply x times $x + 1$, and write the product under the like terms in the dividend.

$$\begin{array}{l} \quad\quad\; x \\ x + 1\overline{)x^2 - x - 2} \\ \quad\quad\; \underline{x^2 + x} \end{array}$$

Now subtract $x^2 + x$ from $x^2 - x$, remembering to change both signs in the polynomial being subtracted.

$$\begin{array}{l} \quad\quad\; x \\ x + 1\overline{)x^2 - x - 2} \\ \quad\; \underline{-\; x^2 + x} \end{array} \qquad\qquad \begin{array}{l} \quad\quad\; x \\ x + 1\overline{)x^2 - x - 2} \\ \quad\; \underline{+ - x^2 - x} \\ \quad\quad\quad\; - 2x \end{array}$$

Bring down the next term, divide x into $- 2x$, and write that value above $- x$. Then multiply and subtract.

$$\begin{array}{l} \quad\quad\quad x \;\; - 2 \\ x + 1\overline{)\;\; x^2 - x - 2} \\ \quad \underline{+ - x^2 - x} \\ \quad\quad\quad - 2x - 2 \\ \quad\quad\; \underline{- - 2x - 2} \end{array} \qquad\qquad \begin{array}{l} \quad\quad\quad x \;\; - 2 \\ x + 1\overline{)\;\; x^2 - x - 2} \\ \quad \underline{+ - x^2 - x} \\ \quad\quad\quad - 2x - 2 \\ \quad\quad\quad \underline{+ + 2x + 2} \\ \quad\quad\quad\quad\quad\quad 0 \end{array}$$

Because there is a zero remainder,
$(x^2 - x - 2) \div (x + 1) = x - 2$

You should check your answer by multiplying the quotient by the divisor and then adding any remainder to that product. The answer should be the dividend of the original problem.

$(x - 2)\,(x + 1) = x^2 + x - 2x - 2 = x^2 - x - 2$

After arranging the polynomial terms in descending order of the exponents, fill in any missing terms with a zero term. The following problem, which is worked in detail, illustrates this process and also involves an answer with a remainder.

EXAMPLE 4 Divide: $(12y^4 + 3y - 4 - 6y^2)$ by $(2 + y)$

Solution Rearrange the terms in both polynomials.

$$y + 2\overline{)12y^4 - 6y^2 + 3y - 4}$$

Fill in the missing y^3 term with $0y^3$.

$$y + 2\overline{)12y^4 + 0y^3 - 6y^2 + 3y - 4}$$

Divide y into $12y^4$ and place the answer above that term. Then multiply $12y^3$ times $y + 2$ and subtract the product (changing the sign of each term).

$$
\begin{array}{r l}
 & \quad 12y^3 \\
y + 2) & \overline{\;12y^4 + 0y^3 - 6y^2 + 3y - 4} \\
 & \ominus \quad \ominus \\
 & \underline{+ 12y^4 + 24y^3} \\
 & \qquad - 24y^3
\end{array}
$$

Bring down the next term, divide y into $-24y^3$, place the answer in the quotient, and then multiply and subtract.

$$
\begin{array}{r l}
 & \quad 12y^3 - 24y^2 \\
y + 2) & \overline{\;12y^4 + 0y^3 - 6y^2 + 3y - 4} \\
 & \ominus \quad \ominus \\
 & \underline{+ 12y^4 + 24y^3} \\
 & \qquad - 24y^3 - 6y^2 \\
 & \qquad \oplus \quad \oplus \\
 & \qquad \underline{- 24y^3 - 48y^2} \\
 & \qquad\qquad + 42y^2
\end{array}
$$

Repeat this sequence of steps until the last term in the dividend is brought down and used.

$$
\begin{array}{r l}
 & \quad 12y^3 - 24y^3 + 42y - 81 \\
y + 2) & \overline{\;12y^4 + 0y^3 + 6y^2 + 3y - 4} \\
 & \ominus \quad \ominus \\
 & \underline{+ 12y^4 + 24y^3} \\
 & \qquad - 24y^3 - 6y^2 \\
 & \qquad \oplus \quad \oplus \\
 & \qquad \underline{- 24y^3 - 48y^2} \\
 & \qquad\qquad + 42y^2 + 3y \\
 & \qquad\qquad \ominus \quad \ominus \\
 & \qquad\qquad \underline{+ 42y^2 + 84y} \\
 & \qquad\qquad\quad - 81y - 4 \\
 & \qquad\qquad\quad \oplus \quad \oplus \\
 & \qquad\qquad\quad \underline{- 81y - 162} \\
 & \qquad\qquad\qquad + 158
\end{array}
$$

Write the remainder as a fraction in the quotient, remainder over divisor. The quotient is

$$12y^3 - 24y^2 + 42y - 81 + \frac{158}{y + 2}$$

Check If you are not very confident with such an answer, check it by multiplying the answer times the divisor and then adding the remainder (+158) to the product.

$$\begin{array}{r} 12y^3 - 24y^2 + 42y - 81 \\ y + 2 \\ \hline + 24y^3 - 48y^2 + 84y - 162 \\ 12y^4 - 24y^3 + 42y^2 - 81y \\ \hline 12y^4 - 6y^2 + 3y - 162 \\ + 158 \\ \hline 12y^4 - 6y^2 + 3y - 4 \end{array}$$

Add the remainder.

The answer does check.

EXAMPLE 5 Find the quotient of $(k^2 - 4)$ and $(k + 2)$.

Solution

$$\begin{array}{r} k - 2 \\ k + 2 \overline{) \; k^2 + 0k - 4} \\ \ominus \; \ominus \\ \underline{k^2 + 2k} \\ -2k - 4 \\ \oplus \; \oplus \\ \underline{-2k - 4} \\ 0 \end{array}$$

Therefore, $(k^2 - 4) \div (k + 2) = k - 2$.

Check $(k - 2)(k + 2) = k^2 + 2k - 2k - 4 = k^2 - 4$

EXAMPLE 6 Divide: $\dfrac{R^2 - 2RT + T^2}{R + T}$

Solution

$$\begin{array}{r} R - 3T + \dfrac{4T^2}{R + T} \\ R + T \overline{) \; R^2 - 2RT + T^2} \\ \ominus \; \ominus \\ \underline{R^2 + RT} \\ -3RT + T^2 \\ \oplus \; \oplus \\ \underline{-3RT - 3T^2} \\ + 4T^2 \end{array}$$

Check $(R - 3T)(R + T) = R^2 + RT - 3RT - 3T^2 =$

$$\begin{array}{r} R^2 - 2RT - 3T^2 \\ \underline{+ 4T^2} \\ R^2 - 2RT + T^2 \end{array}$$

Now add the remainder.

13.7 EXERCISES

A. Divide each polynomial by the monomial. Be sure that all answers are completely simplified.

1. $\frac{39ab - 13a}{13a}$
2. $\frac{12x^2 + 36x}{4x}$
3. $\frac{6x^3 - 12y^2}{-3}$
4. $\frac{8p^2 + 24q^2}{-4}$
5. $\frac{15m^3 - 10m^2 + 25m}{5m}$
6. $\frac{36a^2 - 24a + 12}{-6}$
7. $\frac{30x^2 + 15x + 20}{-10}$
8. $\frac{22d^4 - 33d^3 - 11d^2}{11d}$
9. $\frac{3p - 6q - 9r}{3}$
10. $\frac{7b + 14c - 28}{-7}$
11. $\frac{24v^3w^4x^2 - 18v^2w^2x^3}{6v^2wx}$
12. $\frac{8x^2yz^3 + 12x^2y^2z^2}{-6x^2yz}$
13. $\frac{9a^2b^3c + 15a^3b^2c^3}{6a^2b^2c}$
14. $\frac{27f^3g^2h^2 - 18f^2g^2h^2}{9f^2gh}$
15. $\frac{x^4y^3 + x^2y^2}{-xy}$
16. $\frac{18r^2s^3 - 9r^2s^2 - 27r^3s^3}{9r^2s^2}$

B. If you have studied Section 13.4, simplify the following exercises.

17. $\frac{21a^2b - 14ab^2 - 7a^3b}{7a^2b}$
18. $\frac{p^3t - pt^2}{-p^2t}$
19. $\frac{7r^3 - 5r}{r^2}$
20. $\frac{9t^3 - 5t}{t^2}$
21. $\frac{4a^3b - 3ab^2}{2a^3b^3}$
22. $\frac{24fg^2 - 28f^2g}{3f^2g^2}$
23. $\frac{15x^2yz^3 - 10xy^3z^2 + 5x^2y^2z^2}{10x^3y^2z^2}$
24. $\frac{8m^2n^3p + 16mn^2p^3 - 4m^2n^2p^2}{-4m^2n^3p^2}$

C. Divide these polynomials as indicated.

25. $\frac{A^2 - 5A + 6}{A - 3}$
26. $\frac{T^2 - 7T + 12}{T - 4}$
27. $(n^2 - 4) \div (n + 2)$
28. $(c^2 - 36) \div (c + 6)$
29. $\frac{x^2 - x - 10}{x - 3}$
30. $\frac{y + y^2 - 5}{y + 2}$
31. $\frac{3t + 2t^2 + 1}{t + 1}$
32. $\frac{3P^2 - 2P + 1}{P - 1}$
33. $\frac{8z^4 - 2z^2 + 1}{z^2 + 1}$
34. $\frac{6m^4 + 3m^2 - 2}{m^2 - 1}$
35. $(6x - 2x^2 + 4 - 8x^3) \div (2 + x)$
36. $(7y^2 + 3y^3 + 4 + 6y + y^4) \div (y^2 + y + 1)$
37. $(2 - 8x^2 + 2x^4 - 2x + x^3) \div (x^2 - x - 2)$
38. $(1 - 3m^3 - m + 2m^2) \div (1 - m)$
39. $(a^4 - a^2 - 6) \div (a^2 + 2)$
40. $(c^4 + 3c^2 - 10) \div (c^2 - 2)$

D. Read and respond to the following exercises.

41. Describe in words how the procedures for finding the quotients are similar in:

$27\overline{)8397}$ and $x + 1\overline{)3x^2 + x - 2}$

42. Are $\frac{(24a^2b^2 - 3a^2b + 15ab)}{3ab}$ and $\frac{24a^2b^2}{3ab} - \frac{3a^2b}{3ab} + \frac{15ab}{3ab}$ equivalent expressions? Explain your answer, in words.

13.8 Scientific Notation

One very useful application of the exponent rules occurs in scientific notation. This notation is used extensively in chemistry, physics, electronics, and other technical fields to simplify calculations with very large or very small numbers. Would you want to perform multiplication or division with numbers such as 5,880,000,000,000 (the approximate number of miles in 1 light year) or with 0.00000000000000000000000602 (the number of molecules in a gram molecule)? Scientific notation will transform these quantities into manageable numbers.

From the shortcut for multiplication of decimals (Section 5.3), when 2.5 is multiplied by 1,000 we find the answer by moving the decimal point three places to the right.

$2.5 \times 1{,}000 = 2.500 \times 1{,}000 = 2{,}500$

Because $1{,}000 = 10 \cdot 10 \cdot 10$, it could be expressed as 10^3. Therefore, 2,500 could be written as 2.5×10^3.

2.5×10^3 is scientific notation for 2,500. It expresses the quantity as the product of a number greater than or equal to 1 and less than 10 and the power of 10 that would indicate the movement of the decimal point *back to its original position.*

$10 = 10^1$	$0.1 = 10^{-1}$
$100 = 10^2$	$0.01 = 10^{-2}$
$1{,}000 = 10^3$	$0.001 = 10^{-3}$
$10{,}000 = 10^4$	$0.0001 = 10^{-4}$
and so on	and so on

When 10 has a positive exponent, the decimal point is moved to the *right*, and when the exponent of 10 is negative, the decimal point is moved to the *left*.

To change a number to scientific notation, express the number as a value greater to or equal to 1 and less than 10 times the power of 10 that accounts for the number of places the decimal point would need to be moved—and the direction in which it would need to be moved—to *return to the original number*. For example,

5,880,000,000,000 would be 5.88×10^{12}.

5.88 is greater than 1 and less than 10.

10^{12} indicates that the decimal point would have to be moved 12 places to the right to be returned to the original position

5,880,000,000,000 = 5,880,000,000,000

0.00000000000000000000000602 would be 6.02×10^{-23}.

6.02 is between 1 and 10.

10^{-23} would indicate movement of the decimal point 23 places to the left.

Scientific Notation

If N is a positive number, then $N = n \times 10^c$, where n is a number such that $1 \leq n < 10$ and c is an integer.

To write a positive number in scientific notation:

1. Place the decimal point to the right of the first nonzero digit (make the number greater than or equal to 1 and less than 10).
2. Determine what power of 10 to use by counting the number of places the decimal point in this new number would have to be moved to return it to its original position.
   a. If the decimal point would have to be moved to the right, the exponent is positive.
   b. If the decimal point would have to be moved to the left, the exponent is negative.
   c. If the decimal point would not have to be moved, the exponent is zero.

EXAMPLE 1 Write 558 in scientific notation.

Solution $558 = 5.58 \times 10^2$ because the decimal point would have to move two places to the right to return to its original position.

EXAMPLE 2 Write 0.0065 in scientific notation.

Solution $0.0065 = 6.5 \times 10^{-3}$ because we would have to move the decimal point three places to the left to return to the original number.

EXAMPLE 3 Write 461,300 in scientific notation.

Solution $461{,}300 = 4.613 \times 10^5$

EXAMPLE 4 Change 29.8×10^3 to scientific notation.

Solution 29.8×10^3 is not in scientific notation because 29.8 is not between 1 and 10. 29.8 becomes 2.98×10^1, so $29.8 \times 10^3 = 2.98 \times 10^1 \times 10^3$. The bases are the same, so $10^1 \times 10^3 = 10^4$. Thus

$29.8 \times 10^3 = 2.98 \times 10^4$

EXAMPLE 5 Change each of the following to *ordinary* notation; that is a number written without the power of 10.

a. 1.57×10^{-5} b. 6.9×10^4 c. 4×10^{-3}

Solution a. $1.57 \times 10^{-5} = \mathbf{00001.}57 \times 10^{-5}$
$= 0.0000157$

b. $6.9 \times 10^4 = \mathbf{6.9000} \times 10^4$
$= 69{,}000$

c. $4 \times 10^{-3} = \mathbf{004.} \times 10^{-3}$
$= 0.004$

When you are required to multiply or divide very large or small numbers, first change them to scientific notation. Then multiply or divide the numbers that are less than 10 and apply the rules of exponents to the powers of 10.

Self-Check

1. Write in scientific notation.

   a. 0.0043 b. 572,000

2. Write in ordinary notation.

   a. 1.32×10^{-4} b. 6.7×10^{3}

Answers

1. a. 4.3×10^{-3} b. 5.72×10^{5}

2. a. 0.000132 b. 6,700

EXAMPLE 6 Multiply: $2{,}900{,}000 \times 13{,}000$

Solution

$2{,}900{,}000 \times 13{,}000$

$= 2.9 \times 10^{6} \times 1.3 \times 10^{4}$

$= 2.9 \times 1.3 \times 10^{6} \times 10^{4}$

$= 3.77 \times 10^{10}$

It is permissible either to leave the answer in scientific notation or to write it in ordinary notation as 37,700,000,000.

EXAMPLE 7 Find the quotient of 0.00384 and 3,000.

Solution

$\dfrac{0.00384}{3{,}000}$

$= \dfrac{3.84 \times 10^{-3}}{3 \times 10^{3}}$ Divide 3.84 by 3, and divide 10^{-3} by 10^{3}.

$= 1.28 \times 10^{-3-3} = 1.28 \times 10^{-3+(-3)}$

$= 1.28 \times 10^{-6}$ or 0.00000128

EXAMPLE 8 Simplify and express the answer in scientific notation: $\dfrac{0.0007 \times 0.042}{0.021}$

Solution

$\dfrac{0.0007 \times 0.042}{0.021}$

$= \dfrac{7 \times 10^{-4} \times 4.2 \times 10^{-2}}{2.1 \times 10^{-2}}$ Simplify the numerator.

$= \dfrac{29.4 \times 10^{-6}}{2.1 \times 10^{-2}}$ Divide 29.4 by 2.1, and divide 10^{-6} by 10^{-2}.

$= 14 \times 10^{(-6)-(-2)}$

$= 14 \times 10^{-4}$ Write 14 in scientific notation.

$= 1.4 \times 10^{1} \times 10^{-4}$ Apply exponent rules.

$= 1.4 \times 10^{-3}$

If you are using a calculator in your course work, try doing Examples 6, 7, and 8 with your calculator. Try each example first without using scientific notation, and then try it using the scientific notation keys. Compare the calculator displays and answers with the steps in each example. Consult your calculator manual or your instructor if you need help.

13.8 EXERCISES

A. Write each number in scientific notation.

1. 2,780
2. 0.006
3. 0.00075
4. 18,000,000
5. 0.65
6. 3,000,000
7. 7,885
8. 43.8×10^2
9. 0.431
10. 675×10^{-3}
11. 0.0652
12. 8,841,000

B. Write each number in ordinary notation.

13. 6.5×10^3
14. 7.8×10^2
15. 1.57×10^{-2}
16. 9.8×10^{-3}
17. 9×10^4
18. 1×10^{-5}
19. 7×10^{-4}
20. 5×10^5
21. 3.144×10^{-2}
22. 6.8×10^6
23. 4.7×10^5
24. 4.157×10^{-3}

C. Change each number to scientific notation before doing the calculations. Express your answers in scientific notation.

25. $4{,}600{,}000 \times 9{,}000$
26. 0.0024×0.006
27. 0.009×0.00051
28. $27{,}000 \times 0.003$
29. 0.0064×800
30. $1{,}700{,}000 \times 40{,}000$
31. $\dfrac{6{,}800 \times 400}{40{,}000 \times 2{,}000}$
32. $\dfrac{0.06 \times 7{,}200}{0.008}$
33. $\dfrac{0.0024 \times 6{,}000}{120 \times 40{,}000}$
34. $\dfrac{480 \times 70{,}000}{8{,}000 \times 1{,}400}$
35. $0.135 \div 2{,}700$
36. $0.0185 \div 370$
37. $(0.008)^2$
38. $(20{,}000)^2$
39. $(12{,}000)^2$
40. $(0.0011)^2$

D. Read and respond to the following exercises.

41. Describe, in words, the process you use to change 0.0094 to scientific notation, and then give the answer.
42. Is $1.87 \times 10^3 = 187$? Explain your answer, in words.
43. Do you see any advantage to performing this calculation using scientific notation rather than working it out longhand?

$$\frac{0.068 \times 0.0002}{40{,}000}$$

44. Is 27.3×10^3 scientific notation? In words, explain your answer.
45. Scientific notation isn't used just by scientists. Describe some areas in course work or in your chosen profession where you might need to calculate with very large or very small numbers and where scientific notation might be useful.

13.9 CHAPTER REVIEW

In simplifying monomial and polynomial operations, you must follow these rules.

1. Add or subtract (combine) only *like* terms.

   $3x^2 + 2x + 9x^2 - 5x = 12x^2 - 3x$

2. To multiply quantities with the same base, keep the base and add the exponents.

   $a^2 \cdot a^3 = a^5 \quad (-2y^2)(3y) = -6y^3$

3. A number or variable with no exponent showing is understood to have an exponent of 1.

   $x^2y = x^2y^1$

4. To raise a base in exponent form to a power, keep the base and multiply the exponents.

   $(x^3)^2 = x^6 \qquad (2^4)^2 = 2^8$ or 256

5. To divide quantities with the same base, keep the base and subtract the exponents.

   $\dfrac{m^5}{m^2} = m^3 \qquad \dfrac{3^4}{3} = 3^3$ or 27

6. Any nonzero number raised to an exponent of 0 is equal to 1 and must be simplified.

   $y^0 = 1 \qquad a^2b^0c = a^2 \cdot 1 \cdot c = a^2c$

7. When simplifying polynomials, replace negative exponents with their positive exponent equivalents.

   $2^{-x} = \dfrac{1}{2^x} \qquad a^{-2}b^3 = \dfrac{b^3}{a^2}$

8. Use either the Distributive Principle itself or its long multiplication format to multiply polynomials.

   $a^2b(3a^2 - 4ab + 2b^2) = 3a^4b - 4a^3b^2 + 2a^2b^3$

   $(x^2 + 2x + 4)(2x^2 - x - 6)$

   $= x^2(2x^2 - x - 6) + 2x(2x^2 - x - 6) + 4(2x^2 - x - 6)$

   $= 2x^4 - x^3 - 6x^2 + 4x^3 - 2x^2 - 12x + 8x^2 - 4x - 24$

   $= 2x^4 + 34x^3 - 16x - 24$

   or

$$
\begin{array}{r}
x^2 + 2x + 4 \\
\underline{2x^2 - x - 6} \\
-6x^2 - 12x - 24 \\
-x^3 - 2x^2 - 4x \\
\underline{2x^4 + 4x^3 + 8x^2} \\
2x^4 + 3x^3 + 0x^2 - 16x - 24 = 2x^4 + 3x^3 - 16x - 24
\end{array}
$$

   $(3d + 4)(d - 6)$

   $= 3d(d - 6) + 4(d - 6)$

   $= 3d^2 - 18d + 4d - 24$

   $= 3d^2 - 14d - 24$

9. Binomials can be multiplied using FOIL.

   $(3a + 2)(a - 1)$

F	O	I	L
$(3a)(a)$	$(3a)(-1)$	$(+2)(a)$	$(+2)(-1)$

   $= 3a^2 - 3a + 2a - 2$

   $= 3a^2 - a - 2$

10. In dividing a polynomial by a monomial, write each term over the denominator and then follow the procedure for dividing monomials.

$$\frac{15x^3 - 10x^2 + 20x}{5x}$$
$$= \frac{15x^3}{5x} - \frac{10x^2}{5x} + \frac{20x}{5x}$$
$$= 3x^2 - 2x + 4$$

11. In dividing a polynomial by a polynomial, use the same procedures as in arithmetic long division after arranging the terms in descending order and filling in any missing terms.

12. Very large or very small numbers can be written in scientific notation, with powers of 10 used to represent the decimal places. Once such numbers have been expressed in scientific notation, follow the arithmetic rules for multiplication and division and the exponent rules for the powers of 10.

13. In all operations, the correct signed–number rules must be applied.

REVIEW EXERCISES

A. Perform the indicated operation.

1. $24x - 5x$
2. $14b^2 + 11b^2$
3. $a - 7a$
4. $y - 3y$
5. $8c^2d + 3c^2$
6. $8a - 3a$
7. $p^7 \div p^5$
8. $q^3 \div q^2$
9. $x^4 \div x$
10. $(-2a^2)(4ab^2)$
11. $6mn(-3mn^2)$
12. $b^5 \div b^2$
13. $n^6 \div n^2$
14. $k^5 \div k$
15. $(t^3)^5$
16. $(m^2)^4$
17. $(3b)^2$
18. $(-2a^2)^3$
19. $(5a^3)(-3a^7)$
20. $(-2r^2)(3r^4)$
21. $8m^2n \div 4n$
22. $12c^2d^2 - 4cd^2$
23. $3(2a + b)$
24. $2(3x - 2y)$
25. $2b^{-2}$
26. $a^{-1}b^{-2}c^3$
27. $b^2 \div b^5$
28. $y^3 \div y^4$
29. $(2b)^{-2}$
30. $(ab^2)^{-1}$

B. Simplify each expression by performing the indicated operations and eliminating zero and negative exponents from the answers.

31. $\frac{9t^9}{12t^4}$
32. $\frac{45n^8}{30n^5}$
33. $(2x^2y^4)(7x^5y^2)$
34. $(-6a^8b^3)(-4ab^8)$
35. $\frac{-56a^5b^3}{7a^2b^2}$
36. $\frac{72x^7y^5}{12x^3y^3}$
37. $4pq^4 + p^4q - 7pq^4$
38. $18r^2s - 3rs + 5rs$
39. $(5f + 2g)(4f - g)$
40. $(2z - 3)(5z + 6)$
41. $5x^5 - 8x^5 + 6x^5$
42. $-3y^7 + 9y^7 - y^7$
43. $2rs - 3r^2s$
44. $5p^2q^3 - 2p^2q^2 - 7p^2q^3$
45. $-2a^3b^6 + 8a^7b^3 - 9a^3b^6$
46. $8wz - 11wz^2$
47. $\frac{39t^4}{13t^6}$
48. $\frac{24x^2}{16x^5}$
49. $(3a^{-2}bc^{-1})^{-2}$
50. $(2x^2y^{-3}z^{-1})^{-3}$
51. $2(5a - 7) - 3(4a + 2)$
52. $(-3x^3y)^2(2xy)$
53. $(3c^2d^2)^2(-2cd)$
54. $8(t - 4) - 7(3t - 1)$
55. $(y^2 + 2y - 1)(3y^2 - 5y - 2)$

56. $(2a^2 - a + 7)(a^2 - 4)$

57. $\dfrac{9w^6 + 4w^4 + 2w^3}{w^2}$

58. $\dfrac{7g^5 - 5g^3 + 3g^2}{g^2}$

59. $(5p + 2)^2$

60. $(5x^2 - 6x + 2) - (3x^2 - 4x + 9)$

61. $(y^2 - 6y - 4) - (3y^2 - 8y + 1)$

62. $(7x - 4y)^2$

63. $(5 - 3p)(p + 6)$

64. $(3k - 2) - (3 + 4k)$

65. $(4y + 3) - (9 - 7y)$

66. $(5 - 3z)(7z + 6)$

67. $\dfrac{12b^3 + 16d^2 - 9b}{-6b + 1}$

68. $\dfrac{9d^3 - 15d^2 + 4d}{-3d + 1}$

69. $\dfrac{15x^3y^2z^2 - 25x^2y^2z^2 + 10xy^3z^2}{5xy^2z}$

70. $\dfrac{24a^2b^3c^3 - 18a^2b^2c^2 - 6a^2b^3c^4}{-6ab^2c}$

71. $\dfrac{a^2 + 7a + 10}{a + 5}$

72. $\dfrac{x^2 - 5x + 6}{x - 3}$

73. $(2m^{-2}n^{-3})(-3m^3n^{-2})$

74. $(3m^2 - 1) \div (m - 3)$

75. $(4x^3 - 2x + 1) \div (x - 1)$

76. $(-5r^{-2}s^{-1}t^3)(2rs^{-3}t^{-2})$

77. $\dfrac{(2x^{-1}y^2)^2(x^2y^{-3})^{-1}}{(2^{-3}xy^3)^2}$

78. $(1 + 2t)^3$

79. $(2 + a)^3$

80. $\dfrac{(5m^2n^{-1})^{-1}(2m^2n^2)^{-1}}{(4m^{-3}n^{-2})^{-1}}$

C. Complete the following chart.

Scientific Notation	Ordinary Notation
81.	0.0087
82. 3.44×10^{-3}	
83. 9×10^7	
84.	0.00005
85.	62.3×10^3
86.	0.8×10^{-3}
87. 6.5×10^{-4}	
88. 3.89×104	
89.	6,400,000
90. 1.5×10^{-1}	
91. 9.9×10^3	
92.	2,960,000

D. Perform the indicated operations using scientific notation and the properties of exponents. Express each answer in scientific notation.

93. $0.045 \div 9{,}000$

94. $1{,}114 \times 0.003$

95. 0.0093×0.006

96. $4{,}200{,}000 \div 0.21$

97. $0.007 \times 690 \times 3{,}000$

98. $910 \times 0.008 \times 40{,}000$

99. $\dfrac{64{,}000 \times 0.004}{0.0008 \times 0.02}$

100. $\dfrac{850 \times 0.00001 \times 5{,}000}{17{,}000 \times 0.000005}$

E. In each of the simplifications that follow, an error has occurred. State in words what the error is and give the correct answer.

101. $8m - 11n + 3m + (-n)$

$= 11m - 12n$

$= -1mn$

102. From the sum of $3a^2b$ and $(-5a^2b)$ subtract $(-4a^2b)$.

$3a^2b + (-5a^2b) - (4a^2b)$

$= -6a^2b$

103. $(3x^2)^3$

$= 3x^6$

104. $(a^4b^3)^2$

$= a^6b^5$

105. $6c^2d^2(-2c^3d^4)$

$= -12c^6d^8$

106. $\frac{2^6}{2^4}$

$\frac{\not{2}^6}{\not{2}^4}$

$= 1^2 = 1$

107. $\frac{c^{-2}}{c^5}$

$= c^3$

108. $\frac{14x^2y^3}{6xy^3}$

$= 8x$

109. $-2a^3b^{-4}$

$= \frac{a^3}{2b^4}$

110. $(5x^4y^{-3})^2$

$= 5x^8y^{-6}$

$= \frac{5x^8}{y^6}$

111. $(7c^2 + 5c - 9) - (11c^2 - 4c + 1)$

$= -4c^2 + c - 10$

112. $(x + 2)^2$

$= x^2 + 4$

113. $\frac{9a^2b^2 + 24a^2b}{3ab}$

$= 3ab + 24a^2b$

114. Write 5,120 in scientific notation.

$= 51.2 \times 10^2$

F. Read and respond to the following exercises.

115. Compare $-2a^2$ and $(-2a)^2$.

116. How are addition of monomials and addition of polynomials related?

117. What is the meaning of x^0? Why don't we see this kind of notation very often?

118. Is $(2x - 1)^2 = 4x^2 + 1$? Explain your answer, in words.

119. Describe, in words, how you would apply the Distributive Principle to the following exercise: $(3a - 4)(2a^2 + 5a - 1)$.

120. Describe how to multiply the expression in problem 119 using FOIL method.

G. The following problems may require a little more thought than the previous ones. Simplify completely.

121. $\frac{(2p^{-1}q^{-2}r^2)^{-2}}{(5pq^2r^{-1})(-2p^2)^{-1}}$

122. $\frac{(m^2n^{-3})^2(m^{-1}n^{-2})}{(2m^3n^{-2})(m^{-2}n^{-1})^{-1}}$

123. $\frac{(5x^2y^{-3})(2x^{-1}y^{-2})^2}{(3x^{-1}y^2)^{-1}(x^2y^{-2})}$

124. $(a^2 - 2a + 9)(3a^2 - a - 7)$

125. $(a^2 + b^2)(2a^2 + 3a)$

126. $(x^2 + 2x + 4)(2x^2 - x - 6)$

CHAPTER TEST

In Exercises 1-18, perform the indicated operation(s) and simplify completely, eliminating zero and negative exponents from the answers.

1. $(3a^2b)(-2a^3b^2)$
2. $(y - 4)(2y + 1)$
3. $-8m^2n \div 16mn$
4. $4a^2c + (-6a^2c)$
5. $c^4 \div c^8$
6. $(-2r^2st^3)^2$
7. $(3a^{-2}bc^3)^{-1}$
8. $(-35x^2y^3z) \div (7x^2y)$
9. $-2(5x - 3y)$
10. $F^2 \cdot F \cdot F^3$
11. $(3a^{-2}b^{-3})(-2a^{-3}b^{-2})$
12. $(x + 3)^2$
13. $\left(\dfrac{4x^2}{3y}\right)^2$
14. $\dfrac{56c^2d^2 - 14cd^2 + 21cd^3}{7cd}$
15. From the sum of $(5x)$ and $(-11x)$, subtract $(-3x)$.
16. Subtract: $(5m^2n - 2mn + 7) - (9mn - 11 - 2m^2n)$
17. Multiply: $(a + 2)(3a^2 - 2a + 5)$
18. Divide: $(4y^3 + 3y^2 + 2) \div (y + 1)$
19. Change 0.00067 to scientific notation.
20. Describe, in words, the process of simplifying this expression using scientific notation:

$$\frac{97{,}000 \times 600}{3{,}000{,}000}$$

STUDY SKILLS MODULE

14 Getting Help

Many students, especially those who are new to college, struggle all alone with a course such as math, never realizing that many resources are available to help them. In this Study Skills Module, we will discuss the kinds of resources that are available on most campuses, but you will need to find out what is offered on your own campus.

Your first resource should be your instructor. Write down the location of his or her office, the telephone number, and the office hours. Make an appointment or drop in during one of those scheduled times. If you are not free when office hours are scheduled, then call and ask your instructor to schedule some other time with you. Most full-time instructors are willing to work with their students during office hours or to schedule other times. But some of you instructors may be adjuncts (part-time) and may not have office hours (or even an office). Talk to that instructor to see what help he or she might try to provide.

In addition to help from your instructor, you might try to get tutoring help. Your college may offer group or individual tutoring, and that assistance might be provided by faculty members, professional tutors, or peer tutors (students who have successfully completed the course). It might be regularly scheduled sessions or unscheduled, drop-in help. Services might be free-of-charge. Start by getting information from your instructor; then follow up by contacting the appropriate office on campus (Student Services, Tutoring, Learning Assistance, and so on).

Before going to a help session with your instructor or a tutor, study the material and write down questions or problems that you don't understand. Do not erase a problem that is incorrect, but leave the work so the tutor can see what kind of error you are making. If you come in prepared with questions, your help session will be more beneficial for you. You must attend class and make a serious effort with the material before seeking assistance.

Ask you instructor if there are video- or audio-tapes or computer software supplements for the course. If so, ask where these materials are housed and whether they must be used on campus or if they can be checked out. Look in the library to see if there are other textbooks on the subject that you might understand more easily.

Find someone in your class whom you are comfortable talking to and form a "study-buddy" partnership. Agree to share class notes if one of you is absent. Get together after class and briefly discuss the material covered that day and help each other understand it. Perhaps you can work on homework together in the library or the Learning Center. Exchange phone numbers, so you have a help line when you are at home trying to do homework. Study for tests separately; then get together and test each other on the material.

Don't struggle all alone when there is help available. There are people and resources on campus that are dedicated to helping students succeed. These people want to help you, but you must ask. Make the effort to find out about the resources and then use them.

Your Point of View

1. Go to the campus directory, the college bulletin, or the Office of Student Services and find out if tutoring help is available on your campus. Report what kind of assistance is available, the hours, cost, location, and so on.
2. Write down four advantages to having a study-buddy in your math class. Write down two reasons you would be a good study-buddy.
3. Write a short paragraph explaining what you do when you get stuck with a problem. Be specific about the steps you take.

CHAPTER

Introduction to Factoring

LEARNING OBJECTIVES

When you have completed this chapter, you should be able to:

1. Completely factor expressions containing:

a. greatest common factors

b. the difference of two squares

c. trinomials

2. Check factoring answers by multiplying.

3. Solve quadratic equations by factoring.

In the previous chapters you learned how to multiply monomial and/or polynomial factors together. In this chapter, you will reverse the multiplication process. You will begin with products containing two, three, or four terms and will determine what prime factors were multiplied together to give that answer. This process is called factoring. Some of the factoring techniques may seem complicated at first, but you will come to understand them with practice.

14.1 Greatest Common Factor

Comparing the kinds of problems presented in Chapter 13 and this one reveals that factoring is the reverse of multiplication.

		Chapter 13		Chapter 14
Problem:	Multiply	$3x^2y(-2xy + 5)$	Factor	$14a^2b + 21a^2b^2$
Answer:		$-6x^3y^2 + 15x^2y$		$7a^2b(2 + 3b)$

In the Chapter 14 example, each term in the original expression was examined to see if it contained a factor common to the other. If so, as was the case here, the **greatest common factor** (GCF) was removed from each term and the expression was written as the product of the GCF and the remaining terms. (You may find it helpful to review the arithmetic procedure for finding the GCF in Section 2.7.) The greatest common factor was written in front of a set of parentheses, and the remaining factors of each term were written inside the parentheses. In fact, the Distributive Principle was used to write the factorization. Let's look closely to see how the answer was found.

First, each term was written in prime factored form. (See prime factorization in Section 2.5.)

$14a^2b + 21a^2b^2 = 2 \cdot 7 \cdot a \cdot a \cdot b + 3 \cdot 7 \cdot a \cdot a \cdot b \cdot b$

Note that 7, a, a, and b are in both lists of factors. Consequently, their product, $7a^2b$, was removed from each term.

$= 2 \cdot \cancel{7} \cdot \cancel{a} \cdot \cancel{a} \cdot \cancel{b} + 3 \cdot \cancel{7} \cdot \cancel{a} \cdot \cancel{a} \cdot \cancel{b} \cdot b$

Then $2 + 3b$ remained, so the factorization of $14a^2b + 21a^2b^2$ was written as $7a^2b(2 + 3b)$.

To check your answer, multiply the factors together to see if the answer agrees exactly with the original expression. If the answer and the original expression do agree, your answer is correct, as long as you have done all the factoring that was possible.

Check $7a^2b(2 + 3b) = 14a^2b + 21a^2b^2$, which is the same as the original expression.

EXAMPLE 1 Factor completely: $15m^2n^2 - 20m^2n + 35mn$

Solution $15m^2n^2 - 20m^2n + 35mn$

$5mn$ is in every term and therefore is the GCF. Remove it from every term (divide each term by $5mn$).

$= 3 \cdot \cancel{5} \cdot \cancel{m} \cdot m \cdot \cancel{n} \cdot n - 2 \cdot 2 \cdot \cancel{5} \cdot \cancel{m} \cdot m \cdot \cancel{n} + \cancel{5} \cdot 7 \cdot \cancel{m} \cdot \cancel{n}$

$3mn - 4m + 7$ remains as the quotient. Write the GCF outside the parentheses and the remaining terms inside.

$= 5mn(3mn - 4m + 7)$

Check Multiply the answer:

$$5mn(3mn - 4m + 7) = 15m^2n^2 - 20m^2n + 35mn$$

EXAMPLE 2 Factor completely: $26x^2 + 13x + 39$

Solution $26x^2 + 13x + 39$

13 is the GCF, so factor it out.

$= 13(2x^2 + x + 3)$

Check Multiply the answer:

$13(2x^2 + x + 3) = 26x^2 + 13x + 39$

EXAMPLE 3 Factor completely: $28y^2z^2 + 81y^2z + 15$

Solution $28y^2z^2 + 81y^2z + 15$

$= 2^2 \cdot 7 \cdot y^2z^2 + 3^4y^2z + 3 \cdot 5$

There is no factor common to *all* of the terms. Therefore, this expression cannot be factored by this method.

EXAMPLE 4 Factor completely: $55a^2 + 33a + 121$

Solution You do not need to write a list of factors of each term to see that there is no common literal factor. However, 55, 33, and 121 are all multiples of 11.

$55a^2 + 33a + 121 = 11(5a^2 + 3a + 11)$

Check Multiply the answer:

$11(5a^2 + 3a + 11) = 55a^2 + 33a + 121$

EXAMPLE 5 Factor completely: $-2x^2 + 28$

Solution *When the first term of a polynomial is negative, factor out the common negative factor.*

$-2x^2 + 28 = -2(x^2 - 14)$

Check $-2(x^2 - 14) = -2x^2 + 28$

Looking for the greatest common factor in a polynomial should be the first step in any factoring problem you do. The resulting expression will be much easier to factor further.

Self-Check Factor out any GCF.

a. $12x^3 - 8x^2 + 16x$ b. $-3a^2 + 18$ c. $4m^2 - 22m + 5$

Answers:

a. $4x(3x^2 - 2x + 4)$ b. $-3(a^2 - 6)$ c. $4m^2 - 22m + 5$ (there is no GCF)

Greatest Common Factor

If *all* terms in an expression contain a common factor or factors, remove the greatest common factor from each term and write that GCF and the resulting polynomial as an indicated multiplication.

$14x^2 + 7x + 28 = 7(2x^2 + x + 4)$

$a^2b^2 + a^2b + ab^2 = ab(ab + a + b)$

EXAMPLE 6 Factor completely: $a(x + 2) + b(x + 2)$

Solution Although this doesn't look like the previous examples, there is a common factor in both terms of the expression; it is $x + 2$. Remove that GCF, $x + 2$, from each term, and write it and the resulting polynomial as an indicated multiplication.

$a(x + 2) + b(x + 2) = (x + 2)(a + b)$

Check Multiply the answer:

$(x + 2)(a + b) = ax + bx + 2a + 2b$

Is that the same as $a(x + 2) + b(x + 2)$?

Multiply the original expression.

$a(x + 2) + b(x + 2) = ax + 2a + bx + 2b$

Rearrange the terms, $ax + bx + 2a + 2b$, and you can see that the products are the same.

◀

EXAMPLE 7 Factor completely: $y(y + 5) - 4(y + 5)$

Solution $(y + 5)$ is the common factor.

$y(y + 5) - 4(y + 5)$
$= (y + 5)(y - 4)$

Remove the common factor from both terms, and write the product of the GCF and the remaining terms.

Check $(y + 5)(y - 4)$

$= y^2 - 4y + 5y - 20$

$= y^2 + y - 20$

And $y(y + 5) - 4(y + 5)$

$= y^2 + 5y - 4y - 20$

$= y^2 + y - 20$

◀

EXAMPLE 8 Factor completely: $y^2 + 2y + yz + 2z$

Solution There isn't a common factor in all four terms, but you can factor by taking a y out of the first two terms and a z out of the last two terms.

$y(y + 2) + z(y + 2)$

Now can you see the common factor, $(y + 2)$? Remove $(y + 2)$ from each term.

$y(y + 2) + z(y + 2) = (y + 2)(y + z)$

Check $(y + 2)(y + z) = y^2 + yz + 2y + 2z$

Do you think you would have come up with the same result if you had grouped the terms differently? Let's see:

$y^2 + yz + 2y + 2z$

$= y(y + z) + 2(y + z)$

$= (y + z)(y + 2)$ The factorizations are the same.

The process used in Example 8 is called **factoring by grouping**. If you are to factor an expression with four terms and there is no factor common to all of them, try this method of working with two terms at a time.

EXAMPLE 9 Factor completely: $mn + kn + 2m + 2k$

Solution Use *factoring by grouping* because there are four terms that have no factor in common.

$mn + kn + 2m + 2k$	Factor n from the first two terms, and factor 2 from the last two terms.
$= n(m + k) + 2(m + k)$	Now remove $(m + k)$ from both terms and write the product.
$= (m + k)(n + 2)$	

Check $(m + k)(n + 2) = mn + 2m + kn + 2k$

EXAMPLE 10 Factor completely: $3x^2 + 6y^2 - 2x^2y - 4y$

Solution Try factoring by grouping.

$3x^2 + 6y^2 - 2x^2y - 4y$

$= 3(x^2 + 2y^2) - 2y(x^2 + 2)$

You cannot factor any further because $(x^2 + 2y^2)$ and $(x^2 + 2)$ are not like factors.

EXAMPLE 11 Factor completely: $-12my - 24y + 4m + 8$

Solution

$-12my - 24y + 4m + 8$	Factor out the GCF as -4 so that the first term will be positive.
$= -4(3my + 6y - m - 2)$	Factor a GCF out of each pair of factors.
$= -4[3y(m + 2) - 1(m + 2)]$	Remove $(m + 2)$ from each term.

After factoring $3y$ from the first two terms and getting the factor of $(m + 2)$, examine the last two terms, $-m - 2$, and see if you can factor out any GCF that will also produce an $(m + 2)$ factor. Factoring out a -1 would do that.

$= -4[(m + 2)(3y - 1)]$

$= -4(m + 2)(3y - 1)$

Check Multiply: $-4(m + 2)(3y - 1)$

$= (-4m - 8)(3y - 1)$

$= -4m(3y - 1) - 8(3y - 1)$

$= -12my + 4m - 24y + 8$ Rearrange the terms.

$= -12my - 24y + 4m + 8$

Self-Check Factor by grouping.

a. $3x(x - 2) + 5(x - 2)$ b. $a^2 + 2a + ab + 2b$

Answers:

a. $(x - 2)(3x + 5)$ b. $(a + 2)(a + b)$

14.1 EXERCISES

A. Multiply each set of factors to see if the product that is given is correct.

1. $4(2x + 7)$; product: $8x + 28$
2. $-5(7y^2 - 1)$; product: $-35y^2 - 5$
3. $-8(a^2 + 4b^2)$; product: $8a^2 + 32b^2$
4. $7(m^2n + 2mn)$; product: $7m^2n + 14mn$
5. $c^2b(3b + 5c^2)$; product: $3c^2b + c^4b$
6. $ax^2(-4x + 2a^2)$; product: $-4ax^2 + 2a^3x^2$
7. $9(w^2 + 3w - 2)$; product: $9w^2 + 27w - 18$
8. $2y^3(4y^4 - y^3 + 5y^2 - 1)$; product: $8y^7 - 2y^6 + 10y^5 - 1$
9. $m^5n^6(m^2n^2 - 2mn + 1)$; product: $m^7n^8 - 2m^6n^7 + m^5n^6$
10. $-x^2(x^2 - 2x + 10)$; product: $-x^4 + 2x^3 - 10x^2$

B. Factor out the greatest common factor from each expression. If there is none, write the original expression. Check answers by multiplying.

11. $2x^2 - 4x + 16$
12. $8m^2 - 16m + 24$
13. $5a^2b^2 - 12a^2b + 4ab$
14. $2.8b^3 + 1.4b^2 + 4.2b$
15. $-12r^2 - 12r$
16. $-3y^2 + 9$
17. $1.6m^3 + 6.4m^2 + 2.4m$
18. $-4x^2 + 20$
19. $-a^3 + 2a^2 + 6a$
20. $15d^2 - 10d - 50$
21. $11f^3 + 44g^3$
22. $13c^3 - 39d^3$
23. $-a^3 - a^2 + 8a$
24. $-b^5 - 4b^3 + 2b^2$
25. $64b^4 + 16b^3$
26. $16b^2 - 27$
27. $5a^3 - 15a^2 + 4$
28. $-12m^6 + 36m^5 - 48m^4$
29. $x^5y^4z^2 + 7x^4y^2z^2 - x^2z^2$
30. $2x^2y^2z^3 - 8x^2y^3z^2 + 4x^3y^2z^2$
31. $a(a + 4) - 4(a + 4)$
32. $m(n - 1) + 6(n - 1)$

C. Factor by grouping. Check answers by multiplying.

33. $ax^2 - 7ax + 2x - 14$
34. $8xy - 16x + y^2 - 2y$
35. $18wx + 6x + w^2 + w$
36. $a^2 - 2a - 8ab + 16b$
37. $acx + acy - bcx - bcy$
38. $3rt + 3st + 6r + 6s$
39. $p^2 + 5p + 12pq + 60q$
40. $m^2 - 4m - 9mn + 36n$

D. Factor completely.

41. $3m^2 - 9m$
42. $-15m + 18$
43. $25a^2b + 15b^3 + 35a^2 + 21b^2$
44. $3(x + 5) - 4y(x + 5)$
45. $4a^2b^2 - 12a^2b + 20ab^2$
46. $6b + 12 - 18ab - 36a$
47. $27x^2y + 9xy^2 + 16xy$
48. $-9c^2 - 81c$

E. Read and respond to the following exercises.

49. If you factor $-8a^2 - 24b$ and get $-8(a^2 + 3b)$, how will you check your answer?
50. Describe, in words, how you would factor $5(m + 3n) - 8a(m + 3n)$.
51. What step should be done first in every factoring exercise?
52. What process do you use to check every factoring exercise?

14.2 Difference of Two Squares

The numbers 1, 4, 9, 16, 25, 36, 49, 64, 81, 100, 121, 144, and so on are called **squared numbers**. This means that a rational number was multiplied by itself to produce this product (see Section 2.1).

$1 \times 1 = 1$	$2 \times 2 = 4$	$3 \times 3 = 9$	$4 \times 4 = 16$
$5 \times 5 = 25$	$6 \times 6 = 36$	$7 \times 7 = 49$	$8 \times 8 = 64$
$9 \times 9 = 81$	$10 \times 10 = 100$	$11 \times 11 = 121$	$12 \times 12 = 144$

The quantities x^2, y^2, b^6, and $4a^2$ are also squared quantities. They are the product of a factor times itself.

$x \cdot x = x^2$ $\quad y \cdot y = y^2$

$b^3 \cdot b^3 = b^6$ $\quad 2a \cdot 2a = 4a^2$

Five is said to be the square root of 25 ($\sqrt{25} = 5$). Eleven is the square root of 121 ($\sqrt{121} = 11$), and $2a$ is the square root of $4a^2$ ($\sqrt{4a^2} = 2a$, $a \geq 0$). The **square root** of a quantity is the number that, when multiplied by itself, gives the original quantity. Quantities such as 4, y^2, $121b^2$, and $16y^2$ appear in a special type of factoring problem. When the *difference* of two squared numbers is factored, the result is always the product of the sum and difference of their square roots. This definitely sounds more complicated than it is!

If you are asked to factor $x^2 - 4$, first check for a GCF. There is none. Note that the expression contains only *two terms* and that x^2 and 4 are squared quantities. The factors of their *difference* will always be the sum and the difference of their square roots:

$$x^2 - 4 = (x)^2 - (2)^2 = (x + 2)(x - 2)$$

Check the answer by multiplying the factors together.

$(x + 2)(x - 2)$
$= x^2 - 2x + 2x - 4$

When you combine the like terms, the result is $x^2 + 0x - 4$, which equals $x^2 - 4$. Because this is the original expression, the factors are correct.

EXAMPLE 1 Factor completely: $4b^2 - 9$

Solution There is no GCF, so look to see if this might be the difference of two squared quantities.

$\sqrt{4b^2} = 2b$ and $\sqrt{9} = 3$

Therefore, $4b^2 - 9$ fits the description of the difference of two squared quantities.

$4b^2 - 9 = (2b)^2 - (3)^2 = (2b + 3)(2b - 3)$

Check $(2b + 3)(2b - 3)$
$= 4b^2 - 6b + 6b - 9$
$= 4b^2 - 9$

Multiply the following sets of factors:

$(y + 6)(y - 6)=$

$(2a + 7)(2a - 7) =$

$(b^2 + 5)(b^2 - 5) =$

Did you see the pattern in these products? In each case, the middle terms add to zero. This occurs every time you multiply the sum and the difference of the same two terms.

Would you expect that to occur if you were multiplying two sums of the same terms? How about two differences of the same terms? Let's find out.

$(x + 2)(x + 2)$
$= x^2 + 2x + 2x + 4$
$= x^2 + 4x + 4$

$(2a - 7)(2a - 7)$
$= 4a^2 - 14a - 14a + 49$
$= 4a^2 - 28a + 49$

Apparently, with any combination other than the sum and difference, the middle terms do *not* add to zero and do not disappear.

Check it further by doing these multiplications:

$(y + 6)(y + 6)$ $(3z + 2)(3z - 2)$ $(b^2 - 5)(b^2 - 5)$

Your products should have been $y^2 + 12y + 36$, $9z^2 - 4$, and $b^4 - 10b^2 + 25$. In each case, the *same signs* in the factors *produced a middle term* in the product. But when the *signs in the factors were different*, the *middle terms added to zero*.

Difference of Two Squares

The factors of the difference of two squares are always the sum and difference of the square root of the terms.

$a^2 - b^2 = (a + b)(a - b)$

$16 - x^2 = (4 + x)(4 - x)$

$64m^4 - 9 = (8m^2 + 3)(8m^2 - 3)$

The sum of two squared quantities cannot be factored this way. If you tried to factor $x^2 + y^2$, you might guess $(x + y)(x - y)$, $(x - y)(x - y)$, or $(x + y)(x + y)$ as possible solutions. When those sets of factors are multiplied, they give the products $x^2 - y^2$, $x^2 - 2xy + y^2$, and $x^2 + 2xy + y^2$. None agrees with the original expression. In fact, $x^2 + y^2$ is *prime*; it has *no factors but itself and 1*.

When you begin to factor an expression that you recognize as the difference of two squares, always look first to see if there is a common factor (GCF) that can be removed.

$64x^2 - 16 = 16(4x^2 - 1) = 16(2x + 1)(2x - 1)$

$(8x + 4)(8x - 4)$ would not have been a correct answer, because each factor could be further simplified by removing a common factor of 4. When you are asked to factor an expression, your answer must have each part factored *completely*. Except for the GCF, all other factors must be prime.

Factor Completely

After completing each step of a factoring problem, see if any more factoring can be done.

EXAMPLE 2 Factor completely: $64x^2 - 4$

Solution

$64x^2 - 4$ — First remove the GCF.

$= 4(16x^2 - 1)$ — Factor the difference of two squared quantities.

$= 4(4x + 1)(4x - 1)$

Check

$4(4x + 1)(4x - 1)$

$= (16x + 4)(4x - 1)$

$= 64x^2 - 16x + 16x - 4$

$= 64x^2 - 4$

EXAMPLE 3 Factor completely: $162r^4 - 2$

Solution

$162r^4 - 2$

$= 2(81r^4 - 1)$

$= 2(9r^2 + 1)(9r^2 - 1)$ $\quad$ $9r^2 - 1$ is the difference of two squared quantities.

$= 2(9r^2 + 1)(3r + 1)(3r - 1)$

Check

$2(9r + 1)(3r + 1)(3r - 1)$

$= (18r^2 + 2)(3r + 1)(3r - 1)$

$= (54r^3 + 18r + 6r + 2)(3r - 1)$

$= 162r^4 + 54r^3 + 18r^2 + 6r - 54r^3 - 18r^2 - 6r - 2$

$= 162r^4 - 2$

Self-Check Try these exercises. If you have difficulty, review the steps in the previous examples.

1. $4 - a^2$ 2. $36k^2 - 49$ 3. $3m^3 - 27m$

Answers:

1. $(2 + a)(2 - a)$ 2. $(6k + 7)(6k - 7)$ 3. $3m(m^2 - 9) = 3m(m + 3)(m - 3)$

14.2 EXERCISES

A. Find the product.

1. $(x + 2)(x - 2)$
2. $(b + 6)(b - 6)$
3. $(m + 3)(m + 3)$
4. $(z - 2)(z - 2)$
5. $(2b + 1)(2b - 1)$
6. $(3x + 5)(3x - 5)$
7. $(x + 2z)(x - 2z)$
8. $(m + n)(m - n)$
9. $(1 - 3a)(1 + 3a)$
10. $(6 - 5y)(6 + 5y)$
11. $(3y - 2z)(3y + 2z)$
12. $(7c - 2b)(7c + 2b)$

B. Factor completely. If the expression cannot be factored, write "prime" as your answer. Check your answers by multiplying.

13. $x^2 - 49$
14. $b^2 - 36$
15. $a^2 - 1$
16. $m^2 - 16$
17. $k^2 - 121$
18. $r^2 - 1$
19. $z^2 - 81$
20. $s^2 - 49$
21. $c^2 - 4$
22. $w^2 - 25$
23. $4f^2 - 9$
24. $144d^2 - 1$
25. $25T^2 - 36$
26. $4G^2 - 9$
27. $49J^2 - 64$
28. $121v^2 - 36$
29. $x^2 + y^2$
30. $a^2 - 4b^2$
31. $36 - c^2$
32. $B^2 + 4$
33. $a^2 - 64b^2$
34. $49 - w^2$
35. $16a^2 - 1$
36. $4c^4 - 25c^2$
37. $9b^2c^2 - 16a^2$
38. $36m^2 - 1$
39. $1 - 100x^2$
40. $64x^2y^2 - 49z^2$
41. $6b^3 - 150b$
42. $2a^2 - 32$
43. $x^6 - 81a^2$
44. $c^6 - 16c^2$
45. $3d^4 - 243$
46. $11x^2 - 44y^2$
47. $5m^2 - 45n^2$
48. $9a^2 - 2$
49. $49a^2b^2 - 6$
50. $5z^4 - 80$
51. $a^3 - 9a$
52. $64 - b^4$
53. $4m^2 - 64n^2$
54. $c^4 + c^2d^2$
55. $25x^3 - x$
56. $9m^3 - 36m$

57. $98 + 2a^2$

58. $3a^2 - 27b^2$

59. $27b^2 - 75$

60. $147 - 3y^2$

C. Read and respond to the following exercises.

61. Are $x^2 + 4$ and $x^2 - 4$ factored the same way? Explain your answer, in words.

62. Describe the first step for every factoring exercise.

63. What does "prime" mean when given as an answer to a factoring exercise?

64. Describe, in your own words, how to completely factor this exercise: $a^2 - 25$.

14.3 Factoring Trinomials

When you are asked to factor a polynomial completely, first look to see if either a GCF or difference of two squares applies. Neither apply for the trinomial $x^2 + 13x + 12$, so we need to explore other methods of factoring.

Trial-and-Error Method

Some students and instructors use a trial-and-error approach to finding the factors of trinomials. Such a strategy for the foregoing expressions would follow this reasoning:

$x^2 + 13x + 12$

Factors of x^2 are (x) (x).

Possible factors of +12 are (3)(4), (–3)(–4), (2) (6), (–2) (–6), (1) (12), and (–1) (–12).

Try these combinations to see which produces a middle term of $+13x$.

$(x + 3)(x + 4) = x^2 + 4x + 3x + 12 = x^2 + 7x + 12$

$(x - 3)(x - 4) = x^2 - 4x - 3x + 12 = x^2 - 7x + 12$

$(x + 2)(x + 6) = x^2 + 6x + 2x + 12 = x^2 + 8x + 12$

$(x - 2)(x - 6) = x^2 - 6x - 2x + 12 = x^2 - 8x + 12$

$(x + 1)(x + 12) = x^2 + 12x + x + 12 = x^2 + 13x + 12$

That's the one! $(x + 1)(x + 12)$ has a middle term of $+ 13x$.

Therefore, $x^2 + 13x + 12 = (x + 1)(x + 12)$.

As students use this process, they begin to see clues that enable them to eliminate from consideration some of the factors of the last term without going through the multiplication. For instance, in trinomials where the squared term has a positive coefficient and the last term is positive, both factors will have the same sign. That sign will be the sign of the middle term. And if the last term is negative, one term will have a positive sign and the other a negative sign. The sum of these two terms must equal the middle term.

Apply these clues to the last example, $x^2 + 13x + 12$. Because the last term is positive and the middle term is positive, both of the factors will have + signs:

$(x + \quad)(x + \quad)$

This eliminates the need to look at (–3) (–4), (–2) (–6), and (–1) (–12). And the only factors left that can add to +13 are (+ 1) (+ 12).

Let's look at another trinomial and apply the trial-and-error method to factor it. Factor completely: $m^2 - 12m - 45$

1. The last term is negative, so the factors will have different signs; and their sum must be –12 (the coefficient of the middle term).
2. Possible factors of –45 are (–1) (45), (1) (–45), (–3) (15), (3) (–15), (–5) (9), and (5) (–9). The only pair that would combine to give –12 are (3) and (–15).
3. Try that combination and see if it checks.

$$(m + 3)(m - 15) = m^2 - 15m + 3m - 45$$
$$= m^2 - 12m - 45$$

Let's try to apply these clues as we factor $a^2 - 5a + 6$.

1. The last term is positive, so the factors will have the same sign and their sum will be –5.
2. The middle term is negative, so both of the factors will be negative. The factors of + 6 that are both negative and have a sum of –5 are (–2) and (–3) .
3. Try this combination and see if it checks:

$$(a - 2)(a - 3) = a^2 - 3a - 2a + 6$$
$$= a^2 - 5a + 6$$

This method is more tedious when the coefficient of the squared term is not 1. Factor completely: $12R^2 - 17R - 5$

1. Factors of $12R^2$ must be tested in combination with factors of – 5 to see which products combine to give a middle term of $-17R$.

   Factors of $12R^2$: $(12R)(R)$, $(-12R)(-R)$, $(6R)(2R)$, $(-6R)(-2R)$, $(4R)(3R)$, $(-4R)(-3R)$

   Factors of –5: (1) (–5), (–1) (5)

2. All possible combinations of factors are

$(12R - 5)(R + 1)$	$(6R - 5)(2R + 1)$
$(12R + 5)(R - 1)$	$(6R + 5)(2R - 1)$
$(12R - 1)(R + 5)$	$(6R - 1)(2R + 5)$
$(12R + 1)(R - 5)$	$(6R + 1)(2R - 5)$
$(-12R - 5)(-R + 1)$	$(-6R - 5)(-2R + 1)$
$(-12R + 5)(-R - 1)$	$(-6R + 5)(-2R - 1)$
$(-12R - 1)(-R + 5)$	$(-6R - 1)(-2R + 5)$
$(-12R + 1)(-R - 5)$	$(-6R + 1)(-2R - 5)$
$(4R - 5)(3R + 1)$	$(-4R - 5)(-3R + 1)$
$(4R + 5)(3R - 1)$	$(-4R + 5)(-3R - 1)$
$(4R - 1)(3R + 5)$	$(-4R - 1)(-3R + 5)$
$(4R + 1)(3R - 5)$	$(-4R + 1)(-3R - 5)$

3. After multiplying, look at the sum of the second and third terms to see which one gives $-17R$.

$(4R + 1)(3R - 5)$

$= 12R^2 - 20R + 3R - 5$

$-17R$

4. Therefore, $12R^2 - 17R - 5 = (4R + 1)(3R - 5)$.
5. Check: $(4R + 1)(3R - 5) = 12R^2 - 20R + 3R - 5$

$= 12R^2 - 17R - 5$

Product-Sum Method

For students who prefer a more structured approach to factoring trinomials, the **product-sum method** can be used. *Product* means the product of the coefficient of the first term and the coefficient or constant of the last term. *Sum* means the sum of the factors of the products that add to give the coefficient of the middle term.

In $x^2 + 8x + 12$, the product of the coefficient of the first term and the last term is $+12$ from $(+1)(12)$. Remember that a variable with no coefficient showing is understood to have a coefficient of $+1$. Your job is to find the factors of that product, $+12$, that add together to equal the coefficient of the middle term, $+8$.

$x^2 + 8x + 12$

Product = $(+1)(+12) = +12$

$x^2 + 8x + 12$

Sum = $+8$

Factors of +12:	$(+1)(+12)$	$(+2)(+6)$	$(+3)(+4)$
	$(-1)(-12)$	$(-2)(-6)$	$(-3)(-4)$

The only factors of $+12$ that add to $+8$ are $(+2)$ and $(+6)$. Next, write the original expression in an expanded form, with $+8x$ written as the sum of $+2x$ and $+6x$.

$x^2 + 8x + 12 = x^2 + 2x + 6x + 12$

[Think back to the process of multiplying two binomials, such as $(y + 2)(y - 1)$. The distributive multiplication produces $y^2 - y + 2y - 2$, which is then simplified by combining like terms. The expanded form in this factoring problem is similar to this expression, before the terms are combined.]

From the first two terms, factor their GCF, and then factor a GCF from the third and fourth terms. This is the same kind of factoring you studied in Examples 6 through 10 in Section 14.1.

$x^2 + 2x + 6x + 12$

$= x(x + 2) + 6(x + 2)$

Look at the two terms that were formed. Is there anything common to them? If so, factor it out of both, and write what remains. Because $(x + 2)$ is the common factor, multiply it by the remaining terms.

$x(\underline{x+2}) + 6(\underline{x+2})$

$= (x + 2)(x + 6)$

Therefore, the factorization of the trinomial is

$x^2 + 8x + 12 = (x + 2)(x + 6)$

Check the answer by multiplying the factors together.

$(x + 2)(x + 6) = x^2 + 6x + 2x + 12 = x^2 + 8x + 12$

This whole process seems difficult, but once you see the pattern of steps, you will be able to apply it when factoring any trinomial. Follow the detailed steps in the examples, and then try some on your own.

EXAMPLE 1 Factor completely: $y^2 + 6y + 9$

Solution

1. Product = $(+1)(+9) = +9$ Sum = $+6$

   Factors of $+9$: $(+1)(+9)$

   $(-1)(-9)$

   $(+3)(+3)$

   $(-3)(-3)$

   $(+3)$ and $(+3)$ are the only factors of $+9$ that add to $+6$.

2. Write the original expression in expanded form:

   $y^2 + 3y + 3y + 9$

3. Factor a GCF from the first and second terms, and then a GCF from the third and fourth terms. This is the process you used in factoring by grouping in Section 14.1.

   $y^2 + 3y + 3y + 9$

   $= y(y + 3) + 3(y + 3)$

4. Remove the common factor from both groups and multiply it by the remaining terms.

   $(y + 3)(y + 3)$

   This is factored completely. Therefore,

   $y^2 + 6y + 9 = (y + 3)(y + 3)$

5. Check by multiplication.

   $(y + 3)(y + 3) = y^2 + 3y + 3y + 9$

   $= y^2 + 6y + 9$

EXAMPLE 2 Factor completely: $m^2 - 12m - 45$

Solution

1. Product = $(+1)(-45) = -45$ Sum = -12

   Factors of -45: $(+1)(-45)$

   $(-1)(+45)$

   $(+3)(-15)$

   $(-3)(+15)$

   $(+5)(-9)$

   $(-5)(+9)$

   $(+3)$ and (-15) are the only factors that add to -12.

2. Write in expanded form.

   $m^2 + 3m - 15m - 45$

3. Factor the GCFs from pairs of terms.

   $\underline{m^2 + 3m} - \underline{15m - 45}$

   $= m(m + 3) - 15(m + 3)$

   Note that (-15) was factored from the third and fourth terms. This choice was made so that the terms inside the second parentheses would have the same signs as those inside the first parentheses.

4. Write the common factor times the remaining terms.

   $(m + 3)(m - 15)$

   Therefore, the complete factoring is

   $m^2 - 12m - 45 = (m + 3)(m - 15)$

   In step 2, the order of the middle terms could be reversed and still produce the same answer.

2. Write in expanded form.

   $\underline{m^2 - 15m} + \underline{3m - 45}$

3. Factor the GCFs from the pairs of terms.

   $m(m - 15) + 3(m - 15)$

4. Remove the common factor.

   $(m - 15)(m + 3)$

5. Check:

   $(m + 3)(m - 15)$

   $= m^2 - 15m + 3m - 45$

   $= m^2 - 12m - 45$

◀

EXAMPLE 3 Factor completely: $2T^2 - 8T + 8$

Solution Remove the GCF first; then factor the trinomial.

$2(T^2 - 4T + 4)$

1. Product = (+1) (+4) = +4 Sum = –4

   Factors of +4 (+1) (+4)
   (–1) (–4)
   (+2) (+2)
   (–2) (–2)

   (–2) and (–2) are the only factors of +4 that add to –4.

2. Write the original expression in expanded form.

   $2(T^2 - 2T - 2T + 4)$

3. Factor GCFs from the first and second terms and from the third and fourth terms. The brackets are needed here to keep the first greatest common factor (2) outside the rest of the expression that is now being factored. If the 2 isn't carried along throughout the problem, it might get forgotten in the answer.

   $2[T(T - 2) - 2(T - 2)]$

4. Remove the common factor from both groups and multiply it by what remains.

   $2[(T - 2)(T - 2)]$

   Check to be sure that no more factoring can be done, as is the case here. Therefore,

   $2T^2 - 8T + 8 = 2(T - 2)(T - 2)$

5. Check by multiplication:

   $2(T - 2)(T - 2)$
   $= (2T - 4)(T - 2)$
   $= 2T^2 - 4T - 4T + 8$
   $= 2T^2 - 8T + 8$

EXAMPLE 4 Factor completely: $x^2 - 7xy + 10y^2$

Solution This is the first problem you've seen where the last term also contains a variable. Because y^2 will factor as x^2 does, be sure to include the y with the numerical factors you obtain from the product-sum procedure.

1. Product = (+1) (+10) = +10 Sum = –7

   Factors of + 10: (+ 1) (+ 10)
   (–1) (–10)
   (+2) (+5)
   (–2) (–5)

2. Expanded form:

   $x^2 - 7xy + 10y^2$

   $= x^2 - 2xy - 5xy + 10y^2$

3. Factor GCFs from the pairs of terms:

   $x^2 - 2xy - 5xy + 10y^2$

   $= x(x - 2y) - 5y(x - 2y)$

4. Remove the common factor from both groups:

   $(x - 2y)(x - 5y)$

   This is factored completely. Therefore,

   $x^2 - 7xy + 10y^2 = (x - 2y)(x - 5y)$

5. Check: $(x - 2y)(x - 5y)$

   $= x^2 - 5xy - 2xy + 10y^2$

   $= x^2 - 7xy + 10y^2$

EXAMPLE 5 Factor completely: $12R^2 - 17R - 5$

Solution This is the first example we have done using the product-sum method where the squared term has a coefficient other than 1. You worked this problem earlier in this section using the trial-and-error method. Try it now by following the same steps used in Examples 1 through 4.

1. Product = $(12)(-5) = -60$ Sum = -17

   Factors of −60: $(-1)(+60)$

   $(+1)(-60)$

   $(-2)(+30)$

   $(+2)(-30)$

   $(-3)(20)$

   $(3)(-20)$

   You can stop here, because $(3)(-20)$ are the factors of -60 that have a sum of -17.

2. Expanded form: $12R^2 + 3R - 20R - 5$

3. Factor the GCFs from the pairs of terms:

   $3R(4R + 1) - 5(4R + 1)$

   Have you noticed by now that you get the same factors in both sets of parentheses each time? If that doesn't happen at this step in a problem, then either you have made a mistake somewhere and should start over, or the expression is not factorable.

4. Remove the common factor from both groups:

   $(4R + 1)(3R - 5)$

This is completely factored because the individual factors have no common factors.

$12R^2 - 17R - 5 = (4R + 1)(3R - 5)$

5. Check: $(4R + 1)(3R - 5)$
   $= 12R^2 - 20R + 3R - 5$
   $= 12R^2 - 17R - 5$

EXAMPLE 6 Factor completely: $22X^3 + 77X^2 - 44X$

Solution First take out the greatest common factor:

$11X(2X^2 + 7X - 4)$

1. Product = $(+2)(-4) = -8$ Sum = $+7$
   Factors of -8: $(+1)(-8)$
   $(-1)(+8)$
   No need to list any more factors, because (-1) and $(+8)$ add to $+7$.
2. Expanded form: $11X(2X^2 - 1X + 8X - 4)$
3. Factor GCFs from the pairs of terms:
   $11X[(2X - 1) + 4(2X - 1)]$
   (There's that matching pair again!)
4. Remove the common factor from the groups:
   $11X[(2X - 1)(X + 4)]$
   This is completely factored. Therefore,
   $22X^3 + 77X^2 - 44X = 11X(2X - 1)(X + 4)$
5. Check: $11X(2X - 1)(X + 4)$
   $= (22X^2 - 11X)(X + 4)$
   $= 22X^3 + 88X^2 - 11X^2 - 44X$
   $= 22X^3 + 77X^2 - 44X$

EXAMPLE 7 Factor completely: $3A^2 - 20A - 4$

Solution There is no GCF, so try to factor using the product-sum method.

Product = $(+3)(-4) = -12$ Sum = -20

Factors of -12: $(-1)(+12)$
$(-12)(+1)$
$(-2)(+6)$
$(+2)(-6)$
$(-3)(+4)$
$(+3)(-4)$

No factors of -12 add to -20, so this trinomial cannot be factored. Therefore, $3A^2 - 20A - 4$ is *prime*.

Product-Sum Method

In factoring trinomials of the type $ax^2 + bx + c$, the product-sum method can be used, where *product* means the product of a and c. *Sum* means the sum of the two factors of the product $a \cdot c$, and that sum must equal b.

EXAMPLE 8 Factor completely: $X^2 - 5XY + 6Y^2$

Solution Product = (+1) (+6) = +6 Sum = −5

Factors of +6: (+1) (+6)

(−1) (−6)

(+2) (+3)

(−2) (−3)

$X^2 - 2XY - 3XY + 6Y^2$

$= X(X - 2Y) - 3Y(X - 2Y)$

$= (X - 2Y)(X - 3Y)$

Therefore,

$X^2 - 5XY + 6Y^2 = (X - 2Y)(X - 3Y)$

Check $(X - 2Y)(X - 3Y)$

$= X^2 - 3XY - 2XY + 6Y^2$

$= X^2 - 5XY + 6Y^2$

Self-Check Try these exercises using the trial-and-error method or product-sum method. The answers follow, but you should check your answer by multiplying.

1. $a^2 + 7a + 12$
2. $2r^2 - 5r + 2$
3. $2x^2 - 18x + 36$
4. $6x^2 + 11x - 10$

Answers:

1. $(a + 4)(a + 3)$
2. $(2r - 1)(r - 2)$
3. $2(x - 3)(x - 6)$
4. $(3x - 2)(2x + 5)$

14.3 EXERCISES

A. Factor completely. Check by multiplying.

1. $y^2 + 8y + 15$
2. $x^2 + 4x + 4$
3. $t^2 - 16t + 15$
4. $r^2 - 5r + 4$
5. $a^2 - 7a + 10$
6. $b^2 - 2b - 35$
7. $c^2 + c - 20$
8. $y^2 + 13y + 12$
9. $w^2 + 10w + 25$
10. $d^2 + 11d + 18$
11. $m^2 - 7m + 6$
12. $n^2 - 7n + 12$
13. $x^2 - 10x + 21$
14. $c^2 - c - 12$

15. $r^2 - 2r + 1$
16. $t^2 - 11t + 10$
17. $b^2 + 9b + 20$
18. $m^2 - m - 30$
19. $z^2 - 8z - 33$
20. $x^2 + x - 72$
21. $2a^2 + 3a + 1$
22. $3x^2 + 4x + 1$
23. $3x^2 + 7x + 2$
24. $4y^2 + 8y + 3$
25. $r^2 + 7r - 8$
26. $x^2 - 2x - 8$
27. $3v^2 - 3v - 60$
28. $3R^2 - 21R + 36$
29. $5X^2 + 40X + 75$
30. $2x^2 - 2x - 24$
31. $7t^2 - 42t + 63$
32. $5a^2 + 25a + 30$
33. $m^2 + 6mn + 9n^2$
34. $R^2 - 4RS + 4S^2$
35. $w^2 - 2wz + z^2$
36. $b^2 + 2bc + c^2$
37. $3a^2 + 11ab + 10b^2$
38. $7x^2 + 8xy + y^2$
39. $2x^2 - 11xy + 5y^2$
40. $3m^2 + 11mn + 6n^2$
41. $4a^3 + 8a^2 - 60a$
42. $-3y^2 - 24y^2 + 27y$
43. $-12Y^3 - 22Y^2 + 20Y$
44. $5m^2 + 39m + 72$
45. $4c^2 - 4c - 3$
46. $8t^2 + 14t + 3$
47. $3r^2 + 11r - 20$
48. $2x^2 + 7x + 5$
49. $18x^2 + 3x - 6$
50. $6n^2 + 33n - 18$
51. $7m^2n + 13mn - 2n$
52. $4x^2 + 20x + 25$
53. $16y^2 - 24y + 9$
54. $20 + 3z - 2z^2$
55. $3t^2 - 30t + 63$
56. $21a^2 - 13ab + 2b^2$
57. $20 - 19m + 3m^2$
58. $4x^2 + 4ax - 8a^2$
59. $9a^2b - 12ab + 4b$
60. $m^4 - 2m^2n - 8n^2$

B. Read and respond to the following exercises.

61. What two steps should be done for every factoring exercise?
62. Compare the method used on page 14 to factor $12R^2 - 17R - 5$ with that used in Example 5 on page 19. Which do you prefer and why?

14.4 Factoring Practice

Factoring techniques have been presented one at a time in this chapter. That made it pretty easy to do the homework because you had only one or two types of factoring to worry about at a time. Now that you have practiced all the techniques, however, you should be able to handle any type of problem that you are asked to factor. Any one or more of the following skills will be needed to completely factor an expression: Remove the greatest common factor, factor by grouping, factor the difference of two squares, and factor trinomials by the product-sum method or the trial-and-error method.

Each time you approach a problem, examine it to see which of these skills must be used. Follow the order shown in the list below as you work through each factorization. The examples that follow will require you to use two or three of the skills before the expression is *factored completely*.

Factoring Procedures

1. Look to see if there is a common factor in every term. If there is, *remove the greatest common factor* (GCF) from every term.
2. a. If the expression, with the GCF removed, has only *two* terms, look to see if it is the *diference of two squares*. If so, factor as the product of the sum and the difference of the square roots.

   b. If it is the sum of two squares, $a^2 + b^2$, then the expression is *prime* and cannot be factored.

3. If the expression (with the GCF removed) is a trinomial, use the *product-sum method* or the *trial-and-error method* to factor.
4. If the expression contains four terms, factor by grouping. Remove the GCF from the two pairs of terms; then look to see if a further GCF can be removed.
5. Check all factoring answers by multiplication.

EXAMPLE 1 Factor completely: $25x^2 + 25x + 6$

Solution It is a trinomial and there is no GCF. Using the trial-and-error method:

1. The last sign is positive, so both factors will have the same sign.
2. Because the middle term is positive, both factors will be positive.

   $(+)(+)$

3. Try combinations of the factors of 25 and of 6 to find the one that gives a middle term of $+ 25x$.

   $(5x + 2)(5x + 3)$

Check $(5x + 2)(5x + 3) = 25x^2 + 15x + 10x + 6$

$= 25x^2 + 25x + 6$

EXAMPLE 2 Factor completely: $a^2 - 12a + 36$

Solution Using product-sum: Product = +36; Sum = –12. The factors of +36 that add to –12 are $(-6)(-6)$.

$a^2 - 12a + 36$

$= a^2 - 6a - 6a + 36$

$= a(a - 6) - 6(a - 6)$

$= (a - 6)(a - 6)$

$= (a - 6)^2$

Check $(a - 6)^2$

$= (a - 6)(a - 6)$

$= a^2 - 12a + 36$

EXAMPLE 3 Factor completely: $3x^2 + 18x + 15$

Solution Take out the GCF.

$3x^2 + 18x + 15 = 3(x^2 + 6x + 5)$

The expression is a trinomial, so use the trial-and-error or product-sum method. Product-sum is shown. Product = +5. Sum = +6. The factors are (+1) (+5).

$3(x^2 + 6x + 5) = 3(x^2 + 1x + 5x + 5)$

$= 3[x(x + 1) + 5(x + 1)] = 3(x + 1)(x + 5)$

Check $3(x + 1)(x + 5)$

$= (3x + 3)(x + 5)$

$= 3x^2 + 15x + 3x + 15$

$= 3x^2 + 18x + 15$

EXAMPLE 4 Factor completely: $64t^4 - 4$

Solution $64t^4 - 4$ Take out the GCF.

$= 4(16t^4 - 1)$ Now factor the difference of two squares.

$= 4[(4t^2 + 1)(4t^2 - 1)]$

Because $4t^2 - 1$ is also the difference of two squares, there is still more factoring to be done.

$= 4(4t^2 + 1)(2t + 1)(2t - 1)$

Check $4(4t^2 + 1)(2t + 1)(2t - 1)$

$= (16t^2 + 4)(2t + 1)(2t - 1)$

$= (16t^2 + 4)(4t^2 - 1)$

$= 64t^4 - 16t^2 + 16t^2 - 4$

$= 64t^4 - 4$

EXAMPLE 5 Factor completely: $4a^2 + 28a + 49$

Solution $4a^2 + 28a + 49$

There is no GCF, so try to factor the trinomial.

$4a^2 + 28a + 49$

$= (2a + 7)(2a + 7)$

$= (2a + 7)^2$

Check $(2a + 7)^2$

$= (2a + 7)(2a + 7)$

$= 4a^2 + 14a + 14a + 49$

$= 4a^2 + 28a + 49$

EXAMPLE 6 Factor completely: $AB^2 + 3AB + 6B + 18$

Solution There is no GCF in all four terms, so try to factor by grouping.

$AB^2 + 3AB + 6B + 18$

$= AB(B + 3) + 6(B + 3)$

$= (B + 3)(AB + 6)$

Check $(B + 3)(AB + 6) = AB^2 + 6B + 3AB + 18$

14.4 EXERCISES

A. Factor completely and check the answer by multiplying.

1. $x^2 + 8x + 16$
2. $y^2 - 14y + 49$
3. $a^2 + 16a + 64$
4. $m^2 - 12m + 36$
5. $x^2 - 10x + 21$
6. $c^2 - 8c + 15$
7. $36 - 9x - x^2$
8. $14 - 9n + n^2$
9. $3t^2 + 9t + 6$
10. $5w^2 - 15w - 20$
11. $4y^2 + 4y + 1$
12. $25x^2 - 90x + 81$
13. $16 - 8c + c^2$
14. $4 - 20p + 25p^2$
15. $125a^2 + 64$
16. $9t^2 - 81$
17. $36t^2 + 60tq + 25q^2$
18. $25 - 10a + a^2$
19. $16 + 8m + m^2$
20. $49m^2 + 42mn + 9n^2$
21. $4a^2 - 12ab + 9b^2$
22. $9b^2 + 6b + 1$
23. $3m + 3n - am - an$
24. $2a^3 - 10a^2 - 4a + 20$
25. $y^3 - 4y^2 - 3y + 12$
26. $2ac + 10a + c + 5$
27. $x^4 + 2x^3 - 80x^2$
28. $2z^2 - 28z + 98$
29. $5n^2 - 30n + 45$
30. $b^4 - 12b^3 + 35b^2$
31. $5b^2 - 180$
32. $2x^3 + 3x^2 - 2x - 3$
33. $a^3 + 3a^2 - 4a - 12$
34. $3a^2 - 75$
35. $5ab + bc - 20a - 4c$
36. $2rt + 24s - 16r - 3st$
37. $4t^2 + 14t + 6$
38. $m^3 + 8m^2 + 16m$
39. $3m^4 - 48$
40. $ax - 3x - 2ay + 6y$
41. $ab - ac - 2c + 2b$
42. $2p^4 - 162$
43. $3y^3 - 30y^2 + 75y$
44. $8m^2 + 56m + 98$
45. $20x^4y^2 - 7x^3y^3 - 3x^2y^4$
46. $15a^4b^2 + 2a^3b^3 - 8a^2b^4$

B. Describe, in words, the process used to go from one step to the next.

47. $a^4 - 81$

$(a^2 - 9)(a^2 + 9)$	1.
$(a - 3)(a + 3)(a^2 + 9)$	2.

48. $18m^3n^2 - 8m$

$2m(9m^2n^2 - 4)$	1.
$2m(3mn - 2)(3mn + 2)$	2.

49. $2b - 2c + ab - ac$

$2(b - c) + a(b - c)$	1.
$(b - c)(2 + a)$	2.

50. $3y^2 + 9y - 30$

$3(y^2 + 3y - 10)$	1.
$3(y - 2)(y + 5)$	2.
Check: $(3y - 6)(y + 5)$	3.
$3y^2 + 15y - 6y - 30$	4.
$3y^2 + 9y - 30$	5.

14.5 Solving Quadratic Equations by Factoring

Until now, when asked to solve equations, you have been dealing with first-degree equations (the exponent of the variable was 1). In this section, you will learn to solve second-degree equations—that is, equations with variables that have an exponent of 2. For example, $x^2 - x = 0$ and $y^2 - y = 20$ are second-degree equations.

Before you learn how to solve such equations, you must understand a basic property of the real number system.

Products Equal to Zero

If $ab = 0$, then either $a = 0$ or $b = 0$.

In other words, if you multiply quantities together and their product is zero, then one or more of the quantities must be zero. This situation will arise in the solutions to second-degree equations because your first step is to rearrange the terms so they are all on one side of the equals sign and there is a zero on the other side. Returning to one of our earlier examples, $x^2 - x = 0$ is already in correct form, so factor the polynomial.

$$x^2 - x = 0$$
$$x(x - 1) = 0$$

The product of x and $(x - 1)$ equals zero, so one of the factors must equal zero. Because you don't know which one, set each factor equal to zero and solve the resulting first-degree equation.

$$x(x - 1) = 0$$

If the x factor is zero, then $x = 0$.

If the $x - 1$ factor is zero, then

$$\begin{aligned} x - 1 &= 0 \\ +1 &\quad +1 \\ \hline x &= 1 \end{aligned}$$

The equation has two *possible* answers, so both have to be listed. The solution is given as $x = 0$, $x = 1$. We check the solutions by substituting them into the original equation to see if they produce true statemenst.

Check for $x = 0$

$$\begin{aligned} x^2 - x &= 0 \\ 0^2 - 0 &= 0 \\ 0 - 0 &= 0 \\ 0 &= 0 \quad \text{True.} \end{aligned}$$

Check for $x = 1$

$$\begin{aligned} x^2 - x &= 0 \\ 1^2 - 1 &= 0 \\ 1 - 1 &= 0 \\ 0 &= 0 \quad \text{True.} \end{aligned}$$

Look at the other sample equation, $y^2 - y = 20$. It needs to be rearranged to get all the terms on one side of the equals sign.

$$\begin{aligned} y^2 - y \quad &= 20 \\ -20 &= -20 \\ \hline y^2 - y - 20 &= 0 \end{aligned}$$

Once the equation is set equal to zero, factor the polynomial, if possible.

$y^2 - y - 20 = 0$ $\qquad$ $(y - 5)\ (y + 4) = 0$

Set each factor equal to zero and solve.

$$\begin{aligned} \text{If } y - 5 &= \ \ 0 \\ \underline{+5} \quad &\ \ \underline{+5} \\ y &= \ \ 5 \end{aligned} \qquad \begin{aligned} \text{If } y + 4 &= \ \ 0 \\ \underline{-4} \quad &\ \ \underline{-4} \\ y &= -4 \end{aligned}$$

To check the solutions, $y = 5$ and $y = -4$, substitute each into the original equation.

$$\begin{aligned} y^2 - y &= 20 \\ 5^2 - 5 &= 20 \\ 25 - 5 &= 20 \\ 20 &= 20 \quad \text{True.} \end{aligned} \qquad \begin{aligned} y^2 - y &= 20 \\ (-4)^2 - (-4) &= 20 \\ 16 + (+4) &= 20 \\ 20 &= 20 \quad \text{True.} \end{aligned}$$

In this section, all the second-degree equations, which are called **quadratic equations**, can be solved by factoring. In other courses you will learn techniques for solving quadratic equations that cannot be factored.

Quadratic Equation

A quadratic equation is an equation of the form $ax^2 + bx + c = 0$, where a, b, and c are real numbers and a $\neq 0$. The $ax^2 + bx + c = 0$ form is called the standard form of a quadratic equation.

Study the detailed solutions and checks that follow so that you will be able to do similar steps when you solve the homework exercises.

EXAMPLE 1 Solve and check: $x^2 + 6x + 8 = 0$

Solution The equation is already set equal to zero.

$x^2 + 6x + 8 = 0$ $\qquad$ Factor.

$(x + 4)\ (x + 2) = 0$

Set each factor equal to zero and solve the simple equations.

$$\begin{aligned} x + 4 &= \ \ 0 \\ \underline{\ \ -4} \ &\ \ \underline{-4} \\ x \quad &= -4 \end{aligned} \qquad \begin{aligned} x + 2 &= \ \ 0 \\ \underline{\ \ -2} \ &\ \ \underline{-2} \\ x \quad &= -2 \end{aligned}$$

The solutions are $x = -4$ and $x = -2$.

Check

$$x^2 + 6x + 8 = 0 \qquad x^2 + 6x + 8 = 0$$
$$(-4)^2 + 6(-4) + 8 = 0 \qquad (-2)^2 + 6(-2) + 8 = 0$$
$$(16) + (-24) + 8 = 0 \qquad (4) + (-12) + 8 = 0$$
$$-8 + 8 = 0 \qquad -8 + 8 = 0$$
$$0 = 0 \quad \text{True.} \qquad 0 = 0 \quad \text{True.}$$

EXAMPLE 2 Solve and check: $c^2 - 4c - 45 = 0$

Solution

$c^2 - 4c - 45 = 0$ Factor.

$(c - 9)(c + 5) = 0$ Set each factor equal to zero.

$c - 9 = 0 \qquad c + 5 = 0$ Solve each equation.

$\underline{+9 \quad +9} \qquad \underline{-5 \quad -5}$

$c = 9 \qquad c = -5$

The solutions are $c = 9$ and $c = -5$.

Check

$$c^2 - 4c - 45 = 0 \qquad c^2 - 4c - 45 = 0$$
$$9^2 - 4(9) - 45 = 0 \qquad (-5)^2 - 4(-5) - 45 = 0$$
$$81 - 36 - 45 = 0 \qquad 25 + 20 - 45 = 0$$
$$45 - 45 = 0 \qquad 45 - 45 = 0$$
$$0 = 0 \quad \text{True.} \qquad 0 = 0 \quad \text{True.}$$

EXAMPLE 3 Solve and check: $f^2 + 12f + 36 = 0$

Solution

$f^2 + 12f + 36 = 0$ Factor the trinomial.

$(f + 6)(f + 6) = 0$

Both factors are the same, so there will be only one solution.

$f + 6 = 0$

$f = -6$

When this happens, –6 is called a **double root** or a **double solution**, because every solvable second-degree equation has *two* solutions.

Check

$$f + 12f + 36 = 0$$
$$(-6)^2 + 12(-6) + 36 = 0$$
$$36 + (-72) + 36 = 0$$
$$-36 + 36 = 0$$
$$0 = 0 \quad \text{True.}$$

EXAMPLE 4 Solve and check: $z^2 = 10z$

Solution

$z^2 = 10z$	Rearrange by subtracting $10z$ from both sides of the equals sign.
$z^2 - 10z = 0$	Factor out the GCF.
$z(z - 10) = 0$	Set each factor equal to zero and solve.

$$z = 0 \qquad z - 10 = 0$$
$$z = 10$$

The solutions are $z = 0$ and $z = 10$.

Check

$z^2 = 10z$	$z^2 = 10z$
$0^2 = 10(0)$	$10^2 = 10(10)$
$0 = 0$ True.	$100 = 100$ True.

EXAMPLE 5 Solve and check: $3w^2 - 6w + 3 = 0$

Solution

$3w^2 - 6w + 3 = 0$	Factor out the GCF.
$3(w^2 - 2w + 1) = 0$	Factor the trinomial.
$3(w - 1)(w - 1) = 0$	Set each factor equal to zero and solve.

The constant term, 3, can never be zero and will not affect the solution.

$w - 1 = 0$

$w = 1$ This is a double root.

Check

$$3w^2 - 6w + 3 = 0$$
$$3(1)^2 - 6(1) + 3 = 0$$
$$3 - 6 + 3 = 0$$
$$-3 + 3 = 0$$
$$0 = 0 \quad \text{True.}$$

EXAMPLE 6 Solve and check: $\frac{y^2}{2} + 4y - 10 = 0$

Solution Because you need to have integer coefficients before you can factor, multiply through any fractional equation by the least common denominator (LCD) to eliminate the denominator(s). Remember that you must multiply every term in the equation by the LCD value. (See Section 9.5.)

$$\frac{y^2}{2} + 4y - 10 = 0$$

$\frac{(\not{2}^{1})y^2}{\not{2}_{1}} + (2)4y - (2)10 = (2)0$ Multiply every term by 2.

$$y^2 + 8y - 20 = 0$$

$$(y + 10)(y - 2) = 0$$ Factor the left side.

$$y + 10 = 0 \qquad y - 2 = 0$$ Solve each small equation.

$$y = -10 \qquad y = 2$$

The solutions are $y = -10$, $y = 2$.

Check Check in the original equation.

$$\frac{y^2}{2} + 4y - 10 = 0 \qquad \frac{y^2}{2} + 4y - 10 = 0$$

$$\frac{(-10)^2}{2} + 4(-10) - 10 = 0 \qquad \frac{(2)^2}{2} + 4(2) - 10 = 0$$

$$\frac{100}{2} + (-40) - 10 = 0 \qquad \frac{4}{2} + 8 - 10 = 0$$

$$50 + (-40) - 10 = 0 \qquad 2 + 8 - 10 = 0$$

$$10 - 10 = 0 \qquad 10 - 10 = 0$$

$$0 = 0 \quad \text{True.} \qquad 0 = 0 \quad \text{True.}$$

As you see from these examples, the process of solving quadratic equations by factoring is always the same.

Solving Quadratic Equations by Factoring

1. Rearrange the terms so that zero is alone on one side of the equals sign (standard form).
2. If there are fractional coefficients, multiply every term of the equation by the LCD value.
3. Factor the equation, if possible.
4. Set each nonconstant factor equal to zero.
5. Solve each equation that has been formed,
6. Check each answer by substituting it into the original equation.

EXAMPLE 7 Solve and check: $(x - 3)(x + 4) = 8$

Solution Do not set the factors $(x - 3)$ and $(x + 4)$ equal to zero because the equation must first be arranged in standard form: $ax^2 + bx + c = 0$. Multiply $(x - 3)(x + 4)$ to remove the parentheses.

$$(x - 3)(x + 4) = 8$$

$$x^2 + x - 12 = 8$$ Now rearrange so that zero is alone on one side of the equals sign.

$$\underline{-8} \quad \underline{-8}$$

$$x^2 + x - 20 = 0$$

Then factor the new equation.

$(x + 5)(x - 4) = 0$

$x + 5 = 0 \qquad x - 4 = 0$

$x = -5 \qquad x = 4$

Check

$(x - 3)(x + 4) = 8$	$(x - 3)(x + 4) = 8$
$(-5 - 3)(-5 + 4) = 8$	$(4 - 3)(4 + 4) = 8$
$(-8)(-1) = 8$	$(1)(8) = 8$
$8 = 8$ True.	$8 = 8$ True.

EXAMPLE 8 Solve and check: $4a^2 = 3 - 11a$

Solution

$4a^2 = 3 - 11a$ Rearrange to standard form.

$4a^2 + 11a - 3 = 0$ Factor.

Product = -12 Sum = $+11$

$(+12)(-1)$ add to $+11$.

$4a^2 + 12a - a - 3 = 0$

$4a(a + 3) - 1(a + 3) = 0$

$(a + 3)(4a - 1) = 0$

$a + 3 = 0 \qquad 4a - 1 = 0$

$a = -3 \qquad 4a = 1$

$a = \frac{1}{4}$

Check

$4a^2 = 3 - 11a$	$4a^2 = 3 - 11a$
$4(-3)^2 = 3 - 11(-3)$	$4(1/4)^2 = 3 - 11(1/4)$
$4(9) = 3 - (-33)$	$4(1/16) = 3 - 11/4$
$36 = 3 + 33$	$1/4 = 12/4 - 11/4$
$36 = 36$ True.	$1/4 = 1/4$ True.

Self-Check Solve and check.

a. $x^2 - 8x + 12 = 0$ b. $4y^2 - 16 = 12y$

Answers:

a. $x = 2$	Ck: $0 = 0$	b. $y = 4$	Ck: $48 = 48$
$x = 6$	Ck: $0 = 0$	$y = -1$	Ck: $-12 = -12$

The following application problems require a quadratic equation. All these quadratic equations are solvable by factoring. Study the examples that follow. Use the methods for solving application problems that were shown in Sections 10.2, 10.3, and 12.4.

EXAMPLE 9 One number is four more than twice another number. If their product is 6, what are the two numbers?

List Let n = One number

$2n + 4$ = Second number

Their product is 6.

Equation $n(2n + 4) = 6$

Solution

$2n^2 + 4n = 6$	Rearrange the equation.
$2n^2 + 4n - 6 = 0$	Factor out the GCF.
$2(n^2 + 2n - 3) = 0$	Factor the trinomial.
$2(n + 3)(n - 1) = 0$	Set each non-constant factor equal to zero.
$n + 3 = 0 \quad n - 1 = 0$	Solve the equations.
$n = -3 \quad n = 1$	

Going back to the original list:

When $n = -3$, $2n + 4 = 2(-3) + 4 = -6 + 4 = -2$

When $n = 1$, $2n + 4 = 2(1) + 4 = 6$

Therefore, two sets of numbers satisfy this problem: –3, –2 and 1, 6.

Check The product of both pairs does equal 6 and the condition that one number is four more than twice the other number is also true for these pairs.

EXAMPLE 10 The area of a rectangle ($A = LW$) is 21 square feet. Find the dimensions of the rectangle if the length is 4 feet more than the width (see Figure 14.1).

Sketch Let x = Width

$x + 4$ = Length

$A = LW$

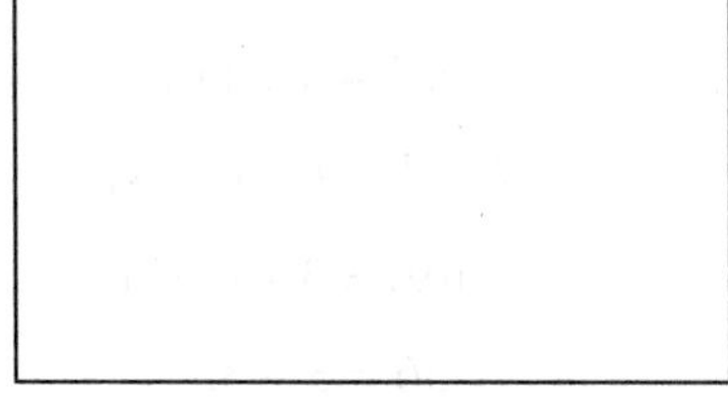

$x + 4$

FIGURE 14.1

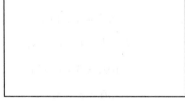

Equation $21 = (x + 4)x$

Solution

$21 = x^2 + 4x$

$0 = x^2 + 4x - 21$

$0 = (x + 7)(x - 3)$

$x + 7 = 0$	$x - 3 = 0$
$x = -7$	$x = 3$
This value must be thrown out because the width must be positive.	This value is the width of the rectangle, 3 feet. The length is $x + 4 = 7$ feet.

Check Seven feet is 4 feet more than the width, and the area of a 3-foot-by-7-foot rectangle is 21 square feet.

14.5 EXERCISES

A. Solve these equations by factoring.

1. $(y + 2)(y - 5) = 0$
2. $(b + 4)(b - 3) = 0$
3. $(a + 8)(a - 9) = 0$
4. $x(x + 5) = 0$
5. $m(3m - 5) = 0$
6. $4a(2a - 3) = 0$
7. $x^2 - 5x = 0$
8. $a^2 - 12a = 0$
9. $b^2 - 49 = 0$
10. $7c^2 = -3c$
11. $4y^2 = 8y$
12. $9y = 36y^2$
13. $3c^3 - 27c = 0$
14. $r^2 - 121 = 0$
15. $5m = -7m^2$
16. $6 - 54t^2 = 0$
17. $t^3 - t = 0$
18. $19x = -x^2$
19. $x^2 + 5x + 6 = 0$
20. $m^2 + 6m + 8 = 0$
21. $n^2 - 3n - 10 = 0$
22. $x^2 - 6x - 40 = 0$
23. $a^2 + a - 72 = 0$
24. $t^2 - t - 56 = 0$
25. $x^2 + 10x = -21$
26. $w^2 = 3w + 40$
27. $4s^2 + 12s = -9$
28. $m^2 - 24m - 25 = 0$
29. $3y^2 - 7y - 6 = 0$
30. $h^2 = 22h - 121$
31. $x^3 - 6x^2 + 9x = 0$
32. $6x^2 + x - 15 = 0$
33. $\frac{R^2}{16} = 1$
34. $V^2 = \frac{1}{16}$
35. $32 - 12x - 2x^2 = 0$
36. $\frac{y^2}{3} - 2y + 3 = 0$
37. $y(y + 3) = 28$
38. $(x + 4)(x - 1) = 6$
39. $\frac{x^2}{3} - 2x - 9 = 0$
40. $\frac{x^2}{2} + 4 = 3x$
41. $(a - 5)(a + 4) = 52$
42. $2m^2 - 5m - 12 = 0$
43. $5a^2 + 13a + 6 = 0$
44. $6y^2 - 13y + 6 = 0$
45. $(m - 4)(m + 6) = -9$
46. $21 - 4x - x^2 = 0$
47. $m^2 - 3m = 10$
48. $a(a - 1) = 20$
49. $3k^2 - 13k + 4 = 0$
50. $(n + 1)(n - 6) = -10$

B. Organize the information, write a quadratic equation, solve it, and give the answers requested.

51. The area of a rectangle ($A = LW$) is 78 square meters. Find the dimensions if the length is 1 meter more than twice the width.
52. The product of two numbers is 100 and one number is four times the other number. Find the numbers.
53. The sum of the squares of two consecutive numbers is 145. What are the numbers?
54. The base of a parallelogram is three times the height, and the area is 108 square yards. Using $A = bh$, find the length of the base.
55. The measure of the base of a triangle is twice the measure of the height and the area is 100 square feet. Using $A = \frac{1}{2}bh$, find the measure of the base.
56. The sum of the squares of two consecutive even numbers is 164. What are the two numbers?
57. The product of two numbers is 60 and one number is seven less than the other number. Find the numbers.
58. The area of a rectangle ($A = LW$) is 27 square inches. Find the dimensions if the width is 6 inches less than the length.

C. Read and respond to the following exercises.

59. How does the Products Equal to Zero rule apply to solving quadratic equations?
60. What does "second-degree" equation mean?
61. In this section you were told that every quadratic equation has two solutions. Why was one of the solutions thrown out in Example 10 but not in Example 9?
62. Describe, in words, the first step that should be performed in solving $10x = x^2 + 24$.
63. Describe, in words, the first step that should be performed in solving $1/2\ x^2 + 4x - 10 = 0$
64. Describe how you check the answers to quadratic equations.

14.6 CHAPTER REVIEW

When you are asked to factor an expression completely, follow these steps:

1. First look to see if there is a common factor in every term. If there is, *remove the greatest common factor* (GCF) from every term.

   $ax^2 + ax + ac = a(x^2 + x + c)$

2. If there are only *two* terms in the expression, look to see if it is the *difference of two squared quantities*. If so, factor as the product of the sum and difference of the square roots.

   $a^2 - b^2 = (a + b)(a - b)$

   If it is the sum of two squared quantities, $a^2 + b^2$, then the expression is *prime* and cannot be factored.

3. If the expression (with the GCF removed) is a trinomial, use the *product-sum* method or the *trial-and-error* method to factor it.

4. If the polynomial has four terms, see if there are common factors in pairs of terms and factor them out. Look again for GCFs.

5. Check all factoring answers by multiplication.

   When solving quadratic equations such as $ax^2 + bx + c = 0$, apply the steps listed on page 598. Check the answers in the original equation.

REVIEW EXERCISES

A. Multiply each of the following expressions. Simplify answers by combining like terms.

1. $3xy(x^2y - 2xy)$
2. $(x + 2)(x - 2)$
3. $(y - 3)(y + 6)$
4. $(8 + b)(8 + b)$
5. $(k - 3)(k + 3)$
6. $5a^2b(3ab + 4)$
7. $3(2c - 1)(c + 3)$
8. $5(m + 4)(m - 2)$
9. $(x + 2)(x + 2)$
10. $(y + 7)(y + 7)$
11. $(a + 3)^2$
12. $(x - 4)^2$

B. Factor these expressions completely. If an expression cannot be factored, write "prime" for the answer. Check all answers by multiplying.

13. $3y^3 + 3y$
14. $m^5 - 3m^3 + 2m$
15. $b^2 - 16$
16. $-6x^2 + 12$
17. $c^2 - 8c + 15$
18. $b^2 - 2b - 35$
19. $x^6 + 2x^4 + x^2$
20. $c^2 + 9$
21. $-t^3 + 2t^2$
22. $10ab - 2a$
23. $4d^2 - 1$
24. $x^2 + 7x - 18$
25. $3m^2n^2 - 5m^2n$
26. $b^2 - 4b - 21$
27. $39 - 16m + m^2$
28. $3y^4 - 81y$
29. $a^2 + 12a + 27$
30. $m^2 + mn - 6n^2$
31. $x^2 + 14x + 49$
32. $72 - 17p + p^2$
33. $35 - 21b$
34. $a^2 + 3ab - 10b^2$
35. $m^2 - 4mn - 45n^2$
36. $10a^3b - 24a^2b^2 + 16ab^3$
37. $4a^3 + 8a^2 - 60a$
38. $-2n^3 - 6n^2 + 20n$
39. $-3y^3 - 24y^2 + 27y$
40. $7 + 14k$
41. $3m^2 - 48$
42. $18x^4 - 8x^2y^2$

43. $9x^3y^3 - 27x^2y^2 + 18xy$
44. $64 - 3x^2$
45. $4a^2 - 64b^2$
46. $25 - 4y^2$
47. $w^4 - 16z^4$
48. $c^4 - 81d^4$
49. $6m^4 - 24m^2n^2 - 30n^4$
50. $a^3bc - a^2b^2c - 30ab^3c$
51. $x^3yz - 2x^2y^2z - 24xy^3z$
52. $4x^4 + 4x^2y^2 - 24y^4$
53. $-8m^4 + 72m^2$
54. $15a^2 - 12a$
55. $-12a^2 + 6a + 6$
56. $-6x^3 + 54x$
57. $x^2 - 49$
58. $a^2 - 64b^2$
59. $10n^2 + 130n + 220$
60. $y^4 - 14y^2 + 45$
61. $a^2 + 4a + 3$
62. $2x^2 + 9x + 7$
63. $x^2 + 11x + 24$
64. $x^2 + 15x + 14$
65. $-4b^2 + 64$
66. $2b^2 - 13b - 24$
67. $2m^2 - 18mn - 20n^2$
68. $4c^2 - 20cd + 16d^2$
69. $3x^2 - 8x - 16$
70. $x^2 - 2x - 15$
71. $x^4yz + x^3yz^2 + x^2y^3z$
72. $a^5b^3c + 2a^3b^3c^3 - 3a^3b^4c^2$
73. $x^2y + xy - 2x - 2$
74. $4ab^2 + 8ab + 8b + 16$
75. $y^3 - 4y^2 - y + 4$
76. $10a + 2ac + 5 + c$
77. $32g^2 - 160gh + 200h^2$
78. $12m^3 + 6m^2 - 30m$
79. $20p^2 + 17p - 24$
80. $5x^3 + 10x^2 - 7x$
81. $18w^2 + 27w - 35$
82. $15b^2 - 65bc + 20c^2$

C. Solve and check.

83. $c^2 - 6c = 0$
84. $\frac{w^2}{4} - 100 = 0$
85. $14x = 7x^2$
86. $a^2 + a - 30 = 0$
87. $6m^2 - 19m + 15 = 0$
88. $y^2 + 21 = -10y$
89. $20 + 3n = \frac{n^2}{2}$
90. $3p^2 - 9p = -6$
91. $3 - 48t^2 = 0$
92. $4x^2 - 12x = 40$
93. $4m^2 - 63 = 4m$
94. $y^2 + 20 = 9y$
95. $(x - 5)(x + 2) = 18$
96. $(m + 4)(m - 7) = 60$

D. Organize the information, write a quadratic equation, solve it, and give the answers requested.

97. The area of a rectangle ($A = LW$) is 120 square yards. Find the dimensions if the length is 1 yard less than twice the width.
98. The product of two consecutive even numbers is 48. Find the two numbers.
99. The sum of the squares of two consecutive numbers is 41. What are the numbers?
100. If six is added to the square of a number, the result is the same as five times the number increased by two. What is the number?
101. If nine is subtracted from the square of a number the result is 16. What is the number?
102. The area of a rectangle ($A = LW$) is 176 square meters. Find the dimensions if the length is 6 meters more than twice the width.
103. The product of two consecutive numbers is 110. Find the numbers.
104. The sum of the squares of two consecutive even numbers is 100. What are the two numbers?

E. An error has occurred in each exercise that follows. State in words what the error was and give the correct answer.

105. $25a^2b^2 - 20a^2b + 35a$
$= 5(a^2b^2 - 4a^2b + 7a)$
106. $-4m^2 + 28$
$= 4(-m^2 + 7)$
107. $c^2 + 2c + cd + 2d$
$= c(c + 2) + d(c + 2)$
108. $X^2 - 4$
$= (X - 2)(X - 2)$

109. $2a^2 + 32$

$= 2(a^2 + 16)$

$= 2(a + 4)(a + 4)$

110. $36m^2 - 4$

$= (6m + 2)(6m - 2)$

111. $y^2 - 6y + 8$

$= (y + 2)(y + 4)$

112. Solve: $3F^2 - 9F = 0$

$3F(F - 3) = 0$

$F - 3 = 0$

$+3 \quad +3$

$F = 3$

F. Read and respond to the following exercises.

113. What two steps are suggested for every factoring exercise?

114. How does looking at the number of terms in a polynomial help with the factoring process?

115. In solving a quadratic equation by factoring, why is each factor set equal to zero?

116. Why are you encouraged to remove any GCF as the first step of a factoring exercise?

117. Compare the factoring of $4a^2 - 1$ and $4a^2 + 1$.

CHAPTER TEST

1. Multiply the expression: $3xy(4x^2y - 7xy + 2)$
2. Multiply: $(4a - 3c)(a + 2c)$

Completely factor each of the following expressions, if possible.

3. $4R^2 - 6R - 40$
4. $c^3 - cd^2$
5. $ax + ay + 5x + 5y$
6. $25 + 40y + 16y^2$
7. $15t^2 - 7t - 2$
8. $2m^2 + 3mn - 2n^2$
9. $9p^2 + 63p + 90$
10. $4w^2 - 16$
11. $x^2 - 5x - 14$
12. $F^2 + G^2$
13. $m^2 - 4m + 6mn - 24n$
14. $2x^2 + 3x + 1$
15. $3a^3 + 21a^2 - 9a$
16. $25 - 9z^2$
17. Describe, in words, how you would completely factor $w^2 - 10w + 21$.

Solve each quadratic equation by factoring.

18. $s^2 = s + 30$
19. $v^2 + 3v - 28 = 0$
20. The product of two consecutive odd numbers is 35. Find the two numbers by writing a quadratic equation and solving it.

STUDY SKILLS MODULE 15

Preparing for the Final Exam

If you have been using good test preparation skills all along in a course, then studying for the final exam will not be a monumental task. If you made topic lists or practice quizzes for previous tests, pull them out and begin reviewing the topics and working through problems. If you have access to the tests you took during the term, rework those problems and be sure you now understand the concepts you missed on each of those tests.

Preparation for the final exam should begin about two weeks ahead of time, if possible. It will be a very busy time for all your classes, so it is imperative that you practice solid time management. You will need to schedule specific time to write papers and to begin preparation for the exam for each class.

Find out from the instructor the type of test, the chapters and topics that will be included, the length of the test, and, of course, the date, time, and place of the exam. Go over topics and chapters that are to be included and list any concepts or topics where you feel you are weak. Schedule time to work on those areas. A detailed calendar for this time of the term is essential. Reread the book and your class notes on those topics. Use other resources such as other textbooks, videotapes, tutors, or computer programs until you feel you have overcome your difficulties.

After doing a thorough job reviewing the topics and concentrating on weaknesses, you are ready to test yourself. You know the number of questions, the type of test, and the topics, so write a practice exam. Use problems from a textbook that has the answers given, so that you will be able to grade the test. Find a quiet place to take the test and don't look up information or get help as you do this practice test. When you finish, grade your exam carefully. Be sure that any variation in an answer is acceptable—and not a variation that may cause you to lose some credit, such as not simplifying completely, not following rounding directions, or not labeling an application problem answer. The errors you made on the practice test should be used as guides to tell you what you still need to study. The results should also tell you how much more studying you need to do.

Perhaps after reviewing the material that is to be included, you could get together with other classmates and quiz one another or discuss the topics that you have studied. You might find out that someone picked up or interpreted something about the exam that you missed. You could exchange the practice tests each of you developed and have additional practice on test taking.

Spread out the entire exam preparation process and try to do only "non-pressure" review the night before or the day of the exam. This should help keep you as relaxed as possible as you go into the exam. This is far healthier (mentally and physically) than cramming all night and becoming extremely tense just before the exam.

By preparing for the exam over a 2-week period, you have *time* to fix problem areas, and, therefore, you should be confident in your math abilities as you undertake the final exam. That positive, confident attitude should contribute to your success on the test.

Your Point of View

1. Make a list of all the information you think you will need to know about the final exam. Have the list available when your instructor announces the final exam schedule, and ask questions to get all the information you have listed.
2. List four or five steps you plan to follow to prepare for the final exam for this course.
3. Make a list of ten topics or sample questions that you need to study to prepare for this exam.
4. Would you prepare differently for a history or psychology exam than you would for the final for this math class? Explain your answer.

CHAPTER

15 Introduction to Graphing

LEARNING OBJECTIVES

When you have completed this chapter, you should be able to:

1. Identify the components of a Cartesian coordinate system (graph).
2. Locate a point on a Cartesian coordinate system and identify that location by an ordered pair of numbers.
3. Produce a table of at least three pairs of values that satisfy the given first-degree equation.
4. Graph the solution of a first-degree linear equation.
5. Write a first-degree linear equation in the form $y = mx + b$.
6. Given two points on a line, find the slope of that line from the graph or by using the two-point formula for slope.
7. Write the equation of a line given its slope and y-intercept.
8. Write the equation of a line given two points on the line.
9. Write the equation of a line given one point on the line and its slope.
10. Graph the solution of a first-degree inequality.

In Chapter 11, on signed numbers, we first looked at the idea of a number line. A number line is a graph that shows what the picture of a set of signed numbers might look like. If we expand that notion to a two-dimensional graph, similar to a map, we graph points whose location is indicated by a *pair* of numbers. In this chapter, we will talk about how to identify the location of points on that graph and about the need for an ordered pair of numbers to indicate right-left position and up-down position. We will also show a picture of the solution set for a first-degree linear equation and a first-degree inequality.

You will need some graph paper of your own to do the work in this chapter. We recommend 1/4-inch graph paper, but a small quantity of any type will do.

15.1 Points on a Graph

Take a few minutes to remember what you learned in Chapter 11 about a number line. It is a picture—a **graph**—of signed numbers that extends indefinitely in both directions. A thermometer is the same type of graph, but it normally is read up and down instead of from side to side like a number line. When two graphs (visual representations) such as a number line and a thermometer are combined, the resulting figure is called a **Cartesian coordinate system**. Figure 15.1 is such a Cartesian coordinate system, a two-dimensional graph.

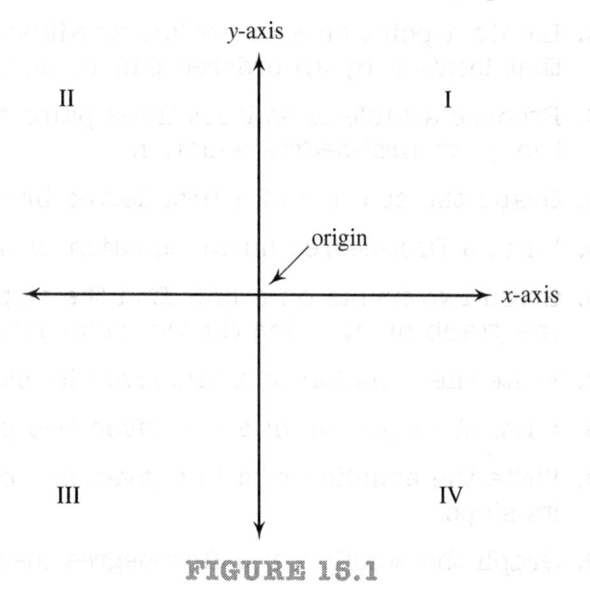

FIGURE 15.1

The vertical scale is called the ***y*-axis**, and the horizontal scale is the ***x*-axis**. The point at which the two scales, or axes, cross is called the **origin**. We measure distance to the *right* of the *y*-axis as *positive* (+) and distance to the *left* as *negative* (–). Distance *up* from the *x*-axis is *positive* (+), and distance *below* the *x*-axis is *negative* (–). We use ordered pairs of numbers, (x, y), to indicate location on the graph. In an ordered pair, the first coordinate, sometimes called the *abscissa*, always measures the distance left or right, and the second coordinate, called the *ordinate*, always measures the distance up or down.

The points that do not lie on the axes are separated into four sections called **quadrants**. The quadrants are designated by roman numerals (I, II, III, and IV) and are numbered counterclockwise as shown in Figure 15.1. Each point on the graph is either right (+) or left (–) of zero and is also either up (+) or down (–) from zero. Pairs of numbers are used to indicate a particular location on the graph.

Look at Figure 15.2. This graph shows the location of several points. For instance, (2, 5) is two units to the right of zero and five units up. (–3, 4) is three units to the left of zero and four units up. (–1, –3) is one unit to the left of zero and three units down. (3, –2 1/2) is three units to the right of zero and 2 1/2 units down. Are the point (–3, 5) and the point (5, –3) in the same place on the graph? Find them. They are different. Even though the integers are the same, –3 and 5, their order in the ordered pair is switched. (–3, 5) is not the same point as (5, –3). The right-left location is always given first in the pair of numbers; the up-down location is always given second.

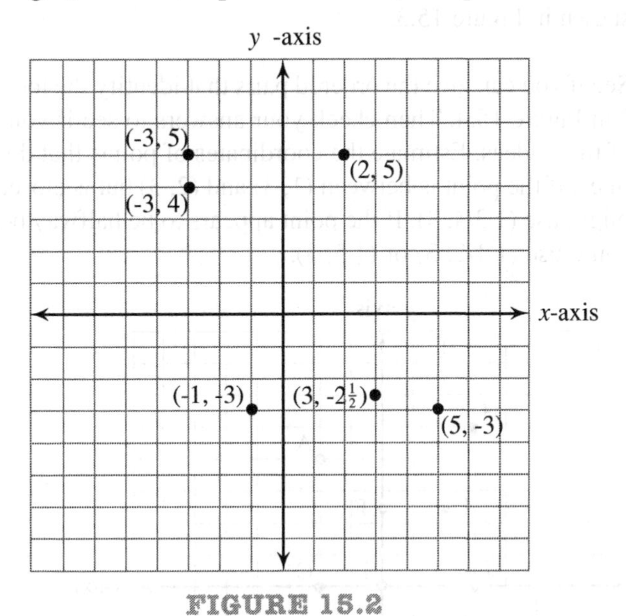

FIGURE 15.2

Look at Figure 15.3. What can you say about the signs of the numbers in the ordered pairs in Quadrant I? All points there are to the right of the origin, (+), and above the *x*-axis, (+). The ordered pairs would therefore look like (+, +). What about points in Quadrant II? In Quadrant III? In Quadrant IV? It will be helpful later on to have figured this out.

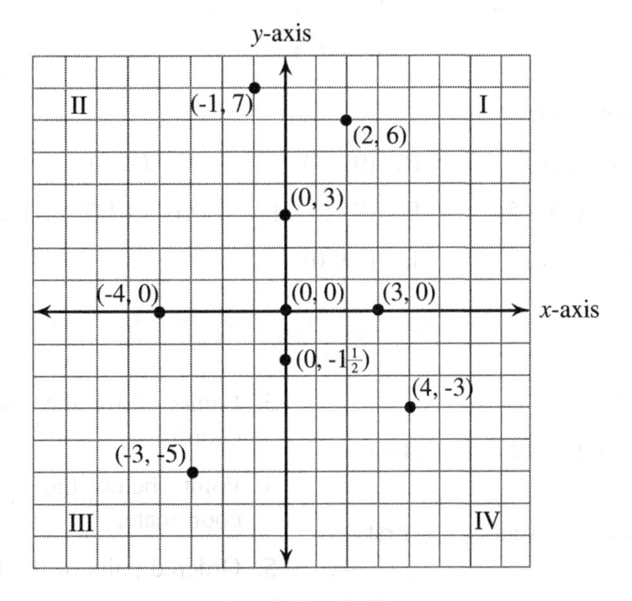

FIGURE 15.3

(+ 3, 0) is three units to the right of zero but zero units up or down. It lies on the horizontal axis, the x-axis. Points that fall on one of the axes have a special feature. Points on the x-axis haven't moved up or down, so they have a y value of 0, and their ordered pairs have the form (left or right value, 0). Points on the y-axis haven't moved left or right, so they have an x value of 0 and take the form (0, up or down value). (0, –1.5) lies on the y-axis, 1.5 units down. (0, 0) is neither right nor left and neither up nor down, so (0, 0) is the location of the origin. (–4, 0) lies on the x-axis. (0, +3) lies on the y-axis. Each of these points is graphed on the coordinate system shown in Figure 15.3.

Self-Check See if you can give the ordered pairs that identify the location of the points A through J in Figure 15.4. Then check your answers to see if you have a good understanding of these ideas. Estimate the coordinates of points that do not fall on two intersecting lines. If the point is between (1, 3) and (2, 3) but is closer to (2, 3) than to (1, 3), you might use (1 3/4, 3). If the point appears to be halfway between (1, 3) and (2, 3), you could use (1 1/2, 3) or (1.5, 3).

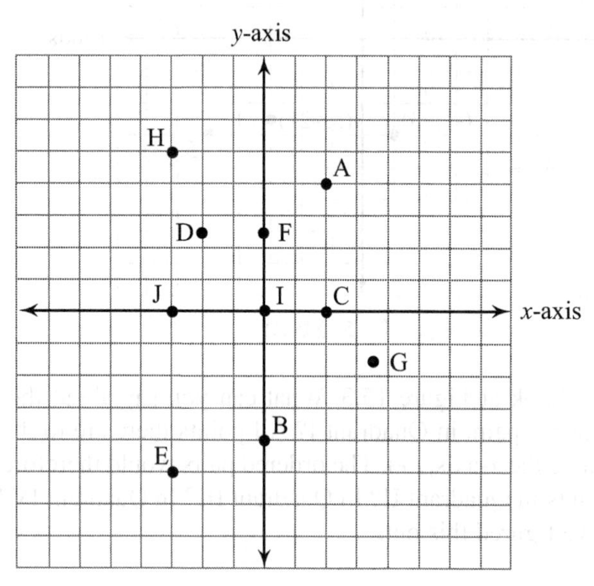

FIGURE 15.4

Answers:

A. (2, 4)	B. (0, –4)	C. (2, 0)	D. (–2, 2 1/2)
E. (–3, –5)	F. (0, 2 1/2)	G. (3 1/2, –1 1/2)	H. (–3, 5)
I. (0, 0)	J. (–3, 0)		

15.1 EXERCISES

A. Complete the following sentences.

1. Points to the right of the y-axis have a _____ first coordinate.
2. Points to the left of the y-axis have a _____ first coordinate.
3. Points above the x-axis have a _____ second coordinate.
4. Points below the x-axis have a _____ second coordinate.
5. Ordered pairs of the form (–, +) lie in Quadrant _____.

6. Ordered pairs of the form (+, –) lie in Quadrant _____.
7. Ordered pairs of the form (0, y) lie on the _____.
8. Ordered pairs of the form (x, 0) lie on the _____.
9. Ordered pairs with two _____ coordinates lie in Quadrant I.
10. Ordered pairs with two _____ coordinates lie in Quadrant III.
11. Two intersecting axes form a _____.
12. The point where the two axes meet is called the _____.

B. Given the coordinates of each point labeled in Figure 15.5.

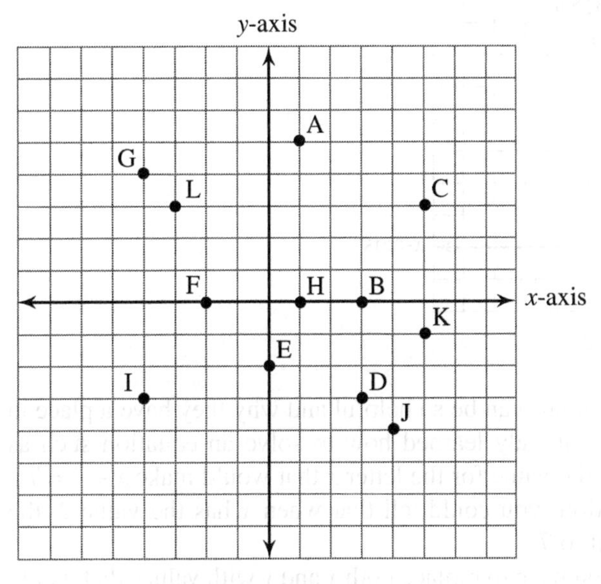

FIGURE 15.5

13. A
14. B
15. C
16. D
17. E
18. F
19. G
20. H
21. I
22. J
23. K
24. L

C. Construct a graph on a sheet of graph paper. Label each of the following items on the graph. Plot each point listed, and label it using its capital-letter designation.

25. x-axis
26. y-axis
27. Quadrant I
28. Quadrant II
29. Quadrant III
30. Quadrant IV
31. origin
32. point A (0, –4)
33. point B (–2, 5)
34. point C (3, 0)
35. point D (3/4, 6)
36. point E (–3, –3)
37. point F (–4, 1)
38. point G (1, –4)
39. point H (5, –2)
40. point I (3/2, 3)

D. Read and respond to the following exercises.

41. Think of a real-life situation where you have seen a coordinate system. Describe that situation.
42. Why do they call (6, 2) an ordered pair?
43. On a globe, what location would be equivalent to the x-axis on a Cartesian coordinate system? Do you know what is comparable to the y-axis? If you can find the answers to these questions, include information about where you found the answers.
44. How can you describe the ordered pair for any point on the x-axis? On the y-axis?

15.2 Graphing Linear Equations by Using Points

In an earlier chapter, we discussed writing a formula from data given in a table. For instance, consider this table of values:

X	3	4	5	6	7
Y	6	7	8	9	10

To write the formula that explains the relationship in the table, we notice that each Y value is 3 greater than its corresponding X value. The formula then would be

$Y = X + 3$ or $X = Y - 3$

Plot these five ordered pairs of values: (3, 6), (4, 7), (5, 8), (6, 9), and (7, 10). Remember that the coordinates are given in (x, y) form and that the x-axis is the horizontal scale and the y-axis is the vertical scale. The result of this graphing is given in Figure 15.6.

The points on the graph appear to fall along a straight line. It appears that the point (2, 5) would also fall on that same line. Is it true that if $Y = X + 3$, then by substituting the values (2, 5), we get a true statement? Is $5 = 2 + 3$ true? Yes, it is. And the same thing would happen for any ordered pair that we found on the graph that falls along that line.

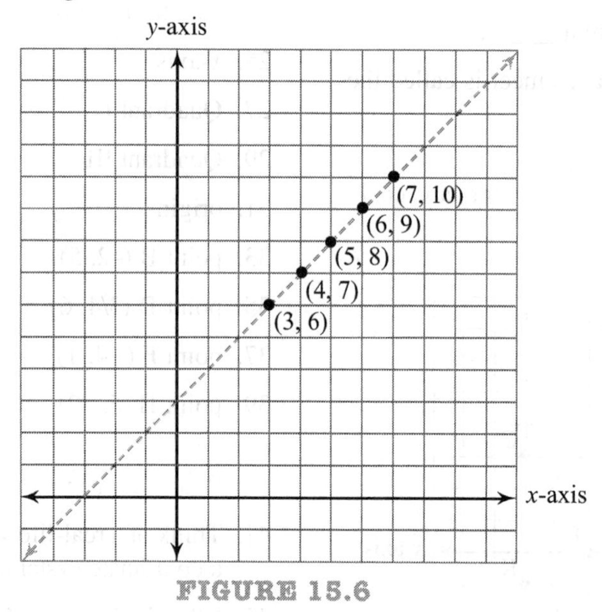

FIGURE 15.6

It is time to discover why graphs can be so helpful and why they have a place in the study of algebra. You have already learned how to solve an equation such as $x + 5 = 7$. Your job was to find the value for the letter x that would make $x + 5 = 7$ a true statement. Just by inspection, you could tell that when x has the value 2, the statement is true. 2 + 5 *is* equal to 7.

In the equation $x + y = 7$, you have to replace both x and y with values that would give a true statement.

If you try $x = 2$ and $y = 5$, you get $x + y = 7 \rightarrow 2 + 5 = 7$. True.

If you try $x = 4$, then y must be 3, because $4 + 3 = 7$ is true.

If you try $x = 6$ and $y = 1$, you get $x + y = 7 \rightarrow 6 + 1 = 7$. True.

Obviously, many pairs of numbers would work. Next, put your findings into an orderly display by using a *table of values* like this:

$x + y = 7$

x	y
2	5
4	3
6	1

Now see if you can find two *more* pairs of numbers that would work. How about (0, 7)? And (3 1/2, 3 1/2)?

There was no special reason for choosing any one of these values; that is, the choice was by chance. Generally, you just pull any value for x out of the air and then see what y must be to give a true statement. If you choose different values from those the book shows, and your values give true statements, then they are also correct.

In the table of values you just created, the first set of numbers ($x = 2$ and $y = 5$) can be represented as an ordered pair (x, y), or (2, 5). Every other point on the table can also be written in that form.

See what happens when you put (or, as we say, *plot*) these ordered pairs of numbers on a graph (Figure 15.7).

EXAMPLE 1 Graph the solution of the equation $x + y = 7$.

x	y
2	5
4	3
6	1
0	7
3 1/2	3 1/2

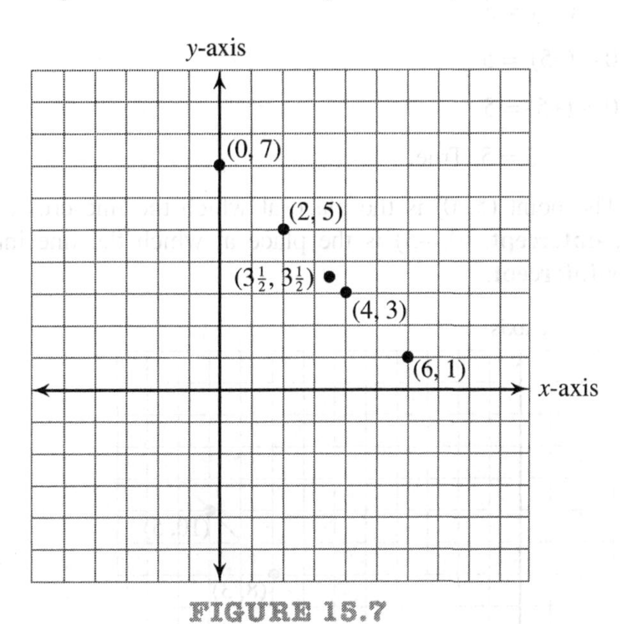

FIGURE 15.7

If the pairs of numbers that have been graphed on a Cartesian coordinate system were connected, they would form a straight line. This line also appears to go through the points (7, 0) and (–1, 8), which are not values on the table but which certainly satisfy the equation $x + y = 7$. In fact, the coordinates of every point on the line satisfy the equation. This line is the picture (graph) of the **solution set** of the equation.

Here are the steps you should follow to graph the solution set for a simple linear equation:

1. Find at least three pairs of values that satisfy the equation. (Two points determine a line, and the third point is for insurance. When all three points fall on the same straight line, you know you have not made any errors) Don't let the values get too large for your graph.
2. List that information in a table.
3. Transfer those ordered pairs to a graph.
4. Connect the points to see if they do form a straight line. If they do not fall in the same straight line, check for an error in your table of values.

EXAMPLE 2 Graph the solution set for the equation $x - y = 5$.

Solution Begin by picking three pairs of numbers that satisfy the equation. Because this is a simple equation, you should be able to find these pairs just by looking at the equation. Figure 15.8 shows three possible choices. After completing a table of values, plot the ordered pairs on the graph and connect the points.

It would appear that the line also goes through the point (5, 0). See if this is true.

$x - y = 0$

$5 - 0 = 5$ True.

Therefore, (5, 0) is also a solution of $x - y = 5$. What about (0, –5)? See if it gives a true statement in the equation.

$x - y = 5$

$0 - (-5) = 5$

$0 + (+5) = 5$

$5 = 5$ True.

The point (5, 0) is the place at which the line crosses the x-axis. 5 is called the ***x*-intercept**. (0, –5) is the place at which the line inters the y-axis so –5 is the ***y*-intercept**.

$x - y = 5$

x	y
6	1
8	3
10	5

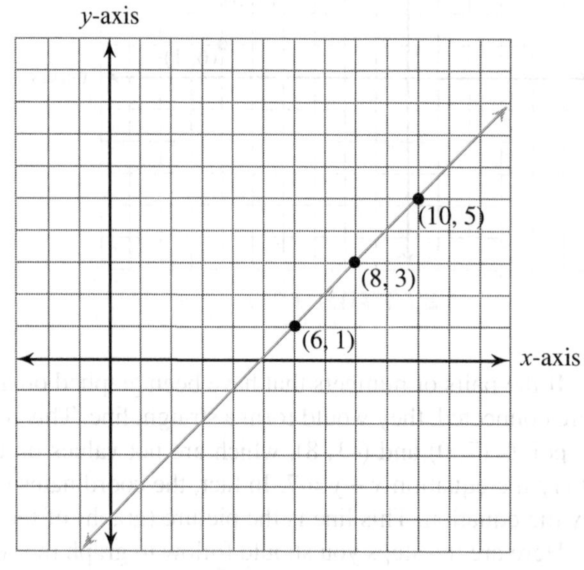

FIGURE 15.8

From Examples 1 and 2, you should see that the coordinates of every point on the line form a solution set for the equation. Also, any ordered pair of values that is a solution of a linear equation falls on the line that is its graph. Thus, unlike the equations you studied earlier (such as $x + 5 = 7$ and $3M + 4 = 5M - 2$), linear equations such as $y + 2x = 5$ don't have just one solution. They have an infinite number of solutions, or values that give a true statement. The graph displays these solutions.

EXAMPLE 3 Graph the line that is the solution set for the equation $2x + y = 8$.

Solution Pick three values for x and determine the corresponding y values.

If $x = 0$, $2(0) + y = 8$; therefore, $y = 8$.

If $x = 2$, $2(2) + y = 8$; therefore, $y = 4$.

If $x = 3$, $2(3) + y = 8$; therefore, $y = 2$.

Put the three ordered pairs in a table.

Plot the three ordered pairs on a graph. (See Figure 15.9.)

x	y
0	8
2	4
3	2

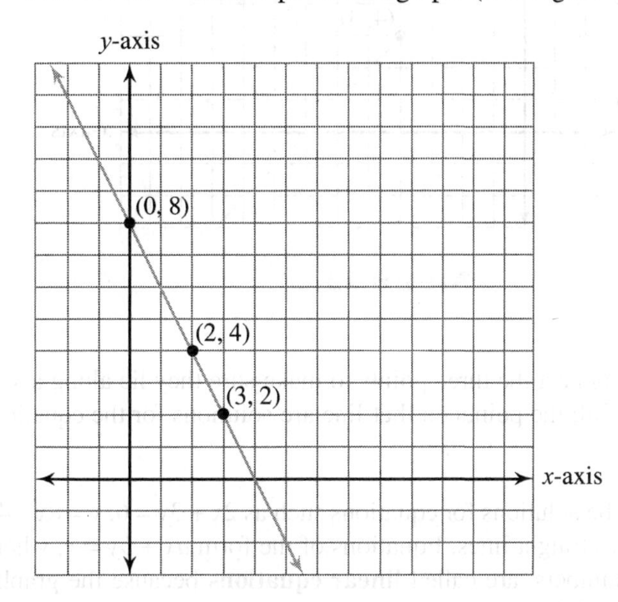

FIGURE 15.9

The points appear to fall along a straight line, and that line also goes through (4, 0). Is (4, 0) a solution of the equation $2x + y = 8$? Substitute the values and see if it is.

$2x + y = 8$

$2(4) + 0 = 8$

$8 + 0 = 8$

$8 = 8$ is true.

EXAMPLE 4 Graph the solution set for $x + y = 7$.

Solution Pick three convenient x values and find the corresponding y values.

If $x = 0$, $0 + y = 7$; therefore, $y = 7$.

If $x = 2$, $2 + y = 7$; therefore, $y = 5$.

If $x = 4$, $4 + y = 7$; therefore, $y = 3$.

Put the values into a table and graph the three ordered pairs. (See Figure 15.10.)

x	y
0	7
2	5
4	3

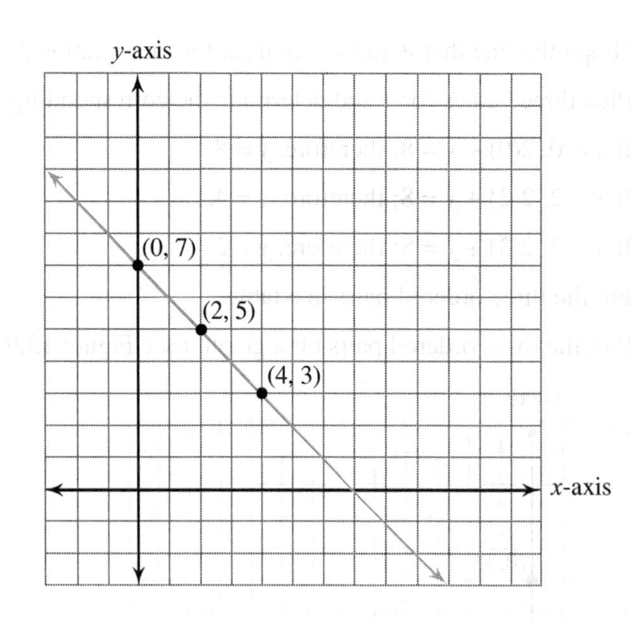

FIGURE 15.10

Connect the three points to make sure they lie along a straight line. The coordinates of all the points on that line are solutions for the equation $x + y = 7$.

◀

The solutions for equations such as $2x + 3y = 6$, $y = 8x - 4$, and $4y - 2x = 11$ all graph as straight lines. Equations of the form $ax + by = c$, where a, b, and c represent real numbers, are called **linear equations** because the graphs of their solutions form a *straight line.*

Self-Check Complete the table of values for these equations.

1. $x - y = 6$

x	y
7	
	0
2	

2. $2x + y = 4$

x	y
–2	
	0
	4

Answers:

1. $y = 1, x = 6, y = -4$
2. $y = 8, x = 2, x = 0$

EXAMPLE 5 Graph the solution set for $3x = 12$.

Solution If we want to put this equation into $ax + by = c$ form, we notice that there is no y term. Therefore, its coefficient must be 0. In linear equation form, this equation is written $3x + 0y = 12$. Upon closer inspection, you can see that y can assume any value because 0 times y will always be 0 and that x must be 4 because 4 is the only value to satisfy $3x = 12$. The three points chosen might be (4, 2), (4, –1), and (4, 3). Any three will work as long as the x value is 4. Make the table of values.

x	y
4	2
4	–1
4	3

Figure 15.11 shows the graph of this line.

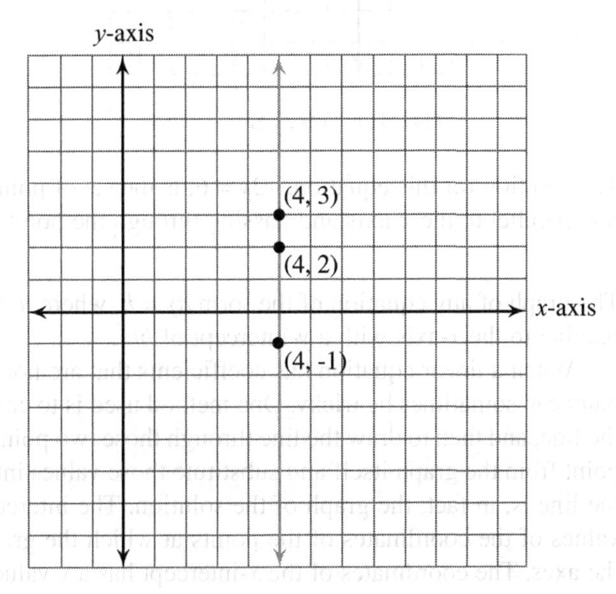

FIGURE 15.11

The graph is a vertical line parallel to the y-axis passing through the point (4, 0).

The graph of any equation of the form $ax = b$, where $a \neq 0$, will be a vertical line parallel to the y-axis with an x-intercept of b/a.

EXAMPLE 6 Graph the solution set for the equation $-2y = 6$.

Solution You notice immediately that there is no x term. In linear equation form, this equation is written $0x - 2y = 6$. The x values can be any number because their coefficient is zero. The only y value to satisfy $-2y = 6$ is $y = -3$. The ordered pairs chosen will all be of the form $(x, -3)$, so you might choose (–2, –3), (0, –3), and (3, –3). Put the ordered pairs in a table of values and plot the points on the graph (Figure 15.12).

x	y
–2	–3
0	–3
3	–3

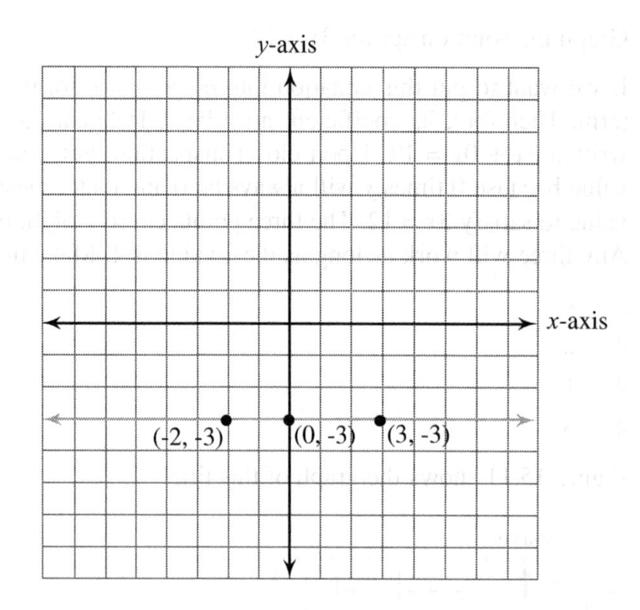

FIGURE 15.12

The solution for this equation, $-2y = 6$, is the set of points that fall on the horizontal line parallel to the x-axis and passing through the point (0, –3).

The graph of any equation of the form $ay = b$, where $a \neq 0$, will be a horizontal line parallel to the x-axis with a y-intercept of b/a.

When a linear equation has coefficients that are not equal to 1, finding the three pairs can sometimes be tricky. One method used is to calculate the two intercepts for the line, and then to draw the line through those two points. Then you can read a third point from the graph itself and substitute those values into the equation to check that the line is, in fact, the graph of the solution. The intercepts are the nonzero x and y values of the coordinates of the points at which the graph crosses, or cuts through, the axes. The coordinates of the x-intercept has a y value of 0. It will be of the form $(x, 0)$. The coordinates of the y-intercept will have an x value of 0. It will look like $(0, y)$. Study the next example to see how this works.

EXAMPLE 7 Graph the solution for $3y = 6x - 12$.

Solution Because this equation has coefficients that are not equal to 1, you might try to graph the line using the intercepts.

First, let $y = 0$ to find out where the line crosses the x-axis.

If $y = 0$, $3(0) = 6x - 12$

$0 = 6x - 12$

$\underline{+12 \qquad +12}$

$+12 = 6x$

$+2 = x$ So $x = 2$ when $y = 0$

The x-intercept is 2 and its coordinates are (2, 0).

Now find the y-intercept. Let $x = 0$.

If $x = 0$, $3y = 6(0) - 12$

$3y = 0 - 12$

$3y = -12$

$y = -4$ So $y = -4$ when $x = 0$.

The y-intercept is -4 and its coordinates are $(0, -4)$.

Now graph the line through the points $(2, 0)$ and $(0, -4)$. (See Figure 15.13.)

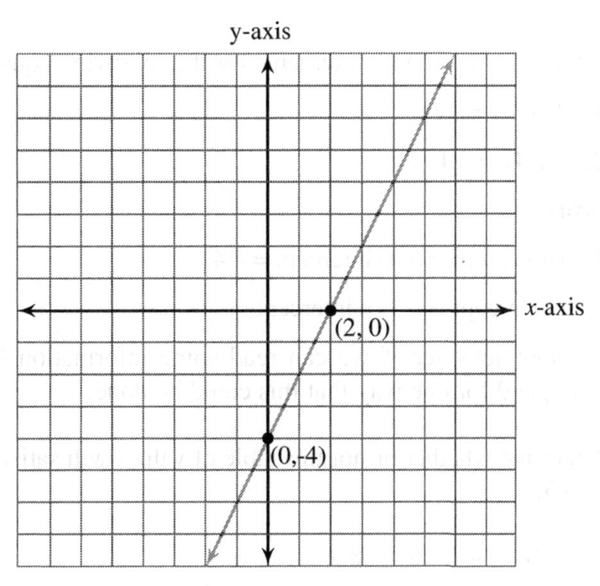

FIGURE 15.13

To check your work, find one more point along that line and see if it is also a solution for the equation. The line goes through $(3, 2)$. If $3y = 6x - 12$, then is $3(2) = 6(3) - 12$ a true statement? Is $6 = 18 - 12$? Yes. The solution has been graphed correctly.

EXAMPLE 8 Graph the solution for $2y + 5x = 5$.

Solution Because this equation has coefficients that are not equal to 1, graph the line using the intercepts.

If $x = 0$, $2y + 5(0) = 5$ and $y = 5/2$.

If $y = 0$, $2(0) + 5x = 5$ and $x = 1$.

Graph the line through the points $(0, 5/2)$ and $(1, 0)$. (See Figure 15.14.)

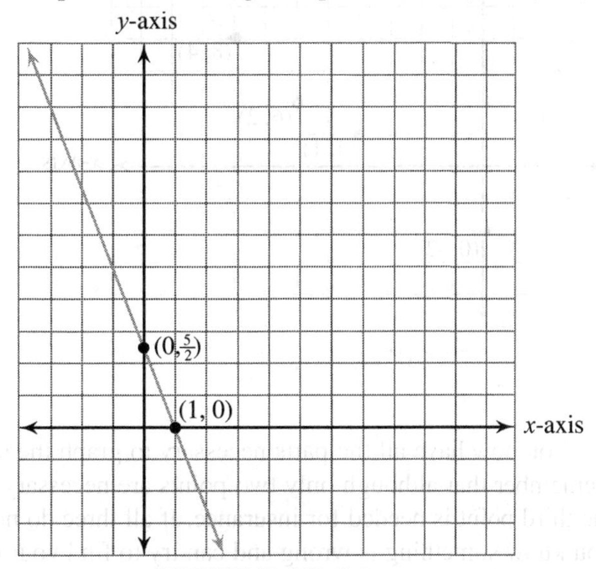

FIGURE 15.14

The line appears to go through the point (–1, 5) and, by substitution, you should see that (–1, 5) satisfies the equation $2y + 5x = 5$. Try it.

◀

Self-Check Determine the x and y intercepts for the following equations.

1. $2x - 3y = 12$
2. $x + 4y = -4$

Answers:

1. x-intercept = 6, y-intercept = –4
2. x-intercept = –4, y-intercept = –1

Now, let's see if we can read some information from the graph itself. Study Example 9 for one way that this could be done.

EXAMPLE 9 Determine whether or not this table of values will satisfy the general linear equation $ax + by = c$.

x	0	4	6	8
y	–2	1	2	4

Solution We could try to figure out what the equation for those values might be (a difficult task at this point). However, if the question is whether or not the values satisfy the general *linear* equation, we need only graph the points and see if they fall on a straight line. As you can see in Figure 15.15, they do not. Therefore, there is no linear equation that could be written from that table of values.

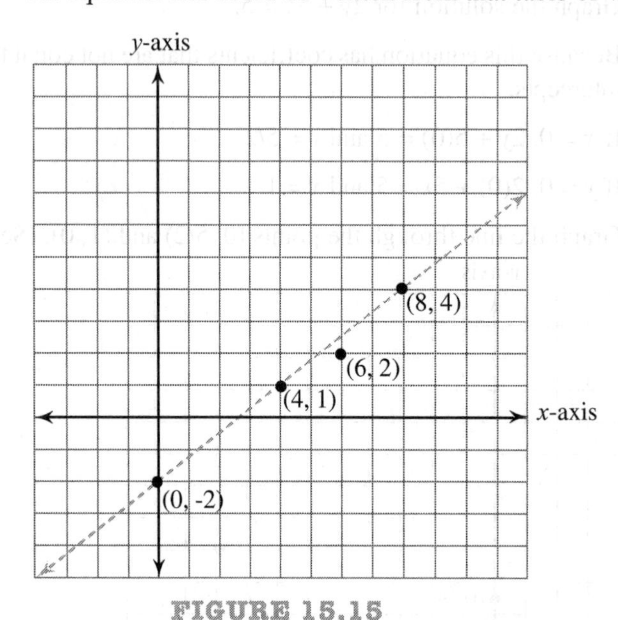

FIGURE 15.15

◀

You now have all the parts necessary to graph the solutions of linear equations. Remember that although only two points are necessary to determine a straight line, the third point is needed for insurance. If all three do not fall on the same line, then you know something is wrong and can try to find your error.

These are the steps you should now follow to graph the solution of a linear equation using points:

1. Pick three x values. Substitute them into the equation and solve for the y values. Use the three ordered pairs to graph the line. Be sure that the line passes through all three points. Label the three points you used so that someone else can follow your work. Check the three pairs of values in the *original equation.*
2. If you choose to graph using the intercepts, let $x = 0$ and find the y value. Let $y = 0$ and find the x value. Plot the points and draw the line through them. It is always a good idea to check one other point by substituting its coordinates into the equation to make sure there are no errors.

15.2 EXERCISES

A. For each of the following equations:

a. Find three ordered pairs of numbers that satisfy the equation, and check those values in the original equation.

b. Graph the solution set for the equation. Be sure to label the three points you used. (When you check your answers against the answer key, your three points may be different from those labeled on the graph shown in the answer key, but the line should look the same.)

1. $x + y = -4$
2. $2x - 4 = y$
3. $-3x + 5 = y$
4. $x + y = 7$
5. $x - y = 3$
6. $3x - y = 4$
7. $2y = -x$
8. $y = 4x$
9. $3y = x + 12$
10. $5x + 2y = 7$
11. $7x - 2y = 6$
12. $y + 4x = -2$
13. $-2x = 6y$
14. $-3y + 2x = -12$
15. $2y - 3x = -4$
16. $3y = x + 9$

B. Graph the solution for each of the following equations by finding the x- and y-intercepts. Be sure to check your work with a third point from the graph.

17. $y + 2 = 3x$
18. $6x = 3y + 12$
19. $2y = 8x - 4$
20. $-2x = 3y - 6$
21. $3x = -4y + 6$
22. $x - y = 5$
23. $x - 2y = 8$
24. $3x - 2y = 0$
25. $2x + y = -6$
26. $3x - y = 12$
27. $-5x + y = 10$
28. $4x + 2y = 0$
29. $-2x + 6y = 0$
30. $-6x + 3y = -12$

C. Read and respond to the following exercises.

31. Explain to a fellow student how you can tell if the line you are looking at is, in fact, the correct graph of the solution for an equation you are solving.
32. If the solution for the equation you are asked to graph goes through (–2, 4), (–1, 0), and (5, 6), is the equation a linear equation? How can you tell?
33. Why do you usually pick three pairs of values to graph a line? Can't you draw a line through any two points? Explain your answer.
34. If you were asked to graph the solution set for $3y + 7 = 2x$, which method would you use and why?
35. What is your first clue that the graph of an equation will be parallel to the x-axis?
36. What is your first clue that the graph of an equation will be parallel to the y-axis?

15.3 Slope

One thing you need to notice when graphing lines on the coordinate system is what the lines look like. Does the line "rise" or "fall"? How steep or flat is the line? Is the line parallel to one of the axes? These are all questions related to the *slope* of the line. If you have ever driven in the mountains, you may have noticed warning signs that caution truck drivers to use a low gear because the next few miles of road have a 5% grade. That means that the road is very steep and drivers need to exercise caution to keep their vehicles under control. That situation is also related to slope.

For purposes of communication (so everyone is talking in the same terms), **slope** is defined as the ratio of vertical to horizontal change between two points on the line. This ratio is usually written in fraction form, and if the slope is, for example, 6/1, it is simply called 6.

Let's go back to the truck example for a minute. Because the caution was to use low gear, the truck was driving down the mountain. A 5% grade could be expressed in fraction form, 5/100, or in reduced fraction form as 1/20. That ratio means that there is a 1 foot vertical drop for every 20 feet traveled, or a 5-foot vertical drop for every 100 feet traveled. The grade of the road and the slope of a line are related.

Look at Figure 15.16 and the three points labeled on the graph.

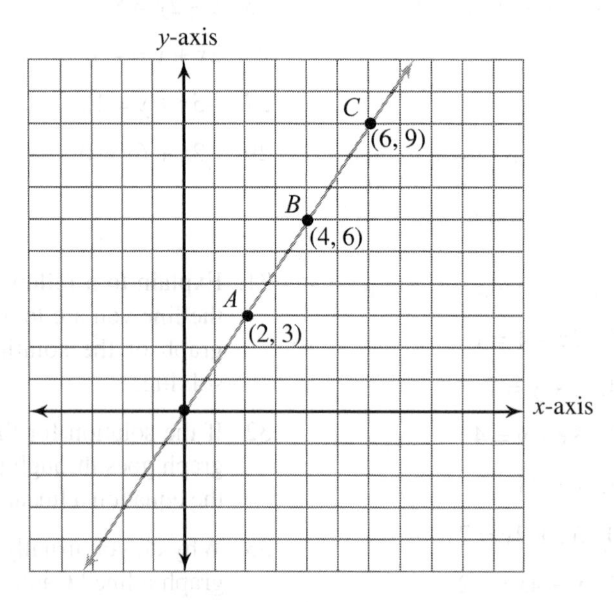

FIGURE 15.16

Slope is defined as the ratio of the vertical change (up-down) to the horizontal change (right-left). A person moving from point *A* to point *B* along the grid lines in Figure 15.16 would go up three units and then to the right two units. Change up is positive and change to the right is positive. The ratio would be + 3/+ 2 or 3/2. A person moving from point *B* to point *C* would go up three units and then to the right two units. That ratio again would be +3/+2 or 3/2. A person moving from point *C* down to point *A* would move down six units and then to the left four units. That ratio would be –6/–4 because movement down is negative and movement to the left is negative. The ratio of –6/–4 would be written in simplest form as 3/2. Notice that the slope of the line among all three points is 3/2. The slope of any nonvertical straight line is a constant value.

EXAMPLE 1 Look at the line in Figure 15.17. There are three points labeled on this graph. Determine the slope of the line by finding the ratio of vertical change (up-down) to horizontal change (right-left).

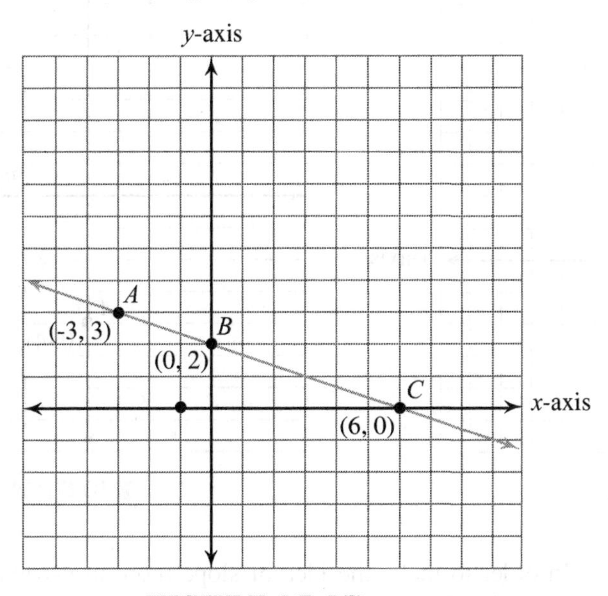

FIGURE 15.17

Solution First consider movement from point A to point B. Be sure to write the ratio of the vertical change to the horizontal change. To go from A to B along the grid lines, a person goes down one unit and then to the right three units. The ratio would be –1/+3 or –1/3. To move from B to C along the grid lines, the person would go down two units and then to the right six units. The ratio would be –2/+6 or –1/3 in simplest reduced form. If you wanted to go from point C to point A, you would move up three units and then to the left nine units. The ratio would be +3/–9 or –1/3 in simplest form. Again, the slope of the line is a constant, –1/3, no matter which two points you use.

EXAMPLE 2 Look at the graph in Figure 15.18 and determine the slope of the line.

Solution This time we will need to pick our own points since none are identified on the graph. Let's use both intercepts and one other point, say (6, 4). Figure 15.19 shows how this might be done.

Moving from A to B would be +4/+3. The slope would be 4/3.

Moving from B to C would be +4/+3, or 4/3.

Moving from A to C would mean going up 8 units, +8, and then to the right six units, +6. The slope would be +8/+6, or 4/3.

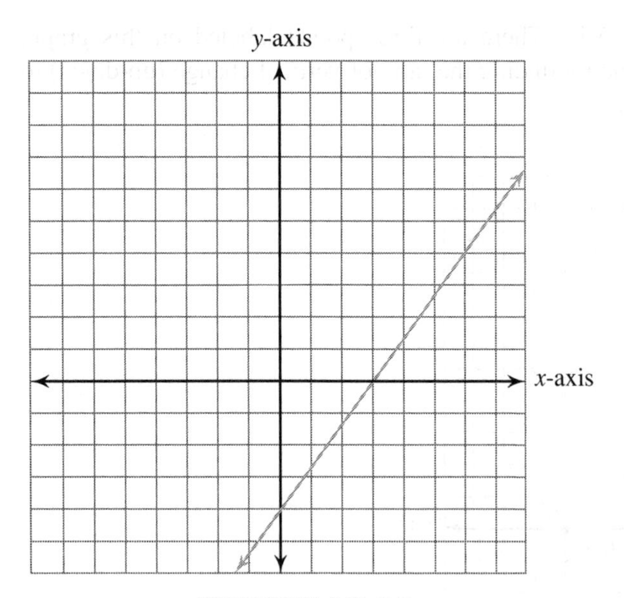

FIGURE 15.18

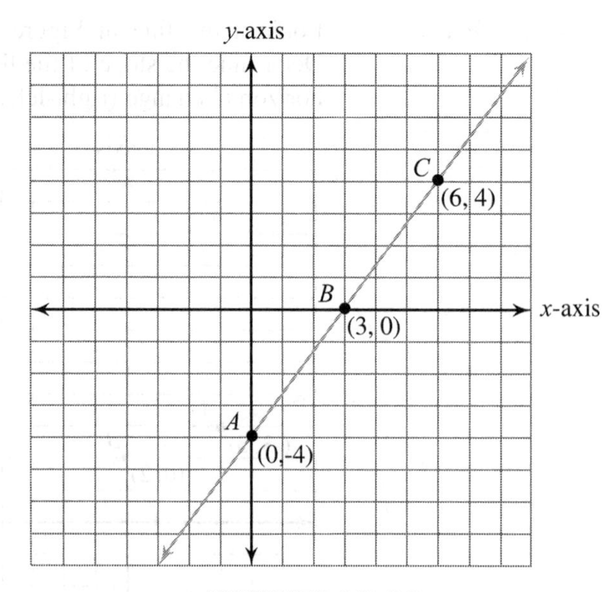

FIGURE 15.19

In order to make the idea of slope more user-friendly, mathematicians discuss slope as "the rise over the run" or rise/run. "Rise," of course, refers to the vertical change, and "run" to the horizontal change.

Notice in Figure 15.16 that the line "rises" or increases as you move along it from left to right and the slope is +3/2. In Figure 15.19 the line also "rises" as you move along it from left to right. Its slope is +4/3. In Figure 15.17 in Example 1, the line decreases or "falls" as you move along it from left to right, and the slope is –1/3. This will always be the case. A line that "rises," or increases, from left to right has a positive slope and a line that "falls," or decreases, from left to right has a negative slope. Conversely, when you are going to graph a line and know that it has a positive slope, you know that the line will rise as it goes to the right. If the slope is known to be negative, the line must decrease, or fall, as you move farther to the right along the line.

Now that you can picture slope as the slant or steepness of a line and know that it equals the ratio rise/run, you can look at slope in more detail. **Slope** can be defined in several ways, but the definition that is the most helpful in understanding the slope of a line is

Definition of Slope

$$\text{Slope} = \frac{\text{Change in } y \text{ values}}{\text{Change in } x \text{ values}} = \frac{\text{Rise}}{\text{Run}}$$

Because slope is used so often, it has been given a special designation, the letter *m*. The definition of slope just given can also be expressed as a formula.

Formula for Slope

Given two points (x_1, y_1) and (x_2, y_2), we can find the slope of the line through those two points by using the formula

$$m = \frac{y_2 - y_1}{x_2 - x_1}$$

This formula may look complicated, but it's very easy to use and it saves time because the line does not have to be graphed to determine its slope. The subscript notation is used to indicate that you are working with two different points but that each of them has two coordinates, x and y. It does not matter which point you call (x_1, y_1), but you must be consistent. For safety's sake, always refer to the first point mentioned as (x_1, y_1) and to the second point as (x_2, y_2). Then you will not get them confused.

Consider two points, (2, 1) and (4, 7). Let (2, 1) be (x_1, y_1) and let (4, 7) be (x_2, y_2). Then, by the definition and formula,

$$m = \frac{y_2 - y_1}{x_2 - x_1} = \frac{7 - 1}{4 - 2} = \frac{6}{2} = \frac{3}{1} = 3$$

Therefore, the slope of the line through (2, 1) and (4, 7) is positive 3, so the line rises to the right. Graph the line through (2, 1) and (4, 7) and you will see that this is correct. Remember also that slope indicates the ratio of the vertical change to the horizontal change. A slope of 3/1 tells you that if you begin at (2, 1) and move up three units (+3) and to the right one unit (+1), you will find another point on the line. Begin at (2, 1) and move up three units and over one unit to (3, 4). Then from (3, 4) move up three units and over one unit, and you come to (4, 7). Check this in Figure 15.20.

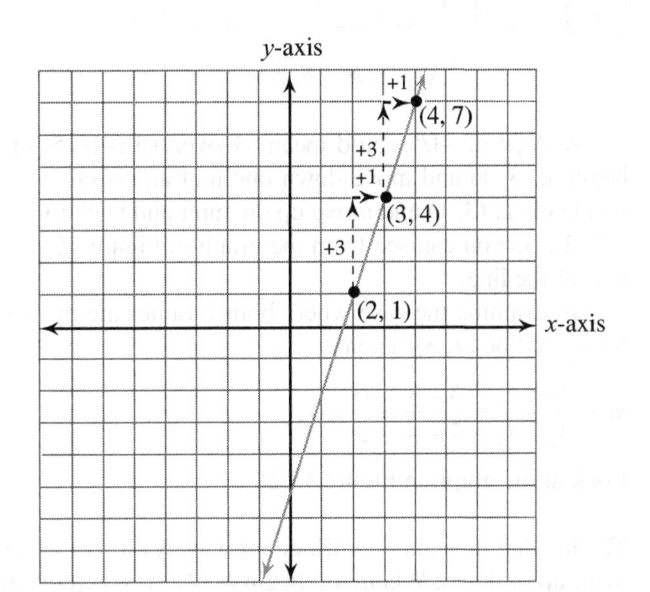

The line is increasing to the right.
Rise/run is +3/+1.

FIGURE 15.20

Actually, it does not make any difference which point is labeled (x_1, y_1). Redo the previous example but let (4, 7) be (x_1, y_1) and (2, 1) be (x_2, y_2). Then by the formula

$$m = \frac{y_2 - y_1}{x_2 - x_1} = \frac{1 - 7}{2 - 4} = \frac{-6}{-2} = +3$$

As you can see, the slope of the line passing through (2, 1) and (4, 7) is +3 in both cases.

Let's try another example. Let (x_1, y_1) be (3, 1), and let (x_2, y_2) be (–2, 2). Then

$$m = \frac{y_2 - y_1}{x_2 - x_1} = \frac{2 - 1}{-2 - 3} = \frac{1}{-5} = \frac{-1}{5}$$

Because the slope is negative, the line should fall; that is, it should slant downward to the right. Look at the graph through (3, 1) and (–2, 2). (See Figure 15.21.) It does, in fact, decrease to the right.

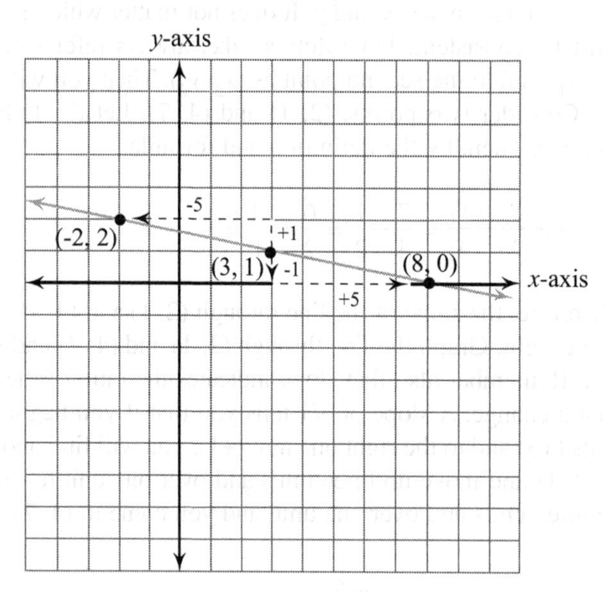

The line is decreasing to the right.
Rise/run is +1/–5
or –1/+5.

FIGURE 15.21

A slope of –1/5 could mean –1 over +5 (–1/+5) or +1 over –5 (+1/–5). If you begin at (3, 1) and move down one unit and over five units, you come to (8, 0). If you begin at (3, 1) and move up one unit and back to the left five units, you come to (–2, 2). As you can see from the graph in Figure 15.21, all three of those points are part of the line.

To examine the case where both *y* values are the same, let (x_1, y_1) be (4, 3) and let (x_2, y_2) be (7, 3). Then

$$m = \frac{y_2 - y_1}{x_2 - x_1} = \frac{3 - 3}{7 - 4} = \frac{0}{3} = 0$$

Look at the graph in Figure 15.22.

The line has no slant—no tilt either way. A ratio of 0/3 means that there is no movement up or down. When you begin at (4, 3), you move (three spaces) to the right or left only. Therefore, *a line parallel to the* x-*axis has slope = 0.* That means that the line will never intersect the *x*-axis.

What happens when both x values are the same? Well, if (x_1, y_1) is (4, 3) and (x_2, y_2) is (4, 9), then

$$m = \frac{y_2 - y_1}{x_2 - x_1} = \frac{9 - 3}{4 - 4} = \frac{6}{0} \qquad \text{Undefined. See Figure 15.23.}$$

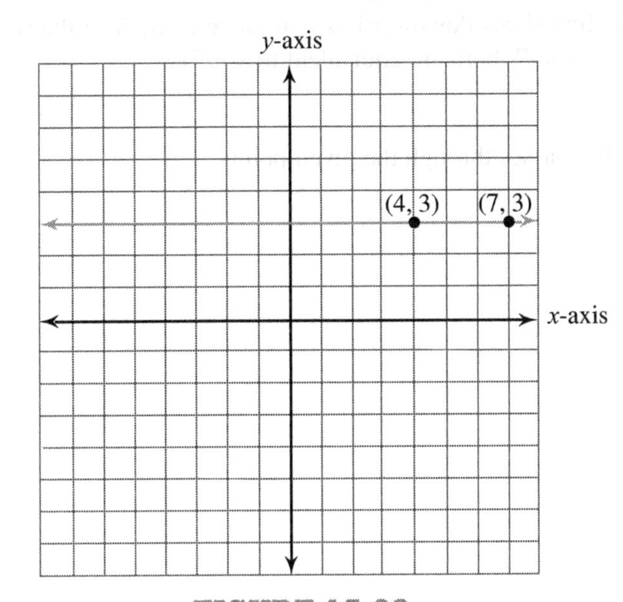

FIGURE 15.22

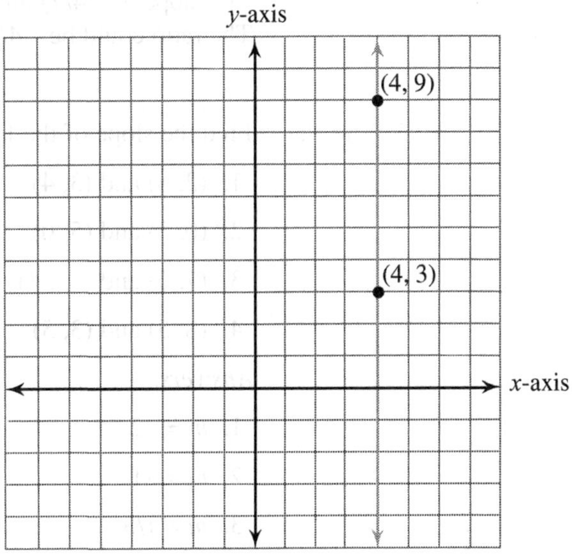

y-axis
(4, 9)
(4, 3)
x-axis

FIGURE 15.23

Because division by zero is impossible in our number system, the slope is said to be *undefined.* A ratio of 6/0 (or a ratio of any other number to zero) means there is no left-right movement between points. Looking at the graph, you can see that the line is straight up and down. Therefore, *a line parallel to the y-axis has an undefined slope.*

When you are given two points, it does not make any difference which point you call (x_1, y_1) and which you call (x_2, y_2). The most important thing to remember in doing these problems is the need to calculate carefully. The two-point formula is used to calculate slope in the following three examples.

EXAMPLE 3 Calculate the slope of the line that passes through the points (9, 4) and (7, –2).

Solution

$$m = \frac{y_2 - y_1}{x_2 - x_1} = \frac{-2 - 4}{7 - 9} = \frac{-6}{-2} = +3$$

The slope is +3, and the line rises to the right. The ratio could be +3/+1 or –3/–1, for example, because both are equivalent to + 3.

EXAMPLE 4 Calculate the slope of the line that passes through the points (–2, –3) and (4, –1).

Solution

$$m = \frac{y_2 - y_1}{x_2 - x_1} = \frac{-1 - (-3)}{4 - (-2)} = -\frac{1 + 3}{4 + 2} = \frac{2}{6} = \frac{1}{3}$$

The slope is +(1/3), and the line rises or increases as you move along it to the right. The ratio could be +1/+3 or –1/–3; both are equivalent to 1/3.

EXAMPLE 5 Calculate the slope of the line that passes through the points (–2, 4) and (5, 0).

Solution

$$m = \frac{y_2 - y_1}{x_2 - x_1} = \frac{0 - (+4)}{5 - (-2)} = \frac{0 - 4}{5 + 2} = \frac{-4}{7} \text{ or } -4/7$$

The slope is –(4/7) and the line slants downward as you move along it to the right. The ratio could be – 4/+7 or + 4/–7; both are equivalent to –(4/7).

Self-Check Find the slope of the line that passes through the given points.

1. (2, 6) and (3, 4)
2. (0, 5) and (5, 0)
3. (7, 6) and (–3, 4)
4. (8, 5) and (3, 5)

Answers:

1. $m = -2$
2. $m = -1$
3. $m = 1/5$
4. $m = 0$

Look at the graph in Figure 15.24. Calculate the slope by using the two points labeled on the graph.

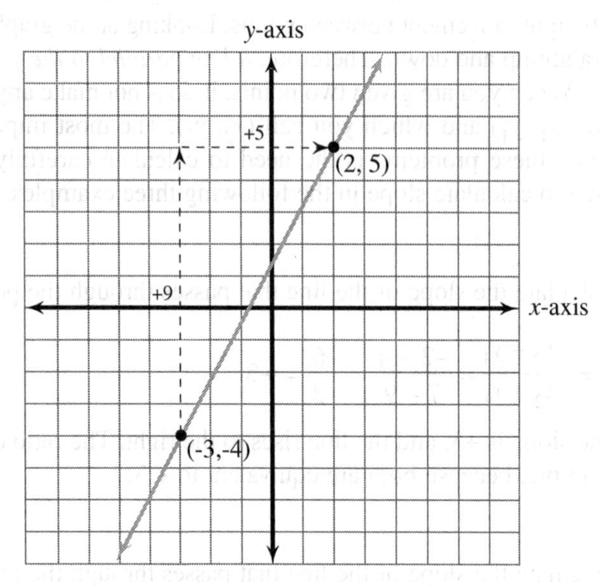

FIGURE 15.24

To move from (–3, –4) to (2, 5), you need to go up nine units and over five units. Therefore, the slope of the line is +9/+5, or 9/5. The slope is positive and the line is increasing to the right. Check the value for the slope by using the formula. Let (x_1, y_1) be (–3, –4) and let (x_2, y_2) be (2, 5). Then

$$m = \frac{y_2 - y_1}{x_2 - x_1} = \frac{5 - (-4)}{2 - (-3)} = \frac{5 + 4}{2 + 3} = \frac{9}{5}$$

The formula value agrees with the value you found for the slope by moving from one point to the other on the graph.

Facts About Slope

Let's summarize what you know now about slope:

1. Slope is the number that describes the steepness or slant of a line.
2. Slope is the ratio of the vertical change to the horizontal change.
3. Slope can be remembered as rise/run.
4. The formula used to calculate slope is

   $$m = \frac{y_2 - y_1}{x_2 - x_1}$$

   where the line passes through (x_1, y_1) and (x_2, y_2).
5. A line with a positive slope increases (rises) from left to right.
6. A line with a negative slope decreases (falls) from left to right.
7. A line with zero slope is horizontal and parallel to the x-axis.
8. A line with undefined (or no) slope is vertical and parallel to the y-axis.

15.3 EXERCISES

A. Study the slope of the line through each pair of points by following the steps listed below.

a. Graph each pair of points and draw the line passing through them.

b. Using the graphs, calculate each slope by studying movement from one point to the other and calculating rise/run.

c. Using the two-point formula for slope, calculate the slope of the line through each pair of points.

d. Make sure that your number answers in parts b and c agree.

e. Compare the numerical result you got in parts b and c with the way the line actually slants—increasing, decreasing, horizontal, or vertical—by inspecting the graphs you drew in part a.

1. (2, 3) and (4, 6)
2. (1, 4) and (–2, –2)
3. (2, 6) and (3, 5)
4. (7, 4) and (3, 5)
5. (0, 4) and (3, 8)
6. (6, 2) and (4, 0)
7. (2, 4) and (2, 8)
8. (3, 2) and (–2, 2)
9. (5, 4) and (2, 6)
10. (0, 3) and (5, 2)
11. (3, 4) and (– 4, 4)
12. (6, 3) and (6, –2)
13. (4, –3) and (2, –5)
14. (–3, 5) and (–2, –3)

B. Read and respond to the following exercises.

15. Describe two situations, besides the steepness of a road, in which discussing slope would be appropriate.

16. How can "rise/run" help you to remember the formula for the slope of a line?

17. How can you tell by looking at the graph of a line whether the slope will be positive or negative?

18. Explain the situation in which the slope of a line is undefined. Why does this happen?

19. The slope of a line passing through (3, 4) is +2. Explain how you could find two other points on the line.

20. Is a line with a slope of +1/2 steeper or flatter than a line with a slope of +2? Explain your answer.

15.4 Slope-Intercept Form

Now we will look at other ways to graph the solutions to linear equations. If the equation is a particularly complex one, finding three points at random may be very time-consuming. Even the intercepts method has its drawbacks. If the line passes through the origin, the x- and y-intercepts are the same point (0, 0) and you still need to find more points in order to graph the line. If the x or y values for the intercepts are fractions, or particularly close to the origin, the intercept method is not very effective.

Another way to graph the equation of a line involves first putting the equation into **slope-intercept form** and then graphing the line with the information you can learn from the rewritten equation. The slope-intercept form of the line is written $y = mx + b$, where y is isolated on one side of the equals sign, m (the coefficient of the x term) represents the slope of the line, and the b value is the second value in the ordered pair $(0, b)$ and indicates where the line intercepts the y-axis. "b" is very commonly referred to as the y-intercept. Use the same steps to rearrange the equation that you used to do formula rearrangement in Chapter 9. To eliminate a term from one side of an equation, add its opposite to both sides of the equation. To eliminate a coefficient from the y term, divide all the terms by that coefficient. Remember that you always want to arrive at a form of the equation where y is isolated on one side of the equals sign and has a coefficient of +1.

EXAMPLE 1 Rearrange $x + y = 7$ into slope-intercept form.

Solution

$$x + y = 7$$
$$\underline{-x} \qquad \underline{-x}$$

To isolate y, you need to subtract the x term from each side.

$$y = 7 - x$$
$$y = -x + 7$$

Rearrange the right side of the equation into the $y = mx + b$ form. The slope of the line is −1 and the value of b is +7. Once the equation is in slope-intercept form, you can graph the line using the y-intercept and the slope. Plot the point $(0, b)$ or (0, 7), in this case, and find two more points on the line using the slope of −1. You can use −1/+1 or, as is shown here, −2/+2. Look at the graph of this line in Figure 15.25.

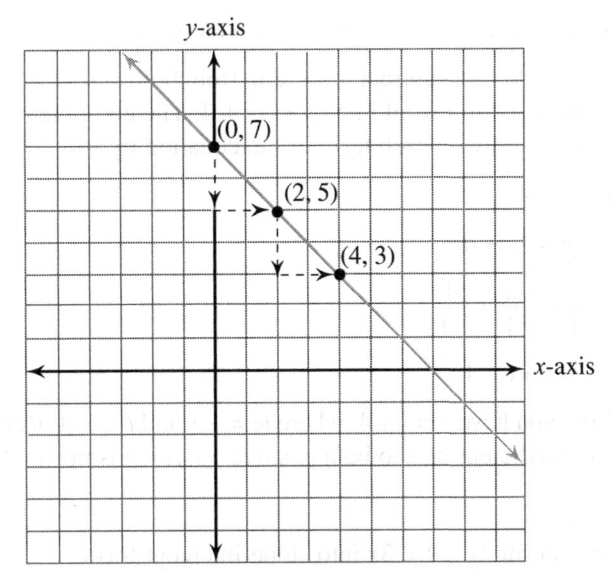

FIGURE 15.25

The line does intercept the y-axis at (0,7); the line decreases from left to right; and you can calculate the slope between two points—for instance, (2, 5) and (4, 3)—and see that the slope is, in fact, –1.

$$m = \frac{y_2 - y_1}{x_2 - x_1} = \frac{3-5}{4-2} = \frac{-2}{+2} = -1$$

Later in this section, you will actually learn to graph the line using the slope and intercept. For now, concentrate on learning to transfer equations into slope-intercept form.

EXAMPLE 2 Transform $y - x = 5$ into slope-intercept form.

Solution

$y - x = 5$

$\underline{+x} \quad \underline{+x}$ To isolate y, you must add x to both sides of the equals sign.

$y = 5 + x$ Rearrange the terms on the right-hand side of the equals sign,

$y = x + 5$ and you have the $y = mx + b$ form.

$y = +1x + 5$ Remember that you may write the coefficient of x alone as +1.

The slope of the line is +1 and the y-intercept is +5.

EXAMPLE 3 Transform $3y = 6x - 15$ into slope-intercept form.

Solution To isolate y, you would divide *every* term in the equation $3y = 6x - 15$ by 3.

$$3y = 6x - 15$$

$$\frac{3y}{3} = \frac{6x}{3} - \frac{15}{3}$$

$$y = 2x - 5$$

Then $y = 2x - 5$ is in $y = mx + b$ form, where $m = +2$ and $b = -5$.

EXAMPLE 4 Transform $x - y = 4$ into slope-intercept form.

Solution You could first subtract the x term from both sides of the equation. The changed form of the equation would be $-y = -x + 4$. The coefficient of y has to be a positive 1, so you must divide each term of the equation by –1.

$$x - y = 4$$

$$-y = -x + 4$$

$$\frac{-y}{-1} = \frac{-x}{-1} + \frac{4}{-1}$$

$$y = +x - 4$$

Now you have $y = x - 4$, where $m = +1$ and $b = -4$. Remember that you always want the coefficient of y to be a positive 1 (which is not written).

EXAMPLE 5 Transform $2y - 4 = 3x$ into slope-intercept form.

Solution Rearrange $2y - 4 = 3x$ into slope-intercept form.

$$2y - 4 = 3x$$

$$\underline{+4} \quad \underline{+4}$$

To isolate the y term, add 4 on *both sides* of the equals sign.

$$2y = 3x + 4$$

$$\frac{2y}{2} = \frac{3}{2}x + \frac{4}{2}$$

To change the coefficient of y to a positive 1, divide *each* term of the equation by +2.

$$y = \frac{3}{2}x + 2$$

In $y = 3/2x + 2$, $m = 3/2$ and $b = +2$.

Self-Check Transform these equations into slope-intercept form and then state the m and b values.

1. $2x + y = 7$
2. $8x - y = -5$
3. $-x + 3y = 6$

Answers:

1. $y = -2x + 7$; $m = -2$ and $b = 7$
2. $y = 8x + 5$; $m = 8$ and $b = 5$
3. $y = 1/3x + 2$; $m = 1 \neq 13$ and $b = 2$

EXAMPLE 6 Write $y = -2x$ in slope-intercept form.

Solution To write the equation in slope-intercept form, all you need to do is add a constant term, zero. $y = -2x + 0$. The slope is –2 and $b = 0$ so the y intercept is 0.

The **slope** in fraction form is the ratio of the vertical change to the horizontal change and is often called "the rise over the run," which means up-down change over right-left change. Consider the graph of the line $y = 2/3x - 3$ shown in Figure 15.26, and study "the rise" and "the run."

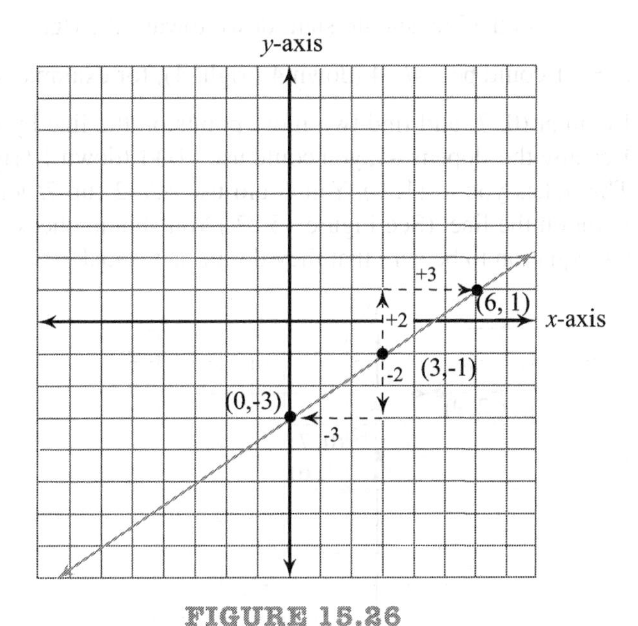

FIGURE 15.26

Because the slope of the graph of $y = 2/3x - 3$ is 2 over 3, the rise is +2 and the run is +3. From any point on the graph, you can rise +2 units and run +3 units and come to another point that lies on the line. Beginning at (3, –1) you would go up two units (rise +2) and go over three units (run + 3) and come to (6, 1), another point on the line. Because –2/–3 has the same value as +2/+3, you could also "rise" –2 and "run" –3 to find another point. Again beginning at (3, –1), you would go down two units (rise –2) and back to the left three units (run –3) and come to (0, –3), which is also a point on the line. Both points that you found using rise and run, (6, 1) and (0, –3), satisfy the equation and therefore lie on the line.

$y = \frac{2}{3}x - 3$	$y = \frac{2}{3}x - 3$
$1 = \frac{2}{3}(6) - 3$	$-3 = \frac{2}{3}(0) - 3$
$1 = 4 - 3$	$-3 = 0 - 3$
$1 = 1$ True.	$-3 = -3$ True.

Look at the following examples of linear equations in slope-intercept form and at their graphs, which have been drawn for you. Note the slopes of the lines, the rise over the run, and the coordinates of the point at which each line intercepts the y-axis $(0, b)$.

EXAMPLE 7 Use the slope and y-intercept to graph the solution set for the equation $y = -x + 7$.

Solution The equation is already in $y = mx + b$ form.

$b = +7$, so the line intercepts the axis at (0, 7).

$m = -1$, so the line should slant down toward the right.

$m = -1$ could be $-4/+4$ (down 4 / right 4), for example, or $+1/-1$ (up 1 / left 1).

Begin at (0, 7) and find two more points on the line by using the rise over the run. Because the slope is –1, you could use –1/+1 (down 1 / right 1) to find another point. That takes you to (1, 6). You could use +2/–2 (up 2/ left 2) to find (–2, 9), a third point on the line. (See Figure 15.27.) You should check all three points in the original equation to be sure that they do, in fact, satisfy the equation.

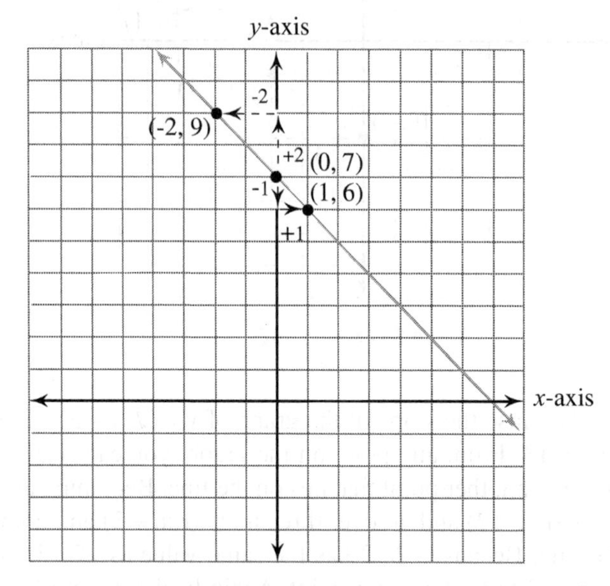

FIGURE 15.27

Check

(–2, 9)	(0, 7)	(1, 6)
$y = -x + 7$	$y = -x + 7$	$y = -x + 7$
$9 = -(-2) + 7$	$7 = -(0) + 7$	$6 = -(1) + 7$
$9 = 2 + 7$	$7 = 0 + 7$	$6 = -1 + 7$
$9 = 9$	$7 = 7$	$6 = 6$

EXAMPLE 8 Graph the solution set for $y = x + 5$ by using the slope and y-intercept.

Solution The equation is already in $y = mx + b$ form. Because $b = +5$, the line goes through the point (0, 5).

$m = +1$ could be +2/+2 (up 2/ right 2) or –1/–1 (down 1 / left 1).

From the graph in Figure 15.28 you can see that +2/+2 takes you to (2, 7) and that, beginning again at (0, 5), –1/–1 takes you to (–1, 4). To check the accuracy of the graphing process, substitute all three points back into the original equation and see if they each result in a true statement.

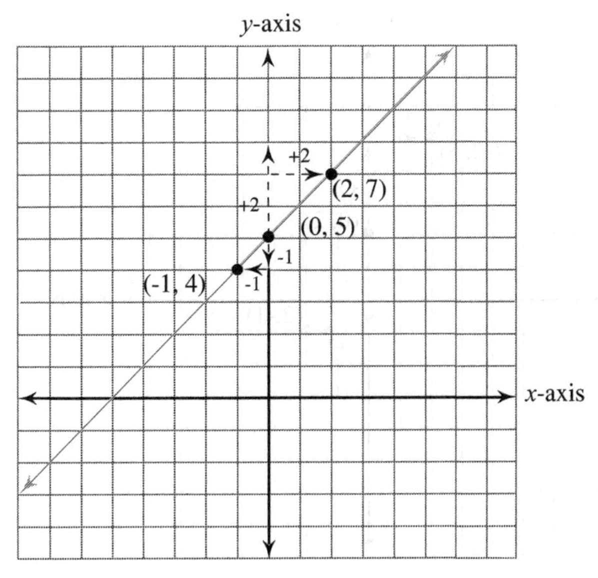

FIGURE 15.28

Check

(2, 7)	(0, 5)	(–1, 4)
$y = x + 5$	$y = x + 5$	$y = x + 5$
$7 = 2 + 5$	$5 = 0 + 5$	$4 = -1 + 5$
$7 = 7$	$5 = 5$	$4 = 4$

EXAMPLE 9 Graph the solution set for $3x + 2y = 4$ by using the slope and y-intercept.

Solution In this problem, you will first need to change $3x + 2y = 4$ into $y = mx + b$ form.

$$3x + 2y = 4$$

$$\underline{-3x} \qquad \underline{\quad -3x}$$

$$2y = 4 - 3x$$

$$\frac{2y}{2} = \frac{-3x}{2} + \frac{4}{2}$$

$$y = -\frac{3}{2}x + 2$$

The slope is –3/2, so the line slants down.

The b value is 2, so the line goes through the point (0, 2).

To graph the line, begin at (0, 2) and use the rise/run ratio to find two more points. –3/2 could mean –3/2 (down 3 / right 2) or 3/–2 (up 3 / left 2). Studying Figure 15.29 reveals that two other points on the line are (–2, 5) and (2, –1). Check the graph by substituting all three points into the original equation to see if they all result in a true statement.

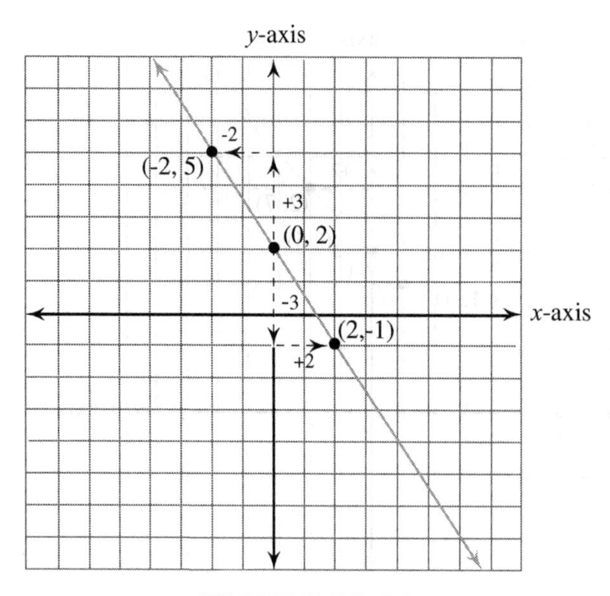

FIGURE 15.29

Check

$(-2, 5)$	$(0, 2)$	$(2, -1)$
$3x + 2y = 4$	$3x + 2y = 4$	$3x + 2y = 4$
$3(-2) + 2(5) = 4$	$3(0) + 2(2) = 4$	$3(2) + 2(-1) = 4$
$-6 + 10 = 4$	$0 + 4 = 4$	$6 + (-2) = 4$
$4 = 4$	$4 = 4$	$4 = 4$

To review: You can graph the solution of any first-degree equation using the y-intercept and slope by following these steps:

1. Write the equation in slope-intercept form, $y = mx + b$.
2. Identify the slope and the y-intercept.
3. Begin by graphing the y-intercept and using the slope (the ratio of rise to run) to find two more points on the line. Draw the straight line through all three points.
4. Check that the graph is correct by substituting all three points back into the original equation to make sure that they all produce true statements.

Here are two more examples worked out completely for you to follow.

EXAMPLE 10 Graph the solution set for $2x + y = 6$.

Solution

1. $2x + y = 6 \rightarrow y = -2x + 6$
2. $b = +6$, so the line goes through $(0, 6)$.
   $m = -2$, so the line decreases to the right.
   $m = -2$ could be $-2/+1$ (down 2 / right 1), for example, or $+2/-1$ (up 2 / left 1).
3. Begin at $(0, 6)$. Go down 2 and right 1; you reach $(1, 4)$. Begin again at $(0, 6)$. Go up 2 and left 1; you reach $(-1, 8)$. (See Figure 15.30.)
4. To check, substitute $(0, 6)$, $(1, 4)$ and $(-1, 8)$ into the original equation.

$2x + y = 6$	$2x + y = 6$	$2x + y = 6$
$2(0) + 6 = 6$	$2(1) + 4 = 6$	$2(-1) + 8 = 6$
$0 + 6 = 6$	$2 + 4 = 6$	$-2 + 8 = 6$
$6 = 6$	$6 = 6$	$6 = 6$

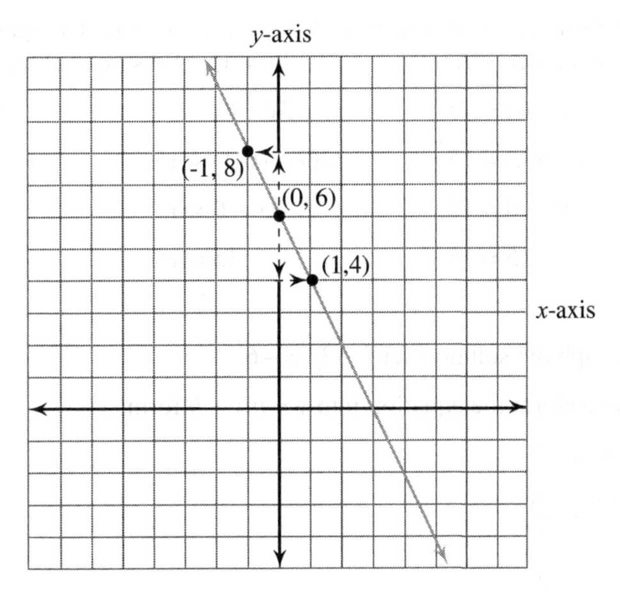

FIGURE 15.30

EXAMPLE 11 Graph the solution set for $3y - 6x = 15$.

Solution Transform the equation into $y = mx + b$ form.

$$\begin{aligned} 3y - 6x &= 15 \\ +6x \quad & \quad +6x \\ 3y &= 15 + 6x \\ \frac{3y}{3} &= \frac{15}{3} + \frac{6x}{3} \\ y &= 5 + 2x \\ y &= 2x + 5 \end{aligned}$$

The b value is +5, so the line goes through the point (0, 5). The slope is +2, so the line rises or increases to the right. To find two more points on the line, begin at (0, 5) and move up 2 and then right 1 (+2/+ 1). Return to (0, 5) and move down 2 and left 1 (–2/–1). You have now located three points on the line. They are (0, 5), (1, 7), and (–1, 3). See them graphed in Figure 15.31.

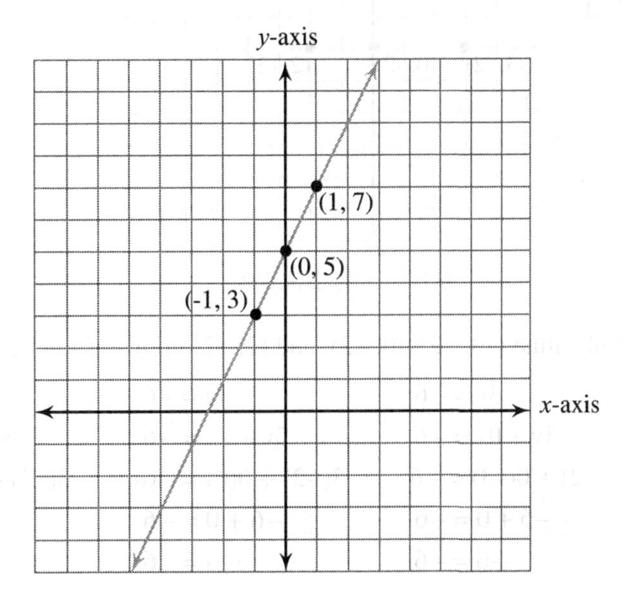

FIGURE 15.31

Check Substitute (0, 5), (1, 7), and (–1, 3) in the original equation to make sure that they are, in fact, solutions for $3y - 6x = 15$. (Be sure to substitute correctly here.)

$3y - 6x = 15$	$3y - 6x = 15$	$3y - 6x = 15$
$3(5) - 6(0) = 15$	$3(7) - 6(1) = 15$	$3(3) - 6(-1) = 15$
$15 - 0 = 15$	$21 - 6 = 15$	$9 + 6 = 15$
$15 = 15$	$15 = 15$	$15 = 15$

EXAMPLE 12 Graph the solution set for $3y = -6$.

Solution Transform the equation into $y = mx + b$ form.

$$3y = -6$$

$$\frac{3y}{3} = \frac{-6}{3}$$

$$y = -2$$

Because there is no x term, the coefficient of x must be 0. Therefore, the equation in slope-intercept form is $y = 0x - 2$. Because the slope is zero, there is no horizontal change between points on the line, and the line is flat. The y-intercept is –2, so the line passes through (0, –2). You can pick any other points as long as the second coordinate is –2. Choose (–3, –2) and 2, –2). Now, graph the line through those three points. (See Figure 15.32.)

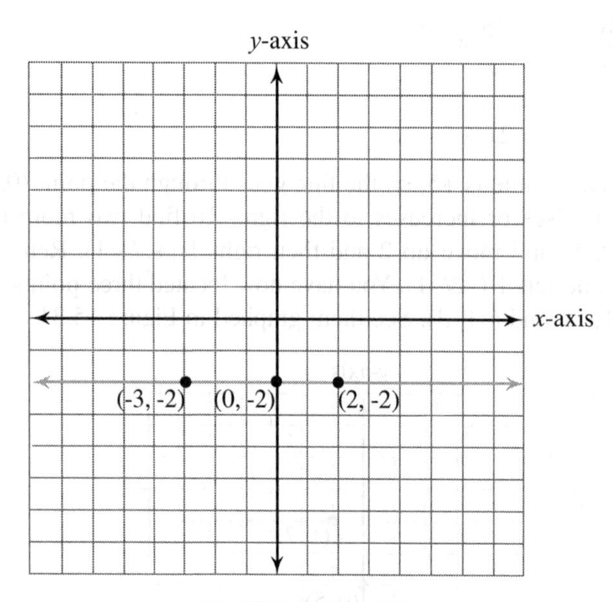

FIGURE 15.32

Check Substitute (–3, –2), (0, –2), and (2, –2) in the original equation.

$3y = -6$	$3y = -6$	$3y = -6$
$3y + 0x = -6$	$3y + 0x = -6$	$3y + 0x = -6$
$3(-2) + 0(-3) = -6$	$3(-2) + 0(0) = -6$	$3(-2) + 0(2) = -6$
$-6 + 0 = -6$	$-6 + 0 = -6$	$-6 + 0 = -6$
$-6 = -6$	$-6 = -6$	$-6 = -6$

EXAMPLE 13 Graph the solution set for $4x = 12$.

Solution When we try to write the equation in slope-intercept form, we notice there is no y term. Solve the equation for x.

$$4x = 12$$

$$\frac{4x}{4} = \frac{12}{4}$$

$$x = 3$$

Because we cannot write the equation in slope-intercept form, we know that there is no y-intercept, that the slope of the line is undefined, and that any point on the line has an x-coordinate of 3. We also know that a line with undefined slope is parallel to the y-axis. We can graph the line from this information. The line will pass through $(3, -4)$, $(3, 0)$, and $(3, 3)$. The line is graphed in Figure 15.33.

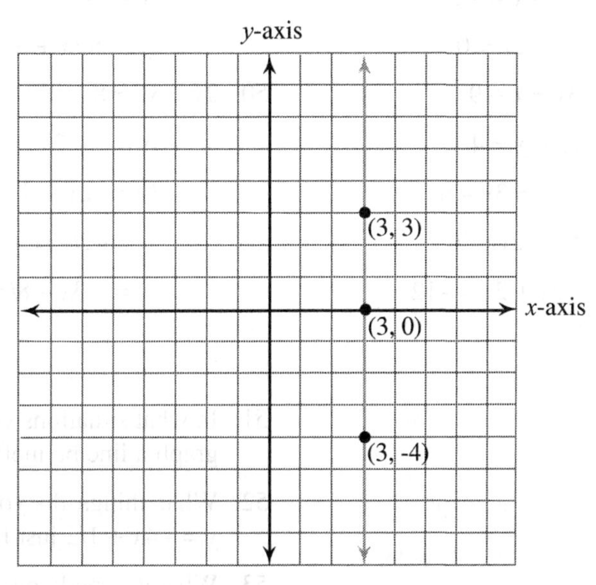

FIGURE 15.33

Check Substitute $(3, -4)$, $(3, 0)$, and $(3, 3)$ in the original equation.

$4x = 12$	$4x = 12$	$4x = 12$
$4x + 0y = 12$	$4x + 0y = 12$	$4x + 0y = 12$
$4(3) + 0(-4) = 12$	$4(3) + 0(0) = 12$	$4(3) + 0(3) = 12$
$12 + 0 = 12$	$12 + 0 = 12$	$12 + 0 = 12$
$12 = 12$	$12 = 12$	$12 = 12$

15.4 EXERCISES

A. Transform each of the following equations into slope-intercept form. Then identify the slope and y-intercept.

1. $x + y = -4$
2. $2x - 4 = y$
3. $-3x + 5 = y$
4. $x + y = 7$
5. $x - y = 3$
6. $6x = 3y + 12$
7. $2y = -x$
8. $y = 4x$
9. $2y = 8x - 4$
10. $x - y = 5$
11. $7x - 2y = 6$
12. $y + 4x = -2$
13. $-2x = 6y$
14. $3x - 2y = 0$
15. $x - 2y = 8$
16. $3y = x + 9$
17. $y + 2 = 3x$
18. $3x - y = 4$
19. $3y = x + 12$
20. $-2x = 3y - 6$
21. $3x = -4y + 6$
22. $5x + 2y = 7$
23. $2y - 3x = -4$
24. $-3y + 2x = -12$

B. For each of the following equations:

a. Change the equation into the $y = mx + b$ form.

b. Identify the slope and the y-intercept.

c. Graph the solution set for the equation by using the y-intercept and the slope.

25. $x + y = -2$
26. $3x + 2 = y$
27. $y = 2x$
28. $-3x = y$
29. $15 = -3y$
30. $2x = 12$
31. $x - y = 1$
32. $x - y = 5$
33. $3y + x = 0$
34. $2y - 4x = 0$
35. $-3x = -6$
36. $8 = -2y$
37. $3y = -9x - 6$
38. $8x = 4y + 4$
39. $3x - y = -1$
40. $2x - 3 = -y$
41. $x - 2y = 6$
42. $2y = x - 8$
43. $y + 4 = -2x$
44. $x - 2y = 3$
45. $3x - 4y = -4$
46. $3y + 6 = 4x$
47. $2y + x = -8$
48. $2x + 5y = -15$

C. In Exercises 49 and 50, describe, in words, the process used to go from one step to the next.

49. $6 = 2y - 3x$

$2y - 3x = 6$	1.
$2y = 6 + 3x$	2.
$2y = 3x + 6$	3.
$y = 3/2x + 3$	4.

50. $2x - 3y = 8$

$-3y = 8 - 2x$	1.
$-3y = -2x + 8$	2.
$y = -2/-3x + 8/-3$	3.
$y = 2/3x - 8/3$	4.

D. Read and respond to the following exercises.

51. In what situations would using only intercepts to graph a line be ineffective?
52. What things do you know about the graph of $y = -4x + 1/2$ just from studying the equation?
53. What do you know about the graph of the line if its equation is $y = -3x - 4$?
54. A friend is having a hard time understanding the idea of a zero coefficient. Explain why you would need to use a zero coefficient to write $7 = 3y$ in slope-intercept form.
55. Explain in your own words what "zero slope" means in terms of "rise over run." What does a line with zero slope look like?
56. Explain in your own words what "undefined slope" means in terms of "rise over run." What does a line with undefined slope look like?

15.5 Writing the Equation of a Line

Reading the Graph

In some situations it is helpful to be able to look at the graph of a line and work in reverse, so to speak, to find the equation of that line.

Consider the graph in Figure 15.34. The line passes through two points, (2, 2) and (0, –2). You know from looking at the graph that its y-intercept is (0, –2). You can calculate the slope of the line by using those points. To move from (0, –2) to (2, 2), you would go up four units (+4) and then over two units (+2). The rise over the run, the slope, is + 4/+2, or +2.

To write the equation of this line, you will use the slope-intercept form for the equation of a line, $y = mx + b$. You know that the slope is +2, so $m = +2$. You can see from the graph that the y-intercept is (0, –2), so $b = -2$. By substituting into $y = mx + b$, you can write the equation of the line $y = +2x - 2$. This is the equation of the line that passes through the points (2, 2) and (0, –2). Whenever you write the equation of a line, you should check to make sure that, in fact, both points do make the equation true.

$y = 2x - 2$	$y = 2x - 2$
Substitute (2, 2).	Substitute (0, –2).
$2 = 2(2) - 2$	$-2 = 2(0) - 2$
$2 = 4 - 2$	$-2 = 0 - 2$
$2 = 2$ True.	$-2 = -2$ True.

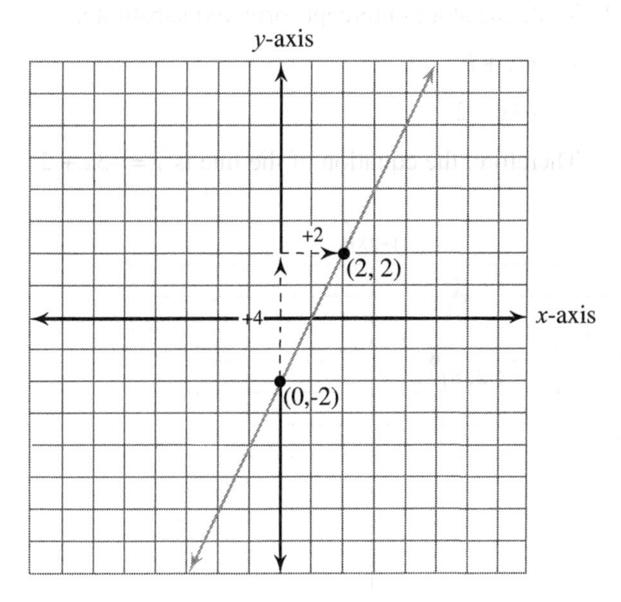

FIGURE 15.34

Slope and y-intercept

Whenever you are given the slope and y-intercept, it is quick to write the equation of the line directly by substituting into the general form, $y = mx + b$. Study Example 1.

EXAMPLE 1 Write the equation of a line whose slope is –5 and whose y-intercept is (0, 1).

Solution You know that the slope is –5 so $m = -5$ and the y-intercept is (0, 1) so $b = +1$. Substitute these values into the slope-intercept form for the equation of a line, $y = mx + b$.

$y = mx + b$

$y = -5x + 1$

Whenever you are given two points and are asked to write the equation, you can graph the line through the two points and see where the line cuts through the y-axis to identify the y-intercept. You can use the graph or the formula to determine the slope and then write the equation, as in Example 1. Study Example 2 to see how you might find the equation by using the graph.

EXAMPLE 2 Write the equation of the line that passes through the points (–2, 8) and (1, –1) by using its graph to determine the slope and y-intercept.

Solution

1. Sketch the graph (see Figure 15.35).
2. You can see that the y-intercept is (0, 2), so $b = +2$.
3. Calculate the slope.

   $m = \text{rise/run} = -9/+3 = -3$

4. Write the slope-intercept form and substitute.

   $y = mx + b$

   $y = -3x + 2$

   Therefore, the equation of the line is $y = -3x + 2$.

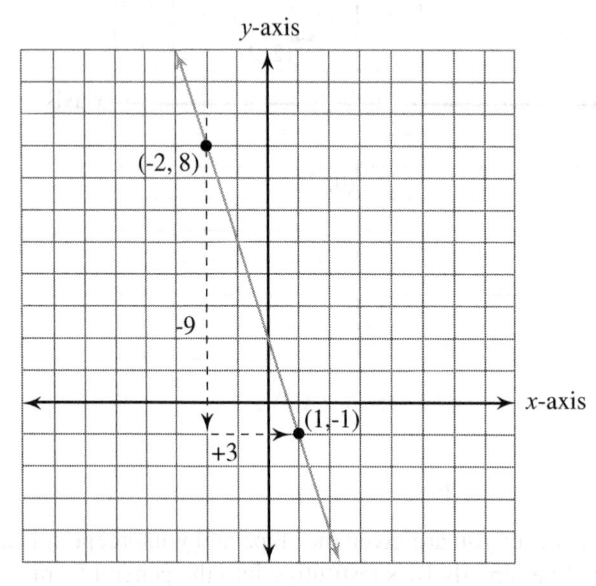

FIGURE 15.35

5. Check by substituting the x and y values for both points into the equation.

$y = -3x + 2$	$y = -3x + 2$
Substitute (–2, 8).	Substitute (1, –1).
$8 = -3(-2) + 2$	$-1 = -3(1) + 2$
$8 = 6 + 2$	$-1 = -3 + 2$
$8 = 8$ True.	$-1 = -1$ True.

Two Points

As you can see from these examples, it is very helpful to be given the y-intercept, as in Example 1, or to be able to read it directly from the graph, as in Example 2. When this is not possible because the y-intercept is not an integer that can readily be determined from the graph, you will need to use another general form of a linear equation. This *two-point formula* can be used when you are given any two points, but it *must* be used when the y-intercept (b value) is not readily available.

When you are given two points and are asked to write the equation of the line through those points, use this two-point formula:

Two-Point Formula

The equation of the line passing through points (x_1, y_1) and (x_2, y_2), is

$$y - y_1 = \left(\frac{y_2 - y_1}{x_2 - x_1}\right)(x - x_1)$$

or

$$y - y_1 = m(x - x_1)$$

Because the second form looks so much easier to use and to remember, it is generally a good idea to calculate the slope (the m value) first and then substitute it into $y - y_1 = m(x - x_1)$.

If you want to write the equation of the line through the points (–4, 6) and (2, 0), use $y - y_1 = m(x - x_1)$. First, find the value for m, letting (–4, 6) be (x_1, y_1) and letting (2, 0) be (x_2, y_2).

$$m = \frac{y_2 - y_1}{x_2 - x_1} = \frac{0 - 6}{2 - (-4)} = \frac{0 - 6}{2 + 4} = \frac{-6}{+6} = -1$$

Then $y - y_1 = m(x - x_1)$.	Substitute (–4, 6) for (x_1, y_1) and –1 for m.
$y - 6 = -1[x - (-4)]$	Use the subtraction rule.
$y - 6 = -1[x + 4]$	Distribute the –1.
$y - 6 = -1x - 4$	Solve for y.
$\underline{+6} \quad \underline{+6}$	
$y = -1x + 2$ or $y = -x + 2$	

Check both points in the equation to make sure that it is, in fact, the equation of the line that passes through the points (– 4, 6) and (2, 0).

$y = -x + 2$	$y = -x + 2$
Substitute (– 4, 6).	Substitute (2, 0).
$6 = -1(-4) + 2$	$0 = -1(2) + 2$
$6 = 4 + 2$	$0 = -2 + 2$
$6 = 6$ True.	$0 = 0$ True.

You would have arrived at the same equation if you had used the longer form. The only difference between the two is that here the slope is calculated separately first and then the simpler formula is used. Try it for yourself.

The only real problem that most students have in writing equations like these is in keeping the points straight. Do not use x_2 when you need x_1, and don't use y_1 when you need y_2! Be sure to label the points when you begin and to substitute carefully.

EXAMPLE 3 Write the equation of the line that passes through the points (–2, 5) and (0, –3).

Solution First calculate the slope. Let (–2, 5) be (x_1, y_1), and let (0, –3) be (x_2, y_2).

$$m = \frac{y_2 - y_1}{x_2 - x_1} = \frac{-3 - 5}{0 - (-2)} = \frac{-3 - 5}{0 + 2} = \frac{-8}{+2} = -4$$

Then $y - y_1 = m(x - x_1)$.

$$y - 5 = -4[x - (-2)]$$

$$y - 5 = -4[x + 2]$$

$$y - 5 = -4x - 8$$

$$\underline{+5} \qquad \underline{+5}$$

$$y \quad = -4x - 3$$

Check

$y = -4x - 3$	$y = -4x - 3$
Substitute (–2, 5).	Substitute (0, –3).
$5 = -4(-2) - 3$	$-3 = -4(0) - 3$
$5 = 8 - 3$	$-3 = 0 - 3$
$5 = 5$ True.	$-3 = -3$ True.

EXAMPLE 4 Write the equation of the line that passes through (3, 2) and (– 4, 0).

Solution First calculate the slope of the line. Let (x_1, y_1) be (3, 2) and (x_2, y_2) be (– 4, 0).

$$m = \frac{y_2 - y_1}{x_2 - x_1} = \frac{0 - 2}{-4 - 3} = \frac{-2}{-7} = \frac{2}{7}$$

Then $y - y_1 = m(x - x_1)$

$$y - 2 = 2/7(x - 3)$$

$$y - 2 = 2/7x - 6/7$$

The preferred form of the equation is one without fraction coefficients.

Multiply each term in the equation by the LCD, 7, to clear the fractions. This gives an equation with **integral coefficients** (coefficients that are integers).

$7(y) - 7(2) = 7(2/7\ x) - 7(6/7)$

$7y - 14 \quad = 2x - 6$

$\underline{\quad + 14} \qquad \underline{\quad + 14}$

$7y \quad = 2x + 8$

Check Check that both points do satisfy the equation.

$7y = 2x + 8$	$7y = 2x + 8$
Substitute (3, 2).	Substitute (– 4, 0).
$7(2) = 2(3) + 8$	$7(0) = 2(-4) + 8$
$14 = 6 + 8$	$0 = -8 + 8$
$14 = 14$	$0 = 0$

EXAMPLE 5 Write the equation of the line that passes through the points (2, 4) and (2, –3).

Solution Let (2, 4) be (x_1, y_1), and let (2, –3) be (x_2, y_2).

Then $m = \dfrac{y_2 - y_1}{x_2 - x_1} = \dfrac{-3 - 4}{2 - 2} = \dfrac{-7}{0}$, undefined.

The slope is undefined, so you know that the line is straight up and down, and because the line passes through (2, 4) and (2, –3), you know that x has the value 2. The first coordinate for every point on that line is 2. Therefore, the equation of the line is $x = 2$.

EXAMPLE 6 Write the equation of the line that passes through the points (–1, –3) and (4, –3).

Solution $m = \dfrac{y_2 - y_1}{x_2 - x_1} = \dfrac{-3 - (-3)}{4 - (-1)} = \dfrac{-3 + 3}{4 + 1} = \dfrac{0}{5} = 0$

The graph of a line with zero slope is horizontal (flat) and is of the form $y = b$. This line goes through two points whose y-coordinates are –3. Therefore, $y = -3$ is the equation of this line.

Check

$y = 0x - 3$	$y = 0x - 3$
Substitute (–1, –3).	Substitute (4, –3).
$-3 = 0(-1) - 3$	$-3 = 0(4) - 3$
$-3 = 0 - 3$	$-3 = 0 - 3$
$-3 = -3$ True.	$-3 = -3$ True.

Slope and One Point

If you are given the slope of a line and only one point, you can still write the equation of the line by using $y = mx + b$. You are given x, y, and m and are missing only b. Therefore, you might substitute the values you have for x, y, and m into $y = mx + b$ and solve the resulting equation for b. Then go back to the general form $y = mx + b$ and substitute the given m value and the b value that you found. Be sure to check the original point, (x, y), in the equation you have written.

EXAMPLE 7 Write the equation of the line that has a slope of –2 and passes through the point (3, –5). Use $y = mx + b$.

Solution Begin with the slope-intercept form of an equation, $y = mx + b$. You know that $m = -2$, and because the point you are given is (3, –5), you also know that $x = 3$ and $y = -5$. Substitute these values and solve to find b.

$y = mx + b$

$-5 = -2(3) + b$

$-5 = -6 + b$

$\underline{+6} \quad \underline{+6}$

$+1 = \quad b$

Go back to $y = mx + b$, and use $m = -2$ and $b = +1$ to write the equation.

$y = mx + b$

$y = -2x + 1$

Check To check to see that $y = -2x + 1$ is the correct equation, substitute (3, –5) and see if you get a true statement.

$y = -2x + 1$

Substitute (3, –5).

$-5 = -2(3) + 1$

$-5 = -6 + 1$

$-5 = -5$ True.

◄

If you are given the slope of a line and only one point, you may also use the simplified version of the Two-Point Formula. You will use $y - y_1 = m(x - x_1)$. It is called the Point-Slope Formula. Normally this approach is simpler than the one used in Example 7 with $y = mx + b$, but the resulting equation will be the same.

Point-Slope Formula

The equation of a line passing through point (x_1, y_1) with a slope of m is $y - y_1 = m(x - x_1)$.

Carefully substitute the x_1 and y_1 values and the slope and then simplify your equation so that it has only integral coefficients. Be sure to check your *original* point in the equation that you have written.

EXAMPLE 8 Write the equation of the line that has a slope of –2 and passes through the point (3, –5). Use $y - y_1 = m(x - x_1)$.

Solution Begin with the Point-Slope form of an equation, $y - y_1 = m(x - x_1)$. You know that $m = -2$, $x_1 = 3$ and $y_1 = -5$.

$y - y_1 = m(x - x_1)$	Make the substitutions.
$y - (-5) = -2(x - 3)$	Apply the subtraction rule.
$y + (+5) = -2(x - 3)$	Distribute the –2.
$y + 5 = -2x + 6$	Solve for y.
$\underline{-5 \qquad -5}$	
$y = -2x + 1$	

This is the same equation that we wrote in Example 7, given the same information, but using $y = mx + b$. You can see that the choice of which formula to use is up to you. The check for Example 8 will be identical to the one in Example 7.

EXAMPLE 9 Write the equation of the line that passes through the point (– 4, 7) and has a slope of –1/2. Use $y - y_1 = m(x - x_1)$.

Solution Begin with the Point-Slope form of an equation, $y - y_1 = m(x - x_1)$. You know that $m = -1/2$, $x_1 = -4$ and $y_1 = 7$.

$y - y_1 = m(x - x_1)$	Make the substitutions.
$y - 7 = -1/2[x - (-4)]$	Apply the subtraction rule.
$y - 7 = -1/2[x + 4]$	Distribute the –1/2.
$y - 7 = -1/2x - 2$	Solve for y.
$\underline{+7 \qquad +7}$	
$y = -1/2x + 5$	Multiply each term by the LCD, 2.

You should multiply through by the common denominator, 2, to clear the fraction and produce an equation with integral coefficients.

$$y = -\frac{1}{2}x + 5$$

$$(2)y = (2)(-1/2)x + (2)(5)$$

$$2y = -x + 10$$

Check Substitute (– 4, 7) to see if it satisfied the equation you have written.

$$2y = -x + 10$$

$$2(7) = -(-4) + 10$$

$$14 = 4 + 10$$

$$14 = 14 \quad \text{True.}$$

To summarize, riting the equation of a line when you are given the slope and the y-intercept is just a matter of substituting into the slope-intercept form of an equation, $y = mx + b$, or the Point-Slope form, $y - y_1 = m(x - x_1)$. Writing the equation from two points is easiest to do by first calculating the slope and then using $y - y_1 = m(x - x_1)$ and simplifying the resulting equation. In equations that have fractions, you should multiply through by the least common denominator to clear the fractions and therefore be able to produce an equation that has all integral coefficients.

Self-Check Write the equations of the lines given the following information. The equations should have only integral coefficients.

1. Passes through (–4, 5) with a slope of –2.
2. Passes through (3, –1) and (2, 3). Use $y = mx + b$.
3. Passes through (1, 3) with a slope of 2/3.

Answers:

1. $y = -2x - 3$
2. $y = -4x + 11$
3. $3y = 2x + 7$

15.5 EXERCISES

A. Write the equation of a line that has the given slope and y-intercept.

1. $m = 3, b = -2$
2. $m = -4, b = +1$
3. $m = 1/2, b = 3$
4. $m = 1/3, b = 2/3$
5. slope = 0, y-intercept = (0, 2)
6. $m = 2, b = -3$
7. $m = -5, b = 6$
8. slope = 0, y-intercept = (0, – 4)
9. slope = –3/4, y-intercept = (0, 2)
10. slope = –1, y-intercept = (0, –1)

B. Write the equation of the line that passes through the given points.

11. (0, 3) and (2, 6)
12. (–3, 4) and (3, –5)
13. (–5, 2) and (–2, 8)
14. (3, 0) and (–1, 4)
15. (3, 5) and (0, 0)
16. (4, 2) and (4, –3)
17. (6, 1) and (6, –2)
18. (–2, –3) and (0, 0)
19. (–2, 3) and (3, 3)
20. (5, –2) and (–1, –2)

C. Write the equation of the line that passes through the given point and has the given slope. Write your final equation with integral coefficients.

21. (6, –1); $m = 3$
22. (–1, –6); $m = 5$
23. (–2, –7); $m = -2$
24. (5, –3); $m = -4$
25. (–3, 0); $m = 1/2$
26. (–2, – 4); $m = 1/3$
27. (5, –2); $m = -1/2$
28. (2, 0); $m = -1/4$
29. (2, –1); $m = 3/4$
30. (3, –2); $m = -2/5$
31. (3, 5); $m = 0$
32. (–1, – 4); undefined slope
33. (2, –3); undefined slope
34. (–2, 6); m = 0

D. Read and respond to the following exercises.

35. Describe the process you use to write the equation of a line given the m and b values.
36. Describe the process you use to write the equation of a line given only two points on the line.
37. Describe the process you use to write the equation of a line given the slope and one point on the line.
38. Describe the process you use to write the equation of a line given that the slope is zero and $b = -6$.

15.6 Graphing Inequalities

You are now going to learn about the graphs of solutions to inequalities with two variables. These graphs will include not only the boundary line (the line that separates the points that satisfy the inequality from the points that don't) but also the shaded portion of the plane or graph that indicates all possible solutions for the inequality.

To find the graphical solution for a first-degree inequality, you follow all of the same steps you used to graph a line, and then add to that graph the inequality portion of the solution. The solution of an inequality may or may not include its boundary line, but it includes all the points on one side of that line.

Remember that when you were working with a number line (Section 12.5), you could graph individual points or whole sections of the number line. When you graphed an inequality, the end point of the graph was either an open circle or a closed (filled-in) circle, depending on whether the inequality included > or < (an open circle), or ≥ or ≤ (a closed circle). Study these simple graphs to refresh your memory.

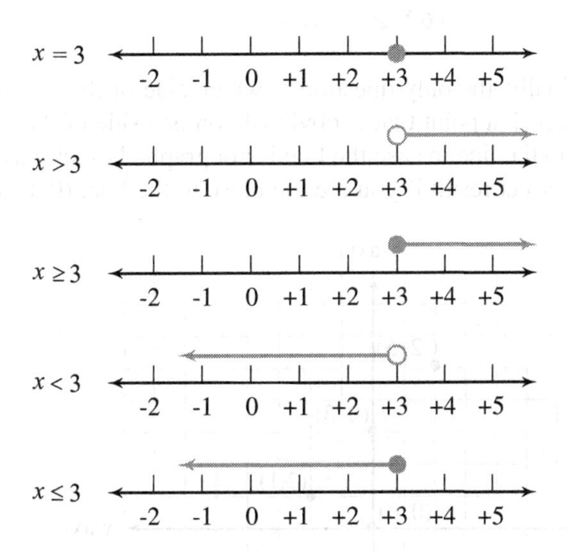

You want to graph the solution set for an inequality that involves two variables. You will have many more values than just the points on the boundary line, if those points are included at all. When graphing the solution to an inequality you will need to indicate a large portion of the graphing plane. Study the steps used in the following sample problem:

Graph the solution set for $x + y > 3$.

First you need to graph the boundary line. You will graph the equation $x + y = 3$. Isolate y, to write the equation in slope-intercept form, obtaining $y = -x + 3$.

The y-intercept is (0, +3) and the slope is –1. The graph of the line is shown in Figure 15.36. Notice that the boundary line is dashed. The inequality is $x + y > 3$, not $x + y \geq 3$. The boundary line will be dashed because the points that are on the line are *not* part of the solution set for $x + y > 3$. The inequality symbol is simply >, not ≥. The dashed line shows that the boundary line *is not* included in the solution set. (See Figure 15.36.)

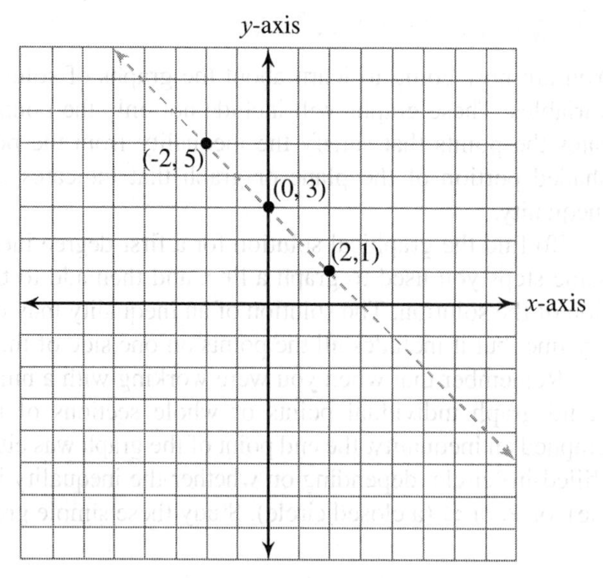

FIGURE 15.36

Finally, the only question is which side of the line to shade. Looking at Figure 15.37, pick a point that is obviously on one side of the boundary line. Don't get too close to the line in case the line is not graphed exactly as it should be. You might get an incorrect result if you are too close to the line. (0, 0) is a convenient choice.

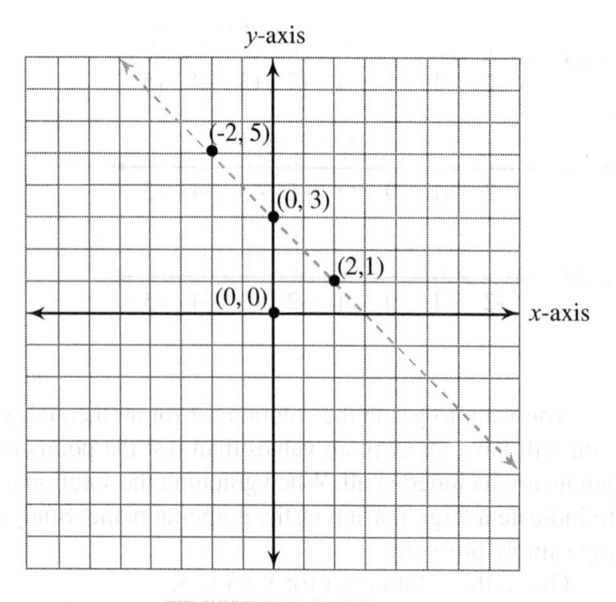

FIGURE 15.37

Substitute that point into the inequality and see if it makes a true statement. If the resulting inequality is true, shade the side of the boundary line that includes (0, 0). If the inequality is not true, shade the other side of the boundary line.

$x + y > 3$
$0 + 0 > 3$
$0 > 3$ False.

Therefore, (0, 0) *is not* part of the solution set, and that side of the line cannot be shaded. If you try the point (5, 2), which is clearly on the other side of the boundary line, the resulting inequality is true.

$x + y > 3$
$5 + 2 > 3$
$7 > 3$ True.

Therefore, you shade the side of the line that includes (5, 2), and the solution set looks like the graph in Figure 15.38.

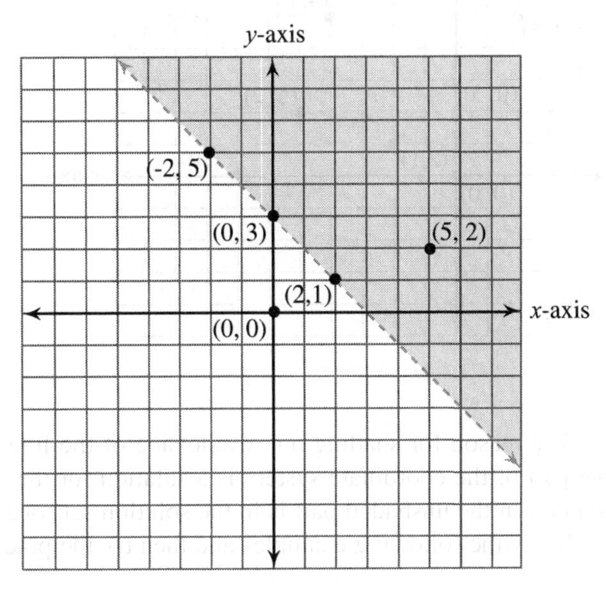

FIGURE 15.38

You could have picked any point on the "top side" of the line and it would have satisfied the inequality. The point (5, 2) was chosen at random. Pick (7, 0), (4, 5), and (10, 10), and you will see that they also satisfy the inequality.

$x + y > 3$	$x + y > 3$	$x + y > 3$
$7 + 0 > 3$	$4 + 5 > 3$	$10 + 10 > 3$
$7 > 3$ True.	$9 > 3$ True.	$20 > 3$ True.

Figure 15.39 shows the location of all four points, and you can see that all four are on the shaded part of the graph.

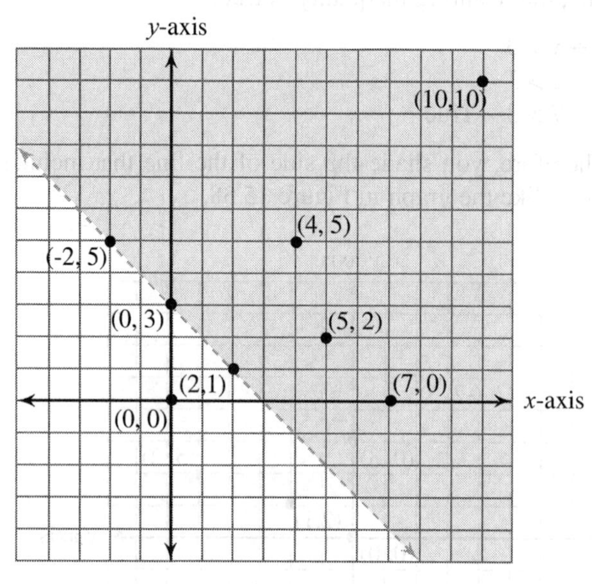

FIGURE 15.39

The reason for shading one whole side of the line is to show that *any point* in that part of the coordinate system is a solution for the inequality. It also shows that *no point* in the unshaded part is in the solution set for $x + y > 3$.

Study the following examples and then try the practice problems.

EXAMPLE 1 Graph the solution set for $x + 2y \leq 4$.

Solution First write the equation of the boundary line by writing the inequality as an equation in slope-intercept form.

$$x + 2y = 4$$

$$\underline{-x \qquad\quad -x}$$

$$2y = 4 - x$$

$$2y = -x + 4$$

$$\frac{2y}{2} = \frac{-1x}{2} + \frac{4}{2}$$

$$y = -\frac{1}{2x} + 2$$

Therefore, the equation of the boundary line is $y = -1/2\, x + 2$.

Graph the boundary line by using the slope and y-intercept. The boundary line will be solid because the points that are on the line are part of the solution set for $x + 2y \leq 4$. (The inequality is less than *or equal to.*) (See Figure 15.40.)

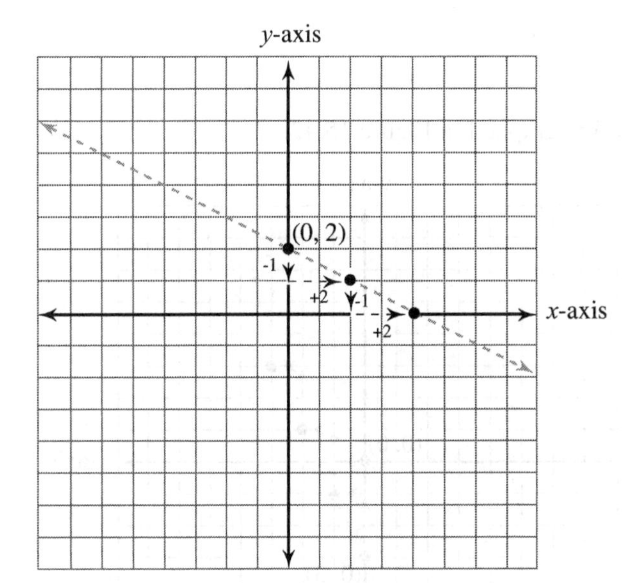

FIGURE 15.40

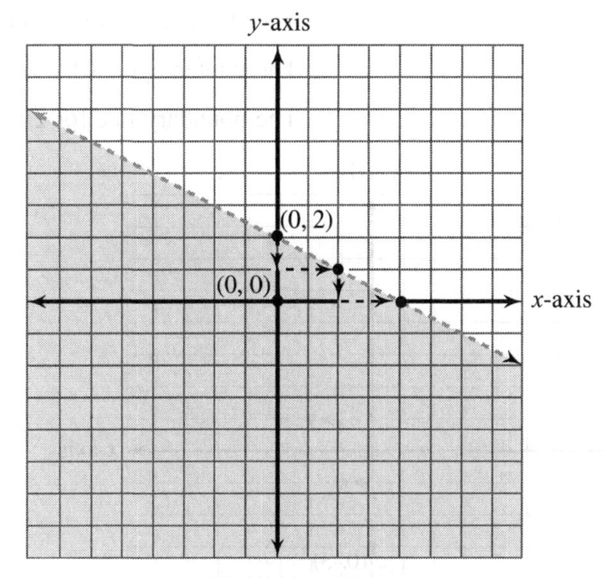

FIGURE 15.41

$b = 2$

$m = -1/2$ (down 1 / right 2)

Pick a point, say (0, 0), and see if it is part of the solution set for the inequality.

$$x + 2y \leq 4$$
$$0 + 2(0) \leq 4$$
$$0 + 0 \leq 4$$
$$0 \leq 4 \quad \text{True.}$$

Therefore, shade the side of the boundary line that includes (0, 0). The graph is shown in Figure 15.41.

EXAMPLE 2 Graph the solution set for the inequality $2x - y < 3$.

Solution First isolate the y in the inequality, and then write the equation for the boundary line in $y = mx + b$ form.

$$2x - y < 3$$
$$\underline{-2x} \qquad \underline{-2x}$$
$$-y < 3 - 2x$$
$$-y < -2x + 3$$
$$\frac{-1y}{-1} > \frac{-2x}{-1} + \frac{3}{-1}$$
$$y > 2x - 3$$

Remember to reverse the direction of the inequality when you multiply or divide by a negative value. (See Section 12.5.)

Therefore, the equation for the boundary line is $y = 2x - 3$. The boundary line will be dashed because the points on the line are not part of the solution set (the inequality is $2x - y < 3$, not $2x - y \leq 3$).

The slope is +2 (+2/+1, up 2 / right 1).

The y-intercept is (0, –3).

The boundary line for $2x - y < 3$ is graphed in Figure 15.42.

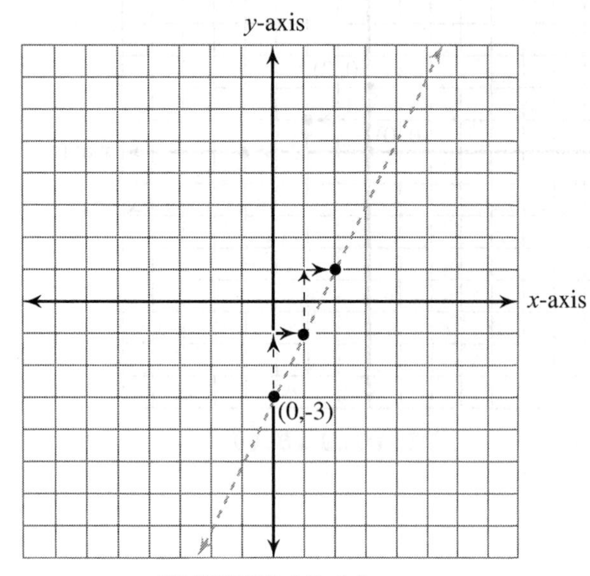

FIGURE 15.42

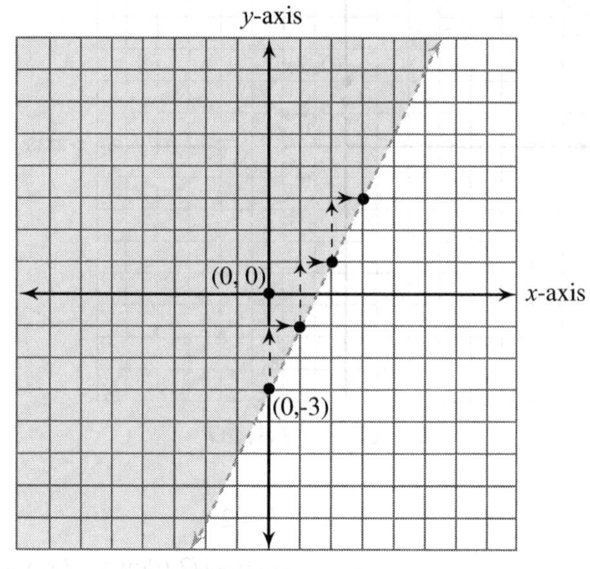

FIGURE 15.43

To determine the shading, try any point on the plane to see if it is part of the solution set. We show the point (0, 0).

$$2x - y < 3$$

$$2(0) - 0 < 3$$

$$0 - 0 < 3$$

$$0 < 3 \quad \text{True.}$$

Shade the portion of the graph that includes (0, 0). (See Figure 15.43.)

EXAMPLE 3 Graph the solution set for the inequality $y < x - 4$.

Solution Write the boundary line in $y = mx + b$ form.

$y = x - 4$

If you so choose, you can graph the boundary line by finding three points that lie on that line.

If $x = 2$, $y = 2 - 4$; therefore, $y = -2$ and the point is (2, –2).

If $x = 0$, $y = 0 - 4$; therefore, $y = -4$ and the point is (0, –4).

If $x = -2$, $y = -2 - 4$; therefore, $y = -6$ and the point is (–2, –6).

The boundary line will be dashed. Because the inequality is $y < x - 4$, not $y \leq x - 4$, the line is not included in the solution set.

Check a point such as (0, 0) to see which side of the line should be shaded.

$y < x - 4$

$0 < 0 - 4$

$0 < -4$ False.

Because (0, 0) does not satisfy the inequality, shade the other side of the boundary line. (See Figure 15.44.)

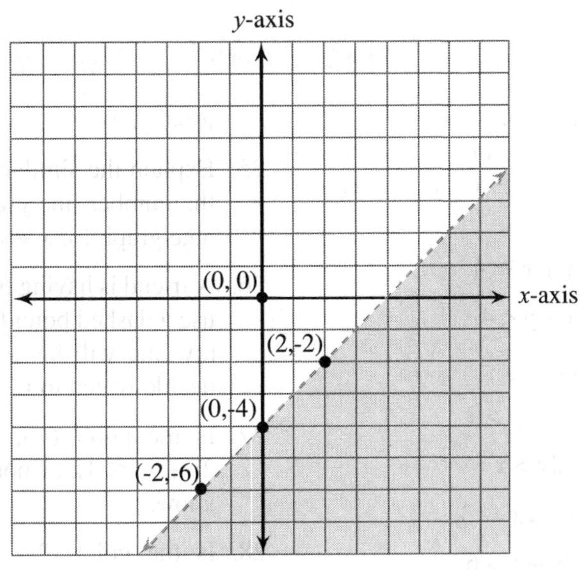

FIGURE 15.44

EXAMPLE 4 Graph the solution set for $x \leq 5$.

Solution There is no y value, so the boundary line is made up of all points whose first coordinate, the x-coordinate, is 5. Therefore, $x = 5$ is the equation for the boundary line. The line will be solid because the inequality states that x is less than *or equal to* 5. When we check the point (0, 0), we find that the inequality is true because $0 \leq 5$. The graph is shown in Figure 15.45.

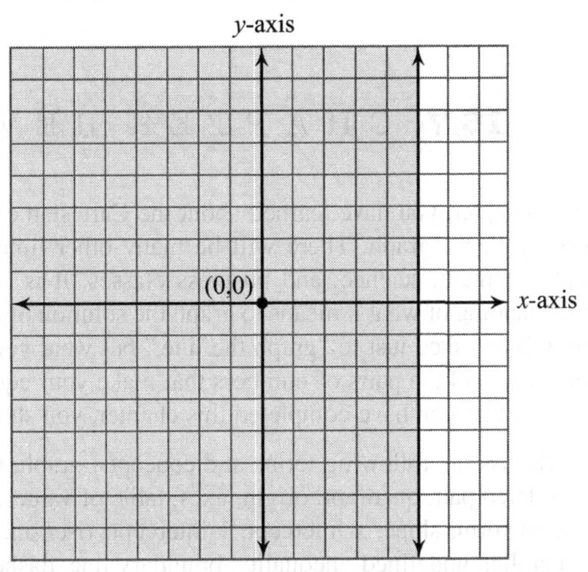

FIGURE 15.45

As you can see from these examples, there are only two new things you need to be careful about when graphing an inequality. Is the boundary line included (solid) or not included (dashed)? And, which side of the boundary line shows the solution set? To answer this question, you pick a point that is obviously on one side of the line, and if that point satisfies the inequality, you shade that side of the boundary line. If the point does not satisfy the inequality, you shade the other side of the boundary line.

15.6 EXERCISES

A. Graph the solution set for each of the following first-degree inequalities. Make sure that your graphs are neatly done and that the boundary line is clearly marked.

1. $x + y < 5$
2. $x + y \geq 4$
3. $x - y \geq 3$
4. $x - y < 4$
5. $2y > x$
6. $3x \leq y$
7. $2x + y > 3$
8. $3x + y > 1$
9. $x \leq -3y$
10. $-2y > x$
11. $y - x \leq 4$
12. $y - 2x \leq 3$
13. $y > -2$
14. $2y + x > 6$
15. $2y - x < 4$
16. $x < -2$
17. $3x - 2y \geq 12$
18. $2x + 3y \geq 6$
19. $x \leq 3$
20. $y \geq -3$
21. $-2y < x + 4$
22. $-y < 2x + 3$
23. $x \geq y - 5$
24. $y - 4 \leq 3x$

B. Read and respond to the following exercises.

25. Explain the similarities and differences between the number line graph for $x < 2$ and the coordinate graph for $x + y \leq 2$.
26. A friend is having trouble understanding when to use a dashed boundary line and when the boundary line will be solid. Explain to her how you decide which to use.
27. Is the point $(-3, 4)$ part of the solution for $2y + x \leq 12$ or not? Explain in words how you know.
28. Is the point $(2, -5)$ part of the solution for $x - 2y < 12$ or not? Explain, in words, how you know.
29. When you are deciding which side of the boundary line to shade, why are you cautioned not to pick a point too close to the line?
30. When you are deciding which side of the boundary line to shade, explain why it doesn't matter which side of the line you choose.

15.7 CHAPTER REVIEW

In this chapter, you have learned about the Cartesian coordinate system, more commonly called a graph. There will be many other times when you will work with graphs in math, science, and business classes. It is crucial that you have a clear understanding of what it means to graph the solution of a linear equation. Most often, you will be asked just to "graph the line," but what you are really doing is drawing a picture of all the pairs of numbers that make your equation or inequality true.

Now that you have completed this chapter, you should be able to:

1. Discuss the following terms and concepts: graph, Cartesian coordinate system, ordered pair, quadrant, origin, axes, table of values, linear equation, slope-intercept form, slope, x-intercept, y-intercept, rise/run, slope = 0, undefined slope, parallel, undefined, inequality, boundary line, dashed or solid line, and shading.

2. Plot ordered pairs of numbers on a graph.
3. Isolate y in order to write any equation or inequality in slope-intercept form.
4. Graph the solution set of a linear equation by following one of these procedures:
   a. Pick three x values and solve for the corresponding y values; then graph the three points and the straight line that passes through them.
   b. Identify both intercepts and draw the line that passes through them.
5. Find the slope from a graph by using rise/run.
6. Find the slope of a line that passes through two given points by using the formula

$$m = \frac{y_2 - y_1}{x_2 - x_1}$$

7. Graph the solution of a linear equation using the y-intercept and the slope by following these steps:
   a. Write the equation in $y = mx + b$ form.
   b. Identify the slope and y-intercept.
   c. Begin with the y-intercept and use the ratio rise/run to find two more points on the line.
   d. Check that the three points satisfy the *original* equation.
8. Write the equation of a line given the slope and y-intercept by substituting into $y = mx + b$.
9. Write the equation of a line given two points by first calculating the slope and then substituting into $y - y_1 = m(x - x_1)$.
10. Write the equation of a line given one point and the slope by substituting into $y = mx + b$ and solving for b. Then substitute again into $y = mx + b$, using the given slope and the newly found b value.
11. Write the equation of a line given one point and the slope by substituting into the Point-Slope form of the line, $y - y_1 = m(x - x_1)$.
12. Write the final form of any equation using only integral coefficients.
13. Graph the solution of a first-degree inequality by following these steps:
    a. Isolate y in the inequality.
    b. Write the equation of the boundary line and determining whether the line is solid (included, $\geq$ or $\leq$) or dashed (not included, $>$ or $<$).
    c. Graph the boundary line.
    d. Pick a point on one side of the boundary line and substitute its coordinates into the inequality to determine which side of the boundary line is to be shaded.

REVIEW EXERCISES

A. Give the location (which quadrant or which axis) of each of the following points.

1. (–3, –1)
2. (5, –2)
3. (0, –4)
4. (–2, 4)
5. (3, –8)
6. (7, 0)
7. (2, 6)
8. (–7, –2)
9. (–3, 0)
10. (0, –11)
11. (–1, 8)
12. (5, 4)

B. In Exercises 13 through 22, graph the line that is the solution for each equation. Be sure to label at least three points on the line.

13. $2y = -6x - 4$
14. $5 - y = 3x$
15. $2x - y = 5$
16. $y + 4 = -x$
17. $x - y = 3$
18. $3y + 9 = 6x$
19. $6 = x$
20. $y = -1$
21. $2y + 3x = 6$
22. $3y + 2x = 6$

C. Find the slope of the line that passes through each pair of points.

23. (4, 5) and (2, 3)
24. (5, 4) and (2, 3)
25. (1, 4) and (3, 5)
26. (3, 2) and (6, 7)
27. (4, 5) and (4, 8)
28. (6, 3) and (6, 7)
29. (–1, –3) and (5, 2)
30. (–2, 4) and (5, –3)

D. Write the equation of a line given the following information. Write the final equations with integral coefficients.

31. slope = –2, y-intercept = (0, 3)
32. passes through (2, 6) and (–3, 2)
33. passes through (2, 5); $m = -3$
34. passes through (–3, 4); $m = -2$
35. passes through (4, 5) and (0, –3)
36. $m = -4$, $b = -2$
37. passes through (1, 3) and (2, 0)
38. passes through (1, 4) and (–2, –5)
39. $m = 2/3$, $b = -1$
40. slope = 3, y-intercept = (0, 2)
41. passes through (–1, –4); $m = -1/2$
42. passes through (5, 6); $m = 1$
43. $m = 0$, $b = -2$
44. $m = -1/4$, $b = 3$
45. passes through (7, 0); $m = 3$
46. passes through (–2, –3); $m = 1/4$
47. slope = –1/2, y-intercept = (0, 3/4)
48. slope = 0, y-intercept = (0, –2/3)
49. passes through (5, 1) and (–2, –3)
50. passes through (3, –1) and (–2, 2)
51. passes through (3, 5); $m = 2/3$
52. passes through (6, 2); $m = 1/3$
53. passes through (–2, 3) and (0, –1)
54. passes through (–1, 2) and (3, 0)
55. passes through (–4, 1); $m = -2/3$
56. passes through (–3, 2); $m = -1/2$
57. $m = -3$, $b = 1/2$
58. $m = -1$, $b = 3/4$
59. passes through (3, –2) and (–4, 3)
60. passes through (2, 2) and (–5, –5)

E. Graph the solution set for each of the following inequalities.

61. $x + y > 6$
62. $x + y \leq 3$
63. $2x - y \leq 3$
64. $-3x + y > 4$
65. $x - 2y < 4$
66. $x + 2y < 6$
67. $3x + 3y \geq 3$
68. $3y - 2x \leq 6$
69. $-2y + x < -4$
70. $2x - y < 5$

F. Each statement that follows has an error. State what the error is; then work the problem and correct the statement.

71. The slope-intercept form of the equation of $2x - 3y = 12$ is $y = -2/3x - 4$.

72. The slope of the line that passes through (–3, 5) and (2, 5) is undefined.
73. The slope of the line that passes through (–2, 3) and (4, 1) is +1/3.
74. The boundary line for the inequality $x + 3y > 6$ is a solid line.
75. The slope of the line through (–1, 0) and 5, –4) is –3/2.
76. To write the equation of the line with a slope of –1/2 that passes through (4, 6) start with $y - 4 = -1/2(x - 6)$.

G. Read and respond to the following exercises.

77. Which method do you prefer to use when graphing the solution to a linear equation? Why do you prefer it?
78. What hints would you give to a classmate who is having trouble remembering how to calculate the slope of a line using the two-point formula?
79. Write your own definition for an intercept. Can you think of another word or situation that could help you remember what an intercept is?
80. Outline the steps you would follow to graph the solution for $2x - 5y > 10$.

CHAPTER TEST

Complete the following sentences.

1. Points of the form (0, y) lie on the _____ axis.
2. The point (–3, 2) is in Quadrant _____.
3. The point (0, 0) is called the _____.
4. What is the slope of the line that is the graph of $2y + x = 5$?
5. Identify the y-intercept from the equation $-3y + 2x = 6$.
6. Write the equation $2x - y = 4$ in slope-intercept form.
7. Find the slope of the line that passes through the points (–2, 5) and (3, –7).
8. Find the slope of the line that passes through the points (–5, 7) and (–5, –4).

Write the equation for a line given the following information. Write the final equations with integral coefficients.

9. slope = –2, y-intercept = (0, 4)
10. slope = 2/3, y-intercept = (0, –3)
11. passes through (–2, 4) and (3, –1)
12. passes through (–3, –2) and (4, –2)
13. passes through (2, –2) and has a slope of –3

Graph the solution set for each of the following equations and inequalities.

14. $2x + y = 5$
15. $-3y = x + 6$
16. $x - y = 4$
17. $-2y + x \leq 4$
18. $2y > 6$
19. $x \geq y - 1$
20. Describe in words the steps used to write the equation of the line passing through (4, 0) and (4, –2).

Cumulative Review, Chapters 1–15

A. Answer each of the following questions. Reduce all fraction answers to lowest terms. In Exercises 1 through 5, add.

1. $1\ 3/8 + 2\ 5/9$
2. $3a + 4a + a + 11a$
3. $(7y^2 - 2y + 3) + (8y - 2y^2 + 5)$
4. $(-69) + (14) + (-8)$
5. $7.6 + 0.09 + 44$

B. In Exercises 6 through 10, subtract.

6. $9 - [6 - 3(3 - 5)]$
7. $0 - (-12) - (3)$
8. $(5a - 2ab + 3b) - (12a + 5ab + 6b)$
9. $11xyz - (-3xyz)$
10. $\frac{5}{9} - \frac{1}{4}$

C. In Exercises 11 through 15, multiply.

11. Find 3/8 of 19.
12. $(-6)(-3)(2)(-1)$
13. $3x^2y(2x^2y^2z)$
14. 0.654×100
15. $(a + 2b)^2$

D. In Exercises 16 through 20, divide.

16. $\frac{4a^2bc^3}{-2abc}$
17. $4\ 7/8 \div 13$
18. $0.06\overline{)44.04}$
19. $\frac{15m^2n^2 - 10m^2n + 5mn^2}{-5mn}$
20. $x + 2\overline{)x^2 - 4x - 12}$

E. In Exercises 21 through 30, answer each question.

21. What is the slope of a line that passes through $(-1, 4)$ and $(-5, -6)$?
22. Change 21% to a decimal.
23. Identify the slope and y-intercept of $7y - 3 = 4x$.
24. Reduce 48/90 to lowest terms.
25. Factor completely: $4x^2 - 20xy + 25y^2$.
26. Perform the indicated operations and express the answer scientific notation: $\frac{0.006 \times 51.6}{0.008}$
27. Evaluate $-4a^2 + 2a - 3$ when $a = -2$.

F. In Exercises 28 through 30, use the following figure.

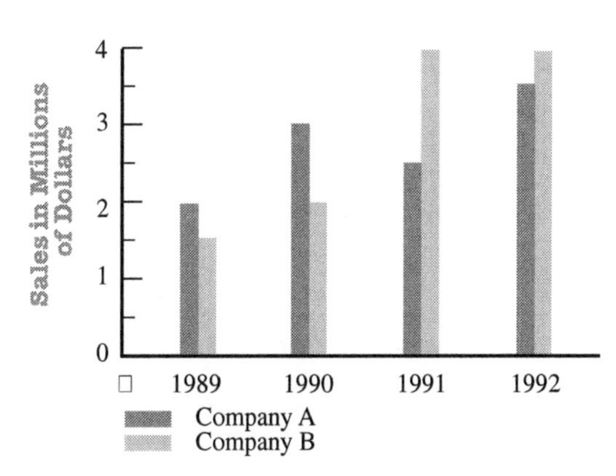

28. Between which two consecutive years did Company B have its greatest increase in sales?

29. How much more sales did Company A have than Company B in 1990?

30. What was the average amount of sales for Company B during the 4-year period?

G. In Exercises 31 through 41, find the answer to each statement. If necessary, round answers to the nearest tenth.

31. $4x - 3 = 15$

32. $1 \le 3x + 4 \le 7$

33. 15% of what number is 16?

34. $x^2 + 6x + 8 = 0$

35. $5(y + 2) = 3y - 4$

36. $\frac{8}{N} = \frac{45}{10}$

37. Solve $Y = a + b + c$ for c.

38. 15 is what percent of 12?

39. $3(2x - 1) \ge 9$

40. $5T - 5(6 - T) = -12$

41. $7Y - \frac{2}{3}(Y - 5) = 2Y - 1$

42. Solve $A = 1/2\ h(b_1 + b_2)$ for h if $b_1 = 9$ feet, $b_2 = 3$ feet, and $A = 42$ square feet.

H. Solve each of the following exercises. You may find it helpful to write an equation or a proportion or to use a formula. When necessary, round answers to the nearest tenth.

43. Six less than twice a number is equal to the product of the number and eight. Find the number.

44. Hani has a total of $2.20 in his pocket, all nickels and quarters. If he has twelve coins, how many are quarters?

45. The daily rainfall during a certain week in Columbus, Ohio, was 0.7 inch on Monday, 0.2 inch on Tuesday, 1.2 inches on Wednesday, 0.9 inch on Thursday, and 1.3 inches on Saturday. On Friday and Sunday, it did not rain. What was the average daily rainfall for the seven days of that week?

46. Using the information from Exercise 45, give the median and range for rainfall during the 7-day period.

47. It took Greg two hours to drive from his home to campus during the morning rush hour. It took him only one hour to go the same distance in the afternoon. If he averaged 20 miles per hour faster on the afternoon trip, how fast did he drive in the morning?

48. Forty-eight percent of the students attending this college are male. If there are 7,872 male students, what is the total enrollment for this term?

49. If 8 more than 5 times an integer is less than 38, what is the maximum value of the integer?

50. The perimeter of a rectangular field is 224 meters. The length is 8 meters less than 3 times the width. Find the dimensions of the field.

APPENDIX

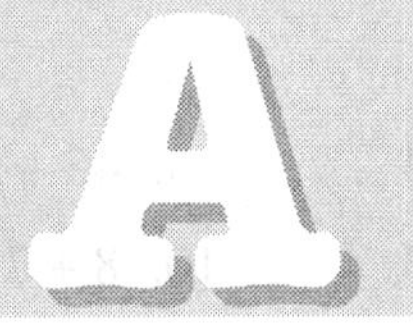

Skills Checks

Overview

There are four parts to Appendix A. Turn to the one that is appropriate to your needs, and use the Skills Checks and materials as directed.

1. The section entitled "Basic Facts" provides a brief test of your knowledge of addition, subtraction, multiplication, and division facts. Activities are given to help you overcome any diagnosed weakness(es).
2. "Addition Facts" contains a chart and activities to help you learn the addition and subtraction facts.
3. "Multiplication Facts" contains a chart and activities to use in memorizing the multiplication and division facts.
4. "Fraction and Decimals" allows you to test your basic abilities in working with these kinds of numbers. References to the textbook sections where these topics are covered enable you to review any topic with which you have trouble.

Basic Facts

Much of the math that you will do in this course is an extension or a review of work that began in elementary school and continued through later grades into high school. This is one of the few subjects that has such a long, ongoing development in your education and builds the new material on the material you have previously learned and used. If you have had difficulty with mathematics, it may be that you never learned early concepts or principles that were essential for success with later topics.

If that is the case, it is important that you stop right now, identify those weaknesses, and take steps to correct them. If you master the basics now, you will be able to move on and succeed with future math courses that are necessary to your educational or career goals.

Many difficulties encountered with arithmetic can be attributed to weakness in the most fundamental step: operations with whole numbers. Many people never learned (memorized) the addition or multiplication facts. Test yourself right now to see if that might be a problem for you.

Find a quiet place where you can work uninterrupted for 3 minutes, and have a watch or clock available. Then do the Arithmetic Facts Check, which follows. Allow yourself only 3 minutes. Go to the test now, cover up the answers, and write your answers on a sheet of paper. Stop working at the end of 3 minutes, whether you are finished or not, and check your answers.

Arithmetic Facts Check

1. $2 + 3$	2. $7 - 4$	3. 6×9	4. $12 \div 4$	5. $45 \div 5$
6. 3×0	7. 7×3	8. $6 + 2$	9. $27 \div 3$	10. $8 + 7$
11. 4×6	12. $14 - 8$	13. $2 + 0$	14. 5×1	15. $72 \div 8$
16. $9 + 9$	17. 4×9	18. $32 \div 8$	19. $13 - 5$	20. 6×3
21. $49 \div 7$	22. $13 - 8$	23. $21 \div 3$	24. 0×5	25. $6 + 7$
26. $42 \div 7$	27. $12 - 3$	28. 2×7	29. $9 + 1$	30. $17 - 9$
31. 5×6	32. 8×3	33. $24 \div 6$	34. $24 \div 8$	35. $12 - 7$
36. $4 - 0$	37. $0 \div 4$	38. $48 \div 8$	39. $3 + 8$	40. $15 - 7$

Answers

1. 5	2. 3	3. 54	4. 3	5. 9
6. 0	7. 21	8. 8	9. 9	10. 15
11. 24	12. 6	13. 2	14. 5	15. 9
16. 18	17. 36	18. 4	19. 8	20. 18
21. 7	22. 5	23. 7	24. 0	25. 13
26. 6	27. 9	28. 14	29. 10	30. 8
31. 30	32. 24	33. 4	34. 3	35. 5
36. 4	37. 0	38. 6	39. 11	40. 8

After checking your answers, answer the following questions.

1. How many problems didn't you have time to do? _____
2. How many errors did you make in addition? _____
3. How many errors did you make in subtraction? _____
4. How many errors did you make in multiplication? _____
5. How many errors did you make in division? _____

If you had fewer than five problems not finished, and had very few errors, then you should be all right in this course. You know your basic facts, and the practice you get doing homework will help you to increase your speed.

If you had more than five problems not completed, then you need to study all the facts so that you really memorize them and don't have to calculate each time, as must be happening now.

If you missed more than three problems in addition and subtraction, then you need to study the addition facts that follow, using the suggestions for learning them.

If you missed more than three problems in multiplication and division, then you need to study and learn the multiplication facts that follow.

After you complete the practice indicated, retake the Arithmetic Facts Check and see what progress you have made. When you can make fewer than three errors in the 3 minutes allotted, you will know the facts very well. If you apply the suggested study methods and still show little progress, then discuss the problem with your instructor.

Addition Facts

If your weakness is in adding and subtracting, work with Table A.1. To use the fact table to find the answer to an addition problem such as 8 + 5, follow the 8 row until you are under the 5 column. The number that appears where that row and column intersect is the answer to the problem: 8 + 5 = 13.

To use the table to practice subtraction, start at a number inside the chart, and trace back and up to the row and column headings. If you start at 11 and trace back and up to the headings 7 and 4, then you know that 11 – 7 = 4 and 11 – 4 = 7.

Besides the facts in the table, remember that zero added to any number does not change the number: 0 + 5 = 5. Similarly, when zero is subtracted from a number, the number does not change: 8 – 0 = 8.

TABLE A.1 Addition Facts

	1	2	3	4	5	6	7	8	9
1	2	3	4	5	6	7	8	9	10
2	3	4	5	6	7	8	9	10	11
3	4	5	6	7	8	9	10	11	12
4	5	6	7	8	9	10	11	12	13
5	6	7	8	9	10	11	12	13	14
6	7	8	9	10	11	12	13	14	15
7	8	9	10	11	12	13	14	15	16
8	9	10	11	12	13	14	15	16	17
9	10	11	12	13	14	15	16	17	18

Here are some other ways of learning the basic addition and subtraction facts.

1. Make flash cards with the "fact" (such as 5 + 3 =) on one side and the answer (8) on the other. Writing the facts with a bright-colored marker may help you "see" them in your memory. Drill with the facts, separating the ones you get correct from the ones you miss. Then work with the incorrect ones until you know them.
2. Punch a fact into a calculator, but guess the answer before the answer is displayed. Check your guess against the display. Record the ones you get wrong so that you can work on them.
3. Record an audiotape, saying the fact and the answer. Listen to the tape during spare moments.
4. Write the facts several times, saying each one to yourself as you write it.
5. If you want some more strategies, see your instructor.

Multiplication Facts

If one of your weaknesses is in multiplying or dividing, use Table A.2 to work on these problems. To use the fact table to find the answer to a multiplication problem such as 8 × 5, follow the 8 row until you are under the 5 column. The number that appears where that row and column intersect is the answer to the problem: 8 × 5 = 40.

To use the table to practice division, start at a number inside the chart, and trace back and up to the row and column headings. If you start at 28 and trace back and up to the headings 7 and 4, then you know that 28 ÷ 7 = 4 and 28 ÷ 4 = 7.

TABLE A.2 Multiplication Facts

	1	2	3	4	5	6	7	8	9
1	1	2	3	4	5	6	7	8	9
2	2	4	6	8	10	12	14	16	18
3	3	6	9	12	15	18	21	24	27
4	4	8	12	16	20	24	28	32	36
5	5	10	15	20	25	30	35	40	45
6	6	12	18	24	30	36	42	48	54
7	7	14	21	28	35	42	49	56	63
8	8	16	24	32	40	48	56	64	72
9	9	18	27	36	45	54	63	72	81

In addition to these facts, you need to learn that zero times any number is equal to zero: 7 × 0 = 0. Also, any number divided *into* zero is equal to zero: 0 ÷ 4 = 0.

But if any number is divided *by* zero, there is no answer; division by zero is said to be *undefined.*

Here are some other ways of learning the basic multiplication and division facts.

1. Make flash cards with the "fact" (such as 7 × 9 =) on one side and the answer (63) on the other. Drill with the facts, separating the ones you get correct from the ones you miss. Then work with the incorrect ones until you know them.
2. Punch a fact into a calculator, but guess the answer before the solution is displayed. Check your guess against the display. Record the ones you get wrong so that you can work on them.
3. Record an audiotape, saying the fact and the answer. Listen to the tape during spare moments.
4. Write the facts several times, saying each to yourself as you write it.
5. If you want some more strategies, see your instructor.

Fractions and Decimals

If you want to "diagnose" weakness in fraction and/or decimal operations, then try the Skills Checks, which follow. These are not timed tests, but you should find a quiet spot to work so that you can concentrate. When you have finished, check your answers and follow the recommendations given.

Fraction Skills Check

Reduce all answers to lowest terms.

1. $\frac{3}{8}+\frac{5}{6}$
2. $3\frac{11}{12}+2\frac{3}{4}$
3. $\frac{7}{9}-\frac{2}{3}$
4. $5\frac{1}{8}-2\frac{3}{4}$
5. $13-4\frac{5}{8}$
6. $\frac{3}{4}\times\frac{8}{15}$
7. $1\frac{3}{5}\times 2\frac{1}{2}$
8. $\frac{5}{8}\div\frac{4}{5}$
9. $\frac{4}{9}\div 8$
10. $\dfrac{\frac{3}{4}}{\frac{2}{3}}$

Check your answers with those that follow. If you made an error, go back to the section in the book that is indicated beside the correct answer. Study that material and try some exercises. When you can do those correctly, go on to other sections that you need to study.

Answers

1. $\frac{29}{24}$ or $1\frac{5}{24}$	2. $5\frac{20}{12}=6\frac{2}{3}$		Sections 4.1 and 4.2
3. $\frac{1}{9}$	4. $2\frac{3}{8}$	5. $8\frac{3}{8}$	Sections 4.1 and 4.2
6. $\frac{24}{60}=\frac{2}{5}$	7. $\frac{40}{10}=4$		Section 3.3
8. $\frac{25}{32}$	9. $\frac{1}{18}$	10. $\frac{9}{8}$ or $1\frac{1}{8}$	Section 3.4

If answers are correct but not reduced, then study Section 3.2.

Decimal Skills Check

1. $0.6 + 2.14$
2. $32.657 + 2.43$
3. $12.009 - 3.597$
4. $9 - 2.48$
5. 9.46×0.38
6. 0.493×1000
7. $3.68 \div 0.04$
8. $\frac{12}{0.8}$
9. $\frac{10.65}{6}$
10. Round 5.693 to the nearest hundredth.

Check your answers with those that follow. If you made an error, go back to the section in the book that is indicated beside the correct answer. Study that material and try some exercises. When you can do those correctly, go on to other sections that you need to study.

Answers

1. 2.74	2. 35.087		Section 5.2
3. 8.412	4. 6.52		Section 5.2
5. 3.5948	6. 493		Section 5.3
7. 92	8. 15	9. 1.775	Section 5.4
10. 5.69			Section 5.1

You may find that you need to go back periodically to these sections to review the facts. That's a good thing to do, because getting a lot of practice is the best way to learn math.

APPENDIX

B Calculator Operations

Introduction

Math educators have many different policies and theories about the use of calculators in basic mathematics courses. For that reason, we have not included any calculator instruction within the regular text. This appendix, however, should help you learn how to use your calculator more efficiently and more effectively.

This material is not written with reference to a particular brand or model, but is directed at scientific calculators that have fraction functions. Most of these calculators cost under $15 and can be found in college bookstores, electronic stores, and discount stores.

On/Off

Battery-powered calculators have keys that allow you to turn the power on and off. Find the key labeled ON or AC and turn your calculator on. Solar-powered calculators will come on when you open the case and expose them to light. Press the AC or ON key on these calculators to clear. Battery calculators will have an OFF key to allow you to turn off the power and conserve the battery, though most of these models automatically shut off after several minutes of nonuse.

+, −, ×, ÷, =

Locate the four basic function keys and input a simple problem like 3 × 4. After you enter the 4, is the answer showing in the display window? The answer to this calculation won't appear until you hit the = key, because the calculator doesn't know that you have keyed in all the numbers and operations until you tell it that you want an answer. The = key requests the answer.

The order that you input numbers for addition or multiplication exercises will not matter, but the order is crucial for subtraction and division. Be sure that the *number being subtracted* or the *number doing the dividing* is entered *after* the operation symbol. Practice inputting these problems, paying close attention to how they are written, and then to how they are entered into the calculator to get the correct answer of 754:

$6\overline{)4524}$ $\qquad \dfrac{4524}{6} \qquad$ $4524 \div 6$

Do the same thing with these problems, given that the correct answer is 553: Find the difference between 729 and 176; subtract 176 from 729; 729 – 176.

Now practice inputting a variety of problems from the Exercises in Sections 1.2 and 1.3. Check your answers in the Answer Key to be sure that you have entered the numbers correctly.

Order of Operations

Input [7] [+] [3] [×] [2] [=]. If the answer shown is 13, then your calculator does have the Order of Operations Agreement programmed into it and will be able to simplify multiple operation arithmetic problems. If your calculator displayed any other answer, then it has not been programmed with the agreement and cannot simplify multiple operation problems correctly. You will need to simplify these problems yourself. Practice inputting Exercises 13–24 from Section 2.2 and check your answers in the answer key.

Grouping Symbols

In the expression 3(5 – 1), by Order of Operations agreement, the grouping symbols indicate that the subtraction should be done before the multiplication. When using a calculator to simplify this expression, you must know if your calculator has Order of Operations Agreement and how the calculator handles multiplication of quantities in parentheses. Try inputting the exercise at the beginning of this paragraph by keying: [3] [(] [5] [–] [1] [)] [=]. If the answer displayed is not 12, then your calculator may need you to input the multiplication operation between [3] and [(]. Try the exercise again, inserting the [×] before [(], [3] [×] [(] [5] [–] [1] [)] [=], and see if your answer is 12.

Input this exercise: (7 – 4)(12 + 2). Did you insert the multiplication symbol between the pairs of parentheses, if your calculator requires them? The answer should be 42.

Order of Operations problems with grouping symbols inside other grouping symbols can be simplified by some calculator models, but not by others. Such problems use parentheses, brackets, and braces to distinguish the different sets of grouping symbols, but some calculators have only parentheses. The parentheses can be used for the brackets and braces. Some calculators limit how many sets of imbedded parentheses you can use in a problem. Practice on the problems that follow and learn the limitations of your calculator model.

Input this exercise:

4 + 2[18 – 3(5 – 2)]

Be sure to insert the multiplication symbol between [2] and [[] and between [3] and [(] if required by your model, and also be sure to close all grouping symbols that you open. Every [(] that you use will need a [)]. The simplification should be 22. If you get some other answer, try inputting the exercise again. If your display shows ERROR, then your calculator cannot handle imbedded grouping symbols. You will first have to simplify within the brackets as a separate problem, 18 – 3(5 – 2) is 9, and then input [4] [+] [2] [×] [(] [9] [)].

Practice using your calculator on a variety of problems from Section 2.2 until you know how to enter the problems and know the limitations for your calculator. Check answers in the Answer Key to be sure that you are simplifying the problems correctly when using your calculator.

x^2, $\sqrt{x}$, 2nd, or SHIFT

To raise a number to the second power (to square it), input the number and then use the x^2 key. The calculator will usually display what x^2 is without using the = key. For example, inputting [5] [x^2] will have 25 show in the display.

In order to include as many operations and functions as possible with only 18 or 20 keys, most keys are given at least two functions. The first function will be what is printed on the key, such as [÷] or [x^2]. The second function is printed above the key in a contrasting color. To have the key do the second rather than the first function, you will need to use the [2nd] or [SHIFT] key (usually in the upper left-hand corner of the key pad) and then the function key. For example, if your calculator has x^2 in contrasting color above a key and you want to calculate 5^2, input [5] [SHIFT] or [5] [2nd] [x^2] and your display will show 25.

The opposite of raising a number to a second power is to find its square root. Find the $\sqrt{\ }$ or $\sqrt{x}$ key. To find the square root of 36, enter [36] [$\sqrt{\ }$]. If 6 does not appear on your display, then try [$\sqrt{\ }$] [36] [=] because some models require the input in this order. You will need to practice to find out how your model works. If your calculator requires the $\sqrt{\ }$ key first, then notice the display. It shows $\sqrt{(}$. After you input the number, you will need to close the parentheses. So the keying sequence should be [$\sqrt{\ }$] [36] [)] [=]. While the calculator would have given the correct answer without closing the parentheses, there are many expressions for which you must close the parentheses to get a correct answer. Try some problems from the exercises in Section 2.1 and check your answers in the Answer Key.

y^x, $\wedge$, $\sqrt[x]{y}$, or $x^{1/y}$

Look to see if your calculator has a key labeled y^x or if on is labeled $\wedge$. These keys are used to raise any number to any power. In either case, input the base number, then the function key, then the exponent, and then the equals key. Try 3^6. [3] [y^x] [6] [=] or [3] [$\wedge$] [6] [=]. Your display should be 729. Some calculators will have y^x as a second function so you would use [2nd] or [SHIFT] before [y^x].

To take a root other than the square root, use $\sqrt[x]{y}$ or $x^{1/y}$. This will probably be a second function. Practice by entering $\sqrt[3]{64}$. The answer should be 4. For most calculators, the keying is [64] [2nd] [$\sqrt[x]{y}$] [3] [=] or [64] [$y^{1/x}$] [3] [=] or [3] [2nd] [$\sqrt[x]{\ }$] [64] [=]. For a few models, it is [3] [2nd] [$\sqrt[x]{y}$] [64] [=]. You will need to determine if your calculator requires the radicand or the index be keyed in first. Practice with some of the problems from the exercises in Section 2.1 and check the answers in the Answer Key.

Fractions *a b/c*

If your calculator has a key labeled *a b/c*, then it can operate with fractions in fraction form. In calculators without fraction capabilities, fractions must first be changed to decimals and then are used as decimals in the calculation.

To enter a proper fraction such as 2/3, enter: [2] [a b/c] [3]. Your display will show 2⌋3 or 2/3 or 2⌈3. Become familiar with the symbol between 2 and 3 because you need to learn to write that notation on paper as $\frac{2}{3}$ or 2/3. The mixed number 5 1/2 is entered as [5] [a b/c] [1] [a b/c] [2] and is displayed as 5 _ 1⌋2 or 5 ⌋1 ⌋ 2 or 5⌈1⌈ 2, depending on the calculator model.

Mixed numbers and improper fractions can be interchanged by using the $^d/c$ function. To change 5 1/2 to an improper fraction, key in [5] [a b/c] [1] [a b/c] [2] [2nd] [d/c] and 11 ⌋ 2 or 11 ⌈ 2 will be displayed.

Some calculators have a F < > D function that changes numbers between fraction and decimal form. If 11⌋ 2 or 11/2 or 11 ⌈ 2 is still displayed, type [2nd] [F < > D] and 5.5 will be displayed. Typing [2nd] [F < > D] again will change the number back to a fraction.

To simplify a fraction to lowest terms, input the fraction using the a b/c key and = . The fraction will appear in lowest terms. Test your calculator to see if it limits the number of digits in the numerator and denominator. Some models are limited to three digits in each part. Input 125/1000. Is your display showing that or 125/100? If 125/100 appears, you will have to reduce by hand any fractions with four or more digits in either the numerator or denominator or divide the numerator by the denominator then change the decimal to a fraction. Try simplifying 125/400 to lowest terms: [125] [a b/c] [400] [=]. Your display should show 5⌋ 16 or 5 ⌈ 16.

Operations with fractions can be put into the calculator just as they appear in the problem. The calculator will know when it needs to use common denominators or improper fractions, and it will give answers in lowest terms.

Practice by inputting problems from the Review Exercises at the end of Chapters 3 and 4. Check your answers in the Answer Key and rework any problems that you miss.

Signed Numbers + / –

All numbers on the calculator with no sign displayed are positive numbers. To indicate that a number is negative, either enter the number first, then use the + / – key or the (+/–) key then the number. The display will show the number as a negative. Key in: (–6) + (–7) –3. The answer should be –16. Practice with problems from the Review Exercises for Chapter 11, and check your answers.

Percent %

The percent function, on most models, will change percent quantities to decimals. On most calculators it is a 2nd function. If you input [20] [%] and 0.2 is displayed, then your calculator has changed the percent to a decimal, and you could use that displayed value in your calculations. Try inputting 12.5% × 20. The answer should be 2.5. If you have keyed in [12.5] [%] [×] [20] [=] and have gotten some other answer, try entering it as [20] [×] [12.5] [%] [=] because some calculators require the base to be entered first.

STO or Min

Your calculator can store or remember numbers until you need them again. Some models have even more than one memory location. Input 231.79 and type STO or Min. If a small M appears in your display, then your number is being held in memory. If your calculator has more than one memory location, then you need to type STO and a number between 1 and 4, and your display will show M1 or M2, and so on. If your calculator shows A B C D E when you hit STO, then move the underline to one of the letters and hit enter. Your value is now saved in that memory location. Turn your calculator off and then on again. If the display shows the M, then that number is saved even when the power is off. Type RCL or MR to recall the number to the display.

To clear the memory when you no longer need the stored number, type [STO] or [Min] [C]. If your calculator has more than one memory location, to clear memory location 1 or A, type [STO] [1] [C] or [STO] [A] [C]. The M should disappear from the display window.

The steps described in this appendix were written to include the three most commonly available models of scientific calculators that have fraction functions. Practice the skills described, and if you have difficulty, consult the manual that came with your calculator. You may be using some other brand or model that differs in format from those described.

Answer Key

EXERCISES—CHAPTER 1

Section 1.1

A. 1. 8,265 3. 6,014,385
5. 48,002,085 7. 8,000,004,007
9. 29,460 11. 46,980,002
13. 72,480,000 15. 8,006,043

B. 17. eight thousand, nine hundred fourteen
19. two thousand, nine
21. seventy thousand, seventy
23. one million, two hundred thirty-four thousand, five hundred sixty-seven
25. fourteen thousand, eight
27. twelve billion, fourteen million, two hundred thirty-five thousand, six
29. thirty-five billion, eight

C. 31. 730 33. 800
35. 2,000 37. 8,500
39. 3,600 41. 14,300
43. 2,000 45. 33,000
47. 42,400 49. 20,000
51. 864,380 53. 864,400
55. 864,000 57. 860,000
59. 900,000 61. 37,070,000
63. 200,000

D. 65. $4,400 67. $36,000
69. $28,000,000

E. 71. Identify digit in thousands place. Look at digit to its right. If this digit is less than 5, leave thousands place as it is and change last three digits to zeros. If the digit is 5 or more, increase thousands digit by 1 and change last three digits to zeros.
73. In 7,863 the 8 becomes a 9 and the 6 and 3 are replaced by 0. In 7,963 increasing the 9 by 1 makes 80 hundred and the 9 and 8 are replaced by zeros.
75. Answers will vary.

Section 1.2

A. 1. 811 3. 272
5. 2,869 7. 651
9. 31,672 11. 174,064
13. 637 15. 1,192
17. 11,443 19. 11,576
21. 3,128 23. 2,110
25. 6,644 27. 1,203
29. 738 31. 5,281
33. 34,510 35. 26,489
37. 1,535 39. 30,186

B. 41. 1,131,391 43. 46,009,775
45. 10,001,277 47. 226,579
49. 16,041,675

C. Estimate answers may vary.

51. Multiply 30 by $400.

53. Multiply $25 by 30 units.

55. Multiply 3 by $10 and 2 by $20 and add the products.

57. Multiply $40 by 12.

59. Add $140 and $90.

61. Answers will vary.

Section 1.3

A. 1. 267 3. 672

5. 6,104 7. 217

9. 609 11. 526 R1

13. 1,033 R2 15. 2,670 R28

17. 691 R8 19. 405 R13

21. 39 R26 23. 124 R8

25. 2,064 27. 308 R12

B. 29. 3,615 31. 17,602

33. 2,003 35. 3,658

37. 368 39. 1,028

C. 41. Divide $368 by 8 vans.

43. Divide $89,635 by 7 coworkers.

45. Multiply the quotient by the divisor and add the remainder to that product.

47. When the remainder is greater than the divisor, increase the quotient by 1 and try again.

49. a. $400,000 and 1,000 citizens are easy to use.

b. $400

c. Answers will vary.

Section 1.4

A. *Note:* Estimate answers may vary somewhat from those shown.

1. Add $2,300, $900, $1,200, $2,000, and $2,400; $8,800

3. Subtract 40 from 70; 30 acres

5. Divide $80,000 by 4; $20,000

7. Multiply $2 increase by 20 days; $40

B. 9. $8,690 11. 28 acres

13. $20,363 15. $42

C. 17. $101 19. 464 miles

21. 86 miles 23. $2,697,367

25. 304 miles 27. $132

D. 29. supplementary 31. an infinite number

33. complementary 35. obtuse

37. vertical 39. protractor

41. degrees 43. intersecting

45. 61° 47. 73°

E. 49. 127° 51. 118°

F. 53. 52° 55. 180°

57. vertical, congruent

G. 59. Answers will vary.

61. Answers will vary

Section 1.5

A. 1. a. 18 hr b. 19 hr

c. 19 hr d. 16 hr

3. a. $193 b. $204

c. $204 d. $100

B. 5. mean: 84, median: 85, mode: none, range: 16

7. Maria and Chris, 16 points; Igor, 8 points

9. Maria 11. 60°

13. 30° 15. July, August, September

17. Game 3 19. 27 more points

21. 27 points 23. 21 points

C. 25. Answers will vary.

27. Answers will vary.

Section 1.6 Review Exercises

A. 1. a. 9 b. 8 c. 2

3. a. 2 b. 1 c. 5

B. 5. 18,004,017 7. 9,042

9. 17,000,009,055 11. 846,000

C. 13. thirty-two thousand, one hundred eighty-nine

15. eight hundred ninety-one thousand, four hundred seven

17. one million, four hundred twenty-four thousand, eight hundred fifty-six

19. seventy-seven thousand, nine hundred twenty-three

21. four hundred ninety-two thousand, one hundred eight

23. twenty-three thousand, eleven

D. 25. 2,660 27. 34,500

29. 185,000 31. 358,900

33. 120,000 35. $92,300

37. $390,000 39. $175,000

E. 41. 45,115 43. 2,367

45. 317 47. 479,136

49. 2,157 51. 127,000

53. 780 55. 44,218

57. 171,342 59. 406

61. 1,376 63. 610,128

F. 65. 92 laps 67. $14 each

69. 535,300 people 71. $25,872

73. 24 pounds 75. 6 pounds/month

77. 18,500 miles

G. 79. acute 81. straight

83. 180º

H. 85. 58º 87. vertical

89. 180º

I. 91. Battery A 93. 25 minutes

J. 95. 9,237.5 points 97. 20 points

99. Friday and Monday

101. Answers will vary.

K. 103. It is sixty-one thousand, not sixteen thousand: eight million, sixty-one thousand, twenty-five

105. Borrowing in the hundreds place is incorrect: 1,503.

107. The problem asks for product which implies multiplication not division: 130,702.

L. 109. Because it is too easy to strike the wrong key and make an error with a calculator.

111. Answers will vary.

M. 113. $1,619,100 115. $10,014

117. $85,820

Test—Chapter 1

1. 47,006,804
2. thirty million, two hundred six thousand, eleven
3. 176,900
4. 300,000
5. 10,063
6. 4,028
7. 12,087
8. 2,088
9. 0
10. Undefined, no solution
11. 65 quarters
12. $7,871
13. $9,900
14. 28 miles per gallon
15. 134º
16. 45º
17. 1,083 – 1,016 = 67 *not* 77; correct answer is 285 R39.
18. mean: 85, median: 84, mode: 76, and 84 range: 24
19. Subtract the given measure from 90º.
20. Multiply 27 by 4 and attach four zeros.

EXERCISES—CHAPTER 2

Section 2.1

A. 1. 27 3. 25

5. 1 7. 256

9. 0

B. 11. 5.916 13. 4.123

15. 6 17. 7.211

19. 4

C. 21. 4 23. 2

25. 1 27. 10
29. 11 31. 8
33. 5 35. 4,796
37. 12 39. 9,165

D. 41. 14 43. 8
45. Yes, 17 × 17 = 289
47. Yes, 7 × 7 × 7 = 343
49. 4

E. 51. 4,096 53. 2
55. 9 57. 1,048,576
59. 100

F. 61. A squared number is the result of multiplying a number by itself. A square root is the opposite of that process. You find the number that, when multiplied by itself, gives the number in the $\sqrt{}$.
63. 1 followed by 9 zeros; 1,000,000,000

Section 2.2

A. 1. 18 3. 2
5. 7 7. 6
9. 28 11. 0
13. 12 15. 2
17. 26 19. 23
21. 44 23. 27
25. 2 27. 10
29. 4 31. 10
33. 29 35. 9

B. 37. 14 39. 12
41. 6 43. 12
45. 12 47. 28
49. 16 51. 0
53. 6 55. 45
57. 154 59. 59

C. 61. 384 miles 63. 18,000 feet

D. 65. 22 67. 18
69. 11 71. 34
73. 13 75. L = 11
77. A = 20 79. M = 15

E. 81. 18.866 83. 225.036
85. 4,915.277 87. 319.77
89. 29.25

F. 91. 7,812 feet 93. $693,960
95. 252 97. 3,517

G. 99. Answers will vary.
101. 35

Section 2.3

A. 1. equilateral or equiangular
3. vertices
5. isosceles
7. pentagon
9. 180°
11. congruent

B. 13. 64° 15. 75°
17. 60°

C. 19. 85 inches 21. 96 centimeters
23. 52 inches

D. 25. 30 yards 27. 42 yards
29. 37°

E. 31. Square, rectangle
33. Square, rectangle, parallelogram, rhombus
35. Square, rectangle, parallelogram, rhombus
37. None

F. 39. 140 square feet
41. 324 square inches
43. 100 feet
45. 396 square inches

G. 47. 216 cubic centimeters
49. 328 square inches

H. 51. 872 square inches
53. 13.74 square meters
55. $11,664

I. 57. Answers will vary.

59. Find the total perimeters and subtract the openings for doorways.

Section 2.4

A. 1. d
3. f
5. h
7. c

B. 9. Answers will vary.
11. Answers will vary.
13. Answers will vary.

C. 15. 112 = 112
17. 42 = 42
19. 48 = 48
21. 56 = 56
23. 154 = 154
25. 2,970 = 2,970

D. 27. Answers will vary.
29. Answers will vary.
31. Answers will vary.

Section 2.5

A. 1. 1, 2, 4, 8
3. 1, 3, 7, 21
5. 1, 7
7. 1, 2, 4, 8, 16, 32
9. 1, 2, 3, 6, 9, 18
11. 1, 3, 5, 9, 15, 45
13. 1, 2, 5, 7, 10, 14, 35, 70
15. 1, 2, 3, 4, 5, 6, 8, 10, 12, 15, 20, 24, 30, 40, 60, 120
17. 1, 2, 3, 5, 6, 10, 15, 25, 30, 50, 75, 150
19. 1, 2, 4, 5, 8, 10, 20, 25, 40, 50, 100, 200

B. 21. No, 1 + 4 = 5
23. Yes, ends in 6
25. No, ends in 6
27. Yes, sum is 12
29. Yes, ends in 5
31. No, ends in 5
33. Yes, ends in 2
35. No, ends in 2
37. Yes, ends in 8
39. Yes, sum is 9

C. 41. $2 \cdot 2 \cdot 3$
43. $2 \cdot 2 \cdot 5$
45. $2 \cdot 2 \cdot 3 \cdot 3$
47. $2 \cdot 3 \cdot 3 \cdot 3$
49. $2 \cdot 2 \cdot 3 \cdot 7$
51. $5 \cdot 17$
53. $2 \cdot 2 \cdot 2 \cdot 3 \cdot 5$
55. $2 \cdot 7 \cdot 11$
57. $2 \cdot 2 \cdot 2 \cdot 2 \cdot 3 \cdot 3$
59. $2 \cdot 11 \cdot 11$

D. 61. $3 \cdot 5 \cdot 11 \cdot 17$
63. $3 \cdot 7 \cdot 11 \cdot 17$
65. $3 \cdot 3 \cdot 3 \cdot 7 \cdot 7 \cdot 11$
67. $3 \cdot 3 \cdot 5 \cdot 7 \cdot 13$
69. $5 \cdot 11 \cdot 13 \cdot 23$
71. $2 \cdot 2 \cdot 5 \cdot 7 \cdot 7 \cdot 11$

E. 73. A number is divisible by 6 if it is an even number and the sum of its digits is divisible by 3.
75. Answers will vary.

Section 2.6

A. 1. 24
3. 84
5. 60
7. 40
9. 60
11. 180
13. 90
15. 60
17. 120
19. 84

B. 21. 180
23. 7,560
25. 1,575
27. 300
29. 1,080

C. 31. Answers will vary.

Section 2.7

A. 1. 2
3. 4
5. 8
7. 1, relatively prime
9. 15
11. 2
13. 7
15. 1, relatively prime
17. 5
19. 6

B. 21. 36
23. 42
25. 72
27. 63
29. 24

C. 31. Answers will vary.

Section 2.8 Review Exercises

A. 1. 125
3. 4
5. 12
7. 2
9. 1

B. 11. 12 13. 4
15. 26 17. 28
19. 7 21. 147
23. 10 25. 2
27. 34 29. 22
31. 36 33. 50

C. 35. 28 37. 28
39. 66 41. 1,232 miles
43. 108 square feet 45. 19 inches
47. 80º 49. 46 yards
51. 24 cubic feet

D. 53. g 55. a
57. b 59. h

E. 61. 72 = 72 True
63. 22 ≠ 110 False
65. 140 = 140 True
67. 102 ≠ 84 False
69. 286 = 286 True

F. 71. 1, 2, 3, 6, 9, 18
73. 1, 2, 3, 5, 6, 10, 15, 30
75. Yes, ends in 6
77. Yes, ends in 0
79. No, sum is 10

G. 81. $2 \cdot 2 \cdot 2 \cdot 2 \cdot 5$ 83. $2 \cdot 3 \cdot 5 \cdot 7$
85. $2 \cdot 3 \cdot 19$ 87. $3 \cdot 5 \cdot 5$
89. $2 \cdot 3 \cdot 11$ 91. $2 \cdot 2 \cdot 3 \cdot 11$

H. 93. 60 95. 120
97. 84 99. 180
101. 210

I. 103. 14 105. 1
107. 6

J. 109. Multiply before subtracting: 8
111. $\sqrt{16}$ is 4 not 8; 15
113. 4 is not a prime number;
$2 \times 2 \times 2 \times 5 \times 5$
115. 1/2 (18)(12) = 9(12), not 9(6); 108 square inches

K. 117. 1. Divide
2. Multiply
3. Multiply
4. Subtract
5. Multiply
6. Subtract

L. 119. Answers will vary.
121. Answers will vary.
123. Answers will vary.

M. 125. 168 square feet
127. 18 2/3 square yards
129. 8 rolls
131. 416 square feet
133. $66

Test—Chapter 2

1. 2, 3, 5, 7, 11, 13, 17, 19, 23
2. Yes, 2 + 7 + 6 = 15 and 15 is divisible by 3.
3. $2 \cdot 3 \cdot 3 \cdot 5 \cdot 7$ or $2 \cdot 3^2 \cdot 5 \cdot 7$
4. 120 5. 6
6. 20 7. 9
8. 47 9. 55
10. undefined
11. $(3 \times 7) \times 6$
12. 1, 2, 5, 10, 25, 50
13. $R = 20$
14. $M = 34$
15. $A = 126$ square feet
16. $P = 50$ feet
17. $D = 301$ miles
18. 112 square yards
19. Find the sum of the three known angles and subtract the sum from 360º. Answer: 45º.
20. 432 square inches

EXERCISES—CHAPTER 3

Section 3.1

A. 1. a. 13/35 b. 22/35

3. a. 13/24 b. 11/24

5. a. 11/20 b. 9/20

B. 7. 4 1/5 9. 3 1/2

11. 1 2/3 13. 2 7/11

15. 2 3/32 17. 1

C. 19. 44/7 21. 58/5

23. 11/8 25. 68/9

27. 18/7

D. 29. 19/4, 8/8, 6/5

31. 1 1/2, 19/4, 6/5, 4 4/9, 25, 3

33. 8/8

35. 1 1/2, 19/4, 6/5, 4 4/9

E. 37. One complete unit shaded and 2 parts shaded out of 9 units in another unit.

Section 3.2

A. 1. Yes, 5/8 = 5/8 3. Yes, 4/5 = 4/5

5. No, 2/3 ≠ 1/2 7. Yes, 12/20 = 12/20

9. No, 10/7 ≠ 15/7

B. 11. 9 13. 15

15. 16 17. 28

19. 3 14/16 or 62/16

21. 52/12 or 4 4/12

C. 23. 5/6 25. 3/5

27. 1/4 29. 3/4

31. 2/3 33. 5/8

35. 8/15 37. 13/7

39. 9/20 41. 49/72

43. 24/31 45. 16/21

47. 3/4 49. 3/7

D. 51. Because 18 ÷ 2 = 9, divide 16 by 2 to find the numerator 8.

53. Yes, because 2 3/5 = 13/5 and if you multiply by 3/3, 13/5 = 39/15.

55. Find the prime factorization of 24 and 39 and divide out the common factor 3. Multiply the remaining factors. 24/39 = 8/13.

Section 3.3

A. 1. 10/27 3. 12/35

5. 27/2 or 13 1/2 7. 1/2

9. 1/5 11. 15

13. 95 15. 22

17. 3 17/27 or 98/27

19. 2000/7 or 285 5/7

21. 497/12 or 41 5/12

23. 315/4 or 78 3/4

B. 25. 6 × 2 = 12

27. 1/2 × 2 = 1

29. 12 × 1/2 = 6

C. 31. 9 students 33. 10 bars

35. 29,037 paperbacks

37. 110 students

39. 33 inches

41. 24 points

43. 1 11/16 boxes

D. 45. 114 2/3 square meters

47. 20 13/24 square inches

E. 49. Multiply 5/8 by 128.

51. It can be used only in division and only after the divisor has been inverted and the operation changed to multiplication. It cannot be used in addition or subtraction.

Section 3.4

A. 1. 3/2 3. 1/4

5. 1/12 7. 7/9

9. 3/13

B. 11. 8/9 13. 5/2 or 2 1/2

15. 48/5 or 9 3/5 17. 5/2 or 2 1/2

19. 7/12 21. 117

23. 32/35 25. 7/9

27. 2/483

C. 29. 1/6 31. 1/6

33. 2/3 35. 7/8

37. 3/2 or 1 1/2

D. 39. 5/16 pounds 41. 36 cars

43. 8 34/45 miles 45. 5 1/2 dollars

47. 10 scarves 49. 54 packages

51. 56 plots 53. 6 servings

E. 55. The reciprocal of a fraction is the fraction that results when the numerator and denominator have been interchanged—that is, when the fraction has been inverted.

57. Write both mixed numbers as improper fractions. Invert the second fraction and change the operation to multiplication. Reduce, if possible, and then multiply.

61. 1 7/9 63. 5/6

F. 65. 66 patties 67. 18 days

69. 800 disks 71. 46 1/2 mph

73. 1 9/16 miles 75. 65 cans

77. 51 9/16 cubic yards of helium

79. Cancellation error: 16 ÷ 4 = 4 not 2. The answer should be 5/12.

81. Multiply 1/2 by 7/8, not divide. 1/2 × 7/8 = 7/16.

83. "Of" means to multiply, not divide. The answer should be 1 9/16.

H. 85. You cannot have zero in the denominator of a fraction because division by zero is undefined.

87. Rewrite as 4/9 ÷ 2/3. Then rewrite as multiplication by the reciprocal of the divisor, 4/9 × 3/2, and simplify to 2/3.

I. 89. Find the area using $A = LW$. Divide the area by 96 square feet. She will need 2 quarts, but will have some paint left over.

Section 3.5 Review Exercises

A. 1. 2 14/15 3. 6 1/3

5. 1 5/6 7. 12 1/5

B. 9. 27/8 11. 38/3

13. 272/5 15. 81/8

C. 17. 3/4 19. 2/5

21. 1/3 23. 2/5

25. 8/13 27. 2/3

29. 16/21 31. 7/10

D. 33. 8 35. 9

37. 12 39. 18

41. 1 43. 4

E. 45. 5/9 47. 3 1/2

49. 6 2/3 51. 10

53. 4 1/2 55. 8/9

57. 1 3/5 59. 4 11/16

Test—Chapter 3

1. 2 5/7 2. 37/8

3. 9/20

4. $\dfrac{2 \times 2 \times 2 \times 2 \times 2 \times 5}{2 \times 2 \times 3 \times 3 \times 7} = \dfrac{40}{63}$

5. 28 6. 5/3

7. 10/21 8. 25/9 or 2 7/9

9. 22/9 or 2 4/9 10. 35

11. 13/5 or 2 3/5 12. 3/2 or 1 1/2

13. 15/8 or 1 7/8 14. $1,980

15. 35 5/8 packages 16. 24 2/7 mpg

17. 39 3/8 ft 18. 71 1/4 sq yd

19. 18 3/8 cu ft

20. 1. Write as improper fractions.

2. Multiply by the reciprocal of the divisor.

3. Divide by common factors, 4 and 5.

4. Multiply.

EXERCISES—CHAPTER 4

Section 4.1

A. 1. 12 3. 8

5. 12 7. 12

9. 120 11. 210

13. 288 15. 720

17. 252 19. 72

21. 288

B. 23. 7/9 25. 3/5

27. 16/15 or 1 1/15 29. 1/2

31. 29/18 or 1 11/18 33. 19/15 or 1 4/15

35. 11/80 37. 23/12 or 1 11/12

39. 13/12 or 1 1/12 41. 9/20

43. 11/36 45. 91/180

47. 101/252 49. 137/72 or 1 65/72

C. 51. 2 1/2 pounds 53. 3/8 pounds

55. 2 1/8 hours 57. 1/8 hour

59. 1 11/12 hours

D. 61. 1 7/8 meters

63. 2 3/4 feet

E. 65. Answers will vary.

67. No. An improper fraction is an acceptable answer in any situation except in response to an application problem, or when the improper fraction is part of a mixed number.

Section 4.2

A. 1. 8 1/8 3. 2 1/6

5. 10 1/12 7. 12 7/36

9. 8 1/6 11. 5 4/15

13. 15 19/28 15. 32 1/40

17. 6 5/9 19. 5/6

21. 143 13/16 23. 4 11/12

25. 4 31/36 27. 42 3/5

29. 29 23/30

B. 31. 20 7/8 33. 38 17/24

35. 16 5/6 pounds 37. 131 3/4 pounds

39. 11 1/4 pounds 41. 5/6 cup

43. 12 7/12 hours 45. 5/6 hour

47. 38 3/8 feet 49. 35 3/8 feet

C. 51. Approximate the measurements to 9 ft by 11 ft and add 9 + 11 + 9 + 11.

53. In the first example, you must borrow 1 from 12 as 8/8 and subtract 5/8 from 8/8 and subtract 7 from 11. In the second, you have no need to borrow. Subtract 0/8 from 5/8 and 7 from 12.

55. Answers will vary.

Section 4.3.

A. 1. 13/12 or 1 1/12 3. 23/30

5. 3/8 7. 1/6

9. 1 5/9 11. 1/96

13. 9/8 or 1 1/8 15. 47/48

17. 59/48 or 1 11/48 19. 1 5/12

21. 6/25 23. 23/40

25. 35/54

B. 27. 1. Multiply

2. Add

3. Reduce

29. 1. Square roots

2. Squaring

3. Least Common Denominator

4. Add and convert

C. 31. 15 1/6 feet 33. 6 8/15

35. $T = 15/64$ 37. $W = \frac{83}{36}$ or $2\frac{11}{36}$

D. 39. No, because $\sqrt{49/4} = \sqrt{49}/\sqrt{4} = 7/2$, not 7/4.

41. No. No, it should be the same.

Section 4.4

A. 1. > 3. >
5. = 7. >
9. < 11. <
13. = 15. >

B. 17. 3/8, 2/3, 5/6 19. 3/4, 7/8, 15/16
21. 2/9, 2/3, 4/5 23. 2/3, 3/4, 15/16
25. 1/2, 5/8, 2/3, 3/4
27. 3/8, 1/2, 4/7, 2/3
29. 7/16, 11/24, 17/32, 5/8

C. 31. 1/12 33. 1/48
35. 1/16 37. 1/48
39. 1/32 41. 1/196

D. 43. Yes. Find the difference between 5/6 and 1/3 and the sum of 1/9 and 8/27. Then write the answers as equivalent fractions with denominators of 54 and see that 27/54 is greater than 22/54.
45. 17/20, 2/3, 7/16. No, because you also need to compare the three fractions in equivalent forms with denominators of 240.

Section 4.5

A. 1. 8 cups 3. 1 1/3 cups
5. 19 3/4 laps 7. 3/4 lap
9. 1 5/6 miles 11. 34 8/11 miles
13. 214 bottles 15. 5/8 hour
17. 11/16 of the class
19. 6 players 21. 7/8 yard
23. 18 1/8 yards 25. 1 13/18 hours
27. 15 volumes 29. 5 pieces

B. 31. 121 1/8 sq ft 33. 356 sq ft
35. 38 1/4 feet; 19 or 20 bricks

Section 4.6 Review Exercises

A. 1. 19/12 or 1 7/12 3. 1/12
5. 4 11/36 7. 4 27/56
9. 3/4 11. 13 33/40
13. 31/28 or 1 3/28 15. 1 7/8
17. 8 1/3 19. 34 23/24
21. 31 4/9

B. 23. 7/8 25. 1/2
27. 0 29. 9/2 or 4 1/2

C. *Note:* Estimate answers may vary somewhat from those shown.
31. 18 – 15 = 3 33. 11 + 9 = 20
35. 1/4 + 1/2 + 1/2 = 1 1/4
37. 3 + 4 × 2 = 11

D. 39. 3/4, 3/8, 5/16 41. 7/9, 2/3, 5/8
43. 5/8, 3/4, 7/12

E. 45. 48 miles 47. 3 5/8 points
49. $69 51. 3 3/4 pounds
53. 20 lots 55. 45 inches
57. 1 4/15 hr.

F. 59. 136 2/3 square meters
61. 54 1/12 meters
63. 7 1/9 square feet; 10 2/3 feet
65. 67 1/2 yards

G. 67. Need to find LCD first; 29 1/8
69. 84 should be 83 3/3; 7 1/3

H. 71. Find the LCD. Write equivalent fractions. Subtract then add.
73. You must have like-size pieces to add or subtract.
75. Answers will vary.
77. Find the areas of those three rectangles. Double those areas because opposite sides are the same size, and then add the areas of the six sides.

I. 79. Area = 233 1/8 sq ft so 1 gallon is needed.
81. 1 gallon, because 74 11/32 sq ft need to be covered.

Test—Chapter 4

1. 120 2. 3/2 or 1 1/2
3. 3/10 4. 21/16 or 1 5/16

5. 6 5/12
6. 1 4/15
7. 3 4/7
8. 47/24 or 1 23/24
9. 14/45
10. 1 7/12
11. 5 4/5
12. 3 2/3
13. 29/18 or 1 11/18
14. 1/2
15. 6 2/5
16. Find the LCD. Write equivalent fractions with the denominator of 16. Then arrange in order from largest to smallest and report fractions in original form.
17. 18 17/24 hours
18. 18 2/3 cubic inches
19. 1 1/6 acres
20. 24 miles on 1 gallon

Cumulative Review 1–4

1. 376
2. 75,330
3. 3/2 or 1 1/2
4. 36/25 or 1 11/25
5. 16 1/18
6. 274
7. 27,039
8. 2/5
9. 2/3
10. 71 5/8
11. 42,042
12. 4,500
13. 27/64
14. 4/9
15. 315/4 or 78 3/4
16. 143
17. 180 R 34 or 180 17/21
18. 5/2 or 2 1/2
19. 175/16 or 10 15/16
20. 30
21. No; yes; no
22. 28
23. 56
24. $3 \cdot 3 \cdot 7$
25. $2 \cdot 2 \cdot 2 \cdot 2 \cdot 3 \cdot 5$
26. 32
27. 9
28. 1, 2, 3, 4, 6, 8, 12, 16, 24, 48
29. 16
30. 5 1/6
31. c
32. a
33. d
34. e
35. b
36. 7
37. 18
38. 13
39. 11
40. 11/12
41. 41/45
42. 13
43. 32
44. 4,800 women
45. 575 miles
46. 14 square inches
47. 19 miles
48. 12 pieces
49. 12 7/24
50. 69

EXERCISES—CHAPTER 5

Section 5.1

A. 1. a. 4 b. 3 c. 6 d. 1
3. a. 7 b. 4 c. 0 d. 3

B. 5. one hundred sixty and six tenths
7. one hundred eighty-two and thirty-four thousandths
9. four hundred fifty-six ten-thousandths
11. three and seventeen thousandths
13. forty-five and forty-five hundredths

C. 15. 6.074
17. 0.107
19. 116,000

D. 21. 0.04
23. 1.5
25. 150
27. 146.3
29. 16.100
31. 16.05
33. 170
35. 0.08
37. 14.264
39. 358.0
41. $286
43. $24.86
45. 268

E. 47. 2.03, 2.12, 2.2
49. 0.635, 0.67, 0.7
51. 4.003, 4.03, 4.304

F. 53. $0.401 > 0.4 > 0.392 > 0.04$
55. $2.015 < 2.123 < 2.146 < 2.15$
57. $2.402 > 2.24 > 2.204 > 2.004$
59. $0.053 < 0.305 < 0.31 < 0.351$

G. 61. Rounding involves identifying the place requested, looking at the digit to its right and either leaving the hundred or hundredths place as it is, or increasing it by 1. Hundreds, however, is the third place to the left of the decimal point; hundredths is the second place to the right of the decimal point.

63. 0.7 = 7/10 and 0.70 = 7/100, which reduces to 7/10. Both 0.7 and 0.70 are equivalent to 7/10 and are, therefore, equivalent to each other.

Section 5.2

A. *Note:* Estimate answers may vary somewhat from those shown.

1. 23 + 15 + 5 = 43
3. 40 – 27 = 13
5. (20,000 + 7,000) – 18,000 = 9,000
7. \$20 – \$8 = \$12

B. 9. 32.74 11. 21.235
13. 3.432 15. 5.778
17. 154.95 19. 2.866
21. 9.677 23. 37.44
25. 41.45 27. \$193.89
29. \$37.27

C. 31. 17.735 33. 147.09
35. 32.66 37. \$12.41
39. 57.68

D. 41. 963.165 43. 3,047.57
45. 165,205.43 47. 854.418
49. 120.790.33 51. 1,220.916
53. 0.01827 55. 0.007809

E. 57. 59.3 inches

59. Add measures of three given angles and subtract the sum from 360º.

Section 5.3

A. *Note:* Estimate answers may vary somewhat from those shown.

1. 25 × 4 = 100 3. 80 × 20 = 1,600
5. 0.002 × 0.005 = 0.000010

B. 7. 48.776 9. 0.0414
11. 0.035 13. 34.8
15. 2,876.8 17. \$1.01
19. 2,300 21. \$404.20
23. 8,848 25. 0.1337
27. 59 29. 0.00189
31. 154.224

C. 33. 4.390864 35. 0.0010602
37. 30,638.1 39. 867.33258
41. 0.0292

D. 43. 36.4 yards
45. 71.92 square yards
47. 47.4 inches
49. 50.4 inches
51. 51.84 square inches
53. 373.248 cubic centimeters
55. 311.04 square centimeters

E. 57. Answers will vary.

59. Entire answer does not appear in the display. Multiply 3,762 by 258 and reposition the decimal point a total of 13 places to the left.

Section 5.4

A. 1. 200 3. 320
5. 0.25 7. 0.0245
9. 400 11. 20
13. 0.265 15. 0.09
17. 0.64 19. 600
21. 0.45 23. 0.1834

B. 25. 21.667 27. 0.867
29. 0.329 31. 5.217
33. 0.026 35. 0.007
37. 0.571 39. 0.271
41. 0.214 43. 28.696
45. 0.059 47. 0.001
49. 0.001

C. 51. 43,104.17 53. 1.4223
55. 36.541 57. 9.7
59. 344.1

D. 61. $K = 1.2$ 63. $K = 3.1$
65. $K = 10.3$ 67. $K = 0.15$
69. $K = 3.85$

E. 71. Answers will vary.
73. Multiply by 1 as 100/100 to clear the decimals.

Section 5.5

A. 1. 7/50 3. 2 3/1000
5. 17 4/5 7. 9/125
9. 36 1/8

B. 11. 0.273 13. 2.833
15. 0.083 17. 10.5
19. 16.211

C. 21. 2.175 or 87/40 or 2 7/40
23. 4.1 or 41/10 or 4 1/10
25. 38.8 or 194/5 or 38 4/5
27. 3.45 or 69/20 or 3 9/20
29. 5.25 or 21/4 or 5 1/4

D. 31. 0.461 33. 3.85
35. 0.280 37. 9.84
39. 0.0350

E. 41. 59.415 43. 2,815.2
45. 3.6291 47. 186.4
49. 2,391.45

F. 51. The calculator rounds the last visible digit.
53. 9 yards times $5 because they are easy to use.

Section 5.6

A. *Note:* Estimate answers may vary somewhat from those shown.
1. 140 – (8 + 62 + 20) = $50
3. 200 × 0.25 - $50
5. 3 × 50 = 150 centimeters

B. 7. $10.12 9. $1.39
11. 6.3 gal 13. 14 vials
15. 6.7 lb 17. $837.72
19. $311.25

C. 21. 25.6 23. 3.927
25. 315.99 27. 18,622 people

D. 29. 4.65 feet 31. 314 square meters
33. 25.12 inches 35. 21.98 feet
37. 254.34 square meters
39. 25.12 feet
41. 100.48 square feet
43. 81.12 feet
45. 104.52 square inches
49. 523.33 cubic inches
51. 120 cubic inches
53. 302.4 cubic inches
55. 1.8 centimeters
57. 2.54 square centimeters

E. 59. Divide $720 by 6 months.
61. To find the circumference, multiply 3.14 by 22.
63. To find two areas multiply $2 \times 3.14 \times 10^2$.

Section 5.7 Review Exercises

A. 1. twenty-four and sixty-three thousandths
3. six and forty-seven thousandths
5. two hundred three and twenty-three thousandths
7. one thousand four hundred thirteen and three thousandths

B. 9. 4.016 11. 409.09
13. 0.0408 15. 12,000.012

C. 17. 4.7 19. 50
21. 92.5 23. 0.101

D. 25. 0.602, 0.609, 0.61
27. 2.002, 2.02, 2.2
29. 0.08, 0.801, 0.81

E. 31. $17.32 > 17.3 > 17.2 > 17.03$

33. $0.006 < 0.06 < 0.066 < 0.606$

F. 35. 90.767 37. 147.886

39. 57.818 41. 2,772.04

43. 1,897.2 45. $53.06

47. 135.48 49. 75.76

51. $45.17 53. 214.93

G. 55. 0.07 57. 2.09

59. 89.47 61. 0.00

63. 80 65. 0.03

67. 0.05 69. 4.97

71. 4.82 73. 0.07

H. 75. 7/25 77. 22 9/20

79. 1/25 81. 45 43/200

I. 83. 0.625 85. 0.778

87. 3.833 89. 0.004

J. 91. $0.57 93. 3.8 yards

95. 0.35 quart 97. $675

99. $7.94

K. 101. 16.61 square inches

103. 37.29 square feet

105. 265.77 square inches

L. 107. Multiply 20 by 26 and divide the product by 25 square feet.

M. 109. 26.300 not 26.003; 18 should be 18.000, not 00.180; 14.37 not 14.037; 59.877

111. Reposition the decimal point two places to the left, not to the right; 1.43261

113. Original decimal point is after the 16 and 1,600 – 1,588 = 12, not 82; 6.32

N. 115. 150.72 square feet

117. Find the area of the pool ($r = 80$ ft); find the area of the apron ($r = 92$ feet). Subtract the area of the pool from the area of the apron. 6,480.96 square feet

119. $c = \pi d$. Therefore, $6{,}400 \div 3.2 = 2{,}000$ km

Test—Chapter 5

1. thirty-two thousandths

2. 16.0406 3. 27.89

4. 0.023, 0.203, 0.23 5. 23.775

6. 7.785 7. 37.166

8. 0.0051 9. $1.06

10. 7.235 11. 0.8

12. 0.000215 13. 23/125

14. 0.917

15. 32.1536 square inches

16. 70.65 cubic inches

17. 7,234.56 cubic inches

18. 15.9 gallons

19. $8.29

20. Add $70 to $220 and subtract $290 from $400.

EXERCISES—CHAPTER 6

Section 6.1

A. 1. 1/3 3. 5/9

5. 40/3 7. 8/1

9. 3/10 11. 4/5

13. 7/27 15. 145/63

17. 24/1

B. 19. 3/2 21. 1/8

23. 2/7 25. 3/5

27. 3/1 29. 4/5

31. 1/4 33. 16/15

35. 3/4 37. 1/12

39. 1/2

C. 41. a. 32/3 b. 15/32 c. 3/32

d. 3/10 e. 16/25

43. a. 21/38 b. 21/17
c. 1/28 d. 3/152

D. 45. \$0.15 per ounce
47. 40/1
49. 2/1
51. 24 miles per gallon
53. 110 miles per hour
55. 35 miles per gallon
57. \$0.995 = \$1.00
59. \$2.50 per bag

E. 61. A, \$0.163 each 63. A, \$0.233 each
65. A, \$0.74 each 67. B, \$0.044 per ounce
69. B, \$0.156 per ounce

F. 71. 16.14 miles per gallon, \$15.03
73. \$0.083 per ounce

G. 75. Change 6 yards to inches by multiplying by 36. Then reduce 16/216 to 2/27.
77. Find the total number 14 + 11 and then write the ratio of 14 to 25.

Section 6.2

A. 1. Yes, 36 = 36
3. No, $225 \neq 315$
5. Yes, 1,372 = 1,372
7. Yes, 1.728 = 1.728
9. Yes, 17.55 = 17.55
11. Yes, 15/8 = 15/8

B. 13. $N = 81$ 15. $R = 3$
17. $A = 40$ 19. $c = 5$
21. $N = 4.2$ 23. $R = 6.67$
25. $A = 16.63$ 27. $B = 189.96$
29. $T = 16$ 31. $A = 160$
33. $W = 2\ 5/8$ 35. $N = 3/5$
37. $A = 10$ 39. $R = 3.17$

C. 41. Yes, 132.066 = 132.066
43. $N = 4.23$
45. $N = 5{,}563.18$

D. 47. The cross products are equal.
49. Make the numbers easier to multiply and divide, such as $c/10 = 20/60$. Then $200 \div 60 \doteq 3$.

Section 6.3

A. 1. 12.5 defective VCRs
3. 265 miles
5. 1.6 inches
7. 2 2/3 houses
9. \$0.80 per pound
11. \$0.59
13. 250 cans
15. 5,500 pounds of gravel
17. 12 hours
19. \$75 interest
21. 100 milligrams
23. 14 feet
25. 471.25 milligrams

B. 27. congruent
29. 1 to 1
31. proportional
33. equiangular, similar

C. 35. $BC = 12$ yards 37. $ML = 9$ yards
39. $XY = 18$ meters 41. $YZ = 36$ meters

D. 43. Write two equal ratios comparing interest to money in the account and solve for the missing interest.
45. Write a proportion comparing pounds to milliliters and then solve for the missing milliliters.
47. Make a sketch of ΔACE and ΔBCD and label the sides. Write a proportion of $5/10 = 4/CE$ and solve.

Section 6.4

A. 1. 4.5 quarts 3. 4,320 minutes
5. 28,800 seconds 7. 0.7 miles per minute

9. 528,000 feet per hour

B. 11. 30 decimeters

13. 15 centimeters

15. 71.1 centimeters

17. 6.4 kilometers per hour

19. 2.1 feet

C. 21. 1.65 decimeters 23. 1,653.5 inches

35. 61.4 meters 27. 22.4 miles per hour

29. 1,234.4 centimeters

D. 31. Two and 3 do not represent the same quantity, but 12 inches and 1 foot are different names for the same thing.

33. In the last ratio, 1 kilometer should be in the denominator.

Section 6.5

A. 1. $\frac{12}{20} = \frac{3}{5}$ 3. $\frac{20}{28} = \frac{5}{7}$

5. 22.25 grams

B. 7. $\frac{10}{6} = \frac{5}{3}$ 9. 1/16

11. 11.5 wins

C. 13. 15/22 15. $15,000

17. $45,000

D. 19. $\frac{30}{55} = \frac{6}{11}$ 21. $\frac{55}{60} = \frac{11}{12}$

23. 39°

E. 25. $\frac{1}{48} = \frac{3}{M}$, 144 miles

F. 27. Answers will vary.

29. Answers will vary.

Section 6.6 Review Exercises

A. 1. 2/9 3. 7 : 2

5. 160 : 3 7. 2 : 3

9. 1 : 9 11. 16 to 5

13. 36 to 5 15. 2 : 5

17. 16 to 7

19. a. 6 : 23 b. 6 : 29 c. 23 : 29

B. 21. $0.13 23. $0.98

25. $0.43 27. $16.59

29. $0.367

C. 31. Yes, 140 = 140

33. No, 156 ≠ 144

35. Yes, 0.144 = 0.144

37. No, 880 ≠ 636

39. No, 75/4 ≠ 90/4

D. 41. $C = 3$ 43. $M = 2.67$

45. $C = 7.2$ 47. $F = 0.47$

49. $K = 48$ 51. $T = 16/15$

E. 53. $3.30 55. 140 miles

57. $7.35

59. 3.75 pizzas, so order 4

61. 3.77 inches

63. 300.8 milligrams

65. 2.4 milliliters

F. 67. $PK = 12.1$ yards

69. $AB = 12$ inches

71. $KN = 25$ yards

73. 3 to 2

G. 75. 1.35 feet 77. 18,000 seconds

79. 64 cups

81. 0.25 miles per minute

83. 152.4 centimeters

H. 85. 2 to 3 87. 60 cars

89. 3 to 10

91. No, it would simplify to the same ratio.

I. 93. Answers will vary.

95. They would be equal.

97. $\frac{5}{A} = \frac{25}{B}$; $5B = 75$ and $25A = 75$. $B = 15$ and $A = 3$, so $\frac{5}{3} = \frac{25}{15}$.

99. $\overline{LM} = 24$ yards

101. 0.01 km/sec

Test—Chapter 6

1. 4/9
2. 60/1
3. 6/5
4. 4/15
5. 60/17
6. $0.205
7. 12 ounce, $0.274
8. 1.3 hours
9. 393.7 inches
10. Compute the cross products. If the results are equal, it is a proportion. If the results are not the same, the original equation is not a proportion. (This is a proportion because both cross products equal 105.)
11. No, $0.128 \neq 0.0144$
12. $W = 63$
13. $C = 25.6$
14. $X = 72$
15. $A = 34.8$
16. $T = 5/4$
17. 20 feet
18. 13.7 gallons
19. $XZ = 60$ meters
    $YZ = 57$ meters
20. a. 3 to 2 b. 17 to 142

EXERCISES—CHAPTER 7

Section 7.1

A. 1. a. 62 b. 12 c. 5

3. a. 89 b. 11 c. 11%

5. 27.4% 7. 8%

9. 18%

B. 11. Subtract 67% from 100%. 100% – 67% = 33% of the students need to improve.

13. Subtract 24% from 100%. 100% – 24% = 76% for household accounts.

Section 7.2

A. 1. 32% 3. 347%

5. 46.4% 7. 1200%

9. 60% 11. 71.4%

13. 225% 15. 350%

B. 17. 0.12 19. 0.009

21. 8.0 or 8 23. 0.105

25. 0.253

C. 27. 19/100 29. 6/25

31. 6 1/2 or 13/2

D.

	Fraction	Decimal	Percent
33.	143/500		28.6%
35.		0.8	80%
37.	1/40	0.025	
39.	3/400	0.008	
41.		0.583	58.3%
43.	3/500		0.6%
45.	2 13/20	2.65	
47.	5/1		500%
49.		0.063	6.3%

E. 51. 0.60 = 60/100 = 60%. 3/5% would be 0.6%, or 0.006.

53. Write 62.4% as a decimal by dividing it by 100% to get 0.624. Write 624 over 1,000, (624/1,000), and simplify to 78/125.

Section 7.3

A. *Note:* Estimate answers may vary somewhat from those shown.

1. 20; 1/4 of 80 = 20

3. 32; 32 is 50% (1/2) of 64

5. 120; 120 is 150% of 80

B. 7. $R = 7\%$ $B = 95$ $A = ?$
$A = 6.65$

9. $R = 124\%$ $B = 70$ $A = ?$
$A = 86.8$

11. $R = 12\ 1/2\%$ $B = 100$ $A = ?$
$A = 12.5$

13. $R = 125\%$ $B = ?$ $A = 30$
$B = 24$

15. $R = 32\%$ $B = ?$ $A = 8$
$B = 25$

17. $R = 12\%$ $B = ?$ $A = 43$
$B = 358.33$

19. $R = ?$ $B = 40$ $A = 4.8$
$R = 12\%$

21. $R = ?$ $B = 30$ $A = 16$
$R = 53.3\%$

23. $R = ?$ $B = 8$ $A = 7$
$R = 87.5\%$

25. $R = 16.5\%$ $B = 42$ $A = ?$
$A = 6.93$

27. $R = 24\%$ $B = ?$ $A = 12$
$B = 50$

29. $R = ?$ $B = 65$ $A = 21$
$R = 32.3\%$

C. 31. $R = 77.3\%$ $B = 92{,}476$ $A = ?$
$A = 71{,}483.95$

33. $R = ?$ $B = 146$ $A = 91.4$
$R = 62.6\%$

35. $R = 29.4\%$ $B = 37{,}645$ $A = ?$
$A = 11{,}067.63$

37. $R = ?$ $B = 8{,}276$ $A = 189.4$
$R = 2.3\%$

39. $R = 38.4\%$ $B = ?$ $A = 127$
$B = 330.73$

41. $R = 82\%$ $B = ?$ $A = 67.4$
$B = 82.20$

43. $R = 46.5\%$ $B = ?$ $A = 35.4$
$B = 76.13$

45. $R = ?$ $B = 8{,}375$ $A = 1{,}276$
$R = 15.2\%$

D. 47. Find 10% of 48, \$4.80. 1/2 of 10% is 5%, so find 1/2 of \$4.80, \$2.40. Add \$4.80 and \$2.40 for the \$7.20 tip and divide \$7.20 by 4 for each person's share.

49. Multiply \$180 by 1/3 and then subtract \$54 from \$180.

Section 7.4

A. 1. $24/B = 15/100$ — $B = 160$

3. $12/40 = P/100$ — $P = 30\%$

5. $A/30 = 94/100$ — $A = 28.2$

7. $6\ 1/2/B = 18/100$ — $B = 36.11$

9. $30/40 = P/100$ — $P = 75\%$

11. $A/40 = 18/100$ — $A = 7.2$

13. $48/B = 8/100$ — $B = 600$

15. $45/80 = P/100$ — $P = 56.3\%$

17. $54/B = 1.8/100$ — $B = 3{,}000$

19. $A/300 = 14.8/100$ — $A = 44.4$

21. $36/B = 3\ 1/2/100$ — $B = 423.53$

23. $52/B = 120/100$ — $B = 43.33$

25. $60/32 = P/100$ — $P = 187.5\%$

27. $45/B = 150/100$ — $B = 30$

29. $21.5/38 = P/100$ — $P = 56.6\%$

B. 31. $18.4/2{,}006 = P/100$ — $P = 0.9\%$

33. $21/B = 43.2/100$ — $B = 48.61$

35. $A/13{,}465 = 67.5/100$ — $A = 9{,}088.88$

37. $623/18{,}540 = P/100$ — $P = 3.4\%$

39. $1{,}765/B = 46.8/100$ — $B = 3{,}771.37$

C. 41. Both P and R represent the percent quantity. P is the actual percent number, but R is the percent value in its decimal form.

43. Write the percent proportion $16/40 = P/100$ and solve to find that P is 40%, or 0.40 in decimal form. 16/40 reduces to 2/5, and 2 divided by 5 is also 0.40. Therefore, they are equivalent.

Section 7.5

A. *Note:* Estimate answers may vary somewhat from those shown.

1. 40% of \$200 is \$80.

3. 7% of \$100 is \$7.

B. 5. \$115.70 7. 12%

9. 500 people 11. \$300

13. 8% 15. \$633.68

17. \$365.88

C. 19. Estimate: $I = 800(10\%)(3) = \$240$

21. Estimate: $I = (600)(8\%)(1) = \$48$; then $600 + 48 = \$648$

23. \$72

25. \$577.50

27. \$845; \$4,095

D. 29. 100% is a form of 1. Multiplying by 1 may change the way a quantity looks, but it does not change its value.

31. 10% of \$18 is \$1.80, so 20% of \$18 is \$3.60.

33. Find the amount of the increase, \$6.50 – \$5.80 = \$0.70. Then solve "0.70 is what percent of 5.80?" Round as requested.

Section 7.6

A. 1. Television

3. \$9,000

5. \$7,200

B. 7. 33%

9. \$1,056

11. \$350

C. 13. \$20.50

15. 5%

17. \$19.70

D. 19. \$1,000

21. 32.5%

23. 17.5%

25. See Figure 7.9 answer.

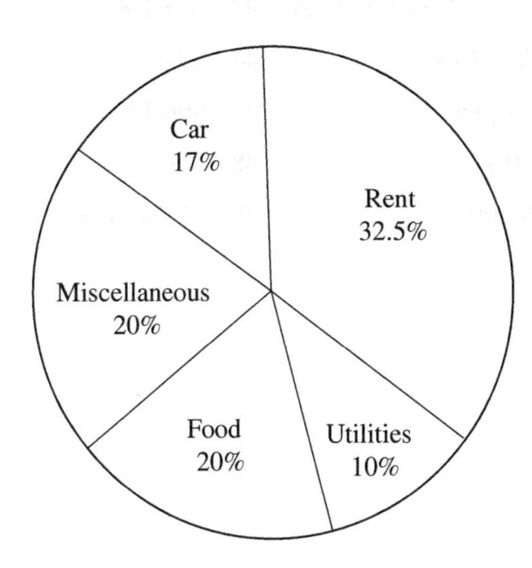

FIGURE 7.9 Answer

E. 27. About 33% (1/3) of \$3,300, which is \$1,100.

29.

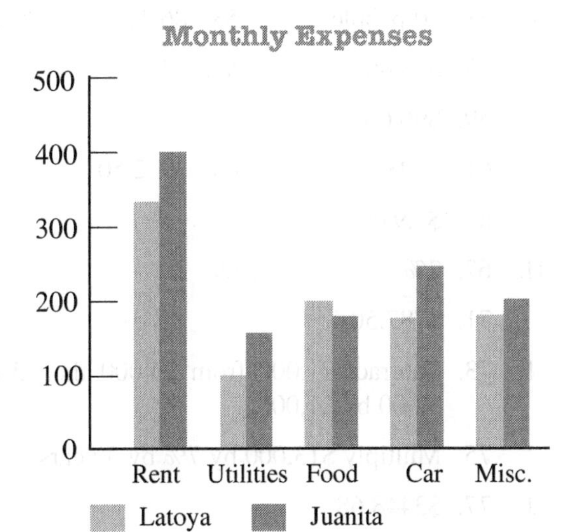

Section 7.7 Review Exercises

A. 1. 38%

3. 15%

B.

	Fraction	Decimal	Percent
5.	4/25		16%
7.		0.25	25%
9.		0.125	12.5%
11.	1 27/50		154%
13.	1 3/25	1.12	
15.		0.083	8.3%

C. *Note:* Estimate answers may vary somewhat from those shown.

17. 20% of 80 is 16.

19. 0.53 is 1% of 53.

21. 160 is 200% of 80.

D. 23. 33.6

25. 60

27. 2,500

29. 42.9%

31. 705.88

33. 2.5%

35. 15.36

37. 718.37

39. 909.09

41. 15%

43. 13.12

E. *Note:* Estimate answers may vary somewhat from those shown.

45. \$75; 30% of \$250 is \$75.

47. \$110; 5% of \$2,200 is \$110.

49. 100; 5% of 2,000 is 100.

F. 51. 40 people
53. 86.4%
55. $48.40
57. 14.1%
59. $96.60

G. 61. $198
63. $312.50
65. $9,975

H. 67. 2%
69. 2%
71. $787.50

I. 73. Subtract 21,000 from 24,000 then divide 3,000 by 21,000.
75. Multiply $13,000 by 7% by 3 years.

J. 77. $3448.68
79. $2,634 + $2,304.75 = $4,938.75

Test—Chapter 7

1. R = 16%, B = 40, A = 6.4
2. 11%
3. 0.7%
4. 0.1675
5. 762.5%
6. 29/200
7. $462
8. You could estimate by asking "18 is what % of $360?" Because 18/360 = 1/20, 18 is about 1/20 or 5% of 360.
9. 17.60 = R × 320; R = 0.055 = 5.5%
10. 17.60/320 = P/100
11. 35
12. 2%
13. 20.3%
14. 70
15. Est.: 10% of 2,800 is $280, so 5% is $140
16. $52.50
17. 12%
18. $20,000
19. $1,166,160
20. Because you are saving 15%, you are paying 100% – 15% or 85% of the original price. Therefore, ask "14,025 is 85% of what amount?" and solve using the percent formula or the percent proportion.

Cumulative Review 1–7

1. 48.9
2. 5/4 or 1 1/4
3. 3,979
4. 10 7/24
5. 26.885
6. 9 1/9
7. 529
8. 55.39
9. 147.7
10. 30 1/5
11. 48
12. 1.1
13. $11,293
14. 1,541/32 or 48 5/32
15. 7.6
16. 1,273.33
17. 372
18. 19/6 or 3 1/6
19. 41.43
20. 33/16 or 2 1/16
21. 49/4
22. 12
23. 13/42
24. 45%
25. 1/8
26. 65.2
27. 0.005
28. 2 3/8 or 19/8
29. 0.42
30. 16/1
31. 4/3
32. 8.556
33. 13.3%
34. 145
35. 8
36. 8/15
37. 0.7476
38. 182
39. 1/2
40. N = 120
41. N = 7 1/3 or N = 7.3
42. N = 1.87
43. N = 14.4
44. $9.54
45. 24 pieces
46. $96.11
47. 60%
48. 3/5
49. $2.00
50. 41 1/4 miles

EXERCISES—CHAPTER 8

Section 8.1

A. 1. $q = 4g$ 3. $M = m/g$

5. $C = 37s$ cents 7. $t = m + f$

9. $f = i/12$

B. 11. $q - 3 = p$ or $p + 3 = q$

13. $b/2 = a$ or $2a = b$

15. $x - 5 = w$ or $w + 5 = x$

C. 17. $A = R - S$; $A = 8$

19. $L = W/Y$; $L = 9$

21. $d = 52h$; $d = 364$ miles

23. $c = rs$; $c = 1,944$ people

25. $L = 32 + 26h$; $110

27. $P = 16.95 + 0.08c$; $19.59

D. 29. Answers will vary.

31. Answers will vary.

Section 8.2

A. 1. a. numerical coefficients b. variables

3. a. 6 b. m and n c. 4, 3, and 8

5. expression

7. numerical coefficient

9. exponent or power

B. 11. $5w$ 13. $12x$

15. $2a^3$ 17. $14P^2 - P$

19. $9c^2$

C. 21. $11y + 7z$ 23. $15f + 9g$

25. $13a + 4b$ 27. $7a^2 + 4b^2$

29. $19b + 5c$

D. 31. Expression: no equals sign; can only be simplified

Equation: equals sign; can be solved

33. Answers will vary.

Section 8.3

A. 1. 31 3. 11

5. 21 7. 8

9. 14 11. 77

13. 18 15. 75

17. 18 19. 31

B. 21. 72 23. 23

25. 108 27. 18

29. 79 31. 11

33. 48 35. 3

37. 31 39. 44

C. 41. 1. Divide

2. Multiply

3. Subtract

4. Add

43. 1. Subtract in ( )

2. Divide in ( )

3. Multiply

45. 1. Exponents and square roots

2. Divide

3. Multiply

4. Add

D. 47. Subtraction is on the left of addition.

49. Answers will vary.

Section 8.4

A. 1. b 3. c

5. g 7. a

9. i 11. k

13. b 15. k

17. k

B. 19. Answers will vary.

21. Answers will vary.

23. Answers will vary.

C. 25. Answers will vary.

27. Answers will vary.

Section 8.5

A. 1. equation 3. solution
5. 9 7. false
9. many 11. Yes
13. Yes

B. 15. $y = 9$ ck: $17 = 17$ 17. $M = 15$ ck: $5 = 5$
19. $T = 17$ ck: $12 = 12$ 21. $N = 12$ ck: $17 = 17$
23. $K = 10$ ck: $11 = 11$ 25. $C = 4$ ck: $12 = 12$

C. 27. $\{x \mid x < 6\}$ 29. $\{y \mid y > 12\}$
31. $\{a \mid a \le 5\}$ 33. $\{a \mid 20 \ge a\}$
35. $\{m \mid m > 10\}$

D. 37. The solution to an equation is a single value, whereas the solution set for an inequality contains an infinite number of values.
39. "$7 < 7$" is false. "$3 \le 3$" is true because 3 is equal to 3.

Section 8.6 Review Exercises

A. 1. $i = 36Y$ 3. $c = 9D$
5. $s = p + a$ 7. $r = d/t$
9. $t + f = V$

B. 11. $K = L + 6$ or $L = K - 6$
13. $A = 3B$ or $B = A/3$
15. $P = R \div 4$ or $R = 4P$
17. $C = D - 2$ or $D = C + 2$
19. $T = V - 5$ or $V = T + 5$

C. 21. x and y 23. 7 and 3
25. p and r 27. 6
29. 5, 3, and 1

D. 31. $2W + 4Y$ 33. $11c$
35. $5a^2$ 37. $5B$
39. $6w - 3x + 2y$ 41. $8y^2$
43. $4a^2$ 45. $4t^3 + 11t^2$

E. 47. 30 49. 12
51. 48 53. 248
55. 8 57. 32
59. 20

F. 61. 28 63. 60
65. 14 67. 96
69. 13

G. 71. $X = 13$ 73. $f = 6$
75. $B = 8$ 77. $m = 15$
79. $y = 4$

H. 81. 9; $\{g \mid g > 9\}$ 83. 18; $\{y \mid y < 18\}$
85. 6; $\{d \mid d \le 6\}$ 87. 30; $\{m \mid 30 > m\}$
89. 4; $\{x \mid x \ge 4\}$

I. 91. Variables in the formula are interchanged; $K - 12 = L$.
93. The coefficient of y is 1.
95. $36 \div 3 = 12$, not $4 \div 3 = 12$; $a = 36$

J. 97. Answers will vary.
99. Answers will vary.

Test—Chapter 8

1. $p = s - a$ 2. $P = c + t$
3. $s = 60m$ 4. $A = P/S$
5. $X = Y + 4$ or $Y = X - 4$
6. $G = H \div 6$ or $G = H/6$ or $H = 6G$
7. $5a$ 8. $11X + 7Y$
9. 78 10. 17
11. 7 12. 16
13. 33 14. 90
15. 5 16. $M = 11$
17. $K = 48$ 18. $\{y \mid y < 8\}$
19. $\{p \mid p \ge 5\}$
20. Wrong values are substituted for y and v; 288.

EXERCISES—CHAPTER 9

Section 9.1

A.

1. $B = 5$
   ck: $9 = 9$
3. $W = 4$
   ck: $12 = 12$
5. $m = 10$
   ck: $7 = 7$
7. $t = 9$
   ck: $6 = 6$
9. $4 = N$
   ck: $12 = 12$
11. $s = 6$
   ck: $30 = 30$
13. $C = 18$
   ck: $6 = 6$
15. $P = 27$
   ck: $9 = 9$
17. $y = 5/4$
   ck: $5/6 = 5/6$
19. $X = 4$
   ck: $7 = 7$
21. $W = 7$
   ck: $21 = 21$
23. $Y = 12$
   ck: $6 = 6$
25. $S = 20$
   ck: $12 = 12$
27. $B = 17$
   ck: $3 = 3$
29. $m = 35/9$
   ck: $5/9 = 5/9$
31. $X = 12$
   ck: $22 = 22$
33. $10/3 = t$
   ck: $10 = 10$
35. $W = 80$
   ck: $20 = 20$
37. $A = 3$
   ck: $14 = 14$
39. $R = 0$
   ck: $8 = 8$

B.

41. Need to multiply both sides by 8 to keep balance—not divide by 8.
43. Answers will vary.

Section 9.2

A.

1. $T = 9$
   ck: $11 = 11$
3. $F = 6$
   ck: $25 = 25$
5. $Y = 6$
   ck: $38 = 38$
7. $M = 8.5$ or $17/2$
   ck: $27 = 27$
9. $W = 3$
   ck: $47 = 47$
11. $M = 1.3$
   ck: $6.65 = 6.65$
13. $A = 50$
   ck: $57 = 57$
15. $T = 0$
   ck: $18 = 18$
17. $P = 8$
   ck: $1.6 = 1.6$
19. $T = 7$
   ck: $41 = 41$
21. $C = 4$
   ck: $48 = 48$
23. $X = 0$
   ck: $3 = 3$
25. $W = 20$
   ck: $2.4 = 2.4$
27. $F = 180$
   ck: $206 = 206$
29. $N = 16$
   ck: $44 = 44$
31. $T = 3$
   ck: $0 = 0$
33. $K = 9/8$
   ck: $13 = 13$
35. $F = 48$
   ck: $17 = 17$
37. $A = 11/15$
   ck: $3 = 3$
39. $W = 0.16$
   ck: $0.8 = 0.8$

B.

41. In the Order of Operations, you first multiply or divide, and then add or subtract. When solving equations, you usually add or subtract, and then multiply or divide.
43. They are solved the same way, subtract from both sides and then divide.
45. In $3m + 6 = 18$, subtract 6 then divide by 3. In $\frac{2}{3}m = \frac{1}{6} = 1\frac{3}{4}$, subtract 1/6 and divide by 2/3. For each, check by substituting your answer for m in the original equation.
47. After subtracting 16, you divide 0 by 8, which equals 0.

Section 9.3

A.

1. $T = 2$
   ck: $16 = 16$
3. $F = 5$
   ck: $13 = 13$
5. $R = 3$
   ck: $43 = 43$
7. $40 = n$
   ck: $240 = 240$
9. $T = 2$
   ck: $22 = 22$
11. $M = 9$
   ck: $18 = 18$
13. $r = 7$
   ck: $21 = 21$
15. $M = 50$
   ck: $25 = 25$
17. $4 = Y$
   ck: $32 = 32$
19. $T = 3$
   ck: $30 = 30$
21. $M = 4$
   ck: $52 = 52$
23. $R = 0$
   ck: $8 = 8$
25. $12 = C$
   ck: $240 = 240$
27. $x = 4$
   ck: $28 = 28$
29. $t = 1$
   ck: $16 = 16$
31. $C = 10$
   ck: $11 = 11$
33. $R = 12$
   ck: $36 = 36$
35. $T = 3/2$
   ck: $15 = 15$

37. $M = 0$
ck: $6 = 6$

39. $N = 0$
ck: $9 = 9$

B. 41. Should not subtract $8C$ twice on one side, but should combine $11C$ and $8\ C$ because they are on the same side of the equals sign.

43. Answers will vary.

Section 9.4

A. 1. $R = 3$
ck: $27 = 27$

3. $t = 1/3$
ck: $2\ 2/3 = 2\ 2/3$

5. $W = 2$
ck: $22 = 22$

7. $1 = k$
ck: $12 = 12$

9. $m = 4$
ck: $24 = 24$

11. $T = 3$
ck: $20 = 20$

13. $w = 3$
ck: $11 = 11$

15. $Y = 3$
ck: $34 = 34$

17. $x = 1.2$
ck: $1.58 = 1.58$

19. $W = 2$
ck: $7 = 7$

21. $x = 3$
ck: $3 = 3$

23. $a = 0$
ck: $7 = 7$

25. $x = 2$
ck: $21 = 21$

27. $C = 3$
ck: $29 = 29$

29. $k = 4$
ck: $24 = 24$

31. $y = 1$
ck: $32 = 32$

33. $W = 10$
ck: $31 = 31$

35. $X = 2$
ck: $5 = 5$

37. $a = 3$
ck: $8 = 8$

39. $c = 6$
ck: $27 = 27$

B. 41. $b = 33$
ck: $27 = 27$

43. $y = 30$
ck: $2 = 2$

45. $a = 7$
ck: $28 = 28$

47. $T = 11$
ck: $8 = 8$

49. $y = 8$
ck: $35 = 35$

51. $c = 5$
ck: $22 = 22$

53. $a = 0$
ck: $7 = 7$

55. $x = 30$
ck: $6 = 6$

57. $Z = 9$
ck: $37 = 37$

59. $R = 2$
ck: $73 = 73$

C. 61. First combine like terms on the same side of the equals sign. Then look for the smaller variable term, and eliminate it by doing its opposite to both sides.

63. Steps 1 through 4 help to isolate the variable on one side of the equals sign.

65. Substitute the answer for M in all places in the original equation. Use the Order of Operations to simplify each side of the equals sign. If you reach an identity statement, you know your solution is correct.

Section 9.5

A. 1. yes

3. no

5. yes

B. 7. $K = 10/9$
ck: $1/2 = 1/2$

9. $N = 3$
ck: $7.12 = 7.12$

11. $W = 20$
ck: $2.4 = 2.4$

13. $A = 175$
ck: $144 = 144$

C. 15. $X = 15/2$
ck: $7/2 = 7/2$

17. $K = 3.3$
ck: $1.12 = 1.12$

19. $N = 2/3$
ck: $15/2 = 15/2$

21. $2.45 = X$
ck: $17.64 = 17.64$

23. $M = 2$
ck: $13/4 = 13/4$

25. $X = 5$
ck: $22.5 = 22.5$

27. $N = 0.25$
ck: $1.35 = 1.35$

29. $A = 4/3$
ck: $6 = 6$

31. $x = 1\ 2/3$
ck: $2\ 1/3 = 2\ 1/3$

33. $c = 8/9$
ck: $10\ 1/3 = 10\ 1/3$

35. $X = 0.8$
ck: $0.897 = 0.897$

37. $c = 2.27$
ck: $3.64 = 3.64$

39. $c = 1/15$
ck: $3/4 = 3/4$

D. 41. $y \doteq 0.31$
ck: $6.12 \doteq 6.35$

43. $t \doteq 0.25$
ck: $0.23 \doteq 0.24$

45. $Y = 1.2$
ck: $2.026 = 2.026$

47. $X = 0.5$
ck: $0.48 = 9.48$

E. 49. $x = 5$
ck: $7 = 7$

51. $a = 5/62$
ck: $5/12 = 5/12$

53. $M = 6\ 2/3$
ck: $6\ 7/12 = 6\ 7/12$

F. 55. Answers will vary.

57. Answers will vary.

59. Did not multiply *all* terms by 8/1; $x = 3\ 1/2$

61. Yes, they are equivalent. If you multiply the fraction equation by 8, it gives $6x + 3 = 16$.

Section 9.6

A. 1. 24 feet
3. 17 centimeters
5. 50 inches
7. \$3,000
9. 119.32 feet
11. 17 1/2 square centimeters
13. 4 years
15. 36 square feet
17. 5 years
19. 1,133.54 square feet
21. \$17,187.50
23. \$157.30
25. 65 square inches

B. 27. 12 centimeters
29. 7 inches
31. 24 inches
33. 25 feet
35. 32
37. 4.5 centimeters
39. 8 kilometers
41. 3 years
43. $S - C = M$
45. $\frac{E}{B} = A$

C. 47. $\frac{3V}{b} = h$
49. $\frac{I}{RT} = P$
51. $\frac{y - b}{m} = x$
53. $\frac{A - P}{PR} = T$
55. $\frac{E}{I} = R$
57. $\frac{3V}{\pi r^2} = h$
59. $r^3 = \frac{3V}{4\pi}$
61. $P - b - c = a$
63. $\frac{2S}{t^2} = g$

D. 65. a. Use Order of Operations on right side because the equation is already solved for P.

b. Use equation-solving steps.

67. Both are equations with two terms on the left side and one term on the right. Both can be solved for a variable—only for one in the first equation, but for three variables in the second formula. First subtract, then divide by 2.

69. When you rearrange the formula, $W = \frac{V}{LH}$.

This means that cubic feet are divided by (feet)(feet). Two pairs of feet will cancel out, leaving feet.

Section 9.7 Review Exercises

A. 1. $X = 4$ ck: $12 = 12$
3. $F = 42$ ck: $23 = 23$
5. $5 = H$ ck: $9 = 9$
7. $A = 6$ ck: $16 = 16$
9. $4 = W$ ck: $11 = 11$
11. $c = 10.81$ ck: $9.44 = 9.44$
13. $Y = 36$ ck: $24 = 24$
15. $2 = M$ ck: $14 = 14$
17. $4 = S$ ck: $32.1 = 32.1$
19. $X = 24$ ck: $24 = 24$
21. $P = 8/3$ ck: $24 = 24$
23. $Y = 3$ ck: $3.6 = 3.6$
25. $C = 4$ ck: $64 = 64$
27. $A = 24/7$ ck: $6 = 6$
29. $S = 0$ ck: $18 = 18$
31. $F = 4$ ck: $7 = 7$
33. $Y = 4$ ck: $16 = 16$
35. $1 = R$ ck: $24 = 24$
37. $y = 54/11$ ck: $45/11 = 45/11$
39. $M = 2$ ck: $2 = 2$
41. $Y = 16/15$ ck: $2/3 = 2/3$
43. $w = 2$ ck: $12 = 12$
45. $R = 4$ ck: $8 = 8$
47. $R = 3$ ck: $39 = 39$
49. $y = 10$ ck: $14 = 14$

B. 51. 7 1/2 hours
53. 70 feet
55. 3 years
57. \$2.34
59. 8 inches
61. 24 cm
63. 4 years
65. $C - A = B$
67. $\frac{W}{N} = M$
69. $\frac{2A}{h} = b$

C. 71. $\frac{P - 2L}{2} = W$
73. $\frac{V}{LW} = H$
75. $x = \frac{36 - 9y}{2}$
77. $a^2 = c^2 - b^2$
79. $\frac{3V}{WH} = L$

D. 81. 1. Divide both sides by 15.

83. 1. Combine $3C$ and C.
2. Combine $17 - 6$.
3. Add 5 to both sides.
4. Divide both sides by 4.
5. Write the original equation.
6. Substitute 4 in place of C's.
7. Use Order of Operations to simplify both sides to an identity.

85. 1. Multiply each term by 8.
2. Subtract 6 from both sides.
3. Divide both sides by 7.

E. 81. The steps are the same because you are isolating a particular variable in both cases.

83. Multiplication Property of Equality
Addition Property of Equality

85. Answers will vary.

87. $12A$ should not be subtracted on the left side. Combine like terms, $18A + 12A$.

89. Every term must be multiplied by 12, even the 8.

91. They did not start in the original equation.

F. 93. Answers will vary.

95. Subtract $2b$ and add $5c$ to both sides. Then divide by 3.

Test—Chapter 9

1. $R = 27$
ck: $81 = 81$

2. $R = 7$
ck: $38 = 38$

3. $C = 14$
ck: $35 = 35$

4. $w = 48$
ck: $12 = 12$

5. $A = 7$
ck: $35 = 35$

6. $m = 5.8$
ck: $14.32 = 14.32$

7. $B = 15/2$
ck: $3/4 = 3/4$

8. $R = 2$
ck: $6 = 6$

9. $w = 1$
ck: $3 = 3$

10. $T = 20/9$
ck: $1 = 1$

11. $A = 15$
ck: $25 = 25$

12. $Y = 3$
ck: $11 = 11$

13. $c = 0$
ck: $17 = 17$

14. $N = 2$
ck: $21 = 21$

15. $T = 32/3$
ck: $4 = 4$

16. $M = 20$
ck: $14.2 = 14.2$

17. $P = 3$
ck: $34 = 34$

18. $PV/nT = R$

19. Combine $7Y$ and $-2Y$—not add $2Y$ twice on the left side.

20. 1. Subtract $9P$ from both sides.
2. Add 2 to both sides.
3. Divide both sides by 3.
4. Write the original equation.
5. Substitute 3 for P's.
6. Use Order of Operations to reach an identity.

EXERCISES—CHAPTER 10

Section 10.1

A. 1. $T + 11$
3. $T - 10$
5. $2T$
7. $17/T$
9. $2/3\ T + 12$
11. $T - 8$
13. $T + 17$
15. $T - 20$
17. $8 - T$
19. $2T - 4$
21. $T + 18$

B. 23. d
25. a
27. j
29. g
31. h

C. 33. $68x$
35. $M - 5$
37. $R/4$
39. $P + 11$
41. $5/6\ S$
43. $\$1.09G$
45. $Y + 4$
47. $\$0.25Q$
49. $D + 6$; $6D$; $D + D + 6 + 6D$, or $8D + 6$

D. 51. Answers will vary.
53. Answers will vary.
55. Answers will vary.

E. 57. The order in a "more than" or "less than" algebraic expression is reversed from the English word order.

59. No, the order is different; $16/N$ and $N/16$.

61. Difference means subtract; *and* indicates where the subtraction symbol is placed.

Section 10.2

A. 1. $N + 17 = 39$; number is 22

3. $11N = 99$; number is 9

5. $N/12 = 4$; number is 48

7. $N - 4 = 13$; number is 17

9. $N + N + 1 + N + 2 + N + 3 = 178$; 43, 44, 45, 46

11. $3N + 12 = 48$; number is 12

13. $2/3\ N - 16 = 56$; number is 108

15. $N + N - 4 = 20$; 12, 8

17. $N + N + 2 = 152$; 75, 77

19. $6N = 2N + 12$; number is 3

21. $N + N + 2 = 146$; 72, 73, 74

23. $B + B - 3 = 113$; Betsy: 58 years Patty: 55 years

25. $2B + 4 - B = 184$; \$180, \$364

27. $3/4\ S = 7{,}200$; 9,600 students

29. $N + 1 + N + 2 = 89$; 43, 44, 45

31. $X + 5X = 12$; 2 feet, 10 feet

33. $1/4\ P = 12$; 48 points

35. $2N + 40 + N = 580$; \$180, \$400

37. $4N - 2N = 28$; 14, 56

39. $N + N + 1 = N + 2$; 1, 2, 3

41. $H + H - 6 = 70$; 32 hours

43. $179 = 47 + 12M$; \$11 per month

45. $10.50d + 12 = 64.50$; 5 days

47. $c + 3c = 780$; chair: \$195, sofa: \$585

49. $N + N + 2 + N + 4 = 57$; 17 yr, 19 yr, 21 yr

51. $4X + 5X = 90$; 40°, 50°

53. $6x + 5x = 77$; 42 and 35

55. $4x + 3x + 2x = 117$; 52, 39, and 26

B. 57. Answers will vary.

59. An equation has an equals sign and can be solved. An expression has no equals sign and can only be simplified.

Section 10.3

A. 1. $7 + 11 + a = 25$; length: 7 feet

3. $s + s + 90 = 200$; equal sides: 55 meters

5. $s + s + s = 135$; 45 feet

7. $6s = 144$; 24 centimeters

9. $x + x + 20 + x + x + 20 = 96$; width: 14 centimeters, length: 34 cm

11. $x + 4x + x + 4x = 40$; 16 inches

13. $x + 3x + 4 + x + 3x + 4 = 88$; width: 10 inches, length: 34 inches

15. $x + x + 2x - 6 = 66$; 18 yd, 18 yd, 30 yd

17. $4x + 3x + 4x + 3x = 252$; 54 meters

19. $9x + 4x + 9x + 4x = 182$; length: 63 in, width = 28 in

21. $3x + 4x + 5x = 60$; 15 yd, 20 yd, 25 yd

B. 23. $62(t) + 56(t) = 531$; 4 1/2 hours

25. $60(t) + 58(t) = 472$; 4 hours

27. $22(t) + 28(t) = 200$; 4 hours

29. $5(r) + 5(2/3\ r) = 500$; 60 mph, 40 mph

31. $64t + 56t = 360$; 3 hours

C. 33. Probably not, because there was nothing to picture. But distance and perimeter could be done by listing information.

Section 10.4

A. 1. $(0.01)N + (0.25)N = 1.56$; 6 pennies, 6 quarters

3. $(0.25)N + (0.10)8N = 210$; 2 quarters, 16 dimes

5. $(2.50)N + (1.50)2N = 269.50$; 49 adults, 98 children

7. $12.95(S) + 18.95(2S) + 7.95(3S) = 149.40$; 2 seniors, 4 adults, 6 children

9. $17{,}520 = 24{,}000(0.12)(4) + X(0.10)(4)$; \$15,000

11. $(1.95)2X + (0.85)X = 9.50$; 2 pens, 4 notebooks

13. $4x(0.06)(5) + x(0.05)(5) = 2900$; \$8,000 at 6%, \$2,000 at 5 %

15. $4X(0.07)(3) + X(0.06)(3) = 2,040$; \$2,000 at 6%, \$8,000 at 7%

17. $1.00(1/4\ X) + 0.25(X) = 14.00$, 7 dollar bills

19. $25(4X) + 15(X) = 46,000$; 1,600 adults

21. $3(4x) + 2x = 630$; 180 adult tickets

23. $6x(0.07)(3) + x(0.05)(3) = 1410$; \$6,000 at 7%, \$1,000 at 5%

B. 25. $2x(3) + x(3) = 270$; 60 mph, 30 mph

27. $45t + 35t = 320$; 4 hours

C. 29. Not really because the chart seems to work best with problems requiring multiplication.

31. Yes, multiply \$0.25 times 9. To find the value of X quarters, you multiply \$0.25 times X. It is the same process.

33. Answers will vary.

Section 10.5 Review Exercises

A. 1. e 3. g

5. b 7. d

B. 9. l 11. n

13. i 15. j

17. r

C. 19. $2n + 7 = 25$; number is 9

21. $n + n + 1 + n + 2 = 87$; 28, 29, and 30

23. $1/2\ n - 4 = 16$; number is 40

25. $x + 12 + x + x + 12 + x = 52$; 7 in., 19 in.

27. $(0.05)N + (0.10)N = 1.95$; 13 nickels, 13 dimes

29. $n + 4n = 35$; 7 and 28

31. $n + n + 4 = 72$; 34, 36, and 38

33. $n + 5 + 3n = 37$; 13

35. $x + 6 + x + x + 6 + x = 48$; 9 in., 15 in.

37. $2x + 3x + 2x + 3x = 180$; 36 yd, 54 yd

39. $81 = 3s$; 27 ft, 27 ft, and 27 ft

41. $3N(0.0625)(3) + N(0.055)(3) = 2,910$; \$12,000

43. $3/4n - 8 = 10$; number is 24

45. $x + 2x + 2x + 1 = 91$; 18 ft, 36 ft, 37 ft

47. $2X(0.07)(4) + X(0.06)(4) = 4,564$; \$11,410

49. $2X + X + 2X + 20 = 180$; 64°, 32°, 84°

51. $3/8\ N + 6 = 12$; number is 16

53. $5x + 9x + 5x + 9x = 112$; width = 20 m, length = 36 m

55. $3x(0.05)(2) + x(0.04(2) = 152$; \$1200 at 5%, \$400 at 4%

57. $3x + x + 3x + x = 72$; 9 meters by 27 meters

59. $X + X + X - 2 = 37$; 13 ft, 13 ft, 11 ft

D. 61. Need 2 lengths and 2 widths; $5x + x + 5x + x = 100$

63. Need to use percent numbers in decimal form; $2X(0.06)(4) + X(0.05)(4) = 1,020$

65. Add only the first and third; $x + x + 4 = 122$

E. 67. Answers will vary.

69. Answers will vary.

F. 71. $x(0.08)(18) + x(0.09)(18) = 76,500$; \$25,000; \$126,500

73. $1\ 1/2\ R + 1\ 1/2(2/3\ R) = 7$; 1 13/15 mph; 4 1/5 miles

75. $X + X + 2X + 8 = 200$; 48 feet by 104 feet

77. $5x(0.07)(4) + 3x(0.06)(4) = 1,060$; \$2,500 at 7%, \$1,500 at 6%

Test—Chapter 10

1. $N - 8$

2. $3N + 4$

3. $4R + 5$

4. $3/4N + 7$

5. $2N - 6 = 20$

6. Number is 13.

7. $X + X + 2 + X + 4 = 27$

8. 7 years, 9 years, 11 years

9. $5/6\ S = 35$

10. 42 students

11. $X + X + 8 + X + X + 8 = 180$

12. 41 meters wide by 49 meters long

13. $0.25(2X) + 1,00(X) = 43.50$

14. 58 quarters

15. $8x + 4x + 3x = 37.5$
16. 20 hours
17. $3(r) + 3(2r) = 1{,}575$
18. prop plane: 175 mph, jet: 350 mph
19. Let X = amount invested at 8%.

    Then $2X$ = Amount invested at 5%.

    The time is 4 years.

    Use the simple interest formula to find expressions for both interests and then combine them to total $2,880. $X(0.08)(4) + 2X(0.05)(4) = 2{,}880$

20. When you have solved the equation for X, you know the amount invested at 8%. Two times that amount, $2X$, is the amount invested at 5%.

EXERCISES—CHAPTER 11

Section 11.1

A. 1. −4 3. +26
5. +276 7. +5,280
9. +75 11. +5
13. −2,200 15. +2,000

B.

-5 +2 +4.2

$-3\frac{1}{3}$ 0

C. 27. $<$ 29. $>$
31. $<$ 33. $>$
35. $>$ 37. $>$
39. $<$ 41. $>$
43. $>$

D. 45. 6 47. 9
49. 0 51. 1.4
53. 8 55. 2
57. 4 59. 24
61. 58

E. 63. $|0|$ $|+2|$ $|-3|$ $|+4|$ $|-7|$
65. $|+3|$ $|-7|$ $|+9|$ $|-10|$ $|-12|$

F. 67. The larger number is on the right.
69. Answers will vary.

Section 11.2

A. 1. −3 3. −11
5. −7 7. +10
9. −5 11. 0
13. −16 15. −6
17. −10 19. +7
21. −7 23. 0
25. +6 27. −4
29. +8

B. 31. $(-14) + (+10) = -4$; lost by 4 points
33. $(-150) + (+250) = +100$; 100 feet above sea level
35. $(+23) + (-40) = -17$; 17 miles west
37. $(-8) + (-8) + (-8) + (+10) + (+10) = -4$; 4 dollar loss
39. $6{,}000 + (-2{,}000) = +4{,}000$; 4,000 feet higher

C. 41. 8 1/3 43. −12
45. −3.0 47. −10 2/3
49. 0

D. 51. 7.177 53. −11.275
55. −92.927

E. 57. Answers will vary.
59. True because $|-3| = 3$ and $|+1| = 1$ and $3 \geq 1$.

Section 11.3

A. 1. −1 3. −3
5. −18 7. +10
9. −6 11. +5
13. −11 15. −9
17. −10.8 19. +13.1

B. 21. +3 23. +2
25. +17 27. −23
29. −17

C. 31. –12 33. +9
35. –19 37. –5
39. –3 1/4 41. –15 1/2
43. –7 45. +4.1

D. 47. (–8) + (+17) – (–3) = +12
49. (–7) – (+12) = –19
51. (–3/8) + (–3/8) + (–3/8) + (–3/8) = –1 1/2; lost 1 1/2 points

E. 53.

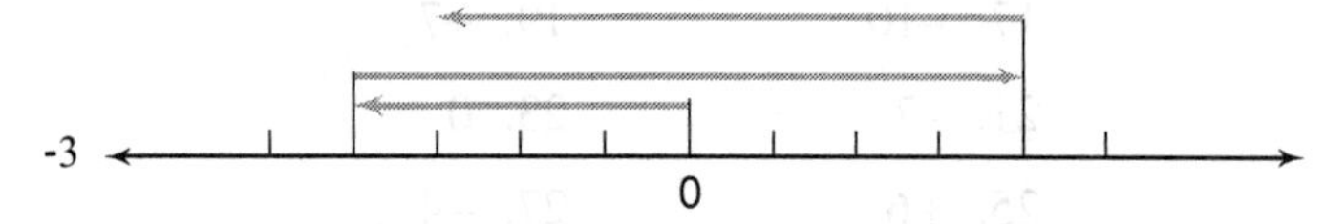

55.

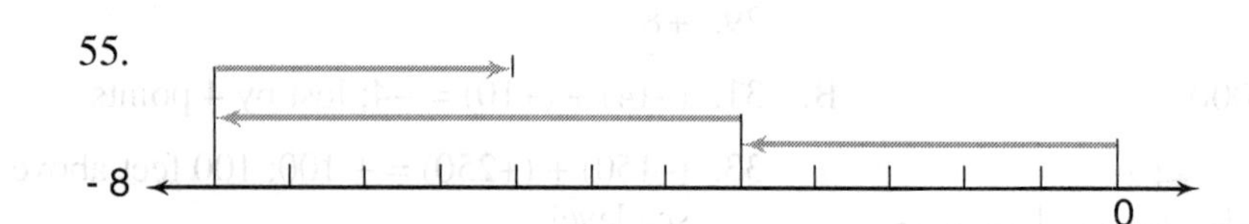

57.

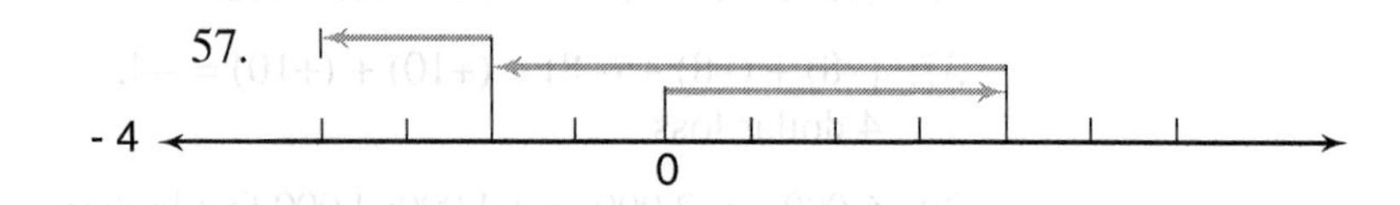

59.

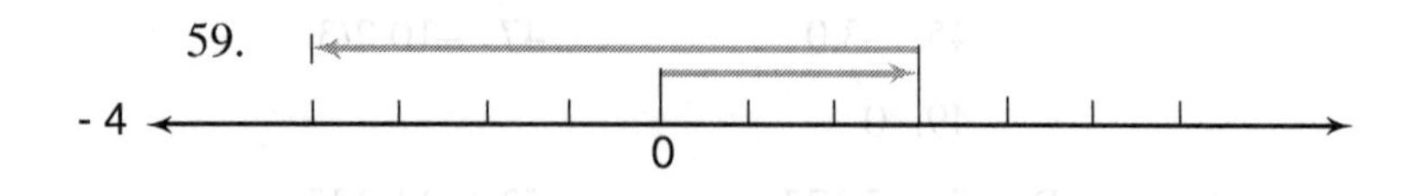

61.

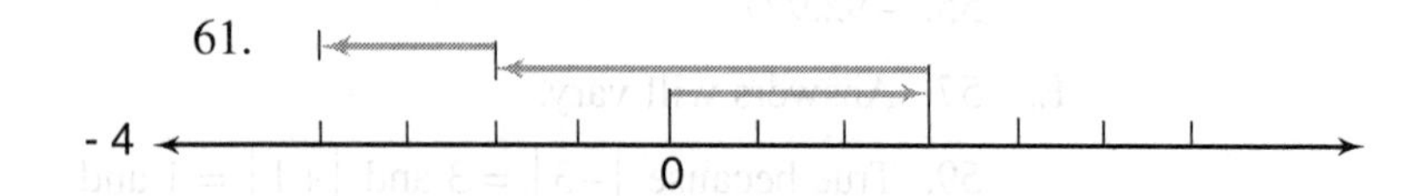

F. 63. –11.682 65. –82.605
67. –17.504

G. 69. 270 – 125 – 82 + 165 – 94 = $134

Section 11.4

A. 1. +55 3. –48
5. +54 7. –36
9. 0 11. –30
13. –42 15. 0
17. +24 19. +24

B. 21. +31 23. –1
25. –24 27. –3
29. +42 31. –4
33. +51 35. +2
37. –20 39. –3

C. 41. –9 43. –64
45. –8 47. +16
49. –4 51. +108
53. –49

D. 55. 13/54 57. –5 1/16

59. –3 13/16

E. 61. –14.71 63. –8,494.26

65. 2,081.51 67. 402.40

69. –518.69 71. 68.81

73. –16.19

F. 75. Three (–2)s means (–2) + (–2) + (–2) = –6.

77. $(-5)^4$ means (–5) (–5) (–5) (–5) = +625, but -5^4 is "the opposite of" 5 to the fourth power, or "the opposite of" +625, which is –625.

Section 11.5

A. 1. –2 3. –4

5. +4 7. –2/5

9. –12/5 11. +9

13. –5 15. –40

17. –2 19. +1

B. 21. +4/3 23. –12/5

25. –12 27. –12

29. 36 31. –62

33. 2 35. –27

37. –1 39. –7/2

41. +1 43. –9

45. –6 47. +31

49. –4

C. 51. (–7) (+6) (–3) – (–9) = +135

53. $(-6) + (+14) - \frac{(-24)}{(-3)} = 0$

55. $\frac{(-8) + (+12)}{(-8) - (+12)} = \frac{+4}{-20} = -\frac{1}{5}$

D. 57. Count the number of negative values in the division problem. An even number of negatives produces a positive quotient. An odd number of negatives produces a negative quotient.

59. $\frac{(-4)^3}{-4^2} = \frac{-64}{-16} = +4$ Raise (–4) to the 3rd. Raise 4 to the 2nd. Opposite of +16 is –16. Divide –64 by –16. Result is +4.

Section 11.6

A. 1. –1 3. +6

5. +24 7. –16

9. +22 11. +7

13. +12 15. –12

17. +45 19. –32

21. –8/3 23. +5/3

25. +12 27. +17

29. –3/5 31. –13

33. +9 35. +5

B. 37. $-\frac{33}{40}$ or –0.825 39. $-3\frac{1}{4}$ or –3.25

41. $+\frac{1}{4}$ or +0.25 43. –8 or –8.0

45. $8\frac{27}{40}$ or 8.675 47. 1/10 or –0.1

49. $-1\frac{1}{10}$ or –1.1

C. 51. 0(0.5) = 0, not 0.05; –30

53. $(+5)^2$ = +25, not +10; +28

55. +(–3) is not +(+3); –14.5 or $14\frac{1}{2}$

57. Sum of two negatives is negative; –9

Section 11.7 Review Exercises

A. 1. distance and direction

3. less than

5. +6

7. add the numbers and use the sign

9. even 11. positive

13. negative 15. negative

17. See Section 11.6. 19. –2

B. 21–29

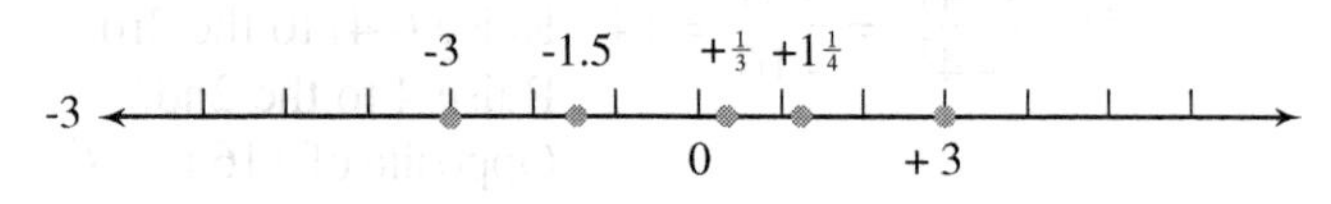

C. 31. +3 33. –15
35. –5 37. –6
39. –9 41. –3
43. –3 45. –16
47. –4 49. –2
51. –4/3 53. –9
55. –16/27 57. –6
59. 1/50 61. 3/16
63. –3 65. –1 3/5
67. –7.224 69. –1

D. 71. 0 73. +7
75. –6 77. +5
79. –36 81. +48
83. +12 85. +3/2
87. +1/5 89. –5/13
91. 3 93. 11
95. –4 97. 6
99. 0

E. 101. –29 103. –0.1854944
105. 0 107. 5.5

F. 109. Sum of two negatives is negative; –15
111. Exponent, not multiplication; –243
113. Positive ÷ Negative = Negative; –4
115. –30 – (–10) = –30 + (+10); –20

G. 117. Answers will vary.
119. $n + n + 1 + n + 2 = -69$
$n = -24$
$n = -24$, $n + 1 = -23$, $n + 2 = -22$
This time, solutions are negative values.

Test—Chapter 11

1. –3 2. 4
3. –24 4. –7
5. –64 6. –28
7. –5 8. –10
9. 8 10. –72
11. 41 12. –28
13. 6 14. 1
15. –3 16. 1
17. 1/5 18. 4
19. (–2) – (–4) ≠ (–2) + (–4)
Also, should multiply [ ] by 3 first, and then subtract product from 14. Simplification is 8.
20. Because both +3 and –3 are three units from zero, by the definition of the absolute value, M could be +3 or –3.

Cumulative Review 1–11

1. 33.36 2. 19 1/9
3. $6x + 11y$ 4. –10
5. 12 6. –3
7. –3 8. 1 5/12
9. 14.44 10. $-9a$
11. 17.6 12. +42
13. 289/24 or 12 1/24 14. 144/5 or 28 4/5
15. +12 16. +4
17. 42 18. –7/8
19. –6 20. 30
21. 7 22. 45/16 or 2 13/16
23. 8 24. –3/4
25. –10 26. $N - 6$
27. $2N + 7$ 28. $N(8)$ or $8N$
29. $15/N = -5$ 30. $C = 5.85N$
31. $Y = +8$
32. $N = 13/5$ or 2 3/5 or 2.6
33. 14.4
34. $x = 5$
35. $M = -80/9$ or –8 8/9
36. $A = 6$
37. $W = +0.05$

38. 20%
39. $X > 7$
40. $N = 7$
41. $W = 21$ inches
42. $M = \$55.70$
43. 5 1/2% or 5.5%
44. 63°
45. –27, –25, –23
46. $14.63
47. 62 miles
48. 4.3 ounces
49. 314 square inches
50. 11 yards by 44 yards

EXERCISES—CHAPTER 12

Section 12.1

A. 1. $-44 = -44$, yes
3. $16 = 16$, yes
5. $2 = 2$, yes
7. $9 \neq 0$, no
9. $0.2 = 0.2$, yes

B. 11. $5K + 10$
13. $-10X - 2$
15. $-4 + 6c$
17. $-28x + 36$
19. $7x + 42$
21. $7M + 35$
23. $3x + 6$
25. $-5m - 20$
27. $6K - 12$
29. $-8Y + 32$
31. $34x - 51 + 68y$
33. $16N + 6.72P$
35. $-36A + 60B - 24C$
37. $9A - 6C$
39. $-4M - 3N$

C. 41. $2.47F - 0.039$
43. $-62.72 + 101.92W - 222.88X$
45. $3{,}199{,}755M + 8{,}753{,}805N$
47. $1\ 75/128Y - 2\ 5/12$
49. $133\ 8/27 + 158\ 4/27T$

D. 51. Answers will vary.

Section 12.2

A. 1. $-1 - 6x$
3. $22 - 5k$
5. $5A - B$
7. $4R + 4$
9. $-10T - 2$
11. $4m + 5$
13. $7B - 2$
15. $-13R + 12$
17. $11 - w$
19. $13 - 4N$
21. $2Y + 6$
23. $24x - 6$
25. $-3 - 5w$
27. $4y - 4$
29. $18y - 38$
31. $-9x + 21$
33. $-17a - 41b$
35. $7H - 8$
37. $18p + 84$
39. $7w - 24$

B. 41. 1. Distribute 2.
2. Distribute 3.
3. Combine variable terms.
4. Combine numerical terms.

43. 1. Distribute –3.
2. Combine like terms.

45. 1. Insert 1.
2. Distribute –1.
3. Combine variable terms.

47. 1. Distribute +2.
2. Distribute –2.
3. Distribute –2.
4. Combine variable terms.

C. 49. Both result in sign changes for all terms.

51. Multiplying by –1 changes the sign; the subtraction operation also uses the opposite of the quantities involved, which in effect changes the sign.

Section 12.3

A. 1. $x = 4$
ck: $18 = 18$

3. $C = 11$
ck: $-4 = -4$

5. $1/5 = P$
ck: $24 = 24$

7. $N = 9$
ck: $38 = 38$

9. $x = -11/4$ or $-2\ 3/4$
ck: $-15/2 = -15/2$

11. $c = -7$
ck: $10 = 10$

13. $B = \frac{7}{2}$
ck: $-\frac{1}{6} = -\frac{1}{6}$

15. $K = -25$
ck: $16 = 16$

17. $r = -7$
ck: $25 = 25$

19. $m = -2$
ck: $-49 = -49$

21. $x = -12$
ck: $-35 = -35$

23. No solution

25. No solution

27. $A = -2$
ck: $21 = 21$

29. $x = 2\ 1/2$
ck: $28 = 28$

31. $-3 = m$
ck: $15 = 15$

33. No solution

35. All real numbers

37. $R = -0.5$
ck: $1.6 = 1.6$

39. No solution

41. All real numbers

B. 43. 1. Distribute 2
2. Add 6 to both sides.
3. Divide both sides by 2.
4. Substitute -3 for A.
5. Combine like terms.
6. Multiply.

45. 1. Distribute 8.
2. Distribute 4.
3. Subtract 8c.
4. State meaning of $16 \neq -12$.

C. 47. Distribute 2/3 and 1/2. Add Y to both sides. Add 6 to both sides. Divide both sides by 5.

49. The variable terms drop out, and the resulting numerical statement is not true. The answer, then, is "no real solution."

51. In both cases, the variable terms drop out, leaving a constant equal to a constant. When that statement is true, the solution is "all real numbers." When that statement is faose, the solution is "no solution."

Section 12.4

A. 1. $2(n) - 3(n - 5) = -7$; sm: 17, lg: 22

3. $J - 19 + J + 2(J - 19) = 163$;
M: \$36, J: \$55, D: \$72

5. $4(n) - (n + 4) = -40$; largest number is -8

7. $6(N - 3) = N - 3$; number is 3

9. $2(N + 4) - 3N = -4$; 12, 14, 16

11. $c + c + 16 + 2(c + 16) = 464$;
Lorraine earned \$240

13. $2(X + 8) + 2(X) = 76$; L: 23 ft, W: 15 ft

15. $2(3X - 20) + 2(X) = 440$; L: 160 ft, W: 60 ft

17. $0.25(2x + 4) + 0.10(x) = 3.40$;
4 dimes, 12 quarters

19. $2.00(X) + 3.50(120 - X) = 360.00$ or
$2.00(120 - X) + 3.50(X) = 360.00$;
40 students, 80 nonstudents

21. $5(x) + 7[2(x + 6)] + 8(x + 6) = 321$;
26 chocolate cakes

23. $6(r) + 6(r + 12) = 732$; 55 mph, 67 mph

25. $3x(0.045)(4) + x(0.035)(4) = 1{,}020$
\$1,500 at 3 1/2, \$4,500 at 4 1/2%

27. $0.08(300 - X) + 0.05(X) = 0.06(300)$ or
$0.08(X) + 0.05(300 - X) = 0.06(300)$;
200 ml of 5%, 100 mL of 8%

29. $1(X) + 5(30 - X) = 50$ or
$1(30 - X) + 5(X) = 50$; 5 5–dollar bills

31. $0.20(X) + 0.15(9) = 0.17(X + 9)$; 6 quarts

33. $0.25(x) + 0.20(20) = 0.23(x + 20)$ 30 liters

35. $3(r) + 3(r + 10) = 330$;
Nan: 60 mph, John: 50 mph

37. $(X + 5{,}000)(0.05)(3) + X(0.04)(3) = 1{,}290$
\$7,000 at 5%, \$2,000 at 4%

39. $2(1/2\,X + 7) + 2\,X = 86$
width = 19 yards

B. 41. Marisol, because she is driving faster.

43. The 10% mixture because it is closer to the 12% mix than is the 15% mixture

Section 12.5

A. 1.

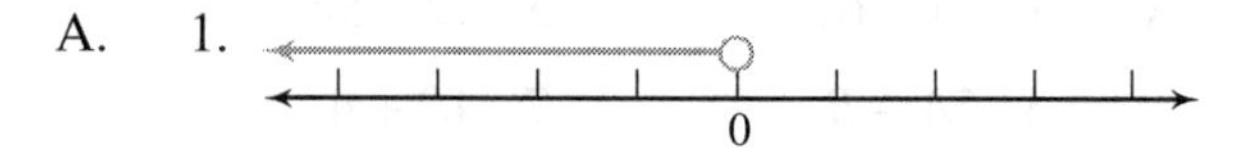

3.

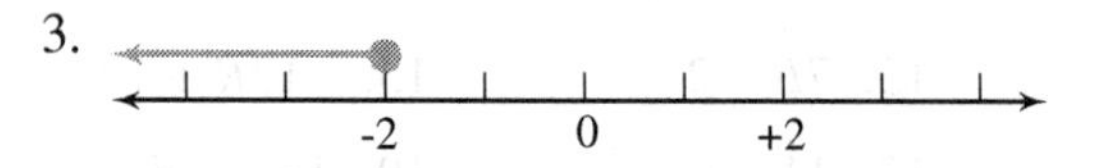

5.

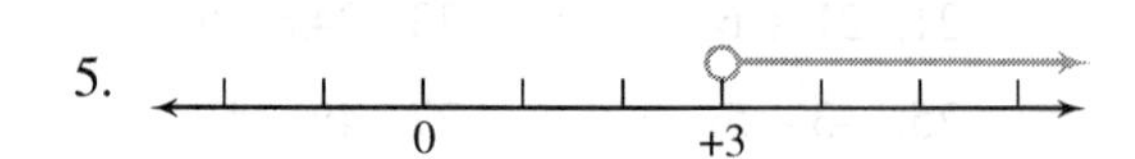

7. 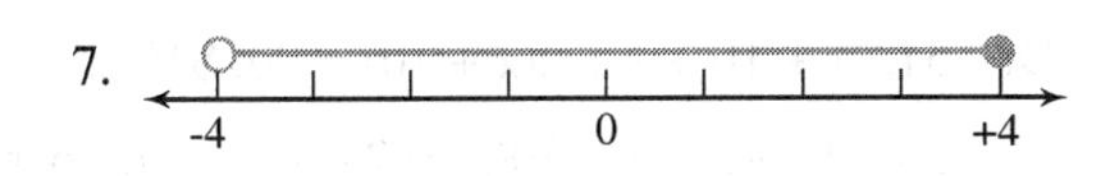

9. 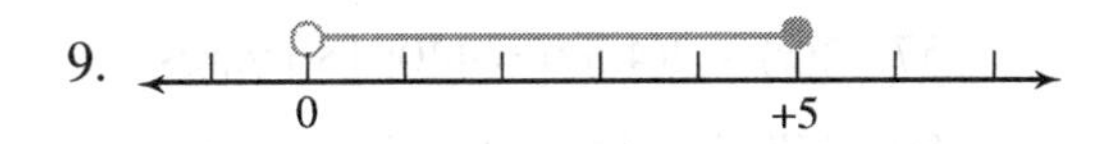

11. 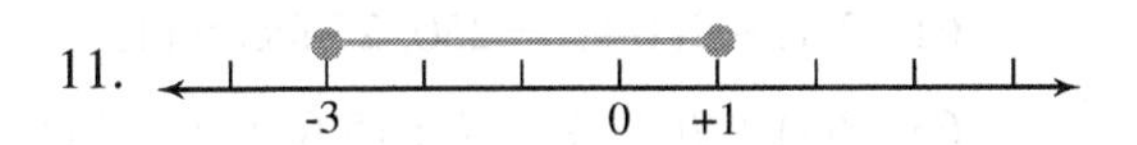

B. 13. $\{x|\ x \leq 7\}$ or $(-\infty, 7]$

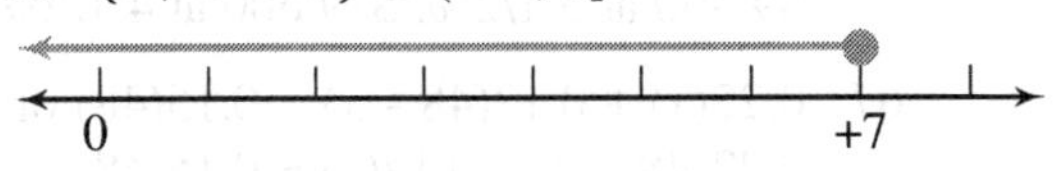

15. $\{x|\ x > 6\}$ or $(6, \infty)$

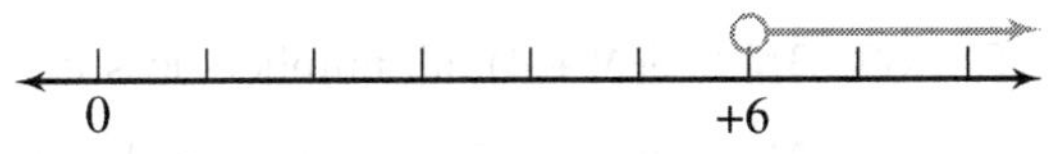

17. $\{x|\ x < 4\}$ or $(-\infty, 4)$

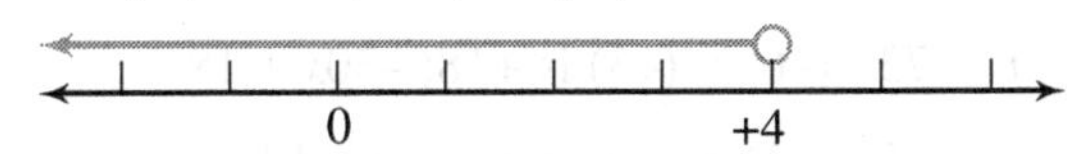

19. $\{x|\ x \leq 3\}$ or $(-\infty, 3]$

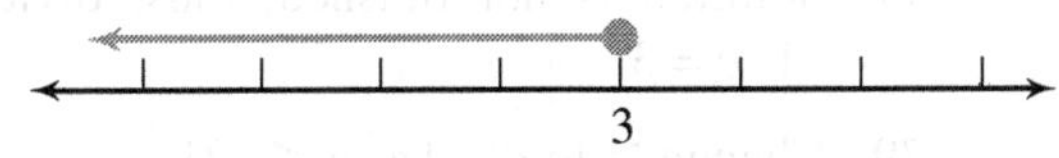

21. $\{x|\ x > 4\}$ or $(4, \infty)$

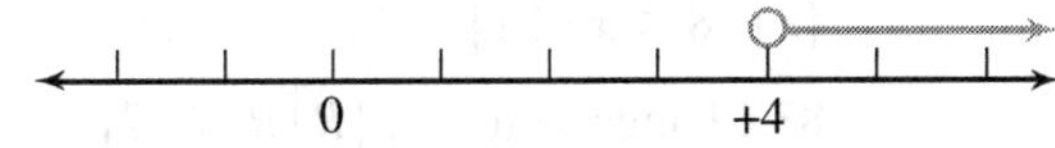

23. $\{x|\ x \leq 2\}$ or $(-\infty, 2]$

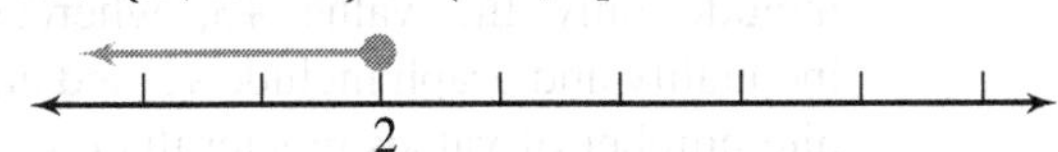

25. $\{x|\ x > 22\}$ or $(22, \infty)$

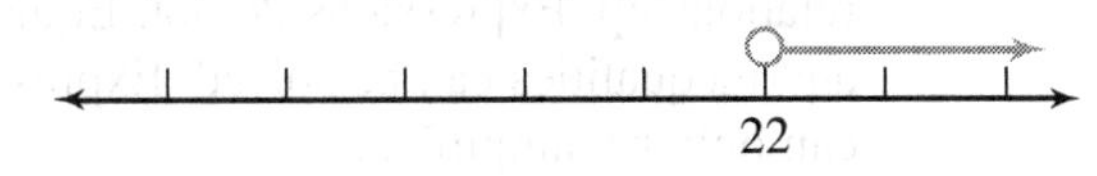

27. $\{x|\ x\ \ 1/2\}$ or $[1/2, \infty)$

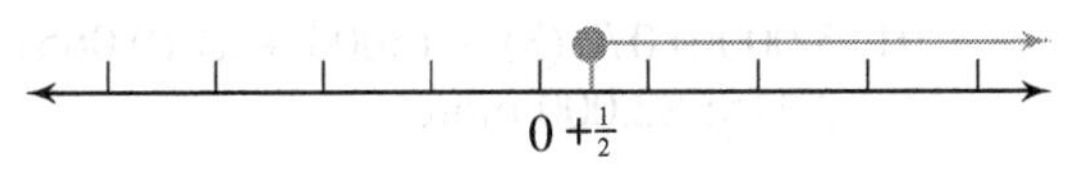

29. $\{x|\ x \geq -9\}$ or $[-9, \infty)$

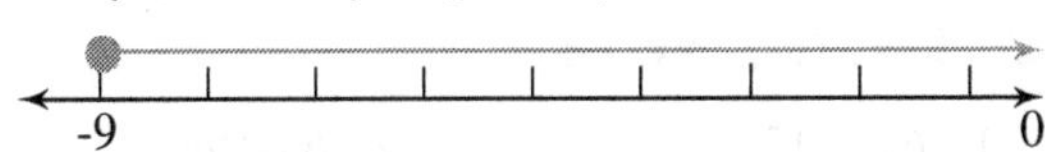

31. $\{x|\ -4 \leq x < -2\}$ or $[-4, -2)$

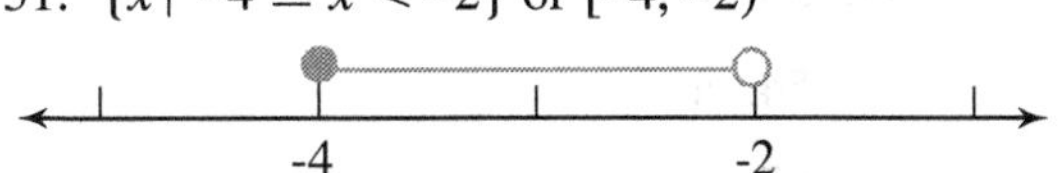

33. $\{x|\, 6 \leq x < 8\}$ or $[6, 8)$

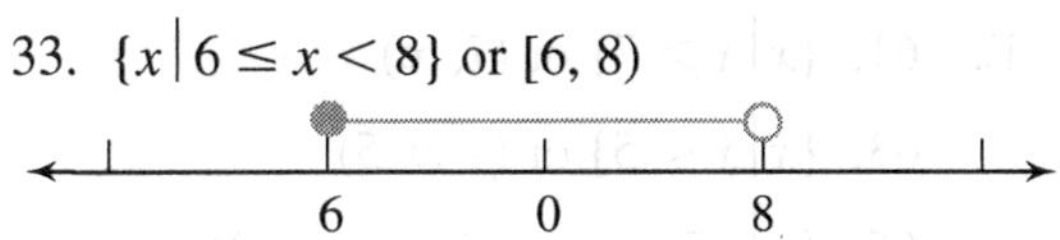

35. $\{x|\, -5 \leq x < 3\}$ or $[-5, 3)$

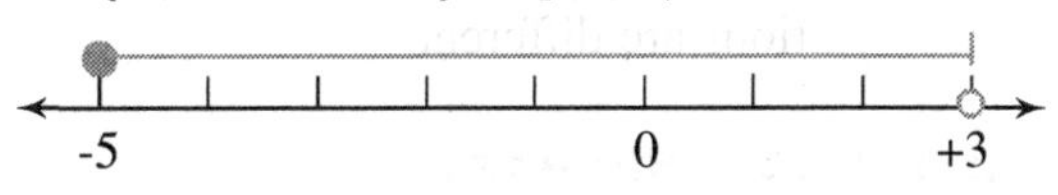

37. $\{x|\, x \geq -3\}$ or $(-3, \infty)$

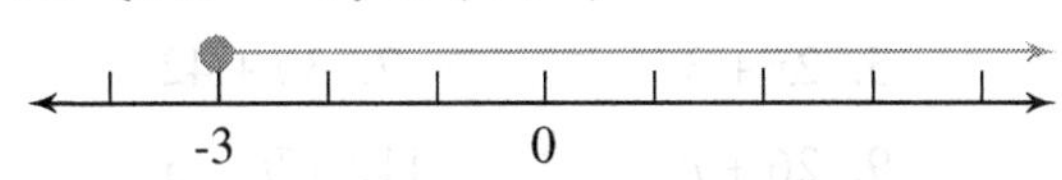

39. $\{x|\, x > -3/2\}$ or $(-3/2, \infty)$

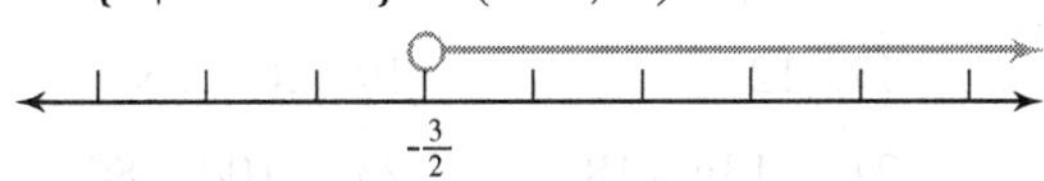

41. $\{x|\, -7/3 < x < 1\}$ or $(-7/3, 1)$

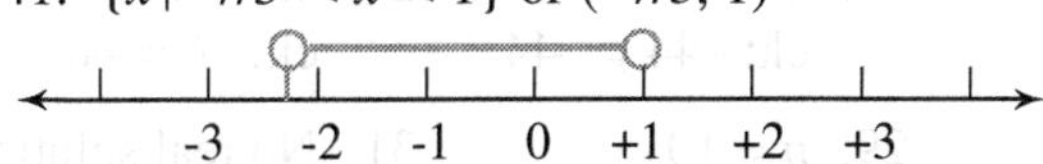

43. $\{x|\, 1 \geq x > -1\}$ or $(-1, 1]$

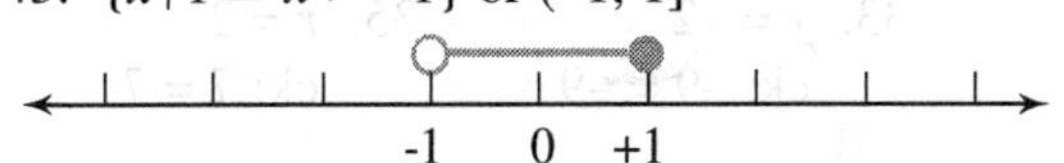

45. $\{x|\, -3 \leq x \leq 9\}$ or $[-3, 9]$

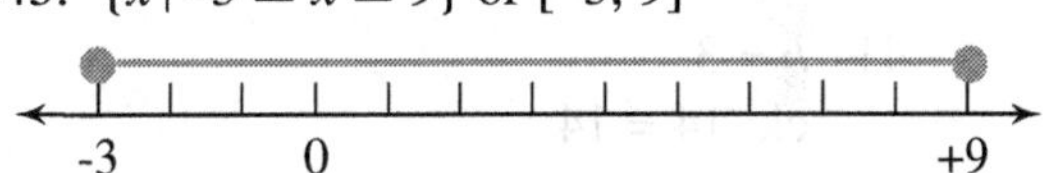

C. 47. $3N - 4 > 5 + 2N$; smallest integer is 10

49. $\dfrac{68 + 82 + X}{3} \geq 80$; at least a 90

51. $2(X + 3) + 2X \leq 72$; length must be at most 19 1/2 feet

D. 53. 1. Add –1 to both sides.

2. Divide both sides by 3.

55. 1. Subtract 8.

2. Divide by –3 and reverse the symbol.

57. 1. Distribute 2.

2. Add 8 to all three parts.

3. Divide all three parts by 2.

59. 53. f

54. a

57. e

E. 61. $\{x \mid x > 7\}$ or $(7, \infty)$

63. $\{x \mid x < 5\}$ or $(-\infty, 5)$

65. $\{x \mid -7 \leq x \leq 3\}$ or $[-7, 3]$

F. 67. The steps to solve are the same. The solutions are different.

Section 12.6 Review Exercises

A. 1. $56 - 21y$ 3. $-3b - 18$

5. $2x + 3$ 7. $3a + 42$

9. $26 + r$ 11. $17n - 5$

13. $30p + 18$ 15. $6x - 12y + 15$

17. 12 19. $20 - 26c$

21. $-13y - 18$ 23. $-10d + 80$

B. 25. $x = -7$ ck: $-44 = -44$ 27. $m = -1$ ck: $-6 = -6$

29. $p = 10$ ck: $45 = 45$ 31. No real solution

33. $y = -2$ ck: $-9 = -9$ 35. $r = 2$ ck: $7 = 7$

37. $5 = x$ ck: $-80 = -80$ 39. $y = 2$ ck: $24 = 24$

41. $b = 3$ ck: $14 = 14$

C. 43. $\{x \mid x \leq 2\}$ or $(-\infty, 2]$

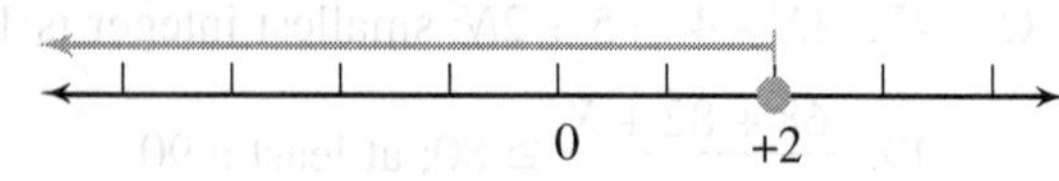

45. $\{x \mid x > 2\}$ or $(2, \infty)$

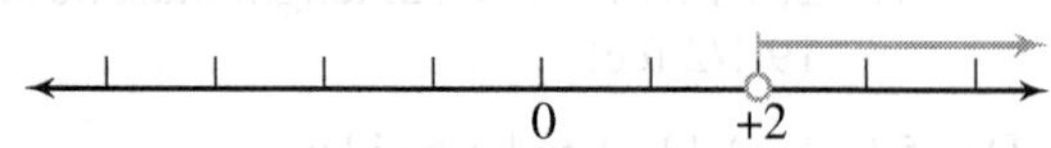

47. $\{x \mid x > 1/7\}$ or $(1/7, \infty)$

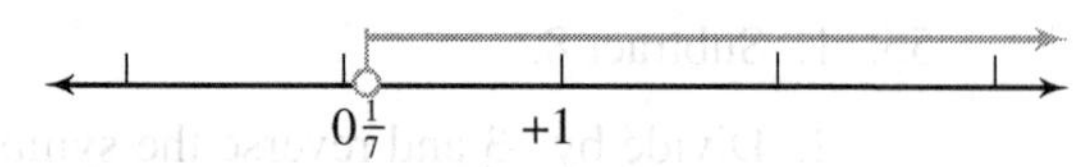

49. $\{x \mid -2 < x < 1\}$ or $(-2, 1)$

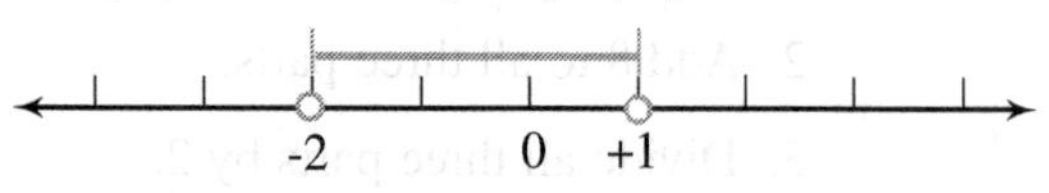

D. 51. $x + 10 + x + 2(x + 10) = 90$; \$15, \$25, \$50

53. $0.45(2x) + 0.30(x) = 2.40$; 4 candy bars

55. $3(n + 4) + 2n = 97$; 17, 19, 21

57. $2(x) + 2(4.5x) = 440$; 180 yards

59. $0.70(X + 1) + 0.95(X) = 4.00$; 2 oz almond beans, 3 oz chocolate beans

61. $2(x) + 2(1/4x) = 250$; 25 feet wide

63. $2(x) + 2(x - 7) = 46$; 15 feet by 8 feet

65. $2X(0.045)(5) + X(0.035)(5) = 6{,}125$; \$9,800 at 3 1/2%, \$19,600 at 4 1/2%

67. $0.22(x) + 0.12(48 - x) = 0.15(48)$ or $0.22(48 - x) + 0.12(x) = 0.15(48)$; 14.4 L of 22% acid, 33.6 L of 12% acid

E. 69. $3N > 5(N + 4)$; all numbers less than -10

71. $2X + 2(20) \leq 72$; maximum length is 16 feet

F. 73. $(-6) \times (-3)$ is $+18$; $-30x + 18$

75. $(-1) \times (-a) = +a$; $8a - 12$

77. Solution is not finished; must divide by -1; $c = 3$

79. Change $>$ to $<$; $\{a \mid a < -2\}$

81. Subtract 2 from all three parts; $\{x \mid -8 < x < 1\}$

83. Change $>$ to $<$; $\{R \mid R < -7\}$

G. 85. Both include +3. Equation and graph include only the value +3, whereas the inequality and graph include +3 and an infinite number of values greater than 3.

87. Equations and inequalities have symbols of relationship. Expressions do not. Equations and inequalities can be solved. Expressions can only be simplified.

89. Answers will vary.

91. $5{,}000(0.07)(3) + (5000 + x)(0.065)(3) = 2{,}415$; \$2,000 more

93. $2(x - 3) + 2(1/2x + 4) = 62$; 20 feet, 10 feet.

Test—Chapter 12

1. $6x + 12$ 2. $6 - 2n$

3. $4y - 8$

4. $x = 10$
ck: $33 = 33$

5. $t = -8$
ck: $13 = 13$

6. $-7 = p$
ck: $24 = 24$

7. No solution
$29 \neq 5$

8. $k = 3$
ck: $-24 = -24$

9. $-9 = T$
ck: $-40 = -40$

10. $11 = c$
ck: $-4 = -4$

11. $y = -12$
ck: $-1 = -1$

12. $5 = w$
ck: $-20 = -20$

13. $\{x \mid x > 1\}$ or $(1, \infty)$

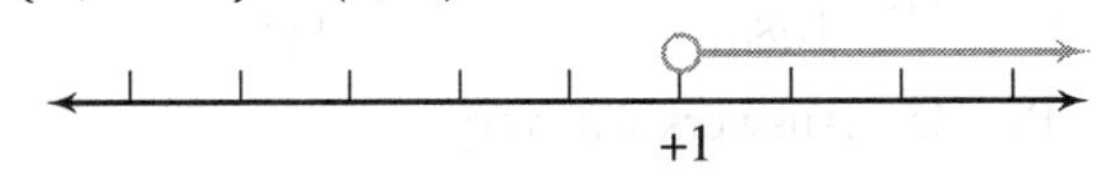

14. $\{x \mid x \geq 4\}$ or $[4, \infty)$

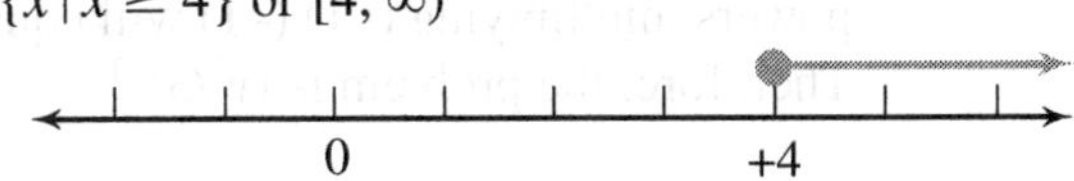

15. $\{x \mid -5 \leq x < 3\}$ or $[-5, 3)$

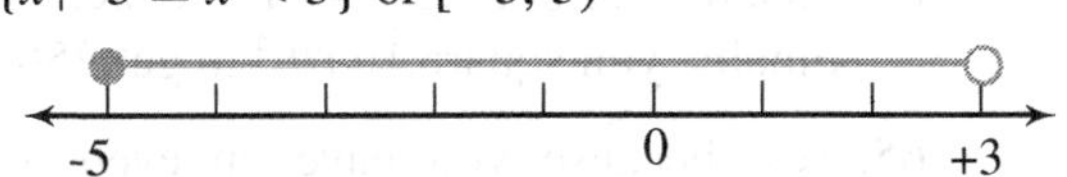

16. $6(r) = 5(r + 10)$

17. 60 miles per hour

18. $0.08(25 - X) + 0.03(X) = 0.05(25)$ or
$0.08(X) + 0.03(25 - X) = 0.05(25)$

19. 10 quarts of 8%, 15 quarts of 3%

20. The expression can only be simplified and it has no symbol of relationship. The equation has a symbol of relationship and has only one solution. The inequality has a symbol of relationship and has an infinite number of solutions.

EXERCISES—CHAPTER 13

Section 13.1

A. 1. $+a$ 3. $-27g$
5. $5r^2s$ 7. $4b^2d^2 + 3b^2d$
9. $5m^2n$

B. 11. $-25k$ 13. $7w^3$
15. $11wxy^2$ 17. $-9b^3c$
19. $45xyz$

C. 21. $14gh$ 23. $6x^2$
25. $-2a^2b^3$ 27. $11mn$
29. $7r^2s$ 31. $272p^3q^2$
33. $-28a^2b^2c^3$ 35. $9c^2d^3$

D. 37. Combining like terms
39. $-3p^2q^2 + 5p^2q^2$ is the sum because they are not like terms and cannot be combined.

Section 13.2

A. 1. y^5 3. b^4
5. a^5 7. z^7
9. 2^5 11. $6a^3b^2$
13. A^6 15. $-y^3$
17. y^7 19. $8x^3$
21. $8m^6$ 23. $-\frac{8}{27}g^6$
25. -64 27. 64
29. -1

B. 31. $-6a^5b^3$ 33. $32x^5z^2$
35. $-a^{15}b^{10}$ 37. x^6y^7
39. $-64a^3b^3c^3$ 41. $-14x^2y^3$
43. $48m^5m^8$ 45. $\frac{9}{16}P^2R^4$
47. $\frac{4m^4n^2}{9}$ 49. $72a^{13}$
51. $\frac{a^9b^6}{8}$ 53. $-6m^4n^3$
55. $\frac{8}{125}R^6S^3$ 57. $-200x^{23}$
59. $64a^4b^6$

C. 61. No, because you add $2 + 3$ in $R^2 \cdot R^3$ and you multiply 2×3 in $(R^2)^3$.

63. Square $-2a^2b$ to get $4a^4b^2$ to get $16a^8b^4$. Finally, you square $16a^8b^4$ to get $256a^{16}b^8$.

65. Yes, because you have an even number of negative factors, so the answer will be positive.

Section 13.3

A.
1. x^3
3. c
5. $4d^2$
7. 1
9. $4g^2$
11. 3^2 or 9
13. 1
15. $4/9x^2$
17. 5/8
19. $4xy^2$
21. $-6b^2cd$
23. $1/3h^2$
25. $-3a^3$
27. $-1/2m$
29. $8c$

B.
31. $8a^2b$
33. $4m$
35. $1/3a^5$
37. $-10n$
39. $9ac$
41. $\frac{5}{3}y^2$
43. $-1/2x$
45. $6ab^3$
47. $-3m^2n^2$
49. $2x/15$

C.
51. No, you keep the base of 2.
53. $(-5x^2y)^2$, because exponents or powers come before multiplying or dividing in the Order of Operations.
55. There are no powers outside the parentheses to consider, so there are three negative factors and the answer will be negative.

Section 13.4

A.
1. $\frac{1}{a^3}$
3. $\frac{4}{R}$
5. $\frac{3}{2M}$
7. $\frac{1}{x^4}$
9. b^4
11. $\frac{a^2}{4}$
13. $\frac{y^2}{4x^4z^6}$
15. $3m^2$
17. x^3
19. $\frac{b^6}{a^4}$
21. $\frac{6x}{y^4z^2}$
23. $\frac{4a^7}{b^3c^3}$
25. $\frac{-2y^3}{3x^5}$
27. $\frac{-m^{15}n^3}{8}$
29. $125a^2b$
31. $-\frac{m}{16}$
33. $-\frac{m^2n^2}{100}$
35. $\frac{b^2}{36a^2}$
37. $\frac{1}{32b^{11}}$
39. $\frac{n}{4m^6}$
41. $\frac{b^{12}c^{11}}{108a}$
43. $\frac{1}{x^6y^4}$

B.
45. Answers will vary.
47. $3/y^2$, because there are three negative powers, multiplying (–1) (–1) will equal –1. Therefore, the problem is $(y^2/3)^{-1}$.

Section 13.5

A.
1. $38x + 6y - 39$
3. $28m^2 - 2m - 9$
5. $7R + S + 3T$

B.
7. $18a + 3ab - 17b$
9. $-8v + 3w - 12x$
11. $18f^2 - 12f + 8h$

C.
13. $-4a + 10b$
15. $-7m + 7$
17. $9x^2 + 9x + 3$
19. $2R - S - 6T$
21. $x^2y - 4xy + 5y^2$
23. $-b^2 - 11b - 13$
25. $7y^3 + 7y^2 - 4y + 2$
27. $-m^3 - m^2 - m - 4$
29. $6x^3 - x^2 + 4x - 6$
31. $7p^2q + 2pq + pq^2$
33. $6x^2 + 3x - 7$
35. $-4x - 3y$
37. $10y^2 + 9y - 5$
39. $m^2n^2 - 2m^2n - mn^2 + 7$

D.
41. Rules must be applied to all terms.
43. Answers will vary.

Section 13.6

A.
1. $3a^3 - 12a$
3. $-56m^3n - 32m^2n^2$

5. $15x^2 + x - 2$

7. $-4p^3 - 35p^2 + 17p - 2$

9. $-3x^4y^3 - 2x^3y^2$

11. $c^3d^2 + 2c^2d^2 - 4cd - 8d$

13. $-5x+^3 - 10x+2+y - 10xy^2$

15. $y^3 + 4y^2 + 8y + 8$

B. 17. $a^4 - 4a^2 - 21$

19. $0.0\text{–}9x^2 - 0.24xy + 0.16y^2$

21. $x^2 - 6xy + 9y^2$

23. $g^2 - 16$

25. $3a^3 - 10a^2b + ab + 8ab^2 - b^2$

27. $-28x^4y^3 + 14x^3y^2 - 21x^2y^3$

29. $-10m^4n + 8m^3n^2 - 4mn^2$

31. $9 + 12w + 4w^2$

33. $y^3 + 3y^2 + 3y + 1$

35. $9y^2 - 1$

37. $x^4 - 9$

39. $x^3y^3 - 5x^2y^2 + 6xy$

C. 41. Answers will vary.

43. No, because you must use the Distributive Principle when multiplying binomials. Also $(5 + 6)^2 = (11)^2 = 121$ and $5^2 + 6^2 = 25 + 36 = 61$ and $121 \neq 61$. In this numerical example, they are obviously not the same.

45. No, because $(m^2 + 2m - 1)^2$ is a trinomial, not a binomial.

Section 13.7

A. 1. $3b - 1$ 3. $-2x^3 + 4y^2$

5. $3m^2 - 2m + 5$ 7. $-3x^2 - 3/2x - 2$

9. $p - 2q - 3r$ 11. $4vw^3x - 3wx^2$

13. $3/2\, b + 5/2\, ac^2$ 15. $-x^3y^2 - xy$

B. 17. $3 - \frac{2b}{a} - a$ 19. $7r - \frac{5}{r}$

21. $\frac{2}{ab^2} - \frac{3}{2a^2b}$ 23. $\frac{3z}{2xy} - \frac{y}{x^2} + \frac{1}{2x}$

C. 25. $A - 2$ 27. $n - 2$

29. $x + 2 - \frac{4}{x - 3}$ 31. $2t + 1$

33. $8z^2 - 10 + \frac{11}{z^2 + 1}$

35. $-8x^2 + 14x - 22 + \frac{48}{x + 2}$

37. $2x^2 + 3x - 1 + \frac{3x}{x^2 - x - 2}$

39. $a^2 - 3$

D. 41. Divide 27 into 83; multiply and then subtract. Bring down the next digit and repeat the process. Divide $(x + 1)$ into $3x^2 + x$; multiply and then subtract. Bring down the next term and repeat the process.

Section 13.8

A. 1. 2.78×10^3 3. 7.5×10^{-4}

5. 6.5×10^{-1} 7. 7.885×10^3

9. 4.31×10^{-1} 11. 6.52×10^{-2}

B. 13. 6,500 15. 0.0157

17. 90,000 19. 0.0007

21. 0.03144 23. 470,000

C. 25. $(4.6 \times 10^6)\,(9 \times 10^3) = 4.14 \times 10^{10}$

27. $(9 \times 10^{-3})\,(5.1 \times 10^{-4}) = 4.59 \times 10^{-6}$

29. $(6.4 \times 10^{-3})\,(8 \times 10^2) = 5.12$

31. $\frac{(6.8 \times 10^3)\,(4 \times 10^2)}{(4 \times 10^4)\,(2 \times 10^3)} = 3.4 \times 10^{-2}$

33. $\frac{(2.4 \times 10^{-3})\,(6 \times 10^3)}{(1.2 \times 10^2)\,(4 \times 10^4)} = 3 \times 10^{-6}$

35. $(1.35 \times 10^{-1})\,(2.7 \times 10^3) = 5 \times 10^{-5}$

37. $(8 \times 10^{-3})^2 = 6.4 \times 10^{-5}$

39. $(12 \times 10^3)^2 = 1.44 \times 10^8$

D. 41. Write 94 as 9.4. Count the number of places between 0.0094 and 9.4, which is 3, and indicate the direction with a negative sign. Then write 9.4×10^{-3}.

43. Answers will vary.

45. Answers will vary.

Section 13.9 Review Exercises

A. 1. $19x$ 3. $-6a$

5. $8c^2d + 3c^2$ 7. p^2

9. x^3 11. $-18m^2n^3$

13. n^4
15. t^{15}
17. $9b^2$
19. $-15a^{10}$
21. $2m^2$
23. $6a + 3b$
25. $2/b^2$
27. $1/b^3$
29. $1/4b^2$

B. 31. $3/4t^5$
33. $14x^7y^6$
35. $-8a^3b$
37. $p^4q - 3pq^4$
39. $20f^2 + 3fg - 2g^2$
41. $3x^5$
43. $2rs - 3r^2s$
45. $8a^7b^3 - 11a^3b^6$
47. $3/7t^2$
49. $a^4c^2/9b^2$
51. $-2a - 20$
53. $-18c^5d^5$
55. $3y^4 + y^3 - 15y^2 + y + 2$
57. $9w^4 + 4w^2 + 2w$
59. $25p^2 + 20p + 4$
61. $-2y^2 + 2y - 5$
63. $-3p^2 - 13p + 30$
65. $11y - 6$
67. $-2b^2 - 3b - \dfrac{6b}{-6b + 1}$
69. $3x^2z - 5xz + 2yz$
71. $a + 2$
73. $\dfrac{-6m}{n^5}$
75. $4x^2 + 4x + 2 + \dfrac{3}{x - 1}$
77. $\dfrac{256y}{x^6}$
79. $8 + 12a + 6a^2 + a^3$

C. 81. 8.7×10^{-3}
83. 90,000,000
85. 6.23×10^4
87. 0.00065
89. 6.4×10^6
91. 9,900

D. 93. 5×10^{-6}
95. 5.58×10^{-5}
97. 1.449×10^4
99. 1.6×10^7

E. 101. Combined unlike terms; $11m - 12n$
103. Did not raise 3 to the third power; $27x^6$
105. Multiplied exponents instead of adding; $-12c^5d^6$
107. Did not subtract exponents correctly;
$$c^{-7} = \frac{1}{c^7}$$
109. Moved a negative number; $\dfrac{-2a^3}{b^4}$
111. Did not change all signs; $-4c^2 + 9c - 10$
113. Divided into only one term; $3ab + 8a$

F. 115. In $-2a^2$, the exponent applies only to the base a. In $(-2a)^2$, $-2a$ is the base.
117. $x^0 = 1$ (for $x \neq 0$), and we don't often see it because when simplifying an answer, x^0 is always replaced by 1.
119. Distribute $3a$ over $(2a^2 + 5a - 1)$ and then distribute -4 over $(2a^2 + 5a - 1)$. Last, combine any like terms.

G. 121. $\dfrac{-p^3q^2}{10r^3}$
123. $\dfrac{60}{x^3y^3}$
125. $2a^4 + 3a^3 + 2a^2b^2 + 3ab^2$

Test—Chapter 13

1. $-6a^5b^3$
2. $2y^2 - 7y - 4$
3. $-\dfrac{m}{2}$
4. $-2a^2c$
5. $1/c^4$
6. $4r^4s^2t^6$
7. $\dfrac{a^2}{3bc^3}$
8. $-5y^2z$
9. $-10x + 6y$
10. F^6
11. $-\dfrac{6}{a^5b}$
12. $x^2 + 6x + 9$
13. $\dfrac{16x^4}{9y^2}$
14. $8cd - 2d + 3d^2$
15. $-3x$
16. $7m^2n - 11mn + 18$

EXERCISES—CHAPTER 14

Section 14.1

A. 1. yes 3. no
5. no 7. yes
9. yes

B. 11. $2(x^2 - 2x + 8)$
13. $ab(5ab - 12a + 4)$
15. $-12r(r + 1)$
17. $0.8m(2m^2 + 8m + 3)$
19. $-a(a^2 - 2a - 6)$
21. $11(f^3 + 4g^3)$
23. $-a(a^2 + a - 8)$
25. $16b^3(4b + 1)$
27. Prime
29. $x^2z^2(5x^3y^4 + 7x^2y^2 - 1)$
31. $(a + 4)(a - 4)$

C. 33. $(x - 7)(ax + 2)$
35. $6x(3w + 1) + w(w + 1)$
37. $c(x + y)(a - b)$
39. $(p + 5)(p + 12q)$

D. 41. $3m(m - 3)$
43. $(5b + 7)(5a^2 + 3b^2)$
45. $4ab(ab - 3a + 5b)$
47. $xy(27x + 9y + 16)$

E. 49. Multiply the factors together to see if you get the original expression.
51. Look for a GCF.

Section 14.2

A. 1. $x^2 - 4$ 3. $m^2 + 6m + 9$
5. $4b^2 - 1$ 7. $x^2 - 4z^2$
9. $1 - 9a^2$ 11. $9y^2 - 4z^2$

B. 13. $(x + 7)(x - 7)$ 15. $(a - 1)(a + 1)$
17. $(k - 11)(k + 11)$ 19. $(z - 9)(z + 9)$
21. $(c + 2)(c - 2)$ 23. $(2f + 3)(2f - 3)$
25. $(5T + 6)(5T - 6)$
27. $(7J + 8)(7J - 8)$
29. Prime
31. $(6 + c)(6 - c)$
33. $(a + 8b)(a - 8b)$
35. $(4a + 1)(4a - 1)$
37. $(3bc + 4a)(3bc - 4a)$
39. $(1 + 10x)(1 - 10x)$
41. $6b(b + 5)(b - 5)$
43. $(x^3 + 9a)(x^3 - 9a)$
45. $3(d^2 + 9)(d + 3)(d - 3)$
47. $5(m + 3n)(m - 3n)$
49. Prime
51. $a(a + 3)(a - 3)$
53. $4(m + 4n)(m - 4n)$
55. $x(5x + 1)(5x - 1)$
57. $2(49 + a^2)$
59. $3(3b + 5)(3b - 5)$

C. 61. $x^2 + 4$ cannot be factored. $x^2 - 4$ is the difference of two squares, so it is factored as $(x - 2)(x + 2)$.
63. It means that the expression cannot be written as the product of anything but 1 and itself.

Section 14.3

A. 1. $(y + 3)(y + 5)$ 3. $(t - 15)(t - 1)$
5. $(a - 5)(a - 2)$ 7. $(c + 5)(c - 4)$
9. $(w + 5)(w + 5)$ 11. $(m - 6)(m - 1)$
13. $(x - 7)(x - 3)$ 15. $(r - 1)(r - 1)$
17. $(b + 5)(b + 4)$ 19. $(z - 11)(z + 3)$
21. $(2a + 1)(a + 1)$ 23. $(3x + 1)(x + 2)$
25. $(r + 8)(r - 1)$ 27. $3(v - 5)(v + 4)$
29. $5(X + 3)(X + 5)$
31. $7(t - 3)(t - 3)$

33. $(m + 3n)(m + 3n)$
35. $(w - z)(w - z)$
37. $(3a + 5b)(a + 2b)$
39. $(2x - y)(x - 5y)$
41. $4a(a + 5)(a - 3)$
43. $-2Y(2Y + 5)(3Y - 2)$
45. $(2c + 1)(2c - 3)$
47. $(3r - 4)(r + 5)$
49. $3(3x - 2)(x + 1)$
51. $n(7m - 1)(m + 2)$
53. $(4y - 3)(4y - 3)$
55. $3(t - 7)(t - 3)$
57. $(5 - m)(4 - 3m)$
59. $b(3a - 2)(3a - 2)$

B. 61. Look first for a GCF and then check the answer by multiplication.

Section 14.4

A. 1. $(x + 4)^2$ 3. $(a + 8)^2$
5. $(x - 7)(x - 3)$ 7. $(3 - x)(12 + x)$
9. $3(t + 1)(t + 2)$ 11. $(2y + 1)^2$
13. $(4 - c)^2$ 15. $(5a + 8)(5a - 8)$
17. $(6t + 5q)^2$ 19. $(4 + m)^2$
21. $(2a - 3b)^2$ 23. $(3 - a)(m + n)$
25. $(y - 4)(y^2 - 3)$ 27. $x^2(x - 8)(x + 10)$
29. $5(n - 3)^2$ 31. $5(b - 6)(b + 6)$
33. $(a + 3)(a + 2)(a - 2)$
35. $(5a + c)(b - 4)$
37. $2(2t + 1)(t + 3)$
39. $3(m^2 + 4)(m + 2)(m - 2)$
41. $(a + 2)(b - c)$
43. $3y(y - 5)^2$
45. $x^2y^2(4x + y)(5x - 3y)$

B. 47. 1. Factored as difference of squares
2. Factored again as difference of squares
49. 1. Factored GCF from 2 pairs of factors
2. Removed the GCF

Section 14.5

A. 1. $y = -2, y = 5$ 3. $a = -8, a = 9$
5. $m = 0, m = 5/3$ 7. $x = 0, x = 5$
9. $b = 7, b = -7$ 11. $y = 0, y = 2$
13. $c = 0, c = 3, c = -3$
15. $m = 0, m = -5/7$
17. $t = 0, t = 1, t = -1$
19. $x = -3, x = -2$
21. $n = 5, n = -2$
23. $a = -9, a = 8$
25. $x = -3, x = -7$ 27. $s = -3/2$
29. $y = -2/3, y = 3$ 31. $x = 0, x = 3$
33. $R = 4, R = -4$ 35. $x = 2, x = -8$
37. $y = -7, y = 4$ 39. $x = -3, x = 9$
41. $a = 9, a = -8$ 43. $a = -3/5, a = -2$
45. $m = -5, m = 3$ 47. $m = 5, m = -2$
49. $k = 1/3, k = 4$

B. 51. $x(2x + 1) = 78$; 6 m by 13 m
53. $n^2 + (n + 1)^2 = 145$; 8, 9, or –9, –8
55. $1/2(x)(2x) = 100$; 10 feet
57. $n(n - 7) = 60$; 12 and 5 or –5 and –12.

C. 59. The factors can each be set equal to zero because their product is zero, and therefore, at least one of them must be zero.
61. You cannot have a negative dimension for length (Example 10), but you can have negative numbers (Example 9).
63. Eliminate the fraction by multiplying all terms by 2.

Section 14.6 Review Exercises

A. 1. $3x^3y^2 - 6x^2y^2$ 3. $y^2 + 3y - 18$
5. $k^2 - 9$ 7. $6c^2 + 15c - 9$
9. $x^2 + 4x + 4$ 11. $a^2 + 6a + 9$

B. 13. $3y(y + 1)(y - 1)$ 15. $(b - 4)(b + 4)$
17. $(c - 3)(c - 5)$ 19. $x^2(x^2 + 1)(x^2 + 1)$
21. $-t^2(t - 2)$ 23. $(2d - 1)(2d + 1)$

25. $m^2n(3n - 5)$
27. $(13 - m)(3 - m)$
29. $(a + 9)(a + 3)$
31. $(x + 7)(x + 7)$
33. $7(5 - 3b)$
35. $(m - 9n)(m + 5n)$
37. $4a(a + 5)(a - 3)$
39. $-3y(y + 9)(y - 1)$
41. $3(m + 4)(m - 4)$
43. $9xy(xy - 2)(xy - 1)$
45. $4(a - 4b)(a + 4b)$
47. $(w^2 + 4z^2)(w + 2z)(w - 2z)$
49. $6(m^2 - 5n^2)(m^2 + n^2)$
51. $xyz(x - 6y)(x + 4y)$
53. $-8m^2(m - 3)(m + 3)$
55. $-6(2a + 1)(a - 1)$
57. Prime
59. $10(n + 11)(n + 2)$
61. $(a + 3)(a + 1)$
63. $(x + 8)(x + 3)$
65. $-4(b - 4)(b + 4)$
67. $2(m - 10n)(m + n)$
69. $(3x + 4)(x - 4)$
71. $x^2yz(x^2 + xz + y^2)$
73. $(x + 1)(xy - 2)$
75. $(y - 4)(y - 1)(y + 1)$
77. $8(2g - 5h)^2$
79. $(5p + 8)(4p - 3)$
81. $(3w + 7)(6w - 5)$

C. 83. $c = 0, c = 6$
85. $x = 0, x = 2$
87. $m = 5/3, m = 3/2$
89. $n = 10, n = -4$
91. $t = 1/4, t = -1/4$
93. $m = -7/2, m = 9/2$
95. $x = 7, x = -4$

D. 97. $x(2x - 1) = 120$; W: 8 yd, L: 15 yd
99. $n^2 + (n + 1)^2 = 41$; 4, 5 or –5, –4
101. $x^2 - 9 = 16$; 5 or –5
103. $n(n + 1) = 110$; –11, –10 or 10, 11

E. 105. The GCF should be $5a$; $5a(5ab^2 - 4ab + 7)$
107. Not finished; $(c + 2)(c + d)$
109. $a^2 + 16$ does not factor; $2(a^2 + 16)$
111. These would not give a middle term of $-6y$; $(y - 4)(y - 2)$

F. 113. Remove the GCF and check the answer by multiplication.
115. Because if the product is zero, at least one on the factors must equal zero.
117. $4a^2 + 1$ cannot be factored. $4a^2 - 1$ is the difference of two squares and factors as $(2a - 1)(2a + 1)$.

Test—Chapter 14

1. $12x^3y^2 - 21x^2y^2 + 6xy$
2. $4a^2 + 5ac - 6c^2$
3. $2(2R + 5)(R - 4)$
4. $c(c - d)(c + d)$
5. $(a + 5)(x + y)$
6. $(5 + 4y)^2$
7. $(5t + 1)(3t - 2)$
8. $(2m - n)(m + 2n)$
9. $9(p + 5)(p + 2)$
10. $4(w - 2)(w + 2)$
11. $(x - 7)(x + 2)$
12. Prime
13. $(m + 6n)(m - 4)$
14. $(2x + 1)(x + 1)$
15. $3a(a^2 + 7a - 3)$
16. $(5 + 3z)(5 - 3z)$
17. Because there is no GCF to remove, and it is a trinomial, use either trial-and-error or product-sum to find $(w - 7)(w - 3)$.
18. $s = 6, s = -5$
19. $v = -7, v = 4$
20. $n(n + 2) = 35$; 5, 7 or –7, –5

EXERCISES—CHAPTER 15

Section 15.1

A. 1. positive 3. positive
5. II 7. y–axis
9. positive 11. Cartesian coordinate system

B. 13. A(+1, 5) 15. C(5, 3)
17. E(0, –2) 19. G(–4, 4)
21. I(–4, –3) 23. K(5, –1)

C.

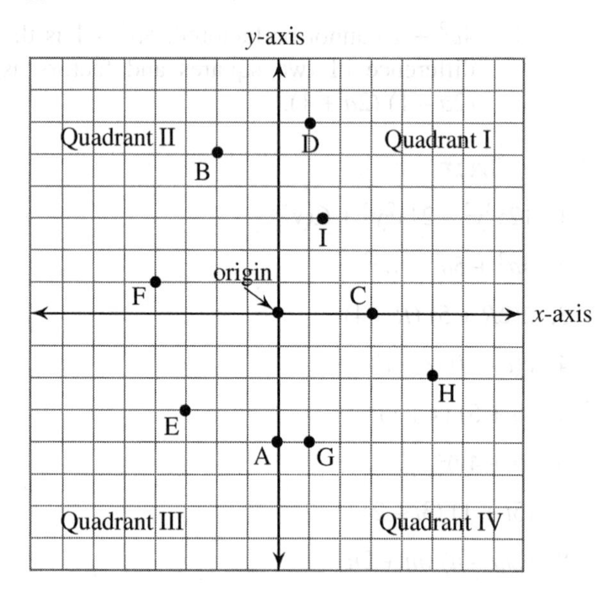

D. 41. Answers will vary.
43. Answers will vary.

Section 15.2

A. and B. The following tables are sample answers and do not contain all possible pairs of numbers.

A. 1.

x	y
0	–4
–4	0
–2	–2

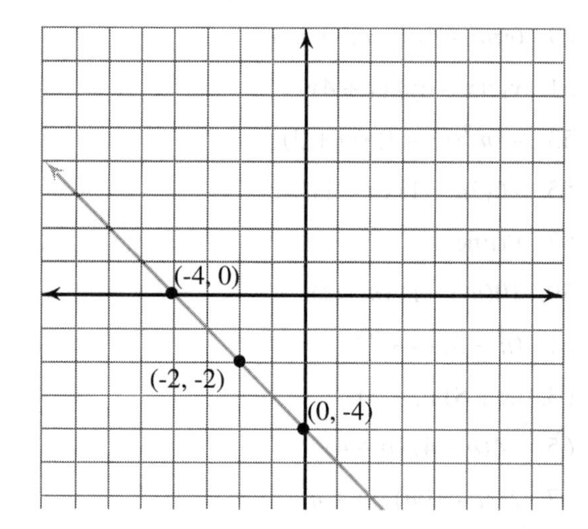

3.

x	y
0	5
5/3	0
3	–4

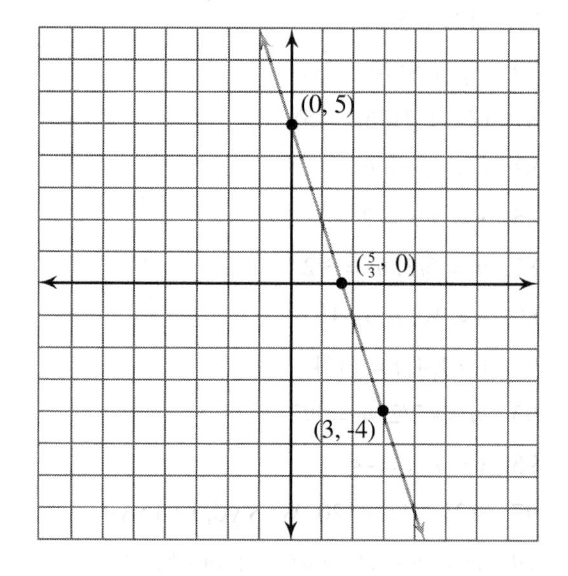

5.

x	y
0	–3
3	0
5	2

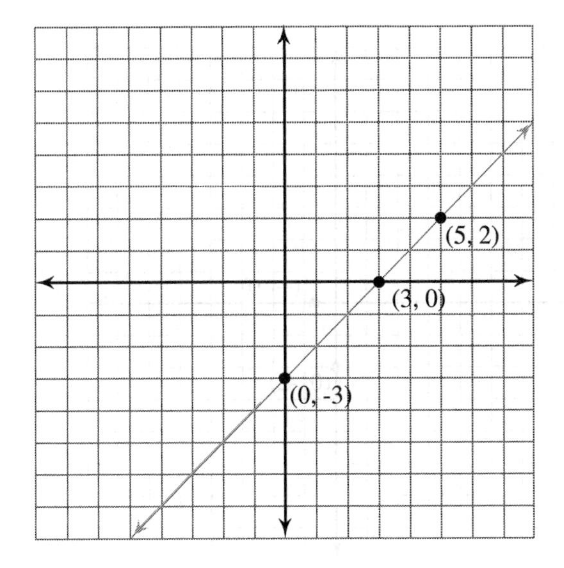

7.

x	y
0	0
2	–1
–2	1

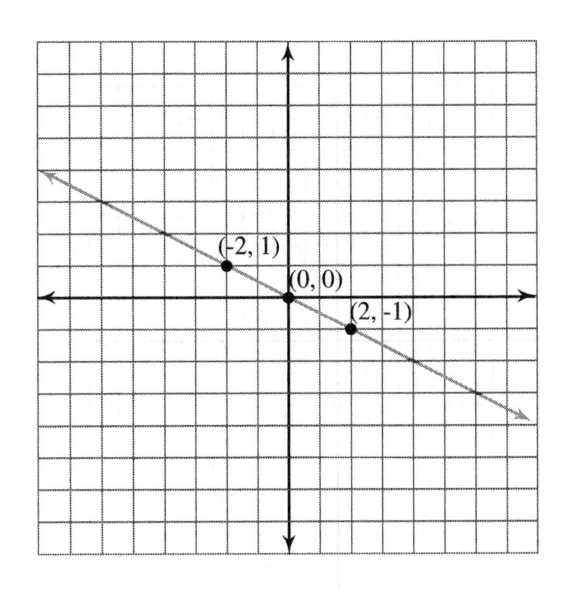

9.

x	y
0	4
3	5
–3	3

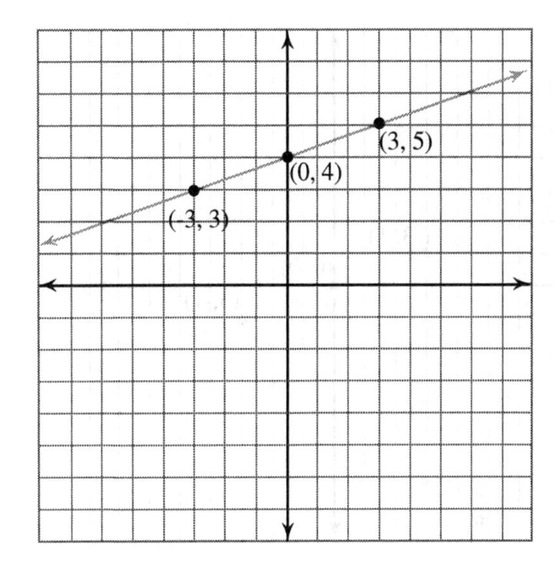

11.

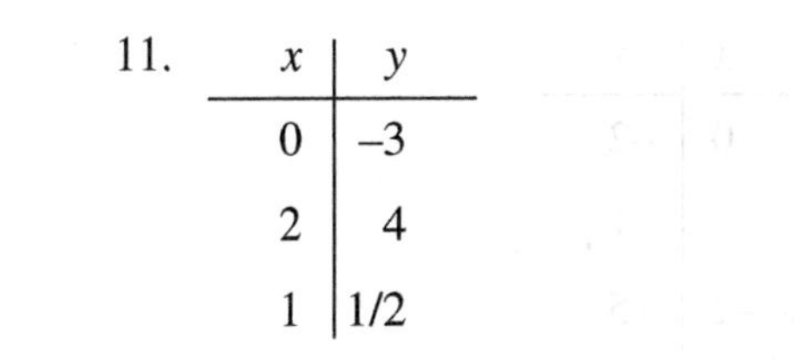

x	y
0	–3
2	4
1	1/2

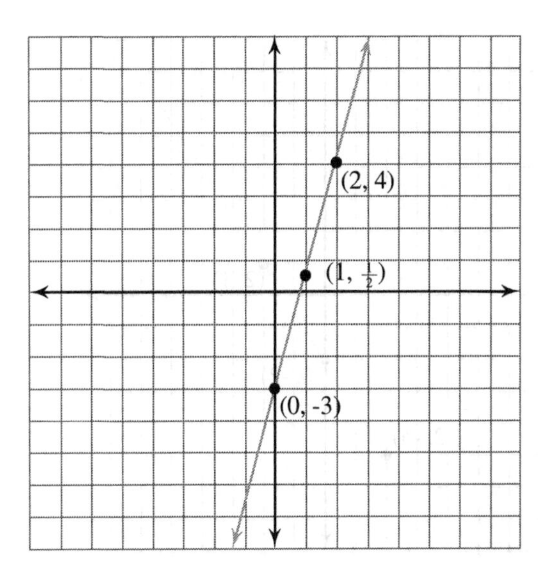

13.

x	y
0	0
3	–1
–3	1

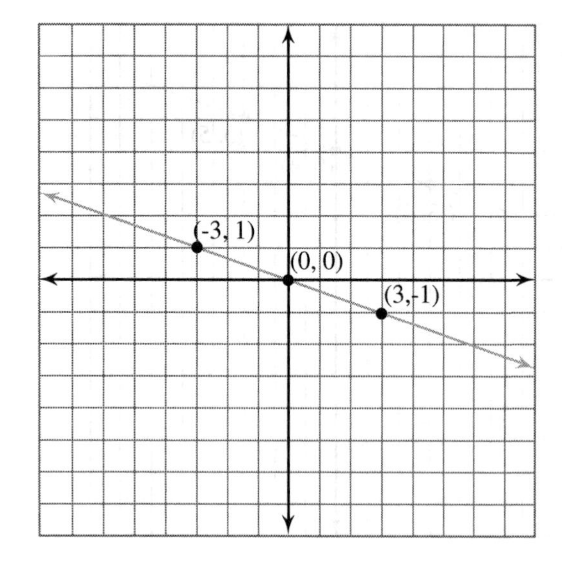

B. 17.

x	y
0	–2
2/3	0

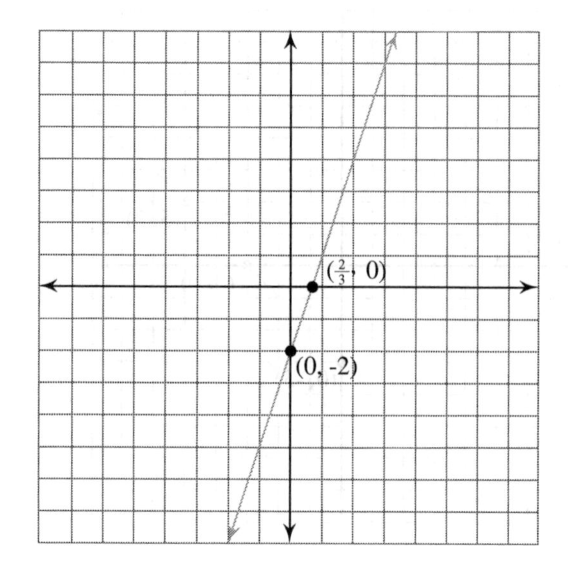

15.

x	y
0	–2
2	1
–2	–5

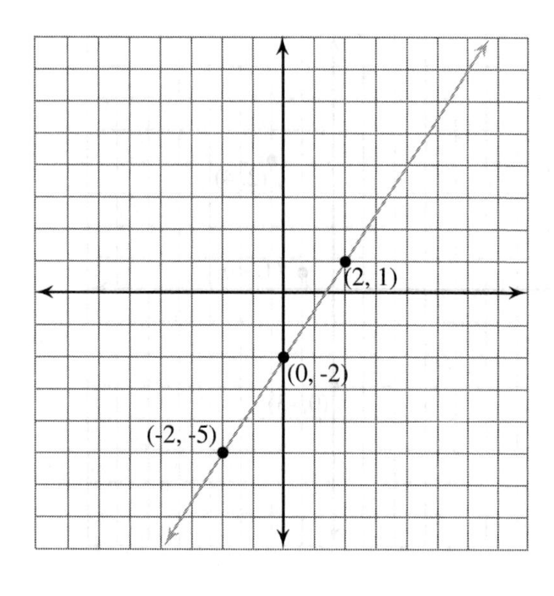

19.

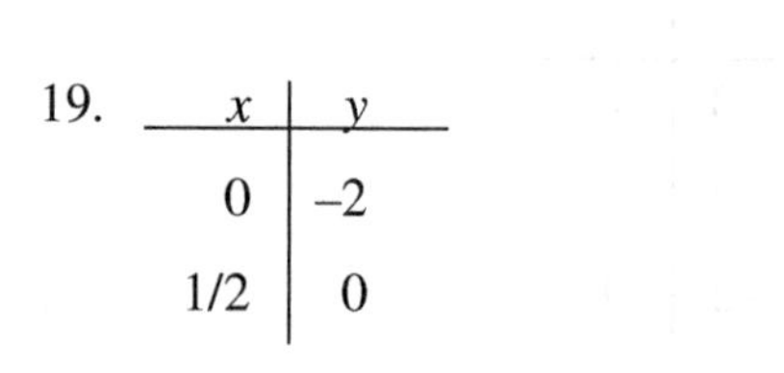

x	y
0	–2
1/2	0

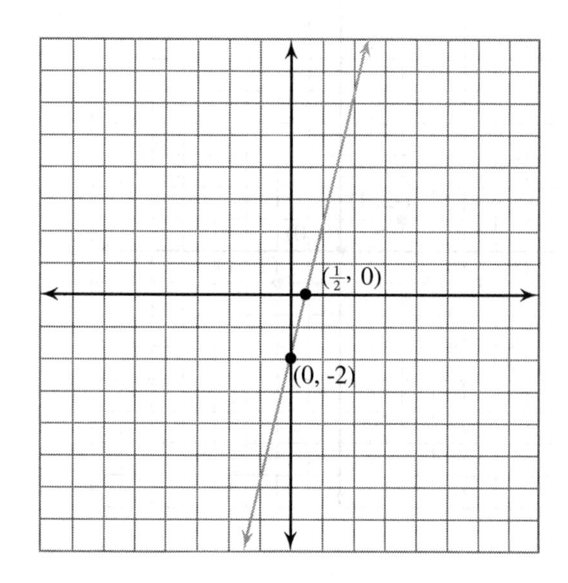

21.

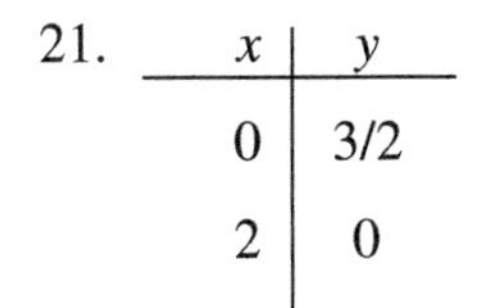

x	y
0	3/2
2	0

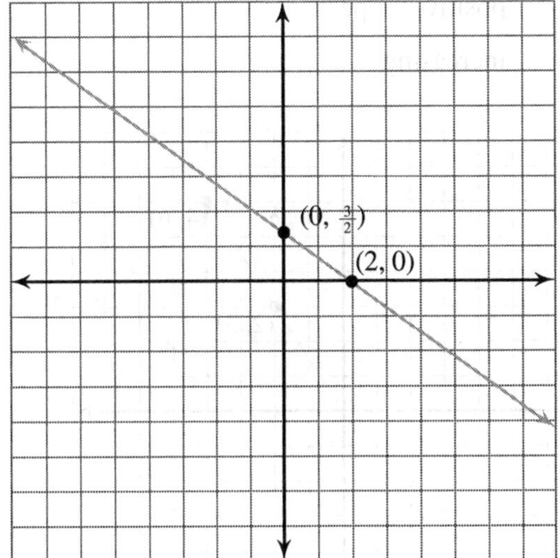

23.

x	y
0	–4
8	0

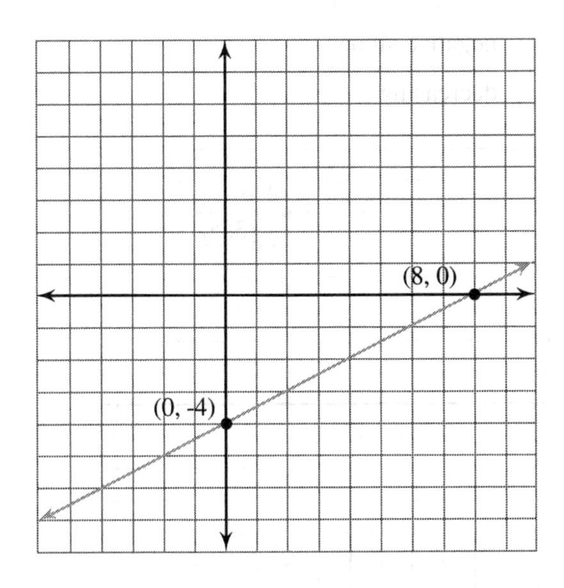

25.

x	y
0	–6
–3	0
–4	2

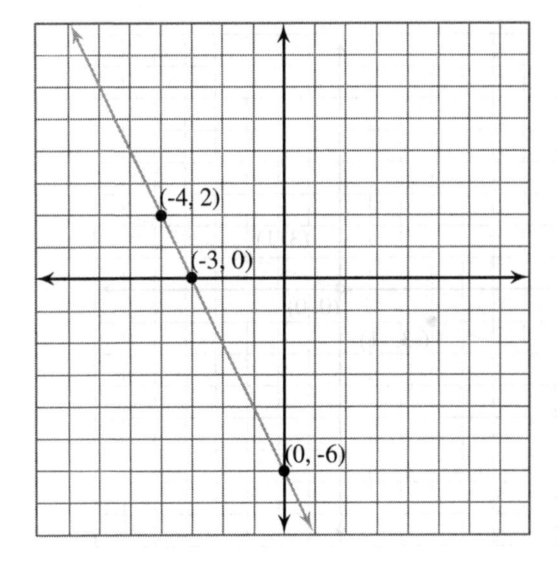

27.

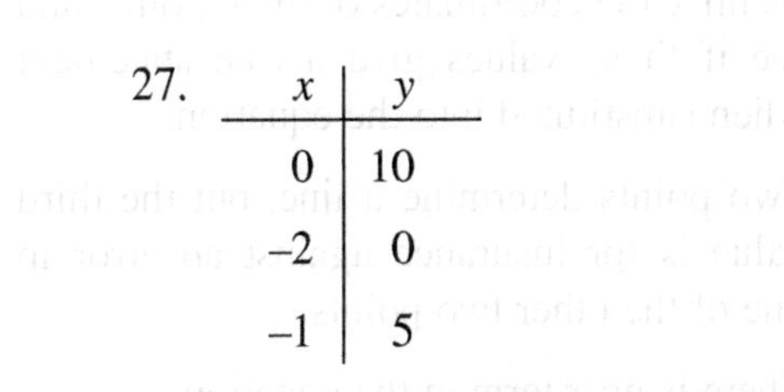

x	y
0	10
–2	0
–1	5

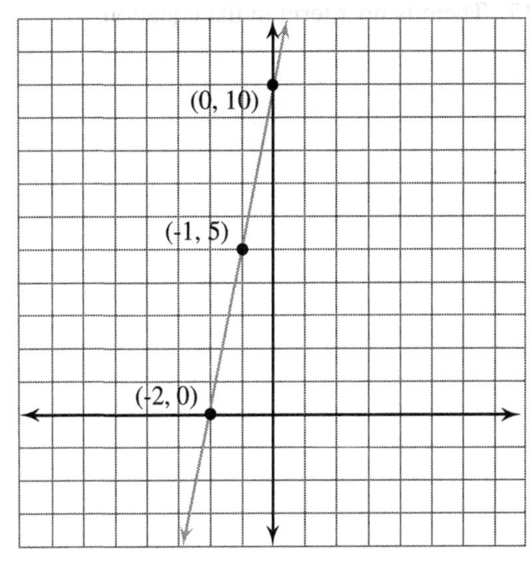

29.

x	y
0	0
3	1
–3	–1

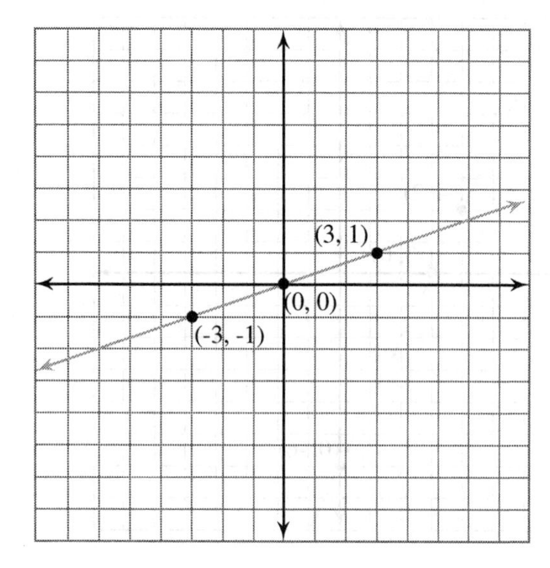

C. 31. Identify the coordinates of three points and see if those values give a true statement when substituted into the equation.

33. Two points determine a line, but the third value is for insurance against an error in one of the other two points.

35. There is no x term in the equation.

Section 15.3

A. 1. $\frac{\text{rise}}{\text{run}} = \frac{+3}{+2}$

$m = \frac{6-3}{4-2} = \frac{3}{2}$

positive slope

increasing

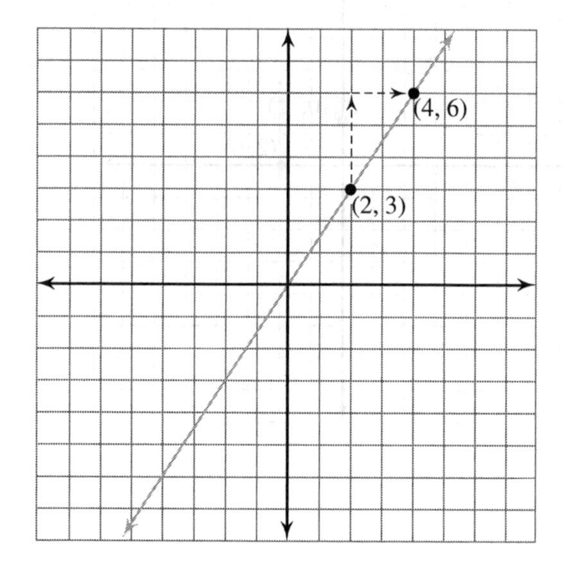

3. $\frac{\text{rise}}{\text{run}} = \frac{-1}{+1}$

$m = \frac{5-6}{3-2} = \frac{-1}{1} = -1$

negative slope

decreasing

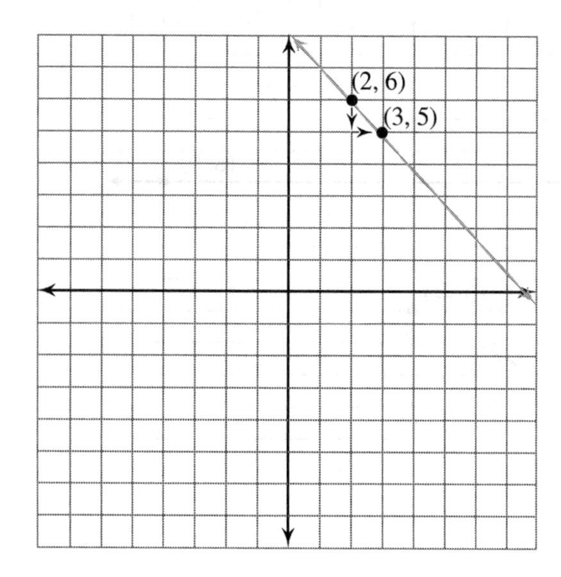

5. $\frac{\text{rise}}{\text{run}} = \frac{+4}{+3}$

$m = \frac{8-4}{3-0} = \frac{4}{3}$

positive slope

increasing

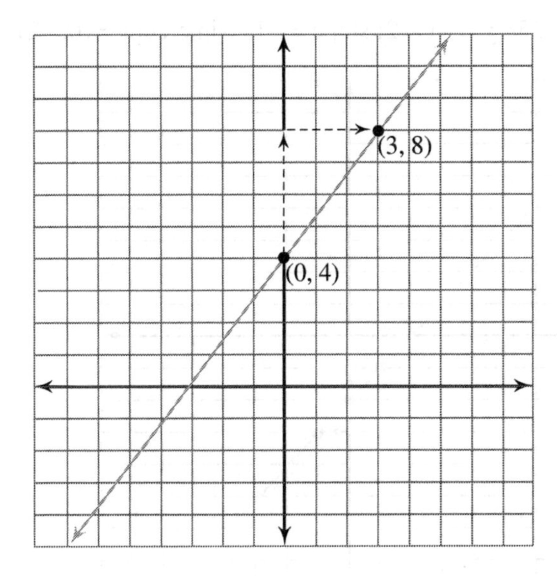

9. $\frac{\text{rise}}{\text{run}} = \frac{+2}{-3}$

$m = \frac{6-4}{2-5} = \frac{+2}{-3} = -2/3$

negative slope

decreasing

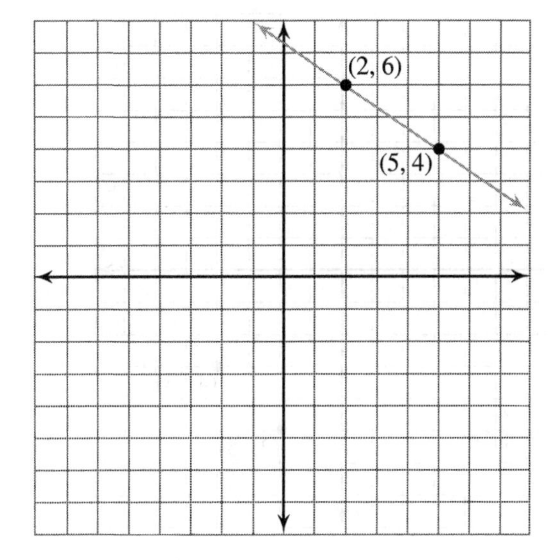

7. $\frac{\text{rise}}{\text{run}} = \frac{+4}{0}$

$m = \frac{8-4}{2-2} = \frac{4}{0} = \text{undefined}$

undefined slope

vertical line

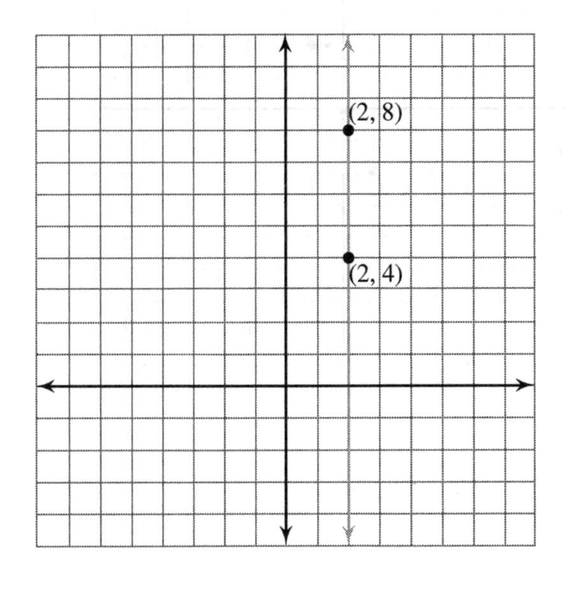

11. $\frac{\text{rise}}{\text{run}} = \frac{0}{-7}$

$m = \frac{4-4}{-4-3} = \frac{0}{-7} = 0$

0 slope

horizontal line

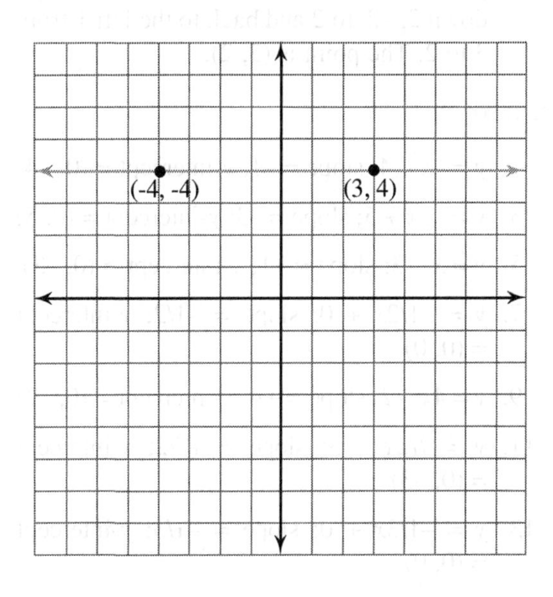

13. $\frac{\text{rise}}{\text{run}} = \frac{-2}{-2}$

$$m = \frac{-5 - (-3)}{2 - 4} = \frac{-2}{-2} = +1$$

positive slope

increasing

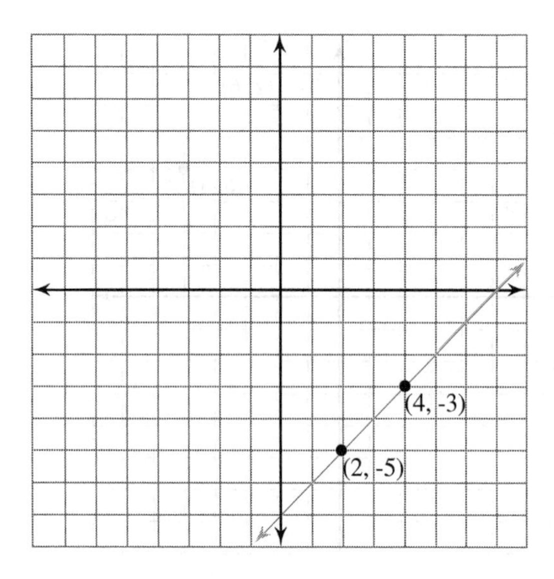

B. 15. Answers will vary.

17. A line with positive slope increases from left to right and a line with negative slope decreases from left to right.

19. Write the slope as +2/+1 and –2/–1. From (3, 4), go up 2 to 6 and right 1 to 4. The point is (4, 6). From (3, 4) using –2/–1, go down 2, –2, to 2 and back to the left 1 from 3 to 2. The point is (2, 2).

Section 15.4

A. 1. $y = -x - 4$; slope = –1, y-intercept = (0, –4)

3. $y = -3x + 5$; slope = –3; y-intercept = (0, 5)

5. $y = x - 3$; slope = +1; y-intercept = (0, –3)

7. $y = -1/2x + 0$; slope = –1/2; y-intercept = (0, 0)

9. $y = 4x - 2$; slope = +4; y-intercept = (0, –2)

11. $y = 7/2x - 3$; slope = +7/2; y-intercept = (0, –3)

13. $y = -1/3x + 0$; slope = –1/3; y-intercept = (0, 0)

15. $y = 1/2x - 4$; slope = 1/2; y-intercept = (0, –4)

17. $y = 3x - 2$; slope = +3; y-intercept = (0, –2)

19. $y = 1/3x + 4$; slope = 1/3; y-intercept = (0, 4)

21. $y = -3/4x + 3/2$; slope = –3/4; y-intercept = (0, 3/2)

23. $y = 3/2x - 2$; slope = 3/2; y-intercept = (0, –2)

B. 25. $y = -x - 2$; slope = –1; y-intercept = (0, –2)

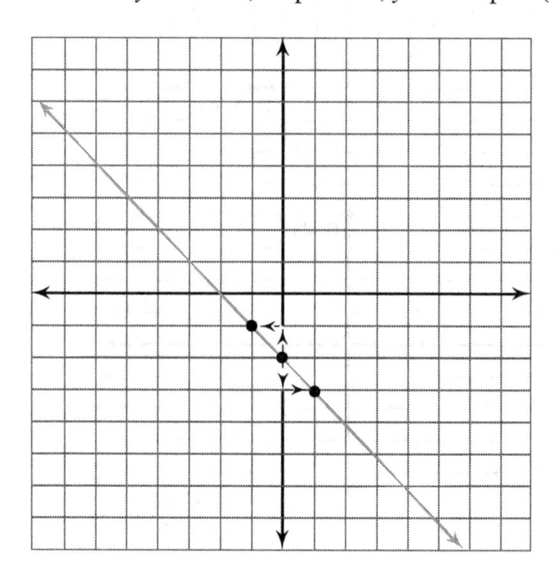

27. $y = 2x + 0$; slope = 2; y-intercept = (0, 0)

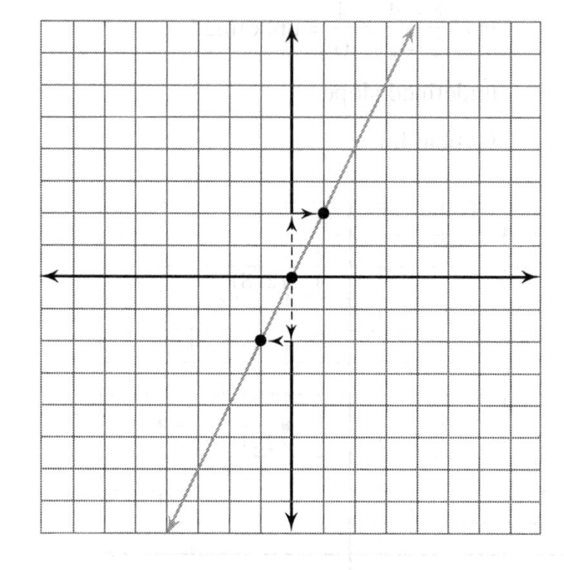

29. $y = 0x - 5$; slope = 0; y-intercept = (0, –5)

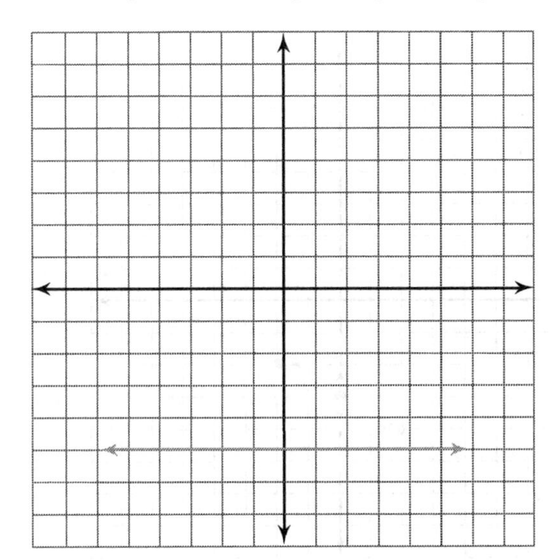

31. $y = x - 1$; slope = +1; y-intercept = (0, –1)

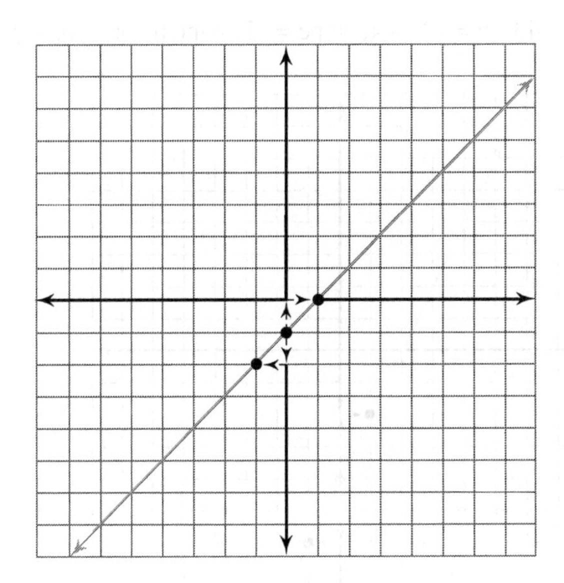

33. $y = -1/3x + 0$; slope = –1/3; y-intercept = (0, 0)

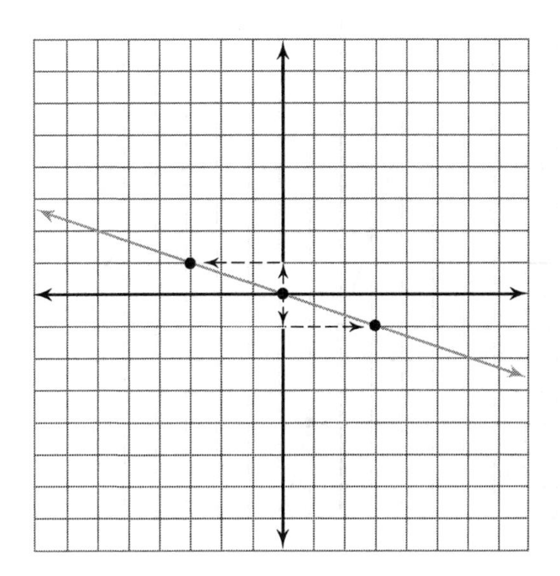

35. Can't do; slope = undefined; no y-intercept

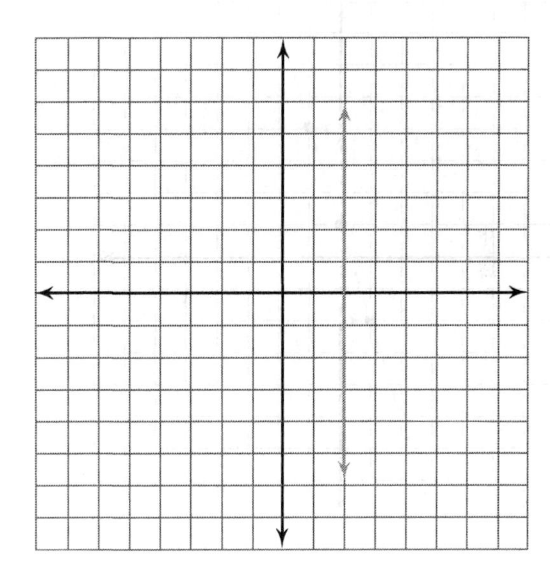

37. $y = -3x - 2$; slope $= -3$; y-intercept $= (0, -2)$

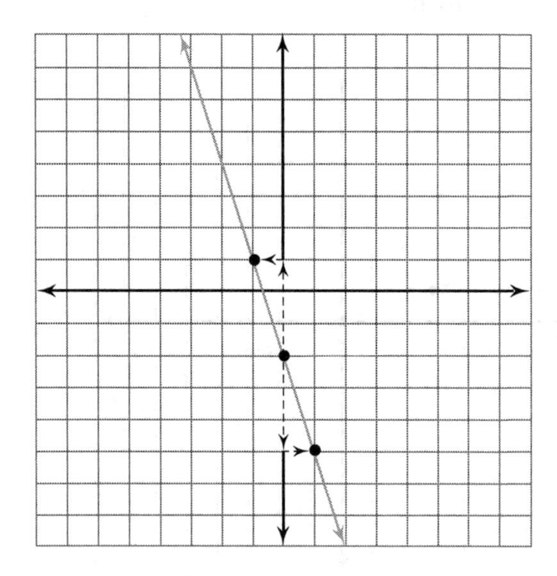

41. $y = 1/2x - 3$; slope $= 1/2$; y-intercept $= (0, -3)$

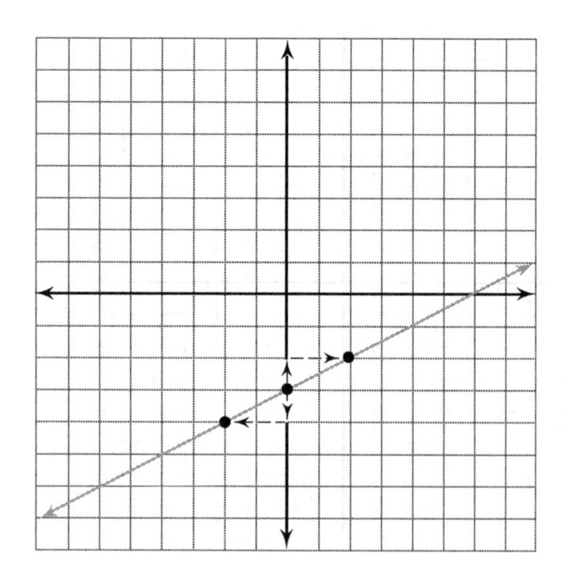

39. $y = 3x + 1$; slope $= +3$; y-intercept $= (0, 1)$

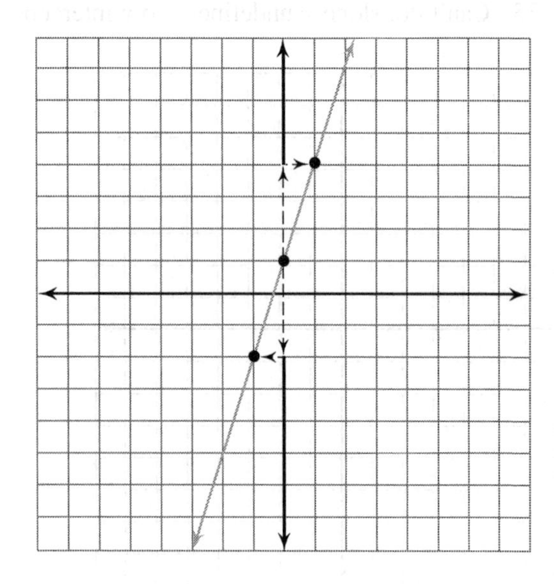

43. $y = -2x - 4$; slope $= -2$; y-intercept $= (0, -4)$

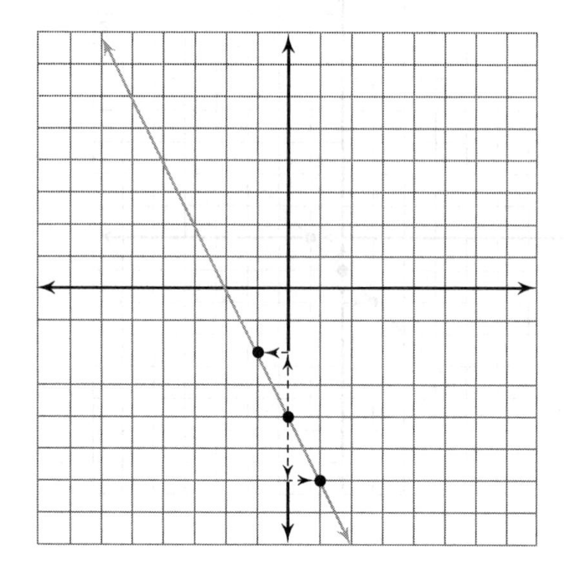

45. $y = 3/4x + 1$; slope = 3/4; y-intercept = (0, 1)

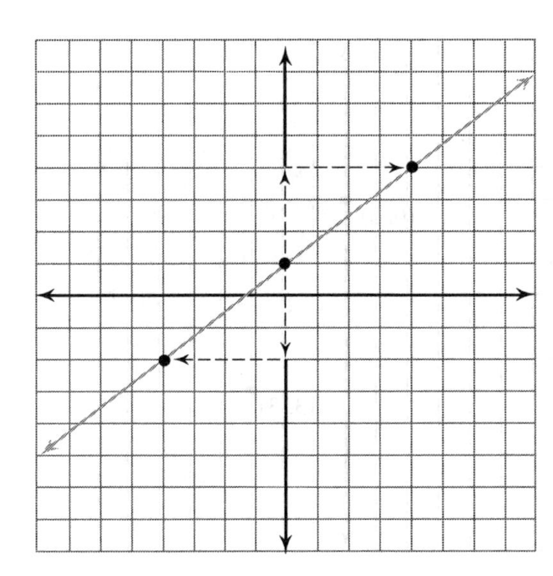

47. $y = -1/2x - 4$; slope = −1/2; y-intercept = (0, −4)

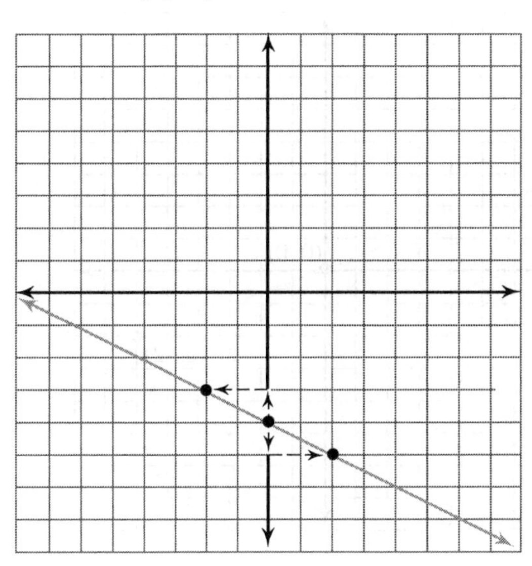

C. 49. 1. Rewrite the equation, y term on left.

2. Add $3x$ to both sides.

3. Rearrange right side to $mx + b$ form.

4. Divide all terms by 2.

D. 51. If the line passes through the origin, both intercepts would be (0, 0) and you would need more points to determine the line.

53. Slope is −3, and it passes through (0, −4).

55. There is no change in the y-coordinates, so the rise is zero, but the difference in the x-coordinates (run) is not zero. So the line is parallel to the x-axis.

Section 15.5

A. 1. $y = 3x - 2$

3. $y = \frac{1}{2}x + 3$

5. $y = 0x + 2$

7. $y = -5x + 6$

9. $y = -\frac{3}{4}x + 2$

B. 11. $y = \frac{3}{2}x + 3$

13. $y = 2x + 12$

15. $y = \frac{5}{3}x$

17. $x = 6$

19. $y = 3$

C. 21. $y = 3x - 19$

23. $y = -2x - 11$

25. $2y = x + 3$

27. $2y = -x + 1$

29. $4y = 3x - 10$

31. $y = 5$

33. $x = 2$

D. 35. Write the slope-intercept formula, $y = mx + b$, and substitute for m and b.

37. Write the slope-intercept formula, $y = mx + b$, and substitute for x, y, and m. Solve the equation for b. Then substitute m and b in $y = mx + b$.

Section 15.6

A. 1.

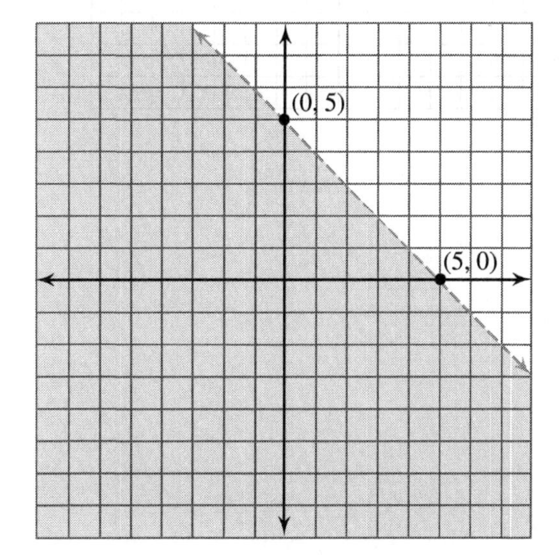

3.

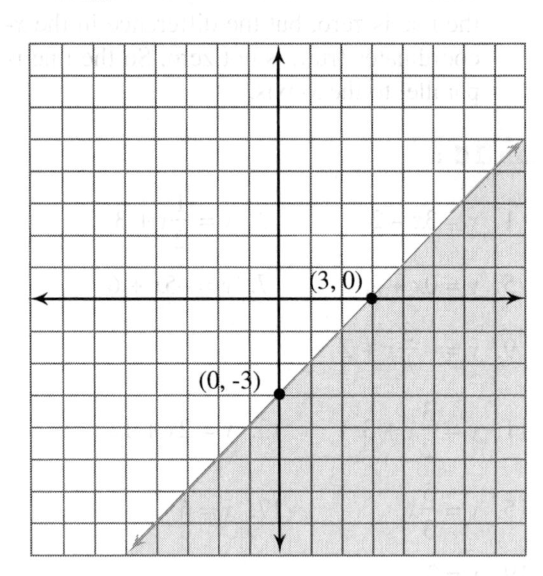

7.

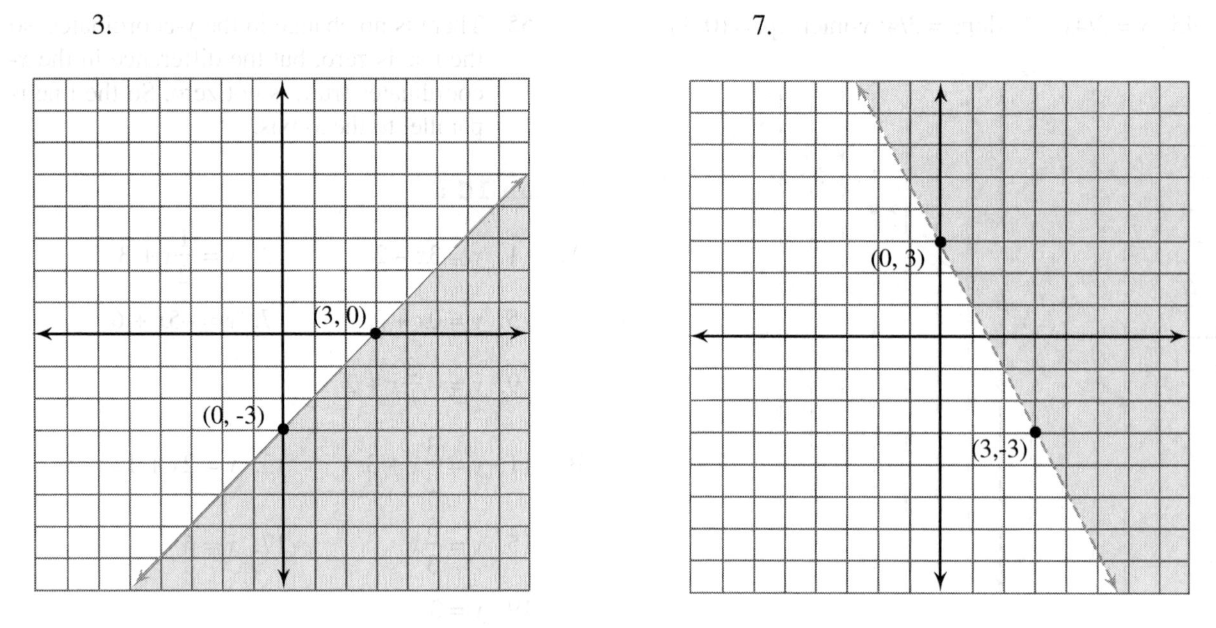

5.

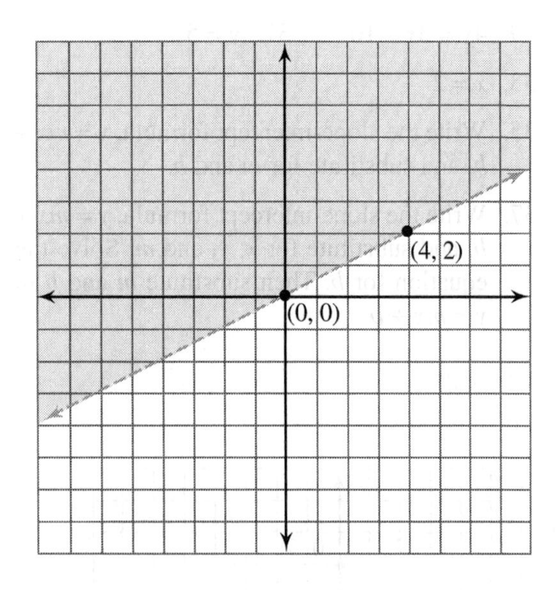

9.

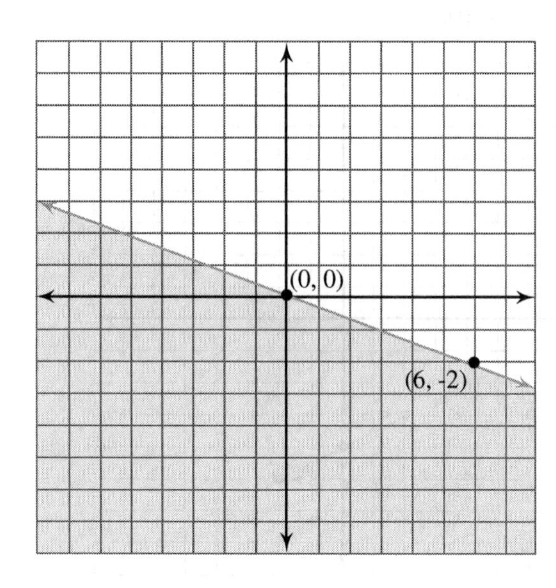

11.

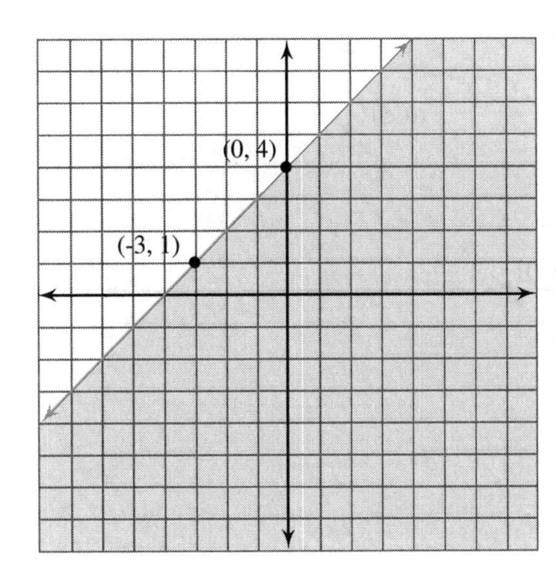

13.

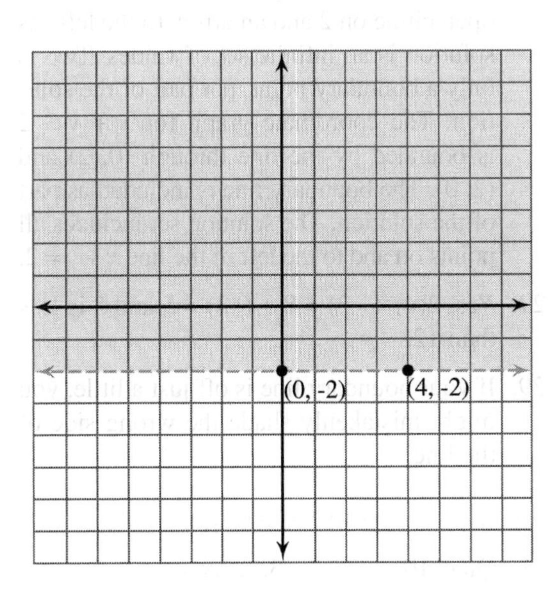

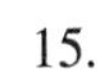

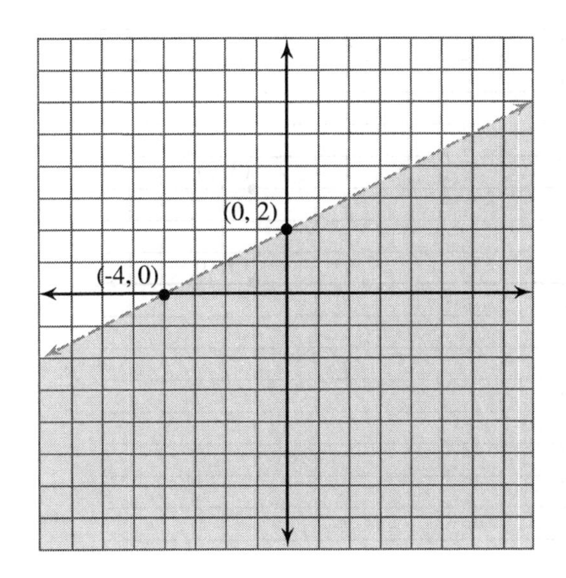

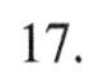

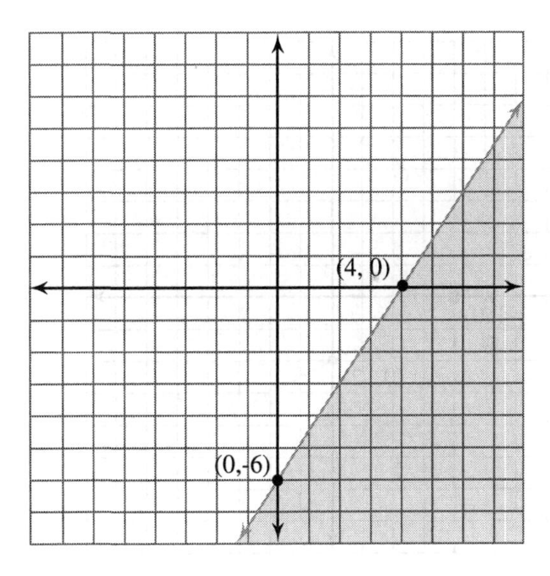

19.

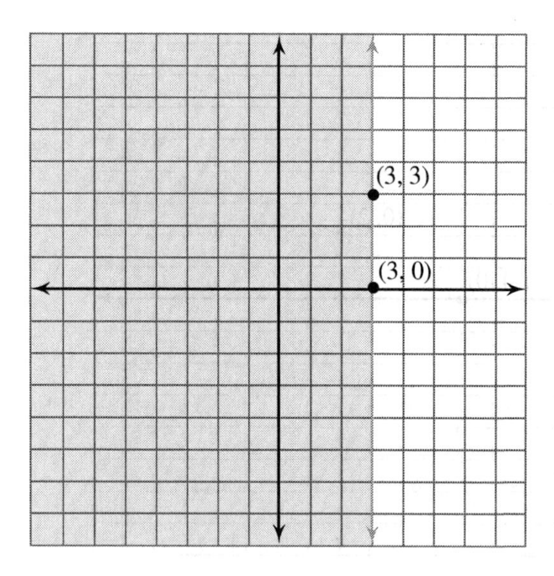

21.

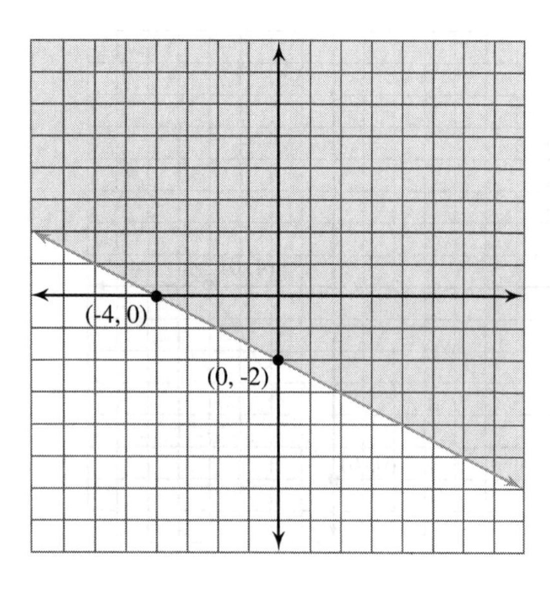

23.

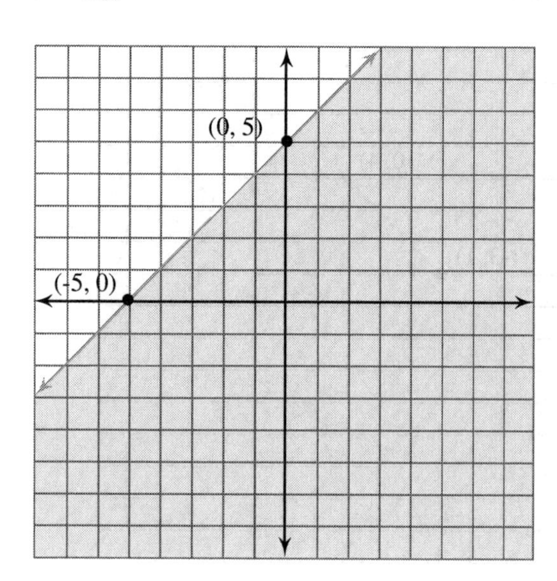

B. 25. The number line graph of $x < 2$ has an open circle on 2 and an arrow to the left. Its solution is an infinite set of values. Two is only a boundary point, not part of the solution. The coordinate graph for $x + y \leq 2$ is bounded by the line through (0, 2) and (2, 0). The boundary line is included as part of the solution. The solution set includes all points on and to the left of the line $x + y = 2$.

27. Yes, $2(4) + -3) = 8 + (-3) = 5$ and 5 is less than 12.

29. If your boundary line is off just a little, you might mistakenly shade the wrong side of the line.

Section 15.7 Review Exercises

A. 1. Quad III
3. y-axis
5. Quad IV
7. Quad I
9. x-axis
11. Quad II

B. The following are sample tables and do not contain all possible pairs of numbers.

13.

x	y
0	−2
−2/3	0
1	−5

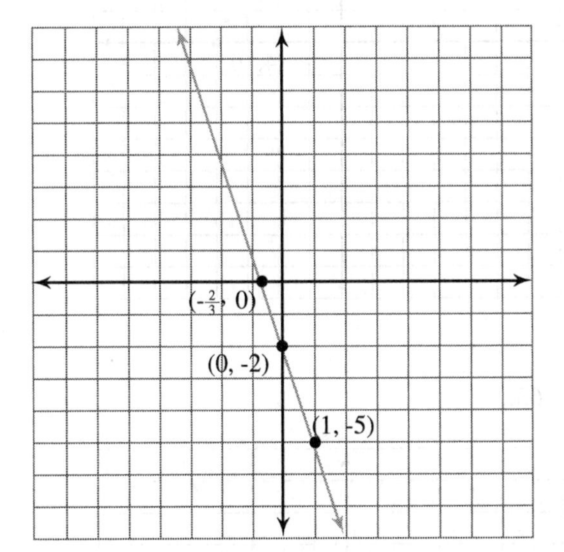

15.

x	y
0	−5
2 1/2	0
−1	−7

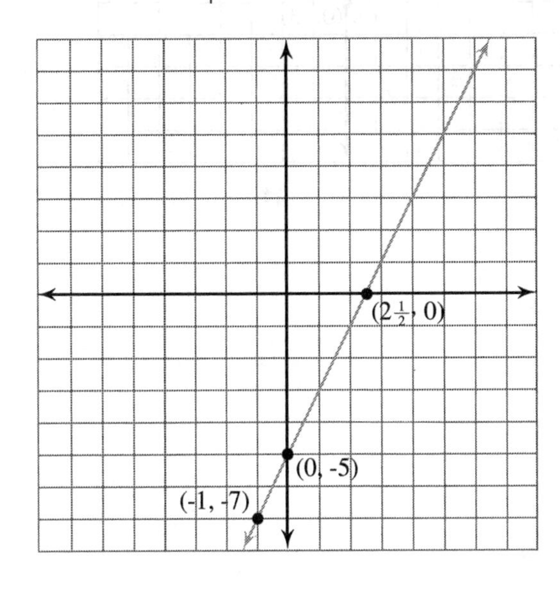

17.

x	y
0	−3
3	0
−1	−4

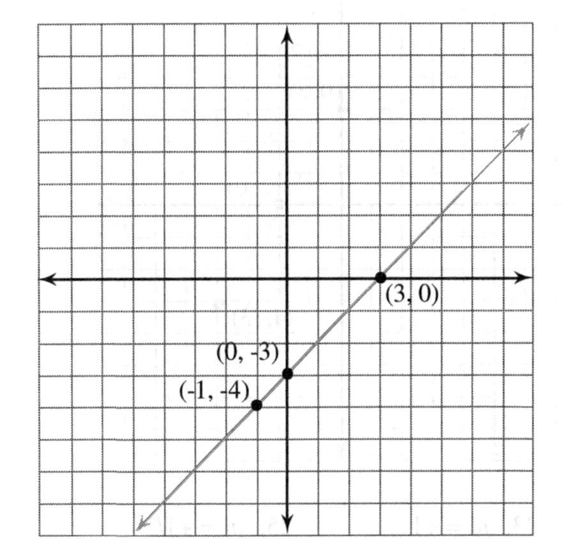

19.

x	y
6	0
6	2
6	−2

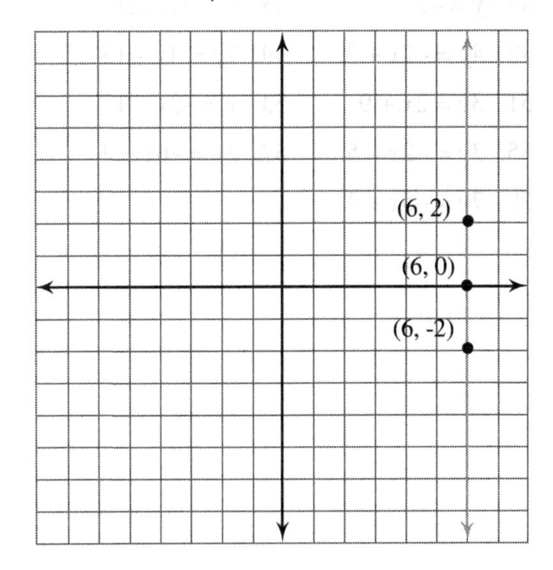

21.

x	y
0	3
2	0
4	–3

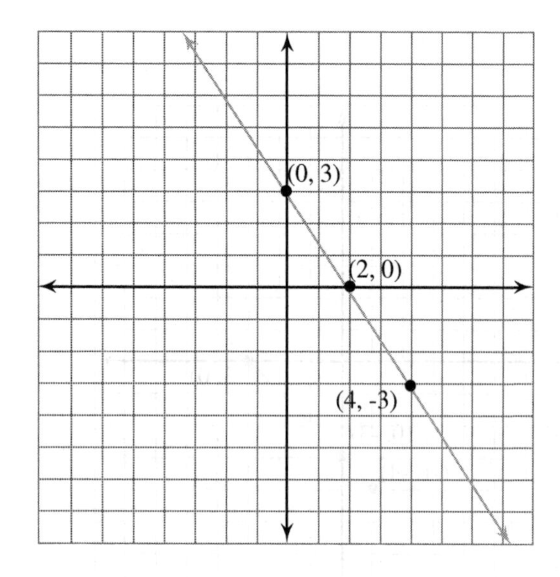

C. 23. $m = +1$

25. $m = +1/2$

27. $m = 3/0$ undefined slope

29. $m = 5/6$

D. 31. $y = -2x + 3$

33. $y = -3x + 11$

35. $y = 2x - 3$

37. $y = -3x + 6$

39. $3y = 2x - 3$

41. $2y = -x - 9$

43. $y = -2$

45. $y = 3x - 21$

47. $4y = -2x + 3$

49. $7y = 4x - 13$

51. $3y = 2x + 9$

53. $y = -2x - 1$

55. $3y = -2x - 5$

57. $2y = -6x + 1$

59. $7y = -5x + 1$

E. 61.

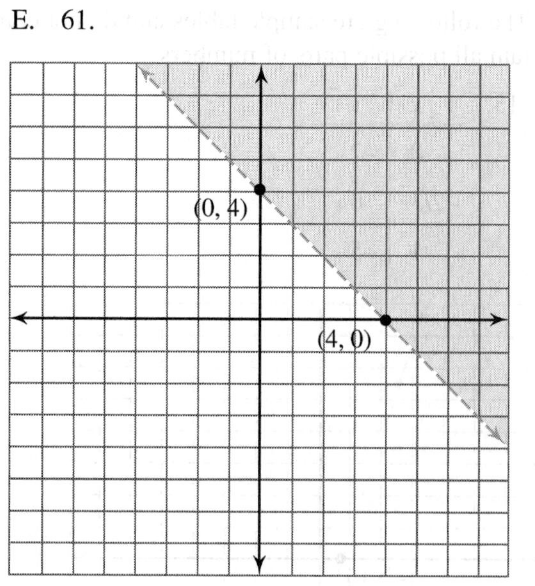

63.

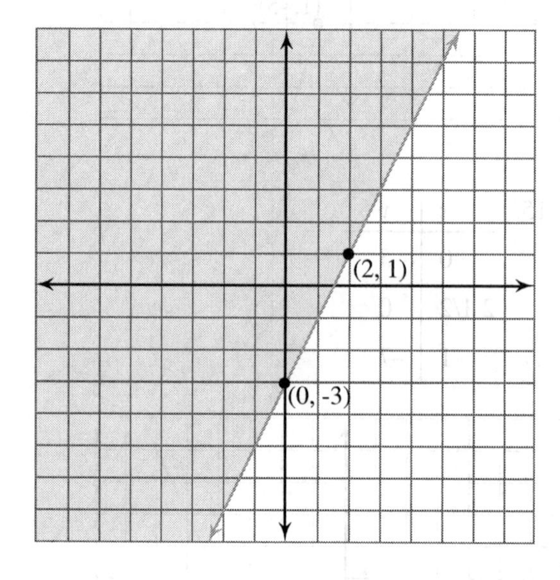

65.

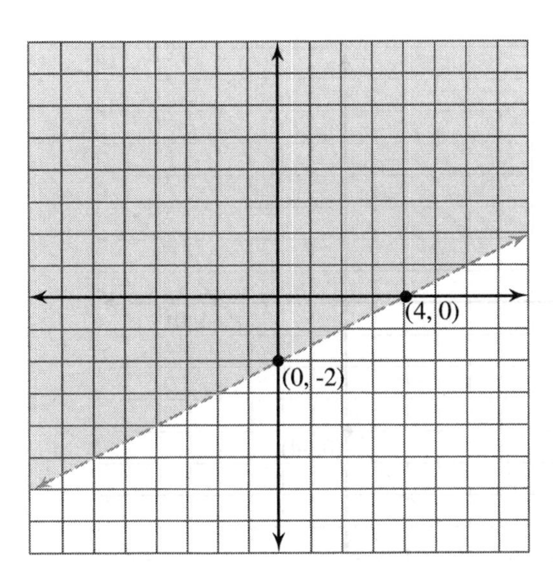

67.

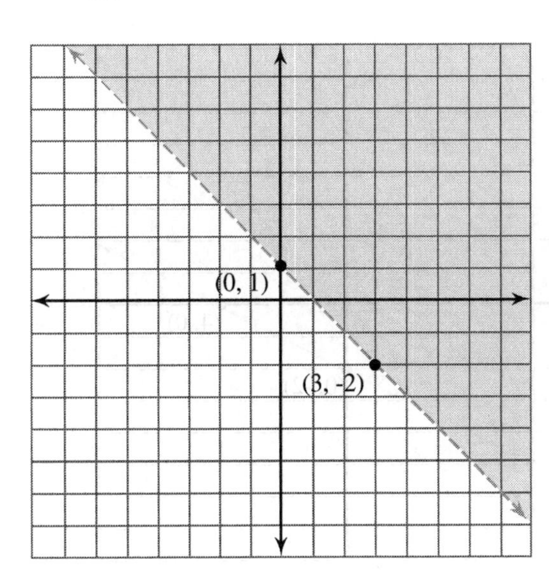

69.

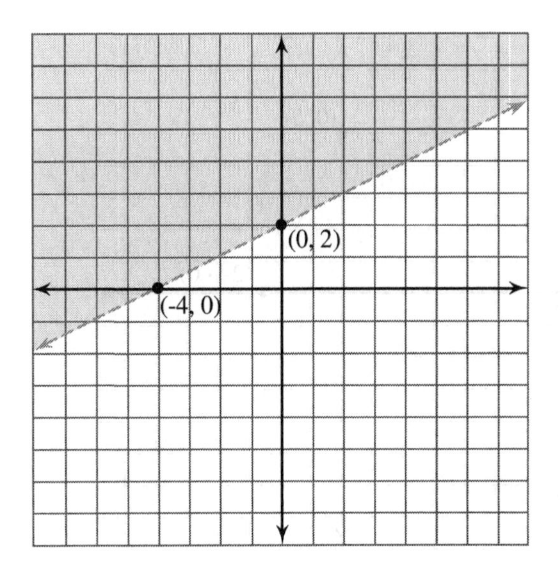

F. 71. $(-2) \div (-3) = +2/3$; equation should be $y = 2/3x - 4$

73. $\dfrac{1-3}{4-(-2)} = \dfrac{-2}{6} = -\dfrac{1}{3}$, not $+\dfrac{1}{3}$

75. $\dfrac{-4-0}{5-(-1)} = \dfrac{-4}{6} = -\dfrac{2}{3}$, not $-\dfrac{3}{2}$

G. 77. Answers will vary.

79. Answers will vary.

Test—Chapter 15

1. y
2. II
3. origin
4. $-1/2$
5. -2
6. $y = 2x - 4$
7. $-12/5$
8. Undefined
9. $y = -2x + 4$
10. $3y = 2x - 9$
11. $y = -x + 2$
12. $y = -2$
13. $y = -3x + 4$

14.

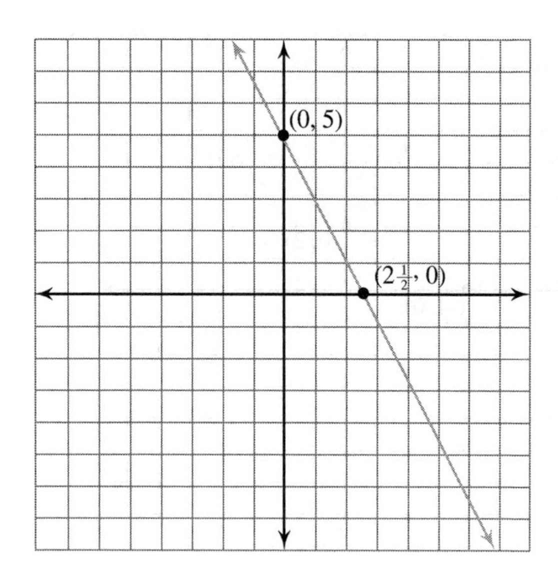

16.

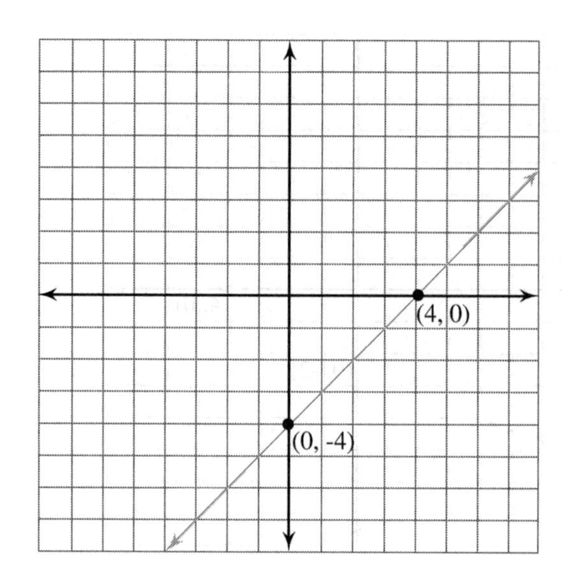

15.

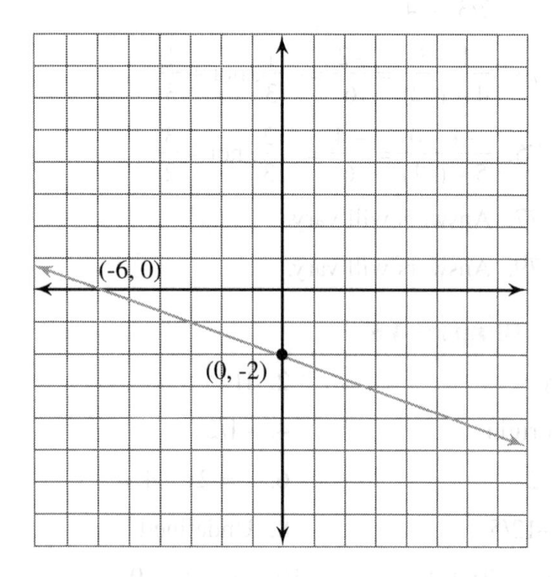

17.

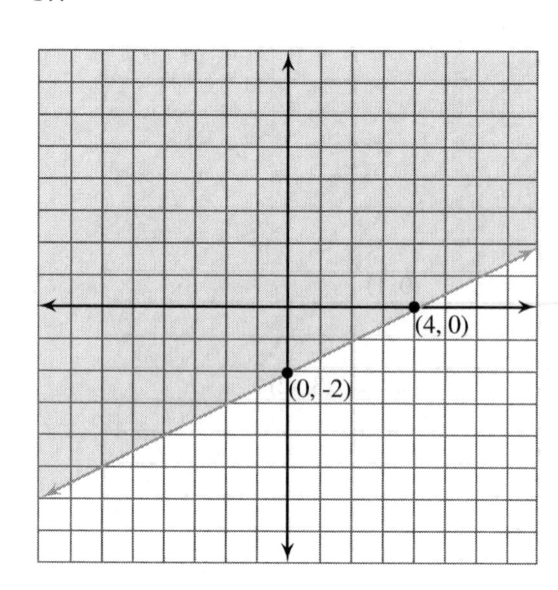

18.

(-3, 3) (0, 3) (4, 3)

19.

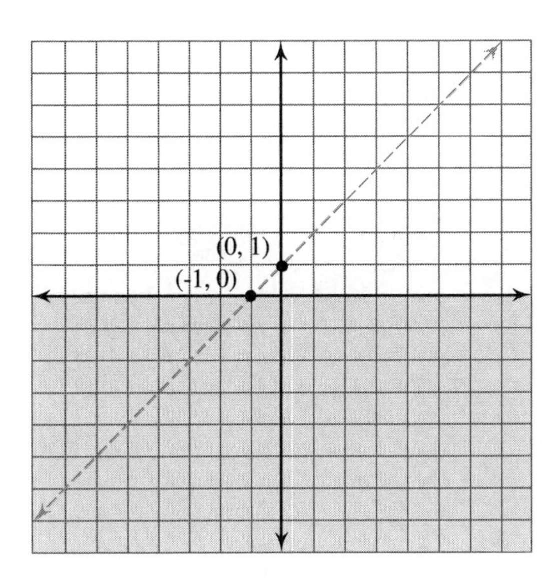

20. Because the x-coordinates are both 4, the slope is undefined, and the equation of the line will be $x = 4$.

Cumulative Review 1–15

1. 3 67/72
2. $19a$
3. $5y^2 + 6y + 8$
4. –63
5. 51.69
6. –3
7. 9
8. $-7a - 7ab - 3b$
9. $14xyz$
10. 11/36
11. 57/8 or 7 1/8
12. –36
13. $6x^4y^3z$
14. 65.4
15. $a^2 + 4ab + 4b^2$
16. $-2ac^2$
17. 3/8
18. 734
19. $-3mn + 2m - n$
20. $x - 6$
21. 5/2
22. 0.21
23. $m = 4/7$, $b = 3/7$
24. 8/15
25. $(2x - 5y)^2$
26. $3.87x^{10}$
27. –23
28. 1990–1991
29. $1,000,000 more
30. $2,875,000
31. $x = 9/2$ or 4 1/2 or 4.5
32. $-1 \leq x \leq 1$
33. 106.7
34. $x = -2$, $x = -4$
35. $y = -7$
36. $N = 1.8$
37. $Y - a - b = c$
38. 125%
39. $x \geq 2$
40. $T = 1.8$
41. $Y = -1$
42. $h = 7$ feet
43. Number is –1
44. 8 quarters
45. 0.6 inches
46. Median: 0.7 inches, range: 1.3 inches
47. 20 miles per hour
48. 16,400 students
49. 5
50. 30 meters by 82 meters

Index

D

E

F